AF880652

Notes on Numerical Fluid Mechanics and Multidisciplinary Design

Volume 151

Founding Editor

Ernst Heinrich Hirschel, Zorneding, Germany

Series Editor

Wolfgang Schröder, Aerodynamisches Institut, RWTH Aachen, Aachen, Germany

Editorial Board Members

Bendiks Jan Boersma, Delft University of Technology, Delft, The Netherlands

Kozo Fujii, Institute of Space & Astronautical Science (ISAS), Sagamihara, Kanagawa, Japan

Werner Haase, Neubiberg, Bayern, Germany

Michael A. Leschziner, Department of Aeronautics, Imperial College, London, UK

Jacques Periaux, Paris, France

Sergio Pirozzoli, Department of Mechanical and Aerospace Engineering, University of Rome 'La Sapienza', Roma, Italy

Arthur Rizzi, Department of Aeronautics, KTH Royal Institute of Technology, Stockholm, Sweden

Bernard Roux, Ecole Supérieure d'Ingénieurs de Marseille, Marseille CX 20, France

Yurii I. Shokin, Siberian Branch of the Russian Academy of Sciences, Novosibirsk, Russia

Managing Editor

Esther Mäteling, RWTH Aachen University, Aachen, Germany

Notes on Numerical Fluid Mechanics and Multidisciplinary Design publishes state-of-art methods (including high performance methods) for numerical fluid mechanics, numerical simulation and multidisciplinary design optimization. The series includes proceedings of specialized conferences and workshops, as well as relevant project reports and monographs.

More information about this series at http://www.springer.com/series/4629

Andreas Dillmann · Gerd Heller ·
Ewald Krämer · Claus Wagner
Editors

New Results in Numerical and Experimental Fluid Mechanics XIII

Contributions to the 22nd STAB/DGLR Symposium

Springer

Editors
Andreas Dillmann
Deutsches Zentrum für Luft und Raumfahrt,
Institut für Aerodynamik und
Strömungstechnik
Göttingen, Niedersachsen, Germany

Ewald Krämer
Institut für Aerodynamik und Gasdynamik
Universität Stuttgart
Stuttgart, Baden-Württemberg, Germany

Gerd Heller
Airbus Deutschland
Bremen, Bremen, Germany

Claus Wagner
Deutsches Zentrum für Luft und Raumfahrt,
Institut für Aerodynamik und
Strömungstechnik
Göttingen, Niedersachsen, Germany

ISSN 1612-2909 ISSN 1860-0824 (electronic)
Notes on Numerical Fluid Mechanics and Multidisciplinary Design
ISBN 978-3-030-79560-3 ISBN 978-3-030-79561-0 (eBook)
https://doi.org/10.1007/978-3-030-79561-0

© The Editor(s) (if applicable) and The Author(s), under exclusive license
to Springer Nature Switzerland AG 2021
This work is subject to copyright. All rights are solely and exclusively licensed by the Publisher, whether the whole or part of the material is concerned, specifically the rights of translation, reprinting, reuse of illustrations, recitation, broadcasting, reproduction on microfilms or in any other physical way, and transmission or information storage and retrieval, electronic adaptation, computer software, or by similar or dissimilar methodology now known or hereafter developed.
The use of general descriptive names, registered names, trademarks, service marks, etc. in this publication does not imply, even in the absence of a specific statement, that such names are exempt from the relevant protective laws and regulations and therefore free for general use.
The publisher, the authors and the editors are safe to assume that the advice and information in this book are believed to be true and accurate at the date of publication. Neither the publisher nor the authors or the editors give a warranty, expressed or implied, with respect to the material contained herein or for any errors or omissions that may have been made. The publisher remains neutral with regard to jurisdictional claims in published maps and institutional affiliations.

This Springer imprint is published by the registered company Springer Nature Switzerland AG
The registered company address is: Gewerbestrasse 11, 6330 Cham, Switzerland

Preface

STAB is the German Aerospace Aerodynamics Association (Deutsche Strömungsmechanische Arbeitsgemeinschaft) founded in the late 1970s, and DGLR is the German Society for Aeronautics and Astronautics (Deutsche Gesellschaft für Luft- und Raumfahrt - Lilienthal Oberth e.V.). The mission of STAB is to foster aerodynamics research and its appreciation in Germany. This is accomplished by creating forums for scientific discussions and by disseminating most recent research results, thereby promoting scientific progress and avoiding unnecessary duplication in research.

STAB brings together German scientists and engineers from universities, research establishments and industry. They present research and project work in numerical and experimental fluid mechanics as well as aerodynamics for a wide variety of fields, such as aeronautics, space, ground transportation, wind turbines and other applications. This format also offers an excellent opportunity for exchange on numerous joint research activities supported by different funding agencies.

Since 1986, the symposium has taken place every two years at different locations in Germany, all having an affinity with fluid mechanics and aerodynamics.

In addition, STAB workshops are regularly held at Deutsches Zentrum für Luft- und Raumfahrt (DLR) in Göttingen in the intermediate years.

Both STAB symposia and workshops provide excellent forums where new research activities can be presented, often resulting in new jointly organized research and technology projects.

In 2020, the 22nd DGLR/STAB Symposium was planned to be held in Berlin. Unfortunately, due to the corona pandemic, the STAB Board had to cancel the conference to protect the health of the participants.

However, it was the Board's strong ambition to give researchers who had planned a presentation for the symposium the opportunity to publish their latest work in an appropriate setting. Thus, the present volume contains the papers originally envisaged for the symposium after having successfully undergone a thorough peer review process.

The review panel for the papers included in this volume consisted of

G. Babij (Göttingen); P. Bahavar (Göttingen); C. Bauer (Göttingen); A. Bauknecht (Braunschweig); P. Bekemeyer (Braunschweig); J. Bell (Göttingen); C. Birkenmaier (Regensburg); K. Bock (Dresden); F. Boden (Göttingen); J. Bosbach (Göttingen); J. Braukmann (Göttingen); C. Breitsamter (München); C. Brückner (Göttingen); M. Cerny (München); M. Cormier (Stuttgart); A. Dannhauer (Göttingen); K. Ehrenfried (Göttingen); M. Ehrle (Stuttgart); S. Fechter (Göttingen); M. Fehrs (Göttingen); J. Feldwisch (Göttingen); U. Fey (Göttingen); J. Franco Sumariva (Göttingen); A. Franco (Braunschweig); M. Franze (Braunschweig); L. Gagnon (Stuttgart); A. Gardner (Göttingen); N. Gauger (Kaiserslautern); F. Genuit (Stuttgart); V. Hannemann (Göttingen); S. Hein (Göttingen); D. Heine (Göttingen); S. Helm (Göttingen); M. Herr (Göttingen); M. Hillebrand (Stuttgart); I. Huismann (Dresden); T. Hüttl (Ismaning); F. Jäckel (Göttingen); S. Jakirlic (Darmstadt); C. Kaiser (Göttingen); M. Keßler (Stuttgart); T. Kirmse (Göttingen); M. Klein (Cottbus); C. Klein (Göttingen); T. Knopp (Göttingen); A. Kohl (Göttingen); N. Krimmelbein (Braunschweig); A. Krumbein (Göttingen); P. Lange (Göttingen); J. Löwe (Göttingen); T. Lutz (Stuttgart); J. Martinez Schramm (Göttingen); E. Mäteling (Aachen); C. Merrem (Braunschweig); O. Meyer (Neubiberg); U. Michel (Berlin); M. Mommert (Göttingen); M. Müller (Göttingen); J. Müller (Stuttgart); K. Niehaus (Göttingen); R. Olmeda (München); R. Pochampalli (Kaiserslautern); A. Probst (Göttingen); D. Puckert (Stuttgart); A. Raichle (Braunschweig); L. Reimer (Braunschweig); M. Rein (Göttingen); C. Reinbold (München); S Risius (Göttingen), U. Rist (Stuttgart); L. Rottmann (Braunschweig); J. Ruhland (München); S. Satcunanathan (Aachen); D. Schiepel (Göttingen); L. Schmidt (Göttingen); M. Schollenberger (Stuttgart); G. Schrauf (Braunschweig); E. Schülein (Göttingen); M. Schulze (Göttingen); D. Schütz (Siegen); A. Shishkin (Göttingen); L. Siegel (Göttingen); D. Simanowitsch (Göttingen); M. Soni (Braunschweig); C. Stemmer (München); L. Streher (Braunschweig); T. Streit (Braunschweig); E. Tangermann (Neubiberg); A. Theiss (Göttingen); G. Voss (Göttingen); A. Waldmann (Stuttgart); V. Wartemann (Braunschweig); K. Weinman (Göttingen); M. Weinschenk (Stuttgart); A. Westhoff (Göttingen); T. Wetzel (Hamburg); H. Wilhelmi (Göttingen); G. Wilke (Braunschweig); C. Wolf (Göttingen); J. Yin (Braunschweig); R. Zahn (München); B. Zhou (Kaiserslautern).

Contents

Rotorcraft

Airplane Aerodynamics/Propulsion Integration

Influence of the Wind Tunnel Model Mounting on the Wake Evolution of the Common Research Model in Post Stall

Maximilian Ehrle(✉), Andreas Waldmann, and Thorsten Lutz

Faculty 6: Aerospace Engineering and Geodesy, Institute of Aerodynamics and Gas Dynamics, University of Stuttgart, Stuttgart, Germany
{ehrle,waldmann,lutz}@iag.uni-stuttgart.de

Abstract. This work describes the influence of the wind tunnel support system on the wake flow and associated forces acting on the CRM aircraft configuration in post-stall conditions at subsonic Mach numbers. Unsteady scale-resolving simulations at $\alpha = 18°$, $M_\infty = 0.25$ and $Re_\infty = 11.6 \cdot 10^6$ were carried out using the CRM with and without a support sting, respectively. A predominantly local effect of the sting was observed at these conditions, causing significant deflection of the local flow direction near the sting. The sting's shape causes downward deflection near the horizontal tailplane, decreasing that surface's effective angle of attack and its lift coefficient. This results in a reduced overall lift coefficient C_L and a significantly reduced nose-down pitching moment C_{My}. The wing pressure distribution is only weakly affected.

Keywords: Transport aircraft · Stall · Hybrid RANS/LES methods

1 Introduction

Validation experiments are of great importance for the development of modern computational methods and for the understanding of physical phenomena. In order to ensure comparability between simulation and experiment, the experimental conditions need to be modeled as accurately as possible in the simulation setup. However, the computational effort increases significantly when the full geometric configuration of a wind tunnel experiment is taken into account. Therefore, simplifications of the setup in combination with correction factors are often used to mimic wind tunnel effects. In the context of aircraft computational fluid dynamics (CFD), it is desirable to handle the isolated aircraft geometry without the added complexity of the surrounding wind tunnel structure.

The physical effects of a mounting system on the flow around the model can be divided in a near-field and far-field effect, as Stojanowski et al. [13] point out. Similarly, Britcher et al. [1] describe three components of interference: overall, local and geometrical modification. The overall effect of the sting on the flow is due to the pressure gradient induced by the sting's physical blockage. The local flow disturbance is caused by interference between support system and

© The Author(s), under exclusive license to Springer Nature Switzerland AG 2021
A. Dillmann et al. (Eds.): STAB/DGLR Symposium 2020, NNFM 151, pp. 3–13, 2021.
https://doi.org/10.1007/978-3-030-79561-0_1

flow in the vicinity of the model, most significantly by altering the development of boundary layers. Finally, the geometrical distortion results from altered model surfaces such as e.g. the interface between sting and model replacing a fuselage surface or a tailcone.

Prior work dealing with flow separation and the wake of the NASA Common Research Model (CRM) [14] in free flight conditions has been carried out in the authors' working group [15,17]. Waldmann et al. [15] showed the necessity of using hybrid RANS/LES methods for the resolution of the complex flow in the separated wake associated with the behavior of the wing leading edge shear layer in post stall conditions. They concluded that the aerodynamic coefficients and the wake propagation can be predicted in good agreement with the experiment by using the Delayed Detached Eddy Simulation method (DDES) [11]. The experimental data used for comparison was taken from the measurement campaign conducted during the ESWI^{RP} project in the European Transonic Wind Tunnel (ETW) [7].

As the simulations in the previously mentioned publications are based on the free flight configuration of the CRM, wind tunnel effects such as wall interference and mounting system are not taken into account. The wind tunnel wall influence is minimized in ETW by the use of slotted walls, and it has been studied in the context of the ESWI^{RP} campaign [3]. However, the wind tunnel model differs from the free flight configuration in the area of the rear fuselage, which includes the interface with the model support system. This leads to inherent differences between wind tunnel and simulation results, and increases uncertainties.

For this reason, different wind tunnel effects such as the wall influence and the impact of the blockage effect of the model support system have been subject of recent research. The influence of the wind tunnel support system on the CRM's flow characteristics has been characterized in the transonic regime [4,6,8] and in the linear range of the angle of attack polar at subsonic conditions [16]. While these studies quantify the influence of the mounting on the aerodynamic coefficients, the wing pressure distribution, and the local flow at the aircraft's tail, its effect on the post stall flow is still insufficiently explored. This flow regime at the edge of the flight envelope is of particular interest because the occurring unsteady aerodynamic loads can cause damage to the aircraft structure and may lead to controllability issues. The present computational study deals with the question of how the mounting of the wind tunnel model affects the propagation of the wake and its interaction with the horizontal tailplane at subsonic flow conditions and at high angles of attack. Knowledge of these effects provides quantitative data on expected differences involved in comparisons of free flight simulations with wind tunnel data. Therefore, more reliable statements towards the development of the turbulent wake can be made.

2 Test Case and CFD Setup

The present work focuses on simulations of the CRM in the post stall regime with the hybrid RANS/LES model DDES. The computational model geometry

Table 1. Flow conditions and geometric parameters

Ma_∞	Re_∞	p_0	T_0	c_{ref}	α
0.25	$11 \cdot 10^6$	303 kPa	115 K	0.189 m	18°

replicates the wind tunnel model as used in ESWI^{RP} [7]. The flow conditions which reproduce run 316 of that campaign as well as the main geometric parameters of the model are listed in Table 1 where p_0 and T_0 denote the total values of pressure and temperature, c_{ref} is the mean aerodynamic chord and α the angle of attack. The data from the ESWI^{RP} [7] campaign include corrections for the wall interference, flow curvature and Mach effects. This study employs the dataset without sting buoyancy correction in order to better isolate the sting's influence on the flow field around the model.

The TAU solver [9] provided by DLR is used for all simulations in the present work. This finite volume solver is second order accurate in space and time using a central differencing scheme for the convective fluxes and an implicit backward Euler scheme with dual time stepping for time discretization. In order to reduce dissipation of small scale turbulent structures and avoid excessive dissipation of small scale instabilities, low dissipation settings are applied incorporating DLR's recommendations for hybrid RANS/LES simulations [2]. More precisely, a matrix-based artificial dissipation approach with a fourth order dissipation coefficient of $k^{(4)} = 1/1024$ is used. A blending between scalar-valued and matrix-valued dissipation models is set up such that matrix-valued dissipation contributes 95% of the total. Furthermore, the mean flow fluxes are discretized using the skew-symmetric scheme by Kok [5]. The dual time stepping scheme employs a physical time step size corresponding to 1% of the convective time unit $t_\infty = c_{\text{ref}}/u_\infty$. This yields a time step of $\Delta t = 3.461 \cdot 10^{-5}$ s. The mean and fluctuating values investigated in this work are obtained from time series spanning $70t_\infty$, which were extracted from longer simulated time series ensuring that no transient effects associated with simulation start-up remain in the solution.

The two investigated configurations represent the CRM with a horizontal tailplane, which is mounted at an incidence of 0° with respect to the fuselage (WBT0 configuration). The wing deformation due to aerodynamic loads in the wind tunnel is taken into account by deforming the grids according to the deformation recorded in the measurement campaign. More details about the deformed shapes and the grid deformation process can be found in Lutz et al. [7] and Waldmann et al. [15]. Waldmann et al. [16] quantified the improvement of computational results, which can be achieved by including the wing deformation. This deformation leads to the WBT0d configuration without wind tunnel support system, which is similar to the grid presented in [15]. The configuration WBT0ssd additionally takes into account the CRM's wind tunnel mounting sting which enters the fuselage in place of a vertical tailplane. The arc sector the sting is attached to in the ETW is not included in the computational model. The sting is cut off in the region where it enters the arc sector after its cylindrical part has reached maximum diameter. The sting's aft end is closed using a hemispherical cap. In the previous investigation of Waldmann et al. [16] this modeled length of the sting was sufficient for capturing the support system's influence on the flow around the CRM. Both configurations are represented in Fig. 1.

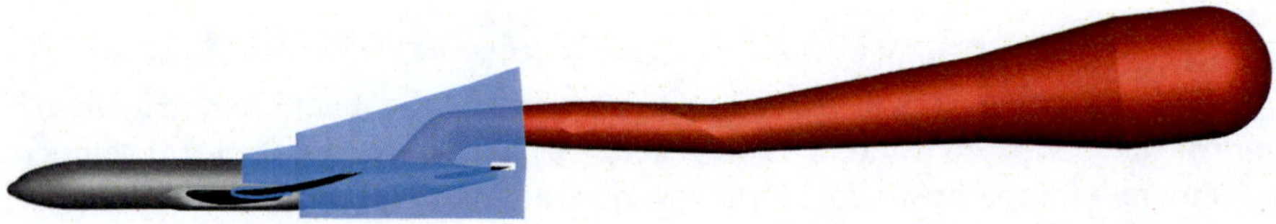

Fig. 1. CRM with and without wind tunnel mounting. Slice through structured wake grid in blue. Sting shown in red.

Two grids created with the CFD mesh generation software Pointwise are used in the present work. They are identical except in the region of the rear fuselage and the sting. The boundary layers are resolved using prism and hexahedron layers, ensuring $y^+ < 1$ everywhere. The area above the wing's suction side and the wake area is discretized using isotropic hexahedral elements in order to ensure a mesh suitable for scale resolving methods in areas of separated flow. A slice through this structured grid area is depicted in blue in Fig. 1. The structured area resolving the separated wake is also identical in both grids.

The junction areas between fuselage and wing, tailplane and wind tunnel mounting system, respectively are meshed using fully structured grid blocks. This enables the resolution of corner flow effects occurring in these areas. Overall, the WBT0d mesh consists of approx. 51 million points, whereas the WBT0ssd grid contains approx. 59 million grid points.

The one-equation Spalart-Allmaras (SA) turbulence model [10] is used as background model for DDES, as in the original model presented by Spalart. SA does not include the f_t and f_{t2} terms in this implementation. All runs are conducted assuming fully turbulent flow due to high Reynolds number and an extreme angle of attack with highly accelerated flow velocities around the leading edge. The hybrid RANS/LES simulations use a filter width based on the maximum edge length of the grid cells. This Δ_{max} filter in combination with the chosen timestep results in a convective CFL number between 1 and 2 in the area of flow separation on the employed grids. Following Spalart's [12] recommendation for DES-type grids, and prior work of Waldmann et al. [15], this is sufficient for resolving the major turbulent scales in the separated wake.

3 Results

The following section comprises studies of local flow effects caused by the support system near the aircraft tail as well as the influence on the wake. In addition to overall lift, drag and pitching moment data, the alteration of the local flow angle in the wake and its impact on the tailplane inflow are quantified and analyzed.

3.1 Near Field Effects

The sting's presence alters the local flow field in its immediate vicinity and forces the incoming flow to move around it instead of remaining largely straight above the fuselage. Figure 2 shows the pressure distribution and surface streamlines in the rear fuselage region of the CRM and on the upstream part of the wind tunnel mounting. The sting forces the flow outward in the spanwise direction. This displacement effect leads to a downward deflection of the surface streamlines upstream of the horizontal tailplane. The flow is locally accelerated around the sting, resulting in the low surface pressure region around the widest area of the vertical portion. Near the tailplane leading edge root and above the upper side of the tailplane, an upward bend and a stronger curvature of the surface streamlines can be observed in the presence of the sting.

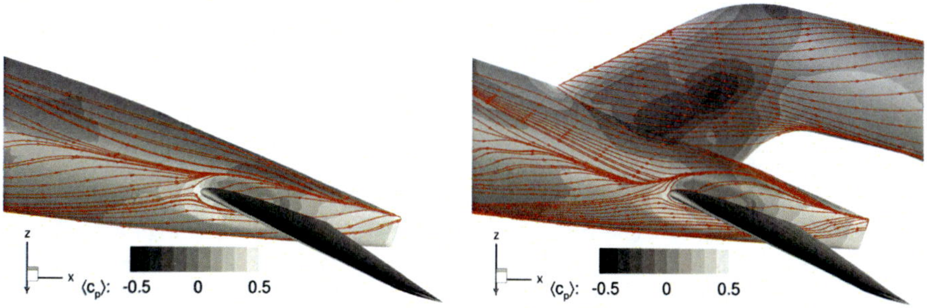

(a) WBT0d configuration (b) WBT0ssd configuration

Fig. 2. c_p and surface streamlines at the rear fuselage of the CRM and the sting.

The difference of the surface pressure coefficient Δc_p on the fuselage caused by the support system is shown in Fig. 3. An increase in local pressure, both upstream and downstream of the area where the sting enters the fuselage is clearly visible. This can be attributed to decelerated flow in these areas. Furthermore, Δc_p on the fuselage's surface is reduced in the region of accelerated flow near the widest point of the sting. The positions of these regions are in agreement with those of Waldmann et al. [16] at lower angles of attack but show a slightly larger magnitude. In addition, the pressure coefficient is increased near the leading edge of the horizontal tailplane when the support system is included. This is due to the reduction of the local flow field angle upstream of the tailplane, which is considered further below.

Figures 4 and 5 visualize the difference of the local flow field angle ξ between the cases with and without the wind tunnel support system. A plane at $\eta = 10.7\%$ of wing semi-span $\eta = y/b$ and a plane upstream of the tailplane and parallel to its leading edge are examined, respectively. The angle ξ is defined with respect to the horizontal inflow direction, i.e. $\xi = \arctan(w/u)$. Generally, Fig. 4 implies that the deflection is directed upward above the sting and downward

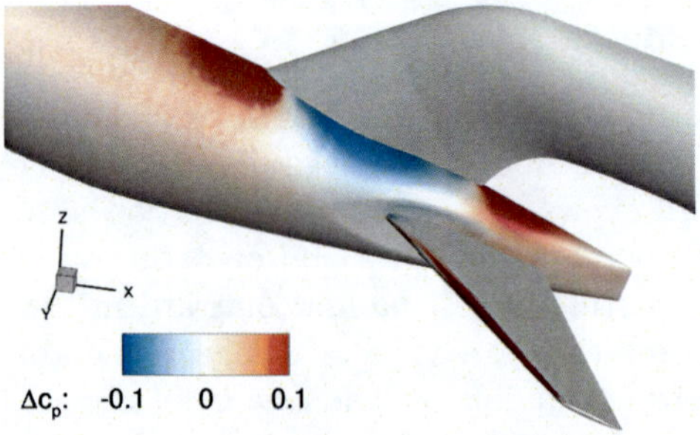

Fig. 3. Pressure difference due to the support system, $\Delta c_p = c_{p,WBT0ssd} - c_{p,WBt0d}$.

below it. This is also indicated by the surface streamlines on the horizontal sting portion in Fig. 2 b). More precisely, the flow displacement due to the sting volume decreases the local flow angle below the sting's spanwise widest point, which includes the flow volume interacting with the tailplane. In the same manner, the flow above the widest point is deflected upwards. ξ is decreased by approximately 2° upstream of the tailplane leading edge in the section shown in Fig. 4. This tendency can be observed over the entire half span in Fig. 5. Except in a small region of positive $\Delta\xi$ at approximately 25% of the halfspan, the tailplane encounters a decrease of the inflow angle. The strongest downward deflection of the flow can be observed near the fuselage surface and, with increasing absolute value, toward the tailplane tip. In the tip region, ξ is decreased by approximately 2° in a large area above the tailplane. From this follows a decrease in the local angle of attack throughout the tailplane halfspan and thus a lower lift coefficient. The flow over the tailplane is attached in both cases except for a small separation bubble near the leading edge in the tip region. This small separation has no significant impact on the tailplane loads. These loads as well as the overall lift and the airplane's pitching moment are examined in Sect. 3.3.

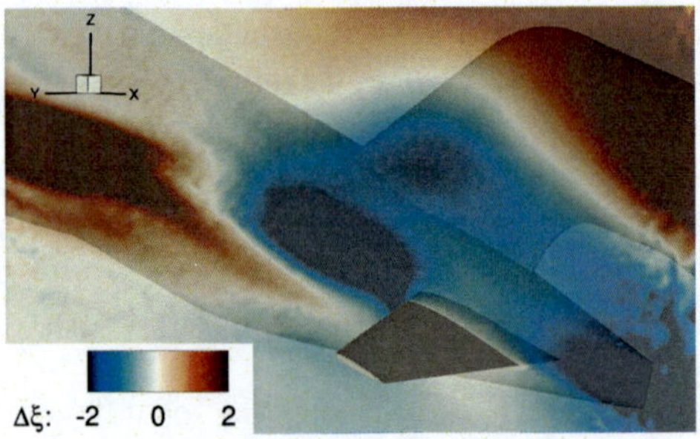

Fig. 4. Increment of the local flow field angle $\Delta\xi = \xi_{WBT0ssd} - \xi_{WBTd}$ at $\eta = 10.7\%$.

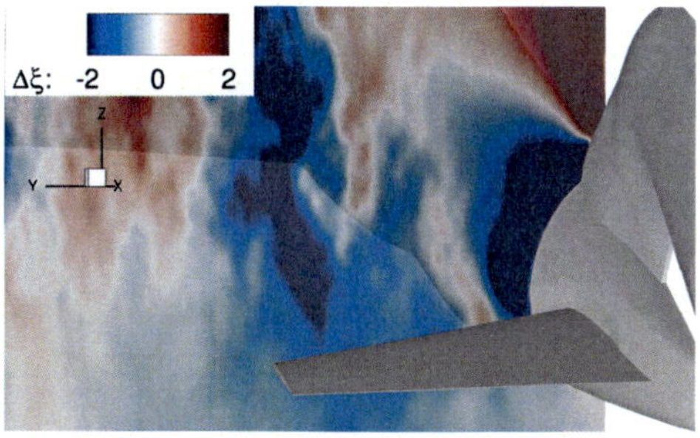

Fig. 5. Increment of the local flow field angle $\Delta\xi = \xi_{WBT0ssd} - \xi_{WBT0d}$ parallel to the htp leading edge.

3.2 Far Field Effects

Figure 6 compares the wake sizes of the two configurations using isolines of the mean streamwise velocity $u/u_\infty = 0.9$. Three planes of constant x/c_{ref} are considered, starting near the trailing edge at $x/c_{ref} = 6$ up to $x/c_{ref} = 9$. The location of these slices is indicated in the bottom center of Fig. 6. The wake extent at $x/c_{\text{ref}} = 6$ in the vicinity of the inboard wing trailing edge shows only minor differences between the two configurations. When progressing to the two downstream planes, the influence of the sting shifts the upper edge of the wake upwards for spanwise positions between $\eta = 0$ and 0.2. In the middle streamwise slice at $x/c_{\text{ref}} = 7.5$, the lower edge of the wake is shifted downwards in the WBT0ssd configuration. This effect is visible up to a spanwise coordinate of $\eta = 0.45$. The most downstream wake plane at $x/c_{\text{ref}} = 9$ shows an increased downward shift of the lower limit of the wake from $\eta = 0$ to $\eta = 0.45$ in comparison to the medium wake position. Further outboard the differences in the wake diminish in all regarded planes. This trend of increasing wake size, especially by shifting the lower wake boundary, is in agreement with the aforementioned far field influence of the sting. While the upper wake edge shows no displacement of the wake except near the sting, the downward deflection of the flow by the wind tunnel support system affects the lower wake edge, which is shifted downwards as the downstream position increases. Effectively, the sting's presence slightly slows down the flow in front of it, thereby widening the wake. The sting's location shifts the overall wake slightly downward. This effect vanishes upstream toward the wing and with increasing spanwise position, i.e. the far field effect is of rather limited spatial extent. Consequently, no such effect can be detected in the vicinity of the wing. The flow separation on the wing and its lift therefore do not show any major influence from the support system.

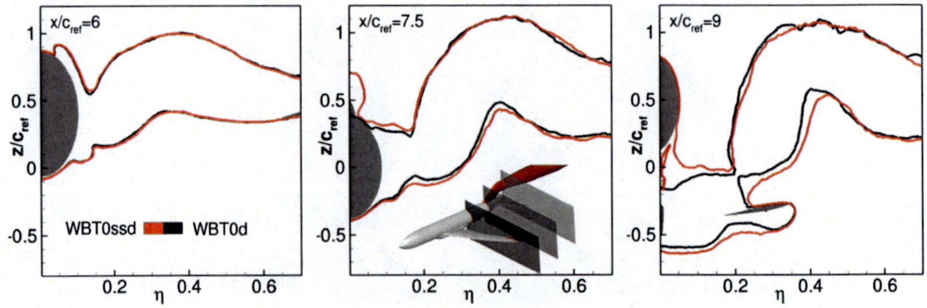

Fig. 6. Wake extent in streamwise planes at $x/c_{\text{ref}} = 6, 7.5, 9$ with lines of $\overline{u}/u_\infty = 0.9$. Fuselage, wing, tailplane and sting in grey.

3.3 Forces and Moments

The local downward deflection caused by the sting in the tailplane's vicinity is associated with decreased lift, as a consequence of the decreased angle of attack of the tailplane. The distributions of C_L over the tailplane span are shown in Fig. 7. The absolute difference between the two cases increases towards the horizontal tailplane's tip. At the same time, the variance of the tailplane lift is consistently higher in the presence of the sting. This results from the downward deflection of the turbulent wake depicted in Fig. 5. As a result, more of the incoming turbulent structures flow toward the tailplane and impinge load. Even though the mean load is decreased by the sting, the fluctuations and therefore the dynamic tailplane loads grow. This is consistent with the observation that load fluctuations are not due to flow separation on the tailplane, but result exclusively from inflow turbulence.

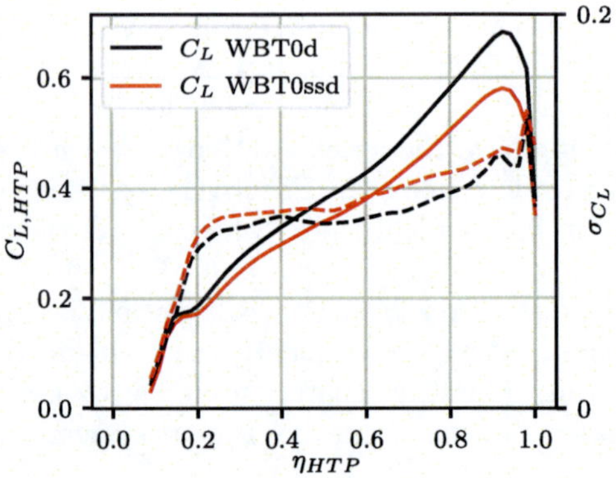

Fig. 7. Spanwise distributions of lift coefficient $c_l(\eta)$ on the horizontal tailplane. Dashed lines denote the variance, corresponding to the right-hand axis.

Time-averaged forces and moments of the two configurations under consideration as well as the experimental values are shown in Table 2. The experimental data does not include forces separated into wing and tailplane contributions, and the total aircraft lift coefficients differ only little. While the WBT0d configuration overpredicts the overall lift C_L, it is underpredicted by the WBT0ssd configuration. The differences in $C_{L,wing}$ are of a comparably small magnitude. The more pronounced difference in $C_{L,HTP}$ cannot be evaluated in terms of accuracy due to lack of experimental data, it is however possible to use the pitching moment C_{My} in its stead. While the WBT0d configuration overpredicts the nose down pitching moment C_{My} by 25%, the WBT0ssd underpredicts it by about 16%. This indicates that slightly better agreement with the experimental values of the pitching moment can be achieved when considering the wind tunnel support system. However, it also indicates that the pitching moment is sensitive to the CFD modeling setup. The drag coefficient C_D decreases slightly when the support system is present, which can be attributed to the decreased lift forces acting on both the wing and the tailplane.

Table 2. Time-averaged forces and moments at $\alpha = 18°$ with and without sting.

	C_L	$C_{L,wing}$	$C_{L,HTP}$	C_D	C_{My}
Experiment	1.078	n/a	n/a	0.3276	−0.159
WBT0d	1.09	0.785	0.321	0.336	−0.199
WBT0ssd	1.07	0.775	0.286	0.331	−0.132

4 Conclusions

The above results and discussion showcase and quantify the influence of the wind tunnel support system on the wake of the CRM in post stall at $\alpha = 18°$. In order to evaluate possible errors when neglecting the wind tunnel mounting in numerical simulations, its effect on the surface pressure coefficient, on the flow in the vicinity of the tailplane and on the entire separated wake emanating from the wing were investigated. Furthermore, the effect of the support system on the forces and moments acting at the airplane as well as the comparison to wind tunnel data is part of this investigation.

The impact in the near field of the wind tunnel support system is due to the displacement effect of the sting which reduces the angle of inflow in the entire tailplane region and alters the pressure distribution on the fuselage surface. This near field effect leads to a decrease in tailplane lift over the entire span. A far field effect of the sting upstream toward the wing cannot be observed as the wake size in the wing region and the wing's lift remains largely unchanged when comparing both configurations. Further downstream, the lower edge of the wake is shifted downwards due to the model mounting, the overall wake extent in vertical direction is increased. This brings about an increase of the tailplane's lift fluctuations due to stronger impingement of the propagated wake turbulence onto its surface. In terms of the overall forces and moments, taking into account the wind tunnel support system leads to a slightly better accordance of all values with the experiment. The overall lift shows a slight decrease when considering the wind tunnel support system, which also leads to a decrease of the overall drag. Furthermore, the decrease of the tailplane lift leads to a lower nose down pitching moment. These small improvements come at the expense of the meshing effort and increase the simulation time due to the rising number of grid points. However, if the flow field near the aircraft's tail is to be investigated, a consideration of the wind tunnel mounting system is recommended because of the sting's deflection of the flow in this area of the wake. A more detailed study on the changing interaction of wake turbulence with the tailplane boundary layer is part of future research.

Acknowledgements. The authors gratefully acknowledge the Deutsche Forschungsgemeinschaft DFG (German Research Foundation) for funding this research in the project LU 809/8. Furthermore, we would like to thank DLR for providing the TAU source code. The computational resources for the simulations were kindly provided by the High Performance Computing Center Stuttgart (HLRS).

References

1. Britcher, C.P., Alcorn, C.W., Kilgore, W.A.: Subsonic sting interference on the drag of a family of slanted-base ogive-cylinders, AIAA 89-2006-CP (1989)
2. Deutsches Zentrum für Luft- und Raumfahrt. DLR TAU-code user guide (2018)
3. Gorbushin, A., et al.: Slotted wall interference investigation in ETW using the NASA CRM model. AIAA 2015-0621 (2015)
4. Kohzai, M., Ueno, M., Koga, S., Sudani, N.: Wall and support interference corrections of NASA common research model wind tunnel tests in JAXA. AIAA 2013-0963 (2013)
5. Kok, J.C.: A high-order low-dispersion symmetry-preserving finite-volume method for compressible flow on curvilinear grids. J. Comput. Phys. **228**(18), 6811–6832 (2009)
6. König, B., Fares, E.: Validation of a transonic Lattice-Boltzmann method on the NASA common research model. In: 54th AIAA Aerospace Sciences Meeting, San Diego, California, USA, AIAA 2016-2023, January 2016
7. Lutz, T., Gansel, P.P., Waldmann, A., Zimmermann, D.-M., Schulte am Hülse, S.A.: Time-resolved prediction and measurement of the wake past the CRM at high reynolds number stall conditions. J. Aircraft **53**(2), 501–514 (2016)
8. Rivers, M., Hunter, C., Campbell, R.: Further investigation of the support system effects and wing twist on the NASA common research model. AIAA 2012-3209 (2012)
9. Schwamborn, D., Gerhold, T., Heinrich, R.: The DLR TAU-Code, recent applications in research and industry. In: European Conference on Computational Fluid Dynamics ECCOMAS CFD 2006, September 2006
10. Spalart, P.R., Allmaras, S.R.: A one-equation turbulence model for aerodynamic flows. Recherche Aerospatiale **1**, 5–21 (1994)

11. Spalart, P.R., Deck, S., Shur, M.L., Squires, K.D., Strelets, M., Travin, A.: A new version of detached-eddy simulation, resistant to ambiguous grid densities. Theoret. comput. Fluid Dyn. **20**(3), 181 (2006)
12. Spalart, P.R., Streett, C.: Young-person's guide to detached-eddy simulation grids. Technical report, NASA CR-2001-211032 (2001)
13. Stojanowski, M., Germain, E.: The falcon 7x from ETW to flight. AIAA 2008-835 (2008)
14. Vassberg, J.C., DeHaan, M.A., Rivers, M.B., Wahls, R.A.: Development of a common research model for applied CFD validation studies. In: 26th AIAA Applied Aerodynamics Conference, August 2008
15. Waldmann, A., Gansel, P.P., Lutz, T., Krämer, E.: Unsteady wake of the NASA common research model in low-speed stall. J. Aircraft **53**(4), 1073–1086 (2016)
16. Waldmann, A., Lutz, T., Krämer, E.: Wind tunnel support system influence on NASA common research model at low speed condition. J. Aircraft **55**(5), 1762–1772 (2018)
17. Zimmermann, D.-M., Waldmann, A., Lutz, T., Krämer, E.: Development of flow structures in the near-field wake region of the common research model. CEAS Aeronautical J. (2016)

Assessment of the Disturbance Velocity Approach to Determine the Gust Impact on Airfoils in Transonic Flow

Jens Müller(✉), Marco Hillebrand, and Thorsten Lutz

Faculty 6: Aerospace Engineering and Geodesy, Institute of Aerodynamics and Gas Dynamics, University of Stuttgart, Pfaffenwaldring 21, 70569 Stuttgart, Germany
jens.mueller@iag.uni-stuttgart.de

Abstract. Unsteady CFD simulations of vertical "1-cos" gusts interacting with the DLR-F15 airfoil at transonic flow conditions are analyzed. The scope of the work is to assess the applicability of the simplified Disturbance Velocity Approach (DVA) for gust simulation, which is implemented in the CFD code TAU. While the DVA covers the interaction of the gust with the airfoil, it neglects the effect of the airfoil and the flow around the airfoil on the gust. Simulations where the gust and all interactions are resolved within the flow field are used to evaluate the accuracy of the DVA. Gust wavelength and angle of attack are varied to ensure a reliable assessment of the DVA. Lift, pitching moment, and shock position can be accurately determined by the DVA. At higher angles of attack and a small gust wavelength, the DVA's drag history shows deviations which should be taken into account. Overall, the DVA provides suitable results with respect to the resolved gust simulations.

Keywords: Aircraft aerodynamics · CFD · Gust loads

1 Introduction

In cruise flight aircraft encounter atmospheric disturbances like gusts which directly affect passenger comfort and structural loads. In order to diminish the negative impact of gusts on the aircraft, the development of gust load alleviation concepts becomes important. This requires accurate and industrially applicable simulation methods to determine the effect of atmospheric gusts on aircraft aerodynamics and structural loads. The accuracy of methods based on potential flow theory, like the unsteady vortex lattice or the doublet-lattice method (DLM) is limited, especially in cruise flight, since they neglect transonic and viscous effects [1]. A transonic correction can be applied to the DLM using specific weighting factors or results based on wind tunnel tests or CFD simulations [2]. Complete coverage of the effects mentioned above is possible when utilizing CFD, which makes it a suitable approach for the analysis of gust interaction. The highest accuracy when simulating gust interaction with CFD is achieved when the gust is resolved and propagated within the flow field. This method is called the

© The Author(s), under exclusive license to Springer Nature Switzerland AG 2021
A. Dillmann et al. (Eds.): STAB/DGLR Symposium 2020, NNFM 151, pp. 14–23, 2021.
https://doi.org/10.1007/978-3-030-79561-0_2

Resolved Atmosphere Approach (RAA) [3]. It covers both the impact of the gust on the aircraft as well as the impact of the aircraft on the gust [4]. The main drawback of this method is the high computational effort due to the fine mesh resolution needed to reduce numerical dissipation and dispersion effects during the propagation of the gust. This limits the industrial application. Therefore, simplified methods for gust interaction simulation in CFD like the split velocity method [5] or the Disturbance Velocity Approach (DVA) [4] were developed where the latter is used within this work. The DVA enables the usage of standard CFD grids without any special refinement, but covers only the influence of the gust on the aircraft. Different to the RAA, the effect of the aircraft on the gust is neglected within the DVA. Hence, the question arises to what extend the DVA is applicable for gust interaction simulation.

Both RAA and DVA were implemented into the flow solver TAU [6] by Heinrich and Reimer [4]. They provide a first assessment of the accuracy of the DVA compared to the RAA by analyzing the lift history of two consecutive NACA 0012 airfoils and an aircraft configuration encountering vertical "1-cos" gusts of different wavelengths at sub- and transonic speeds [3,4]. For subsonic flow conditions these investigations were extended in [7] where vertical "1-cos" gusts interact with a single airfoil. Gust wavelength, angle of attack, and airfoil shape were varied. These studies show that for gust wavelengths larger than the chord length the DVA provides suitable results. At subsonic speeds the pitching moment turned out to be the most critical parameter regarding the accuracy of the DVA. The DVA is also applicable to airfoils or wings interacting with broadband atmospheric turbulence at subsonic speeds, as shown in [8,9]. The scope of the work at hand is to extend the investigations in [3,4,7] to a detailed assessment of the applicability of the DVA to supercritical airfoils at transonic speeds. Compared to the NACA 0012 airfoil investigated in [4], the DLR-F15 airfoil [10] used within this work is representative of supercritical airfoils designed for transonic flow conditions. The utilization of an airfoil compared to a full aircraft configuration enables the assessment of specific effects occurring at a wing section which influence the accuracy of the DVA. Lift, drag, and pitching moment history resulting from CFD simulations of the DLR-F15 airfoil encountering a vertical "1-cos" gust are evaluated. For different gust wavelengths and angles of attack the results of the DVA are compared to the RAA results in order to assess the applicability of the DVA.

2 Numerical Methods and Setup

The numerical approach, as well as mesh generation, follows Müller et al. [7]. Further details can be found there. The finite volume Navier-Stokes solver TAU [6] is applied for the CFD simulations within this work. The solver developed by the German Aerospace Center (DLR) provides a second order accuracy in time and space. In order to represent atmospheric gusts in CFD simulation the Resolved Atmosphere Approach and the Disturbance Velocity Approach are used. A detailed description of RAA and DVA can be found in [3,4]. The RAA represents the physically accurate reference for gust interaction simulation in CFD.

Here, modified unsteady far field boundary conditions are used to feed the gust into the flow field. At these boundary conditions the far field velocity at each cell is dependent on time and space, with the disturbance velocities $\mathbf{u}'$ of the gust added to the mean velocity of the flow [4]. After being fed into the flow field the gust is fully resolved and propagated within the flow field.

Within the DVA the time dependent disturbance velocities $\mathbf{u}'$ are treated similar to grid velocities. At each cell $\mathbf{u}'$ is added to the flux balance by superposition. With the convection across the cell interfaces changed from $\mathbf{u} - \mathbf{u}_b$ to $\mathbf{u} - \mathbf{u}_b - \mathbf{u}'$, where $\mathbf{u}_b$ is the velocity of the boundary of a control volume [4], the continuity equation changes to

$$\frac{d}{dt}\int_V \rho \, dV + \oint_S \rho(\mathbf{u} - \mathbf{u}_b - \mathbf{u}') \cdot \mathbf{n} \, dS = 0. \tag{1}$$

Within this work the supercritical DLR-F15 airfoil [10] interacting with a vertical "1-cos" gust is evaluated. The flow conditions correspond to transonic cruise flight at 35,000 feet with Mach number $M = 0.74$, angle of attack $\alpha = 1.1°$, and Reynolds number $Re = 20.2 \cdot 10^6$ with respect to the chord length c of the airfoil. The amplitude of the vertical "1-cos" gust is set to $A_{gust} = 10\text{m/s}$. The resulting vertical gust velocity w, as a function of the gust wavelength λ, is defined according to the EASA certification specification CS 25.341 as

$$w(x) = \frac{A_{gust}}{2}\left[1 - \cos\left(\frac{2\pi x}{\lambda}\right)\right] \quad \text{with} \quad 0 \leq x \leq \lambda. \tag{2}$$

In order to evaluate the accuracy of the DVA at transonic cruise flight conditions two parameters are varied within this work. First, the gust wavelength λ is varied at a constant angle of attack of $\alpha = 1.1°$. λ is set to $\lambda = c$, $\lambda = 2c$, and $\lambda = 4c$. Second, the angle of attack is changed to $\alpha = 0°$ and $\alpha = 2.2°$ with a constant gust wavelength $\lambda = c$. At all cases unsteady RANS simulations are performed. The Spalart-Allmaras turbulence model is applied. Within the dual time-stepping approach of TAU, 250 inner iterations per time step are used with a physical time step size of $\Delta t/t_{ref} = 0.005$. The reference time $t_{ref} = c/u_\infty$ is the time a gust needs to travel a distance of one chord length. u_∞ denotes the freestream velocity. Similarly to [7], the Chimera overset grid method is used to represent different angles of attack within the RAA and DVA simulations. This results in a background grid and a separate airfoil grid. The airfoil grid is fully structured and identical for both RAA and DVA simulation in order to avoid any numerical differences. Due to the transonic flow regime a shock is present on the upper side of the airfoil. To ensure the correct resolution of the shock the number of points on the airfoil surface is increased in this region. The chordwise position of the refined shock region is adapted to the local shock position for each individual angle of attack. It is also ensured that the shift in the shock position due to the gust interaction is covered by the refined region. The boundary layer is resolved and $y^+ < 1$ is ensured.

The resulting airfoil grid is shown in blue in Fig. 1.

The background grid differs for RAA and DVA simulations. Both background grids are similar to the grids used in [7] for the subsonic investigations. As mentioned above, the RAA requires a fine mesh resolution in the propagation region of the gust. In order to reduce the numerical losses during the propagation of the gust, an overset gust transport grid is used according to the specifications in [4] with $dx = dz = 0.01c$. This is similar to the subsonic investigations in [7]. The width of the transport grid is set to 2λ in streamwise direction. The background grid of the RAA simulations is fully structured in the gust propagation region to avoid interpolation errors between gust transport and background grid. Further downstream from the airfoil the grid becomes unstructured to reduce the total number of grid cells, see also [7]. At the start of the simulation the gust transport grid is located at the most upstream position where the gust is fed into the flow field. The initial position of the transport grid is shown in yellow in Fig. 1. When the gust is located in the center of the gust transport grid this grid starts to move with the freestream velocity u_∞. Since there is no relative movement between gust and transport grid, no numerical dissipation and dispersion occur during this propagation. The gust transport grid is stopped close to the airfoil. Its final position is shown by the dashed yellow lines in Fig. 1. The gust then leaves the transport grid and interacts with the airfoil. In contrast to the RAA, the DVA background grid is mainly unstructured with a small structured region in the vicinity of the airfoil. This ensures that there are no differences in the Chimera interpolation between airfoil grid and background grid compared to the RAA. Since the DVA does not require a fine resolution in the background grid, the grid resolution is coarsened with increasing distance from the airfoil. More details regarding RAA transport grid as well as the background grid of RAA and DVA can be found in [7]. The benefit of the DVA becomes visible when comparing the total grid sizes for both approaches. While the RAA grid consists of about 50 million points, the DVA reduces the necessary grid size to 0.2 million points.

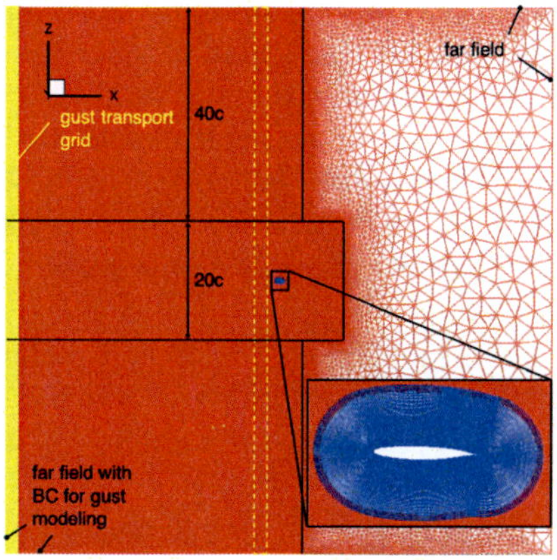

Fig. 1. RAA grid setup with background grid (red), airfoil grid (blue), and gust transport grid (yellow). The dashed yellow line represents the final position of the gust transport grid. Picture adapted from [7].

3 Results of the Gust Wavelength Variation

As described in Sect. 2, the flow regime is transonic at $M = 0.74$ and $\alpha = 1.1°$. The selected gust wavelengths $\lambda/c = 1$, $\lambda/c = 2$, and $\lambda/c = 4$ are similar to the ones used for the subsonic single airfoil investigations in [7] and the sub- and transonic lift evaluation using two consecutive NACA 0012 airfoils in [4]. These previous investigations revealed that gust wavelengths in the order of the chord length lead to noticeable deviations between RAA and DVA results. Hence, the selected wavelengths enable a good assessment of the applicability of the DVA for the current investigations and allow a good comparison to the results of the previous investigations. Due to the transonic flow regime a shock is present at 61.4%c on the upper side of the supercritical DLR-F15 airfoil in steady flow conditions. The lift (c_l), drag (c_d) and pitching moment coefficient (c_m) history of the airfoil interacting with the different vertical "1-cos" gusts is shown in Fig. 2 for RAA and DVA simulations. c_m is evaluated around the quarter chord of the airfoil. While the gust wavelength varies, the gust amplitude is kept constant at $A_{gust} = 10\text{m/s}$. The reference time $t_{ref} = c/u_\infty$ the gust needs to travel a distance of one chord length is used to normalize the time axis. At $t/t_{ref} = 0$ the front of the gust is located at the leading edge of the airfoil. In line with the results presented in [7], the differences between RAA and DVA at selected points are shown in Table 1 with

$$err_{c_{l,max}} = \frac{|c_{l,max,RAA} - c_{l,max,DVA}|}{|c_{l,max,RAA}|} \quad \text{and} \quad \Delta c_i = c_{i,RAA} - c_{i,DVA}. \tag{3}$$

Comparing the c_l history of RAA and DVA in Fig. 2a, it can be seen that the shape of the lift coefficient curve can be adequately reproduced by the DVA. Only at the point of maximum lift $c_{l,max}$ the DVA results show lower $c_{l,max}$ values than the RAA, especially for $\lambda/c = 1$. While the relative DVA error of 1.69% at the smallest gust wavelength should not be neglected, it decreases with increasing gust wavelength to 0.32% for $\lambda/c = 4$. The relative error at $c_{l,max}$ agrees with the subsonic investigations of Müller et al. in [7] where the

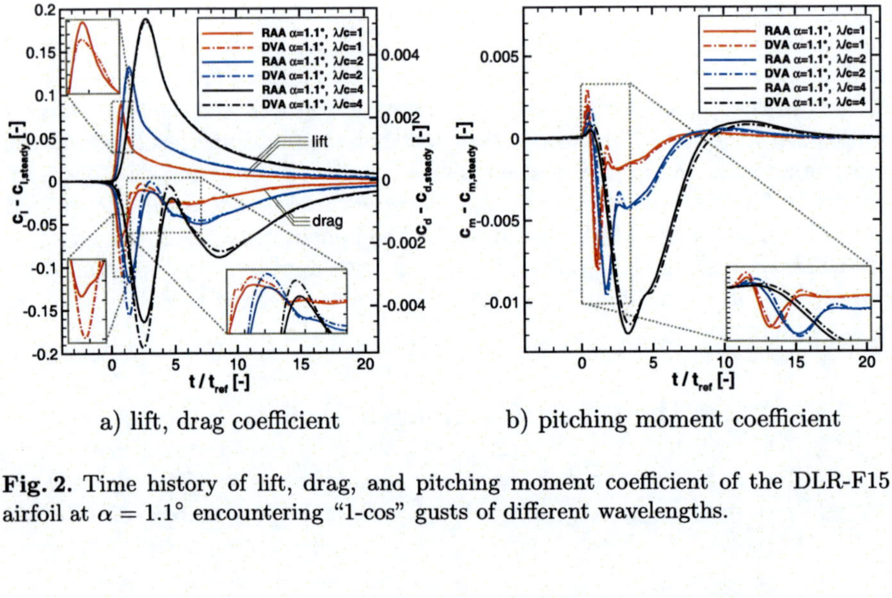

a) lift, drag coefficient b) pitching moment coefficient

Fig. 2. Time history of lift, drag, and pitching moment coefficient of the DLR-F15 airfoil at $\alpha = 1.1°$ encountering "1-cos" gusts of different wavelengths.

Table 1. Deviations between the RAA and DVA results at the extreme points of the global load response for different gust wavelengths.

	$\Delta c_{l,max}$	$\mathrm{err}_{c_{l,max}}$	$\Delta c_{d,min}$	$\Delta c_{m,min}$	$\frac{\Delta t_{c_{l,max}}}{t_{ref}}$	$\frac{\Delta t_{c_{d,min}}}{t_{ref}}$	$\frac{\Delta t_{c_{m,min}}}{t_{ref}}$
$\lambda/c = 1$	0.01299	1.69%	0.001051	0.000243	0.020	−0.125	−0.130
$\lambda/c = 2$	0.00513	0.63%	0.001062	0.000511	−0.025	0.030	−0.040
$\lambda/c = 4$	0.00274	0.32%	0.000769	−0.000512	0.020	0.060	−0.110

$err_{c_{l,max}}$ is in the same order of magnitude. The c_p distribution at $c_{l,max}$ reveals that for $\lambda/c = 1$ the absolute c_p value of the RAA directly upstream from the shock position is higher than the one of the DVA. This difference decreases with increasing λ reducing the lift difference between RAA and DVA. The main reason for the differences between RAA and DVA is that the RAA covers the changes of the gust due to the interaction with the airfoil, whereas the DVA neglects this influence. Caused by the flow around the airfoil, the gust is accelerated above and decelerated below the airfoil, leading to an offset in streamwise gust position compared to the analytical gust position used by the DVA as illustrated in [7]. Within the DVA the gust always travels with u_∞ in streamwise direction. In the vicinity of the airfoil the gust amplitude is reduced in the RAA simulation, which is also not covered by the DVA. With increasing gust wavelength the changes in the gust shape are less important for the gust interaction. The area of influence of the gust increases and the velocity gradients of the gust decrease [7]. This leads to a better agreement between RAA and DVA. Compared to the subsonic investigations of Müller et al. in [7], the relative DVA error at $c_{l,max}$ is in the same order of magnitude. In contrast, the transonic simulations with two consecutive NACA 0012 airfoils in [4] lead to $err_{c_{l,max}} = 11.08\%$ for $\lambda/c = 1$ which is significantly higher than the results presented here. The main reason is most likely the supercritical DLR-F15 airfoil used within this work, whereas the NACA 0012 airfoil in [4] is not suited for transonic flow conditions. The additional oscillation in the DVA c_l course for $\lambda/c = 1$, when the gust is located at the rear part of the airfoil, present in the subsonic investigations in [7] as well as in the sub- and transonic investigations in [4] does not occur in the results presented here leading to a better agreement in lift between RAA and DVA for the DLR-F15 airfoil.

Different to the subsonic investigations, the time history of the drag coefficient shows noticeable differences compared to the time history of the lift coefficient. It has to be mentioned that the coordinate system for lift and drag evaluation is aligned with the inflow velocity u_∞ and the global angle of attack. Hence, it is not adjusted with the local angle of attack which changes during gust interaction. The differences between RAA and DVA at $c_{d,min}$ presented in Table 1 are in the same order as for the subsonic investigations in [7]. The difference in the RAA gust position above and below the airfoil described above leads to an earlier occurrence of $c_{d,min}$ for $\lambda/c = 1$. In transonic flow conditions the wave drag induced by the shock has a major contribution to the total drag. This leads to a local drag maximum between $t/t_{ref} = 2.4$ and $t/t_{ref} = 4.7$ not present in the subsonic gust interaction studies in [7]. Different to $c_{l,max}$ and $c_{d,min}$, the difference of the local drag maximum between RAA and DVA slightly increases with increasing λ/c. The higher λ/c, the stronger the shock on the upper side of the airfoil at the local drag maximum and the closer the shock moves towards the trailing edge compared to the steady solution.

The pitching moment evaluated around the quarter chord of the airfoil shown in Fig. 2b reveals that the gust interaction leads to a slight increase of the pitching moment compared to the steady pitching moment when the gust starts to interact with the airfoil. At $c_{m,max}$, the center of the gust is still located in front of the airfoil. The suction peak is more pronounced compared to the steady solution while the c_p distribution in the vicinity of the shock, the strength of the shock, and the shock position are not affected. The increase in the suction peak is reduced for higher λ/c which leads to a reduction of this pitching moment extreme point as visible in Fig. 2b. The DVA suction peak is higher than the one of the RAA which leads to a higher $c_{m,max}$. $c_{m,max}$ is followed by a significant decrease of the pitching moment relative to the steady reference at all wavelengths for both RAA and DVA results. At $c_{m,min}$ for $\lambda/c = 1$ the gust increases the absolute c_p values on the upper side of the airfoil upstream from the shock. Downstream from the shock the c_p distribution is nearly unaffected. The shock position is also comparable to the steady solution. On the lower side of the airfoil the surface pressure is increased which increases the pressure difference between upper and lower side of the airfoil compared to the steady solution, especially upstream from the shock. This leads to a global minimum of c_m. The same effects at $c_{m,min}$ occur for $\lambda/c = 2$ but the pressure difference is further increased. At $\lambda/c = 4$ the position of the shock is additionally shifted towards the trailing edge combined with a further increase of the pressure difference between upper and lower surface. At all wavelenghts, a time shift is present between RAA and DVA results. As illustrated in [7], the pitching moment is prone to differences in the gust position above and below the airfoil covered by the RAA but neglected by the DVA. Due to the interaction with the airfoil the gust velocity in the vicinity of the airfoil is decreased as well for the RAA simulation as shown in [7]. This is also not covered by the DVA leading to the more pronounced c_m turning points especially at small wavelengths in the DVA results. Compared to the subsonic investigations in [7] there is a better agreement between the RAA and DVA pitching moment for $\lambda/c = 1$.

4 Results of the Angle of Attack Variation

The results presented in Sect. 3 show that also for the transonic flow regime, the smallest wavelength $\lambda/c = 1$ is the most critical regarding the accuracy of the DVA compared to the RAA. Even at transonic speeds, an aircraft does not always operate at design conditions. Hence, angle of attack variations of the DLR-F15 airfoil encountering a vertical "1-cos" gust with $\lambda/c = 1$ are considered and its effect on the agreement of the DVA to the RAA is evaluated. The investigations in [7] show that for subsonic flight conditions high angles of attack decrease the agreement of drag and pitching moment history between RAA and DVA. In addition to $\alpha = 1.1°$ presented in Sect. 3, $\alpha = 0°$ and $\alpha = 2.2°$ are considered. At $\alpha = 0°$ without the gust present, the shock is relatively weak and is located at 59.3%c. With increasing α the strength of the shock increases and the position of the shock is shifted towards the trailing edge being at 61.4%c at $\alpha = 1.1°$ and at 64.1%c at $\alpha = 2.2°$. At $\alpha = 2.2°$ a small shock induced separation occurs covering a distance of 2.36%c in the steady reference case. The RAA and DVA results of the gust interaction at different angles of attack are shown in Fig. 3. The differences between RAA and DVA at the global extreme points are listed in Table 2. The c_l history in Fig. 3a shows that the only effect of the α variation on the agreement between RAA and DVA occurs at $c_{l,max}$. Here, the accuracy of the DVA marginally increases with increasing α. The rest of the c_l history shows a very good agreement between RAA and DVA during the entire gust interaction. The impact of the α variation on the accuracy of the DVA's c_m history is also relatively small. The deviations at the extreme points are in the same order of magnitude at all considered angles of attack. With increasing α the difference in the RAA's gust position above and below the airfoil increases due to the increased circulation around the airfoil. Accordingly, the RAA's gust amplitude in the vicinity of the airfoil reduces the higher α. These effects are not covered by the DVA. These differences in the gust shape variations due to different angles of attack seem to be only of minor influence on the accuracy of the DVA's pitching moment. The time difference between RAA and DVA due to the difference in the RAA's gust position above and below the airfoil is also not affected by the different angles of attack. This leads to a consistent agreement of the approaches at all angles of attack considered. A different situation occurs when looking at

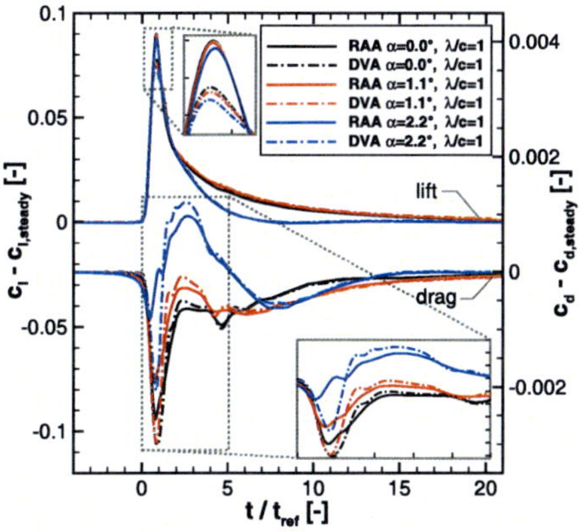

a) lift, drag coefficient

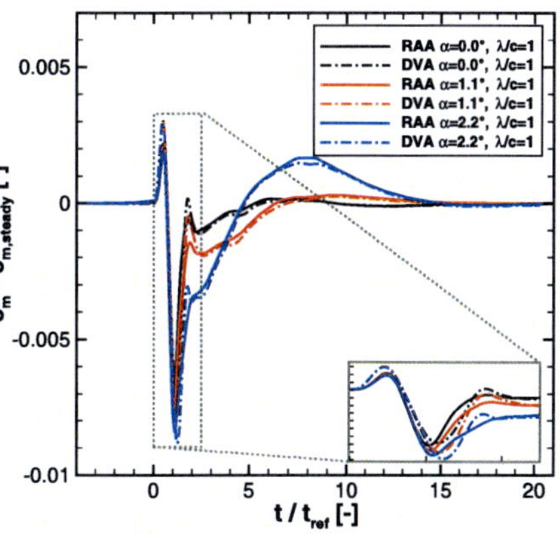

b) pitching moment coefficient

Fig. 3. Time history of lift, drag, and pitching moment coefficient of the DLR-F15 airfoil at $\alpha = 0°$, $\alpha = 1.1°$, and $\alpha = 2.2°$ encountering "1-cos" gusts with $\lambda/c = 1$.

Table 2. Deviations between the RAA and DVA results at the extreme points of the global load response for different angles of attack.

	$\Delta c_{l,max}$	$\mathrm{err}_{c_{l,max}}$	$\Delta c_{d,min}$	$\Delta c_{m,min}$	$\frac{\Delta t_{c_{l,max}}}{t_{ref}}$	$\frac{\Delta t_{c_{d,min}}}{t_{ref}}$	$\frac{\Delta t_{c_{m,min}}}{t_{ref}}$
$\alpha = 0.0°$	0.01226	2.27%	0.000455	−0.000336	0.025	−0.110	−0.120
$\alpha = 1.1°$	0.01299	1.69%	0.001051	0.000243	0.020	−0.125	−0.130
$\alpha = 2.2°$	0.01375	1.41%	0.001219	0.000460	0.045	−0.380	−0.130

the c_d history. Subsonic investigations in [7] show that an increased α mainly influences the DVA's accuracy at the drag minimum. The higher α, the higher the differences between RAA and DVA at this point. This trend is confirmed by the transonic results presented in Fig. 3a and Table 2. But these results also show differences in the shape of the c_d curve between RAA and DVA, which are not present in the subsonic investigations. They mainly occur between the global drag minimum, when the gust starts to interact with the airfoil, and the drag maximum. They diminish shortly after the occurrence of the drag maximum between $t/t_{ref} = 2.3$ and $t/t_{ref} = 4.7$. While the differences between the c_d time history of RAA and DVA are comparable for $\alpha = 0°$ and $\alpha = 1.1°$, the curves at $\alpha = 2.2°$ show larger differences between $c_{d,min}$ and $c_{d,max}$. Here, $c_{d,min}$ of the DVA is considerably more delayed than at smaller α. Due to the higher acceleration of the gust above the airfoil with increasing α, $c_{d,min}$ in the RAA simulations occurs earlier the higher α. At this point, the center of the gust is located in the area of the leading edge. Thus, it mainly influences the c_p distribution upstream from the shock. Here, the DVA's c_p level downstream from the suction peak is increased compared to the RAA, until the shock is reached. This leads to the different shapes of the c_d curve between RAA and DVA close to $c_{d,min}$ at $\alpha = 2.2°$. It is also the case at $c_{d,max}$, where the higher c_p level of the DVA downstream from the suction peak leads to a slight increase in the shock strength. In general, the global c_d maximum at $\alpha = 2.2°$ compared to the only local maxima for the smaller α occurs due to the strongest shock at this α and the significant increase in size of the shock induced separation. While it covers a distance of 2.36%c at steady flow conditions the separation at $c_{d,max}$ covers 7.27%c (RAA) respectively 6.75%c (DVA).

5 Conclusion

CFD simulations of the DLR-F15 airfoil encountering a discrete vertical gust at $M = 0.74$ and $Re = 20.2 \cdot 10^6$ were analyzed. The applicability of the simplified Disturbance Velocity Approach (DVA) for gust representation in CFD was evaluated by comparing it to the more accurate Resolved Atmosphere Approach (RAA). While the DVA neglects all changes in the gust shape due to the interaction with the airfoil, the RAA covers all mutual interactions between gust and airfoil. To enable a reliable assessment of the DVA's capability, gust wavelength and angle of attack were varied. It can be concluded that the DVA is capable of covering the main effects of gust interaction in transonic flow regimes. The lift can be accurately simulated using the DVA at all wavelengths and angles of attack considered within this work. An adequate determination of the pitching moment history and the shock position is also possible. In contrast to the subsonic investigations in [7], the most significant deviations occur in the drag coefficient history due to the complex interaction of the gust with the shock and the shock induced separation at high angles of attack. Especially at high angles of attack, these differences should be considered when aircraft performance is evaluated. Since lift and pitching moment are most important when it comes to the design of gust load alleviation systems, whereas the drag is only of minor importance in this context, the DVA is a suitable method for vertical gust interaction simulation in transonic flow regimes.

Acknowledgements. The authors gratefully acknowledge the Federal Ministry for Economic Affairs and Energy, which funded the work presented in this report as part of the LuFo project VitAM-Turbulence. Also, we acknowledge the High Performance Computing Center Stuttgart (HLRS) for the provision of computational resources.

References

1. Wales, C., Valente, C., Cook, R., Gaitonde, A., Jones, D., Cooper, J.E.: The future of non-linear modelling of aeroelastic gust interaction. In: 2018 Applied Aerodynamics Conference (2018)
2. Thormann, R., Dimitrov, D.: Correction of aerodynamic influence matrices for transonic flow. CEAS Aeronaut. J. **5**(4), 435–446 (2014)
3. Heinrich, R., Reimer, L.: Comparison of different approaches for modeling of atmospheric effects like gusts and wake-vortices in the CFD code TAU. In: International Forum on Aeroelasticity & Structural Dynamics (2017)
4. Heinrich, R., Reimer, L.: Comparison of different approaches for gust modeling in the CFD code TAU. In: International Forum on Aeroelasticity & Structural Dynamics (2013)
5. Wales, C., Jones, D., Gaitonde, A.: Prescribed velocity method for simulation of aerofoil gust responses. J. Aircraft **52**(1), 64–76 (2015)
6. Schwamborn, D., Gerhold, T., Heinrich, R.: The DLR TAU-Code, recent applications in research and industry. In: European Conference on Computational Fluid Dynamics ECCOMAS CFD (2006)
7. Müller, J., Hillebrand, M., Schollenberger, M., Ehrle, M., Lutz, T., Krämer, E.: Studies on the applicability of a simplified gust simulation approach in the CFD Code TAU. In: High Performance Computing in Science and Engineering '20. Springer, Cham (2021, submitted)
8. Müller, J., Ehrle, M., Lutz, T., Krämer, E.: Numerical simulation of the FNG wing section in turbulent inflow. In: High Performance Computing in Science and Engineering '19. Springer, Cham (2021)
9. Müller, J., Lutz, T., Krämer, E.: Numerical simulation of the swept FNG wing in atmospheric turbulence. In: AIAA Aviation 2020 Forum (2020)
10. Wild, J.: Mach and reynolds number dependencies of the stall behavior of high-lift wing-sections. J. Aircraft **50**(4), 1202–1216 (2013)

Check for updates

Comparison of Different Methods for the Extraction of Airfoil Characteristics of Propeller Blades as Input for Propeller Models in CFD

Michael Schollenberger(✉) and Thorsten Lutz

Faculty of Aerospace Engineering and Geodesy,
Institute of Aerodynamics and Gas Dynamics (IAG), University of Stuttgart,
Pfaffenwaldring 21, 70569 Stuttgart, Germany
schollenberger@iag.uni-stuttgart.de
https://www.iag.uni-stuttgart.de/en/

Abstract. Four different methods for the generation of input data for propeller models (Actuator Disk, Actuator Line) in CFD simulations are described and their results are compared with an unsteady fully resolved simulation of the propeller geometry. It is shown that the neglect of the rotation effects with a 2d method overestimates the propeller thrust and the axial velocity in the wake. If the rotation effects are instead considered with a 3d or a new 2.5d approach or via a recovery from the slipstream data, there is better agreement regarding thrust and slipstream velocities. Thereby, the 2.5d method shows a good compromise between computational effort and the quality of the results.

Keywords: Propeller blade aerodynamics · Actuator Disk · Actuator Line

1 Introduction

The CFD simulation of propeller flows is of increasing relevance due to research into new configurations such as distributed propulsion, wingtip mounted propellers, etc., which make use of positive aerodynamic interactions between the propellers and the other components of the aircraft. To reduce calculation time, there are different models to simulate aircraft propellers within CFD instead of fully resolving the flow around the propeller blades. In the DLR TAU-Code, a steady-state Actuator Disk (ACD) method is available, see Raichle [1], and in addition an unsteady Actuator Line (ACL) method has been implemented by the presenting authors [2]. These models have in common that either the airfoil characteristics of the propeller blades in the terms of lift and drag polars (c_l, c_d) or the direct force distributions (f_n, f_t) are required as input data. Since the quality of the models depends on the accuracy of the input data, it is essential that these are represented in a physically correct manner. Himmelskamp found

© The Author(s), under exclusive license to Springer Nature Switzerland AG 2021
A. Dillmann et al. (Eds.): STAB/DGLR Symposium 2020, NNFM 151, pp. 24–34, 2021.
https://doi.org/10.1007/978-3-030-79561-0_3

in experimental investigations that the lift and drag distributions of propeller blades are influenced by rotation effects (centrifugal and Coriolis forces), thus differ from those of airfoils with parallel inflow, see Raichle [1]. If the input data are not known, they can be generated based on CFD or even experimental data. In the research area of wind turbine simulation different methods for extracting airfoil characteristics from CFD data have been developed [3]. Since other inflow conditions exist in the area of aviation propellers due to significantly higher rotational speeds and significantly smaller diameters, the aim of the present paper is to compare and evaluate the suitability of different methods on the example of a two bladed aircraft propeller. Four different methods with different computational effort and degree of simplification are described and their results are compared, a 3d, a 2.5d and a 2d method to determine the polars, as well as a simple force recovery from the slipstream data.

2 Methodology

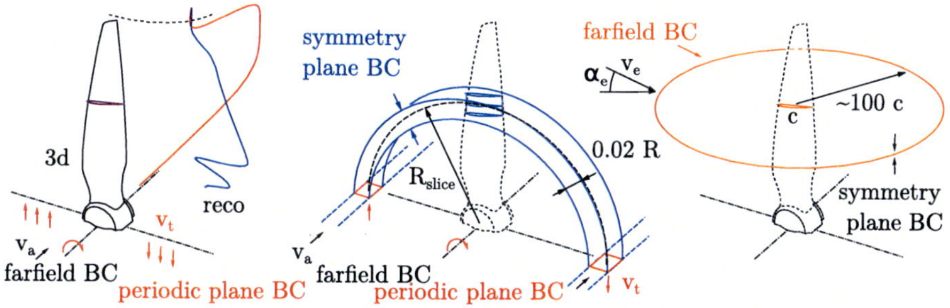

(a) 3d and (d) reco (b) 2.5d (c) 2d

Fig. 1. Propeller blade input data methods

The flow around a propeller blade is characterized by the rotation effect as well as the effect of the blade tip vortex. Different methods are possible to extract polars from simulations of the propeller geometry. The principles of the methods are shown in Fig. 1. The highest level of detail is thus expected with the simulation of a 3d rotating propeller blade, where the rotation and tip effects are both present in the CFD simulation - method (a). To reduce the computational time method (b) was developed, where blade segments are rotated on a cylindrical grid, thus 2.5d, simulating the rotation effects and neglecting the tip effects. A further simplification is the 2d simulation of airfoils, neglecting both effects - method (c). The greatest simplification is achieved by not using the propeller blade geometry at all. Instead, the normal and tangential force distributions are directly recovered from existing slipstream data - method (d).

2.1 Method (a) - 3d Propeller Blade Polars

In order to reduce the number of cells and enable a steady-state calculation, the axial symmetry of the two bladed propeller is utilized by applying periodic boundary conditions (BC). Instead of rotating the entire propeller in a fixed grid, only a one-half grid including one propeller blade is rotated. Periodic BC allow a steady rotational in- and outflow through the computational domain, see Fig. 1a. In case of a three-bladed propeller, a one-third grid would be used instead, etc. The tangential velocity is generated by the rotation of the grid and the axial velocity by the farfield BC. There are two possible ways to vary the angle of attack at the propeller blade, either by changing the global blade pitch β or by changing the inflow conditions. Since in the present study a fixed pitch propeller is analysed, the inflow conditions are varied. To change the inflow conditions there are again two possibilities, either the variation of the propeller rotation ω at constant inflow velocity v_∞ or the variation of the inflow velocity at constant rotation. The latter is used here to consider aerodynamic interactions.

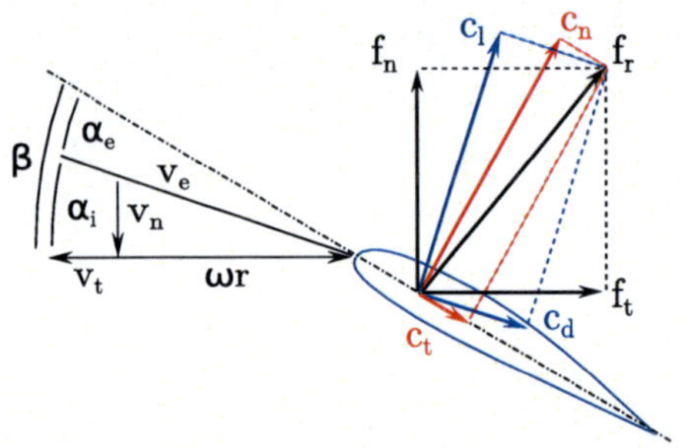

(a) Blade velocity and force triangle

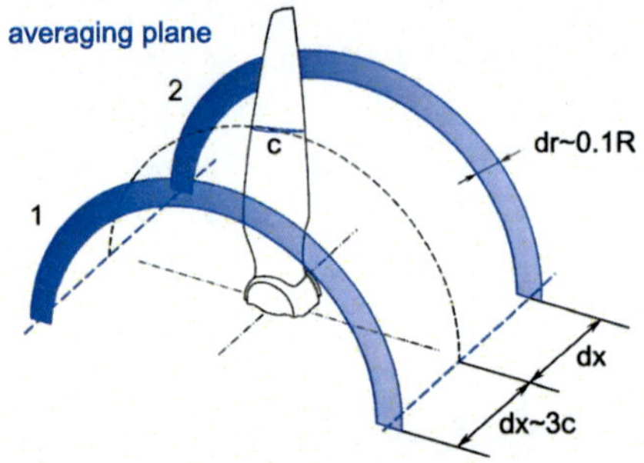

(b) RAV method

Fig. 2. Propeller blade forces and angle of attack

After the simulations, the coefficients c_l and c_d have to be determined for each inflow case at different sections of the propeller blade. They are obtained from the normal and tangential force coefficients c_n, c_t and the effective angle of attack α_e, see Fig. 2a. Thereby the friction drag $c_{d,f}$ has to added to the pressure drag. The normal and tangential force coefficients are calculated from the difference in pressure coefficient c_p between upper and lower side of the respective blade section by integration in normal and tangential direction and the friction drag $c_{d,f}$ by integrating the skin friction coefficient c_f. The local induced angle of attack α_i is extracted from the CFD data using the reduced axial velocity technique (RAV). This method was proposed by Hansen et al. to determine α_i at wind turbine blades from CFD data, taking into account the disturbed axial Δv_a and tangential Δv_t velocity, due to the flow around the blade, see Johansen et al. [3]. For this purpose, two planes are shaped as a circular ring sector in front of and behind the blade section, on which the flow data are averaged, see Fig. 2b. According to Johansen et al. [3], the correct velocity is theoretically obtained by traversing the plane through the propeller disk and can be estimated by linear interpolation between the planes. With α_i and the known blade pitch β, the effective angle of attack α_e of the blade section is given and all variables are known to determine the coefficients:

$$c_l = c_n\ cos(\alpha_e)\ -\ c_t\ sin(\alpha_e), \tag{1}$$

$$c_d = c_n\ sin(\alpha_e)\ +\ c_t\ cos(\alpha_e)\ +\ c_{d,f}, \tag{2}$$

$$\alpha_e = \beta - \alpha_i, \qquad \alpha_i = \arctan\left(-v_n/(v_t - \omega\ r)\right). \tag{3}$$

2.2 Method (b) - 2.5d Blade Element Polars

Since method (a) requires the simulation of the whole propeller blade and a large farfield distance, although coefficients are determined only at certain blade elements, the idea behind method (b) is to reduce the computational effort by means of a well chosen set of further BC, see Fig. 1b. Instead of the entire blade, the individual sections are used to create blade elements by an extrusion of 0.01 R in both radial directions. The blade elements are intersected with a cylindrical grid. Periodic BC are used as well to allow the rotational inflow and outflow through the computational domain. The inner and outer grid boundary in form of cylindrical surfaces are set as symmetrical BC to generate geometrically an infinite blade and prevent a crossflow over the boundary. The symmetrical BC in TAU does not act with a flat symmetry plane, instead it is formed for each cell which enables the cylindric shape. It is therefore analogous to an Euler wall. Since the periodic BC causes the tangential velocity to vary in radial direction over the height $dz = 0.02\ R$, it can be described as a 2.5d approach. The simulation process and the determination of the coefficients is analogous to the 3d method. The angle of attack is also determined by the RAV method, but in contrast to the 3d method, no induced angle of attack is included. According to the authors' knowledge, a similar approach to generate blade polars has not yet been published.

2.3 Method (c) - 2d Airfoil Polars

In the purely 2d calculation, no axial and tangential velocity is varied, since the angle of attack on the non-twisted airfoil ($\beta = 0°$) can be set directly, see Raichle [1]. The simple grid topology is an airfoil surrounded by a circular farfield boundary at which the inflow is specified, see Fig. 1c. For each blade section a Mach and Reynolds number corresponding to the radius is given for $v_e = \sqrt{v_\infty^2 + (\omega\ r)^2}$ via:

$$M(r, v) = \frac{v_e}{\sqrt{\kappa R T}}, \qquad Re = \frac{v_e\ c}{\nu}. \tag{4}$$

Since in the 2d method the rotation effects are not captured, the use of rotation corrections models known from the application of wind turbines is available to correct to the lift coefficients via $c_{l,rot} = c_{l,2d}\ +\ f\ \Delta c_l$. They are based on the difference $\Delta c_l = c_{l,2d} - c_{l,inv}$ to inviscid lift without separations and a factor f. The factor f varies for the different models, for example by Snel, Lindberg and Chaviaropoulos & Hansen, see Bak [7]:

$$f_s = 3 \left(\frac{c}{r}\right)^2, \qquad f_l = 3.1 \left(\frac{c}{r}\right)^2 \left(\frac{\omega\, r}{v_e}\right), \qquad f_{ch} = a\ (c_r)^h\ cos(\alpha_e)^n. \tag{5}$$

2.4 Method (d) - Slipstream Recovered Forces

Method (d) has the approach, to determine the thrust and torque distributions from the slipstream's velocity and pressure data, thus recovering them, see Fig. 1d. This method is especially intended if slipstream data are already available, e.g. from experiments or CFD but no propeller geometry. Thrust and torque distribution can be recovered from the slipstream data by integrating the total pressure and velocity distributions over the slipstream area. This approach was already described by Samuelsson [4], in order to check whether measured slipstream data are reasonable. Samuelsson also provided a way to derive the torque distribution from the slipstream data. Following the actuator disk theory, the thrust T produced by the propeller is equal to the total pressure increase in the slipstream

$$T = \int \Delta p_t\ dA = \int (p_t(r) - p_{t,\infty})\ 2\pi\ r\ dr, \tag{6}$$

and following the conservation of the angular momentum, the toque is given as

$$Q = \int \rho\ u_x\ u_t\ r\ dA = \int \rho\ u_x\ u_t\ 2\pi\ r^2\ dr, \tag{7}$$

assuming an axisymmetrical flow $dA = 2\pi\ r\ dr$. The normal force distribution f_n of the propeller blade is equal to the differential thrust dT/dr and the tangential force distribution f_t is related to the differential torque $dQ/dr = dF_t/dr\ r$:

$$f_n = \frac{dT}{dr} = 2\pi\ r\ \Delta p_t, \qquad f_t = \frac{dF_t}{dr} = \frac{dQ}{r\ dr} = 2\pi\ \rho\ r\ (u_x\ u_t). \tag{8}$$

In the present study the slipstream data are taken from fully resolved simulations. Because the forces were recovered from the position of $x/R = 1.18$, a scaling of the radius has been carried out to compensate for the slipstream contraction.

2.5 Tip Loss Correction

Since the methods (b) and (c) neglect the influence of the tip vortex on the propeller blade, a correction must be applied to take the tip loss effects into account. The first tip loss correction approach was proposed by Prandtl [5], in which the circulation is corrected via a factor F

$$F = \frac{2}{\pi}\ exp(-f). \tag{9}$$

Different correction models, for example from Prandtl, Glauert and Shen [5,6] are present, which are formed using either the local $\lambda_l = v_n/v_t$ or the global $\lambda_g = v_\infty/(\omega\ R)$ advance ratio:

$$f_p = \frac{b}{2}\left(1-\frac{r}{R}\right)\frac{1}{\lambda_g}, \quad f_g = \frac{b}{2}\left(1-\frac{r}{R}\right)\frac{R}{r}\frac{1}{\lambda_l}, \quad f_s = g\,\frac{b}{2}\left(\left(1-\frac{r}{R}\right)\frac{R}{r}\right)^n \frac{1}{\lambda_l} \tag{10}$$

with $g = exp\left(\frac{-c_1(b-c_2)}{(1-2\ \min(dc/dr)}\right) + 0.1$ and $n = 1+1/2\ \min(dc/dr)$. The parameters of the Shen model were calibrated on the present 3d results to $c_1 = 0.05,\ c_2 = 21$.

3 Numerical Setup

The study consists of two parts, the application of the methods (a)-(d) and their validation. For the validation, three propeller settings ($\lambda_g = 0.11,\ 0.19,\ 0.26$) were simulated via an ACD propeller model with the different input data (2d, 2.5d, 3d, reco) as well as via an unsteady fully resolved simulation of the propeller geometry with the rotation realized by the Chimera technique as assumption for correct results. Thereby, the entire propeller with both blades is simulated in contrast to the 3d method. For the different methods three different grid types are required. In addition to a grid for the validations, the study contains the four grid types, sketched in Fig. 3. The half-circular grid of the 3d method, with periodic BC in the symmetry plane and a refined area in the wake, consists of ~ 3 million cells. The entire circle is used for the validation, whereby the propeller is either fully resolved and rotated by chimera technique or replaced by a circular propeller disk (ACD). The 2.5d method uses a cylindrical grid for each element with $\sim 100k$ cells each, see also Fig. 1b. The grid consists of three cells in radial direction. Nine blade element were used, see Fig. 4. The 2d method is carried out on a circular structured grid with a farfield distance to the airfoil of $\sim 100c$ and consists of $\sim 10k$ cells, see also Fig. 1c. For comparability reason, the propeller blade and the airfoils were each equally meshed regarding number of cells and boundary layer (bl) structure.

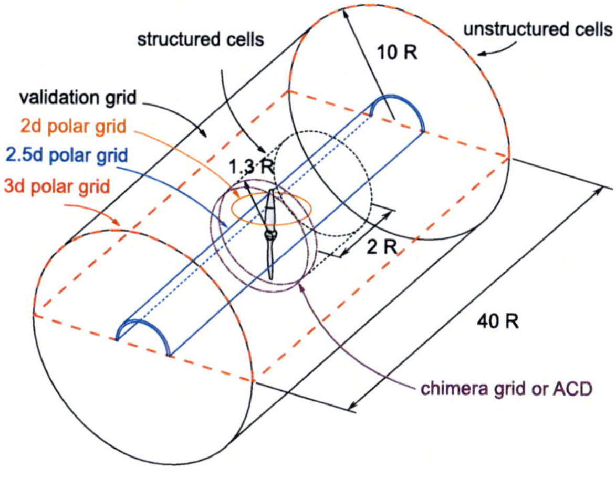

Fig. 3. Computational Grids

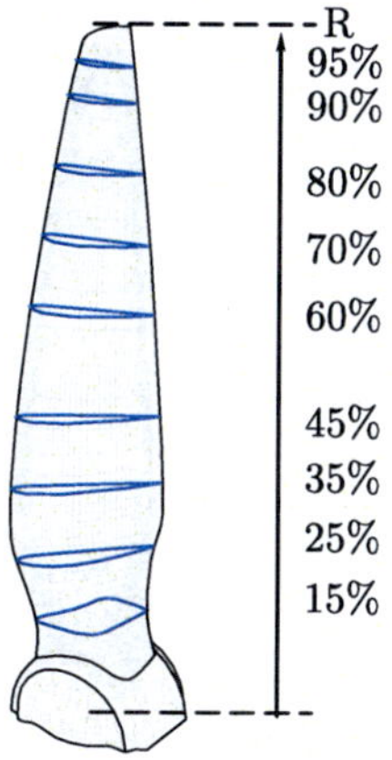

Fig. 4. Blade elements

The research was preceded by a study of the spatial discretization of a single propeller blade with the 3d grid arrangement. The results are shown in Fig. 5. The number of cells in chordwise and radial direction as well as the number of cells across the bl were varied.

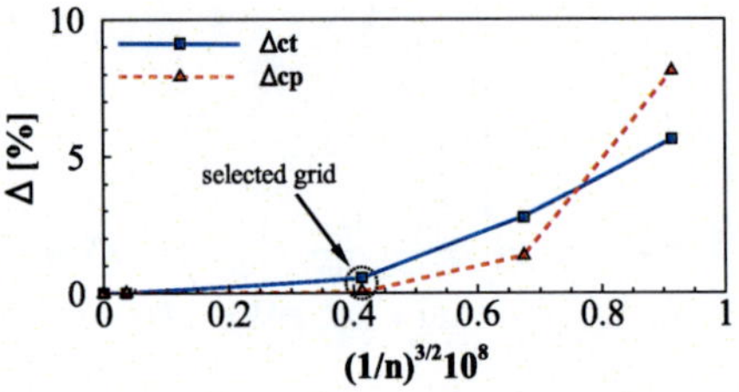

Fig. 5. Spatial discretization

The selected grid has a deviation from the Richardson-extrapolated value of $\Delta c_t = 0.5\%$ and $\Delta c_p = 0.1\%$. The airfoil is discretized by 116 points, the blade by 62 points in radial direction, excluding the hub, and the bl by 24 structured cells and $y^+ < 1$. The propeller geometry has a radius of 0.127 m and the rotational speed was 10,000 RPM. For the 3d and 2.5d methods the axial velocity was varied between 15–45 m/s and set to $v_\infty = 25$ m/s for the 2d method. For all calculations the second order central scheme was used for spatial discretization and the implicit Backward Euler for time discretization. The turbulence was modeled by the one-equation Spalart-Allmaras model, no transition was considered. The calculation time of the fully resolved simulations covered five propeller revolutions. The physical timestep corresponds to 4° rotation per timestep for the first four and 1° for the last revolution over which an averaging was performed. The 3d method was computed in parallel with 72 processors, whereas the 2d and 2.5d ones with one processor. The computational effort for a polar corresponds to: ∼1.4 cpuh (2d), ∼24 cpuh (2.5d), $\sim 4.1 \cdot 72 = 295$ cpuh (3d).

4 Results and Discussion

4.1 Airfoil Characteristics

Figure 6 shows the polars obtained from the different methods (2d, 2.5d, 3d) exemplary for one section in the area of the blade root ($r/R = 0.35$), two sections in the area of the main thrust ($r/R = 0.60$, 0.80) and one section in the area of the blade tip ($r/R = 0.95$). The zero lift angle is predicted similar by all methods. However, the 2d method overestimates the lift gradient, thus for increasing angles of attack the lift is too high. This was also observed by Raichle [1]. The rotation correction models (2d cor.) do not result in a correct lift gradient. The reason for this is, that the models are based on the difference to the inviscid (inv.) gradient. However, the 3d gradient is already lower before the onset of separations. As expected, the deviation decreases with the radius, because the rotation effect decreases and the blade inflow condition approaches a parallel flow. At $r/R =$ 0.80 the lift gradient is almost reached until the tip effect gets dominant. At $r/R = 0.35$ and $r/R = 0.60$, the lift gradient is much better reproduced by the 2.5d method, which indicates that the rotation effect is basically captured with the new method. The drag, which is generally underestimated by the 2d method, is also better reproduced by the 2.5d method. At $r/R = 0.95$ the effect of the blade tip vortex is evident, leading to a decrease in lift and an increase in drag, which is not captured by the uncorrected 2d and 2.5d results shown here. In Fig. 7 the lift and drag coefficients are plotted over the propeller radius for the inflow case of $\lambda_g = 0.19$. The corresponding angles of attack were taken from the 3d results for all methods. The overestimation in lift in the inner area by the 2d method is visible as well, while the 2.5d method corresponds to the 3d results. To account for the influence of the blade tip, the Shen tip loss model was used for the 2d and 2.5d methods. For comparison, the uncorrected results and the overestimating correction by Glauert are also shown for the 2.5d method.

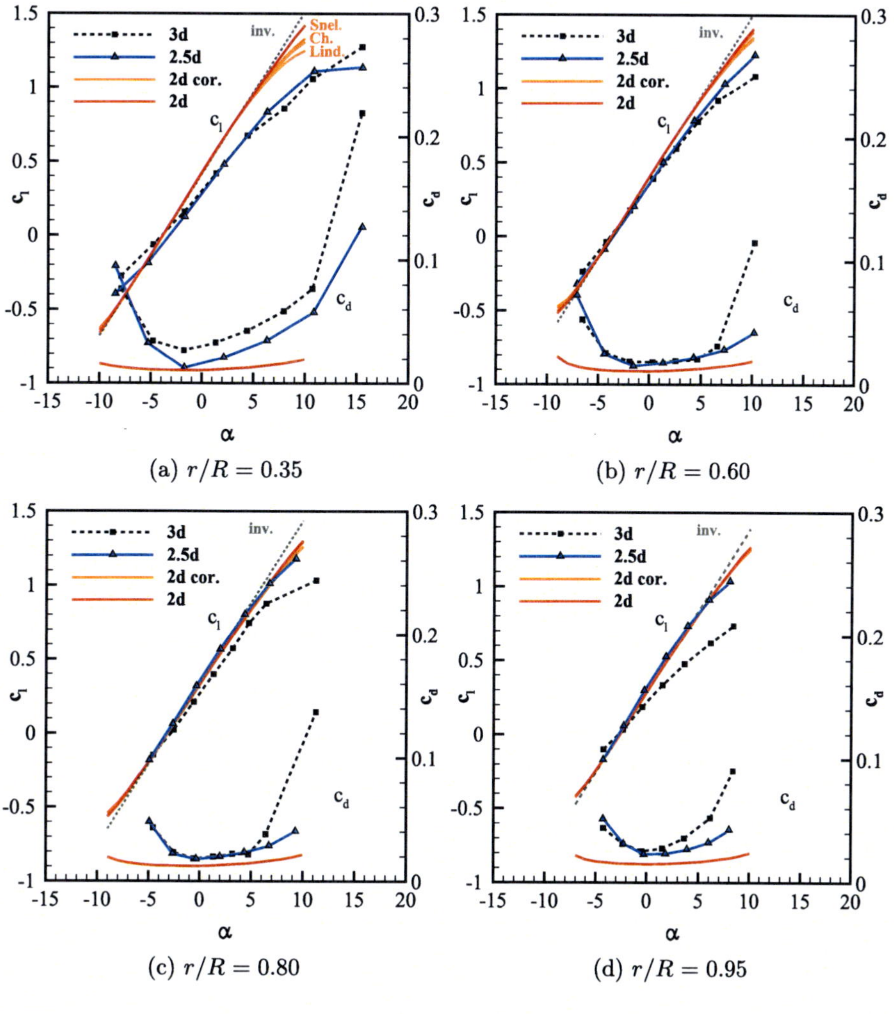

Fig. 6. Airfoil characteristics at different blade sections

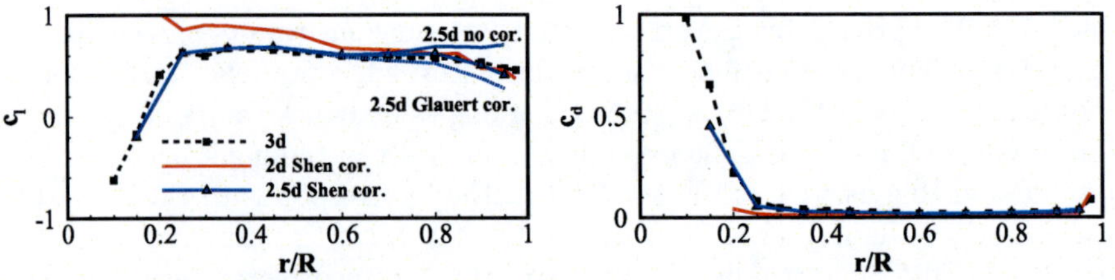

(a) lift coefficient (b) drag coefficient

Fig. 7. Radial distribution of airfoil coefficients for $\lambda_g = 0.19$

4.2 Validation

Figure 8 shows the forces along the radius for the different input data (2d, 2.5d, 3d, reco) as well as the fully resolved results (full). The ACD results with the 3d input data show a very close match. Only for $\lambda_g = 0.11$ the forces are slightly underestimated. As a consequence of the too high lift with the 2d method, the forces are overestimated in all cases, especially in the inner blade area. Again the 2.5 method shows an improved agreement toward the 3d data. The results of the forces recovered from the slipstream with method (d) show a good reproduction of the thrust for $\lambda_g = 0.26$ and $\lambda_g = 0.19$, with a slight overestimation of the torque at $\lambda_g = 0.19$. At $\lambda_g = 0.11$ the peak of the thrust is too high, but the inner area coincides well. In contrast the torque is by far overestimated. The differences are manifested in the integral propeller coefficients for thrust and power as well, see Fig. 9. The thrust by the 2d method generally exceeds the fully resolved results, while the 2.5d and 3d methods are more accurate, except of $\lambda_g = 0.11$, where the thrust is underestimated. The thrust is captured most accurately by method (d), but the power deviates with decreasing advance ratio λ_g. For example, in the case $\lambda_g = 0.19$ the deviations with ACD in the thrust coefficient are: 1.36% (reco), 2.68% (3d), 1.28% (2.5d), 7.84% (2d).

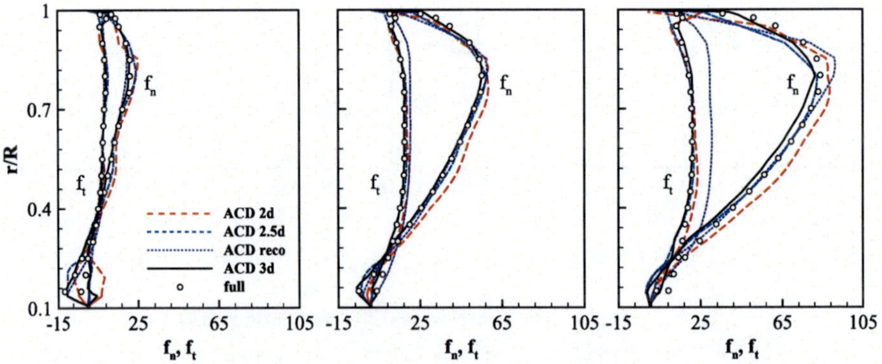

(a) $\lambda_g = 0.26$ (b) $\lambda_g = 0.19$ (c) $\lambda_g = 0.11$

Fig. 8. Radial force distributions at the propeller disk

Figure 10 shows the axial and tangential velocity in the slipstream, in a section $x/R = 1.18$ behind the propeller disk. A correct simulation of the wake is essential for the investigation of the aerodynamic interactions. The close match of the fully resolved results with the 3d and the 2.5d method is shown as well as the overestimation by the 2d method. The method (d) results reproduce the velocities with the exception of the too high tangential velocity at $\lambda_g = 0.11$.

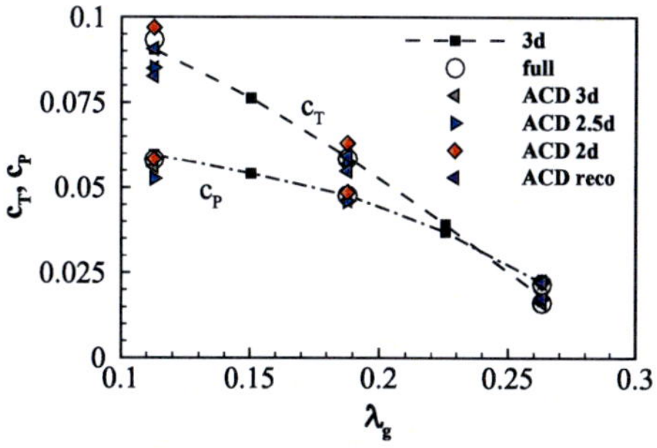

Fig. 9. Global propeller coefficients

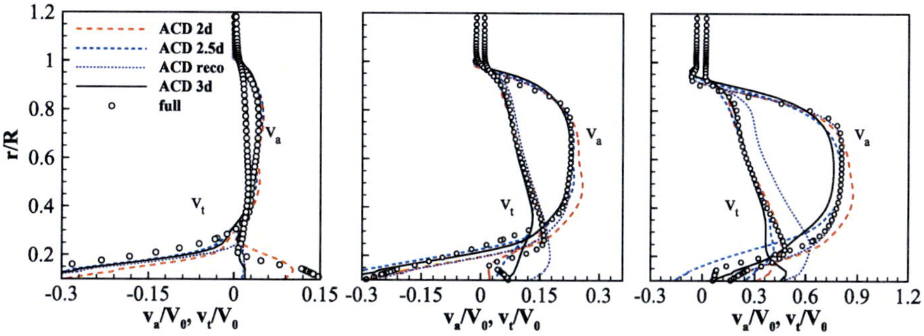

(a) $\lambda_g = 0.26$ (b) $\lambda_g = 0.19$ (c) $\lambda_g = 0.11$

Fig. 10. Radial velocity distributions in the slipstream

5 Conclusions

The neglect of the rotation effects in 2d polars as input for propeller models (ACD, ACL) leads to an overestimated propeller thrust and a too high axial velocity in the wake. Correction models, used in the area of wind turbines, failed to correct the lift gradient. The 3d simulation of an entire propeller blade captures the rotation effects as well as the tip effects and shows a close agreement with the results of fully resolved simulations. However, the computational effort to generate the input data is considerably higher. As a compromise, a new 2.5d approach was developed, calculating different blade sections with a rotating cylindrical grid to take into account the rotation effects. After a tip loss correction, the results are almost similar to the 3d results, regarding force distributions, propeller coefficients and wake velocities.

Acknowledgements. The authors gratefully acknowledge the Federal Ministry for Economic Affairs and Energy (BMWi) for funding this work inside the research project ELFLEAN. Special thanks is directed to Jan Denzel and Dominique Bergmann from the Institute of Aircraft Design of the University of Stuttgart for providing the scanned propeller geometry.

References

1. Raichle, A.: Flusskonservative Diskretisierung des Wirkscheibenmodells als Unstetigkeitsfläche. Ph.D. thesis, DLR (2017)
2. Schollenberger, M., Lutz, T., Krämer, E.: Boundary condition based actuator line model to simulate the aerodynamic interactions at wingtip mounted propellers. In: Dillmann, A., Heller, G., Krämer, E., Wagner, C., Tropea, C., Jakirlić, S. (eds.) New Results in Numerical and Experimental Fluid Mechanics XII. Notes on Numerical Fluid Mechanics and Multidisciplinary Design, vol. 142, pp. 608–618. Springer, Cham (2020). https://doi.org/10.1007/978-3-030-25253-3_58
3. Johansen, J., Sørensen, N.N.: Aerofoil characteristics from 3D CFD rotor computations. Wind Energy **7**(4), 283–294 (2004). https://doi.org/10.1002/we.127
4. Samuelsson, I.: Experimental Investigation of low speed model propeller slipstream aerodynamic characteristics including flow field surveys and nacelle/wing static pressure measurements, ICAS-90-3.1.3 (1990)
5. Prandtl, L., Betz, A.: Vier Abhandlungen zur Hydrodynamik und Aerodynamik. Universitätsverlag Göttingen, vol. 3 (Reprint 2010)
6. Shen, W.Z., Zhu, W.J., Sørensen, J.N.: Study of tip loss corrections using CFD rotor computations. J. Phys. **555**, 012094 (2014)
7. Bak, C., Johansen, J., Andersen, P.: Three-dimensional corrections of airfoil characteristics based on pressure distributions. In: EWEC (2006)

Turbulence and Transition

Check for updates

Effect of the Motion Pattern on the Turbulence Generated by an Active Grid

Meike Herbert[1(✉)], Tanja Skeledzic[1], Hermann Lienhart[1], and Özgür Ertunc[2]

[1] Friedrich-Alexander-Universität Erlangen-Nürnberg (FAU), Egerlandstraße 7-9, 91058 Erlangen, Germany
meike.herbert@fau.de

[2] Özyeğin University, Çekmeköy Campus Nişantepe District, Orman Street, 34794 Çekmeköy - İstanbul, Turkey

Abstract. In the field of turbulence research, many experiments are performed using turbulence generated under reproducible conditions. In the past mostly static grids having a fixed blockage were adopted, but in order to generate turbulence with high intensity and high turbulence Reynolds numbers, so called active grids can be used. Therefore, an experimental study has been carried out to investigate the effect of the motion pattern on the turbulence generated by an active grid with individually controllable paddles in the wind tunnel. Measurements were performed using the hot-wire technique. In comparison to former experiments, modifications of the active grid had been made concerning the arrangement of the paddles. The results show that the homogeneity of the turbulence could be improved using the modified grid. Turbulence Reynolds numbers could reach values up to 540. In order to produce even higher integral time scales or higher turbulence Reynolds numbers, synchronous movement of adjacent paddles was tested but did not show the intended improvements. In this paper, the results of the experimental study are presented in more detail.

Keywords: Turbulence · Active grid · Experimental study

1 Introduction

Almost all technical flows such as the flows in pipelines, around cars or aircraft, in reactors or mixers, as well as flows occurring in nature such as wind in the atmosphere, river or ocean flows are of turbulent nature. For describing and/or modelling turbulent flows most commonly Kolmogorov's assumption of local isotropy [1, 2] is adopted. This assumption presumes that the motion of the smallest eddies in which the turbulent energy is finally dissipated by the action of viscosity is nearly isotropic and is only weakly related to the motion of the anisotropic and inhomogeneous large eddies where the production of the turbulent energy occurs. The energy is transferred to the smaller scales through the eddy cascade in the inertial range of the velocity spectrum. This spectral separation in the scales develops with increasing turbulence Reynolds number

$$\mathrm{Re}_\lambda = \frac{\sqrt{\overline{u^2}}\lambda}{\upsilon}$$

© The Author(s), under exclusive license to Springer Nature Switzerland AG 2021
A. Dillmann et al. (Eds.): STAB/DGLR Symposium 2020, NNFM 151, pp. 37–46, 2021.
https://doi.org/10.1007/978-3-030-79561-0_4

based on the Taylor micro scale λ, which is a measure of dissipation length scale, and the energy of turbulence represented by the r.m.s. of the longitudinal fluctuations, υ is kinematic viscosity of the fluid.

Based on these theoretical considerations many experiments have been undertaken to study turbulence in the Re_λ range where Kolmogorov's assumption holds, i.e. at high values of Re_λ. As Comte-Bellot and Corrsin [3] in 1971 many of these studies considered the wake flow behind static grids in order to generate approximately homogeneous isotropic turbulence. However, the turbulence intensities and turbulence Reynolds numbers occurring in the wake of static grids were very much limited which lead to the development of so-called active grids as proposed in 1991 by Makita [4]. Similar experiments with active grids were performed by Mydlarski and Warhaft [5], Kang et al. [6] and Cekli & van de Water [7]. More recently, Bewley et al. [8] and Griffin et al. [9, 10] introduced a refined active grid layout with independently rotating paddles allowing to control the large scales of turbulence.

Such an active grid with individually controllable paddles was built at Friedrich-Alexander University Erlangen-Nürnberg (FAU) aiming at the generation of strong turbulence. The results of a former experimental study presented in [11] showed that the characteristic values such as integral scale, turbulence intensity and turbulence Reynolds number could be increased significantly in comparison to the usage of a static grid. The generated turbulence was characterized by intensities up to 20%, turbulence Reynolds numbers up to 520 and integral time scales up to 0.1 s.

Further experiments were carried out with the same test-setup but a modified active grid, which will be described in the following section. The aim was to improve the homogeneity and further investigate the effect of different motion sequences.

2 Experimental Setup

2.1 Modified Active Grid in the Wind Tunnel

The test-setup (see Fig. 1) consisted of a closed test section inserted in the wind tunnel at FAU, which is a closed-loop type wind tunnel with a maximum speed of 60 m/s and a free stream turbulence intensity of less than 0.3%. In the present study, mean velocity was kept constant at about 4.5 m/s for all experiments. Some more notes on this specific setup and the boundary conditions adopted are given in [11]. Measurements were performed using a hot-wire probe mounted on a three-dimensional traversing unit to position the sensor in the test section. The reference coordinate system was placed in the center of the active grid with the x-axis pointing in flow direction and the z-axis pointing upwards as shown in Fig. 1.

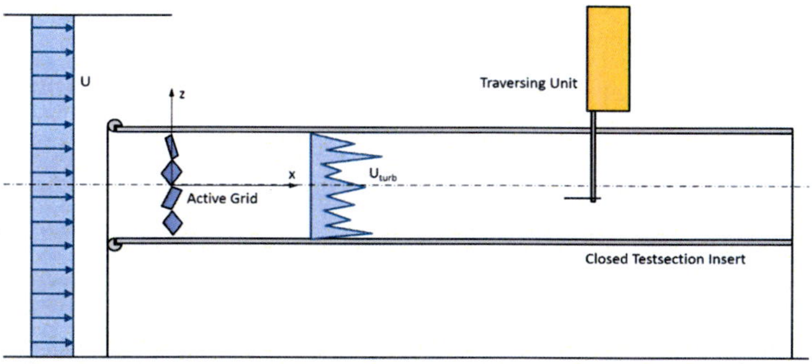

Fig. 1. Schematic illustration of the test-setup at FAU [11]

The active grid consisted of 24 paddles made of stainless steel (see Fig. 2 left). The paddles were controlled by servomotors so that the rotational position of each paddle could be adjusted individually between 0 degree (closed) and 90 degree (open), corresponding to an estimated blockage ratio of about 60% and 30%, respectively. In the closed position, the tunnel was not completely blocked due to openings and gaps between the paddles. The total cross section of the closed test section insert was 0.6 m × 0.45 m. Figure 2 shows the active grid (left) and the experimental setup (right).

Fig. 2. Active grid (left) and experimental setup with the active grid mounted in the closed test section in the wind tunnel at FAU (right)

In comparison to the grid used in [11], modifications were made concerning the geometry of the paddles and the arrangement of the paddles relative to each other. In the former grid, four neighboring paddles pointed into or out of a mesh (see Fig. 3 left). In the modified grid, two opposite paddles moved into one mesh, while at the same time the other two paddles pointed out of this mesh (see Fig. 3 right). The major aim of this modification in the motion pattern was to improve the homogeneity of the generated turbulence in the wake of the active grid.

Fig. 3. Arrangement of the paddles of the former active grid (left) and the modified active grid (right), rotational axis is along the diagonal of the quadratic panel

2.2 Experiments and Data Acquisition

The focus of the experimental study was to investigate the effect of different motion sequences of the paddles. To characterize the turbulence in the wake of the active grid, measurements were performed using constant temperature hot-wire anemometry with a single normal wire probe (Dantec 55P11). Based on preliminary tests the sampling rate was set at 6 kHz while the sampling time was chosen at 60 s which corresponded to about 1500 integral time scales. This resulted in velocity time traces of 360000 data points for each measurement position which were processed to derive the statistical quantities of the velocity fluctuations: mean value U_{mean}, standard deviation σ and their dimensionless quotient turbulence intensity Tu, integral time scale L_t, Taylor micro scale λ_t and turbulence Reynolds number Re_λ. More details on the measurement technique adopted and the data processing applied are described in Skeledzic et al. [11].

Measurements were performed at different positions in the wake of the active grid while horizontal as well as vertical profiles were considered. In horizontal direction traversing was made along the center line (z = 0 m) and along a parallel line at z = −0.1125 m, which is a quarter of the grids height below the center line. Measurement positions are set starting at 0.4 m behind the grid and ending at 2.0 m in steps of 0.2 m each. In vertical direction traversing was made along a line at 0.6 m behind the grid and at a further distance of 1.6 m to the grid while measurement positions are beginning at z = −0.2 m to z = 0.2 m in steps of 0.025 m each.

Different temporal motion sequences with random angular positions of the paddles were applied. The clock rate (period of time after which a new angular position is transmitted) was set at 160 ms for all experiments as it did not show a significant influence on the generated turbulence in [11]. The motion sequences featured a Gaussian probability density distribution (random Gaussian) or a uniform distribution (random uniform). The probability density functions for both cases are shown in Fig. 4.

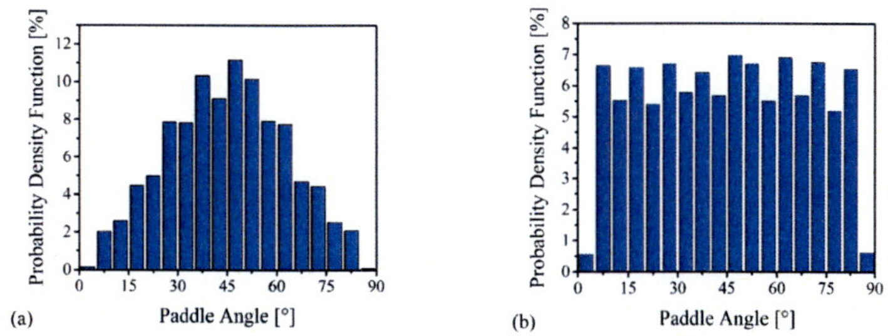

Fig. 4. Probability density functions, random Gaussian (a) and random uniform (b) [11]

Besides the individual movement of the paddles, the influence of synchronous movement of adjacent paddles was examined. For this purpose, groups of eight paddles (three blocks) and four paddles (six blocks), respectively, were formed as shown in Fig. 5. The paddles in one group moved synchronously, but again with random angular positions using Random Gaussian probability density distribution. The intention for these test runs was to investigate whether in this way even higher values of turbulence intensity and integral time scale might be achieved.

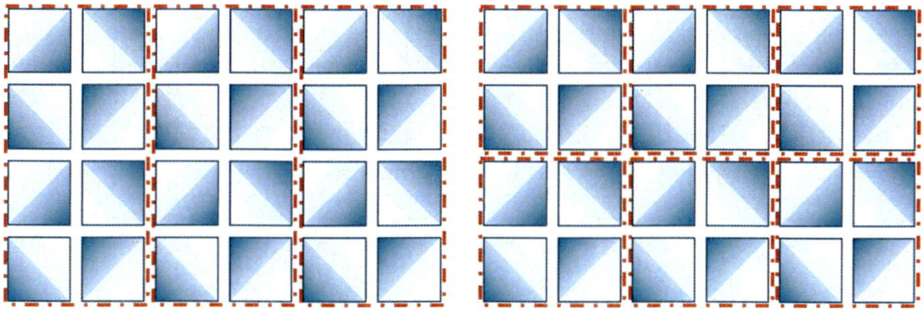

Fig. 5. Synchronous movements of paddles, 3 × 8 (left) and 6 × 4 (right)

Table 1 gives a summary of the motion sequences applied to the active grid. Measurements include the horizontal as well as the vertical profile, respectively. The results of the experimental study will be presented and discussed in the following section.

Table 1. Overview of motion sequences

Motion sequence	Probability density function	Paddle movements
RU	Random uniform	Individual
RG	Random Gaussian	Individual
RG 6 × 4	Random Gaussian	Synchronous 6 × 4
RG 3 × 8	Random Gaussian	Synchronous 3 × 8

3 Results and Discussion

The following diagrams show the development of mean velocity, standard deviation, integral time scale, Taylor micro scale, turbulence intensity and turbulence Reynolds number measured along horizontal or vertical lines in the wake of the active grid while using different motion patterns of the servomotors.

Figure 6 compares the results of the modified grid with the former grid for measurements on centre line and off centre using a random uniform motion sequence. The mean velocity shows differences between centre line and off centre, especially for the former grid at 0.4 m to 0.8 m. The graphs almost converge with increasing distance to the grid. Using the modified grid, a more homogenous profile also close to the grid could be achieved. The standard deviations were higher with the modified grid in comparison to the former grid; in both cases the measured values decreased downstream. The integral time scale showed fluctuations for the former grid and increased with increasing distance, while for the modified grid it was smoother and a trend could not be noted. The Taylor micro scale increased downstream as expected; the values of the modified grid were slightly lower than those of the former grid. The turbulence intensities *Tu* showed a similar behavior as described for the mean velocities and the values declined with an increasing distance to the grid. Using the modified grid, higher turbulence intensities could be achieved at almost all measurement positions; *Tu* is about 25% at x = 0.4 m to 12% at a greater distance to the grid. Turbulence Reynolds numbers of the modified grid decreased downstream starting at about 400; the values on centre line and off centre were closer together than those of the former grid. Overall, a more homogenous profile and also higher standard deviations and higher turbulence intensities could be achieved using the modified grid in comparison to the former grid.

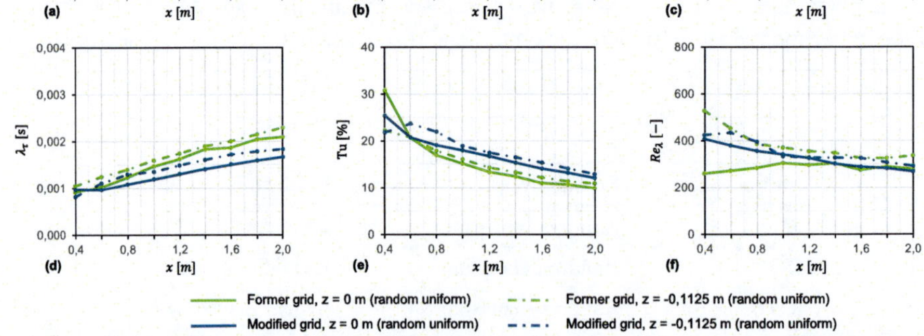

Fig. 6. Development of mean velocity (a), standard deviation (b), integral time scale (c), Taylor micro scale (d), turbulence intensity (e) and turbulence Reynolds number (f) measured on centre line and off centre using the modified grid in comparison to the former grid with a random uniform distribution

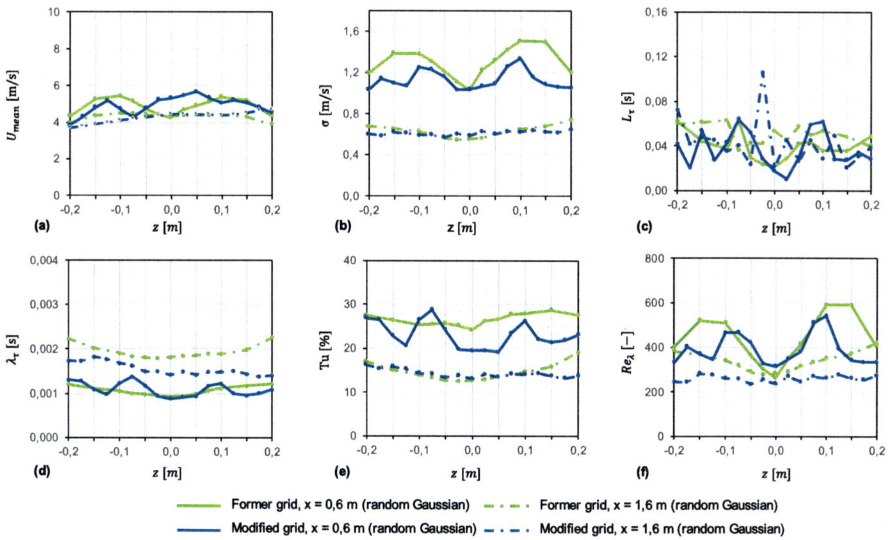

Fig. 7. Development of mean velocity (a), standard deviation (b), integral time scale (c), Taylor micro scale (d), turbulence intensity (e) and turbulence Reynolds number (f) measured in vertical direction at two distances to the grid using the modified grid in comparison to the former grid with a random Gaussian distribution

For further investigation of the homogeneity of the flow field, measurements were performed with traversing in vertical direction at two distances of 0.6 m and 1.6 m to the grid. Figure 7 shows the results for the vertical profile with random Gaussian motion sequence using the former grid in comparison to the modified grid. The influence of the grid can be recognized from the distinctive maxima of standard deviations, turbulence intensities and turbulence Reynolds numbers at x = 0.6 m. At a distance of 1.6 m to the grid the graphs are flattened.

The modification of the grid had an apparent influence on the generated turbulence according to the vertical profile. Comparing the standard deviation of the former grid to the modified grid, the diagram indicates that there is less influence of the z-position for the modified grid. The graph of the standard deviations at x = 0.6 m still shows maxima, but less distinctive in comparison to the former grid. As a negative effect of the modifications made to the grid, turbulence Reynolds numbers were slightly lower. Turbulence intensities reached values exceeding 20% at x = 0.6 m so it has to be noted that the results of the measurements close to the grid are not fully dependable because the range for reliable single hot-wire measurements is exceeded. For future investigation it should be considered to use a different measurement technique such as Laser-Doppler Anemometry (LDA).

Figure 8 shows the results for the vertical profiles measured at two distances to the grid comparing random Gaussian and random uniform motion sequences. Measurements were performed using the modified grid. The graphs are very similar for both motion sequences. Turbulence Reynolds numbers are slightly higher for the random uniform motion sequence with Re_λ up to 540, but the effect is very small. In general, the motion sequence random Gaussian or random uniform did not have a great influence on the generated turbulence.

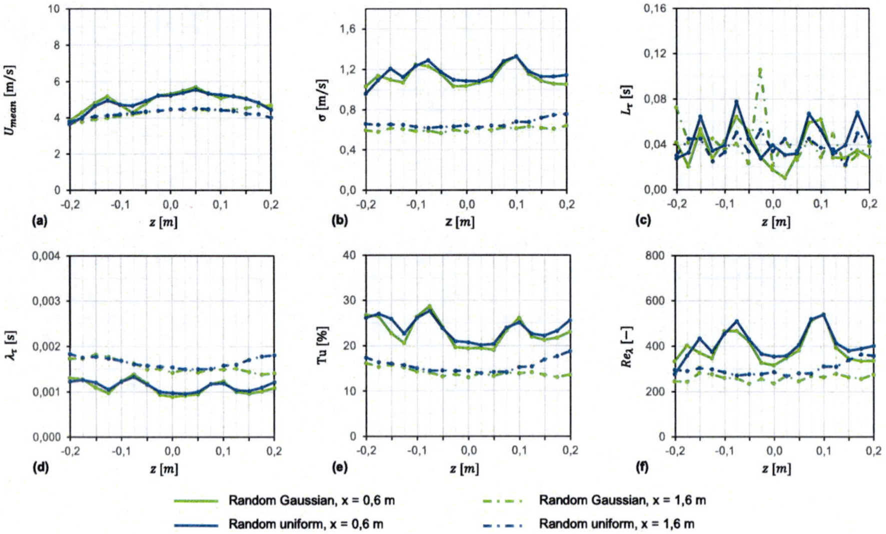

Fig. 8. Development of mean velocity (a), standard deviation (b), integral time scale (c), Taylor micro scale (d), turbulence intensity (e) and turbulence Reynolds number (f) measured in vertical direction at two distances to the grid with random Gaussian and random uniform distribution

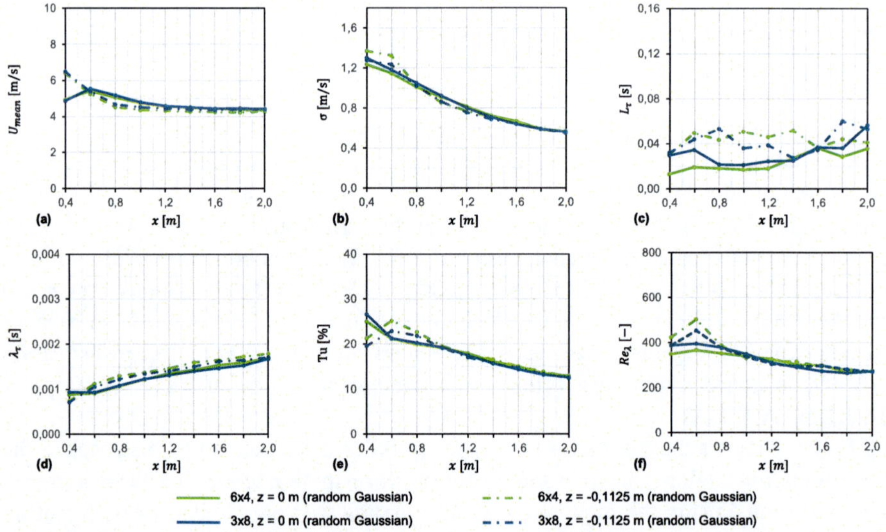

Fig. 9. Development of mean velocity (a), standard deviation (b), integral time scale (c), Taylor micro scale (d), turbulence intensity (e) and turbulence Reynolds number (f) measured on centre line and off centre with synchronous paddle movements (6 × 4 and 3 × 8)

Furthermore, the effect of synchronously moving paddles in groups of four and eight was investigated. The results shown in Fig. 9 are quite similar to those of the motion sequences with individual paddle movements. Contrary to supposition, it was not possible to achieve higher integral time scales and higher turbulence Reynolds numbers while as expected, the homogeneity of the flow field behind the grid was deteriorated. From these results it also might be concluded that the number of paddles in the present active grid is about at the lower end for producing homogenous turbulence.

4 Conclusions

An active grid with individually controllable paddles was used to generate turbulence under repeatable conditions in the wind tunnel. Measurements were performed using the hot-wire technique. In an experimental study, the effect of the motion pattern on the generated turbulence in the wake of the grid was investigated. In comparison to former experiments, modifications of the active grid have been made concerning the geometry and the arrangement of the paddles. Different motion sequences with random angular positions of the paddles were applied, that featured a Gaussian probability density distribution or a uniform distribution, respectively. Traversing in the wake of the active grid was made in horizontal as well as in vertical direction. The homogeneity of generated turbulence could be improved using the modified grid in comparison to the former grid. Turbulence with Re_λ up to 540 could be generated with the random uniform motion sequence. The synchronous movement of adjacent paddles with random sequences did not increase the scales of the generated turbulence. The integral time scale and turbulence Reynolds numbers remained at almost the same level compared to the turbulence generated with individual paddle movements.

References

1. Kolmogorov, A.N.: Equations of turbulent motion in an incompressible fluid. Izv. Akad. Nauk. SSSR Serr. Fiz. 6, 56–58 (Imperial College Report No. N/6–1968) (1942)
2. Kolmogorov, A.N.: A refinement of previous hypotheses concerning the local structure of turbulence in a viscous incompressible fluid at high Reynolds number. J. Fluid Mech. **13**, 82–85 (1962)
3. Comte-Bellot, G., Corrsin, S.: Simple Eulerian time correlation of full- and narrow-band velocity signals in grid-generated "'isotropic"' turbulence. J. Fluid Mech. **48**, 273–337 (1971)
4. Makita, H.: Realization of a large-scale turbulence field in a small wind tunnel. Fluid Dyn. Res. **8**, 53–64 (1991)
5. Mydlarski, L., Warhaft, Z.: On the onset of high-Reynolds-number grid-generated wind tunnel turbulence. J. Fluid Mech. **320**, 331–368 (1996)
6. Kang, H.S., Chester, S., Meneveau, C.: Decaying turbulence in an active-grid-generated flow and comparison with large-eddy simulation. J. Fluid Mech. **480**, 129–160 (2003)
7. Cekli, H.E., van de Water, W.: Tailoring turbulence with an active grid. Exp. Fluids **49**, 409–416 (2010)
8. Bewley, G.P., Kassel, J., Bodenschatz, E.: Control of turbulence with a high degree-of-freedom active grid. In: 14th European Turbulence Conference (2013)
9. Griffin, K., Wei, N., Hultmark, M., Bodenschatz, E., Bewley, G.P.: Correlated forcing with an active grid to control turbulence. In: XXIV ICTAM Montreal (2016)

10. Griffin, K.P., Wei, N.J., Bodenschatz, E., Bewley, G.P.: Control of long-range correlations in turbulence. Exp. Fluids **60**(4), 1–14 (2019). https://doi.org/10.1007/s00348-019-2698-1
11. Skeledzic, T., Krauss, J., Lienhart, H., Ertunc, Ö., Jovanovic, J.: Characterization of turbulence generated by an active grid with individually controllable paddles. In: Dillmann, A., Heller, G., Krämer, E., Wagner, C., Bansmer, S., Radespiel, R., Semaan, R. (eds.) New Results in Numerical and Experimental Fluid Mechanics XI, pp. 105–114. Springer International Publishing, Cham (2018). https://doi.org/10.1007/978-3-319-64519-3_10

Check for updates

Stochastic Modeling of Passive Scalars in Turbulent Channel Flows: Predictive Capabilities of One-Dimensional Turbulence

Marten Klein(✉) and Heiko Schmidt

Lehrstuhl Numerische Strömungs- und Gasdynamik,
Brandenburgische Technische Universität Cottbus-Senftenberg,
Siemens-Halske-Ring 15A, 03046 Cottbus, Germany
marten.klein@b-tu.de
https://www.b-tu.de/fg-stroemungsmodellierung

Abstract. Numerical simulations of passive scalars in turbulent channel flows up to friction Reynolds number $Re_\tau = 5{,}200$ and Schmidt number $Sc = 2{,}000$ are performed by utilizing the stochastic one-dimensional turbulence (ODT) model as stand-alone tool. The model is calibrated once for the turbulent velocity boundary layer at $Re_\tau = 5{,}200$ so that the passive scalar is a model prediction. ODT is able to reproduce with reasonable accuracy the scaling regimes of the scalar transfer and locally resolve the boundary layer structure. Albeit the model is unable to capture the emerging dissimilarity of near-wall scalar and momentum transport for high Sc, it can economically and accurately represent fluctuating wall-normal fluxes.

Keywords: Boundary layers · One-dimensional turbulence · Passive scalars

MSC codes: 76F25 · 76F25 · 80A20 · 82C31 · 82C70

1 Introduction

Numerical modeling of scalar transport in turbulent boundary layers is a standing challenge that is relevant for applications from the technical to the atmospheric scales. Key problems are related to small-scale correlations, scale interactions, and counter-gradient fluxes (e.g. [1,2,13]). Due to the latter, all relevant scales of a turbulent flow have to be resolved for robust numerical predictions. Direct numerical simulation (DNS) would be the ideal tool, but it is of limited applicability due to the resolution requirements imposed by the Kolmogorov and Batchelor scales (e.g. [1,3,16]).

In this paper, we address the standing numerical challenge of small-scale resolution with the stochastic one-dimensional turbulence (ODT) model [5].

© The Author(s), under exclusive license to Springer Nature Switzerland AG 2021
A. Dillmann et al. (Eds.): STAB/DGLR Symposium 2020, NNFM 151, pp. 47–57, 2021.
https://doi.org/10.1007/978-3-030-79561-0_5

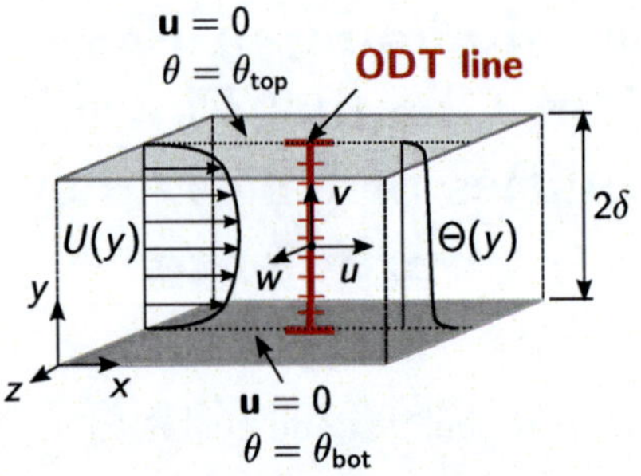

Fig. 1. Schematic of the channel flow set-up investigated

The numerical efficiency is further improved by utilizing a dynamically-adaptive finite-volume grid [10]. Here we apply the model to canonical channel flow and investigate the Schmidt and Reynolds number dependence of the wall-normal transport of a passive scalar that has negligible effect on the mass, momentum, and energy balances. Figure 1 shows a sketch of the set-up investigated.

The rest of this paper is organized as follows. In Sect. 2 we give an overview of the model and its application to channel flow. In Sect. 3 we discuss the model results for the passive scalar. Last, in Sect. 4, we give some concluding remarks.

2 ODT Model Formulation, Application, and Calibration

2.1 Key Idea of ODT Modeling

Kerstein's [5,6] one-dimensional turbulence (ODT) model aims to resolve all relevant scales of a turbulent flow. This is made feasible for high Schmidt and Reynolds number flows by modeling the effects of turbulent eddies by a stochastic process. The model directly resolves molecular-diffusive transport processes and distinguishes them from turbulent-advective ones for a notional line-of-sight through the turbulent flow that is denoted as 'ODT line' in Fig. 1. Note that no closure and no closure modeling is involved in the present ODT application.

2.2 Governing Equations

Here we limit our attention to constant-density flows with a passive scalar that has the same mass density as the surrounding fluid. The lower-order stochastic equations that are numerically solved in ODT may be written as [8]

$$\frac{\partial \mathbf{u}}{\partial t} + \mathcal{E}_{\mathbf{u}}(\alpha) = \nu \frac{\partial^2 \mathbf{u}}{\partial y^2} - \frac{1}{\varrho}\frac{\mathrm{d}P}{\mathrm{d}x}\mathbf{e}_x \,, \qquad \frac{\partial \theta}{\partial t} + \mathcal{E}_\theta = \Gamma \frac{\partial^2 \theta}{\partial y^2} \,, \tag{1a, b}$$

where $\mathbf{u} = (u, v, w)^{\mathrm{T}}$ denotes the velocity vector with Cartesian components, θ the scalar concentration, t the time, x the streamwise and y the vertical coordinate, ϱ and ν the fluid's density and kinematic viscosity, Γ the scalar diffusivity,

$\mathrm{d}P/\mathrm{d}x$ the prescribed mean pressure gradient, $\mathbf{e}_x$ the unit vector in streamwise direction, and $\mathcal{E}$ the stochastically sampled eddy events that are detailed in [5]. $\mathcal{E}_{\mathbf{u}}(\alpha)$ and $\mathcal{E}_\theta$ are coupled, but the latter only models the effects of turbulent advection, whereas the former also accounts for fluctuating pressure-gradient forces that redistribute kinetic energy among u, v, and w. The model parameter $0 \le \alpha \le 1$ can be used to modify the efficiency of this mechanism [6]. The rate at which the discrete eddy events occur in $\mathcal{E}$ is controlled by the nondimensional rate parameter C [5]. The sampling procedure is energetically constrained such that unphysically small eddy events are suppressed through the model parameter $Z \simeq Re_{\mathrm{crit}}^2$ that represents the square of a critical Reynolds number [5]. Here, eddy events are bounded from above by the channel half-height δ [15]. No-slip Dirichlet wall-boundary conditions are prescribed as indicated in Fig. 1.

A nondimensional form of the governing Eq. (1a,b) may be obtained with the aid of the mean wall-stress balance and the friction concentration,

$$u_\tau^2 = \nu \left|\frac{\mathrm{d}U}{\mathrm{d}y}\right|_{\mathrm{w}} = \frac{\delta}{\varrho}\left|\frac{\mathrm{d}P}{\mathrm{d}x}\right|, \qquad \theta_\tau = \frac{\Gamma}{u_\tau}\left|\frac{\mathrm{d}\Theta}{\mathrm{d}y}\right|_{\mathrm{w}}, \tag{2a, b}$$

where the upper-case symbols $U(y) = \overline{u(y,t)}$ and $\Theta(y) = \overline{\theta(y,t)}$ denote Reynolds-averaged variables, u_τ and θ_τ the friction velocity and concentration, respectively, and the subscript 'w' evaluation at the wall. The scaled (superscript '+') variables are $y^+ = yu_\tau/\nu$, $t^+ = tu_\tau/\delta$, $u^+ = u/u_\tau$, and $\theta^+ = (\theta - \theta_{\mathrm{w}})/\theta_\tau$, where θ_{w} is a reference (wall) concentration. The similarity parameters are given by the Schmidt (Prandtl), $Sc = \nu/\Gamma$, and friction Reynolds, $Re_\tau = \delta u_\tau/\nu$, number.

2.3 Remarks on Computational Resources Used

Stand-alone ODT simulations need to sufficiently sample the state space. We therefore ran long-time simulations for $t^- = t/t_\nu \gg 1$, where $t_\nu = u_\tau^2/\nu$ is the viscous time scale, with the minimum grid cell size, $\Delta y_{\min}$, below the Kolmogorov and Batchelor scales. Further details are summarized in Table 1, where CPU hours were measured on Intel® Xeon® E5-2630 (2.40 GHz) processors.

Table 1. ODT computational resources for representative cases

Sc	Re_τ	$\Delta y^+_{\min}$	$t^-/10^6$	CPU-h	Re_τ	$\Delta y^+_{\min}$	$t^-/10^6$	CPU-h
0.025	180	0.2	10.8	22	5,200	0.2	10.0	124
1	180	0.2	10.8	23	5,200	0.2	10.0	130
1,000	180	0.02	1.1	280	5,200	0.10	2.5	130

2.4 Model Application and Calibration

Equations (1a,b) are numerically integrated, which yields a temporal sequence of synthetic but statistically representative wall-normal flow profiles. Conventional and conditional statistics [5,8] are performed on these profiles and gathered on

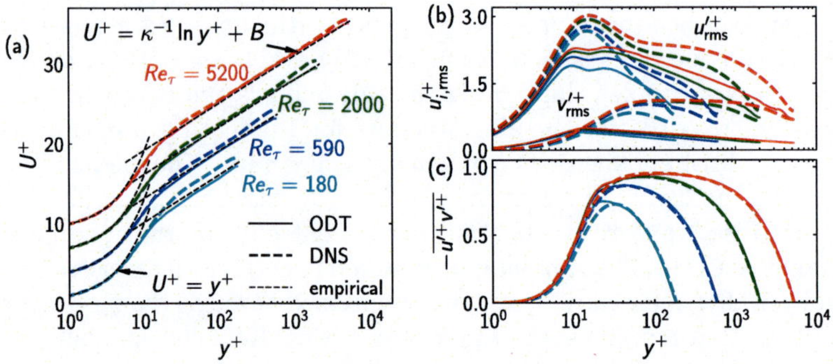

Fig. 2. (a) Mean streamwise velocity, U^+ (offset by $\Delta U^+ = 3$ for visibility), (b) streamwise, $u'^+_{\rm rms}$, and wall-normal, $v'^+_{\rm rms}$, r.m.s. fluctuation velocities, and (c) Reynolds stress, $\overline{u'^+v'^+}$, for various Re_τ numbers. Reference DNS results are from [9,12] and the empirical mean profile with coefficients $\kappa = 0.389$ and $B = 4.23$ from [11]

a predefined post-processing grid for the fully-adaptive solver [10] used. While cumulative statistics are straightforward, the computation of the ODT-resolved turbulent fluxes is conditional on the eddy events and described, for example, in [8] (appendix). Note that 'pre-simulations' *without* passive scalars were conducted for model calibration. Following [14,15], we obtain the model parameters as $C = 6$, $Z = 300$, and $\alpha = 1/6$ in the usual nomenclature from [6].

Figure 2(a) shows normalized mean velocity profiles, U^+, for various Re_τ numbers for the calibrated model. Simulated mean velocity profiles are consistent with the empirical law-of-the-wall of high asymptotic Re_τ numbers [11]. The ODT model accurately captures the linear and the log layer, but not fully the buffer and outer layers of the reference DNS [9,12].

Figure 2(b) shows the normalized root-mean-square (r.m.s.) fluctuation velocity components, $u'^+_{i,\rm rms} = (\overline{u_i^2} - U_i^2)^{1/2}/u_\tau$, for $i = 1$ (u') and $i = 2$ (v'). In ODT, the fluctuation velocity is generally underestimated, which is a known modeling artifact (e.g. [5,6,14,15]). The inner peak of $u'^+_{\rm rms}$ is degraded due to the microstructure modeling in the discrete eddy events [10]. The outer peak in $u'^+_{\rm rms}$ is only weakly captured by ODT [14]. Reasonable agreement with reference DNS is obtained, however, for both $u'^+_{\rm rms}$ and $v'^+_{\rm rms}$ in the vicinity of the wall.

Figure 2(c) shows the normalized Reynolds stress component $\overline{u'^+v'^+}$, which is in very good agreement with the reference DNS [9,12].

These pre-simulation results demonstrate that ODT captures the mean and wall-normal flux rather than the fluctuation variance which is a relevant model property for transport modeling. We remind that turbulent fluxes are obtained diagnostically by conditional eddy-event statistics [8] (appendix).

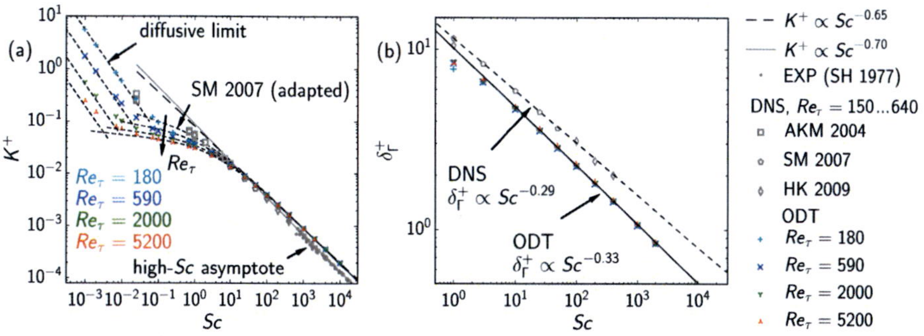

Fig. 3. (a) Scalar transfer coefficient, K^+, and (b) diffusive sublayer thickness, δ_Γ^+, for various Sc and Re_τ numbers. δ_Γ^+ is given by the intersection of the linear and log layer for $Sc \geq 1$ (compare with Fig. 4(b) below). Broken and solid lines indicate empirical scaling laws. High Re_τ and Sc number reference measurements are from [17]. Low Re_τ and moderate Sc number reference DNS are from [1,3,16]

3 Results and Discussion

3.1 Scaling Regimes of the Scalar Transfer to the Wall

Scaling regimes of the scalar transfer to a wall (among other integral properties) is of particular interest for various applications. It is quantified by the scalar transfer coefficient, K^+, that is defined as (e.g. [16–18])

$$K^+ = Sh/(\gamma\; Sc\, Re_\tau) = \theta_\tau/\Delta\theta\,, \qquad (3a, b)$$

where Sh denotes the Sherwood (Nusselt) number, γ a geometrical proportionality constant, θ_τ the friction concentration from Eq. (2b), and $\Delta\theta = |\theta_\mathrm{b} - \theta_\mathrm{w}| = |\theta_\mathrm{top} - \theta_\mathrm{bot}|/2$ the bulk-wall concentration difference. The Re_τ and Sc dependence of K^+ is shown in Fig. 3(a) and discussed in the following.

For low asymptotic Sc numbers, molecular-diffusive processes dominate over turbulent advection so that $Sh \simeq 1$. Equation (3a) in turn yields $K^+ \propto Sc^{-1} Re_\tau^{-1}$ for the diffusive limit. ODT exhibits this limit exactly, whereas reference DNS data [1] for $Sc = 0.025$ with $Re_\tau = 395$ and 640 exhibit larger K^+ values than ODT. These reference DNS appear to be somewhat less sensitive to Re_τ presumably due to details of the scalar forcing term employed there.

For high asymptotic Sc numbers, asymptotic one-dimensional theory for the mean scalar conservation equation is used with neglect of the thin asymptotic diffusive surface layer. This yields $K^+ \propto Sc^{-(n-1)/n}$ with theoretically estimated $n = 3$ or 4 [17,18]. Available measurement data for $700 \leq Sc \leq 37{,}000$ are consistent with the effective scaling $K^+ \propto Sc^{-0.70}$, that is, $n = 3.38$ [17]. Reference DNS for up to $Sc = 400$ also seem to approach this limit [3]. Present ODT results reach up to $Sc = 2{,}000$ and these results are well described by $K^+ \propto Sc^{-0.65}$, that is, $n = 2.85$, for $Sc > 100$ investigated. We will come back to this below in Sect. 3.4 for the turbulent eddy viscosity.

For intermediate Sc numbers, an empirical equation has been derived that takes into account the linear and log laws of the scalar boundary layer that are discussed below. The empirical relation reads [16]

$$K^+(Sc, Re_\tau) = \left[\kappa_\theta^{-1}\ln(Re_\tau) + \xi\, Sc^{1-r} + r\,\kappa_\theta^{-1}\ln(Sc) - \kappa_\theta^{-1}\ln(\xi)\right]^{-1}, \quad (4)$$

where κ_θ denotes the scalar von Kármán constant, ξ and r are fit parameters of the high Sc asymptote, $\delta_\Gamma^+ \simeq \xi\, Sc^{-r}$, that is shown in Fig. 3(b), and δ_Γ^+ the diffusive boundary-layer thickness that is defined by the intersection of the linear and log law of the scalar boundary layer (see Fig. 4(b) below). ODT simulations for $Re_\tau = 180$ with $Sc \leq 2{,}000$ yield $\kappa_\theta = 0.27$, $\xi = 9.7$, and $r = 0.32$. These coefficients are used for the curved dashed lines in Fig. 3(a). The values are close to DNS [16], which yields $\kappa_\theta = 0.27$, $\xi = 11.5$, and $r = 0.29$ (not shown). Note that Eq. (4) is strictly valid only for $Sc \geq 1$, but we use it as bound for the Re_τ dependence of K^+. In fact, ODT results for moderate $Sc < 1$ are reasonably well described by the empirical formula. So, the main difference between ODT and DNS is seen in the exponent r of the diffusive sublayer scaling that governs the high Sc asymptote of K^+.

3.2 Mean Scalar Concentration and Boundary-Layer Structure

Figures 4(a) and 4(b) show wall-normal profiles of the mean scalar concentration, $\Theta^+ = (\bar{\theta} - \theta_\mathrm{w})/\theta_\tau$, for various Re_τ and Sc numbers separated into $Sc < 1$ and $Sc > 1$, respectively. In Fig. 4(a), the diffusive limit is approximately realized for the ODT simulations with $Sc = 0.025$ and $Re_\tau \leq 590$ which is notable by the absence of the log layer for these cases. The defect law obtained by ODT is consistent with available reference DNS [1,4] for $Sc = 0.7$ and 0.025, respectively. In these DNS, the scalar is prescribed by a forcing term due to which the functional dependence $\Theta^+(y^+)$ takes a different form in the outer layer with $y^+ \geq O(Re_\tau/2)$ [13].

For all other cases shown in Figs. 4(a,b), the turbulent boundary layer exhibits a defect law. The linear and log fits shown (dashed black lines) have been obtained for $Re_\tau = 5{,}200$ in the case of ODT, but for $Re_\tau = 180$ in the case of DNS [7]. For $Sc \geq 10$ shown in Fig. 4(b), the boundary-layer structure predicted by ODT exhibits qualitative but not quantitative agreement with the available reference DNS from [3,16]. ODT overestimates K^+ which is, due to Eq. (3b), the consequence of overestimated θ_τ. This manifests itself exactly by the underestimation of the additive constant, B_θ, indicated in Fig. 4(b). For low $Re_\tau = 180$, the scalar von Kármán constant, $\kappa_\theta = 0.24 \pm 0.03$, predicted by ODT [7] matches locally with the value obtained for available reference DNS [3,16]. For large $Re_\tau = 5{,}200$, present ODT results exhibit an increase of the scalar von Kármán constant to $\kappa_\theta = 0.34 \pm 0.01$, which is closer to $\kappa \approx 0.4$ of the velocity boundary layer [11].

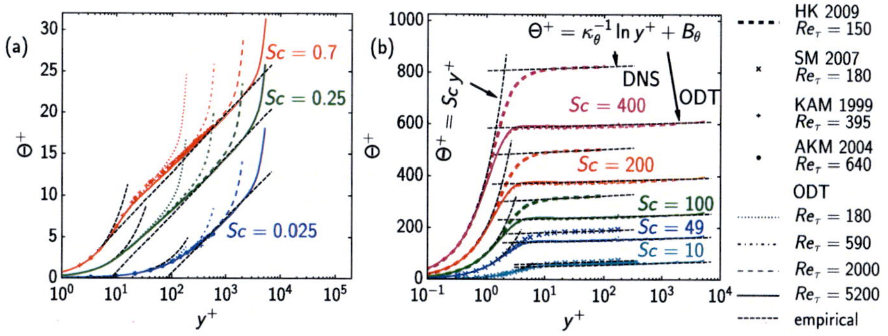

Fig. 4. Profiles of the normalized mean scalar concentration, Θ^+, for various Sc and Re_τ numbers. (a) Diffusion-dominated regime with $Sc < 1$, and (b) convection-dominated regime with $Sc > 1$. Reference DNS results are from [1,3,4,16]

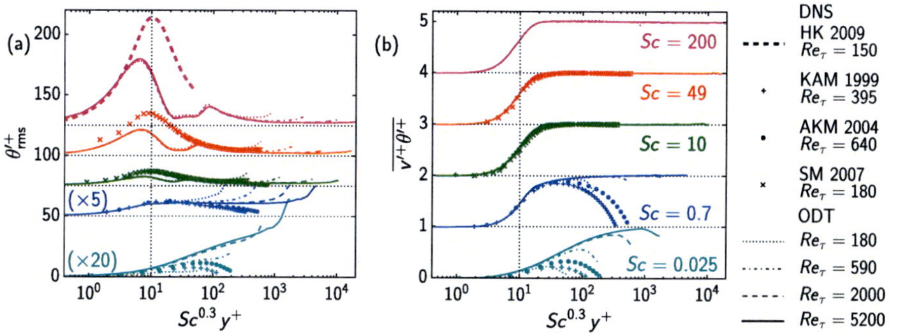

Fig. 5. (a) Root-mean-square (r.m.s.) scalar fluctuation concentration, θ'^+_{rms}, and (b) wall-normal turbulent scalar flux, $\overline{v'^+\theta'^+}$ for various Sc and Re_τ numbers. Both θ'^+_{rms} and $\overline{v'^+\theta'^+}$ are shifted for visibility, where dashed horizontal lines mark the corresponding baselines. A dashed vertical line marks the notional maximum location of θ'^+_{rms} that corresponds with the inflection point in $\overline{v'^+\theta'^+}$. Reference DNS data are from [1,3,4,16]

3.3 Scalar Fluctuation Variance and Turbulent Fluxes

We use ODT to estimate the variability of all flow variables without additional closure modeling. This variability is addressed in the following for the passive scalar.

Figure 5(a) shows wall-normal profiles of the normalized r.m.s. fluctuation concentration, $\theta'^+_{\mathrm{rms}} = (\overline{\theta^2} - \Theta^2)^{1/2}/\theta_\tau$, for various Re_τ and Sc numbers alongside with available reference DNS results. The r.m.s. fluctuations notably increase with Sc. For $Sc < 1$, θ'^+_{rms} predicted by ODT is consistent with available reference DNS [1,4] in the linear and log layers. For increasing $Sc > 1$, a near-wall fluctuation peak develops at nominal location $Sc^{0.3}y^+ \approx 10$ [3,16]. This peak obeys inner scaling but its shape and magnitude is degraded in the ODT results, which is a modeling artifact. The origin of the latter lies in the prescribed micro-structure of the eddy events. This is discussed in [10] for the inner u'^+_{rms} peak

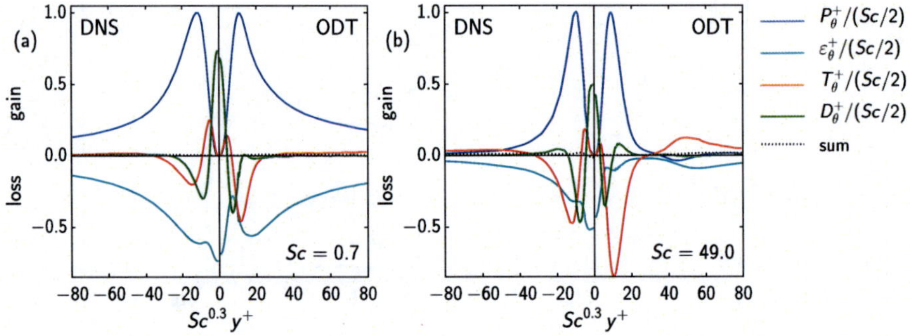

Fig. 6. Scalar variance budget for (a) $Sc = 0.7$ with $Re_\tau = 590$ (ODT, $y^+ \geq 0$) and 640 (DNS [4], $y^+ \leq 0$), and (b) $Sc = 49$ with $Re_\tau = 180$ (ODT and DNS [16])

shown in the above Fig. 2(b). By analogy, the artifact manifests itself also in the scalar fluctuations that are governed by the velocity boundary layer for $Sc > 1$.

Figure 5(b) shows the turbulent scalar flux, $\overline{v'^+\theta'^+}$, which is estimated directly from the ensemble of discrete eddy events as described, for example, in [8]. ODT accurately captures the turbulent scalar flux for high Sc numbers in agreement with reference DNS [16]. The inflection point is located at $Sc^{0.3}y^+ \approx 10$, which is consistent with the location of the peak in θ'^+_{rms} even though ODT does not fully capture the latter peak. For low Sc numbers, there is agreement between ODT and reference DNS [1,4] only for the universal near-wall region, but not for the outer layer, which depends on the case set-up.

Next, we investigate the budget of the scalar variance in order to clarify the missing scalar fluctuations at high Sc numbers. This budget reads (e.g. [16])

$$\frac{\partial \overline{\theta'^2}}{\partial t} + \mathbf{U} \cdot \nabla \overline{\theta'^2} = P_\theta + \varepsilon_\theta + D_\theta + T_\theta \, , \tag{5}$$

$$P_\theta = -2\overline{v'\theta'}\frac{\mathrm{d}\Theta}{\mathrm{d}y} \, , \quad \varepsilon_\theta = -2\Gamma\overline{\frac{\partial\theta'}{\partial y}\frac{\partial\theta'}{\partial y}} \, , \quad D_\theta = \Gamma\frac{\mathrm{d}^2\overline{\theta'^2}}{\mathrm{d}y^2} \, , \quad T_\theta = -\frac{\mathrm{d}\,\overline{v'\theta'^2}}{\mathrm{d}y} \, , \tag{6a, b, c, d}$$

where $\overline{\theta'^2} = \theta'^2_{\mathrm{rms}}$ denotes the scalar variance, P_θ the production, ε_θ the dissipation, D_θ the diffusive transport, and T_θ the turbulent transport of the scalar variance. The left-hand side vanishes for statistically stationary channel flow, and the terms on the right-hand side only retain their wall-normal contributions within ODT. Corresponding nondimensional expressions for all terms are obtained by division with $\theta_\tau^2 u_\tau^2/\nu$ which is indicated by the super-script '+'.

Figures 6(a) and 6(b) show the normalized scalar variance budget balance for both ODT and reference DNS for $Sc = 0.7$ and $Sc = 49$, respectively. The nondimensional terms are additionally normalized with the peak production, $P^+_{\theta,\max} = Sc/2$ [16]. ODT exhibits a balance of the scalar variance budget as the right-hand-side terms of Eq. (5) sum to zero. The gross structure of all terms is captured by ODT, but somewhat better for the lower $Sc = 0.7$ than for the higher $Sc = 49$. The results obtained suggest that the turbulent production peak

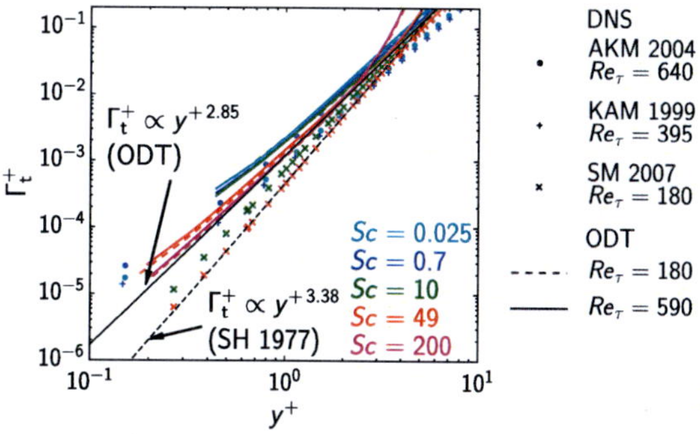

Fig. 7. Normalized turbulent eddy diffusivity, $\Gamma_t^+ = \Gamma_t/\nu$, in the vicinity of the wall for various Sc and Re_τ numbers and its limiting behavior. Only data for $Re_\tau \leq 640$ investigated are shown to aid visibility. Reference DNS results are from [1,4,16]

is correctly predicted, except for its tail at high Sc numbers. In addition, dissipation is increasingly underestimated for large Sc numbers so that compensating transport maintains the balance. The compensation is primarily achieved by turbulent and only weakly influenced by diffusive transport. This is consistent with the previously discussed modeling artifacts in the r.m.s. fluctuation.

3.4 Eddy Diffusivity and Its Relation to Asymptotic Scalar Transfer

We noted in Sect. 3.1 that the thin diffusive surface layer can be neglected for $Sc \gg O(1)$. Therefore, the scalar transfer coefficient, $K^+ \propto Sc^{-(n-1)/n}$, is governed by the near-wall structure of the turbulent eddy diffusivity,

$$\Gamma_t^+ = \Gamma_t/\nu \propto y^{+n}\,, \qquad \Gamma_t = -\overline{v'\theta'}/(\mathrm{d}\Theta/\mathrm{d}y)\,, \tag{7a, b}$$

that obeys inner scaling [18] and diagnostically parametrizes the ensemble effect of turbulent advection, $\overline{v'\theta'}$, by the mean scalar gradient, $\mathrm{d}\Theta/\mathrm{d}y$.

Figure 7 shows the normalized eddy diffusivity, $\Gamma_t^+ = \Gamma_t/\nu$, in the vicinity of the wall. All results shown exhibit a limiting behavior of the form $\Gamma_t^+ \propto y^{+n}$ for high asymptotic Sc numbers. Reference data [3,16,17] for such high Sc numbers is consistent with $n = 3.38$, that is, $K^+ \propto Sc^{-0.70}$ that is shown in Fig. 3(a). Present ODT results for $100 \leq Sc \leq 2{,}000$ and $Re_\tau \leq 5{,}200$ yield the limiting exponent $n = 2.85 \pm 0.05$, which is consistent with $K^+ \propto Sc^{-0.65}$. Latter is closer to the reference DNS [1,4] with $Sc < 1$ and very close to asymptotic one-dimensional theory [17] with $n = 3$. This suggests similarity of the mean scalar and momentum transfer in ODT and is consistent with the mapping rule used.

4 Concluding Remarks

Economical numerical simulations of passive scalars in turbulent channel flows have been performed with the stochastic one-dimensional turbulence (ODT) model as stand-alone tool. ODT is able to accurately capture wall-normal turbulent fluxes and the general turbulent boundary-layer structure, but not for the total fluctuation variance. Furthermore, ODT correctly predicts various scaling regimes of the scalar transfer to a wall, but it does not capture the near-wall dissimilarity of the scalar and momentum transport that emerges for high asymptotic Sc numbers. These limitations, however, would be less of concern when ODT is used as wall model in LES to locally improve the bulk-wall coupling on small scales.

References

1. Abe, H., Kawamura, H.: Surface heat-flux fluctuations in a turbulent channel flow up to $Re_\tau = 1020$ with $Pr = 0.025$ and 0.71. Int. J. Heat Fluid Flow **25**(3), 404–419 (2004)
2. Deardorff, J.W.: The counter-gradient heat flux in the lower atmosphere and in the laboratory. J. Atmos. Sci. **23**(9), 503–506 (1966)
3. Hasegawa, Y., Kasagi, N.: Low-pass filtering effects of viscous sublayer on high Schmidt number mass transfer close to a solid wall. Int. J. Heat Fluid Flow **30**(3), 525–533 (2009)
4. Kawamura, H., Abe, H., Matsuo, Y.: DNS of turbulent heat transfer in channel flow with respect to Reynolds and Prandtl number effects. Int. J. Heat Fluid Flow **20**(3), 196–207 (1999)
5. Kerstein, A.R.: One-dimensional turbulence: model formulation and application to homogeneous turbulence, shear flows, and buoyant stratified flows. J. Fluid Mech. **392**, 277–334 (1999)
6. Kerstein, A.R., Ashurst, W.T., Wunsch, S., Nilsen, V.: One-dimensional turbulence: vector formulation and application to free shear flows. J. Fluid Mech. **447**, 85–109 (2001)
7. Klein, M., Schmidt, H.: Investigating the Reynolds number dependency of the scalar transfer to a wall using a stochastic turbulence model. Proc. Appl. Math. Mech. **18**(1), e201800238 (2018)
8. Klein, M., Zenker, C., Schmidt, H.: Small-scale resolving simulations of the turbulent mixing in confined planar jets using one-dimensional turbulence. Chem. Eng. Sci. **204**, 186–202 (2019)
9. Lee, M., Moser, R.D.: Direct numerical simulation of turbulent channel flow up to $Re_\tau \approx 5200$. J. Fluid Mech. **774**, 395–415 (2015)
10. Lignell, D.O., Kerstein, A.R., Sun, G., Monson, E.I.: Mesh adaption for efficient multiscale implementation of one-dimensional turbulence. Theor. Comput. Fluid Dyn. **27**(3), 273–295 (2013)
11. Marusic, I., McKeon, B.J., Monkewitz, P.A., Nagib, H.M., Smits, A.J., Sreenivasan, K.R.: Wall-bounded turbulent flows at high Reynolds numbers: Recent advances and key issues. Phys. Fluids **22**, 065103 (2010)
12. Moser, R.D., Kim, J., Mansour, N.N.: Direct numerical simulation of turbulent channel flow up to $Re_\tau \approx 590$. Phys. Fluids **11**(4), 943–945 (1999)
13. Pirozzoli, S., Bernardini, M., Orlandi, P.: Passive scalars in turbulent channel flow at high Reynolds number. J. Fluid Mech. **788**(2), 614–639 (2016)
14. Rakhi, Klein, M., Medina M., J.A., Schmidt, H.: One-dimensional turbulence modelling of incompressible temporally developing turbulent boundary layers with comparison to DNS. J. Turbul. **20**(8), 506–543 (2019)

15. Schmidt, R.C., Kerstein, A.R., Wunsch, S., Nilsen, V.: Near-wall LES closure based on one-dimensional turbulence modeling. J. Comput. Phys. **186**, 317–355 (2003)
16. Schwertfirm, F., Manhart, M.: DNS of passive scalar transport in turbulent channel flow at high Schmidt numbers. Int. J. Heat Fluid Flow **28**(6), 1204–1214 (2007)
17. Shaw, D.A., Hanratty, T.J.: Turbulent mass transfer rates to a wall for large Schmidt numbers. AIChE J. **23**(1), 28–37 (1977)
18. Son, J.S., Hanratty, T.J.: Limiting relation for the eddy diffusivity close to a wall. AIChE J. **13**(4), 689–696 (1967)

Check for updates

Study of RANS Turbulence Models for Turbulent Wake Flow in Adverse Pressure Gradient

Tobias Knopp[1(✉)], Paul Korsmeier[1], Mikael Strelets[2], and Ekaterina Guseva[2]

[1] DLR, Institut für Aerodynamik und Strömungstechnik, Bunsenstr. 10, 37073 Göttingen, Germany
{tobias.knopp,paul.korsmeier}@dlr.de
[2] Saint Petersburg Polytechnic University, Saint Petersburg, Russia

Abstract. We present an analysis of RANS turbulence models for a new turbulent wake flow case in adverse pressure gradient. The reference data are obtained from an LES. RANS simulations using the Spalart-Allmaras model, the SST model and the SSG/LRR-ω model show a tendency to underpredict the onset of flow reversal in the wake. The comparison with the LES data reveals that the SSG/LRR-ω model underpredicts the turbulent transport of the Reynolds stresses and overpredicts the ratio of production to dissipation. Therefore two modifications of the SSG/LRR-ω model are studied, i.e., the modification of the coefficient of the gradient diffusion hypothesis for the turbulent transport of the Reynolds stresses and the sensitization of the dissipation rate to irrotational strains. Both modifications improve the agreement with the reference data.

1 Introduction

The aerodynamic performance of a multi-element high-lift wing depends not only on the separation behaviour of the boundary layers, but also on the behaviour of the turbulent wake flows generated by the slat and the main wing element when subjected to a strong adverse pressure gradient, see [13]. The wake flow can be prone to flow reversal, which can lead to wing stall even if the boundary layer flow remains attached. Several investigations in the past showed that standard RANS models are not able to predict this flow phenomenon accurately. The SST model was found to underpredict the tendency for flow reversal for the test-case by Driver & Mateer, see [3], and this was attributed to the tendency to overpredict the Reynolds stresses. Similar observations were found in [2] for the SST model and for the SSG/LRR-ω model for the test-case by Driver & Mateer, and by Tummers et al. for their own experiment [12].

Albeit both experiments by Driver & Mateer and by Tummers et al. provide detailed data, the reasons for the deviation between the RANS simulations and the experiments cannot be fully explained, for which different reasons are possible. The first reason is that the boundary conditions in the experiments are

© The Author(s), under exclusive license to Springer Nature Switzerland AG 2021
A. Dillmann et al. (Eds.): STAB/DGLR Symposium 2020, NNFM 151, pp. 58–67, 2021.
https://doi.org/10.1007/978-3-030-79561-0_6

not defined exactly. In both cases flow control was applied on the wind tunnel walls, which generate the adverse pressure gradient, but the information given by the authors is not sufficient to reproduce the boundary conditions in RANS simulations. The second issue is that using experimental techniques alone, it is not possible to measure all terms of the turbulent kinetic energy equation and of the Reynolds stress equation. The third issue is that the experiments are likely to be influenced by side-wall effects, especially when flow reversal takes place. The displacement effect of the side-walls causes a mean flow acceleration in the centerplane, and this cannot be accounted for in a 2D simulation. On the other hand, in 3D simulations of the wind-tunnel experiments of wake-flow in adverse pressure gradient cited above (despite the lack of detailed information about the flow actuation), the onset of flow separation of the side-wall boundary layers in adverse pressure gradient plays a significant role and can hardly be distinguished from the onset of flow reversal in the centerline of the wake.

For these reasons, a joint German-Russian research project "Wake Flows in Adverse Pressure Gradient", supported by DFG and RBRF, was started in 2017. A new test-case was designed with exactly specified boundary conditions. The test-case was studied experimentally [1] and numerically (using Large-Eddy Simulation (LES)) [6]. The comparison between experiment and LES will be described in a separate paper. The LES provides all terms of the Reynolds stress equation and of the turbulent kinetic energy equation. In particular, unique data for the turbulent transport term and for the dissipation ϵ were obtained. Moreover the numerical experiment provides a well-defined test case with precisely known boundary conditions. The flow of the numerical experiment is statistically two-dimensional and free of side-wall effects.

2 Test Case Description

The test-case was designed to achieve a strong adverse pressure gradient in the wake with a significant reduction of the centerline velocity and a small region of flow-reversal. At the same time, the test-case should be suitable for RANS simulations, i.e., the boundary conditions should be well-defined and without the use of active flow control, e.g., synthetic jets, to avoid flow separation on the geometry parts which generate the adverse pressure gradient. Moreover, the aim was to achieve a macroscopically statistically steady flow, which was presumably not fully met, for example, for the experiment by Driver and Mateer, see [6].

Regarding the geometry of the test case in Fig. 1, a flat plate (FP) of length $L = 1.058\,\text{m}$ and thickness of $0.02\,\text{m}$ with a rounded leading edge and a linearly tapered trailing edge was used as a wake generator. On both sides of the symmetry plane, two thin liner foils (LF1 and LF2) act as a diffusor-stabilizer arrangement to impose the APG and avoid unsteady vortex shedding. The slot between the first and second liner foil serves as a passive mechanism to prevent flow separation on the liner foil. Active boundary layer control techniques were not necessary. This yields well-defined and stationary boundary conditions for the simulations. The origin of the coordinate system is located at the trailing edge of the splitter plate. The vertical distance h of the liner foil leading edge to the symmetry plane was varied to change the strength of the pressure gradient. The present publication shows the results for $h = 0.07\,\mathrm{m}$ (moderately strong APG, wake at the verge of flow reversal) and for $h = 0.05\,\mathrm{m}$ (strong APG, flow reversal in the wake) for $Re_L = 3.2 \times 10^6$ based on the onflow velocity U_{ref} and the length of the splitter plate L.

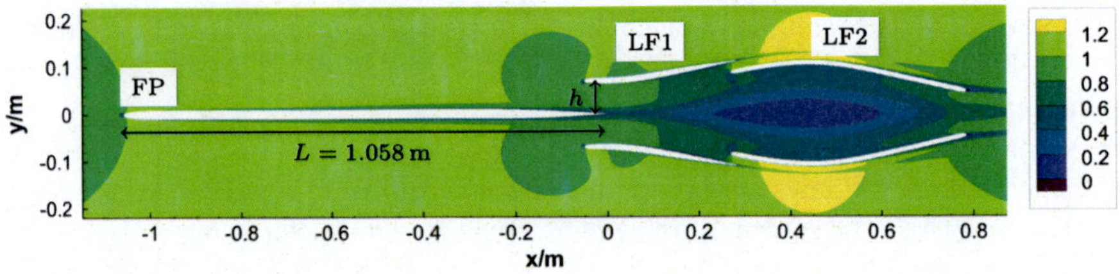

Fig. 1. Geometric setup and ratio of local to onflow velocity U/U_{ref} for $h = 0.07\,\mathrm{m}$

The reference data were obtained with the use of a scale-resolving simulation of the considered flow. Specifically, a zonal approach to turbulence representation was employed (see Fig. 2). It includes SST RANS computation of the initial part of the flat plate boundary layer and Improved Delayed Detached-Eddy Simulation (IDDES) [11] of the rest part of the flow. In the downstream part of the plate the IDDES functions as an LES with wall modelling (WMLES) while in the wake it performs as a pure LES. For triggering the WMLES branch of IDDES, the Volume Synthetic Turbulence Generator (VSTG) [10] is used, in order to create artificial turbulence at the inlet of the IDDES sub-domain needed for rapid transition from fully-modeled turbulence in the RANS sub-domain to mostly-resolved turbulence in the IDDES sub-domain, see Fig. 3. The simulation was performed with the use of the in-house high-order finite-volume structured multi-block code NTS [9]. The grid used in the simulation has about 50 million cells and satisfies commonly accepted requirements for RANS-LES computations. Its sufficiency is supported by the grid-sensitivity study.

For the specifics of the wind-tunnel experiment we refer to [1].

3 Numerical Setup for RANS Simulations

The present simulations were performed using the DLR TAU code, a compressible finite volume solver for unstructured and hybrid meshes, using a second-order accurate central scheme for the spatial discretisation and a local time stepping. Low Mach number preconditioning of the Rossow-Swanson type was employed.

At the inflow boundary, the reservoir pressure inflow condition was used. The reference values in the TAU code were set to $Re_L = 3.2 \times 10^6$ based on the onflow (reference) velocity $U_{\mathrm{ref}} = 50\,\mathrm{m/s}$, and the reference temperature was

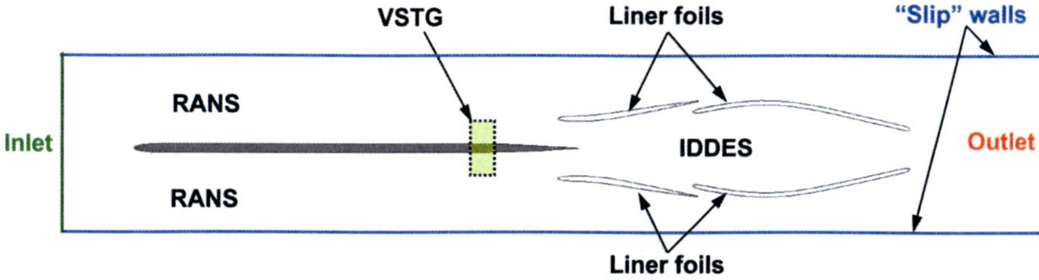

Fig. 2. Schematic of the zonal RANS-IDDES computational setup.

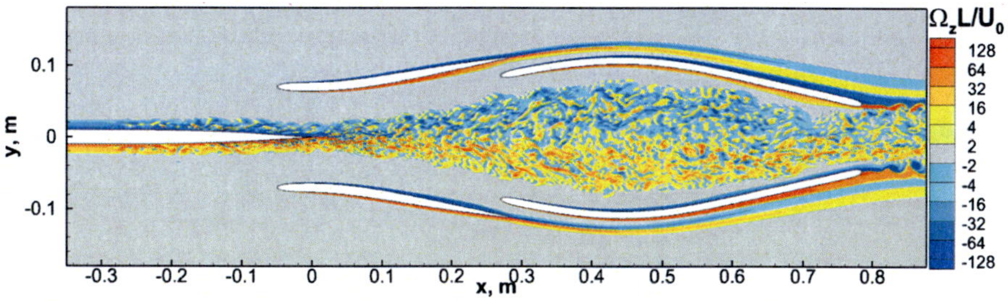

Fig. 3. Snapshot of the out-of-plane component Ω_z of the vorticity vector.

$T_{\mathrm{ref}} = 310.15\,\mathrm{K}$. The exit-pressure outflow condition was used at the outflow boundary, and the Euler wall condition at the upper and lower wind tunnel walls. The viscous surfaces of the flat plate and of all airfoils were fully turbulent.

The RANS meshes were generated using the Pointwise mesh generator. The spacing in streamwise direction is $\Delta x/L = 0.006$ on the flat plate and $\Delta x/L \leq 0.002$ in the wake. The wall-distance of the first grid point above the viscous walls y^+ are below one on the flat plate and on the liner foils.

4 RANS Turbulence Modelling

For the RANS simulations, both eddy-viscosity models and differential Reynolds stress models were used. Regarding the turbulent viscosity models, we used the Spalart-Allmaras (SA) and the Menter Shear-Stress Transport model (SST) both without and with the extensions for rotation and curvature (RC) and the quadratic constitutive relation (QCR). For the differential Reynolds stress models we used the SSG/LRR-ω model. The transport equation for the Reynolds stress tensor reads as

$$\frac{\partial(\overline{\varrho}R_{ij})}{\partial t} + \frac{\partial(\overline{\varrho}R_{ij}U_k)}{\partial x_k} = \overline{\varrho}(P_{ij} + \Pi_{ij} - \epsilon_{ij} + D_{ij})\,. \qquad (1)$$

The arising terms and their modelling are defined in Table 1 for the SSG/LRR-ω model. A single blending function (borrowed from the Menter SST model) is used

Table 1. The terms of the SSG/LRR-ω model and their modelling.

Name	Term	Model
Production	$\overline{\varrho}P_{ij}$	Unmodeled
Dissipation	$\overline{\varrho}\epsilon_{ij}$	Menter baseline for $\overline{\varrho}\omega$; $\epsilon = C_\mu k\omega$; $\overline{\varrho}\epsilon_{ij} = 2\overline{\varrho}\epsilon\delta_{ij}/3$
Pressure-strain	$\overline{\varrho}\Pi_{ij}$	SSG in outer part of boundary layers, LRR in inner wall layer
Diffusion	$\overline{\varrho}D_{ij}$	Simple or generalised gradient diffusion hypothesis

for all coefficients. For the turbulent transport (part of $\overline{\varrho}D_{ij}$), the generalised gradient diffusion hypothesis (GGDH) is used:

$$\overline{\varrho}D_{ij} = \frac{\partial}{\partial x_m}\left[\left(\overline{\mu}\delta_{mn} + C_{\text{GGDH}}\frac{\overline{\varrho}kR_{mn}}{\epsilon}\right)\frac{\partial R_{ij}}{\partial x_n}\right] \tag{2}$$

For more details, see [4] and references therein.

4.1 Sensitization to Pressure Gradients

In [7], the sensitization of the dissipation equation to irrotational strains by means of an additional term in the ϵ equation was proposed. The present work uses the corresponding term

$$S_{\omega,4} = -C^*_{\epsilon,4}\frac{\omega}{k}\left(\overline{u_s u_s}\frac{\partial U_s}{\partial x_s} + \overline{u_{n,1}u_{n,1}}\frac{\partial U_{n,1}}{\partial x_{n,1}} + \overline{u_{n,2}u_{n,2}}\frac{\partial U_{n,2}}{\partial x_{n,2}}\right) \tag{3}$$

which is based on the works [5,8]. Here, $C^*_{\epsilon,4} = 1.16$ and x_s is the streamline direction that forms an orthonormal frame with the coordinate directions $x_{n,1}$ and $x_{n,2}$.

4.2 Calibration of the GGDH Using the LES Data

The LES data provide all terms of the TKE balance, in particular the exact data for the turbulent transport term D_k^t from the triple correlations. We recall that the gradient diffusion hypothesis (GDH) for the k-equation considers the diffusive transport term $D_k = D_k^t + D_k^p + D_k^\nu$ with the turbulent part D_k^t, the fluctuating pressure part D_k^p, and the viscous part D_k^ν given by

$$D_k^t = -\frac{\partial}{\partial x_i}\frac{\overline{u_i'u_j'u_j'}}{2}, \qquad D_k^p = -\frac{\partial}{\partial x_i}\frac{\overline{u_i'p'}}{\varrho}, \qquad D_k^\nu := \nu\frac{\partial}{\partial x_i}\frac{\partial k}{\partial x_i} \tag{4}$$

and models the sum of the three terms as

$$D_k = \frac{\partial}{\partial x_i}\left((\nu + C_k\nu_t)\frac{\partial k}{\partial x_i}\right), \tag{5}$$

In the standard models like the SST model, $C_k = 1$ and ν_t is the eddy viscosity. In this work the calibration for C_k is revisited for wake flows at APG. From the LES data we evaluate D_k from (5). Here the eddy viscosity is computed from the relation

$$\nu_t = \frac{-\overline{u'v'}}{\frac{\partial U}{\partial y} + \frac{\partial V}{\partial x} + 10^{-22}}, \tag{6}$$

The comparison at $x = 0.2\,\text{m}$ yielded that $C_k = 2$ is more suitable than $C_k = 1$ which is shown in Fig. 4. The analogous modification of the generalized gradient diffusion hypothesis coefficent C_{GGDH} (cf. (2)) in the SSG/LRR-ω model was thus implemented and activated for $x \geq 0.1\,\text{m}$. Those results are denoted throughout the rest of this paper as "GGDH mod fac 2.0".

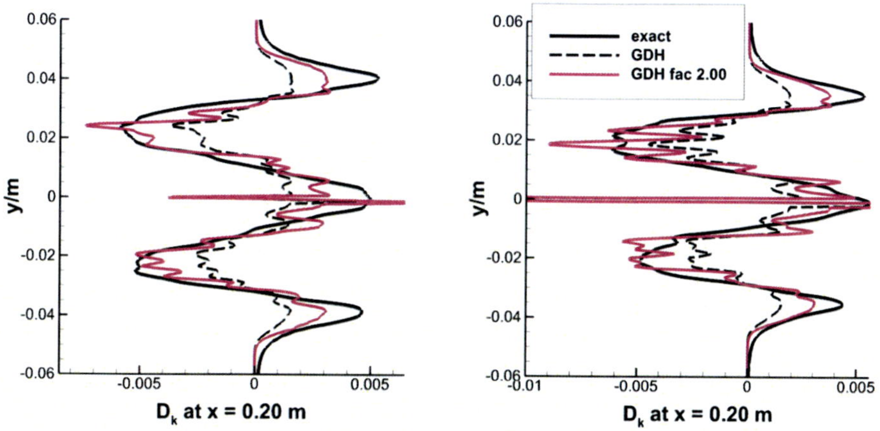

Fig. 4. Comparison of 'exact' TKE transport term D_k (in [m^2s^{-3}]) from the LES data and its modelings according to (5) with ν_t defined in (6) and either $C_k = 1$ (standard GDH) or $C_k = 2$ (modified GDH); for $h = 5\,\text{cm}$ (left) and $h = 7\,\text{cm}$ (right)

5 Results

5.1 Mean Velocity

First the mean velocity U_{CL} on the centerline is considered and shown in Fig. 5. In the LES, the case $h = 5\,\text{cm}$ produces an APG strong enough to cause mean flow reversal in the wake, while the case $h = 7\,\text{cm}$ is close to flow reversal, but U_{CL} remains positive.

In the very near wake region, i.e., in the region $x \approx 0.04\,\text{m}$ downstream of the trailing edge of the flat plate, the local maximum of U_{CL} obtained from the LES is significantly underpredicted by all RANS models. In this region, the flow is subjected to a local acceleration and is in non-equilibrium, which is hardly resolved RANS models. This needs to be analysed in future work. Further downstream, for $x > 0.1\,\mathrm{m}$, the streamwise decrease of U_{CL} for the standard RANS models is smaller than for the LES. This becomes even more clear just upstream of the minimum of U_{CL} for $x > 0.3\,\mathrm{m}$. The modification of the SSG/LRR-ω model using the $S_{\epsilon 4}$-term causes a significant reduction of U_{CL}, and brings the results closer to the LES data. The GGDH modification for the SSG/LRR-ω model also causes a reduction of U_{CL}. The RC-QCR-extension causes a significant increase of the centerline velocity for the SA model, and a small decrease for the SST model. For the case $h = 5\,\mathrm{cm}$, the magnitude of negative velocity in the region of flow reversal cannot be reached by any of the RANS models. Only the SSG/LRR-ω model with $S_{\epsilon 4}$-term predicts a small region of reverse flow. For the case $h = 7\,\mathrm{cm}$, the minimum centerline velocity predicted by the SSG/LRR-ω model with $S_{\epsilon 4}$-term is close to the LES data, albeit even a little lower.

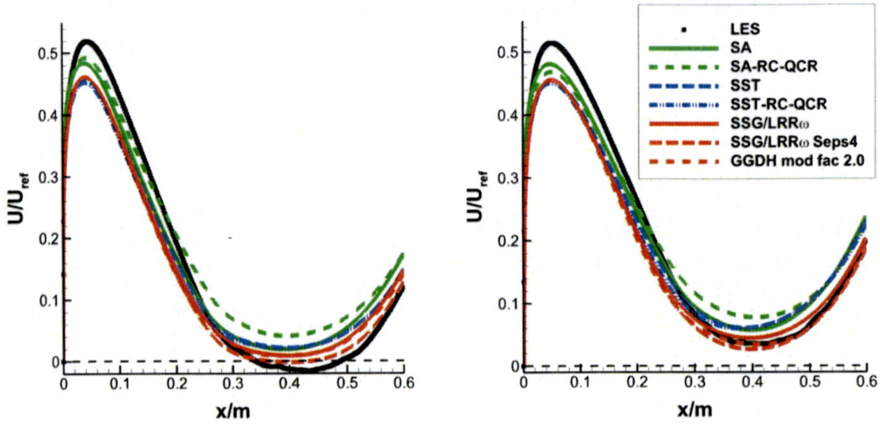

Fig. 5. Centerline velocity for $h = 5\,\mathrm{cm}$ (left) and $h = 7\,\mathrm{cm}$ (right)

5.2 Reynolds Stresses

Figure 6 shows the profiles of the Reynolds stress $\overline{u'v'}$ in y-direction at two different streamwise positions $x = 0.2\,\mathrm{m}$ and $x = 0.4\,\mathrm{m}$. The eddy-viscosity based RANS models tend to overestimate the magnitude of $\overline{u'v'}$ more than the SSG/LRR-ω model. The RC-QCR-extension for SA and SST increases the maximum of $\overline{u'v'}$ compared to the original model at $x = 0.2\,\mathrm{m}$. The modification of the SSG/LRR-ω using the $S_{\epsilon 4}$-term causes a significant overall reduction of $\overline{u'v'}$, except near the outer edge of the wake. The GGDH modification causes a similar reduction of $\overline{u'v'}$ and increases $\overline{u'v'}$ near the outer edge of the wake. This improves the agreement with the LES data near the outer edge of the wake.

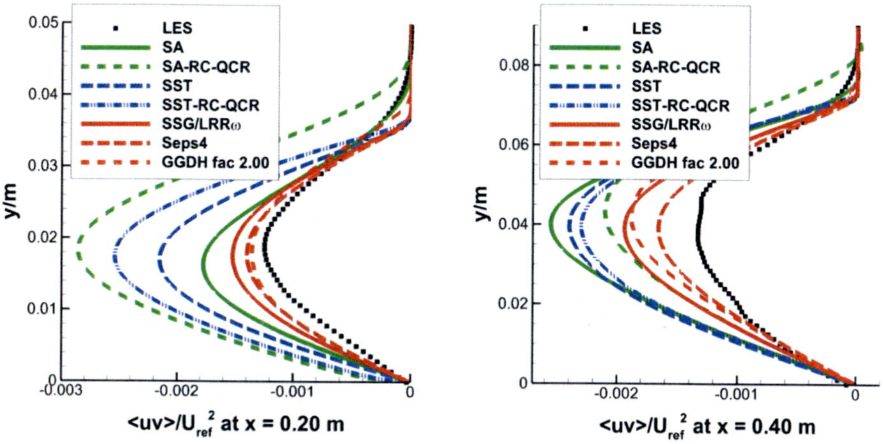

Fig. 6. Profiles of Reynolds stress $\overline{u'v'}$ for $h = 7\,\text{cm}$ at $x = 0.2\,\text{m}$ (left) and $x = 0.4\,\text{m}$ (right).

5.3 Ratio of Production to Dissipation of Turbulence

The LES data reveal that the most relevant terms in the balance equation for the turbulent kinetic energy are production P_k and the dissipation ϵ_k. The ratio of P_k/ϵ_k plays a central role in the streamwise evolution of the Reynolds stresses, and helps to gain further understanding in the effect of the $S_{\epsilon 4}$-term. The results are shown in Fig. 7.

Three regions of the wake (in wall-normal direction) can be distinguished for the comparison between the SSG/LRR-ω results and the LES data. The first region is the centerline region ($y < 0.005\,\text{m}$). The second region is the so-called shear center region of the largest mean velocity gradient in the middle part of the wake ($0.02\,\text{m} < y < 0.03\,\text{m}$). The third region is the outer edge of the wake. Here the y-values of the different regions are for $x = 0.2\,\text{m}$, see Fig. 7 (left).

At $x = 0.2\,\text{m}$, the SSG/LRR-ω model significantly overpredicts P_k/ϵ_k both in the centerline region and in the shear center region. For the SSG/LRR-ω with $S_{\epsilon 4}$-term, P_k/ϵ_k is significantly reduced both in the centerline region and in the shear-center region at $x = 0.2\,\text{m}$. Further downstream, at $x = 0.3\,\text{m}$, P_k/ϵ_k is still a little overpredicted in the shear center region. Hence, overall, the $S_{\epsilon 4}$ modification reduces P_k/ϵ_k and brings the prediction closer to the LES data.

An additional indication that it is the shear center region which is of crucial importance can be inferred from P_k/ϵ_k for the SST model. The SST model overpredicts P_k/ϵ_k in the shear center region, whereas SST-RC-QCR is closer to the LES. This is in agreement with the higher values for U_{CL} for the SST model compared to SST-RC-QCR in Fig. 5 (left) for $0.2\,\text{m} < x < 0.3\,\text{m}$. The centerline velocity remains much higher for both models at $x = 0.4\,\text{m}$ and is even larger than for the SSG/LRR-ω model, see Fig. 5 (right).

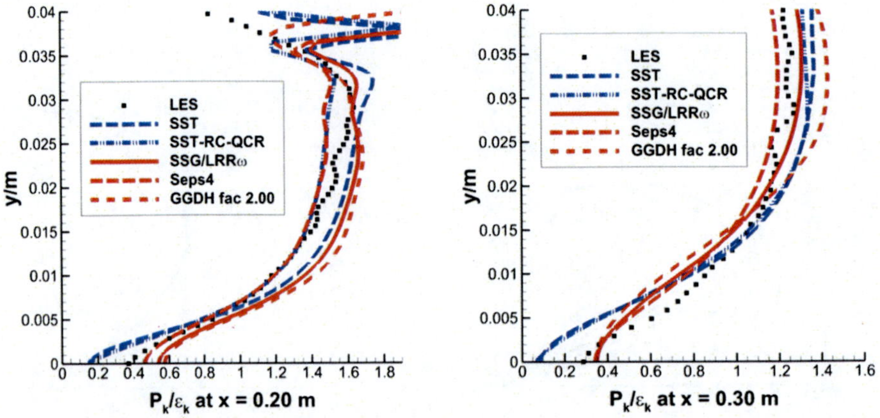

Fig. 7. Profiles of the production to dissipation ratio P_k/ϵ_k of turbulent kinetic energy at $x = 0.2\,\mathrm{m}$ (left) and $x = 0.4\,\mathrm{m}$ (right) for the case $h = 7\,\mathrm{cm}$.

6 Conclusion

The present work describes the results of RANS computations of a turbulent wake flow in adverse pressure gradient with the use of the SA, SST and SSG/LRR-ω turbulence models. The models were found to underpredict the onset of flow reversal in the wake. The comparison with the LES data revealed two shortcomings of the SSG/LRR-ω model. First, it underpredicts the turbulent transport of the Reynolds stresses. Second, it overpredicts the ratio of production to dissipation, and this becomes most important in the so-called shear center region of the largest mean velocity gradient. Therefore two modifications of the SSG/LRR-ω were studied, i.e., the modification of the coefficient of the gradient diffusion hypothesis for the turbulent transport of the Reynolds stresses and the sensitization of the dissipation rate to irrotational strains. Both modifications reduce the predicted Reynolds stresses in the shear center region and improve the agreement with the LES data. The results indicate that possibly a single modification term cannot eliminate the differences between the SSG/LRR-ω model and the LES reference data.

In the very near wake region just downstream of the trailing edge, the strong mean flow acceleration observed in the LES is significantly underpredicted by all RANS models. This issue needs to be studied in future research.

Acknowledgement. The funding of the project "Wake Flows in Adverse Pressure Gradient" by DFG and RBRF (Grants No. RA 595/26-1, 17-58-12002 and KN 888/3-1) is gratefully acknowledged.

References

1. Breitenstein, W., et al.: A wind tunnel experiment for symmetric wakes in adverse pressure gradients. In: AIAA Paper 2019–1875 (2019)
2. Burnazzi, M., et al.: Characterization of off-surface separation caused by adverse pressure gradient in a turbulent wake. In: 21th STAB/DGLR Symposium on New Results in Numerical and Experimental Fluid Mechanics, 21, STAB/DGLR Symposium 2018, 6–8 November 2018, Darmstadt, Deutschland (2018)
3. Driver, D.M., Mateer, G.G.: Wake flow in adverse pressure gradient. Int. J. Heat Fluid Flow **23**, 564–571 (2002)
4. Eisfeld, B., Rumsey, C., Togiti, V.: Verification and validation of a second-moment-closure model. AAIA J. **54**, 1524–1541 (2016)
5. Grenson, P., Probst, A.: Implementation and validation of a near-wall reynolds-stress model based on the homogeneous specific dissipation rate for application to aerodynamic flows. In: Proceedings of the 10th PEGASUS-AIAA Student Conference (2014)
6. Guseva, E., et al.: Experimental/numerical study of turbulent wake in adverse pressure gradient. In: Hoarau, Y., Peng, S.H., Schwamborn, D., Revell, A., Mockett, C. (eds.) Progress in Hybrid RANS-LES Modelling. Notes on Numerical Fluid Mechanics and Multidisciplinary Design, vol. 143, pp. 401–412. Springer, Cham (2020). https://doi.org/10.1007/978-3-030-27607-2_33
7. Hanjalic, K., Launder, B.E.: Sensitizing the dissipation equation to irrotational strains. J. Fluid Eng. T. ASME **102**, 34–40 (1980)
8. Probst, A., Radespiel, R.: Implementation and extension of a near-wall Reynolds-Stress Model for application to aerodynamic flows on unstructured meshes. In: AIAA 2008-770 (2008)
9. Shur, M., Strelets, M., Travin, A.: High-order implicit multi-block Navier-stokes code: ten-years experience of application to RANS/DES/LES/DNS of turbulent flows (2004). In: 7th Symposium on Overset Composite Grids and Solution Technology, Invited Lecture, Huntington Beach, CA, October (2004). http://cfd.spbstu.ru/agarbaruk/c/document_library/DLFE-42505.pdf
10. Shur, M., Strelets, M., Travin, A.: Improved embedded approaches. Notes Numer. Fluid Mech. Multidiscip. Des. **134**, 51–87 (2017)
11. Shur, M.L., Spalart, P.R., Strelets, M.K., Travin, A.: A hybrid RANS-LES approach with delayed-DES and wall-modelled LES capabilities. Int. J. Heat Fluid Flow **29**, 1638–1649 (2008)
12. Tummers, M.J., Hanjalic, K., Passchier, D.M., Henkes, R.A.W.M.: Computations of a turbulent wake in a strong adverse pressure gradient. Int. J. Heat Fluid Flow **28**, 418–428 (2007)
13. Ying, S.X., Spaid, F.W., McGinley, C.B., Rumsey, C.L.: Investigation of confluent boundary layers in high-lift flows. J. Aircraft **36**(3), 550–562 (1996)

Study on Large-Scale Amplitude Modulation of Near-Wall Small-Scale Structures in Turbulent Wall-Bounded Flows

Esther Mäteling(✉), Michael Klaas, and Wolfgang Schröder

Chair of Fluid Mechanics and Institute of Aerodynamics, Wüllnerstraße 5a, 52062 Aachen, Germany
e.maeteling@aia.rwth-aachen.de

Abstract. Recently, a novel approach to study the inner-outer interactions in turbulent wall-bounded flows has been presented by Mäteling et al. [8], which tackles the combined effects of superposition, bursting events, and amplitude modulation based on the time-dependent streamwise velocity fluctuations. This approach is extended in the current study to all three velocity fluctuation components. Due to the incorporation of the wall-normal velocity fluctuations, the bursting events can be study in more detail and independently of the superposition. By using a multivariate empirical mode decomposition for the scale separation, all velocity components at both wall-normal positions can be decomposed simultaneously to ensure proper scale alignment between the resulting modal representations. The results show that the superposition effect and the amplitude modulation exist in all three velocity components but are most intense for the streamwise fluctuations. A high level of coherence of the sweeps and ejections is also detected by conventional cross-correlation.

Keywords: Amplitude modulation · Scale interaction mechanisms · Turbulent wall-bounded flow

1 Introduction

The interaction between outer-layer large-scale structures and near-wall turbulence has been a subject of turbulence research for the past decades. Aiming at a deeper understanding of the internal structure of turbulent wall-bounded flows, the inner-outer coherence seems to play a fundamental part in the energy and momentum exchange of flow structures of different scales. Yet, four phenomena have been assigned to this structural interaction, which are superposition, bursting events, amplitude modulation, and frequency modulation. The interested reader is referred to the excellent reviews of [4,7] concerning this topic.
In a recent study, Mäteling et al. [8] introduced a novel approach to investigate the combined effects of superposition, bursting events, and amplitude modulation. They used time-dependent one-dimensional (1D) streamwise velocity

© The Author(s), under exclusive license to Springer Nature Switzerland AG 2021
A. Dillmann et al. (Eds.): STAB/DGLR Symposium 2020, NNFM 151, pp. 68–76, 2021.
https://doi.org/10.1007/978-3-030-79561-0_7

fluctuations at an inner- and an outer-layer position of a direct numerical simulation (DNS) of a turbulent channel flow (TCF) at a friction Reynolds number $Re_\tau \approx 1000$. By removing the influences of superposition and bursting events prior to performing the analysis of amplitude modulation, the latter could be revealed in a very clear manner. This is additionally achieved by a modified algorithm to calculate the large-scale amplitudes. They are directly obtained from the large scales after performing a second scale separation instead of the conventionally used large-scale trend of the small-scale amplitudes [1,9].

The current study extends the approach of Mäteling et al. [8] to all three velocity fluctuation components. This additionally enables a separate, in-depth investigation of superposition and bursts since the bursting events can now be defined based on the signs of streamwise and wall-normal fluctuations using quadrant analysis. The authors of [8] showed that a scale separation by empirical mode decomposition (EMD) and by spectral filtering yields similar outcomes of the interaction analysis. In the present analysis, an extension of the univariate EMD to multiple variates by Rehman & Mandic [11] is applied. This multivariate EMD (MEMD) enables a simultaneous decomposition of all velocity components at both wall-normal positions, which ensures a proper scale alignment between individual modes. The spectral alignment guarantees that the same scales are assigned to the large scales at the near-wall and the outer-layer position, which increases the precision of the subsequent inner-outer interaction analysis.

In the following, the DNS data of the TCF are presented in Sect. 2. Then, the methods for scale separation and for revealing the interaction mechanisms between the outer-layer and the near-wall signals are outlined in Sect. 3. The results are shown in Sect. 4 and conclusions are drawn in Sect. 5.

2 Data Set

The present study uses DNS data of a TCF at $Re_\tau = \frac{u_\tau h}{\nu} \approx 1000$, where u_τ is the friction velocity, h the channel half height, and ν the kinematic viscosity. The simulation data are provided by the *Johns Hopkins Turbulence Database* [2] and comprise a domain of $(L_x \times L_y \times L_z) = (8\pi h \times 2h \times 3\pi h)$ with $h = 1$ and $(2048 \times 512 \times 1536)$ nodes. The cell sizes measure $(\Delta x^+ \times \Delta y^+ \times \Delta z^+) = (12.2639 \times 1.65199 \cdot 10^{-2}$ (at the wall) -6.15507 (at the centerline) $\times 6.13196)$.

For the current investigation, 1D time-dependent signals at a fixed streamwise and spanwise position are extracted at two wall-normal locations. The near-wall velocity fluctuations are taken at $y^+_{NW} = 1$ and the outer-layer fluctuations are extracted at $y^+_{OL} = 123$, where the large-scale motions are most energetic according to $y^+_{OL} = 3.9\sqrt{Re_\tau}$ [6]. In total, 4000 time steps with a spacing of $\Delta t^+ = 0.325$ are available, which results in a total temporal length of $t^+ \approx 1300$. The temporal resolution is comparable to hot-wire anemometry data that is often used to study inner-outer interactions, e.g., [10]. For more information about the DNS data set, the reader is referred to [2].

Throughout this paper, the superscript + indicates quantities scaled by inner units, i.e., by u_τ and ν. The index LS indicates large scales. A near-wall signal is denoted by the index NW, an outer-layer location is defined by OL, and the symbol ^ marks an amplitude.

3 Methodical Approach

3.1 Multivariate Empirical Mode Decomposition

When applying EMD for scale separation in the inner-outer interaction analysis, past studies, e.g., [1,8], used its univariate form originally introduced by Huang et al. [3]. This data-driven algorithm decomposes a non-linear, time-dependent signal into Intrinsic Mode Functions (IMFs), which are solely based on the signal's characteristics. Further information of its features and application to fluid flow data can be found in, e.g., [8]. However, this method has some disadvantages, which certainly influence the precision of the interaction analysis. Since velocity signals at different wall-normal locations usually possess individual characteristics, a varying number of decomposition levels, i.e., a different number of IMFs, is obtained when applying the univariate EMD. In addition, mode mixing occurs, which is a consequence of signal intermittency causing an overlapping of the mode spectra. These effects lead to improper mode alignment and complicate a comparison of individual IMFs at the considered near-wall and outer-layer location as it is done to study the inner-outer interactions.

To overcome these drawbacks, which also play a major role in various physics-related applications of EMD, Rehman & Mandic developed a multivariate extension (MEMD), which simultaneously decomposes multi-channel temporal signals [11]. The authors utilized real-valued signal projections to tackle the challenging definition of local extrema in multivariate data, which is a key step to determine the local mean during the sifting process. The main steps of the algorithm are outlined in the following based on the description found in [11].

1. Selection of a suitable low-discrepancy point set, e.g., a Hammersley sequence, for sampling on an $(n-1)$ sphere, where n denotes the number of variates. The point set defines the angles φ_k, with k being the projection directions, from which the unit projection vectors $p^{\varphi_k}(t)$ are obtained.
2. Projection of the multivariate input signal $I(t)$ on $p^{\varphi_k}(t)$, which is denoted by $P^{\varphi_k}(t)$.
3. Determination of the locations of the local extrema of each projection $P^{\varphi_k}(t)$.
4. Interpolation of the original signal $I(t)$ at these extrema locations yields the mean envelopes $e^{\varphi_k}(t)$.
5. Averaging the envelopes $e^{\varphi_k}(t)$ over all projection directions gives the local mean envelope $m(t)$.
6. Extraction of the detail $d(t)$ using $I(t)$ - $m(t)$. If the detail fulfills the stoppage criterion for a multivariate IMF [11], the first IMF equals $d(t)$ and the above procedure is applied to $I(t)$ - $d(t)$. Otherwise, the sifting is proceeded with $d(t)$.

The interested reader is referred to [11] for more details about the MEMD algorithm. For the current analysis, six variates (u', v', w' at y^+_{NW} and at y^+_{OL}) are simultaneously decomposed using 22 projection directions. The first six IMFs are assigned to the small scales and the remaining modes to the large scales.

3.2 Inner-Outer Interaction Analysis

The three phenomena of superposition, bursting activity, and amplitude modulation are investigated with respect to the inner-outer interaction. The procedure described in [8] is adapted to all three velocity components and due to the lack of space, only a rough outline and the variations to the original approach are given in the following. The reader is referred to [8] for in-depth physical and mathematical foundations of the analysis.

Besides the application of the MEMD, a main difference of the current study is the separate investigation of superposition and bursts. Therefore, the large-scale components of all three velocity fluctuations at y^+_{NW} and y^+_{OL} - obtained by MEMD - are correlated by standard cross-correlation (Eq. 1 with $q' = \{u', v', w'\}$) to analyze the superposition. Additionally, based on quadrant analysis, the large-scale streamwise and wall-normal components that either indicate sweeps or ejections are separately correlated (Eq. 1 with $q' = \{q'_{sw}, q'_{ej}\}$). Therefore, the streamwise and the wall-normal fluctuations of both events are combined to complex numbers prior to performing the cross-correlation.

$$R_i\left(\Delta t^+\right) = \frac{\overline{q'_{NW,LS}\left(t^+\right)} \cdot \overline{q'_{OL,LS}\left(t^+ + \Delta t^+\right)}}{\sqrt{\overline{q'^2_{NW,LS}\left(t^+\right)}} \cdot \sqrt{\overline{q'^2_{OL,LS}\left(t^+ + \Delta t^+\right)}}} \tag{1}$$

The outer-layer large-scale influence is subsequently removed from all three near-wall velocity components. The event-specific strength of correlation $R_{i,max}$ as well as the inclination angle

$$\Theta_i = tan^{-1}\left(\frac{\Delta y^+_{NW-OL}}{\Delta x^+_i}\right) \tag{2}$$

are thereby considered. The difference Δy^+_{NW-OL} corresponds to the wall-normal distance between inner and outer layer, while Δx^+_i is the spatial shift obtained from Δt^+_i via Taylor's hypothesis. The remaining near-wall signals with $q' = \{u', v', w'\}$ are calculated by

$$\begin{aligned} q'_{NW,woSB}(t^+) = q'_{NW}(t^+) &- R_{S,q',max} \cdot q'_{OL,LS}(t^+, \Theta_{S,q'}) \\ &- R_{sw,max} \cdot q'_{OL,LS,sw}(t^+, \Theta_{sw}) \\ &- R_{ej,max} \cdot q'_{OL,LS,ej}(t^+, \Theta_{ej}). \end{aligned} \tag{3}$$

With respect to the bursts, only the streamwise (wall-normal) component is subtracted from the streamwise (wall-normal) near-wall signal and the spanwise near-wall fluctuations are certainly not affected by bursting activity.

A second MEMD is applied to the remaining near-wall signals and the unchanged outer-layer fluctuations. Using the Hilbert transform with $q' = \{u', v', w'\}_{NW,OL}$

$$q'_{HT,LS2}\left(t^{+}\right) = \frac{P}{\pi}\int_{-\infty}^{\infty} \frac{q'_{LS2}\left(t^{\star}\right)}{t - t^{\star}} dt^{\star}, \tag{4}$$

imaginary components are determined to build an analytic signal from each large-scale component, which enables the calculation of the large-scale amplitudes via

$$\hat{q}'_{LS2}\left(t^{+}\right) = \sqrt{q'^{2}_{LS2}\left(t^{+}\right) + q'^{2}_{HT,LS2}\left(t^{+}\right)}. \tag{5}$$

They are correlated for the amplitude modulation analysis by standard cross-correlation (Eq. 1 with $q' = \{\hat{u}', \hat{v}', \hat{w}'\}$).

4 Results

The large-scale components obtained by MEMD are shown in Fig. 1 for all three velocity components and the bursting events. They are scaled by the corresponding rms values to obtain comparable fluctuation values. All distributions possess a similar trend between the near-wall and the outer-layer scales. Local extrema of the outer large scales are often found in the near-wall signals but with a temporal shift and a varying strength. Particularly for the wall-normal fluctuations (Fig. 1b), an offset occurs between near-wall and outer-layer distribution. The near-wall bursting events shown in Fig. 1d often possess a shorter lifetime compared to their outer-layer counterparts.

To quantify the degree of inner-outer interaction, the large-scale components shown in Fig. 1 are cross-correlated. The resulting correlation functions R_i are given in Fig. 2 as a function of the time shift Δt^{+}. The permitted shifts along the positive and the negative time axis comprise inclination angles in the range $\Theta_i \geq \pm 1.27°$. From Fig. 2a it is obvious that the superposition of the streamwise fluctuations is most intense with a maximum correlation coefficient of $R_{S,u,max} = 0.69$. The value is similar to the coefficients reported in the literature [5, 10]. The wall-normal correlation is quite low with $R_{S,v,max} = 0.16$ and the spanwise correlation features two distinct peaks of which the slightly higher peak indicates $R_{S,w,max} = 0.28$. This somehow shows the limitations of classical cross-correlation when applied to fluctuation values. Although an inner-outer coherence is visible in Fig. 1 with the naked eye for all velocity components, the correlation does not reflect this effect. Particularly for the wall-normal fluctuations, this is caused by the offset between inner and outer distribution, which often leads to opposed signs that falsify the correlation results. Hence, alternative approaches should be developed in the future for a more accurate determination of the inner-outer correlation.

Figure 2b shows high maximum correlation values for the bursting events. The sweeps' peak correlation measures $R_{sw,max} = 0.68$ and the value of the ejections is slightly lower with $R_{ej,max} = 0.63$. These values are thus comparable

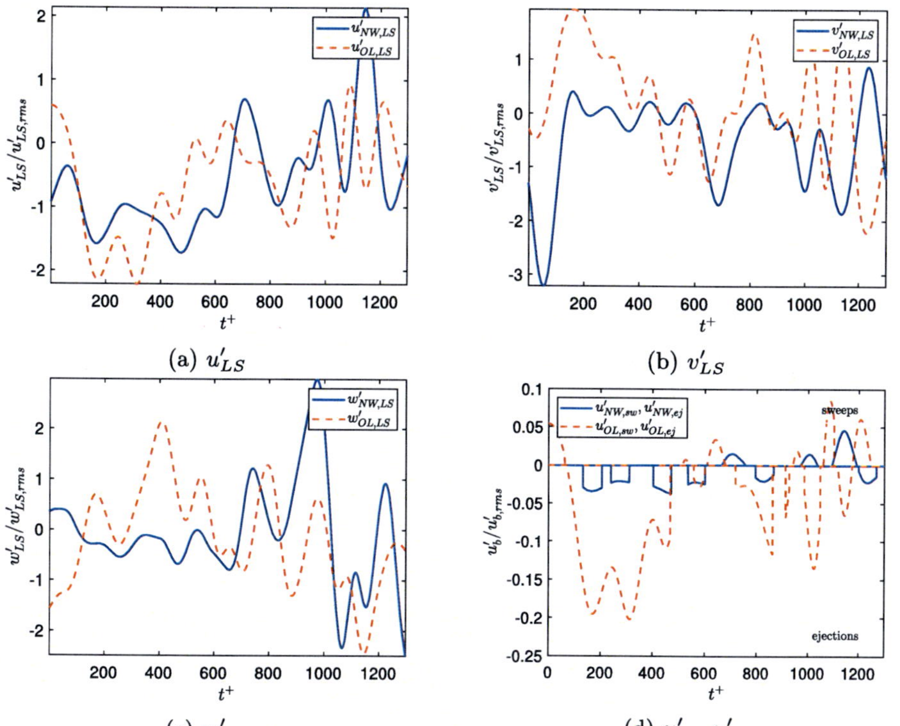

(a) u'_{LS} (b) v'_{LS} (c) w'_{LS} (d) u'_{sw}, u'_{ej}

Fig. 1. Near-wall $\{u, u_b, v, w\}'_{NW,LS}$ and outer-layer large-scale signals $\{u, u_b, v, w\}'_{OL,LS}$.

to the superposition of the streamwise fluctuations. Since the signs of near-wall and outer-layer features are always the same, the above addressed drawback of the cross-correlation method does not appear for the bursts.

All maximum correlation values and the corresponding inclination angles are summarized in Table 1. While the superposition of streamwise and wall-normal fluctuations indicates a similar inclination angle around $\Theta_S \approx 8°$, the spanwise correlation has a lower angle of $\Theta_{S,w} = 2.4°$. A more surprising deviation occurs for the bursting events. The sweeping interaction possesses a negative angle, which indicates leading near-wall activity, and the ejections are characterized by a very high angle, i.e., only a small shift between inner and outer ejections. This demands for further analysis with an extended data range that is beyond the scope of the current study.

In a subsequent step, the outer-layer influences are removed from the near-wall signals. Afterwards, the remaining near-wall and the unchanged outer-layer signals are decomposed by MEMD yielding "updated" large scales. The amplitudes of these large scales are determined from the analytic signals and are

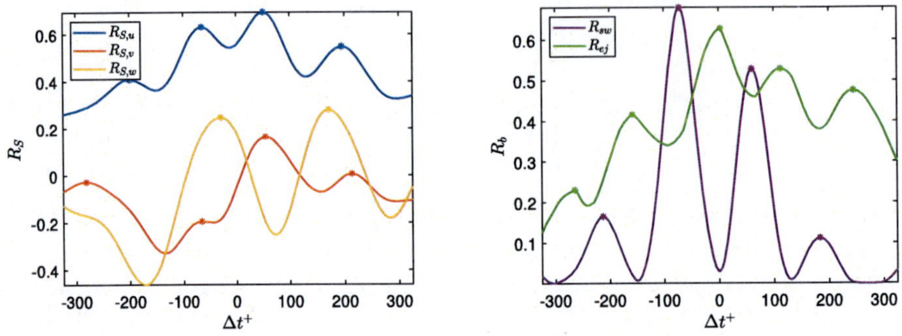

(a) R_S (b) R_b

Fig. 2. Correlation coefficients of large-scale features shown in Fig. 1.

Table 1. Correlation coefficients and inclination angles of all phenomena.

	$R_{S,max}$	Θ_S	$R_{sw,max}$	Θ_{sw}	$R_{ej,max}$	Θ_{ej}	$R_{AM,max}$	Θ_{AM}
u'	0.69	8.08°	0.68	−5.62°	0.63	65.71°	0.96	16.25°
v'	0.16	7.47°					0.86	90°
w'	0.28	2.40°	-	-	-	-	0.87	10.55°

shown in Fig. 3. A very high coherence between inner and outer distributions is observed, which is also reflected in the correlation coefficients pictured in Fig. 3d. With a maximum value of $R_{AM,u,max} = 0.96$, the streamwise amplitudes possess the highest correlation. Wall-normal and spanwise amplitudes behave very similar with $R_{AM,max} \approx 0.87$. However, the spanwise correlation function shows two distinct peaks as it was also observed with respect to the superposition. The periodicity of alternating near-wall streaks might cause this effect. Using the peak closer to a zero shift, its inclination angle is in a similar range as for the streamwise amplitudes with $\Theta_{AM,u} = 16.25°$. The near-wall wall-normal amplitudes are not shifted with respect to their outer-layer counterparts.

A study of the amplitude modulation of all three velocity components has been previously performed by Talluru et al. [12]. It must be noted that the approach to investigate the amplitude modulation is the same as reported in [9] and it was already shown in [8] that the currently used, novel approach gives a much clearer picture of the phenomenon for the streamwise components. Thus, it is not surprising that the detected degree of modulation is also higher for the two remaining velocity components. However, the study of [12] did not use near-wall components as close to the wall as in the present analysis due to experimental restrictions, which additionally hampers a precise comparison. Therefore, we applied their method to the DNS data used in the present study. The results reveal maximum correlation coefficients close to zero or even negative values. Hence, it is fair to state that the novel approach for the amplitude modulation analysis is superior in revealing this phenomenon for all three velocity components.

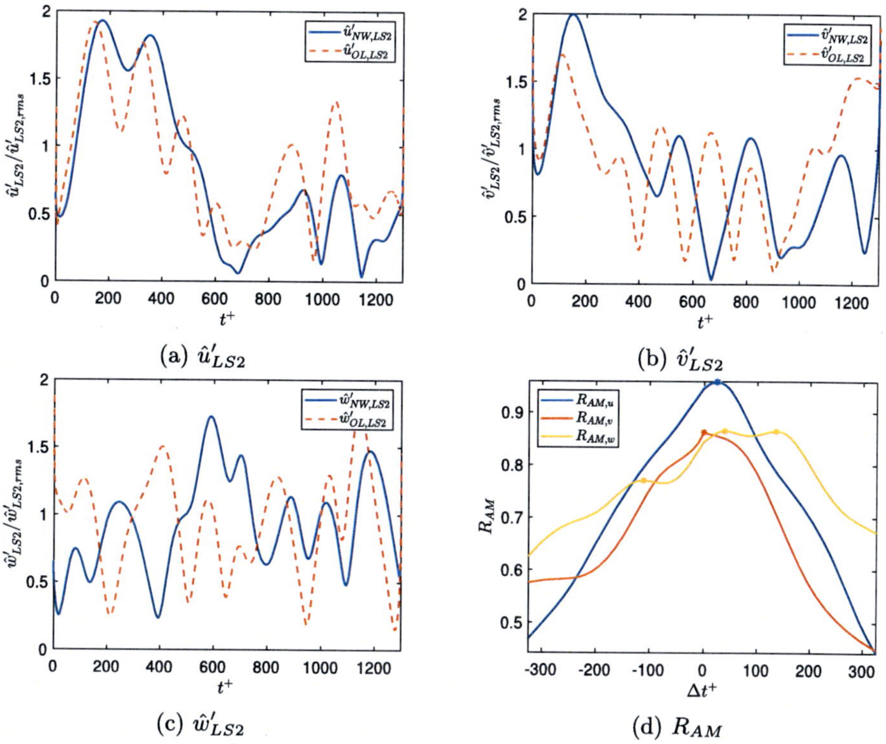

Fig. 3. Near-wall $\{\hat{u}, \hat{v}, \hat{w}\}'_{NW,LS2}$ and outer-layer large-scale amplitudes $\{\hat{u}, \hat{v}, \hat{w}\}'_{OL,LS2}$ and their correlation R_{AM}.

5 Conclusions

The inner-outer interaction is studied for a turbulent channel flow at $Re_\tau \approx 1000$. All three velocity components at the near-wall location $y^+_{NW} = 1$ and at the outer-layer location $y^+_{OL} = 123$ are considered in time-dependent, one-dimensional form. To obtain the large-scale components, the velocity fluctuations are simultaneously decomposed by a multivariate empirical mode decomposition, which ensures the alignment of scales in the individual modes. The subsequent analysis of superposition and bursting activity based on standard cross-correlation reveals a high inner-outer correlation between the large-scale streamwise components, the sweeps, and the ejections. The wall-normal and the spanwise large scales also possess a coherence between both locations but this is mathematically lost to a great extent in the applied cross-correlation due to opposite signs of the components. Hence, alternative methods are to be investigated to quantify the degree of inner-outer coherence.

Removing the influences of the superposition and bursting activity from the near-wall signals enables the analysis of the amplitude modulation. With $R_{AM} \geq 0.86$, the correlation is very high in all three directions but the streamwise amplitudes correlate the most. While the inclination angles of streamwise and spanwise amplitude modulation are similar and comparable to the effects of superposition, the spanwise modulation indicates a zero shift between inner and outer amplitudes. The high degree of amplitude modulation across all velocity components deviates from the findings in other studies and reveals the superior performance of the methodical approach introduced by Mäteling et al. [8].

References

1. Altıntaş, A., Davidson, L., Peng, S.: A new approximation to modulation-effect analysis based on empirical mode decomposition. Phys. Fluids **31**(2), 025117 (2019)
2. Graham, J., Kanov, K., Yang, X., Lee, M., Malaya, N., Lalescu, C., Burns, R., Eyink, G., Szalay, A., Moser, R., et al.: A web services accessible database of turbulent channel flow and its use for testing a new integral wall model for LES. J. Turbul. **17**(2), 181–215 (2016)
3. Huang, N.E., Shen, Z., Long, S.R., Wu, M.C., Shih, H.H., Zheng, Q., Yen, N.C., Tung, C.C., Liu, H.H.: The empirical mode decomposition and the Hilbert spectrum for nonlinear and non-stationary time series analysis. Proc. R. Soc. London Ser. A Math. Phys. Eng. Sci. **454**(1971), 903–995 (1998)
4. Jiménez, J.: Cascades in wall-bounded turbulence. Ann. Rev. Fluid Mech. **44**, 27–45 (2012)
5. Liu, H.Y., Bo, T.L., Liang, Y.R.: The variation of large-scale structure inclination angles in high Reynolds number atmospheric surface layers. Phys. Fluids **29**(3), 035104 (2017)
6. Marusic, I., Mathis, R., Hutchins, N.: High Reynolds number effects in wall turbulence. Int. J. Heat Fluid Flow **31**, 418–428 (2010)
7. Marusic, I., Baars, W.J., Hutchins, N.: Scaling of the streamwise turbulence intensity in the context of inner-outer interactions in wall turbulence. Phys. Rev. Fluids **2**(10), 100502 (2017)
8. Mäteling, E., Klaas, M., Schröder, W.: Detection of small-scale/large-scale interactions in turbulent wall-bounded flows. Phys. Rev. Fluids **5**(11), 114610 (2020)
9. Mathis, R., Hutchins, N., Marusic, I.: Large-scale amplitude modulation of the small-scale structures in turbulent boundary layers. J. Fluid Mech. **628**, 311–337 (2009)
10. Mathis, R., Hutchins, N., Marusic, I.: A predictive inner-outer model for streamwise turbulence statistics in wall-bounded flows. J. Fluid Mech. **681**, 537–566 (2011)
11. Rehman, N., Mandic, D.P.: Multivariate empirical mode decomposition. Proc. R. Soc. A Math. Phys. Eng. Sci. **466**(2117), 1291–1302 (2010)
12. Talluru, K., Baidya, R., Hutchins, N., Marusic, I.: Amplitude modulation of all three velocity components in turbulent boundary layers. J. Fluid Mech. **746**, 1–11 (2014)

Application of Multitask Learning for Enhancement of Spalart-Allmaras Turbulence Model

Rohit Pochampalli(✉), Emre Özkaya, Beckett Y. Zhou, Guillermo Suarez, and Nicolas R. Gauger

Chair for Scientific Computing, TU Kaiserslautern, Bldg/Geb 34-36, Paul-Ehrlich-Strasse, 67663 Kaiserslautern, Germany
rohit.pochampalli@scicomp.uni-kl.de

Abstract. Correcting functional terms using data driven techniques is one recent approach to improve the predictive accuracy of turbulence models. We present a multitask learning framework that can be used to generalize functional corrections of turbulence models to different flow conditions and test cases. The approach is developed with a view to generalize the correction of turbulence models using data from different flow conditions. The machine learning model first learns a compressed latent representation of flow variables that are most correlated with the correction term and has enhanced predictive accuracy for new test cases. The approach is shown to improve the performance of Spalart-Allmaras turbulence model in massively separated flow regimes.

1 Introduction

Reynolds averaged Navier Stokes (RANS) based turbulence models are the predominant approach to turbulence modeling by virtue of their low computational cost. Turbulence models, typically based on the Boussinesq hypothesis, introduce one or more transport (or algebraic) equations, that are used to model the unknown Reynolds stress terms in the RANS equations using known mean flow quantities. In general, turbulence models are developed for specific applications and perform poorly if the flow conditions are drastically different from those for which it was calibrated. On the other hand, high fidelity simulations such as direct numerical simulation (DNS) and large eddy simulation (LES) methods are infeasible at high Reynolds numbers owing to their high computational cost. Hybrid approaches such as Detached Eddy Simulation (DES) [1] and Delayed Detached Eddy Simulation (DDES) [2] improve on the fidelity over RANS based simulations [3] typically by employing a redefinition of the length scales used in the turbulence models. In this work, we address the effects of introducing additional complexity into the model on its fidelity. We show that the predictive accuracy of a specific turbulence model, in this case, the accuracy of lift coefficients of airfoils as estimated by the Spalart - Allmaras turbulence model [4] is improved by introducing an additional term into the model. The unknown term

© The Author(s), under exclusive license to Springer Nature Switzerland AG 2021
A. Dillmann et al. (Eds.): STAB/DGLR Symposium 2020, NNFM 151, pp. 77–86, 2021.
https://doi.org/10.1007/978-3-030-79561-0_8

is then modeled using a data driven approach. The development of a machine learning model in a multitask learning framework is the main contribution of this paper. We note that although the experimental investigations are limited to the SA model, the approach is easily extended to other models.

Recent years have seen a resurgence in turbulence modeling, specifically data driven approaches, which utilize data from experiments or high fidelity simulations to estimate and subsequently correct functional inaccuracies [5,6] in turbulence models. Several papers [7,8] also propose approaches that use machine learning to model Reynolds stress tensors. In this paper, we propose a multitask learning framework to improve the functional inaccuracies in the turbulence models. For numerical investigation, we consider the Spalart - Allmaras turbulence model and model a correction functional against its production term in order to improve the model's fidelity. An optimal estimate of the correction function is thus required to correct the inaccuracy of the model. This estimation is carried out using the framework introduced by [6]. The correction function is found for different flow conditions of two airfoils (NACA0012 and NACA0021) at different angles of attack. By different flow conditions, we mean that the Reynolds number and Mach for the NACA0012 cases are different from those for the NACA0021. Inverse problems are formed for each configuration, with the objective of minimizing the discrepancy between experimentally measured lift coefficient and its numerical approximation. These inverse problems are solved with respect to the correction function, producing optimal estimates that can be used to train a machine learning model.

A brief overview of the training process follows along with the contribution of this paper. The data from the inverse problems is divided into three subcollections – two for training and validating the machine learning model and the last for testing. Each dataset contains a selection of flow variables and correction function values. Due to the differences in flow conditions and domains, the distributions associated with the correction function vary from one configuration to another. If the correction functional is treated as a random variable, each configuration corresponds to a unique random variable, with a different probability distribution function. To our knowledge, this issue has been overlooked in the literature on data driven turbulence modeling. Instead, a mapping f, that takes some selection of the flow variables, from all the configurations (thus variables from multiple distributions) as input and produces a corresponding correction term for the SA model is sought. The framework proposed in this paper explicitly addresses this issue. We argue that acknowledging the differences in the data arising from their respective domains and flow conditions would lead to a model that is better in terms of generalization ability. In order to construct a common mapping f as described above, a framework that can treat multiple non-identically distributed random variables is necessary, which is the main contribution of this paper. Our framework simplifies the inductive transfer of relevant information from different domains into a common domain and as a consequence improves the machine learning model's generalization ability. The machine learning model is designed to learn a compressed latent representation of the data from multiple distributions. These *latent variables* are the result of a non-linear mapping from the flow variables to a low dimensional space, that represent the information most correlated with the optimal correction field. Subsequently, these latent variables form the input to a feed-forward neural network that maps to the correction field.

A more detailed summary of the machine learning model is presented in Sect. 2.2. The trained model is then embedded into a CFD solver (we use SU2 [9]) to predict the correction term for the SA model. At each iteration, the machine learning model queries the current state of the flow variables and estimates the correction term. The results of this model on the test datasets are presented in Sect. 3.

2 Numerical Method

The focus of this work is on improving the predictive accuracy of turbulence models using machine learning. Although our presentation is limited to the Spalart - Allmaras model, the general technique is easily extended to other turbulence models. The Spalart - Allmaras turbulence model is a one equation model that defines a conservation equation for the modified eddy viscosity $\tilde{\nu}$ [4]. Owing to its use of a single closure equation, the model is cost-effective and accurate for applications involving flows with mild separation and attached wall-bounded flows. The model was developed for use in the aerospace industry and suffers from limitations to its general applicability.

We specifically look at a modification of the production term P of the SA turbulence model by introducing the correction function as a function of the space variable x as follows

$$\tilde{P} := \beta(x)P = \beta(x)c_{b1}(1 - f_{v2})\tilde{S}\tilde{\nu}. \quad (1)$$

We refer to [4] for the SA turbulence model and the description of the terms in Eq. (1). In this paper, we are only concerned with steady flows, and consequently the correction is just a function of the flow domain. Estimation of the optimal correction function entails the formulation of several inverse problems as described in Sect. 2.1. Optimal estimates of the correction function are used, along with flow variables, to train a machine learning model that can approximate the correction function for new flow conditions. Subsequently, the machine learning model is embedded into the CFD solver and to produce real-time corrections of the source term in the SA model.

2.1 Inverse Problem Formulation

The unknown correction function $\beta(x)$ is henceforth treated as an unknown high dimensional parameter, that assumes a value at each discretization point. The modified production term Eq. (1) is equivalent to the production term of the SA model, when the functional is a constant with $\beta(x) = 1.0$. We consider several configurations as described below and formulate an inverse problem for each of these with respect to the correction function. By optimizing the accuracy of the SA model, optimal correction functions for each configuration are obtained.

The configurations considered come from two airfoils, the NACA0012 and NACA0021 at different angles of attack. The NACA0012 configurations are run with Reynolds number 6×10^6, Mach 0.15. An unstructured grid with 897×257 grid points is used. We consider several different angles of attack between 2 and 19 degrees for this airfoil. The second is the NACA0021 airfoil at Reynolds number 1.5×10^6, Mach 0.2. A hybrid (structured boundary layer and rest of the domain unstructured) grid is used with 143,757 cells, triangular and quadrilateral. The angles of attack considered are between 6 and 20 degrees. As mentioned in Sect. 1, the set of configurations is divided into training, validation and testing subsets. Only those cases marked for training are subjected to inverse design and thus produce optimal correction function estimates. So we reserve the NACA0012 airfoil at angles of attack $8°, 14°, 19°$ and the NACA0021 airfoil at $12°$ for testing. The remaining cases, which account for training and validation can be seen in Fig. 1. An inverse problem is formulated for each of these configurations, with respect to the target functional

$$J(U(\beta), \beta) = (c_l - c_l^*)^2 + \lambda \Sigma_{j=1}^{N} (\beta_j - 1)^2. \tag{2}$$

Here c_l is the lift coefficient as estimated by the numerical solution U, c_l^* is the experimental measurement for the lift coefficient for that configuration, β_i is the value of the correction function at the ith grid point and λ is a constant. Also, N represents the total number of grid points. The second term represents a Tikhonov regularization on the correction function, biasing it towards the SA model. For each configuration the inverse design can be formulated as an optimization problem:

$$\underset{\beta}{\text{argmin}} \quad J(U(\beta), \beta), \tag{3}$$

$$\text{s. t.} \quad \mathcal{R}(U, \beta) = 0. \tag{4}$$

The constraint Eq. (4) represents the numerical solution residual of the compressible Reynolds averaged Navier-Stokes equations along with the Spalart-Allmaras turbulence model. The numerical simulations are carried out in SU2, using a compressible solver. Each of the inverse problems is solved using the L-BFGS algorithm [10]. The latter is a quasi-Newton algorithm that computes an approximation B_k of the true Hessian $\nabla^2 J_k$, at the iteration k of the optimization algorithm, using gradients of the previous iterations. The BFGS algorithm is characterized by a recursive update to the approximate Hessian given by

$$B_{k+1} = B_k - \frac{B_k s_k s_k^T B_k}{s_k^T B_k s_k} + \frac{y_k y_k^T}{y_k^T s_k}.$$

Here $y_k = \nabla J_{k+1} - \nabla J_k$ and $s_k = x_{k+1} - x_k$. The limited memory version mainly differs from BFGS in the storage of gradients and calculation of line search directions. The gradients are evaluated using the discrete adjoint framework

[11]. A comparison of lift coefficient evaluated using the usual SA model in SU2 against experiment and subsequently after inverse design, for the cases used for training are shown in Table 1. For the NACA0012 set of configurations, data was obtained from Ladson [12] and for NACA0021 configurations, the data is taken from [13].

Table 1. Comparison of lift coefficients and errors for training data.

Case	AoA	c_l	c_l	Rel. error	c_l	Rel. error
NACA		Exp.	SA	SA vs Exp.	Inv.	Inv.
0012	2.05	0.21	0.23	9.1	0.22	1.18
	4.04	0.43	0.45	5.52	0.44	1.71
	6.09	0.65	0.68	4.28	0.66	0.79
	10.12	1.07	1.11	3.76	1.09	1.80
	11.13	1.17	1.21	3.72	1.18	0.98
	12.12	1.26	1.31	3.72	1.29	1.94
	13.08	1.35	1.39	3.7	1.36	1.08
	15.26	1.51	1.57	3.88	1.52	0.47
	16.3	1.57	1.63	3.92	1.61	2.29
	17.13	1.61	1.66	3.43	1.65	2.38
	18.02	1.00	1.65	65.43	1.08	9.30
0021	6	0.52	0.60	14.36	0.53	1.14
	16	1.07	1.19	11.59	1.07	0.18
	20	0.98	0.91	−7.13	0.98	−0.17

2.2 Machine Learning

The inverse problems formulated in the previous section result in a collection of datasets containing the flow variables for each grid point. A selection of flow variables is made from the set of all flow quantities, with the constraint that the selection is comprised of those quantities that are not linearly correlated. These selected flow variables will be used as the inputs of the machine learning model. For example, the quantities that we have used are the eddy viscosity, wall distance, vorticity, velocity and pressure gradients. However, this selection is based on judgement and a more rigorous way of selecting features is a possible open question. The selected flow variables are represented as vectors q in some space $\mathbb{R}^n$, where n is the dimension of the vector of the selected flow variables. These form inputs to the machine learning model, whereas the corresponding correction function estimates β, are the training targets. For notational simplicity, we consider the union of the set of vectors q, from all test configurations, and call it $\Phi \subset \mathbb{R}^n$. This set Φ along with the correction function estimates forms the data to train the machine learning model.

The data is divided further into two subsets which will be called the training and validation sets with the latter set containing a third of the entire data. The training set is used to calculate gradients, and consequently, parameter updates for the machine learning model and the validation set is used to find the set of parameters with least generalization error. Note that we consider fewer cases of the second airfoil than those in the first deliberately as a means to verify the machine learning model's generalization ability. The domains, flow conditions, and consequently the estimated correction functions, are expected to be different for each configuration and therefore, not directly comparable. The multi-task learning framework addresses these issues.

In principle, one can model β as a functional, $g\colon \Phi \to \mathbb{R}$ and subsequently use this *direct model* to predict the correction function for new configurations. Appealing again to the discussion at the end of Sect. 1, the dissimilarities in configurations renders these datasets incomparable and direct modeling ignores these dissimilarities. In more rigorous terms, the data Φ, represented as a union of their original domains, are not independent samples and moreover, the data for each single configuration, are heteroscedastic and geometrically correlated. This is a significant factor in the modeling of data and ill-specified models could be severely inconsistent or biased [14].

Consequently, an additional strategy is required to find a meaningful approximation of the correction function β, that is independent of differences in the domains. We introduce two non-linear mappings $\mathcal{H}\colon \Phi \to Z$, and $\mathcal{F}\colon Z \to \Phi$ where $Z \subset \mathbb{R}^s$ is a set of latent variables, with the condition that $s < n$. The function $\mathcal{H}$ forms compressed representations of the data points in the latent variable space. Whereas the function $\mathcal{F}$, can be thought of as the reconstruction of data Φ from the latent variables in Z. These mappings are also known as encoders and decoders, respectively [15]. Finally, a mapping $\mathcal{G}\colon Z \to Y$ can then be formed that predicts the correction term, irrespective of the domain of the data point. The reasons for using this scheme are twofold. Firstly, the treatment of data from several different configurations under one machine learning model is made possible by the mapping $\mathcal{H}$, which takes data from related, yet dissimilar domains onto a latent variable domain. Secondly, the issues due to heteroscedasticity and geometric correlations in the dataset Φ, can be resolved using a regularization that is described below.

Several machine learning frameworks are available for modeling non-linear mappings. We use neural networks for all three mappings $\mathcal{F}, \mathcal{G}$ and $\mathcal{H}$. If $\mathbf{x}_j \in \mathbb{R}^{n_j}$ is the vector representing the layer number j, then a neural network computes $\mathbf{x}_{j+1} = \sigma(W_j \mathbf{x}_j + \mathbf{b}_j)$, for the vector representing the next layer with a total of L layers and W_j, b_j, called the weight and bias of layer j, represent unknown parameters. The networks for $\mathcal{H}, \mathcal{F}$ are an encoder and decoder respectively with one hidden layer with five neurons, and the network used for $\mathcal{G}$ is a residual feed-forward network with five hidden layers and thirty neurons each. Hinge activation functions are used for all the networks. By training the model, we mean the optimization problem that minimizes the functional

$$J(\mathbf{x}, y^*) := \|\mathcal{G}(\mathbf{z}; \theta) - y^*\|_{\mathcal{L}^2} + \kappa \mathrm{D}(\mathbf{x}, \mathcal{F}(\mathcal{H}(\mathbf{x}; \theta'); \theta'')). \tag{5}$$

with respect to the parameters θ, θ' and θ''. Here, $\mathbf{x}$ is a subset of the data Φ, $\mathbf{z}$ is the latent representation of $\mathbf{x}$, and $\theta, \theta', \theta''$ are parameters of the networks $\mathcal{G}, \mathcal{H}$ and $\mathcal{F}$, respectively. The output of the model, $y = \mathcal{F}(\mathbf{x}; \theta'')$ is the model's estimate of the field parameter for input $\mathbf{x}$ whereas y^* is the value of the correction parameter estimated from the inverse problem Eq. (3). The parameters are updated by calculating the gradient of J using the adjoint method, usually called the backpropagation algorithm in machine learning. A modified version of the ADAM optimization algorithm is used to train the network [16].

Once the optimal parameters $\mathbf{W}^*$ are determined, the model, along with the parameters are embedded into the CFD solver such that at each iteration, a new value of the correction parameter β is determined using the flow variables, that are computed by the CFD solver. Subsequently we run this machine learning coupled solver on the testing data set and compare the predicted lift coefficient against experiment. The computation of β from an input q by the trained machine learning model requires, on average, a few milli-seconds. There is a small marginal increase in computation time resulting from the coupled machine learning solver. The implementations were made in Pytorch [17].

3 Results

As outlined in Sect. 2.1, the inverse design of parameters β is carried out based on the optimization problem Eq. (3). The CFD flow solver performs adjoint based optimization to estimate the optimal parameters for the training set of data. The regularization parameters λ and κ was chosen from a set of values ranging from 10^{-6} to 10^{-2}, based on the least error value obtained. The value of λ used is 10^{-5} and that for κ is 10^{-4}. A more rigorous selection of these parameters is considered in future work, for example based on generalized cross validation and near optimal methods.

Table 2. Lift coefficients and errors in the test cases.

Case	AoA	c_l Exp.	c_l SA	c_l ML+SA	Rel. error SA vs Exp.	Rel. error ML+SA vs Exp.
NACA0012	8.30	0.89	0.92	0.91	3.85	3.35
	14.22	1.44	1.49	1.48	3.91	2.93
	19.08	1.13	1.38	1.17	21.44	2.70
NACA0021	12.00	1.02	1.11	1.05	8.44	3.22

The trained machine learning model is embedded into SU2 in order to predict the values of the correction field at each iteration. The model is tested at the angles of attack 8.3°, 14.22° and 19.08° for the NACA0012 airfoil and at angles of attack 12° for the NACA0021 airfoil, by comparing the lift coefficients against the baseline SA and experimental values. No data pertaining to these cases is

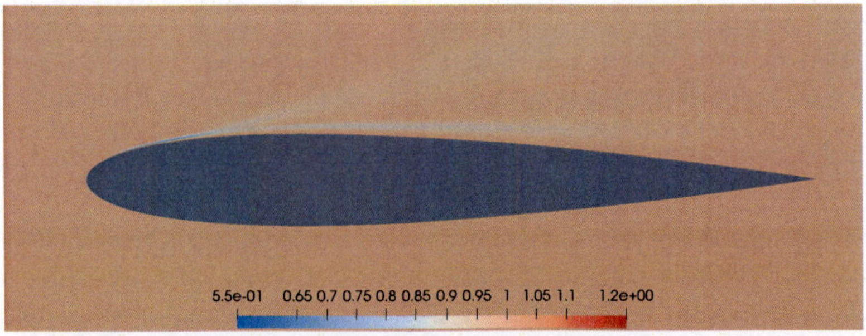

(a) Field inversion, NACA 0012, $\alpha = 18.02°$

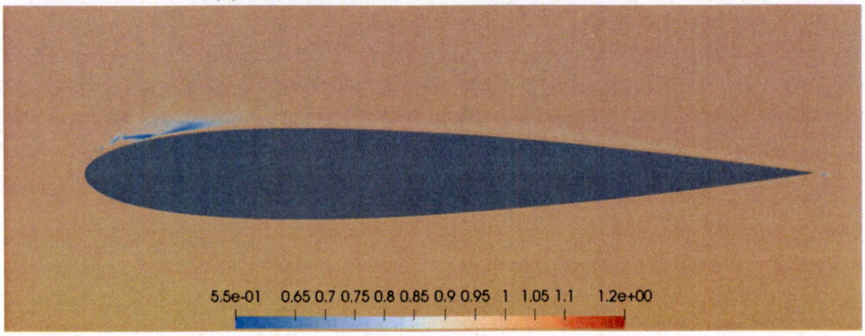

(b) Machine learning, NACA 0012, $\alpha = 19.08°$

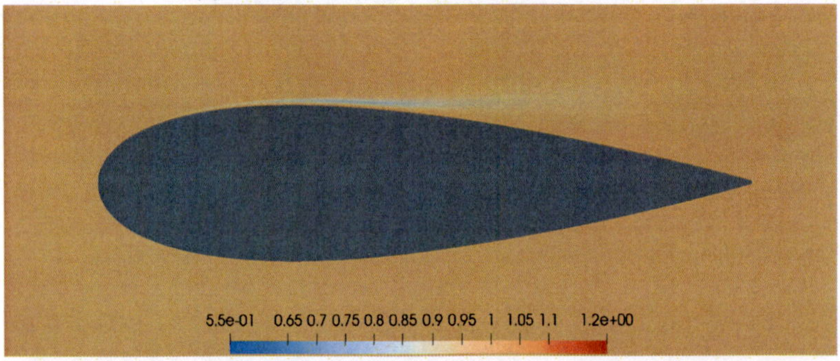

(c) Field inversion NACA 0021, $\alpha = 16.00°$

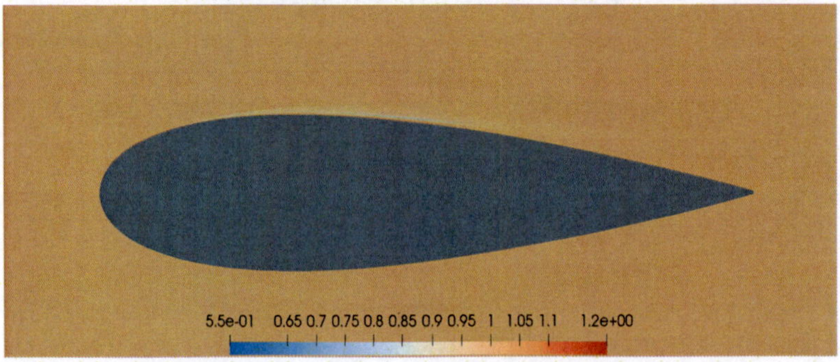

(d) Machine learning, NACA 0021, $\alpha = 12.00°$

Fig. 1. Correction terms obtained from inverse design and machine learning.

used in training the machine learning model. The results of inverse design as well as the predictions of correction function by the machine learning model are shown in Fig. 1, the various measurements involved are outlined in Table 2. Each of the cases have shown improvement over the baseline SA turbulence model. Higher angles of attack have shown better improvement, possibly due to the larger discrepancy of the baseline SA model. Note that fewer NACA0021 airfoil configurations were used to train the model. Notwithstanding this limitation, the correction function prediction from the machine learning model is consistent with expected results.

4 Conclusion

We presented a multitask learning framework for correcting functional errors in turbulence models. The method is illustrated on the production term of the Spalart Allmaras turbulence model. The SA model is modified by introducing a function β that corrects the functional inaccuracy of the model's production term. The correction function is estimated at different angles of attack for the NACA0012 and NACA0021 airfoils by solving inverse problems based on experimental lift coefficient. A machine learning model which learns to map a subset of flow variables to the estimated correction functions, is trained using data obtained from the inverse problems. The trained machine learning model is then used simultaneously with the SA model to predict correction functions in real time. The machine learning enhanced SA model is shown to improve the lift coefficients on four new test cases.

The selection of the optimal regularization parameters for both the inverse design phase and the machine learning phase are important issues marked for future work. The selection of features used in this work was made based on judgement. Other approaches have made use of hill-climbing algorithms for feature selection [6], but this approach is not suitable for models that require a lot of computational power, and when the size of the data is large. We consider this an open problem that requires further attention. A related issue is that of the design of the size of latent variables in the model used. In order to obtain a compressed representation of the input, it is sufficient to ensure that the latent variables are lower dimensional random variables, compared to the input features. However, an optimal dimension of the latent variables if known beforehand, would expedite the learning process. Future work includes determining the optimal size of latent variables for such models.

Acknowledgments. The authors gratefully acknowledge financial support by the Federal Ministry of Economic Affairs and Energy of Germany (BMWi) in the project DIGIfly (FKZ: 20X1909G) and the computational resources provided by the RHRK high performance computing center via the 'Elwetritsch' high performance cluster at the TU Kaiserslautern.

References

1. Spalart, P.R.: Comments on the feasibility of LES for wings, and on a hybrid RANS/LES approach. In: Proceedings of first AFOSR international conference on DNS/LES. Greyden Press (1997)
2. Spalart, P.R., Deck, S., Shur, M.L., Squires, K.D., Strelets, M.K., Travin, A.: A new version of detached-eddy simulation, resistant to ambiguous grid densities. Theoret. Comput. Fluid Dyn. **20**(3), 181 (2006)
3. Shur, M., Spalart, P.R., Strelets, M., Travin, A.: Detached-eddy simulation of an airfoil at high angle of attack. In: Engineering Turbulence Modelling and Experiments, vol. 4, pp. 669–678. Elsevier (1999)
4. Spalart, P., Allmaras, S.: A one-equation turbulence model for aerodynamic flows. In: 30th Aerospace Sciences Meeting and Exhibit, p. 439 (1992)
5. Tracey, B.D., Duraisamy, K., Alonso, J.J.: A machine learning strategy to assist turbulence model development. In: 53rd AIAA Aerospace Sciences Meeting, p. 1287 (2015)
6. Duraisamy, K., Zhang, Z.J., Singh, A.P.: New approaches in turbulence and transition modeling using data-driven techniques. In: 53rd AIAA Aerospace Sciences Meeting, p. 1284 (2015)
7. Ling, J., Kurzawski, A., Templeton, J.: Reynolds averaged turbulence modelling using deep neural networks with embedded invariance. J. Fluid Mech. **807**, 155–166 (2016)
8. Sotgiu, C., Weigand, B., Semmler, K., Wellinger, P.: Towards a general data-driven explicit algebraic reynolds stress prediction framework. Int. J. Heat Fluid Flow **79** (2019)
9. Economon, T.D., Palacios, F., Copeland, S.R., Lukaczyk, T.W., Alonso, J.J.: SU2: an open-source suite for multiphysics simulation and design. AIAA J. **54**(3), 828–846 (2015). https://doi.org/10.2514/1.J053813
10. Nocedal, J., Wright, S.: Numerical Optimization. Springer (2006)
11. Albring, T., Sagebaum, M., Gauger, N.R.: A consistent and robust discrete adjoint solver for the SU2 framework - validation and application. Notes on Numerical Fluid Mechanics and Multidisciplinary Design, vol. 132, pp. 77–86 (2016)
12. Ladson, C.L.: Effects of independent variation of Mach and Reynolds numbers on the low-speed aerodynamic characteristics of the NACA 0012 airfoil section, volume 4074. National Aeronautics and Space Administration, Scientific and Technical ... (1988)
13. Sheldahl, R.E., Klimas, P.C.: Aerodynamic characteristics of seven symmetrical airfoil sections through 180-degree angle of attack for use in aerodynamic analysis of vertical axis wind turbines. Technical report, Sandia National Labs., Albuquerque, NM (USA) (1981)
14. White, H., Domowitz, I.: Nonlinear regression with dependent observations. Econometrica J. Econ. Soc. 143–161 (1984)
15. Kramer, M.A.: Nonlinear principal component analysis using autoassociative neural networks. AIChE J. **37**(2), 233–243 (1991)
16. Kingma, D.P., Ba, J.: Adam: a method for stochastic optimization. arXiv preprint arXiv:1412.6980 (2014)
17. Paszke, A., et al.: Pytorch: an imperative style, high-performance deep learning library. In: Advances in Neural Information Processing Systems, pp. 8024–8035 (2019)

Check for updates

Investigation of Coherent Motions in a Flat-Plate Turbulent Boundary Layer with Adverse Pressure Gradient

Matthias Weinschenk(✉), Ulrich Rist, and Christoph Wenzel

Institute of Aerodynamics and Gas Dynamics, University of Stuttgart, Pfaffenwaldring 21, 70550 Stuttgart, Germany
weinschenk@iag.uni-stuttgart.de

Abstract. Convection velocities of small- and large-scale patterns in a turbulent boundary layer obtained via Direct Numerical Simulation are computed by correlating finite-size interrogation windows between consecutive time steps, which is common practice in Particle-Image Velocimetry. A particular region in the logarithmic layer and the lower defect layer of the mean flow is identified where different patterns belonging to different flow variables are convected downstream by the local fluid material. The corresponding structures are unique and match with the well-known high- and low-speed streaks of the streamwise velocity component. Outside of the region where these structures are observed, the convection of local disturbance patterns exhibits wave character, i.e., they travel faster than the mean flow close to the wall and slower in the outer defect layer and at the edge of the boundary layer. This further confirms our interpretation of the presence of dominant material-bound coherent structures in the intersection zone of the log layer with the defect layer.

Keywords: Coherent motions · Turbulent boundary layer · Convection velocity

1 Introduction

Turbulent boundary layer flows contain many small- and large-scale coherent patterns, cf. [9,12,13], etc. A proven way to investigate these structures in a turbulent boundary layer is the evaluation of their convection velocities. Earlier attempts are summarized by [3,5], for instance. Krogstad *et al.* [10] have observed large variations of the convection velocities with wall distance and scale of events [using two-point correlations of the measured streamwise velocity component]. For all scales and positions, the convection velocity for ejections is distinctly lower than that for sweeps, for instance. Typically, the largest-scale motions of the order of the boundary-layer thickness are convected at velocities close to the local mean and the velocities drop significantly as the scales are reduced for all types of events studied. Hence, it appears that the convection velocity varies in

© The Author(s), under exclusive license to Springer Nature Switzerland AG 2021
A. Dillmann et al. (Eds.): STAB/DGLR Symposium 2020, NNFM 151, pp. 87–96, 2021.
https://doi.org/10.1007/978-3-030-79561-0_9

space and time. Dennis & Nickels [4] consider convection velocities in the log-layer in connection with Taylor's hypothesis, also based on experimental data for the streamwise velocity component. They find that the motion of large-scale structures can be traced much longer with acceptable accuracy than that of small-scale structures. Low-pass filtering was shown to be helpful if not applied in an excessive manner. In a more recent study by [11] convection velocities at three wall distances have been obtained from the spatio-temporal spectra of PIV data. These authors find a variation with wall distance in agreement with [4] and [10]. However, how are the structures of the individual flow variables (u, v, w, ρ, T, p) related to each other and how do they behave in a turbulent boundary layer with adverse pressure gradient?

The purpose of the present contribution is to apply a simple method for estimating convection velocities in a turbulent flat-plate boundary layer with adverse pressure gradient, based on recent DNS data by [15]. The present work complements a much earlier investigation which has discovered a very similar behaviour of convection velocities for the first time in turbulent channel flow [8]. The paper is organized as follows: Sect. 2 shortly presents the underlying DNS data used for the present investigation. Exemplary visualisations are shown to demonstrate the presence of well-known streaky structures and their length scales to confirm the validity of the present data. The correlation technique is also presented in that section. This is followed in Sect. 3 by the outcome of the technique. Conclusions in Sect. 4 summarize the results.

2 DNS Data Base and Correlation Technique

For the present investigation the DNS data computed by [15] for a turbulent flat-plate boundary layer with streamwise pressure gradient are used. The data itself, its statistics and its scaling with respect to Reynolds number, Mach number and pressure gradient are presented and discussed in full detail in [15]. The present investigation is based on a data set with an inflow Mach number $Ma = 0.5$ and an adverse pressure gradient with constant Rotta-Clauser parameter $\beta_k = 1.0$. The Reynolds number based on the momentum thickness covers the range $1925 \leq Re_\theta \leq 3209$ with $499 \leq Re_\tau \leq 766$. 200 snapshots with a temporal distance of $\Delta t \cdot U_\infty/\delta_0 = 8$ are available for post processing [15]. The reference length δ_0 is the boundary-layer thickness at the inflow of the integration domain at $Re_\theta = 287, Re_\tau = 127$. The choice of using an adverse streamwise pressure gradient is motivated by the expectation that this leads to a faster thickening of the boundary layer with Re_θ which in turn is necessary to observe a second peak in the rms-spectra due to the occurrence of large-scale motions.

Typical spanwise and streamwise pre-multiplied spectra of streamwise velocity fluctuations for the streamwise position $Re_\theta = 2622$ are shown in Fig. 1. They indicate the presence of two regions with increased turbulent activity, one close to the wall and the other further away in the log layer of the mean flow at $100 < y^+ < 500$. The latter contains large-scale structures with a dominant spanwise length scale of $\lambda_z^+ \approx 500 - 600$ and a streamwise scale $\lambda_x^+ \approx 1000$.

Another maximum with typical spacing of $\lambda_z^+ \approx 120$ and $\lambda_x^+ \approx 500 - 1000$ is observed near $y^+ \approx 10$.

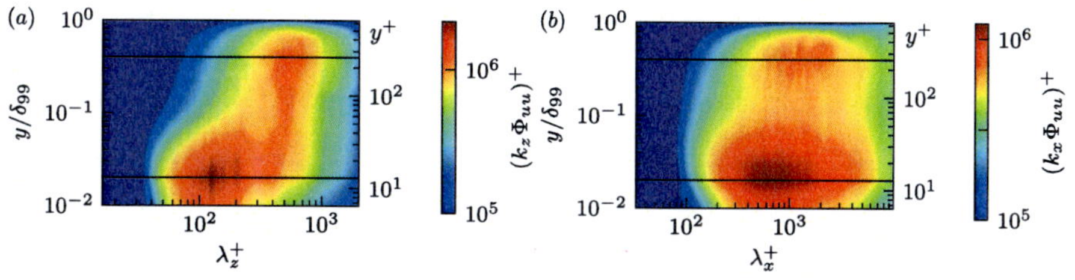

Fig. 1. Pre-multiplied spanwise (a) and streamwise (b) u'-velocity spectra at $Re_\theta = 2622$, $Re_\tau = 642$.

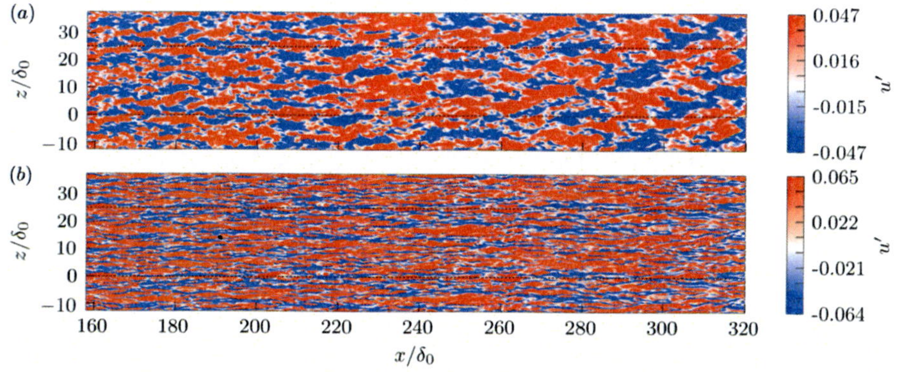

Fig. 2. Snapshots of streamwise velocity disturbance in planes corresponding to: (a) $y/\delta_{99} \approx 0.4$, ($b$) $y/\delta_{99} \approx 0.02$ (2 spanwise periods shown).

Snapshots of the streamwise velocity disturbance $u'(x, z)$ are shown in Fig. 2 for the two y-positions indicated by the black lines in Fig. 1. Note that the spanwise coordinate covers two periods of the integration domain. High- and low-speed streaks (red and blue, respectively) can be identified and a streamwise evolution from smaller to larger scales is especially apparent at the larger wall distance $y/\delta_{99} \approx 0.4$. The two different kinds of structural patterns corresponding to the two local extrema in Fig. 1 stand out clearly: Narrow streamwise streaks close to the wall and large-scale structures at the larger wall distance.

As the focus of the present contribution will be on the latter these are visualized in Fig. 3 which is obtained from the corresponding slice in Fig. 2(a) by application of a Gaussian low-pass filter with a filter-size 35×35 and a standard deviation $\sigma = 7$; both in grid points. The filter size corresponds to $6.80 \times 3.34\ \delta_0$ in x- and z-direction, respectively. This choice is motivated by the observation that the correlation used further down has been performed with these widths. Compared to the spectra in Fig. 1 it turns out that the corresponding widths in x

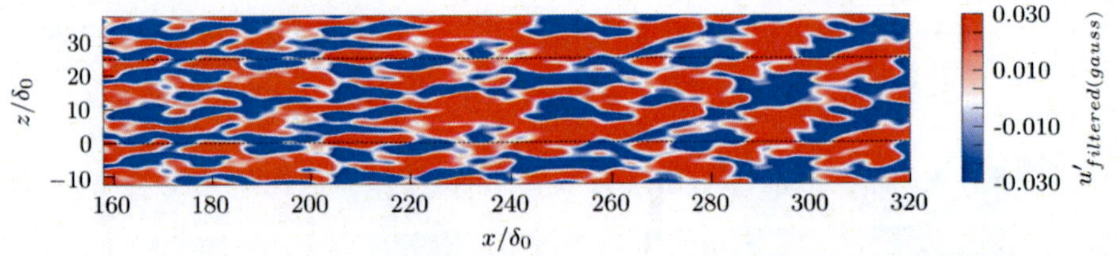

Fig. 3. Low-pass filtered streamwise velocity disturbance corresponding to Fig. 2(a) at $y/\delta_{99} \approx 0.4$.

and z direction correspond to $\lambda_x^+ \approx 568.47$ and $\lambda_z^+ \approx 276.75$, respectively, which are near the amplitude maxima observed there. Streamwise and obliquely connected regions of high-speed flow are the dominant patterns and there is a clear trend that these evolve as they travel downstream from smaller, streamwise oriented structures into larger oblique wave fronts with an angle of approximately 30 degrees relative to the streamwise axis x.

What has been shown for the streamwise velocity component u, so far, can be repeated for any other quantity; especially as such data is readily available from DNS. As an example, the two other velocity components are presented together with the corresponding instantaneous density and pressure disturbances in Fig. 4 for comparison with the streamwise velocity component in Fig. 2 (top). Note that a corresponding slice for the temperature is not shown because its structures T' are just the negative of the density fluctuations ρ', arising from the fact that p' is small. It is difficult to find any obvious correlation between the different fields in Fig. 4, apart from some local, possibly more fortuitous resemblance. Each data variable seems to have its own structure and, especially, there are less streamwise streaks in contrast to the streamwise velocity component. This is in agreement with correlation measurements in earlier DNS data, where other flow-field variables, like the two other velocity components, have been considered as well, e.g. [7], for instance. These yield different length scales for different variables, i.e., there is not the same structure as in the streamwise component.

Looking at the present DNS results in Fig. 4 it is obvious that all available 200 equidistant time steps for each flow variable contain quasi-random patterns. Due to their resemblance with particle images in PIV (numbers in the present case instead of grey-scale values in PIV) it is straightforward to estimate the convection velocity of these disturbance patterns in a PIV-like manner, see [16]. The correlation technique applied is based on Pearson's correlation-coefficient:

$$r_{\Phi\Phi}(\Delta x) = \frac{\sum_{i=1}^{N}(\Phi_i^t - \overline{\Phi^t})(\Phi_i^{t+\Delta t}(\Delta x) - \overline{\Phi^{t+\Delta t}(\Delta x)})}{\sqrt{\sum_{i=1}^{N}(\Phi_i^t - \overline{\Phi^t})^2} \cdot \sqrt{\sum_{i=1}^{N}(\Phi_i^{t+\Delta t}(\Delta x) - \overline{\Phi^{t+\Delta t}(\Delta x)})^2}} \; . \qquad (1)$$

Computation of this coefficient is performed with the help of so-called "interrogation windows" of size $N = n \times m$ grid points. Φ stands for any of the flow variables (u, v, w, ρ, T, p) and overbars indicate the averages of all N data values in each window. For estimating the streamwise displacement of data patterns in

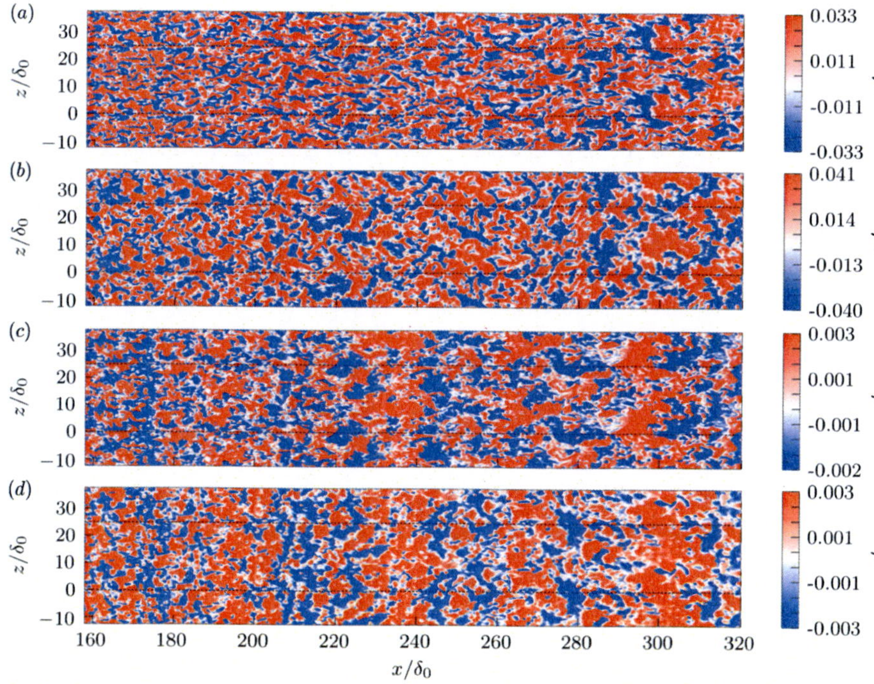

Fig. 4. Comparison of instantaneous disturbance patterns at the same wall-normal positions $y/\delta_{99} \approx 0.4$ and the same snapshot in time.(a) wall-normal velocity component v', (b) spanwise velocity component w', (c) density ρ', (d) pressure p'.

a flow variable Φ^t at time t in the first window to time $t + \Delta t$ in the second, i.e., from Φ^t to $\Phi^{t+\Delta t}$, the second window is shifted in x-direction by $j \cdot \Delta x$. Where Δx is the grid-step size in the DNS data and $1 < j < J_{max}$. In the present case $J_{max} = 45$ suffices to capture the correlation maximum. This leads to a discrete representation of $r_{\Phi\Phi}$ versus x whose maximum $x_{max(r_{\Phi\Phi})}$ is then interpolated via a third-order polynomial, for obtaining "sub-grid" accuracy. Dividing the displacement of the maximal streamwise correlation $x_{max(r_{\Phi\Phi})}$ by the time step Δt, the instantaneous, local convection velocity is obtained: $c_\Phi = x_{max(r_{\Phi\Phi})}/\Delta t$. The interpolated maximal value of the correlation coefficient is also stored for later evaluations of the method's accuracy for each estimated data point. This procedure is applied to each grid point of the snapshot in wall-parallel planes at time t. Applying this technique to all available unfiltered snapshots of the selected flow variable, $c_\Phi(x, y, z, t)$ is obtained, and from this the average convection velocity $\overline{c}_\Phi(x, y)$, by averaging with respect to time and spanwise coordinate z.

3 Results

The interrogation window size has been varied in the range 15×15 to 65×65 (grid points) with an increment of 10 grid points from one case to the next. It turned out that the estimated mean convection velocity versus wall-normal coordinate is least dependent on the interrogation window size in the range 15 × 15 to 45 × 45. Therefore, an interrogation window size with 35 × 35 grid points has been used for computing the maximal streamwise correlation $x_{max(r_{\Phi\Phi})}$ for each of the 900 × 256 ($x \times z$) input data points of two consecutive snapshots.

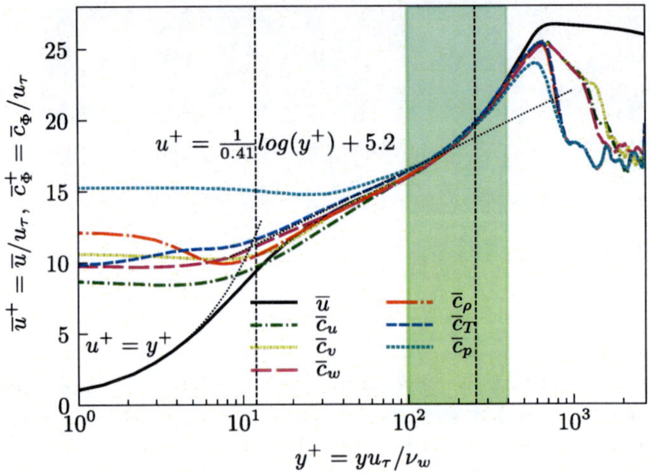

Fig. 5. Comparison of mean convection velocity $\overline{c}_\Phi$ for different flow variables i with mean-flow profile $\overline{u}$ in wall units at $Re_\theta = 2622$.

Figure 5 shows the comparison of the mean convection velocities of all flow variables among each other and with respect to the mean flow $\overline{u}$ of the turbulent boundary layer at $x = const.$ The average correlation maxima for these data varied in the range 0.65 with 1.6% outliers in the T-field and 0.78 with 0.08% outliers in the u-field; data points where the correlation maximum falls below 0.5 are considered as 'outliers'. The time-averaged convection velocities $\overline{c}_\Phi$ not only vary with wall distance but also closely follow the time-averaged streamwise velocity $\overline{u}$ in the overlap region between the logarithmic layer and the wake of the mean-flow profile (highlighted by a green band in the figure). This is identical to the observation in [8] for a turbulent channel flow. A comparison with the spectra in Fig. 1 using the two y-positions marked in black show that the coincidence of the mean convection velocities of all flow variables with the mean flow corresponds to the area where large-scale structures have been detected in the spectra. The observation that apparently different patterns which have been exemplarily shown in Fig. 4 for the wall distance $y^+ \approx 255$, are traveling at least in the mean at the local mean-flow speed in the area $100 < y^+ < 400$ means that all patterns or structures in this region are bound to the motion of the fluid material.

Interestingly, closer to the wall ($y^+ < 100$) and further away ($y^+ > 400$) the different convection velocities begin to diverge such that the disturbance motion can no longer be bound to material moving at the mean-flow velocity. In both limits, $y \to 0$ and $y \to \infty$, the disturbance patterns thus either move faster or slower than the fluid particles, which means that they have wave character. In both areas, the pressure convection velocity $\overline{c}_p$ is the first and strongest to deviate from the rest. This can be explained by the fact that pressure decays at a much lower rate compared to other variables with distance from a source. In addition, pressure signals can travel in a wave-like, mean-flow independent fashion through the fluid. In the present context, this means that the pressure signals observed towards the wall must be the footprints of the strongest pressure-inducing structures further above. Apparently, the source for such pressure signals lies near the lower border of the green area where the fluid has almost the same velocity as $\overline{c}_p$ below it. On the other hand, the higher-than-mean-flow value of $\overline{c}_p$ at $y^+ < 40$ means that pressure signals belonging to the coherent motions further above run in a wave-like manner over the slower-moving fluid close to the wall. This corroborates the interpretation that there is a coherent pressure-fluctuation-inducing and dominating structure in the green area. One may even hypothesize that the wave-like character of all flow variables at $\overline{c_\Phi/U_e} \approx 0.3 - 0.35$ is an indication for "continuous laminar-turbulent transition" in the viscous sublayer or at its edge as a source for continuous turbulence production that balances turbulent dissipation above.

Towards the edge of the boundary layer and outside of it, the convection velocities are smaller than the local mean-flow speed $\overline{u}$ because the present DNS have been performed with zero free-stream turbulence, such that there is no externally forced turbulence convecting with the free-stream speed outside the boundary layer. Instead, an intermittent laminar-turbulent interface at the boundary-layer edge appears and it is in this interface where the average convection velocities $\overline{c}_\Phi$ approach their free-stream values. This free-stream value at 60–70% of the mean-flow is related to very small disturbances which have wave-like character, supposedly linked to pressure waves.

Kim & Hussain [8] in their evaluation of convection velocities using all three velocity and vorticity components, as well as pressure in an incompressible plane channel flow at $Re_h = 3300$ have observed a very similar wave-character of disturbances near the wall with the pressure wave moving much faster than all other components. However, the corresponding near-wall phase speeds in their data is much higher at 55 to $75\% U_c$, where U_c is the channel center-line velocity. Another difference is the absence of waves towards the center of the channel, but this is simply due to the presence of fully developed turbulence there in contrast to the present investigation with a laminar free-stream which allows observing waves.

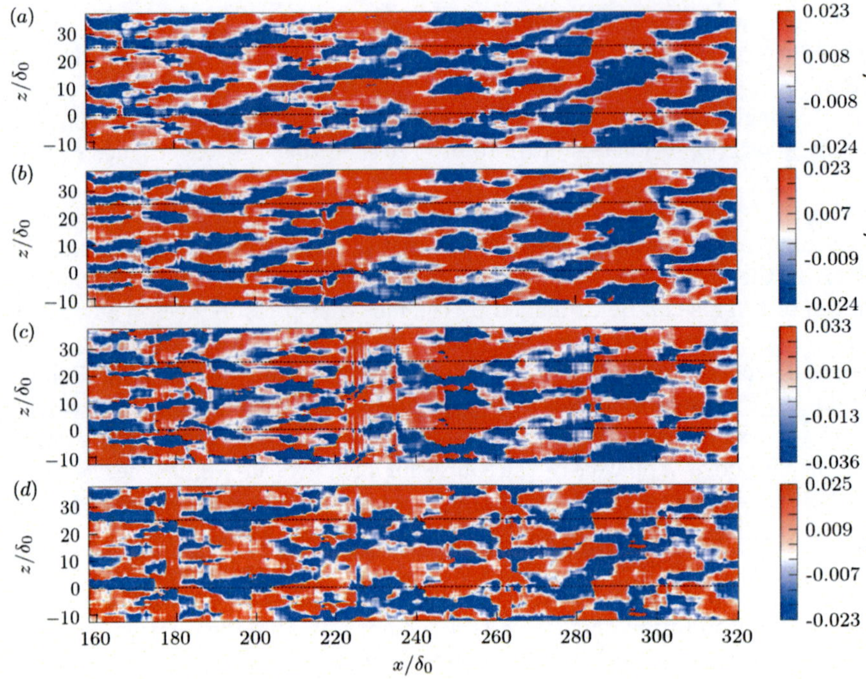

Fig. 6. Comparison of instantaneous convection-velocity disturbance patterns at the same wall-normal positions $y/\delta_{99} \approx 0.4$ and the same snapshot in time as in Fig. 4.(*a*) wall-normal velocity component v', (*b*) spanwise velocity component w', (*c*) density ρ', (*d*) pressure p'.

Following the above reasoning, the present data contain material-bound coherent motion in the area $100 < y^+ < 400$. But what is its spatial structure? Is it universal or variable-specific, because Fig. 2 and Fig. 4 show that each flow variable has different instantaneous patterns? Subtracting the time average from the instantaneous convection velocity leads to the convection velocity disturbance $c'_\Phi = c_\Phi - \overline{c}_\Phi$, for which Fig. 6 shows examples corresponding to the snapshots shown in Fig. 4. Interestingly, all instantaneous disturbance fields contain high and low convection-velocity streaks which appear as spatially connected patterns with either positive or negative c'_Φ. There is an unexpected difference to the patterns observed in v', w', ρ', p' in Fig. 4. At the same time, there is a considerable resemblance of all c'_Φ with the filtered u' snapshot in Fig. 3.

Apparently all these flow-variable-dependent disturbance patterns in Fig. 4 simply mark the fluid, like tracer particles in PIV experiments. For the state variables ρ, p and T this appears obvious as these are bound to the local gas state, i.e., to the local thermodynamic properties of the fluid. Applied to ρ or T, for instance, our PIV-like tracking algorithm has tracked the downstream displacement of the low- and high-density and -temperature regions of the fluid, as if these were particle patterns. On the other hand, the streamwise fluid motion (which dominates over all other directions by large) is already quantified by the streamwise velocity component u in the DNS data. Therefore, it can be expected that the high- and low-speed patterns in c'_Φ must correspond to those in u', apart from those areas where the spatio-temporal pattern correlation technique produces erroneous estimates, so called 'outliers'. The average maximum correlation values of (u, v, w, ρ, p, T) in the data set in Fig. 6 range between 0.646 in the T-field and 0.78 in the u-field, which confirms the presence of erroneous estimates. In standard PIV estimation such outliers are usually corrected, e.g. by interpolation of surrounding credible data points. This step has not been performed here. Averaging over time and spanwise coordinate has reduced the influence of 'outliers' on the mean value below an acceptable level, but the c'_Φ snapshots still contain them. As the local correlation maxima have been stored together with the estimated convection velocities, we are able to identify these outliers. The corresponding maximum number of outliers in this data set amounts to 13.6% in the T-field but only to 0.5% in the u-field.

4 Conclusions

The comparison of estimated average convection velocities with the mean-flow velocity in Fig. 5 has led to the identification of the region $100 < y^+ < 400$ where all flow patterns are bound to the fluid material. Physically, this would mean that there are structures in the Lagrangian sense, presumably related to so-called Lagrangian Coherent Structures (LCS), a concept from mathematics, see [6]. The region identified here is in the overlap between log-layer and wake of the turbulent mean-flow profile and corresponds to the region where Large-Scale Motions (LSMs) are identified in the pre-multiplied velocity spectra (Fig. 1). New results presented in Fig. 6 together with the according reasoning show that their spatial structure corresponds to the high- and low-speed streaks in the streamwise velocity component u' in that area.

In contrast to these, the unsteady flow patterns in other areas, especially those in the viscous sublayer, have wave character. These waves originate in the LSM area and travel faster than the local fluid near the wall. They may cause reproduction of turbulence from the viscous sublayer, but this process must be investigated further in following work. A practical consequence of the present observation that instantaneous high- and low-speed streaks correspond to LSMs is that LSMs can most easily be extracted by low-pass filtering of the streamwise velocity component instead of correlating snapshots of any other flow variable.

References

1. Adrian, R.J.: Twenty years of particle image velocimetry. Exp. Fluids **39**(2), 159–169 (2005)
2. Balakumar, B.J., Adrian, R.J.: Large-and very-large-scale motions in channel and boundary-layer flows. Phil. Trans. R. Soc. A Math. Phys. Eng. Sci. **365**(1852), 665–681 (2007)

3. del Alamo, J.C., Jiménez, J.: Estimation of turbulent convection velocities and corrections to Taylor's approximation. J. Fluid Mech. **640**, 5–26 (2009)
4. Dennis, D.J.C., Nickels, T.B.: On the limitations of Taylor's hypothesis in constructing long structures in a turbulent boundary layer. J. Fluid Mech. **614**, 197–206 (2008)
5. Goldschmidt, V.W., Young, M.F., Ott, E.S.: Turbulent convective velocities (broadband and wavenumber dependent) in a plane jet. J. Fluid Mech. **105**, 327–345 (1981)
6. Haller, G.: Lagrangian coherent structures. Annu. Rev. Fluid Mech. **47**, 137–162 (2015)
7. Handler, R.A., Swearingen, J.D., Swean Jr., T.F., Leighton, R.I.: Length scales of turbulence near a free surface. In: 22nd Fluid Dynamics, Plasma Dynamics and Lasers Conference, p. 1775 (1991)
8. Kim, J., Hussain, F.: Propagation veloctiy of perturbations in turbulent channel flow. Phys. Fluids A **5**, 695–706 (1993)
9. Kline, S.J., Reynolds, W.C., Schraub, F.A., Runstadler, P.W.: The structure of turbulent boundary layers. J. Fluid Mech. **30**(4), 741–773 (1967)
10. Krogstad, P.Å., Kaspersen, J.H., Rimestad, S.: Convection velocities in a turbulent boundary layer. Phys. Fluids **10**(4), 949–957 (1998)
11. LeHew, J., Guala, M., McKeon, B. J.: A study of convection velocities in a zero pressure gradient turbulent boundary layer. In: 40th Fluid Dynamics Conference and Exhibit, p. 4474 (2010)
12. Ringuette, M.J., Wu, M., Martin, M.P.: Coherent structures in direct numerical simulation of turbulent boundary layers at Mach 3. J. Fluid Mech. **594**, 59–69 (2008)
13. Robinson, S.K.: Coherent motions in the turbulent boundary layer. Annu. Rev. Fluid Mech. **23**(1), 601–639 (1991)
14. Smits, A.J., McKeon, B.J., Marusic, I.: High-Reynolds number wall turbulence. Annu. Rev. Fluid Mech. **43** (2011)
15. Wenzel, C., Gibis, T., Kloker, M., Rist, U.: Self-similar compressible turbulent boundary layers with pressure gradients. Part 1. Direct numerical simulation and assessment of Morkovin's hypothesis. J. Fluid Mech. **880**, 239–283 (2019)
16. Willert, C.E., Gharib, M.: Digital particle image velocimetry. Exp. Fluids **10**, 181–193 (1991)

Hypersonic Aerothermodynamics

HyperCODA – Extension of Flow Solver CODA Towards Hypersonic Flows

Immo Huismann[1(✉)], Stefan Fechter[2], and Tobias Leicht[2]

[1] German Aerospace Center (DLR), Institute of Software Methods for Product Virtualization, Dresden, Germany
Immo.Huismann@dlr.de

[2] German Aerospace Center (DLR), Institute of Aerodynamics and Flow Technology, Braunschweig and Göttingen, Göttingen, Germany

Abstract. This paper presents HyperCODA, the hypersonics extension to the flow solver CODA (CFD for ONERA, DLR and Airbus). Similar to the spacecraft extensions of TAU, HyperCODA extends CODA for applications at high Mach numbers, including non-ideal gas thermodynamics, gas mixtures, and chemistry. The paper discusses simulations demonstrating numerical convergence and stability properties, as well as a supersonic rocket retropropulsion maneuver. Cross-code comparisons are made against the established DLR TAU code.

Keywords: CFD · Hypersonic flow · Finite volume method · Numerical software

1 Introduction

Modern numerical methods are required to complete an expanding array of tasks which can lead to significant code complexity. Furthermore, modern hardware incorporates new performance enhancements, such as multiple cores per socket, deep cache hierarchies with non-uniform memory access, and accelerator cards. Implementation of changes on the flow solver side, as well as the need to take advantage of hardware optimizations, can significantly influence the design of a code.

Rather than retrofitting large-scale legacy code bases such as TAU, a new modular infrastructure has been devised and is commonly developed as the CODA CFD software by various European partners, around a nucleus of ONERA, DLR and Airbus. As developed for TAU in [10] and further published in [9], CODA requires extensions for high-enthalpy flows and this paper presents the hypersonics extension to CODA: HyperCODA.

2 Flow Solver CODA

2.1 Motivation and Background

In 2015, Airbus expressed its interest for a New Generation CFD to be co-developed by ONERA and DLR. In 2016 and 2017, a series of Steering

© The Author(s), under exclusive license to Springer Nature Switzerland AG 2021
A. Dillmann et al. (Eds.): STAB/DGLR Symposium 2020, NNFM 151, pp. 99–109, 2021.
https://doi.org/10.1007/978-3-030-79561-0_10

Committees involving Airbus, DLR and ONERA took place to further drive this cooperation. The Steering Committee was fed by a Technical Working Group looking at the framework architecture for such a CFD platform and a Governance Working Group. In the May 2017 Steering Committee meeting all three parties reached an agreement based on the industrial needs and constraints and decided to pursue the way proposed by Airbus. It has been decided to base the common framework and architecture on the prototype code Flucs, previously developed by DLR [13].

The resulting CFD flow solver is now being jointly developed by the aforementioned as well as further partners under the name "CFD for ONERA, DLR and Airbus", for short: CODA. This common development focuses on subsonic and transonic flows in aviation context and individual extensions are possible. For instance, HyperCODA is currently a DLR-only activity.

2.2 Solver Capabilities

The CFD solver CODA is a framework for steady and transient partial differential equations (PDEs) comprising first-order spatial derivative terms for convection, second-order derivative terms for diffusion, as well as generic source terms. Each solver component consists of a choice of PDE to solve, algebraic relations for the closure, e.g. thermodynamic relations or equations of state, and then the choice of temporal and spatial discretization as well as iterative solver. This modularity allows for using PDEs in conjunction with any combination of implemented spatial and temporal discretization schemes.

For spatial discretization, CODA implements finite volume (FV) and discontinuities GALERKIN (DG) techniques. A cell-centered scheme has been chosen for the former, whereas for the latter modal basis functions are implemented, allowing for fixed order simulations as well as spatially varying orders. For both discretization techniques, unstructured curvilinear high-order meshes are supported. For the diffusion terms, a central scheme averages the flux contributions from the left and right element, with the gradients either augmented in face-normal direction or in direction of the cell-mid-point connecting edge. The evaluation of the convective fluxes requires more care to achieve stability. Here, an upwind treatment is used, with the ROE flux including an enforceable minimal dissipation (entropy fix), a local LAX-FRIEDRICH flux, and the simple low dissipation AUSM variant SLAU2 from [11] being implemented. For each of these, linear reconstruction can be used to increase the consistency order to, ideally, two with non-linear limiters available for increased stability.

For temporal discretization, CODA implements time-accurate schemes and fully implicit schemes for computing the steady-state solution. The former contain explicit multi-stage RUNGE-KUTTA methods with definable Butcher tableaus as well as diagonally implicit RUNGE-KUTTA techniques. Both allow for spatially uniform time step sizes, as well as acceleration techniques via local time steps. A NEWTON-style method implements the implicit scheme for solving for the steady-state. It uses automatic differentiation (AD) of the residual evaluation to generate the Jacobian as the system matrix, allowing for one Jacobian computation working for all PDEs.

At the time of writing, CODA supports the EULER equations for inviscid flow, the NAVIER-STOKES equations applicable for laminar viscous flows, and the REYNOLDS-FAVRE-averaged NAVIER-STOKES equations for time-averaged turbulent flows. For the latter, multiple turbulence models are supported: The SPALART-ALLMARAS one-equation model in negative formulation [1], a parametrized two-equation model [2] which, with specific choices of the parameters, yields either the k-ω model [18] or the MENTER-SST model [14]. Furthermore, scale-resolving simulations are supported via DNS, implicit LES, mainly intended for the high-order DG mode, and traditional LES via the WALE subgrid-scale model [5]. Ongoing work consists of hybrid RANS/LES methods and an incompressible version, including stabilization in the spatial discretization.

2.3 Infrastructure and Implementation

The CFD solver CODA was designed as part of an eco-system, fully concentrating on the core CFD capabilities while reusing already existing infrastructure as much as possible. This leads to a separate infrastructure layer for data structures called Flis which handles parallelism, AD support, and interfacing to the outside world, with the latter being implemented via the FlowSimulator DataManager (FSDM) [16]. Linear problems arising from implicit schemes and linearized equations are solved using DLR's linear solver library Spliss, as described in [12].

For exploiting HPC hardware to its fullest extent, Flis provides a coarse-level domain decomposition with a ghost cell layer implemented via message passing. This first level harnesses either MPI or GASPI and exploits one-sided and asynchronous communication, including overlap of communication and computation. To further capitalize on shared-memory architectures, a fine-level parallelization implements a shared-memory layer. Compared to the message passing layer, the second level parallelization uses fixed subdomains without any data duplication, such that data locality can be exploited. This two-layer parallelization allows to exploit data parallelism on either node, socket, or L3-cache level.

The software development process for CODA focuses on extensibility and maintainability: For the core flow solver, an abstract formulation is used, such that PDEs, their closures, the convection schemes, and spatial discretizations are specified separately and in a modular fashion. This allows to test them separately and interchange them. Furthermore, a continuous integration process is in place, including unit testing and code reviews. This core of CODA is implemented using C++ 14 for the flow solver. This allows to leverage template metaprogramming: One can compile for one specific combination of PDE, closure, convection scheme, and spatial discretization, i.e. fully adapted to the choice of the user, leading to automatic, syntax-safe generation of efficient machine code. However, the end user will never have to cope with the C++ interface itself. As done for FSDM and TAU, the CODA user interface consists of Python code. These range from simple IO and then starting the flow solver, over scenario scripts, to full MDA/O couplings [16].

3 HyperCODA – Hypersonics Extension to CODA

3.1 Governing Equations

The flow solver CODA targets subsonic and transonic flows, explicitly excluding hypersonic ones. These flows come with their own difficulties: Strong shocks are prevalent, leading to large differences in temperatures and pressures. These, in turn, cause chemical reactions and the fluid separating into its constituents, i.e. the approximation of a single fluid not being accurate anymore. Multi-component fluid models are required to simulate this regime, and are the explicit target of HyperCODA.

At the time of writing, HyperCODA extends CODA's implementation of the EULER, NAVIER-STOKES and RANS equations from the assumption of a (single) perfect gas to the consideration of gas mixtures at thermodynamic equilibrium. In CODA terms, implementing these equations consists of providing the state variables and the corresponding convection and diffusion terms, e.g. for the EULER equations the state consists of a vector containing the partial densities $\rho_i = \rho Y_i$ of the N species, where the Y_i denote the mass fraction of the species, the momentum $\rho\mathrm{v}$, and the stagnation energy E times density:

$$\partial_t \begin{pmatrix} \rho_1 \\ \vdots \\ \rho_N \\ \rho\mathrm{v} \\ \rho E \end{pmatrix} + \nabla \cdot \begin{pmatrix} \mathrm{v}\rho_1 \\ \vdots \\ \mathrm{v}\rho_\mathrm{N} \\ \mathrm{v} \otimes \rho\mathrm{v} + \mathrm{p} \\ \mathrm{v}\left(\rho\mathrm{E} + \mathrm{p}\right) \end{pmatrix} = 0 \quad , \tag{1}$$

with the ideal gas closure connecting pressure p, temperature T and the state:

$$p = \rho R T \tag{2}$$

$$R = \sum_{i=1}^{N} R_i Y_i, \quad c_\mathrm{v} = \sum_{i=1}^{N} c_{\mathrm{v},i} Y_i, \quad T = \frac{1}{c_\mathrm{v}} \left(E - \frac{1}{2} \mathrm{v} \cdot \mathrm{v} \right) \quad . \tag{3}$$

Here, each R_i denotes the species gas constants for species i resulting from the universal gas constant R_u and the molecular weights M_i via $R_i = R_\mathrm{u}/M_i$. Furthermore, the $c_{\mathrm{v},i}$ denote the specific heat at constant volume for the species resulting as $c_{\mathrm{v},i} = R_i/(\gamma_i - 1)$.

For the NAVIER-STOKES equations for gas mixes, three diffusion terms occur on the right-hand side, one due to the shear stresses, one due to the varying species concentrations and one due to the enthalpies of the species [9]. Lastly, the RANS implementation consists of adding further variables to the state and reusing the same terms as used in CODA.

3.2 Infrastructure, Spatial, and Temporal Discretization

The main benefit of CODA lies in its modularity, which allows to define the PDE, its closure and flux treatment separately from the spatial and temporal discretization. This modularity allows HyperCODA to seamlessly integrate into CODA, reusing all of the discretization while requiring solely a specification of the additional PDEs, their closures, and boundary conditions. At the time of writing, this amounts to less than 3000 lines of code. However, this low number stems from most infrastructural changes and features, such as stabilization techniques, transport terms, and convection fluxes, being implemented directly in CODA and, therefore, benefiting both activities.

HyperCODA can focus on the implementation of equations, and specialities for the discretization thereof. Due to the strong shocks occurring in high-enthalpy flows, shock-stable convection fluxes are required. As a first step, the HARTEN-LAX-VAN LEER with contact (HLLC) [17] flux using the HARTEN-LAX-VAN LEER-EINFELDT wave estimates [7] were implemented. These serve as an accurate baseline and have a higher resistance to the Carbuncle phenomenon compared to the ROE flux. Thereafter, shock indicators and Carbuncle fixes based on these were investigated, specifically: an added shear viscosity term for the ROE flux [4] and the AUSM+M flux [3].

4 Numerical Results

4.1 Species Transport for a Multi-species Gas

For a two-species gas in a periodic domain $\Omega = [0, 1)$, the initial conditions are:

$$\left(\rho_1 \; \rho_2\right)^T = \left((1 + \sin(2\pi x_1)) \quad 2\,(1 + \cos(2\pi x_1))\right)^T \quad , \tag{4}$$

with $v_1 = 1.5$ and $p = \pi$. The specific heat ratios and molecular weights are chosen as $\gamma_1 = 1.4, M_1 = 1.0$ and $\gamma_2 = 1.13, M_2 = 1.0$.

HyperCODA simulated the advection problem using linear reconstruction in conjunction with the HLLC flux, a two-stage explicit RUNGE-KUTTA method and periodic boundary conditions. This choice of parameters allows for second order convergence in time and space. For a fixed CFL number of 0.8 the number of elements was scaled from 32 to $32 \cdot 2^8$ and each simulation carried out until $t_{\text{end}} = 0.5$.

Figure 1 depicts the solution for ρE at t_{end} on the coarsest grid, which suffices to accurately resolve the advection problem. The grid refinement study proves consistency of second order in both ρ and ρE. While only these two quantities are shown, the behaviour is present in all quantities of HyperCODA's augmented state.

4.2 SOD Problem with Multiple Species

This section considers the SOD shock tube problem as defined in [17] in a domain $\Omega = (-1, +1)$ for a two-species gas with $\gamma_1 = 1.4, M_1 = 1.0$, $\gamma_2 =$

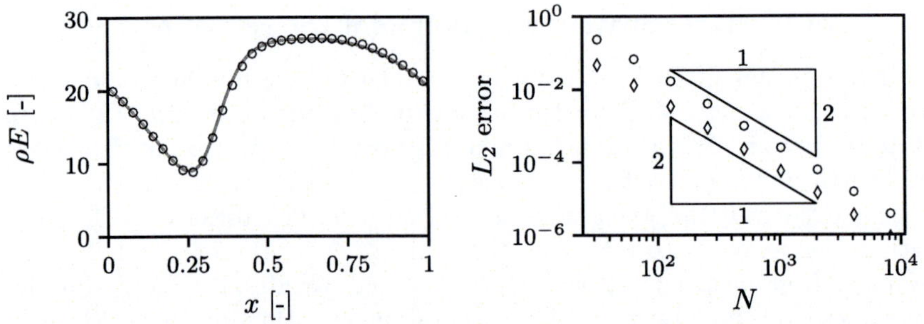

Fig. 1. Results for the convection problem. Left: ρE for $N = 32$ elements, comparing the analytical solution (grey line) to the result computed by HyperCODA (no line and circles). Right: L_2 errors for the density ρ (diamonds) and the stagnation energy times density ρE (circles). (Color figure online)

$1.13, M_2 = 1.0$ with the first species residing on the left and the second one on the right of the discontinuity, which initially resides at $x = 0$. The boundary conditions on the left and right boundaries of the domain match the left and right initial state, respectively. The test problem is computed until $t_{\text{end}} = 0.25$ using the HLLC flux, constant reconstruction, and an explicit fourth-order RUNGE-KUTTA method.

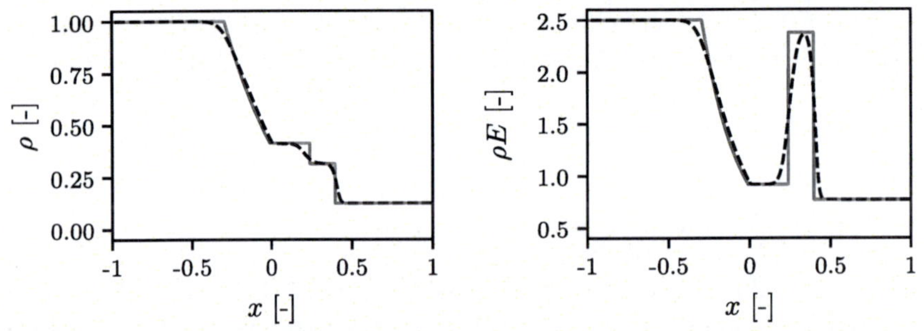

Fig. 2. Results for the RIEMANN problem when using $N = 256$ elements, comparing the analytical solution (grey) to the result computed by HyperCODA (black). Left: Density ρ, right: stagnation energy times density ρE.

Figure 2 compares the solution computed by HyperCODA using $N = 256$ elements to the analytical solution of the RIEMANN problem. The overall solution is captured, with the correct values in ρ and ρE being captured albeit with high numerical diffusivity involved, which will diminish with higher resolution [8]. This inaccuracy is to be expected and stems from the approximate flux calculation: The HLLC flux estimates the wave speed of the contact discontinuity from the speeds computed for the left- and right-most wave, i. e. the wave speed of the middle wave is only approximative based on integral conservation while the outer wave speeds are exact.

4.3 Blunt Body Test Case

This section investigates the numerical stability of the flow solver for high-enthalpy flows utilizing the blunt body test problem from, e.g., [15]. The geometry in Fig. 3 with $r = 1$ and $R = 3$ is used. While the half-cylinder serves as a wall boundary, the left bound serves as inlet with $Ma_\infty = 15$ and $\rho_\infty = 1$, whereas the right side serves as outlet with $p_\infty = 1$. This domain was discretized using a structured mesh using $N_r = 160$ grid points in the radial direction and $N_\phi = 480$ in the azimuthal one.

For a single-species gas with $\gamma = 1.4$, the problem was simulated four times: First, a numerical reference solution is generated with DLR TAU, using the AUSM-DV flux including the Carbuncle fix proposed in [10]. Second, with CODA using standard configuration: A Roe flux including a Harten-like entropy fix, with the fix fraction set to 0.2. Third, with HyperCODA using the shear-viscosity fix for the Roe flux from [4] and, lastly, with HyperCODA using AUSM+M [3]. In all cases, constant reconstruction was employed in conjunction with an explicit time-stepping scheme and the simulations ran until their density residual was reduced by a factor of 10^{-10}.

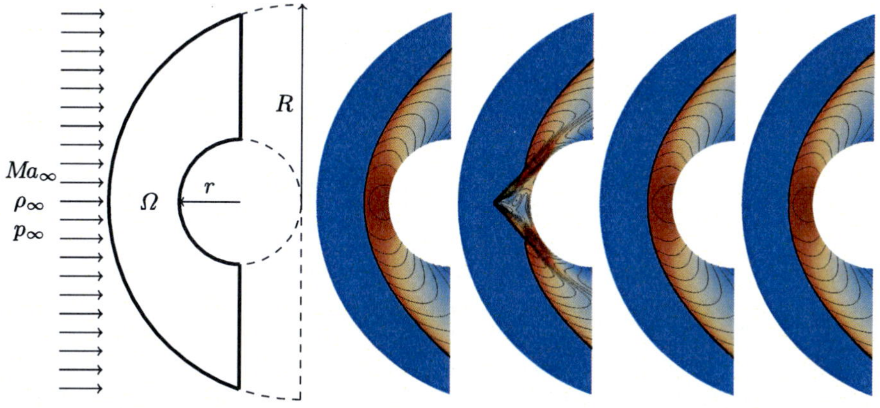

Fig. 3. Left: Setup for the blunt body testcase. Bold lines bound the computational domain Ω, dashed lines are used for construction of it. Right: Simulation results for the blunt body test case at $Ma = 15$. The color corresponds to the density ρ, with blue denoting the upstream density ρ_∞ and red the stagnation density $6.2\rho_\infty$. The corresponding temperature isolines are plotted in black. From left to right: DLR TAU using AUSMDV with Carbuncle fix, CODA using standard configuration of an entropy-fixed Roe flux, HyperCODA using Roe flux with an additional shear viscosity, HyperCODA using AUSM+M.

Figure 3 depicts the results of the simulations. TAU computes the physically correct solution: A sharp, bow-shaped shock surrounds the cylinder. For CODA, however, a completely different solution results, with a stem-like feature in front of the cylinder leading to a wedge-shaped shock and a qualitatively as well as quantitatively incorrect solution. It is to be noted, that this is not due to an erroneous implementation as the results exactly match those from, e.g., [11]. This is called the Carbuncle phenomenon and is dependent on the flux and numerical diffusivity. The Carbuncle phenomenon can significantly influence solution accuracy and is not trivial to detect in complex 3D flows.

For HyperCODA adding the shock-indicator-based shear viscosity from [4] suffices to suppress the Carbuncle. However, the temperature gradient near the stagnation point differs from that of TAU, leading to different isolines which are, moreover, not as smooth. The AUSM+M flux from [3] is able to regain the smoothness and yields the same isolines as the AUSMDV from TAU. In addition to preventing the Carbuncle, the AUSM+M flux allows to compute this testcase using limited linear reconstruction, which was not possible with any previously implemented flux in CODA.

Table 1. Thermodynamic properties of the fluids

Fluid	Air		Exhaust			
Species composition	N_2	O_2	CO	CO_2	H_2	H_2O
Mass fractions Y [-]	0.76	0.233	0.39	0.10	0.15	0.31
Mixture molar mass M $\left[\frac{kg}{mol}\right]$	0.02884		0.0234			
Heat capacity ratio γ [-]	1.4		1.267			

4.4 Application to a Rocket Example: Falcon9 Retropropulsion

This section performs a cross-code comparison between DLR TAU and the HyperCODA solver framework with the retropropulsion test case described in [6]: A simplified, descending Falcon 9 rocket. A cylinder with an attached 9-engine configuration represents the rocket and only three of the nine Merlin 1D+ engines are fired during the reentry maneuver. The simulation tries to recapture the trajectory point 5 in [6], corresponding to an altitude of 25 km and a free-stream Mach number of $Ma = 2$, using the SPALART-ALLMARAS turbulence model. First results for the effect of retropropulsion on the flow features are shown.

A two-fluid physical model is used: air and exhaust (both modeled as perfect gases) with the assumption of frozen heat capacity ratios γ for both. The composition and perfect gas properties are evaluated at free-stream conditions for the air and at the nozzle exit for the kerosene exhaust fluid, with the properties thereof being calculated from the mean components in the nozzle based on a separate nozzle calculation described in [6]. Table 1 lists the constituents, mass fractions, and fluid properties.

Compared to the detailed study in [6], the following simplifications were introduced for the first simulation: First, the rocket walls are modelled as adiabatic walls. Second, the profile at the nozzle exit is assumed to be a block profile and, third, constant fluid properties are assumed for all gas constituents.

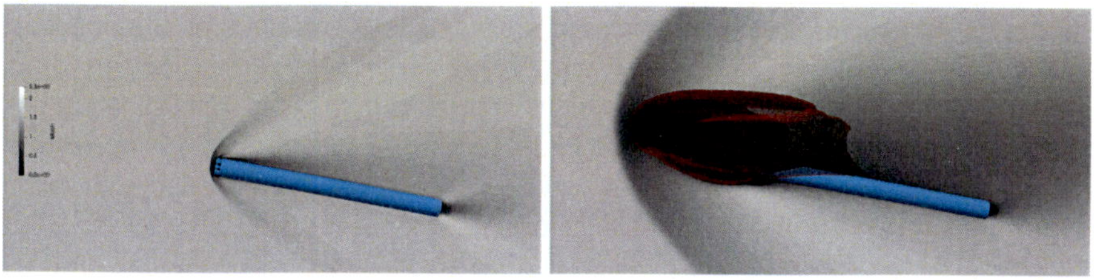

Fig. 4. Comparison of Mach number distribution with engine off and engine on at $\alpha = 10$ deg, $h = 25$ km, and $M_\infty = 2.0$. The rocket body is highlighted in blue. Left: Simulation with engine off. Right: Simulation with engine on with visualization of the plume isosurface with levels of 25, 50, and 75 % exhaust mass fractions (from inside to outside) in reddish colors. (Color figure online)

Figure 4 visualizes the effect of the hot exhaust plume onto the Mach number distribution. Without retropropulsion, a thin shock lies very close to the rocket and an expansion wave is present on the upper side. With activated retropropulsion, the exhaust moves the shock further away from the rocket while making it stronger. Due to the effect of the plume, the expansion wave on the upper side is not present in the case of engine on. Similar findings were reported in [6], using a more elaborate gas model implemented in TAU. With HyperCODA, the main flow features could be reproduced for the advanced use case of rocket propulsion. Further work will consist of implementing models for non-perfect gases, allowing for quantitative comparison to [6].

5 Planned Features for HyperCODA

For accurately modelling and simulating gases in high-enthalpy flows, HyperCODA will require the following features: Chemical reactions including ionization, high-temperature equations of state, i.e. non-ideal gas descriptions, and, lastly, multi-temperature models for chemical high-temperature modes. This will include tabulated reactions as well as equations of state.

Furthermore, cryogenic flows are of interest for, e.g., simulating combustion chambers, where the fluid is close to its critical point. For these, non-ideal equations of states for fluids and their chemical reaction kinetics are required.

6 Conclusions

This paper presented HyperCODA, the extension of DLR's flow solver CODA to high-enthalpy flows and multi-species fluids. First, the joint infrastructure was presented and the currently solved equations were discussed as well as their discretization. Then, HyperCODA was verified: First, second-order convergence was proven in a one-dimensional setting, as was the correctness of the approximate RIEMANN solvers. Then, the Carbuncle fixes were presented for the blunt body testcase in two dimensions. Lastly, a three-dimensional test case constituted a first application of HyperCODA to three-dimensional flows. The results are plausible in comparison with TAU while the code base is modular, more easily verifiable and testable, and offering novel temporal and spatial discretization schemes.

The next steps consist of further validation and a first extension to non-perfect gases and reactions.

References

1. Allmaras, S.R., Johnson, F.T.: Modifications and clarifications for the implementation of the Spalart-Allmaras turbulence model. In: Seventh International Conference on Computational Fluid Dynamics (ICCFD7) (2012)
2. Bassi, F., Crivellini, A., Rebay, S., Savini, M.: Discontinuous Galerkin solution of the reynolds-averaged Navier-Stokes and k-ω turbulence model equations. Comput. Fluids **34**(4–5), 507–540 (2005)
3. Chen, S.-S., Cai, F.-J., Xue, H.-C., Wang, N., Yan, C.: An improved AUSM-family scheme with robustness and accuracy for all mach number flows. Appl. Math. Model. **77**, 1065–1081 (2020)
4. Chen, S.-S., Yan, C., Lin, B.-X., Liu, L.-Y., Yu, J.: Affordable shock-stable item for Godunov-type schemes against carbuncle phenomenon. J. Comput. Phys. **373**, 662–672 (2018)
5. Ducros, F., Nicoud, F., Poinsot, T.: Wall-adapting local eddy-viscosity models for simulations in complex geometries. Numer. Meth. Fluid Dyn. 293–299 (1998)
6. Ecker, T., Karl, S., Dumont, E., Stappert, S., Krause, D.: Numerical study on the thermal loads during a supersonic rocket retropropulsion maneuver. J. Spacecraft Rockets **57**(1), 131–146 (2020)
7. Einfeldt, B.: On Godunov-type methods for gas dynamics. SIAM J. Numer. Anal. **25**(2), 294–318 (1988)
8. Ferziger, J.H., Perić, M., Street, R.L.: Computational Methods for Fluid Dynamics, vol. 3. Springer (2002)
9. Hannemann, K., Schramm, J.M., Wagner, A., Karl, S., Hannemann, V.: A closely coupled experimental and numerical approach for hypersonic and high enthalpy flow investigations utilising the HEG shock tunnel and the DLR TAU code. Technical report, German Aerospace Center, Braunschweig (Germany), Institute of Aerodynamics and Flow Technology (2010)
10. Hannemann, V.: Numerische Simulation von Stoß-Stoß-Wechselwirkungen unter Berücksichtigung von chemischen und thermischen Nichtgleichgewichtseffekten. PhD thesis, Dt. Forschungsanstalt für Luft-und Raumfahrt, Abt. Operative Planung (1997). in German

11. Kitamura, K., Shima, E.: Towards shock-stable and accurate hypersonic heating computations: A new pressure flux for AUSM-family schemes. J. Comput. Phys. **245**, 62–83 (2013)
12. Krzikalla, O., Rempke, A., Bleh, A., Wagner, M., Gerhold, T.: Spliss: a sparse linear system solver for transparent integration of emerging HPC technologies into CFD solvers and applications (2020). submitted
13. Leicht, T., et al. DLR-project Digital-X – next generation CFD solver Flucs (2016)
14. Menter, F.R., Kuntz, M., Langtry, R.: Ten years of industrial experience with the SST turbulence model. Turbul. Heat Mass Transfer **4**(1), 625–632 (2003)
15. Quirk, J.J.: A contribution to the great Riemann solver debate. In: Upwind and High-Resolution Schemes, pp. 550–569. Springer (1997)
16. Reimer, L.: The flowsimulator – a software framework for CFD-related multidisciplinary simulations. In: NAFEMS European Conference: Computational Fluid Dynamics (CFD) – Beyond the Solve (2015)
17. Toro, E.F.: Riemann Solvers and Numerical Methods for Fluid Dynamics: a Practical Introduction. Springer Science & Business Media (2013)
18. Wilcox, D.C.: Reassessment of the scale-determining equation for advanced turbulence models. AIAA J. **26**(11), 1299–1310 (1988)

Experimental Approach on Concentration Measurements of NO in Hydrogen Combustion Based on Heterodyne Laser Absorption Spectroscopy Using Quantum Cascade Lasers

Jan Martinez Schramm(✉) and Diana Luís

Department Spacecraft, German Aerospace Center, Institute of Aerodynamics and Flow Technology, Bunsenstraße 10, 37073 Göttingen, Germany
{Jan.Martinez,Diana.Luis}@dlr.de

Abstract. Experimental determination of NO and H_2O production during hydrogen combustion applying laser absorption spectroscopy will be performed in the High Enthalpy Shock Tunnel Göttingen of the German Aerospace Center. The combustion process forming these species within the propulsion unit of a small scale wind tunnel model is to be studied. A novel experimental approach to obtain absorption spectra of both species in the infrared region within μ-second range is reported. This technique is based on the use of frequency combs generated with quantum cascade lasers. This method enables multi-species measurements of absorption spectra from which concentrations can be inferred. The generated frequency combs are within the $1760 \pm 30\,\text{cm}^{-1}$ wavenumber region which features strong NO and H_2O spectral lines allowing characterization of these species. The technique, the application in short duration ground testing and its calibration procedure will be reported.

Keywords: HEG · H2020 · STRATOFLY · Laser absorption spectroscopy · Hydrogen combustion · SCRAMJET · Dual comb frequency · Quantum cascade lasers

1 Introduction

A worldwide growing willingness to dramatically reduce the environmental impact of civil aviation within the next decades is noticeable. At the same time the need to improve the current civil transport aircraft performance is urging for the development of new high speed vehicles. The H2020 STRATOFLY (Stratospheric Flying Opportunities for High Speed Propulsion Concepts) project funded by the European Commission under the European Union's Horizon 2020 research and innovation programme was set up to study the feasibility of high speed passenger stratospheric flight. One main goal is to drastically decrease the transfer time of long range civil flights which implies performing an evaluation of

© The Author(s), under exclusive license to Springer Nature Switzerland AG 2021
A. Dillmann et al. (Eds.): STAB/DGLR Symposium 2020, NNFM 151, pp. 110–120, 2021.
https://doi.org/10.1007/978-3-030-79561-0_11

the economic sustainability of the future operation of hypersonic vehicles. The use of hydrogen combustion based scramjets as propulsion units is one possibility to drive hypersonic passenger transport. It is scientifically well established that cruise NO emissions cause a significant part of the current total climate impact of aviation [1]. The experimental determination of NO and H_2O production in hydrogen combustion is part of the contribution of the German Aerospace Center (DLR) to the STRATOFLY project. The measurements are performed in the High Enthalpy Shock Tunnel Göttingen (HEG), one of the major European hypersonic test facilities. The test time for experiments in HEG is typically in the ms range, which requires diagnostic techniques with high acquisition rates. The combustion process within the propulsion unit of a small scale flight experiment wind tunnel model will be studied employing heterodyne dual frequency comb based absorption spectroscopy with quantum cascade lasers (QCLs) in the infrared region within μ-second range. The paper reports on this novel approach, describes the technical and experimental prerequisites, and demonstrates the ability of this technique in the context of the measurement task.

2 Dual Comb Laser Absorption Spectroscopy

The process which led to the selection of the laser absorption spectroscopy based on the heterodyne processing of dual comb laser beating signals is explained in this section. The theoretical absorption spectra expected based on numerical flow simulations are shortly introduced. From here, the heterodyne technique allowing the detection of IR wavelength absorption spectra in the μs range and the QCL (quantum cascade laser) hardware enabling this are discussed.

Figure 1 (left) presents a broad NO and H_2O absorption spectra for the hydrogen combustion flow case to be studied (p = 4.14 bar, T = 2320 K). More details about the flow case are given in Sect. 3.2. The strongest NO absorption lines expected in the planned combustion experiment occur in a $1500\,\mathrm{cm}^{-1} - 2000\,\mathrm{cm}^{-1}$ region. The QCLs chosen for this work have spectral coverage of $1730\,\mathrm{cm}^{-1}$-$1790\,\mathrm{cm}^{-1}$. Figure 1 (right) shows a feature of the expected absorption spectra in the hydrogen combustion flow for NO and H_2O in the QCL emission region. The simulated spectra are overlaid with symbols to indicate the spectral resolution of the spectrometer. It is concluded that the absorption of H_2O and NO can be separated and the reconstruction of the state of both species can be performed with one measurement. The challenge is to obtain IR-absorption spectra at high repetition rates. The constant flow state of the planned experiment will be in the ms range and thus the acquisition rates of the spectrometer have to be in the μs range. The optical frequency comb technique allows for that.

A significant challenge in the field of optical frequency measurement is to obtain values with high accuracy, as light frequencies oscillate 10^5 times faster than state of the art digital electronics can resolve [3]. Up to the millennium, methods to access an optical frequency relied on wavelength determination and on the knowledge of the speed of light [3], which lead to poor precision. Around

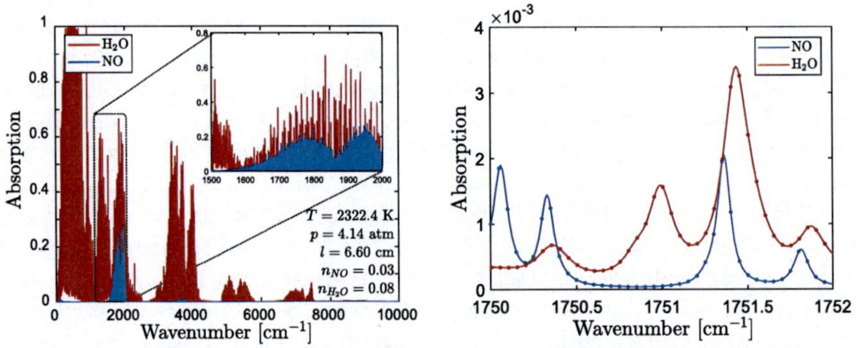

Fig. 1. Simulated NO and H_2O absorption spectra for a representative gas state during hydrogen combustion ($p = 4.14$ bar, $T = 2320$ K) obtained using HITRAN/HAPI [2]. Left: Broad spectra. Right: Feature of the expected spectra with points corresponding to the wavenumber resolution of the spectrometer.

2000, John L. Hall and Theodor W. Hänsch developed the frequency comb technique in which laser light with a series of equidistant frequencies is used to measure frequencies with great precision. For this accomplishment, the two scientists were awarded the Nobel Prize of Physics in 2005 for "their contributions to the development of laser-based precision spectroscopy, including the optical frequency comb technique" [4].

The application of this technique allows the determination of optical difference frequencies with good accuracy. These difference frequencies in the infrared light region can be used because they fall within the bandwidth limit of precision frequency counters (<10 GHz) [3]. Consider as a simplified example, two optical frequency comb carriers with slightly different repetition frequencies $f_{rep,1}$ and $f_{rep,2}$, shown by the blue and red lines in Fig. 2 on the left. The combination of these combs can be treated as an optical carrier with an amplitude modulation at the frequency difference, $\Delta f_{rep} = f_{rep,2} - f_{rep,1}$. When these laser signals are aligned on a photo detector, the detector produces a voltage proportional to the amplitude modulation, referred as heterodyne beating frequency. This heterodyne beating signal occurs in a much smaller frequency region allowing to capture its frequency contents by analyzing the photo detector signals (green line in Fig. 2 left). A single optical frequency comb consists of a very regular pattern of sharp spectral lines in the frequency domain. The comb frequency lines [5] at frequencies f_n, given by

$$f_n = nf_{rep} + f_o, \tag{1}$$

are spaced by the repetition frequency f_{rep} and shifted by the offset frequency f_o, satisfying $f_o < f_{rep}$. Both f_{rep} and f_o are in the radio frequency (RF) range, while the mode number n is large enough to scale f_n to optical frequencies [6].

Optical frequency combs are conventionally generated at the output of a mode-locked laser that emit short pulses. A combination of successive

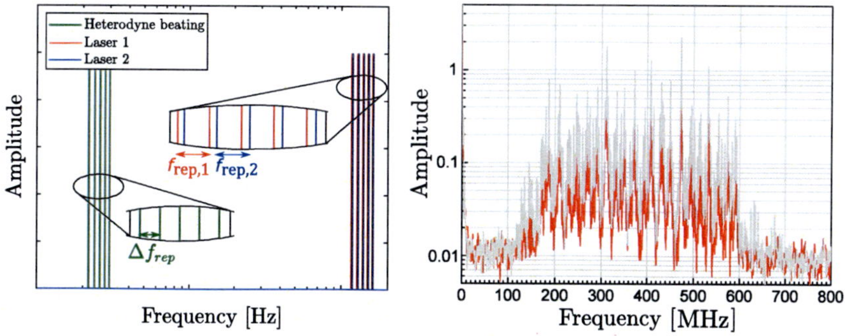

Fig. 2. Left: Two frequency combs with slightly different repetition frequencies (blue and red lines) and the resulting heterodyne beating frequency (green lines) observed when both frequency combs are optically aligned on the same detector. Right: Realistic beating signals from the IR-lasers used. The signals show the beating of two aligned beams carrying an absorption information (in red) and two beams propagating through a N_2 purged path (in grey). (Color figure online)

longitudinal laser modes gives rise to pulses every τ_{rep} if they have the same phase. The fixed phase relationship between the optical modes is needed for the formation of the optical comb spectrum described in Eq. 1. This relationship is obtained by having a single pulse propagating in the laser cavity. The round trip time τ_{rep} is the inverse to the repetition frequency f_{rep} of the emitted pulse train. After each round-trip, the temporal pulse is exactly restored which results in a stable pulse envelope at the laser output.

The generation of optical frequency combs with QCLs lies on the intrinsic physical properties of these laser sources. QCLs are semiconductors injection lasers emitting throughout the mid-infrared (416 − 3333 cm^{-1}) and THz (40 − 200 cm^{-1}) regions of the electromagnetic spectrum. A state-of-art infrared (IR) QCL laser source installed in a spectrometer (IRis-F1, IRsweep AG, Switzerland) is chosen to conduct the absorption measurements, to ultimately infer concentrations of NO and H_2O in a supersonic hydrogen combustion flow case. The IRis-F1 spectrometer uses a four wave mixing of phase-locked QCL to fabricate stable and repeatable IR frequency combs [7]. The comb is formed from an approximately 6-mm-long indium gallium arsenide/indium aluminum arsenide (InGaAs/InAlAs) QCL cavity and has approximately 200 comb lines with 7.5 GHz (0.25 cm^{-1}) comb-line spacing [8]. A heterodyne beating signal obtained from this device is shown in Fig. 2 on the right. The position of the frequency comb shown in Fig. 2 on the right with respect to the base frequency depends mainly on the length of the laser cavity. The length of the cavity depends on its temperature and the temperature can be controlled by the current applied. Changing the current can be used to tune the measurement to a specific absorption line. The line spacing within the comb is only weakly dependent of the varying length of the cavity.

3 Experimental Apparatus

3.1 Spectrometer

The IRis-F1 spectrometer has two QCLs installed, with a spectral coverage of 30 cm^{-1} centered at 1760 cm^{-1}. Details on the spectrometer can be found in Ref. [9]. The beam of each QCLs is attenuated with 25% transmission (0.6 optical density) neutral density (ND) filters (Thorlabs ND06A) and combined through a 50:50 CaF_2 beam splitter (BS) inside the laser module as shown in Fig. 3. The beam splitter divides the light into a sample beam path which reaches the sample detector (D1), and a reference beam path which is aligned onto the reference detector (D2). The detectors are a high-bandwidth (1 GHz) AC-coupled HgCdTe (Vigo PV-4TE-10.6), and are reached by the light beams through a parabolic mirror (Thorlabs MPD019-M01). Each of the detectors yield a heterodyne beating signal (grey and red signals in Fig. 2 on the right). When the sample beam passes through the combustion chamber of the model, specific frequencies of the laser are absorbed leading to a decrease of the intensity of the light beam and, consequently, the baseline heterodyne signal. To align the sample beam out and back inside the spectrometer through the combustion chamber, additional optics are used. The optical setup coupling the sample beam out of the spectrometer consists of a polarizer (Thorlabs PS810) combined with a lens which focuses the light onto an optical fiber with an FC connector at Brewster angle (IRflex IRF-Se-12). The Brewster angle minimizes back reflections and maximizes the power being coupled into the fiber. The other end of the fiber transporting the light to the model has a SMA connector inserted on a fiber coupler (Thorlabs RC02SMA-P01). A second coupler receives the beam after it passed the combustion chamber and feeds it to a second fiber, which is then coupled again on a third coupler and aligned with the parabolic mirror inside of the spectrometer. An external electronic box (current driver) from IRsweep is used to control the current driving the lasers.

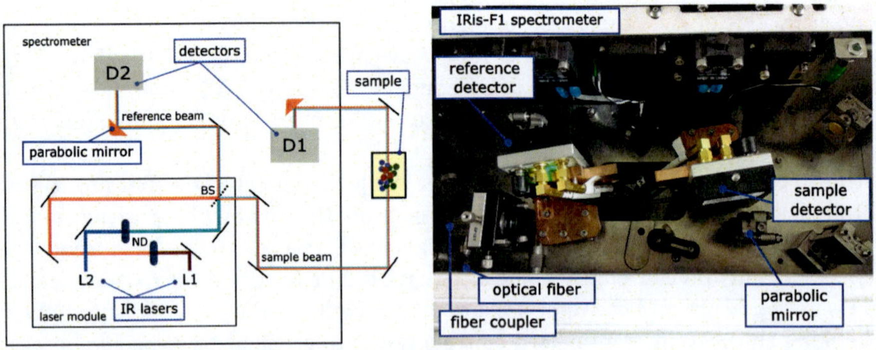

Fig. 3. Left: Simple schematic of the beam paths. Right: Actual setup inside of the spectrometer for detection of the laser beams.

The spectrometer allows three different types of processing modes: *static*, *time resolved* and *sweeping*. In *static* mode, a single spectrum is obtained from averaging multiple number of measurements; in *time resolved* mode, the multiple measurements are taken and saved, yielding a spectrum for each time instant. The *static* and *time resolved* modes leave the lasers current constant throughout the acquisition allowing time resolutions at 4 μs [9] and a spectral point spacing of 0.25 cm^{-1}. In *sweeping* mode the current of the lasers is periodically modulated in a saw-tooth wave (Fig. 4) at a rate of 5 mA/V, by a wave generator. Each ramp is set to increase from 0 V to approximately 6 V and takes about 1 ms to sweep through. The exact maximum voltage value is chosen by fine tuning so that the heterodyne beating signal is stable and well defined. The ramps are recorded and thus, the frequency changes can be precisely coordinated in the post processing. This method decreases the time resolution of the measurements, but increases the spectral resolution by a factor of almost 250. This is a theoretical value when sampling with 4 μs during 1 ms that is checked experimentally as well.

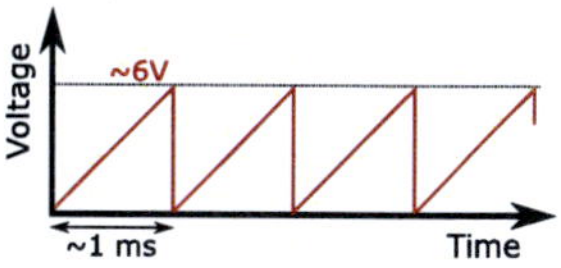

Fig. 4. Saw-tooth wave

To perform calibration in the tunnel, the sample beam is aligned through the model, the spectrometer is purged with N_2, the test section is in vacuum, and the lasers conditions and instrument parameters are the same as during the experimental test. Ahead of the measurements, transfer functions are taken with the lasers light blocked. These functions are used as background and help eliminating the electronic noise of the system when there is no light on the detectors. During calibration, compressed room air is injected by the hydrogen supply system into the combustion chamber. The measurement is done in *time resolved* processing mode and triggered by the hydrogen injection trigger line. The pre-trigger time is used as a background measurement as the test section and, consequently, the combustion chamber, are in vacuum. The comparison between the spectra before and after the trigger allows observing the H_2O content on the compressed air. Fitting the H_2O spectrum obtained experimentally with the theoretical water spectrum generated with HAPI [2] allows calibration of the wavenumber axis (see Sect. 3.4).

3.2 HEG and SSFE Wind Tunnel Model

Experiments will be conducted in the High Enthalpy Shock Tunnel (HEG) of the German Aerospace Center, DLR. HEG is designed on the working principle of a free piston driven shock tube (Stalker tube [10]) and is one of the major European hypersonic test facilities of this kind. Additional information about the working principle of the facility, its capabilities and former use in combustion research can be found in Ref. [11,12]. The operating condition that will be used for the experimental campaign is condition H3.3R3.7 (former condition XIII). Table 1 lists the nominal reservoir conditions for the nozzle, the Pitot pressure p_{t2} and stagnation point heat flux q_{t2}, as well the complete set of free stream parameters. The subscript 0 refers to the reservoir conditions and $t2$ to the total conditions behind the normal shock. The other variables p yield pressure, T temperature, h specific enthalpy, u velocity, Ma Mach number, Re

Reynolds number, ρ density, and m_{O_2} mass flow rate of oxygen. The total available test time is approximately 5 ms. Based on the nozzle reservoir conditions, the free stream is determined by numerical computation of the nozzle flow [12]. The nozzle reservoir gas temperature is quite low, when the total simulation range of HEG is considered. Typical problems in these facilities related to the development of sooth in the freestream leading to dimming of light traversing through the gas are not expected for the planned experiments. A final check has still to be done when the experiments are performed. Another typical and challenging problem for optical measurement techniques in shock tunnels is the free gas radiation (self-illumination of the flow). Pre-tests did not show any emission in the wavenumber region worth considering for the laser absorption technique.

Table 1. HEG operating conditions

p_0 [MPa]	T_0 [K]	h_0 [MJ/kg]		p_{t2} [kPa]	q_{t2} [MW/m^2]	
170	2740	3.3		146.53	3.71	
p [Pa]	T [K]	ρ [g/m^3]	u [m/s]	m_{O_2} [g/m^3]	Ma [-]	Re [1/m]
1990	266	25.9	2398	6.34	7.4	$3.7 \cdot 10^{-6}$

The supersonic hydrogen combustion flow to be studied will be generated in the propulsion unit of the SFFE wind tunnel model. SSFE stands for small scale flight experiment. This configuration is based on the full scale LAPCAT (Long-Term Advanced Propulsion Concepts and Technologies) II Mach 8 MR 2 vehicle proposed by ESA-ESTEC as described in Ref. [13]. The MR 2 vehicle is characterized by the combination of an aerodynamic efficient wave rider configuration and the dorsal integration of a high performance propulsion unit whose cruise and aero-propulsive performance was numerically assessed [14]. The SSFE configuration was derived from the MR2 as reported in Ref. [15]. The combustor layout is shown in Fig. 5. Two semi strut and one full strut injector expand the hydrogen into the combustor. The generation of NO in the combustion process is visualised by iso surfaces from the numerical rebuilding of the combustor flow. Additionally the lines of sight for the laser absorption spectroscopy measurements are shown. Further details about the model design are given in Ref. [16].

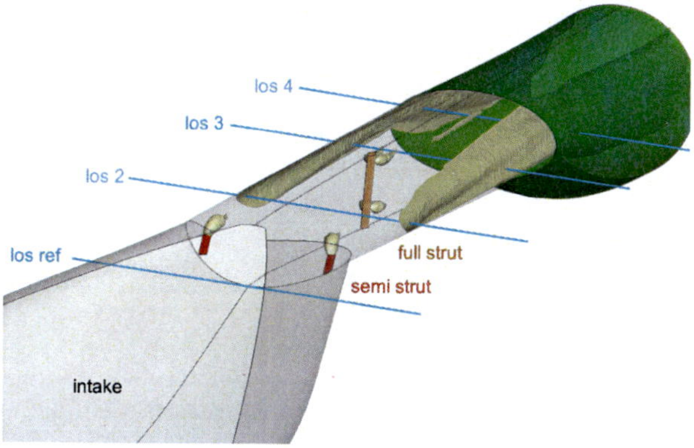

Fig. 5. CFD simulation showing iso surface for NO (mass fraction of 0.1).

To couple the laser into the model and to allow measurements along the lines of sight, eight optical ports were manufactured into the model. Each of these ports contains a Zinc-Selenide (ZnSe) window. The optical parts coupling the infrared light into the combustion chamber are mounted on and supported by an additional optical rail fixed outside to the model as pictured in Fig. 6.

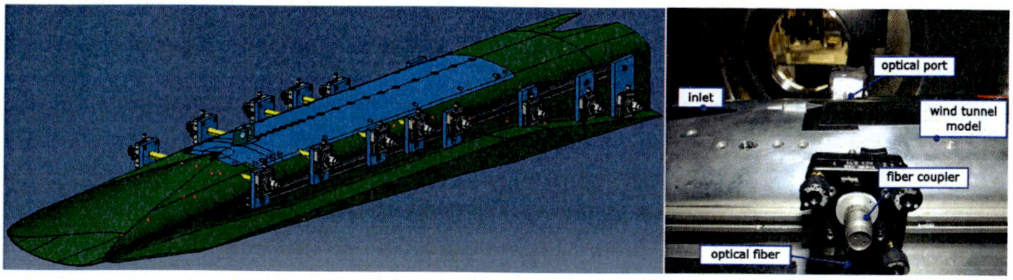

Fig. 6. Left: Schematic of the SSFE model with the optical couplers installed. Right: Photograph of an optical coupler installed on the model.

3.3 Flow State of the SSFE Propulsion Unit

The numerical flow properties along the $los\,2$ coordinate Y are given in Fig. 7 on the left. The distribution of temperature T, pressure p and the mass fractions for H_2O and NO are shown. Two different procedures have been pursuit to obtain the resulting NO and H_2O absorption spectra using HITRAN/HAPI [2]. The first assumes that the gas properties at the symmetry axis are constant throughout the line of sight; the latter uses a more sophisticated determination procedure. Here, individual spectra are obtained using the varying gas properties along the line of sight and the distances between the individual points are considered as integration length. All spectra are then integrated to give a resulting spectrum. The comparison between both procedures is shown in Fig. 7 on the right. The solid lines denote the results for the first procedure and the dotted lines the results for the latter. There is a notable difference between both procedures and thus the integration through the complete flow field is necessary, due to the large variation of the gas properties perpendicularly to the flow direction.

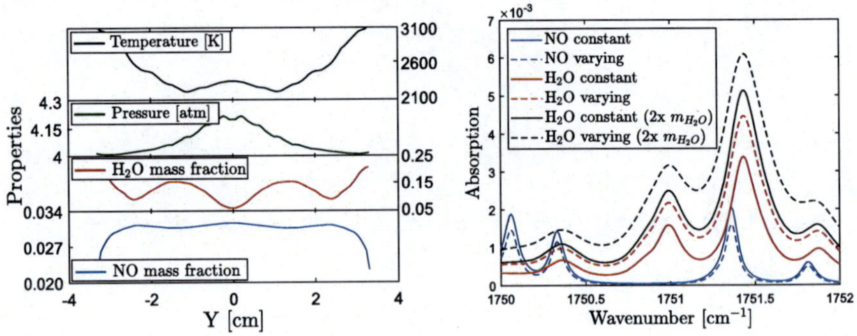

Fig. 7. Left: Flow properties (temperature, pressure, and NO and H_2O mass fractions) along the *los* 2 extracted from the numerical solution. Right: Simulated NO and H_2O absorption spectra using constant properties from the axis symmetry (solid lines) and properties varying (dashed lines) both along the *los* 2.

3.4 Wavenumber Calibration

Some preliminary calibration measurements are conducted under lab conditions to test the technique, prior to the experimental campaign at HEG. Measurements are taken with atmospheric air introduced into the sample path after obtaining the background measurements. Figure 8 presents some calibrated experimental points within a water feature (black dots) using *sweeping* mode. The calibrated wavenumber is obtained by moving the data along the wavenumber axis to fit the theoretical results obtained with HAPI [2]. In Fig. 8 (left), the spectral resolution in *time resolved* mode is highlighted and the number of points measured in between a $0.25\,\text{cm}^{-1}$ interval is visible. As previously mentioned, the spectral resolution of the measurements increases when using the *sweeping* mode.

A parametric study is conducted to analyse the variation of the theoretical spectrum with pressure (Fig. 8 left), temperature (Fig. 8 center), and H_2O concentration (Fig. 8 right). It is observed that small changes in temperature do not influence significantly the transmission spectrum. The most significant parameter is the concentration. To obtain an impression of the flow case investigated, an increase by double of the concentration of H_2O is shown in Fig. 7 on the right (black lines). A change in roughly 25% is expected in the absorption signal. Additional and optimized measurements that will access other absorption lines of water and subsequently NO will be performed as the establishment of a well defined calibration method is one of the most important prerequisites before the experimental campaign at HEG.

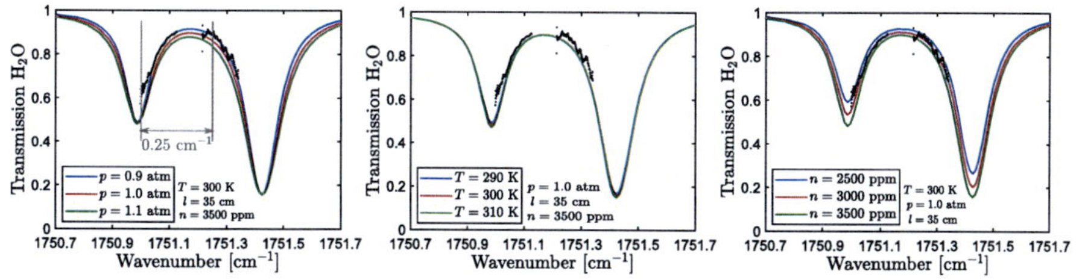

Fig. 8. Feature of experimental data using *sweeping* mode when measuring water (black dots) and fit with numerical results from HAPI [2]. Left: Variation of the theoretical H_2O transmission spectrum with pressure. Detail on the spectral resolution with *time resolved* mode. Center: Variation of the spectrum with temperature. Right: Variation of the spectrum with H_2O mole fraction.

4 Conclusions

Experimental determination of NO and H_2O production during hydrogen combustion is to be performed by employing laser absorption spectroscopy. This novel experimental approach to obtain absorption spectra of both species in the infrared region within μ-second range is reported. This technique is based on the use of frequency combs generated with quantum cascade lasers. It has been demonstrated that this technique enables multi-species measurements at rapid acquisition rates. The analysis of numerical simulations of the supersonic hydrogen combustion flow with respect to the absorption spectral distribution lead to the selection of frequency combs within the $1760 \pm 30\,\text{cm}^{-1}$ wavenumber region which contains strong NO and H_2O spectral lines. A method to obtain integrated line of sight spectral absorption spectra was created. This method allows to include the boundary layer effects in the analysis and it was demonstrated to be necessary for a more exact comparison. A calibration method was reported and its importance was shown with the help of parametric studies of the absorption line spectra.

References

1. Faber, J., et al.: Lower NOx at Higher Altitudes Policies to Reduce the Climate Impact of Aviation NOx Emission. Technical report, CE Delft (2008)
2. Kochanov, R.V., et al.: HITRAN application programming interface (HAPI): a comprehensive approach to working with spectroscopic data. J. Quant. Spectrosc. Radiat. Transfer **177**, 15–30 (2016)
3. Fortier, T., Baumann, E.: 20 years of developments in optical frequency comb technology and applications. Commun. Phys. **2**(153) (2019)
4. The Nobel Prize in Physics 2005. Nobel Media AB 2020. https://www.nobelprize.org/prizes/physics/2005/summary/. Accessed 19 Oct 2020
5. Faist, J., et al.: Quantum cascade laser frequency combs. Nanophotonics **5**(2), 272–291 (2016)

6. Probst, R.A.: Laser Frequency Combs for Astronomy. PhD thesis, Fakultät für Physik der Ludwig-Maximilians-Universität, München, 8 May 2015
7. Hugi, A., et al.: Mid-infrared spectrometer featuring μ-second time resolution based on dual-comb quantum cascade laser frequency combs. In: 2017 Conference on Lasers and Electro-Optics (CLEO), pp. 14–19. IEEE, May 2017
8. Overton, G.: Spectrometers: Mid-IR spectrometer uses quantum-cascade laser frequency combs, 10 May 2017. https://www.laserfocusworld.com/test-measurement/spectroscopy/article/16548279/spectrometers-midir-spectrometer-uses-quantumcascade-laser-frequency-combs. Accessed 5 Oct 2020
9. IRsweep. IRis-F1 spectrometer ASC/PSC variants, 2 edition, September 2019. https://s3-eu-west-1.amazonaws.com/mediahomepage.irsweep.com/resources/manuals/IRis-F1_UserManual_v2.pdf. Accessed 28 Aug 2020
10. Stalker, R.J.: A study of the free-piston shock tunnel. AIAA J. **5**(12) (1967)
11. Hannemann, K., et al.: The high enthalpy shock tunnel Göttingen of the German aerospace center (DLR). J. Large-scale Res. Facil. **4**(A133) (2018)
12. Hannemann, K., Martinez Schramm, J.: High enthalpy, high pressure short duration testing of hypersonic flows. Springer Handbook of Experimental Fluid Mechanics. Springer-Verlag, Berlin Heidelberg, January 2007
13. Steelant, J., Langener, T.: The LAPCAT-MR2 hypersonic cruiser concept. In: ICAS-2014-0428, 29th Congress of the International Council of the Aeronautical Sciences, St. Petersburg, pp. 7–12 September 2014
14. Roncioni, P., et al.: Numerical simulations and performance assessment of a scramjet powered cruise vehicle at mach 8. J. Aerospace Sci. Technol. **42**, 218–228 (2015)
15. Langener, T., et al.: Layout and design verification of a small scale scramjet combustion chamber. In: Proceedings 21st International Symposium on Airbreathing Engines, Busan, Korea, 9–13 September 2013
16. Martinez Schramm, J., et al.: Ground testing synthesis of the LAPCAT II small scale flight experiment configuration scramjet flow path. In: 20th AIAA International Space Planes and Hypersonic Systems and Technologies Conferences, Glasgow, Scotland, 6–9 July 2015

Internal Application of Ultra-Fast Temperature Sensitive Paint to Hydrogen Combustion Flow

Jan Martinez Schramm(✉) and Leni Schmidt

Department Spacecraft, German Aerospace Center, Institute for Aerodynamics and Flow Technology, Bunsenstr. 10, 37073 Göttingen, Germany
{Jan.Martinez,Leni.Schmidt}@dlr.de

Abstract. The paper will discuss an experimental approach to determine wall surface heat flux in a hydrogen combustion chamber using a temperature sensitive paint. This measurement technique represents a non-intrusive method to determine wall surface heat flux with a high spatial resolution. The application of this method to an internal flow, generated in a hydrogen fueled combustion wind tunnel model, is a new approach in the application of the measurement technique. The wind tunnel model represents the small scaled flight experiment configuration of the MR2 vehicle of the EU co-funded Long-Term Advanced Propulsion Concepts and Technologies project. The paper will discuss the special demands and prerequisites needed to apply TSP in high temperature flows. The experimental setup will be described and the feasibility of the application of TSP to internal flows and the resulting technical demands will be reviewed. The obtained result will be discussed with respect to the validity by comparison to a numerical simulation of the combustor flow.

Keywords: High enthalpy · Shock tube · Heat flux · Internal flow · High temperature hypersonic flow · Hydrogen combustion · HEG · TSP · SSFE · LAPCAT

1 Introduction

The precise determination of the surface heat flux is mandatory for experiments in high enthalpy impulse facilities [1]. The usage of surface-mounted temperature sensors like thermocouples or thin film gauges is common in these facilities. The High Enthalpy Shock Tunnel Göttingen, HEG, is one of the major European hypersonic test facilities of this kind [2]. Thermocouples and thin film gauges guarantee fast response times in the order of a few microseconds and therefore a high temporal resolution. But the spatial resolution is limited and can only be increased by increasing the number of sensors installed in a wind tunnel model. Additionally, it is challenging to attach sensors on complex or sharp-edged models. A non-intrusive option is the use of the temperature-sensitive paint technique

© The Author(s), under exclusive license to Springer Nature Switzerland AG 2021
A. Dillmann et al. (Eds.): STAB/DGLR Symposium 2020, NNFM 151, pp. 121–131, 2021.
https://doi.org/10.1007/978-3-030-79561-0_12

(TSP). The application of TSP enables the measurement of spatially resolved surface temperatures which allow for the subsequent calculation of surface heat fluxes. The application of the technique to an internal flow will be presented in this paper. The measurement technique will be used to determine wall surface heat flux in the propulsion unit of the small scale flight experiment configuration (SSFE) of the Long-Term Advanced Propulsion Concepts and Technologies (LAPCAT) MR2 vehicle [3] in order to validate the numerical modeling of the flow [4,5]. TSP has been successfully applied in HEG [6] and to the external flow in the SSFE intake [7].

2 Theoretical Background

A TSP typically consists of a polymer binder, luminescent molecules called luminophores and a solvent. Here, the used TSP was a mixture of Dichlorotris(1, 10-phenanthroline) ruthenium (Ru(phen)) serving as luminophore dissolved in the polyamide AQ Nylon (P-70) and ethanol. The binder has to be oxygen impermeable in chemically reacting hypersonic flows to prevent recombination processes in the boundary layer and on the surface. In addition, the impermeability is mandatory to prevent any pressure dependence of the TSP. Figure 1a shows a schematic of a typical TSP setup. The paint is illuminated by light emitting diodes (LEDs) having a specific wavelength λ_{ex} in order to excite the luminophores of the paint, while a light sensitive device detects the luminescence signals emitted at a wavelength λ_{em} from the TSP. The luminophores are exited to higher electronic states by the absorption of photons. The return to the ground state occurs through different mechanisms, either non-radiative relaxations or radiative decay [8]. One of those is the deactivation by the emission of a photon (photoluminescence). The other predominant mechanism is the thermal quenching which is a radiation-less process due to particle collisions. This process is temperature dependent, which means a rise in temperature increases the probability of radiation-less transitions. As a result, there are less emissions and thus the luminescence of the paint decreases. The energy loss due to the radiation-less transitions causes a shift in the wavelength. As shown in Fig. 1b, the spectrum of the excitation light source (LED) differs from the detected emission spectrum of the TSP. The maximum in wavelength of the emitted luminescence radiation is increased in comparison to the excitation wavelength. The difference in wavelength between both maxima is defined as the Stokes shift. A large Stokes shift is advantageous, because it facilitates the mandatory separation of luminescent emission from excitation radiation before the detection. For this purpose, a filter is placed in front of the detector to suppress light of shorter and longer wavelengths. For the application in hypersonic flows and under ultra-fast conditions, there are some special requirements for the paint due to the limited test time [9]. As already discussed, a temperature increase in the TSP leads to a lower intensity of the emitted light. This means the detectable temperature rise is limited; the signal becomes zero at an upper temperature limit. The temperature sensitivity of the TSP has to be adapted to the measurement situation.

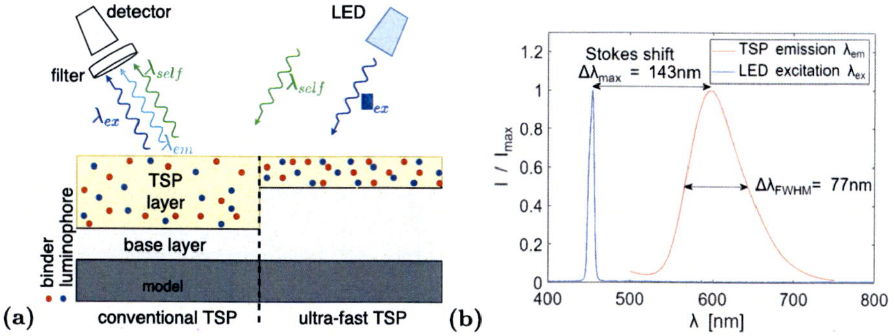

Fig. 1. (a): Schematic of typical and high-speed TSP set-up. (b): Emission and excitation spectrum of the selected TSP under excitation with blue LED ($\lambda_{ex} = 460$ nm).

Further, the luminescent lifetime of the luminophore of the TSP has to be significantly shorter than the test time and thus, in our case, it is in the order of less than 1 microsecond. When the TSP method is used in short duration flows, the detection instrument needs to be set up for short exposure times to resolve the fast changes appropriately. But these short exposure times require a fast response and high intensity luminescence signals of the paint [9]. The utilisation of a reflective base layer helps to increase the intensity of the luminescence signal and is essential in our case. Normally, a thick TSP layer causes an increase of the signal. But the paint's response time is proportional to the square of the paint thickness, so a rather thin layer in order of a few micrometers is required for the application in ultra-fast context [8]. As illustrated in Fig. 1a, the TSP layer for ultra-fast TSP is much thinner than the base layer in comparison to a conventional TSP setup. If long time scales are considered, temperature equilibrium between TSP and base layer can be assumed and the luminescence can directly be related to the temperature of both, the TSP and the base layer. In ultra-fast applications, the TSP layer needs to be thin enough to assume a negligible effect of it on the heat conduction process. The heat conduction process has to be considered with respect to time. To do so, we can treat the base layer as a uniform semi-infinite medium; this allows to follow to the analysis for thin-film gauges and other conventional sensors, as described by Schultz and Jones [10]. Under these presuppositions, it is possible to recover the heat transfer $\dot{q}_s$ from time dependent temperature profile in a discrete form [11]:

$$\dot{q}_s\left(t_n\right) = 2\frac{\sqrt{\rho c k}}{\pi}\sum_{i=1}^{n}\frac{T(t_i)-T(t_{i-1})}{\sqrt{t_n-t_i}+\sqrt{t_n-t_{i-1}}}. \tag{1}$$

$T(t_i)$ is the surface temperature obtained at time t_i. The heat flux depends on the thermal properties $\sqrt{\rho c k}$ with the density ρ, specific heat capacity c and the heat conductivity k of the base layer material. For the special case of a constant heat flux, the surface temperature can be expressed as [10]:

$$T(t) = \frac{2\dot{q}_s}{\sqrt{\pi}}\sqrt{\frac{t}{\rho c k}}. \tag{2}$$

Additionally, the implementation of the TSP technique in high enthalpy test facilities like the HEG creates some other challenges due to the special test conditions [12]. In hypersonic high speed flows, a self-illumination of the test gas may occur because of the arising high temperatures, this leads to a superimposition with the signals from the TSP luminescence.

3 Experimental Apparatus and Data Processing

The layout of the hydrogen-fueled vehicle is characterized by an aerodynamically ecient wave rider design and the integration of a propulsion unit [13]. A schematic of the 1.5-m-long SSFE configuration installed in the HEG test section is shown in Fig. 2a. The model is held by two feet that are fixed on the test section rail system. The vehicle's inward turning intake was designed based on a modied Busemann template ow eld and has an elliptical capture area [13]. The intake leads to an elliptical combustion chamber in which a two-stage multi-strut injector system is installed. Directly behind the intake, an upstream stage consisting of two semi-struts is injecting the hydrogen almost normal to the flow while downstream a full strut injects in the direction of the flow from a central position. The nozzle merges the elliptical combustor cross section into a circular shape and then expands. The hydrogen is supplied through a valve into a pipe that attaches to the model and supplies two steel bottles, one for each injector. Once the system is filled to the required pressure, the valve is closed and the hydrogen is injected into the combustor a few milliseconds before the run. The model is instrumented with 44 pressure transducers. The transducers in the combustor are located at the sidewall, while more transducers cover the region near the intake and in the nozzle. Their transducer signals are amplified

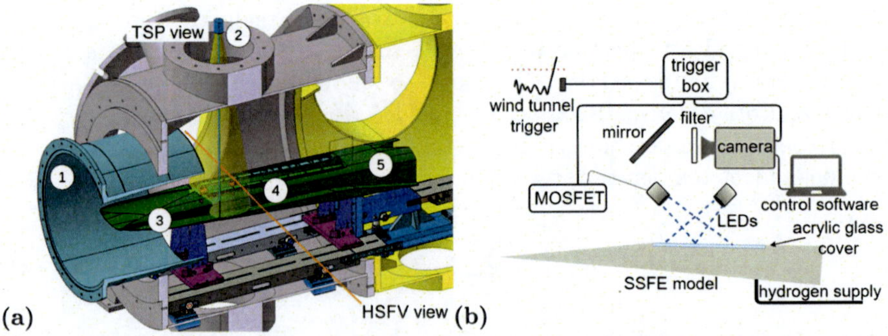

Fig. 2. (a) Schematic of the model installation in the HEG test section. 1: HEG nozzle; 2: LED; 3: intake; 4: combustor; 5: thrust nozzle. (b) Schematic of the TSP setup.

to increase the signals before they are recorded by the data acquisition system. A high speed camera is used in conjunction within a Schlieren setup to visualize the flow through the optical windows perpendicular to the flow direction. The high speed flow visualisation (HSFV) system allows to visualize the outer flow, ensure proper flow start up and can indicate unwanted movement of the model. A transient recorder is used for data acquisition of the wind tunnel instrumentation, the model instrumentation and all relevant signals of the TSP and Schlieren system. For the TSP measurements a top view has been realized. A schematic of the TSP setup is shown in Fig. 2b. Ten LEDs with each emitting at a wavelength of 460 nm illuminate the combustion chamber wall through the upper optical window of the shock tube. They are aligned in such a way that an optimal homogeneous illumination of the combustion chamber wall is attained. To optimize the power output of the LEDs, the currents used have to be maximized. This can only be achieved for short amounts of time. Therefore, the LEDs are loaded with high currents using a metal oxide semiconductor field-effect transistor (MOSFET) for approximately 8 ms. We use $I = 64.7$ A and $U = 51$ V to drive this pulse, which results in a total power of 3299.7 W. The complete acquisition chain is triggered by the tunnel signals. Once the trigger is active, the high-speed high-resolution uesd to visualize the TSP starts the recording. The camera records at a frame rate of 60 kHz and an exposure time of 2.5 µs. The image of the combustion chamber wall is mapped onto the camera chip through a mirror and a lens system. The inclination of the mirror can be changed easily to obtain an optimal image view. To allow optical access to the internal flow, the combustion chamber is closed with an acrylic glass cover. This causes an optical distortion in the image between the view without (Fig. 3a) and with cover (Fig. 3b). Due to that, the measurements have to be corrected. Therefore, points in a grid are marked on the layer which enable the comparison of the view with and without the cover. White light recordings of the view with and without the cover have been done to make the determination of the distortion possible. Moreover, a coordinate measuring arm was used. This device has a special needle tip. This tip can be pointed to each grid marker and stores the 3D coordinates in relation to a defined coordinate system. The acquired data set of the grid can be used later to build a 3D model of the combustion chamber and thus reconstruct the 3D internal heat flux from the TSP recording. The operating condition used in HEG for the experimental investigation was condition XIII [2]. To transfer the measured TSP intensities and obtain results to compare to the numerical modeling, the data need to be processed. The first step is to normalize the acquired intensity data. This correction can be achieved by taking the ratio I/I_0 of each shot image to the corresponding reference image for each pixel. As described above, an optical distortion occurs caused by the acrylic glass cover. Therefore, points in a grid were marked on the foil before the test to identify the distortion. As illustrated in Fig. 3c, the points are particularly distorted along the y-direction, while the displacement in x-direction is rather negligible small. The DLR software `ToPas` enables the warping of a picture. It compares the grid markers on the images with and without cover and creates a projection matrix

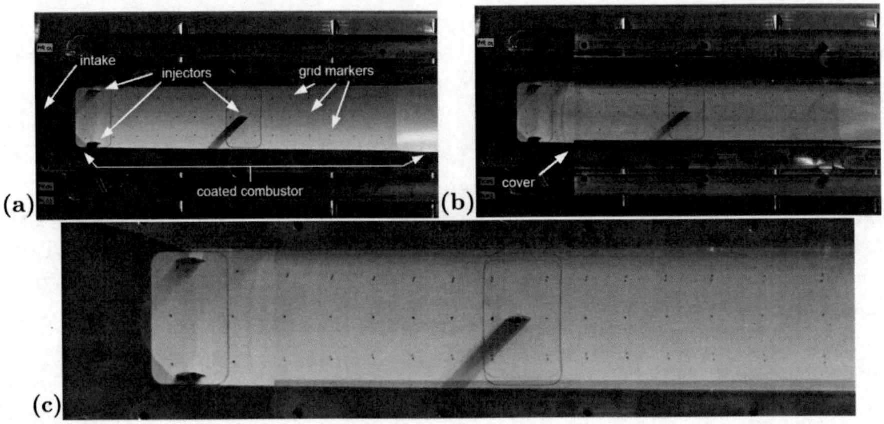

Fig. 3. Before the experiment white light shots were used to compare the distortions caused by the cover. a) without cover. b) with cover. c) overlay of both images.

between the grid markers. The area around the grid is interpolated. With the help of this projection matrix, each distorted image can now be corrected. The TSP technique is based only on the reduction of the luminescence intensity. But the hydrogen burning causes a self-illumination of the flow and thus leads to an increase in intensity. In order to separate the TSP signals, a mask was overlaid for pixels whose intensity deviates upwards more than 8% from a defined reference image at the beginning. This reference frame was chosen to the point when the LED excitation stabilized and thus initial fluctuations in the intensity stopped. The results are smoothed using a moving average filter to minimize the noise signals. Next, the intensity values need to be transferred to temperatures. Using the time dependent temperatures, the heat flux can be determined using Eq. 1.

4 Results

We will discuss and present the data we have been able to obtain from one experiment, being performed as a feasibility study. The measurement technique itself, even though it is quite simply from theory and can be described in a straight forward manner, is quite complex when real experimental data has to be analyzed. The first thing to note here is that the outer regions at the walls of the combustion chamber have to be excluded from the analysis. The reason for that is the extreme gradient of distortion caused by the strong bending of the cover in that area. We excluded this region in the analysis. Figure 4a shows a pressure recording of a sensor in the combustion chamber as a blue line and the pressure which is recorded in the reservoir of the hydrogen combustion unit as a red line. Please note that the units for the hydrogen reservoir pressure are given in arbitrary units. The time axis is normalized to the instant in time, when

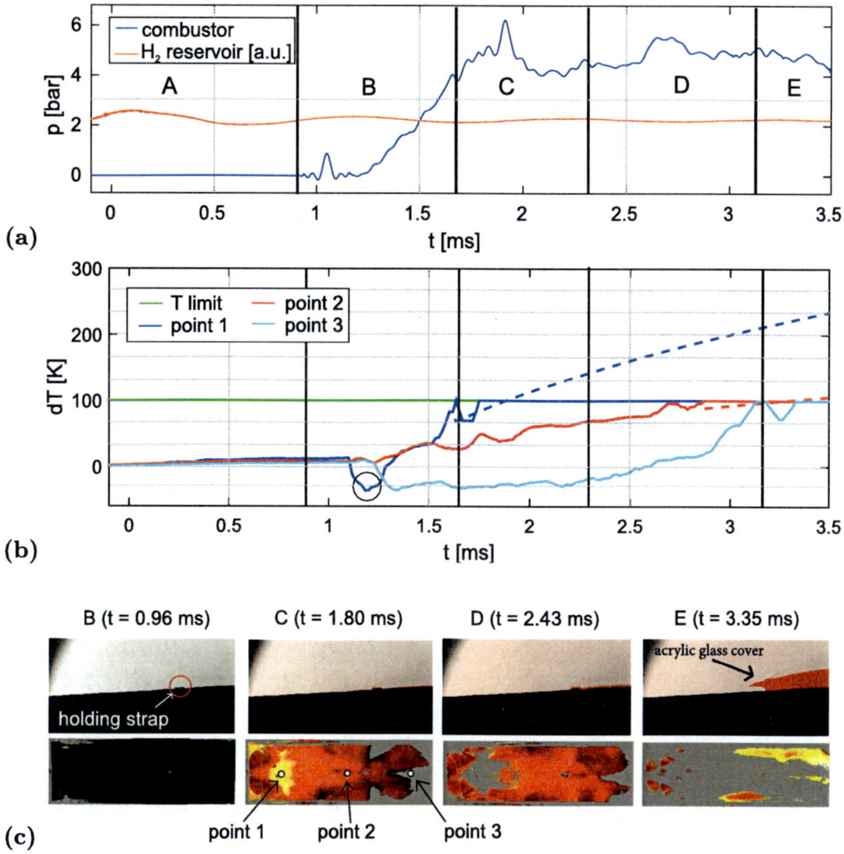

Fig. 4. a) Pressure trace of sensor in the combustion chamber and pressure in hydrogen reservoir. b) Time-dependent temperature profile for three points with extrapolations. c) Selected TSP heat flux distributions and false-color HSFV images (acrylic glass cover in red).

the reservoir of the hypersonic nozzle of HEG starts ($t = 0$ ms). The flow needs roughly 1 ms before it reaches the SSFE model. The time needed to obtain a constant mass flow rate through the injectors is larger than the flow start time in the combustion chamber, therefore it is mandatory to start the hydrogen injection process prior to the flow start. This process can be seen in Fig. 4a. The injection process is already steady at roughly 0.7 ms before the arrival of the test gas air in the combustion chamber, which happens at 1 ms. To discuss the evaluation process, we divide the experiment in five phases denoted by the letters A through E as given in Fig. 4a. Phase A is the observation phase prior to the actual experiment and before the flow start in the combustor. During phase A, the hydrogen is floating into the combustion chamber and a small increase in surface wall temperature is notable already in the time interval $t = 0 - 0.7$ ms as given in Fig. 4b. This data shown in Fig. 4b originate from three specific points (1), (2) and (3) on the TSP layer and the corresponding locations are shown in Fig. 4c. The first point located just behind the axial position of the two semi struts and in the symmetry plane shows the temperature development in a high temperature region, which can also be identified in the heat flux distributions given in Fig. 4c. We can observe from the temperature development with respect to time, that the measurement range for most of the points in the combustor's front part is exceeded at approximately $t = 2.43$ ms (the limit is marked with the green line). Some individual points like point (1) exceed the limit temperature already before at about $t = 1.8 - 1.9$ ms. After this time $t = 2.43$ ms, the heat flux integration using Eq. 1 is not valid anymore and the evaluation must stop. The gray areas in Fig. 4c mark the regions, where the heat flux determination had to be stopped due the above described effect. However to investigate this region, the assumption of a constant wall surface heat flux is made and an extrapolation of the time behaviour based on Eq. 2 can be performed. The extrapolation of the data is given by the dotted blue line in Fig. 4b. A second effect that has to be considered in the data is the illumination of parts of the CCD chip which originates from the hydrogen combustion during the startup and could not be suppressed for the run. This disturbing light which will increase the intensity on the chip will force the evaluation into a negative temperature and therefore a negative heat flux reconstruction. Regions with this behaviour are completely masked out in the evaluation and are given as grey regions in Fig. 4c. The effect can be seen in the signal around $t = 1.2$ ms and is marked with a black circle. The overshot in intensity will lead to a negative temperature undershoot in the evaluation, this has to be considered when extrapolating the data. A lot of TSP readings fall in phase B, which covers the time during the flow start up in the combustor. The Schlieren images given in Fig. 4c (top row) allow to define a third phase, which we denote with phase C. The acrylic glass insert forming to top wall of the combustor was not able to withstand the large forces at some point in time and started to bend, this can be observed in the Schlieren images in Fig. 4c. The acrylic glass cover, separating from the model, has been colored in red for better visibility. At $t = 2.43$ ms, the covers starts to deform slightly and this ends the phase C at which the experiment remains unchanged. The temperature development for the point (3) in Fig. 4b shows a negative development of the wall surface heating. There is no physical process possible that would allow for a negative wall surface heating as we observe it. The reason behind this behavior is the same as for the negative overshoot, we explained earlier: the camera sees intensities originating from the hydrogen combustion process during the experiment. The locations that show this behaviour can not be treated and are also marked in grey (see Fig. 4c). The phase D is not considered for the evaluation anymore, here the cover deforms already and it can not be assured that the flow will not be influenced. The phase E shows the phase when the cover lifts of the model. It is interesting to note that the pressure is not strongly influenced by this effect. To show the flow processes we can observe by the TSP measurement, the inter- and extrapolated temperature

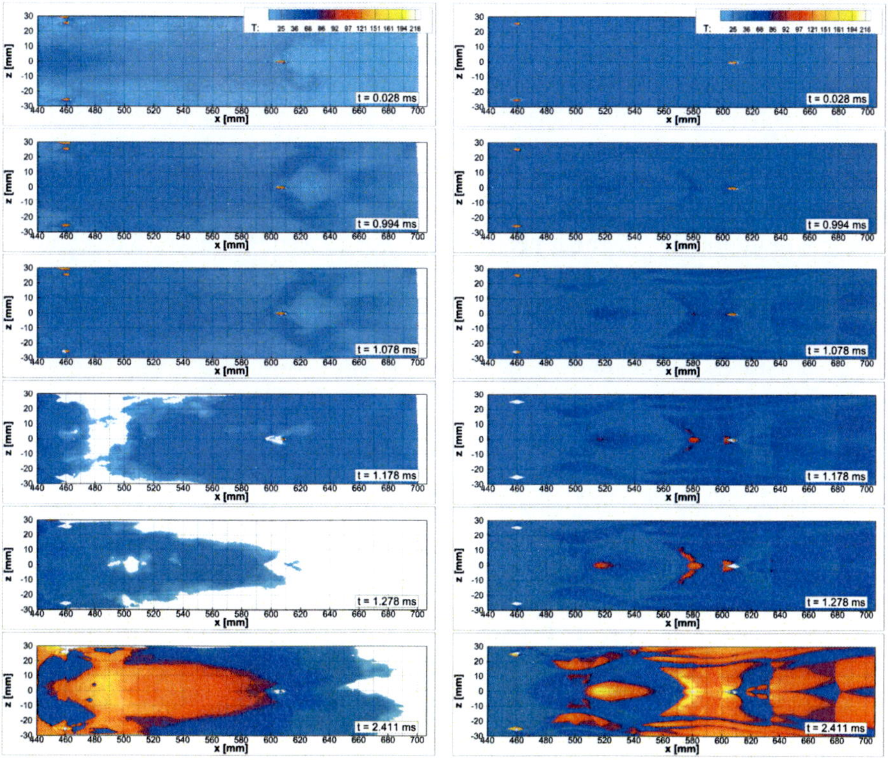

Fig. 5. Reconstructed temperature distribution at different times.

data is given in Fig. 5 for different time during the flow startup. The left column shows the experimental temperature evolution and corresponds to Fig. 4. We show the time $t = 0.028$ ms which is recorded prior to the arrival of the test gas expanding from the HEG nozzle at the model. It can be seen that the cooling of the wall due to the hydrogen, which is injected into the combustion chamber prior to this time instant, is picked up by the TSP. We then present a sequence ($t = 1.111$ ms, $t = 1.178$ ms, $t = 1.278$ ms) showing that the hydrogen, which is already existing in the combustor, starts to burn at about $x = 400$ mm and is then blown downstream by the incoming flow. The main part of the combustion anchors at late times at a position of approximately $x = 660$ mm. The right column of Fig. 5 shows the numerical calculation of the developing temperature. To obtain this time resolved numerical temperature data from the steady flow field calculation, the heat flux data from the solution has been transformed by the application of Eq. 2, here t is set according to the experimental time. The numerical temperature distribution shows a temperature development for the assumption of a constant heat flux, which is not necessarily the case in the experiment, but is a quite fair assumption. Discussing the comparison, we can show that the overall agreement of the individual flow features is reasonable to some point, but certain specific features differ. The high local heating point on the symmetry axis of the combustor located at $x = 520\,\text{mm}$ in the numerical solution appears more upstream in the experimental data around $x = 490\,\text{mm}$. The characteristic footprint of the hydrogen-jet emerging from the two first stage semi strut injectors are better visible in the numerical data than in the experiment but the pattern is quite similar. The strong shock in the numerical data almost perpendicular to the flow direction at $x = 490\,\text{mm}$ is not visible in the experimental data and the high heating zone not as well. We have to be careful, with a more extended analysis, since the flow might not have reached a steady state at the early time, where the analysis has to stop. Usually we regard the flow to be stable after $t = 3.00\,\text{ms}$. A modified version of the acrylic glass cover will allow for extended experiments and these will be reported in the future.

5 Summary, Conclusion and Outlook

The feasibility of the application of TSP to internal flows has potential but comes with high demands. This experimental investigation has outlined several technical improvements for further experiments. A modified version of the acrylic glass cover will prevent the unwanted movement. To scope with the high heat loads during the experiment, the TSP has to be adapted with regard to the temperature sensitivity. The strong illumination of the hydrogen burning zones has to be suppressed. Emission spectroscopy to evaluate the emission wavelengths will be applied.

References

1. Hannemann, K., Martinez Schramm, J.: High enthalpy, high pressure short duration testing of hypersonic flows. In: Springer Handbook of Experimental Fluid Mechanics, pp. 1081–1125. Springer, Berlin Heidelberg (2007)
2. Deutsches Zentrum für Luft - und Raumfahrt (DLR). The high enthalpy shock tunnel Göttingen of the German aerospace center (DLR). J. large-scale Res. Facilit. (2018). https://doi.org/10.17815/jlsrf-4-168
3. Hannemann, K., Martinez Schramm, J., Laurence, S., Karl, S., Langener, T., Steelant, J.: Experimental and numerical analysis of the small scale LAPCAT II scramjet flow path in high enthalpy shock tunnel conditions (2014). In: Proceedings Space Propulsion 2014, SP2014-2969350
4. Karl, S., Martinez Schramm, J., Hannemann, K.: Post-test analysis of the LAPCAT-II subscale scramjet. In: HiSST: International Conference on High-Speed Vehicle Science Technology (2018)
5. Karl, S., Martinez Schramm, J., Hannemann, K.: Post-test analysis of the LAPCAT-II subscale scramjet. CEAS Space J. **12**(3), 385–395 (2020). https://doi.org/10.1007/s12567-020-00307-7
6. Martinez Schramm, J., Edzards, F., Hannemann, K.: Calibration of fast-response temperature sensitive paints for their application in hypersonic high enthalpy flows. In: New Results in Numerical and Experimental Fluid Mechanics XI, Notes on Numerical Fluid Mechanics and Multidisciplinary Design. Springer (2017). https://doi.org/10.1007/978-3-319-64519-3_13

7. J Martinez Schramm, S Karl, K Hannemann, and H Ozawa. Ultra-fast Temperature Sensitive Paint Shock Tunnel Heat Flux Measurements on the Intake of the LAPCAT II Small Scale Flight Experiment Configuration., (2018). HiSST: International Conference on High-Speed Vehicle Science Technology
8. Liu, T., Sullivan, J.P.: Pressure and Temperature Sensitive Paints. Springer-Verlag, Berlin Heidelberg (2005)
9. Martinez Schramm, J., Edzards, F., Hannemann, K.: Development of ultra-fast temperature sensitive paints for hypersonic high speed flows. In: 21st AIAA International Space Planes and Hypersonics Technologies Conference (2017). https://doi.org/10.2514/6.2017-2211
10. Schultz, D.L., Jones, T.V.: Heat transfer measurements in short-duration hypersonic facilities. AGARD-AG-165 (1973)
11. Cook, W.J., Felderman, E.J.: Reduction of data from thin-film heat transfer gage: a concise numerical technique. AIAA J. **4**(3), 561–562 (1966)
12. Martinez Schramm, J., Hannemann, K., Ozawa, H., Beck, W., Klein, C.: Development of temperature sensitive paints for the high enthalpy shock tunnel Göttingen HEG. In: 8th European Symposium on Aerothermodynamics for Space Vehicles (2015)
13. Steelant, J.: Sustained hypersonic flight in Europe: technology drivers for LAPCAT II. In: AIAA 2009–7240, 16th AIAA/DLR/DGLR International Space Planes and Hypersonic System Technologies Conference (2009)

Aerodynamic Data Set Generation for the Experimental Vehicle ReFEx

Clemens Merrem(✉), Viola Wartemann, and Thino Eggers

German Aerospace Center (DLR), Institute for Aerodynamics and Flow Technology, Spacecraft, Lilienthalplatz 7, 38108 Braunschweig, Germany
{Clemens.Merrem,Viola.Wartemann,Thino.Eggers}@dlr.de

Abstract. Aerodynamic characterization of vehicles is mandatory to identify their flight envelope, their controllability behavior and ultimately their shape. During the design process one goal is to efficiently determine the aerodynamic properties for every possible state during the flight. Therefore, the aerodynamic force and moment coefficients need to be known for all combinations that are expected to occur of velocity, orientation, altitude, control surface deflections and dynamic values such as rotation rates. The amount of conditions that need to be evaluated can be reduced if an independence within these variation parameters is proven. For an aerodynamic design process via CFD this paper purposes an additional method to reduce the amount of data points. Through the use of the DLR TAU code with an overlapping meshes feature, envelope datasets can be produced for the momentum free trimmed state in each flight point. With this starting set, several subsets with parameter variations can be computed. This reduces the amount of simulations needed by 65% for the investigated case, the flight experiment ReFEx.

Keywords: Aerodynamic design · CDF · ReFEx

1 Introduction

The Reusability Flight Experiment (ReFEx) [1] is a flight experiment to gain knowledge on reusable rocket first stages or booster stages. The approach is to use a winged vehicle for a horizontal landing after launching vertically (VTHL). During the autonomous reentry flight, the Mach number range is $0.5 < M < 5.5$ and the angle of attack range is $-45° < AoA < 10°$. One challenge of the aerodynamic design process is to efficiently determine the needed aerodynamic coefficients within this envelope. ReFEx has a cylindrical fuselage with a flattened ogive tip and small rearward wings. There are three control surfaces: two canards and one flap in the vertical fin (see Fig. 1 - left).

The reference trajectory starts belly-up with high negative angles of attack for the initial part of the re-entry. This allows for an early deceleration which reduces the mechanical and thermal loads. After decelerating to Mach numbers lower than 2, action must be taken, as there is no controllable flight path to finish the flight in the belly-up orientation. Therefore, a roll maneuver is conducted to change to the nominal belly-down orientation. During this maneuver, the angle of attack changes from negative to positive

© The Author(s), under exclusive license to Springer Nature Switzerland AG 2021
A. Dillmann et al. (Eds.): STAB/DGLR Symposium 2020, NNFM 151, pp. 132–140, 2021.
https://doi.org/10.1007/978-3-030-79561-0_13

values. The remainder of the flight is then flown in the nominal orientation at moderate positive angles of attack for a maximum lift-to-drag ratio (see Fig. 1 - right).

The aerodynamic design process consists of iteratively calculating partial datasets of different geometries and checking in with all involved parties [2]. As the final shape is found alongside with a reference trajectory, a full set of aerodynamic data containing all parameter variations for all flight conditions needs to be created.

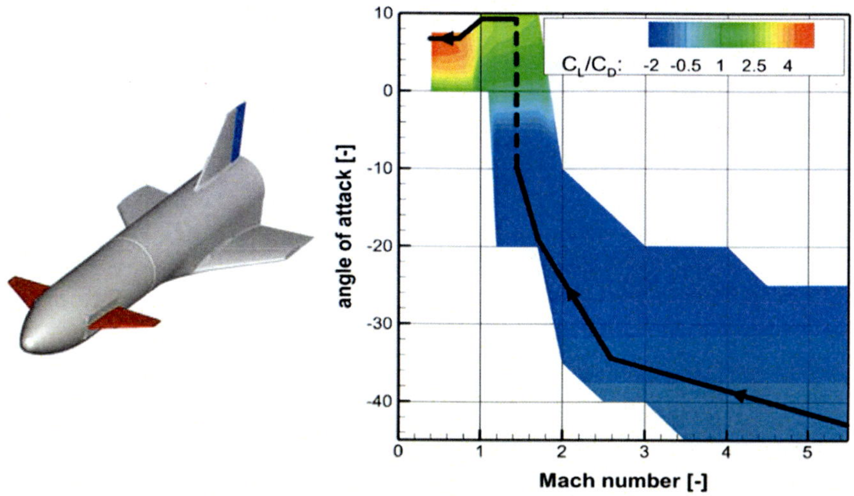

Fig. 1. Left: Final outer shape of ReFEx for CFD simulations with canards (red) and rudder (blue). Right: Flight envelope with CL/CD as contour and a model trajectory.

2 Aerodynamic Characterization

The aerodynamic dataset needs to provide information for the flight controller to adequately command the control surfaces to adjust the vehicle's orientation. This information must contain the aerodynamic coefficients and the effect of any control surface deflection on the aerodynamic coefficients. Additionally, the effect of dynamic parameters such as the vehicle's rotation rates further completes the aerodynamic characterization. Each flight state of the envelope needs to be covered.

To cover all possible flight conditions a grid of simulation points is spanned throughout the entire envelope. In this grid the aerodynamic coefficients for each possible combination of parameters can be interpolated. For ReFEx the list of parameters that describes the flight condition consists of Mach number (M), angle of attack (AoA), sideslip angle (AoS), altitude (A), canard deflections left and right (η_l, η_r), rudder deflection (ζ) and rotation rates around vehicle fixed axes (p,q,r). Hereby AoS, η_l, η_r and ζ variations consist of mirrored positive and negative values. Without any reduction strategies, a fully filled 10-dimensional grid would be necessary. The amount of grid points would be the product of the amount of variations of every parameter, e.g. for ReFEx the grid would contain 16 different Mach numbers. The total number of points for ReFEx would multiply up to more than 44 million (see Table 1). For every point a simulation would be necessary to obtain the aerodynamic coefficients, therefore measures to reduce this number are imperative. The following strategies are described with respect to ReFEx.

Table 1. Parameter space of ReFEx and the respective product without any reduction strategies.

Parameter	M	AoA	η_l	η_r	AoS	ζ	A	p	q	r	$\prod$
All variations	16	21	7	7	5	5	4	3	3	3	44.5 M

2.1 Reduction of Calculation Points by Traditional Means

The first approach would lead to a multidimensional, entirely filled box. However, a lot of the points are outside of the envelope, as there is no set of control surface deflections that enables the vehicle to attain these flight conditions. To exclude a lot of M-AoA combinations that will not be reached during the flight the dataset was cut into three sections: re-entry before the roll maneuver, during the roll maneuver and afterwards. For each section, only a subset of angles of attack is expected and included in the section. This results in a reduction from 336 to 114 M-AoA combinations. A similar strategy was used for the SPACE SHUTTLE Orbiter [2]. Also, a reduction of the range of flap deflections further reduces the total number. As for ReFEx this reduction is different for each box, here it will be estimated with a reduction from 7 to 5 possible canard deflections which equals a total of 25 combinations of which 5 are symmetrical (η_{Sym}) and 20 are asymmetrical (η_{Asm}). Respectively, the number of possible altitudes is reduced from 4 to 2. The resulting dataset after this step consists of three filled 10-dimensional boxes, with a total of 7.7 million points.

The next step includes symmetry. Through mirroring a lot of conditions can be covered. These are all conditions that can be reached through switching the sign of all parameters that can change either the flow conditions or the geometry from symmetrical to asymmetrical: AoS, η and ζ. For one of these parameters, only simulations for half of the non-zero values have to be conducted, as the remainder can be generated through mirroring. For the others, the full set needs to be computed, except for conditions at which the first parameter is symmetrical. Overall the canards each have 5 designated deflections, which gives 25 total combinations. Of these 5 are symmetrical and 20 are asymmetrical combinations. Half of the 20 asymmetrical combinations can be mirrored and therefore do not have to be computed, leaving 15 canard combinations. Adding all possible ζ and AoS to the remaining canard deflections, the combined factor of these 4 parameters is 375. Additional simulations can be cut out through the reduction of ζ and the AoS to non-negative values for symmetrical canard deflections. For both, two less possibilities remain that do not have to be simulated with the 4 symmetrical canard deflections. This reduces the factor to $375 - (2 \times 2 \times 5) = 355$. Overall, taking symmetry conditions into account reduces the total points to 4.37 M.

A further reduction can be achieved through the assumption of linear behavior. However, this assumption needs to be proven. For ReFEx, this has been done for the rotation rates [4] so only one non-zero value instead of two was considered. The total points are reduced to 1.30 M.

The final step that was carried out were assumptions regarding superimposing effects. The question that has to be answered is, whether the effect of two or more parameter variations is the same as the effect of the same variations done separately and their effect added up afterwards. Usually, assuming superposition is valid and used for re-entry vehicles [4]. The reason being that the overall characteristic of any parameter effect will not be changed. Even if small differences of the effectiveness occur, the flight controller will be able to compensate. For ReFEx, or probably for any vehicle with canards, this assumption needs to be investigated thoroughly, because changing the deflection of the canards will affect the downstream flow around the vehicle. The currently available aerodynamic data for ReFEx suggests, that, with one exception, all parameters other than M and AoA are sufficiently independent from one another and therefore can be superimposed. The exception being η_{Sym} and η_{Asm} (see Chapter 3). This assumption allows to neglect sign changes in AoS and ζ, reducing these parameters from 5 to 3 due to symmetry. The resulting multiplication factor of the parameters besides M and AoA sums up to 27 of which the share of canard combinations is 15. The total number of simulations that remains is 3.1 K (Table 2).

Table 2. Parameter space of ReFEx and the respective product with successive reduction strategies.

Parameter	M	AoA	η_{Sym}	η_{Asm}	AoS	ζ	A	p	q	r	Π
All variations	16	21	49		5	5	4	3	3	3	44.5M
Boxes	114		25		5	5	2	3	3	3	7.70M
Symmetry	114		355				2	3	3	3	4.37M
Linearity	114		355				2	2	2	2	1.30M
Superposition	114		27								3.1K

2.2 Aerodynamic Dataset Structure with Trimmed Sets

For a winged re-entry vehicle, no highly dynamic maneuvers are expected. Therefore, the state of the vehicle will most certainly be close to a momentum free state. For ReFEx, this state in which $p = q = r = 0$ is utilized as reference point and referred to as trimmed point. As the geometry is symmetrical in its x-z-plane, $p = r = 0$ is given for AoS = 0, symmetric canard deflection and zero rudder deflection. Therefore, with a set of all relevant M-AoA states combined with different symmetric canard deflections, an initial dataset can be produced to interpolate the canard deflection angle for each M-AoA point at which $q = 0$. If no such point is found, the M-AoA combination is not a state in which the vehicle can fly in a steady state and is therefore discarded. As a reference trajectory is usually present in the early design phase, each Mach number is calculated with the expected atmosphere properties to set a baseline for the Reynolds number dependency. To calculate the initial set the trajectory was divided into three sections as described above. 456 simulations are necessary to create the initial set. With this initial set, 106 M-AoA combinations were determined, for which a reference canard deflection (rotation rate free state) can be found. For each combination, the respective reference canard deflection is set and simulated with movable mesh blocks. The edges of the mesh blocks and the canard deflection mapping are shown in Fig. 2.

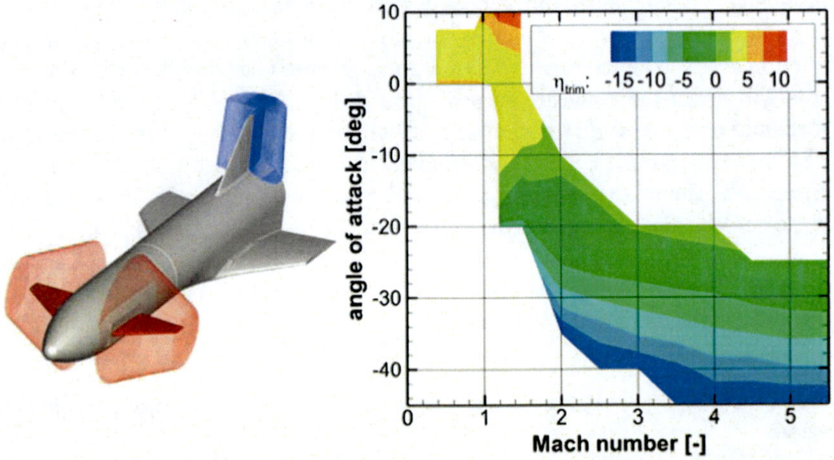

Fig. 2. Left: Layout of movable mesh blocks for the control surfaces. Right: Mapping of canard trim deflections.

The 106 points of this reference set are the baseline for all further parameter variations. Assuming linear behavior except for the AoS (two angles of sideslip are calculated up to this point) and superposition, every parameter other than AoS adds one set of 106 simulations. Assuming these conditions, the total number of simulations is 1060 (see Table 3). The trimmed set data set generation method allows reducing η_{Sym} and η_{Asm} to only one deflection off the momentum free state which for ReFEx reduces the number of simulations in the dataset by more than 65% compared to the previously available reduction strategies.

Table 3. Parameter space of ReFEx and the respective product utilizing reduction strategies of the trimmed set.

Parameter	M	AoA	η_{Sym}	η_{Asm}	AoS	ζ	A	p	q	r	Π
All variations	16	21	49		5	5	4	3	3	3	44.5 M
Final set	106		10								1060

2.3 Obtaining Aerodynamic Coefficients from the Trimmed Aerodynamic Dataset

The aerodynamic properties in this sort of set are given with respect to the reference state. Starting from this state, one point per parameter is simulated which enables calculating the gradient of all aerodynamic coefficients for this parameter. As every varied parameter potentially influences every aerodynamic coefficient, every coefficient can be formulated as the Taylor series

$$C(M, AoA) = C_{ref}(M, AoA) + \sum\left(\frac{\partial C}{\partial X}(M, AoA)\, dX\right) \quad (1)$$

at all points of the M-AoA grid. In Eq. 1, C is the aerodynamic coefficient and C_{ref} is the respective coefficient that is obtained through interpolation within the reference data set in the M-AoA grid. $\partial C/\partial X$ is the gradient of the coefficient with respect to the parameter X and dX is the difference of X from the reference state. If parameters are not superimposable, X has to be extended to include combinations of these parameters.

For a M-AoA combination, that is not part of the grid but within the grid's boundaries, the proposed method to obtain the aerodynamic coefficients consists of two steps. First, a new set of reference information needs to be calculated through 2D interpolation between the closest grid points of the reference set. For ReFEx, this reference data must not only contain the aerodynamic coefficients, but also the interpolated canard deflection angle and the interpolated parameter gradients. For linear behavior, the calculated reference state is a momentum free state. The second step is the incorporation of all parameters to this state. Their influence can be calculated with their difference to the newly found reference state and the respective interpolated gradients.

If the M-AoA combination lies outside of the provided grid, the reference information can only be obtained through extrapolation. Alternatively, additional points could be calculated at the corners of the spanned M-AoA range to perform an interpolation. If corner points are involved, a weighted interpolation should be used, giving significantly more weight to the gradients at the point of the mesh rather than the corner point. This can be crosschecked using the initial set.

3 Superposition of Parameter Effects

As described in Sect. 2.1, the assumption that parameter variations do not affect one another is reasonable for most parameters. However, this chapter will give an example for the ReFEx vehicle for which a superposition is not valid: the symmetrical and asymmetrical canard deflection.

The greatest concern for ReFEx regarding controllability is the rolling motion. The expectation during the early design phases to use the canards asymmetrically to control rolling turned out to be too optimistic. The left diagram of Fig. 3 shows the rolling coefficient for an asymmetric canard deflection of $\pm 5°$ at $M = 1.5$ for different angles of attack and two mean deflection angles, 0° and 5°. It reveals, that not only the sign of the produced rolling moment changes, which equals a rudder reversal, but also the coefficient depends on the mean deflection angle. For example, at $AoA = -15°$, the sign of the rolling coefficient is different for both shown mean canard deflections.

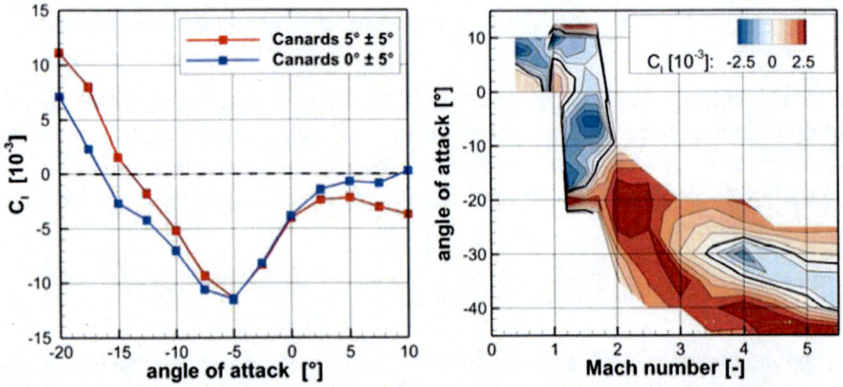

Fig. 3. Rolling coefficient C_l of the ReFEx for asymmetric canard deflections. Left: Variation of mean canard deflection angle at Mach 1.5 with $\pm 5°$. Right: Characterization of the flight envelope with $\pm 1°$ deflection around the trim deflection.

As the knowledge of the rudder reversal areas in the envelope is needed to ensure controllability, their characterization in the data set must be as precise as possible. In the right diagram of Fig. 3 this is done with a trimmed set from the aerodynamic dataset described in Sect. 2.2. For each flight point, the symmetric canard deflection of a momentum free state with an asymmetric delta deflection of $\pm 1°$ is simulated. The shown contour is the resulting rolling coefficient. The broader lines are the rudder reversal lines with respect to the rolling motion.

4 Interpolation Quality and Valid Deflections

In the trimmed dataset for all linearly behaving parameters only one additional simulation is carried out per M-AoA combination. Together with the reference state, the respective gradient can be calculated. The deflection effects have maximum accuracy at the reference state, but the question arises how far off this state this type of set still gives accurate aerodynamic coefficients. The nature of the trimmed dataset is to calculate the gradient close to the trimmed state. However, if the points are too close together, the share of the numerical error will become problematic. Therefore, the crosscheck with the initial dataset also serves as control for this problem.

Figure 4 shows data points from the trimmed set (black) and the initial set (red) that are built as described in Sect. 2.2. The left diagram is representative for lower velocities with $M = 0.8$ and $AoA = 5°$, the right diagram represents high Mach numbers ($M = 4.0$ and $AoA = -35.0°$). The deflection angles of the final set are not exactly at the position of the linearly interpolated data points from the initial set, because two iterations to find the momentum state were carried out to increase accuracy.

For both points, a mismatch of the C_m value can be observed. For the subsonic point, it occurs due to the lack of calculation points in the initial set. For the supersonic point, the reason is, that the initial set was calculated with inviscid simulations.

To characterize the range of validity of the trimmed set, the gradient which represents the efficiency is more important. Defining a valid region depends on various constraints

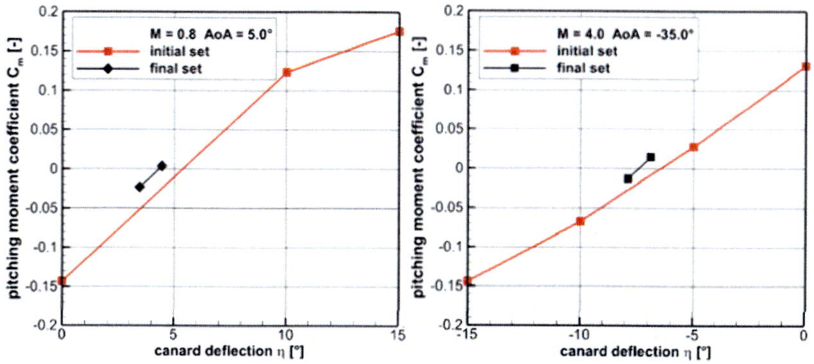

Fig. 4. Pitching moment coefficient of simulations from the initial set and the final data set. Left: Subsonic point with matching derivative. Right: Supersonic point with expected mismatch for gradient and values due to inviscid/viscous comparison.

that ultimately will lead to an allowed mismatch between the gradients of the final set and respective validation simulations. This process is still ongoing for ReFEx.

For the subsonic point the gradient of the final and the initial set match very well. The valid canard deflection range was preliminary set from 0° to 10° in which the trimmed set is valid. For the supersonic point, there is a mismatch between the gradients. But as the gradient of the initial set does net change much, the preliminary validity range is even larger, ranging from −15° to 0°.

5 Conclusion

A new method to create aerodynamic data sets for vehicles is introduced: the trimmed data set. For this set, first a reference subset is created that contains the momentum free state for each M-AoA combination of the envelope. This is achieved through the usage of moving mesh blocks, to precisely deflect all control surfaces at each point. With respect to this reference set, small variations of all parameters are introduced and simulated. These sets yield the influence of the respective parameters on the aerodynamic coefficients and therefore their gradients. With this information, the aerodynamic coefficients at any flight conditions can be interpolated.

For ReFEx, symmetrical and asymmetrical canard deflections cannot be superimposed. However, with the trimmed dataset the symmetric canard deflection to obtain a momentum free state is set for each flight point. Therefore, only one symmetric and asymmetric canard deflection around this deflection needs to be simulated, which saves a significant amount of otherwise necessary simulations.

The introduction of trimmed data sets comes with some great benefits but also has a few downsides. Put in a nutshell, the positive features of this type of dataset are:

- Reduction of the number of meshes needed due to the moving mesh approach
- Reduction of the number of necessary simulations
- Increased accuracy at the reference state

Negative aspects include the following:

- Increased complexity and number of points in the used CFD meshes
- Reduced accuracy at flight conditions that differ from the reference state

This type of set is recommended to use for vehicles with a wide variety of flow conditions, that are not expected to conduct highly dynamic maneuvers. This includes winged re-entry vehicles.

For ReFEx the estimated overall reduction of simulation points by incorporating trimmed sets was from 3038 to 1060, or a factor of about 0.34. Creating meshes with rotating blocks is more time consuming than the regular meshes, but therefore less meshed are required. The meshes with moving blocks have additional mesh points, hence additional computation time per simulation. For ReFEx this factor is in the order of 1.5. Together with the reduced amount of computations, an overall reduction of raw calculation time of around 50% is achieved.

References

1. Rickmers, P., et al.: Refex: reusability flight experiment a flight experiment to demonstrate controlled aerodynamic flight from hypersonic to subsonic velocities with a winged rlv. In: 7th European Conference for Aeronautics and Space Sciences (EUCASS). https://doi.org/10.13009/EUCASS2017-644 (2017)
2. Merrem, C., Wartemann, V., Eggers, T.: Preliminary aerodynamic design of a reusable booster flight experiment. CEAS Space J. **12**(3), 429–439 (2020). https://doi.org/10.1007/s12567-020-00313-9
3. NN Aerodynamic Design Data Book – Orbiter Vehicle STS-1. Rockwell International, USA (1980)
4. Merrem, C., Kiehn, D., Wartemann, V., Eggers, T.: Aerodynamic design of a reusable booster stage flight experiment. In: 8th European Conference for Aeronautics and Space Sciences (EUCASS). https://doi.org/10.13009/EUCASS2019-644 (July 2019)
5. Weiland, C.: Aerodynamic Data of Space Vehicles. Springer Science & Business Media (2014)

Contribution of Numerical and Experimental Flow Simulations to the Aerodynamic Data Base of the DLR Reusable Flight Experiment ReFEx

Viola Wartemann[1(✉)], Ntarinai Konosidou[1], Andreas Flock[2], and Clemens Merrem[1]

[1] Department Spacecraft, German Aerospace Center (DLR), Institute for Aerodynamics and Flow Technology, Lilienthalplatz 7, 38108 Braunschweig, Germany
{Viola.Wartemann,Clemens.Merrem}@dlr.de

[2] Department Supersonic and Hypersonic Technologies, German Aerospace Center (DLR), Institute for Aerodynamics and Flow Technology, 51147 Linder Höhe, Köln, Germany
Andreas.Flock@dlr.de

Abstract. ReFEx is a Reusable Flight Experiment of the DLR. For the layout of the aerodynamic shape of the vehicle, the identification of a flight trajectory and finally the control systems, extensive data sets are necessary. The focus of this paper lays on the comparison of calculated versus measured aerodynamic coefficients. All simulations were carried out with the DRL TAU Code. The measurements were performed in the trisonic wind tunnel (TMK) of the DLR. One aerodynamic flight control element of ReFEx are the canards. The here presented analysis delivers a good agreement comparing the calculated and measured coefficients, looking at the symmetrical canard deflections. The occurred differences of the asymmetrical canard deflection are discussed in detail, including the numerical as well as experimental uncertainties.

Keywords: ReFEx · CFD · TMK · Aerodynamic coefficients

1 Introduction

The German Aerospace Center (DLR) is currently developing the Reusable Flight Experiment ReFEx as a technology demonstrator for a reusable, Vertical Take-off and Horizontal Landing (VTHL) booster stage [1]. A two-staged VSB-30 sounding rocket transports ReFEx to an altitude of about 130 km. After the separation from the carrier rocket, an autonomous re-entry is performed with a maximum Mach number of approx. 5. The first part of the following re-entry will be flown upside down at high, negative angles of attack α. This initial deceleration keeps the structural and aerothermal loads within acceptable limits. At

© The Author(s), under exclusive license to Springer Nature Switzerland AG 2021
A. Dillmann et al. (Eds.): STAB/DGLR Symposium 2020, NNFM 151, pp. 141–150, 2021.
https://doi.org/10.1007/978-3-030-79561-0_14

$M_\infty \approx 1.5$, a 180° roll manoeuvre will be performed to reach the nominal orientation and an angle of attack close to $\alpha = 10°$ for a maximum lift to drag ratio.

One major task covered by the Department Spacecraft of the DLR Institute of Aerodynamics and Flow Technology in this project is the aerodynamic design of ReFEx [2–4]. Extensive CFD data are needed to lay out the vehicle itself, the flight envelope for the trajectory and the control systems. The numerical generation of these aerodynamic data sets, including the overall and separated aerodynamic coefficients for the whole flight range, are the second big task. The data bases are supported by wind tunnel experiments, performed by the Department Supersonic and Hypersonic Technologies in Cologne. Especially, the planned measurements of the dynamic aerodynamic coefficients are of interest for the comparison. The first wind tunnel test campaign, which is the focus of the current paper, includes measurements of the static coefficients of the belly-up flight phase. Consequently, the investigations of the current paper are concentrated on this flight phase with higher Mach numbers and high negative angles of attack. The geometry with orientation of the belly-up flight phase is visualised in Fig. 1.

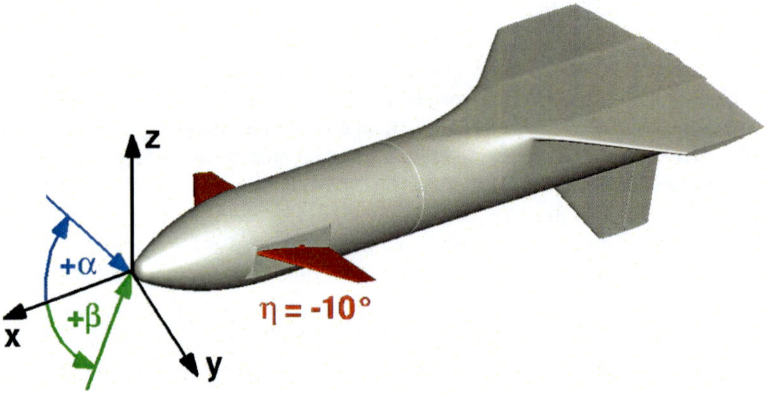

Fig. 1. ReFEx geometry in belly-up flight orientation

The first part of the results of this paper is based on calculations with symmetrical canard deflections and summarizes the numerical study of different turbulence models and small deviations of the applied free stream conditions (Sect. 4.1). For the comparison with the experimental aerodynamic coefficients, additionally the sting effect and the Reynolds Number effect are analysed (Sect. 4.2). The second part of the results is focused on the investigations with asymmetrical canard deflections (Sect. 5).

2 Wind Tunnel

The experiments were performed in the trisonic wind tunnel (TMK), which is a constant flow, blow down wind tunnel. The TMK covers a broad range of Mach and Reynolds numbers and allows testing at low hypersonic, supersonic, trans- and subsonic velocities. A scaled model of ReFEx (1:13) was designed to fit into the TMK test section. It enables canard angles of $0°$, $\pm 5°$, $\pm 10°$ and $\pm 15°$ and rudder deflections of $0°$, $5°$ and $10°$. The aerodynamic coefficients are measured with a six-degree-of-freedom balance system: A 3/4 in. forces and moments balance from the TASK corporation. The balance main axis is aligned with the ReFEx wind tunnel model and mounted onto a sting. With the balance system and a calibration, the forces and moments in all three directions can be measured. The wind tunnel model has a length of $L_{\text{ref}} = 0.209$ m and a base area of $A_{\text{ref}} = 0.686 \times 10^{-4}$ m^2. For more details about the model and the wind tunnel experiments itself, see Flock et al. [5].

3 Numerical Methods

For the numerical analysis presented in the current paper, the DLR Tau code, see e.g. Schwamborn et al. [6], is applied. The Tau code is a finite volume scheme, solving for this analysis the Reynolds-averaged Navier-Stokes (RANS) equations, using tetrahedrons primarily in inviscid flow layer regions and prismatic elements for boundary layer regions. Different turbulence models are applied (see Sect. 4.1). Due to the short experiment duration, the temperature of the model wall is assumed with a constant isothermal boundary condition with $T_{\text{w}} = 293$ K. The grid is refined at the nose, the canards, the leading and trailing edges of the wings, the fin and the rudder. The grid study delivers a final grid with 12.5 – 15.5 million grid points, depending on whether a model with or without sting is selected.

4 Results for Symmetrical Canard Deflection

Section 4 summarizes the results with symmetrical canard deflection. For the investigated belly-up flight phase, a canard deflection of around −10 is necessary, due to the aerodynamic stability and trimmability (see e.g. Merrem et al. [3]). Consequently, a symmetrical canard deflection of $\eta = -10$ is chosen for this chapter, see also Fig. 1. A symmetrical canard deflection, without rudder deflection and side-slip angle, which are not part of this paper, delivers $\text{C}_{\text{y}} = 0$ (coefficient of side force), $\text{C}_{\text{l}} = 0$ (coefficient of rolling moment) and $\text{C}_{\text{n}} = 0$ (coefficient of yawing moment). Thus the investigations for the symmetrical canards are focused on C_{x} (coefficient of axial force), C_{z} (coefficient of normal force) and C_{m} (coefficient of pitching moment). The calculations are performed for a free stream Mach number $M_\infty = 4.5$ with a unit Reynolds number $Re_{\text{u}} = 73.90 \times 10^6$/m and a free stream Mach number $M_\infty = 2.5$ with a unit Reynolds number $Re_{\text{u}} = 37.21 \times 10^6$/m (see also Table 1).

Table 1. Overview of the flow conditions

M_∞	Re_u [10^6/m]	T_∞ [K]	p_∞ [N/m^2]
4.5	73.9	58.5	7300
2.5	37.2	129.0	21600

4.1 Uncertainties of the Calculated Aerodynamic Coefficients

The chosen turbulence model is one influence factor on the calculated coefficients. Three different turbulence models are applied: the one equation Spalart-Allmaras model, SA-neg [7], in the below figures marked by squared black symbols, and the following *kw* - two equation models: Menter's Shear Stress Transport turbulence model, SST [8], marked as green circle symbols, and the Wilcox *kw* model [9], which is illustrated with blue delta symbols. All simulations include the wind tunnel sting. Applying the three turbulence models, the C_x coefficient delivers the highest difference, compared to the other coefficients. Figure 2.a shows the C_x coefficient as a function of the angle of attack for $M_\infty = 4.5$ (lines). The influence of the chosen turbulence model depends on the selected Mach number, thus Fig. 2.a additionally includes $M_\infty = 2.5$ (dotted lines). To extract the uncertainties due to the selected turbulence model, the differences between the SA and SST model are used, because the highest differences appear between these two models. $M_\infty = 4.5$ delivers a maximum difference of about 1.5%. With a decrease of the Mach number the differences increase to about 2%. For the investigated cases, the respective uncertainty for C_z and C_m is lower. Including the performed grid study the numerical error results in $\Delta C_x < \pm 3\%$ and less for C_z as well as C_m. For all following results, the SA model is applied.

Beside the turbulence model and the grid influence, small deviations in the applied free stream conditions are investigated. The uncertainty of the selected free stream Mach number is given by $\Delta M_\infty < \pm 0.5\%$. The measurement uncertainties of the reservoir temperature and reservoir pressure are $\Delta T_0 = \pm 1.5\,^\circ\mathrm{C}$ and $\Delta p_0 = \pm 0.1\% \times 30$ bar for $M_\infty = 4.5$. These parameters are used to calculate the maximum possible deviations of the applied free stream conditions, using the following two equations:

$$\Delta T_\infty = \left|\frac{\partial T_\infty}{\partial T_0}\right| \Delta T_0 + \left|\frac{\partial T_\infty}{\partial M_\infty}\right| \Delta M_\infty \tag{1}$$

$$\Delta p_\infty = \left|\frac{\partial p_\infty}{\partial p_0}\right| \Delta p_0 + \left|\frac{\partial p_\infty}{\partial M_\infty}\right| \Delta M_\infty \tag{2}$$

As for the turbulence model error, the C_x coefficients deliver the highest differences compared to the other coefficients and is consequently illustrated in Fig. 2.b. Two different combinations are investigation: The combination $\Delta M_\infty = +0.5\%$, $\Delta T_0 = +1.5\,^\circ\mathrm{C}$ and $\Delta p_0 = +0.1\% \times 30$ bar, which is marked as + max. uncert. in Fig. 2.b, whereas the combination $\Delta M_\infty = -0.5\%$, $\Delta T_0 = -1.5\,^\circ\mathrm{C}$ and $\Delta p_0 = -0.1\% \times 30$ bar are marked as - max. uncert.. The investigated free stream uncertainties have a minor effect on the aerodynamic coefficient and can be declared with a maximum deviation of $\pm 0.3\%$ for C_x and less for C_z as well as C_m.

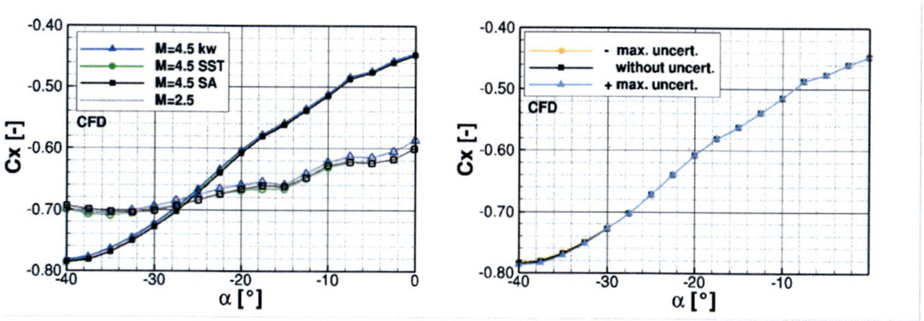

(a) (b)

Fig. 2. CFD comparison of C_x: (a) influence of different turbulence models and (b) influence of the deviations of the used free stream conditions

One additional uncertainty is the angle of attack α, which is given by $\Delta\alpha < \pm 0.25\%$. This uncertainty can be read out directly of all shown diagrams. Additionally, in Fig. 3.a (black symbols) the accumulated, absolute uncertainties, including the angle of attack uncertainties, is given. The side-slip angle β has the same uncertainty of $\Delta\beta < \pm 0.25\%$ and is investigated in Sect. 5.

Please note, usually the deviations/errors are not given as percental values, because this approach does not work, if coefficients approach small values. However, for the presented cases the percental approach is possible and hence applied.

4.2 Further Influence on the Aerodynamic Coefficients

The previous sections cover mainly the numerical uncertainties. This subsection shows the comparison of the calculated and measured aerodynamic coefficients, the influence on the measured coefficients induced by the sting and deals with the Reynolds number effect.

Sting Effect. The wind tunnel model is mounted onto a sting. The effect of this mounting system is in this subsection in the focus of the analysis. The numerical calculations, which are performed without sting, marked by blue color, and with sting, illustrated by black color, in Fig. 3.a, which shows C_x as a function of the angle of attack. Due to the different wake flows clearly visible differences appear between the coefficients of axial force for $M_\infty = 4.5$ (lines) as well as $M_\infty = 2.5$ (dotted lines) and increase for the region with high angle of attack, since the wake flow interaction with the sting increases. For the data bases, which are prepared for the flight experiment, the sting effect has to be considered, especially for lower Mach numbers ($\Delta C_{x,\mathrm{with/without\ sting}}$ increase to 13% for $M_\infty = 2.5$).

The parallel offset between the calculations with and without sting can be explained with the influence of the wake flow on the rear area. For a better overview in Fig. 3.b only the higher Mach number 4.5 is summarized. Comparing the blue line, which illustrates the calculated results without sting, with the dashed blue line, which shows the same simulation, but with subtracted rear area, the effect of the area is clearly visible. Nevertheless, the measurements consider this rear area, although this area is covered by the sting, which lessens the base area pressure drop. However, a more accurate way for the comparison between measurements and calculations, is to perform the simulations with a small gap of a few millimeters between model and sting to include the effect of sting as well as the rear area. This results in a slightly better agreement between the measured C_x (red line) and calculated C_x (green line) at high angles of attack. Applying the measured uncertainties to the maximal error of the balance system calibration of $\Delta C_x \leq \pm 0.0128$ and considering the numerical error of Sect. 4.1, a good agreement can be observed, including the high angle of attack regime.

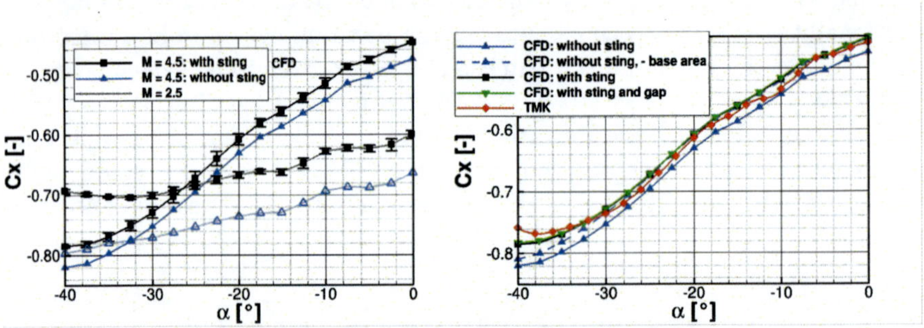

(a) (b)

Fig. 3. Sting effect on the axial force coefficient C_x

As expected, for the other coefficients C_z and C_m the sting effect is less. In Fig. 4.a the C_z coefficient as function of the angle of attack is given, with the same colors as before in Fig. 3. The sting effect is neglectable and a good agreement between the calculated and measured values, especially for the smaller angles of attack, can be seen. The increasing calculated and measured difference with increase of the angle of attack is still under investigation. In the experiments, continuous measurements were performed for various angles of attack. That means, during the experiment, the angle of attack is increased from 0° to

40 ° and then gradually decreased again to 0 °. Thus, an increasing measurement error during the experiment can be excluded. However, the maximum deviation between the calculated and measured coefficients is still in good agreement as it is less than 5% for all angles of attack.

To assess the static longitudinal stability C_m is investigated. The static requirement of the longitudinal stability is defined as $\partial C_m/\partial\alpha < 0$. The effect of the sting on C_m is neglectable, see Fig. 4.b. The mean deviation of the simulation compared to the measurements is about 4%. However, for the crucial pitch stability analysis, both calculation methods are in agreement with the measurements: the pitch stability is given for $M_\infty = 4.5$ from $-40.0\,° < \alpha < -17.5\,°$ and for $M_\infty = 2.5$ from $-40.0\,° < \alpha < -5.0\,°$.

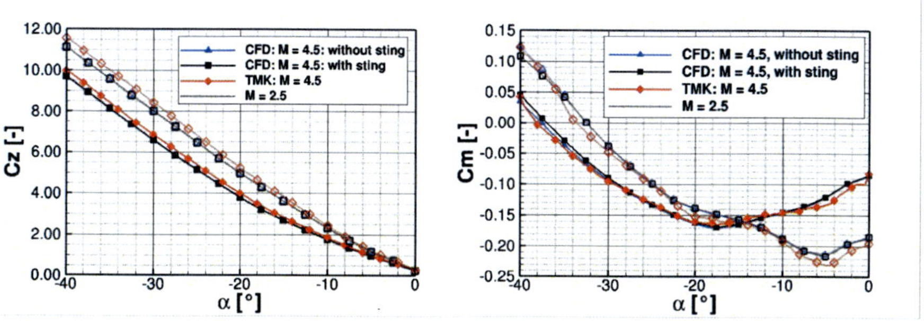

Fig. 4. Sting effect on C_z (a) and C_m (b)

Reynolds Number Effect. From the wind tunnel's unit Reynolds number $Re_{u,m}$ and the reference length of the wind tunnel model $L_{ref,model} = 0.209$m, a model Reynolds number Re_m is calculated. This model Reynolds number is then used to calculate a corresponding flight altitude (h^*), which is finally applied to rerun the numerical simulations, using the corresponding free stream conditions and original dimensions of ReFEx ($L_{ref,ori} = 2.72$ m), see also Table 2:

$$Re_m = Re_{u,m} L_{ref,m} \rightarrow h^* \rightarrow Re_{u,ori} \rightarrow Re_m(\text{build with } L_{ref,ori}) \qquad (3)$$

To ensure the comparability the same wall temperature condition as before is chosen. In summary, that means, the unit Reynolds numbers differ, whereas the model Reynolds numbers Re_m are the same. The effect of the different dimensions is known as Reynolds number effect. This effect has an influence on the boundary layer transition, the separation as well as the reattachment (see e.g. Blackwell et al. [10] or Song et al. [11]) and consequently can influences the forces and moments exerted on the body. Comparing in Fig. 5 the black lines (calculations based on TMK condition, TMK model dimensions) versus the blue lines (calculations based on h^* and ori. ReFEx dimensions), differences are visible, depending on the Mach number. For the data bases, which are prepared for the flight experiment, the Reynolds number effect should be considered for lower Mach numbers ($\Delta C^*_{x,TMK/h}$ increase to 4% for $M_\infty = 2.5$). For $M_\infty = 4.5$, h^* is on the currently planned ReFEx flight trajectory. For $M_\infty = 2.5$, h^* is about 5 km higher than the altitude of the planned flight trajectory. Thus Fig. 5.b additionally includes the calculations for the planned altitude during the flight h_f, marked by lavender color (compare blue versus lavender line, to see the influence of the altitude on C_x).

Table 2. Reynolds numbers

Ma	$Re_{u,m}$	Re_m	$Re_{u,ori}$	h^*
[-]	[10^6/m]	[10^6]	[10^6/m]	[km]
4.5	73.9	15.4	5.6	22.3
2.5	37.2	7.8	2.9	22.9

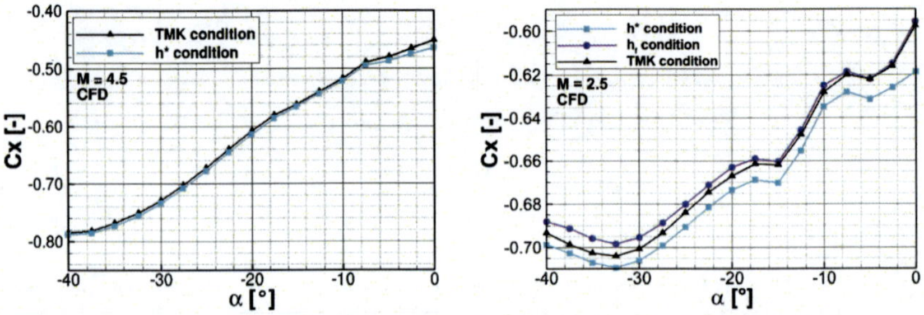

(a) (b)

Fig. 5. CFD analyses of the Reynolds number effect: (a) $M_\infty = 4.5$, (b) $M_\infty = 2.5$

5 Results for Asymmetrical Canard Deflection

Former analyses of the numerical data sets with asymmetrical canard deflections detected a region of rudder reversal at higher Mach numbers. Thus for this section a Mach number of 4.5 is selected. In the context of this paper rudder reversal means, that the same asymmetrical canard deflection induces roll moment coefficients C_l with different signs, depending on the Mach number and the angle of attack. An asymmetrical canard deflection of $0\,°/-10\,°$ is chosen for this analysis. Figure 6.a summarized the measured and calculated roll moment coefficients C_l as a function of the angle of attack. The maximum uncertainty of the measured C_l coefficients is included as error bars, which seems disproportionally high. Please notice the extremely small values compared to the previously shown C_x values. Beside the already discussed uncertainties, for this subsection an additional uncertainty parameter has to be considered. The angle of attack and the side-slip angle have the same uncertainty of $\beta = \pm 0.25\,°$. Thus Fig. 6.b includes the effect of the side-slip angle. Beginning with Fig. 6.a, in contrast to the CFD data, the measurements show a region of rudder reversal, marked by negative C_l values. A side-slip angle increases this effect (see the numerical comparison in Fig. 6.b). Nevertheless, the principal trend is similar. Thus for the detection of eventual further critical rudder reversal regions, the measurements can be utilized as an auxiliary input for the data bases of the flight experiment.

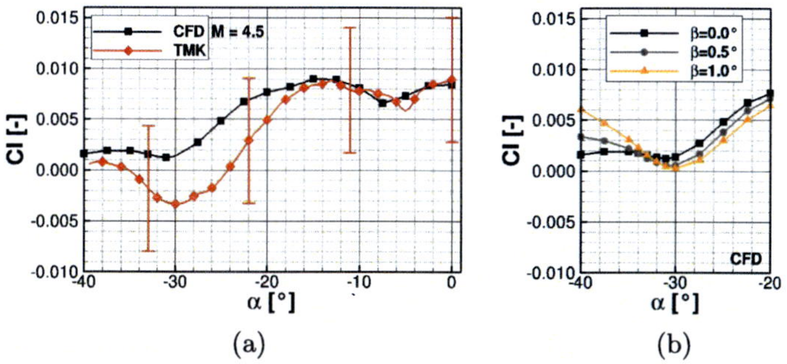

Fig. 6. Asymmetrical canard deflections: (a) Comparison of meas./calc. C_l, (b) CFD analysis of side-slip angle influence on C_l

6 Conclusion

For the layout of the ReFEx vehicle, the identification of a flight trajectory and finally the control systems, extensive data sets are necessary, which include aerodynamic coefficients of the forces and moments for the whole flight range. The numerically generated data bases are supported by wind tunnel experiments. The comparison of the calculated versus measured aerodynamic coefficient is the focus of this paper. As measured data the first wind tunnel test campaign of the TMK is used, which includes static measurements of the belly-up flight phase. Consequently, the investigations of the current paper is concentrated on this flight phase with higher Mach numbers and negative angles of attack.

For the investigated free stream condition the numerical uncertainties resulting from the grid, the chosen turbulence model as well as possible inaccuracies in the selected free stream conditions, can be summarizes as small (Sect. 4.1).

A good agreement of the calculated and measured coefficients is demonstrated, applying symmetrical canard deflections (see Sect. 4.2). To support the

CFD data sets of ReFEx with the wind tunnel measurements, the influence of the sting and the Reynolds number effect on the axial coefficient C_x have to be taken into account. However, the numerical longitudinal stability predictions based on C_m are in agreement with the measurements.

The investigations of the asymmetrical canard deflections show, despite the small C_l values in combination with the maximal possible error of the six-degree-of-freedom balance system, the measured versus calculated C_l coefficients deliver the same trend. Hence, the measurements can be used to detect eventual further critical rudder reversal regions in the aerodynamic parameter space of ReFEx.

Acknowledgements. The presented results are partly performed in the framework of the ReFEx project. Hence the authors would like to acknowledge the support of the ReFEx project management Dr. P. Rickmers.

References

1. Rickmers, P.: ReFEx: reusability flight experiment. A flight experiment to demonstrate controlled aerodynamic flight from hypersonic to subsonic velocities with a winged RLV. In: EUCASS 2017–644 (2017). https://doi.org/10.13009
2. Merrem, C., Wartemann, V., Eggers, T.: Preliminary aerodynamic design of a re-usable booster flight experiment. In: International Conference on High-Speed Vehicle Science Technology (HiSST), pp. 26–29 November 2018. Russia, Moscow (2018)
3. Merrem, C., Kiehn, D., Wartemann, V., Eggers, T.: Aerodynamic design of a reusable booster stage flight experiment. In: 8th European Conference for Aeronautics and Space Sciences (EUCASS), vol. 1, no. 04, July 2019
4. Merrem, C., Wartemann, V., Eggers, T.: Preliminary aerodynamic design of a reusable booster flight experiment. CEAS Space J. (2020). https://doi.org/10.1007/s12567-020-00313-9
5. Flock, A., Thiele, T., Neeb, D., Gülhan, A.: Planned wind tunnel experiments at DLR Köln for the reusability flight experiment (ReFEx). In: International Conference on Flight Vehicles, Aerothermodynamics and Re-entry Missions & Engineering, Monopoli, Italien (2019)
6. Schwamborn, D., Gerhold, T., Heinrich, R.: The DLR Tau-code: Recent Applications in Research and Industry, European Conference on Computational Fluid Dynamics ECCOMAS CFD (2006)
7. Allmaras, S., Johnson, F., Spalart, P.: Modifications and clarifications for the implementation of the spalart-Allmaras turbulence model, ICCFD7-1902. In: 7th International Conference on Computational Fluid Dynamics (2012)
8. Menter, F.: Two-equation eddy-viscosity turbulence models for engineering applications. AIAA J. **32**(8) (1994)
9. Wilcox, D.: Formulation of the k-omega turbulence model revisited. AIAA J. **46**(11), 2823–2838 (2008)
10. Blackwell, J., James, A: Preliminary study of effects of Reynolds number and boundary-layer transition location on shock-induced separation, NASA Technical Note D-5003 (1969)
11. Song, S., Eaton, J.: Reynolds number effects on a turbulent boundary layer with separation, reattachment, and recovery. Exper. Fluids **36**(2), 246–258 (2004)

Flow Control

Unsteady Characterization of Fluidic Flow Control Devices for Gust Load Alleviation

Salvatore Asaro[1,2(✉)], Khalid Khalil[1,2], and André Bauknecht[1,2]

[1] Cluster of Excellence SE2A–Sustainable and Energy-Efficient Aviation, Technische Universität Braunschweig, Braunschweig, Germany
[2] Institute of Fluid Mechanics, Technische Universität Braunschweig, Hermann-Blenk-Str. 37, 38108 Braunschweig, Germany
s.asaro@tu-braunschweig.de

Abstract. The lift reduction potential of fluidic flow control devices is investigated. The devices target unsteady gust load alleviation to achieve a reduction in overall wing weight and fuel consumption. The study focuses on a jet normal to the surface and jet over a Coanda surface installed at the trailing edge. The fluidic actuators are compared to a trailing edge flap for reference using unsteady Reynolds-averaged Navier–Stokes simulations. Unsteady interactions of the airfoil with different gusts are simulated to estimate the required load change and response time of the control devices. To reduce computational costs, a surrogate model based on precomputed aerodynamic polars is developed to estimate gust-induced load changes. The effect of jet temperature for the fluidic actuators and dual blowing for Coanda actuator are studied.

Keywords: Load alleviation · Gusts · Unsteady actuation

1 Introduction

To fulfil the goal of a 75% CO_2 reduction per passenger-kilometer of the European Aviation Roadmap described in the European Commission's Flightpath 2050 report [1], new technologies have to be identified to reduce the airplane structural weight, drag, and fuel consumption. Today, conventional control surfaces such as flaps or ailerons are used to reduce unsteady gust and maneuver loads. Further reduction in wing weight appears feasible if dedicated flow control devices are applied instead. A number of previous studies have investigated the steady performance of active flow control, e.g. with fluidic and micro-mechanical flow actuators. Some promising actuator designs for load alleviation on a passenger aircraft wing were compared in [2]. It was found that a jet perpendicular to the surface placed in the rear part of the suction side (Normal jet) and a tangential jet over a Coanda surface inserted at the trailing edge (Coanda jet) perform similar to a conventional trailing edge flap (Flap) and show potential for faster actuation over a range of different altitudes and flight Mach numbers.

© The Author(s), under exclusive license to Springer Nature Switzerland AG 2021
A. Dillmann et al. (Eds.): STAB/DGLR Symposium 2020, NNFM 151, pp. 153–163, 2021.
https://doi.org/10.1007/978-3-030-79561-0_15

The present study investigates the suitability of a Normal and Coanda jet in comparison to a Flap for compensating typical gust peak loads in the time frames they occur. The gust time frames and actuator responses are determined using unsteady Reynolds-averaged Navier–Stokes (RANS) equations. To accelerate the investigation of the gust effects, a surrogate model based on quasi-steady airfoil polars is proposed. The outputs of the RANS gust simulations and surrogate model are airfoil peak load increments and corresponding temporal response times, which serve as target values for unsteady actuator simulations. The results allow for a quantitative assessment of actuator control authority and response time for realistic operating conditions. Parametric studies are carried out to answer the questions how baseline airfoil performance can be restored for the Coanda airfoil design, what the effect of jet temperature is on the fluidic actuator performance, and whether dual blowing can improve Coanda load alleviation.

2 Numerical Approach

The unsteady 2D RANS simulations are conducted with the TAU code from the German Aerospace Center (DLR). The Spalart-Allmaras negative formulation is chosen as turbulence model with rotation and curvature corrections for the Coanda actuator [3], and a 2nd-order central scheme for spatial discretization. The appropriate size of the mesh elements is identified with a mesh independence study, resulting in $y^+(1) \leq 1$ and a total number of elements of ca. 1 million, including local refinement around the actuator, as shown in Fig. 1. The DLR-F15 airfoil with a relative maximum thickness of 12.5% is used in clean cruise configuration. This supercritical airfoil is considered relevant for transport aircraft application. Simulations are performed at take-off without high-lift system (altitude $h = 0\,\text{m}$, Mach number $M_\infty = 0.37$), climb ($h = 5239\,\text{m}$, $M_\infty = 0.59$) and cruise ($h = 10668\,\text{m}$, $M_\infty = 0.78$). Conditions of a trimmed medium-range aircraft are assumed. The chord Reynolds number varies in the range of $\text{Re}_c = 1.3\text{–}1.8 \cdot 10^7$. The airfoil corresponds to a mean aerodynamic chord of 2.4 m at 40% of the wing span. It is scaled according to infinite swept wing theory to obtain representative operating conditions with a sweep angle $\phi = 17°$, shortened chord c, reduced effective M, and increased angle of attack α.

2.1 Unsteady Simulations

The unsteady simulations are started from steady converged solutions without actuation. A Courant–Friedrichs–Lewy condition of $CFL = 1$ is chosen and the time step size Δt is estimated as the Kolmogorov timescale, $4.5 \cdot 10^{-6}$ s at take-off, $3 \cdot 10^{-6}$ s at climb and $3 \cdot 10^{-6}$ s at design altitude. For the Coanda actuator Δt is reduced by a factor of 10 because of the relatively small lip height that has an impact on the smallest turbulence scales. Unsteady fluidic actuation is achieved through time-dependent modification of an "actuation" boundary condition, whereas flap deflections are realized through a grid deformation approach.

The gust interactions with the airfoil are simulated applying the Disturbance Velocity Approach (DVA), i.e. a disturbance velocity due to the gust is added as a source term to the Navier-Stokes equations [4]. In this analysis the airfoil is considered rigid, i.e. no aeroelastic deformations or body motion are allowed.

3 Actuator Definitions

The most promising active flow control devices identified in [2] are:

- **Trailing edge flap**: the virtual hinge line of the hinge-less flap is located at $x/c = 0.8$ and the deflection angle δ is $0°$ when the Flap coincides with the original airfoil, as shown in Fig. 1a. Upwards deflections (negative δ) are considered in this study to compensate the gust loads.
- **Normal jet**: air jet that is locally perpendicular to the airfoil's surface and located at $x/c = 0.9$, with a slot height of $h/c = 1\%$ (Fig. 1b).
- **Coanda jet**: air jet that is locally parallel to a Coanda surface installed at the trailing edge (Fig. 1c). Based on a parametric study, not presented here for brevity, an elliptical Coanda surface was selected with the major axis aligned with the airfoil camber line and 1.5 aspect ratio. The airfoil thickness is increased aft of $x/c = 0.7$ to implement the Coanda actuator and to obtain a trailing edge thickness and ellipse minor axis of $t_{TE}/c = 1.5\%$. The Coanda radius calculated with the major ellipse semi-axis is $r/c = 1.2\%$, slot height $h/c = 0.05\%$, and lip height is half the slot height based on the findings in [5]. The jet slots are located forward of the ellipse minor semi-axis to reduce their impact on airfoil baseline performance and improve jet attachment to the Coanda surface.

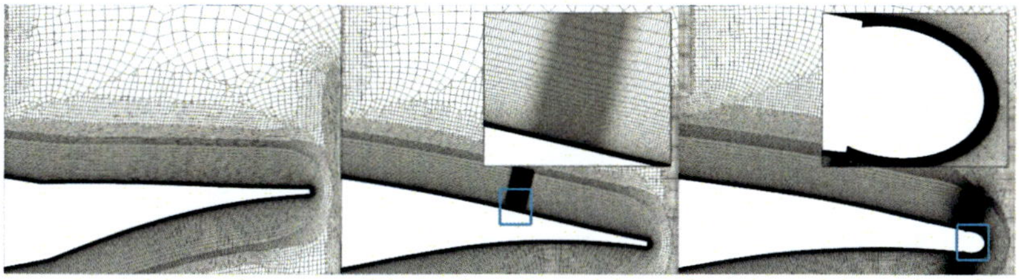

(a) Flap (b) Normal jet (c) Coanda jet

Fig. 1. Detail view of actuators and mesh in trailing edge region of airfoil.

An important parameter of the fluidic actuators is the momentum coefficient:

$$c_\mu = \frac{\dot{m} \cdot v_{jet}}{\frac{1}{2}\rho_\infty U_\infty^2 S_{ref}} \tag{1}$$

with mass flow $\dot{m}$, jet velocity v_{jet}, reference density ρ_∞, free stream velocity U_∞ and reference surface $S_{ref} = c_{mean} \cdot b$, with unity span $b = 1\,\text{m}$ for this 2D study. Some important aspects of the Coanda jet and Normal jet actuator design are discussed next.

3.1 Performance Recovery for Coanda Baseline Airfoil

The Coanda surface replaces the sharp trailing edge of the airfoil, leading to performance degradation compared to the original airfoil. There are several ways to restore the original performance, for example increasing airfoil camber in the trailing edge area, increasing angle of attack, or using continuous and unequal blowing from both upper and lower slots (dual blowing). The latter two correction methods are investigated in this study. Table 1 shows the effect of the angle of attack correction $\Delta\alpha$ and of dual blowing on lift (c_L) and pitching moment coefficient (c_M). Values are normalized with the unmodified and unactuated baseline airfoil values (index "bl"). Dual blowing is realized with $c_{\mu,up}/c_\mu$ $=99\%, 90\%, 93\%$, $c_{\mu,low}/c_\mu = 1\%, 10\%, 7\%$ at take-off, climb, and cruise respectively.

The values in Table 1 indicate the angle of attack correction leads to a pitching moment reduction, whereas dual blowing is able to get very close to the original pitching moment values. The unsteady behavior of the angle of attack and dual blowing corrections is compared in Sect. 5. The current results show that steady dual blowing is a promising approach for restoring baseline airfoil performance with small impact on airfoil drag.

Table 1. Baseline performance recovery by increased α or dual blowing.

Case	Altitude	Adapted angle of attack			Dual blowing		
	h, m	$\Delta\alpha$	$\frac{c_{L,\Delta\alpha}}{c_{L,bl}}$	$\frac{c_{M,\Delta\alpha}}{c_{M,bl}}$	c_μ	$\frac{c_{L,dual}}{c_{L,bl}}$	$\frac{c_{M,dual}}{c_{M,bl}}$
Take-off	0	0.53°	1	0.81	0.0543%	1	0.97
Climb	5239	0.44°	1	0.85	0.0479%	1	0.98
Cruise	10668	0.32°	1	0.84	0.0398%	1	0.95

3.2 Jet Temperature Effect

For reduced temperature of the supplied pressurized air and fixed c_μ, the required mass flow increases and resulting jet velocity decreases. Table 2 shows the lift change Δc_L for air jets of different temperature at the three altitudes with constant $c_\mu = 0.105\%$ for both Normal and Coanda jet. Lift change $\Delta c_L(T_\infty)$ with jet air at ambient temperature $T_\infty = 288.15$ K (take-off), $T_\infty = 254$ K (climb) and $T_\infty = 218$ K (cruise) is compared to the lift change for hot engine bleed air $\Delta c_L(T_{eng} = 473K)$ and for an intermediate temperature $T_{avg} = 380$ K (take-off), $T_{avg} = 363$ K (climb), and $T_{avg} = 346$ K (cruise) relative to $\Delta c_L(T_{eng})$. In addition, the required power for the compression process of the jet air is estimated from [8]:

$$P = \dot{m}_{jet} \cdot (c_v T_{jet} + \frac{1}{2} v_{jet}^2 + \frac{p_{jet}}{\rho_{jet}}) - \dot{m}_a \cdot (c_v T_a + \frac{1}{2} v_a^2 + \frac{p_a}{\rho_a}) \quad (2)$$

with index "a" denoting the condition upstream the compression process and "jet" at jet nozzle, mass flow $\dot{m}_{jet} = \dot{m}_a$, specific heat capacity at constant volume c_v, and pressure p. The conditions at the jet exit and upstream of the compression are calculated with isentropic relations.

Table 2. Impact of reduced jet temperature on lift change relative to hot engine bleed air.

Case	Altitude	c_μ	Normal jet		Coanda jet	
	h, m		$\frac{\Delta c_L(T_\infty)}{\Delta c_L(T_{eng})}$	$\frac{\Delta c_L(T_{avg})}{\Delta c_L(T_{eng})}$	$\frac{\Delta c_L(T_\infty)}{\Delta c_L(T_{eng})}$	$\frac{\Delta c_L(T_{avg})}{\Delta c_L(T_{eng})}$
Take-off	0	0.105%	0.86	0.94	1.11	1.08
Climb	5239	0.105%	0.87	0.95	1.06	1.05
Cruise	10668	0.105%	0.90	0.96	1.28	1.16

The values in Table 2 show that higher temperature improves the Normal jet lift reduction, whereas lower jet temperature improves the Coanda actuator lift reduction. According to Eq. 2, higher jet temperature requires increased power, whereas lower jet temperature increases mass flow. For example, $\frac{\Delta c_L(T_{avg})}{\Delta c_L(T_{eng})} = 1.16$ is obtained at cruise condition for the Coanda actuator with a 17% increase in mass flow, but a 58% decrease in power for a jet at T_{avg} compared to T_{eng}. Depending on the other airplane subsystem requirements, the jet air could be provided by the low pressure compressor of the engine or by a dedicated compressor system with a heat exchanger or mixing with uncompressed air. Based on these results, T_{avg} is chosen as total temperature for the unsteady simulations for both Normal and Coanda jet.

3.3 Dual Blowing for Load Alleviation

The potential benefits of operating both upper and lower Coanda slots at the same time with different c_μ for lift reduction are studied. Table 3 shows the lift change obtained with dual blowing $\Delta c_{L,dual}(T_{eng})$ normalized with only lower blowing with $\Delta c_{L,low}(T_{eng})$. For the dual blowing case, the c_μ of the upper jet is 2%, 5%, 10% of the total c_μ listed in Table 3 and the c_μ for the lower jet is 98%, 95%, 90% of the total c_μ, respectively.

Table 3 shows that steady blowing with reduced lower slot c_μ and minimal upper slot blowing generally increases actuator control authority, with highest improvements of around 10–13% for $c_{\mu,up} = 5\% c_\mu$.

Table 3. Improvement in lift reduction for dual blowing vs. lower slot blowing.

Case	Altitude	c_μ	$\Delta c_{L,dual}/\Delta c_{L,low}$		
	h, m		$c_{\mu,up} = 2\% c_\mu$	$c_{\mu,up} = 5\% c_\mu$	$c_{\mu,up} = 10\% c_\mu$
Take-off	0	0.105%	1.06	1.13	1.07
Climb	5239	0.105%	1.10	1.11	1.05
Cruise	10668	0.105%	1.07	1.10	0.97

Figure 2 shows the skin friction coefficient c_f in the trailing edge region. The c_f distribution shows that for the lower jet configuration, an area of separated flow develops in the region downstream of the upper jet location and the outer flow reattachment occurs at the point of maximum thickness of the ellipse. For the dual jet configuration, the upper slot jet reduces the size of the separated flow region and leads to an earlier reattachment and subsequent separation of the external flow compared to the lower jet configuration, marked by circles in the plot. The described effect has an impact on the c_p distribution on the entire airfoil. As the gains through dual blowing are promising, the impact of $c_{\mu,up}$=2%c_μ is studied on unsteady Coanda actuation in Sect. 5.

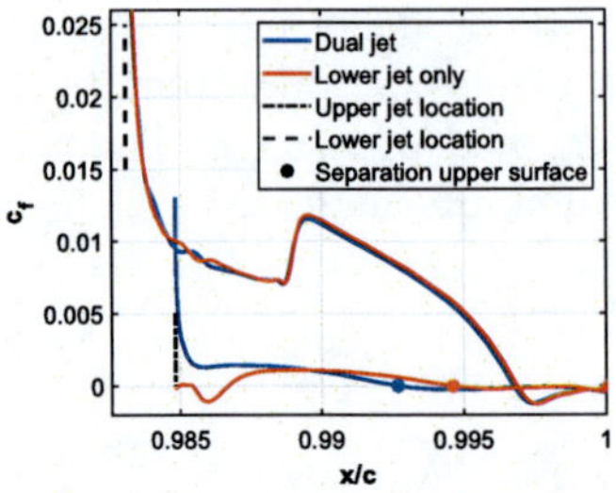

Fig. 2. Dual blowing effect on c_f over Coanda surface with $c_{\mu,up} = 5\%c_\mu$ at Climb.

4 Gust Definition

Based on aircraft certification requirements [6], three upward-directed 1-cos gusts are studied with half gust lengths of $H = 9$ m, 58 m, and 107 m. The gust amplitudes are calculated as a function of altitude and airplane properties:

$$U_{gust} = \frac{U_{ds}}{2} \cdot (1 - cos(\frac{\pi \cdot s}{H})) \tag{3}$$

where U_{ds} is the maximum gust velocity, depending on H, altitude and airplane weight, and $s = 0$–$2H$ is the gust penetration distance. For the selected gusts, U_{ds} varies between 9.9 m/s and 16 m/s True Air Speed.

In the surrogate model the lift change Δc_L due to the gust-airfoil interaction is based on precomputed steady airfoil polars. For the three altitudes, steady simulations at five different α are conducted. The values of c_L and α are interpolated with two curves, a linear polynomial from 0 to the last α in the linear region and a quadratic polynomial for the remaining points. For a given gust-induced angle of attack change $\alpha_{gust} = \arctan(U_{gust}/U_\infty)$, the corresponding lift coefficient is interpolated from the steady lift polars. This is also calculated for quasi-steady points s along the gust profile, resulting in a time history for c_L. Unsteady aerodynamic effects are relevant for the shortest gusts at take-off and climb altitude, and c_L is therefore scaled using the Küssner function [9].

Figure 3 shows the c_L changes as a function of time t at the three altitudes and gust lengths, comparing the DVA and the surrogate approach. The surrogate model exhibits good overall agreement with the CFD results, with deviation in peak-load below 7%. The most significant differences occur at high c_L due to non-linear effects such as airfoil dynamic stall and unsteady vortex shedding [7].

Table 4 lists corresponding gust data for each altitude and H, the time at which the peak load occurs (t_{peak}), gust-induced lift change Δc_L relative to the

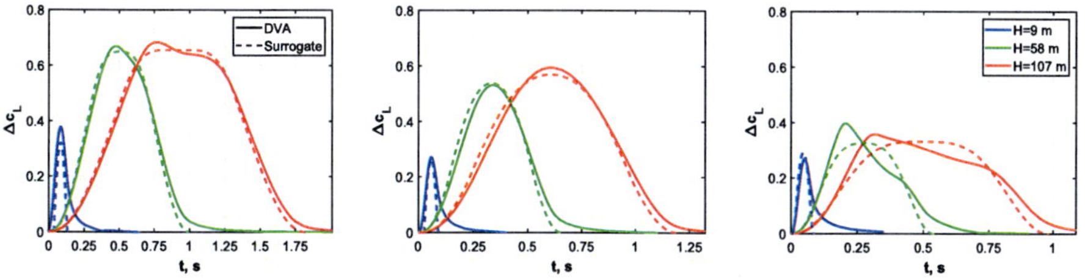

(a) Take-off (h=0 m) (b) Climb (h=5239 m) (c) Cruise (h=10668 m)

Fig. 3. c_L change due to gust airfoil interaction.

undisturbed case, and percentage difference of the peak load between the DVA ($c_{L,dva}$) and surrogate approach ($c_{L,su}$), divided by the DVA peak load. Gust rise times vary between 0.052–0.77 s with gust-load changes Δc_L between 0.27–0.68. The surrogate model can be useful for rough estimation of gust-load changes, not only in terms of lift coefficient but also drag and pitching moment coefficients. The Δc_L and t_{peak} values predicted by DVA are used as load alleviation targets for the following unsteady simulations.

Table 4. Gust rise time t_{peak}, induced load change Δc_L, and surrogate model error for different half gust lengths H and altitudes h.

H, m	Take-off			Climb			Cruise		
	t_{peak}, s	Δc_L	$\frac{c_{Lsu}-c_{Ldva}}{c_{Ldva}}$	t_{peak}, s	Δc_L	$\frac{c_{Lsu}-c_{Ldva}}{c_{Ldva}}$	t_{peak}, s	Δc_L	$\frac{c_{Lsu}-c_{Ldva}}{c_{Ldva}}$
9	0.0874	0.379	−5.2%	0.0594	0.272	−1.3%	0.0525	0.269	2.2%
58	0.478	0.669	−1.3%	0.342	0.531	0.62%	0.208	0.397	−6.7%
107	0.767	0.683	−2%	0.606	0.595	−2.3%	0.316	0.357	−2.5%

5 Unsteady Behavior of Active Flow Control

Unsteady simulations are conducted with the Flap, Normal jet, and Coanda jet to compensate the previously defined gust-load changes Δc_L within the gust rise time t_{peak} for H = 9 m and 58 m. Based on commercial aileron slew rates of 35–40°/s [10], the flap slew rate is set to a constant value of 45°/s for H = 9 m, and 22.5°/s for H = 58 m. For the fluidic actuators, the jet velocity is impulsively increased to the steady target values. For H = 9 m at climb and cruise conditions the Coanda jet simulations are not directly started with the target c_μ in

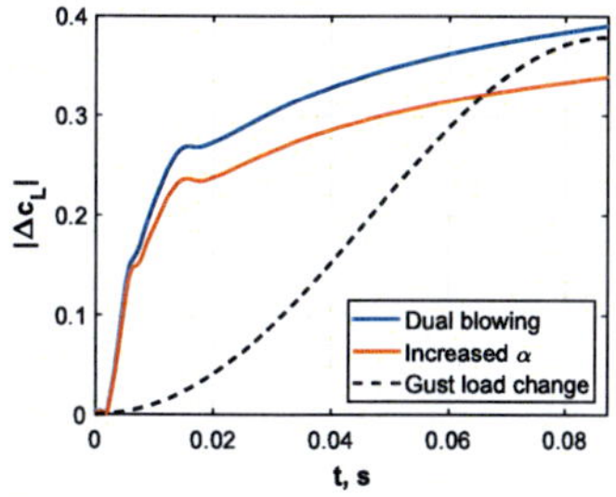

Fig. 4. c_L-change for Coanda actuator starting from dual blowing and increased α to compensate short gust at take-off.

order not to exceed a jet Mach number $M_{jet} = 0.9$ initially at the throat. After reaching $20\% t_{peak}$ for $H = 9$ m and $5\% t_{peak}$ for $H = 58$ m, c_μ is slightly adjusted to reach $M_{jet} = 0.97$. The target c_μ of Coanda jet at take-off and Normal jet at the three conditions is increased at $20\% t_{peak}$ for $H = 58$ m.

Section 3.1 shows how the airfoil baseline performance with the Coanda geometry can be restored through increased angle of attack or dual blowing. Figure 4 compares the unsteady behavior of the two configurations with the same $c_\mu = 0.47\%$ at take-off. Both cases feature a step change in blowing at $t = 0$ s. The case with increased angle of attack therefore features a jet startup at both slots from zero to peak flow rate $c_{\mu,up} = 2\% c_\mu$ and $c_{\mu,low} = 98\% c_\mu$, whereas for the dual blowing the blowing rate is adjusted from baseline performance values. The case with increased α would require a higher c_μ to reach the same load changes as for dual blowing. Dual blowing is therefore selected for the following cases.

5.1 Actuator Comparison

Table 5 shows the target Δc_L and t_{peak} for $H = 9$ m and 58 m at the three altitudes, as well as the time-averaged c_μ for the fluidic actuators and reached δ for the flap at the gust peak time.

Table 5. Required δ for flap and c_μ for fluidic actuators to alleviate gusts.

Case	Altitude	Gust length	Rise time		Flap	Normal jet	Coanda jet
	h, m	H, m	t_{peak}, s	Δc_L	δ	c_μ	c_μ
Take-off	0	9	0.0874	0.379	−3.93°	1.92%	0.47%
		58	0.478	0.669	−10.7°	4.40%	0.78%
Climb	5239	9	0.0594	0.272	−2.67°	0.71%	0.32%
		58	0.342	0.531	−7.69°	2.81%	0.33%
Cruise	10668	9	0.0525	0.269	−2.36°	0.45%	0.19%
		58	0.208	0.397	−4.68°	0.56%	0.19%

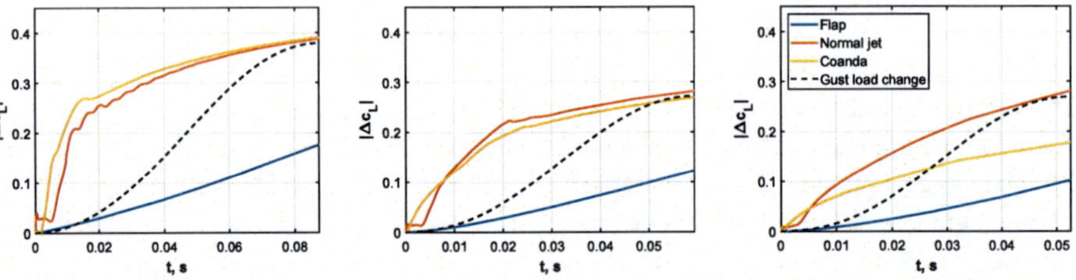

(a) Take-off altitude (b) Climb altitude (c) Cruise altitude

Fig. 5. Actuator-induced changes in c_L to compensate $H = 9$ m gust.

Figure 5 shows a comparison of the unsteady c_L behavior of the three actuators at the three selected operating conditions with $H = 9$ m, as well as the gust-induced loads for reference. When comparing actuator response with the gust, it is apparent that the flap does not react fast enough, only compensating 47%, 45%, and 38% of the gust-load change at take-off, climb, and cruise condition, respectively. The fluidic actuators react significantly faster and are able to compensate the short gusts. Only for cruise condition, the Coanda jet compensates just 66% of the gust load due to an imposed limitation on c_μ to maintain subcritical jet velocities. The initial lift plateau and consequent asymptotic change in lift in the time course signal of the Normal jet are linked to the actuator-induced formation of a recirculation region and its upstream propagation, as well as unsteady vortex shedding proportional to the decline in bound circulation. The effects are less pronounced at cruise condition.

The actuator-induced load changes at $H = 58\,\text{m}$ are similar to $H = 9\,\text{m}$, as Fig. 6 shows at cruise condition. The flap slew rate of 22.5° is sufficient to compensate the gust-induced lift change. The Normal jet and Coanda jet show faster reaction time compared to the Flap. As for $H = 9\,\text{m}$, the limitation on c_μ has a strong impact on the load-change capability of the Coanda jet at cruise.

For impulsive actuation, the lift change rates due to the fluidic actuators exceed the actual gust-induced changes, leading to a reduction of effective angle of attack prior to reaching peak gust load. The fluidic actuator speeds exceed the gust requirements in certifications, rendering them especially useful for load alleviation systems with increased delay times due to robust gust detection and controller response times. A final load alleviation system would require a more tailored activation profile to avoid overcompensating occurring gust loads.

Figure 7 shows the unsteady c_M behavior compared to the gust-induced loads. The actuator-induced pitching moment changes all far exceed the gust load at the three operating conditions. For the fluidic actuators at take-off and climb altitude the pitching moment changes occur faster than lift changes and level off well before peak lift changes are reached, whereas the Flap-induced pitching moment is proportional to lift change.

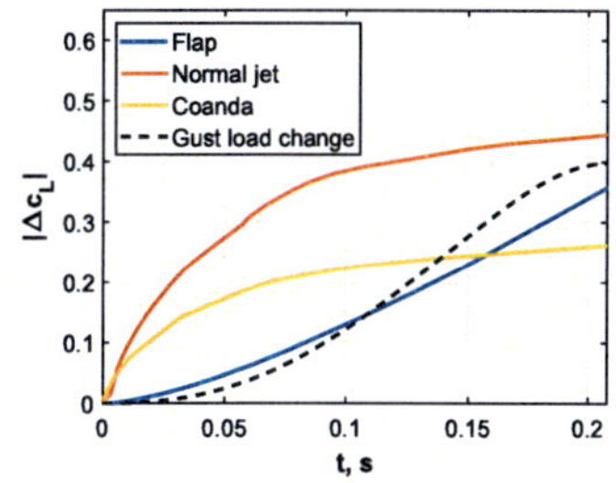

Fig. 6. Actuator-induced changes in c_M to compensate $H = 58\,\text{m}$ gust during cruise.

The present results show that the studied fluidic actuator feature initial change rates of lift and pitching moment that surpass a typical trailing edge flap system. A flap system would have to feature a significantly higher slew rate to compensate the same short gusts. These increased actuation speeds come at the price of additional system weight and energy consumption for providing the required air flow. Still, reduced onset times for the load alleviation might prove critical for countering short gusts and compensating delays in the gust detection system for gusts of all lengths, leading to more complete load alleviation and additional improvements in passenger comfort. A more detailed investigation is underway to evaluate the performance of the entire load alleviation system against classical approaches. Normal and Coanda jets exhibit similar load alleviation potential. As Table 5 shows, the Coanda jet requires significantly lower c_μ

(down to 1/8 of Normal jet), whereas the Normal jet has an expected reduced system complexity.

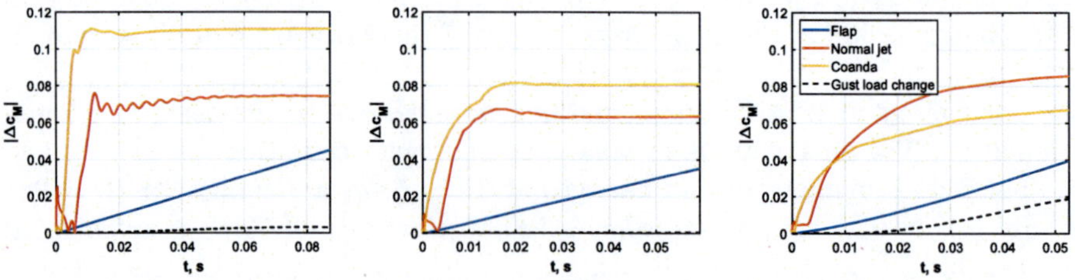

(a) Take-off altitude (b) Climb altitude (c) Cruise altitude

Fig. 7. Actuator-induced changes in c_M to compensate $H = 9\,\text{m}$ gust.

6 Conclusions

The present work investigates the unsteady response of a trailing edge flap, a jet normal to the airfoil, and a jet over a Coanda surface for compensating the load change due to a vertical gust on a supercritical airfoil representative of a medium-range airplane. An investigation of dual blowing from both upper and lower slots over a Coanda surface shows that baseline performance and peak load alleviation could be improved with minimal blowing from the upper slot. Normal and Coanda jet further exhibit opposite performance trends with varying jet temperature, as for the former higher temperature and for the latter lower temperature increases load alleviation performance. Airfoil-gust interactions are simulated by means of a URANS method with a Disturbance Velocity Approach. A surrogate approach, based on steady aerodynamic polars, is developed to estimate the peak load change due to the gust, leading to prediction errors below 7%. The unsteady actuator simulations show that the studied fluidic actuators exhibit a faster initial load response than a classical trailing edge flap over a comprehensive range of flight conditions, which could be crucial for countering short gusts or compensating a significant delay in the gust detection system.

Acknowledgment. We acknowledge the funding by the Deutsche Forschungsgemeinschaft (DFG, German Research Foundation) under Germany's Excellence Strategy-EXC 2163/1-Sustainable and Energy Efficient Aviation – Project-ID 390881007.

References

1. Krein, A., Williams, G.: Flightpath 2050: Europe's vision for aeronautics. In: Proceedings of the Sixth European Aeronautics Days Innovation for Sustainable Aviation in a Global Environment, p. 63 (2012)

2. Khalil, K., Asaro, S., Bauknecht, A.: Active flow control devices for wing load alleviation. In: Proceedings of the AIAA Aviation Forum 2020, p. 2940, online (2020)
3. Spalart, P. R., Shur, M.: On the sensitization of turbulence models to rotation and curvature. Aerospace Sci. Technol. **1**(5), 297–302 (1997)
4. Heinrich, R., Reimer, L.: Comparison of different approaches for gust modeling in the CFD Code TAU. In: International Forum on Aeroelasticity & Structural Dynamics, Bristol, United Kingdom (2013)
5. Nishino, T., Karim, S.: Effect of jet nozzle lip momentum loss on circulation control airfoil performance. AIAA J. **50**(3), 551–558 (2012)
6. EASA: Certification Specifications and Acceptable Means of Compliance for Large Aeroplanes CS-25. Technical report, Amendment 20, European Aviation Safety Agency, Europe (2007)
7. Mittal, S., Saxena, P.: Hysteresis in flow past a NACA 0012 airfoil. Comput. Meth. Appl. Mech. Eng. **191**(19–20), 2207–2217 (2002)
8. Radespiel, R., et al.: Active flow control for high lift with steady blowing. Aeronaut. J. **120**(1223), 171–200 (2016)
9. Cebeci, T., et al.: Analysis of Low-Speed Unsteady Airfoil Flows. Springer, Berlin Heidelberg (2005)
10. National Research Council: Aviation Safety and Pilot Control: Understanding and Preventing Unfavorable Pilot-vehicle Interactions. The National Academies Press, Washington, DC (1997)

Aeroelastic Behavior of a Laminar Wing in Transonic Flow

Michael Fehrs(✉), Sebastian Helm, Christoph Kaiser, and Thomas Klimmek

DLR, Institute of Aeroelasticity, 37073 Göttingen, Germany
michael.fehrs@dlr.de,
https://www.dlr.de/ae/

Abstract. A laminar wing test case for the flutter behavior in transonic flow with free boundary layer transition is presented. The DLR-AE-L1 airfoil and wing are designed to serve as an aeroelastic testbed to develop a better understanding for the effect of boundary layer transition on flutter stability. Coupled CFD-CSM simulations are used to determine the flutter onset in a transitional and a fully turbulent flow. It is found that flutter occurs at lower stagnation pressures for the transitional flow, which is in accordance with earlier research on laminar airfoils.

Keywords: Laminar wing · Boundary layer transition · CFD-CSM coupling · Transonic flutter

1 Introduction

Natural laminar flow airfoils and wings allow more ecologic and economic transport aircraft by reducing the friction drag during cruise flight significantly. Laminar flow control by active and passive means is well understood from an aerodynamic point of view, but aeroelastic effects have not been considered sufficiently yet [1]. The partly laminar boundary layer flow alters the lift and moment acting on the wing as the displacement thickness of the boundary layer is changed. In transonic flow, changes in the shock location and strength have the potential to introduce shock-induced boundary layer separations that can affect the flutter behavior directly. In addition, any partial or complete loss of laminarity introduces aerodynamic non-linearities, which in turn might affect the dynamic aeroelastic stability.

The objective of this paper is to introduce an airfoil and wing geometry as an aerodynamic and aeroelastic testbed that can be distributed without any restrictions[1]. The airfoil is mainly designed to attenuate Tollmien-Schlichting (TS) transition. A small leading-edge sweep angle is used to prevent crossflow transition. There is no intention to design a state-of-the-art laminar airfoil and wing that serves all needs of a modern transport aircraft design as it is done by other research groups [2,3].

[1] Please contact first author.

© The Author(s), under exclusive license to Springer Nature Switzerland AG 2021
A. Dillmann et al. (Eds.): STAB/DGLR Symposium 2020, NNFM 151, pp. 164–173, 2021.
https://doi.org/10.1007/978-3-030-79561-0_16

2 Methods

2.1 ModGen: Geometry and Structural Model

The wing geometry and structural model are created with the DLR computer program ModGen. ModGen is mainly designed to generate finite element models for the load carrying structure of wing-like and fuselage components of an aircraft [4]. As ModGen is setting up parametrically defined geometry models as a prerequisite for the finite element model generation, these geometry models of the outer wing geometry can also be used for the CFD mesh generation.

2.2 DLR TAU-Code: Transition Prediction

The flow field is computed with the DLR TAU-Code [5]. Boundary layer transition is considered based on the correlation-based γ transition model [6] and linear stability theory with the DLR transition module [7,8]. The γ transition model is used for most of the steady and unsteady computations. It is advantageous for unsteady computations as the transition prediction is performed at each inner iteration of each physical time step and on each computational grid point.

As the γ model is not able to predict crossflow transition, the e^N method of the DLR transition module is used to confirm the transition prediction by the γ transition model for the steady flow computations. The SST $k-\omega$ turbulence model is used for all fully turbulent computations. It is also used as the underlying turbulence model for both transition prediction methods.

2.3 FlowSimulator: CFD-CSM Coupling

The CFD-CSM coupling is performed within the FlowSimulator environment which is designed to enhance CFD-related multidisciplinary analysis [9]. The time-integration is conducted in a co-simulation exchanging structural deformations and aerodynamic forces at each time step. The mode shapes are mapped onto the CFD grid to account for displacements of the structure by mesh deformation and the structural displacements are obtained by employing the Newmark-beta method [10].

3 Results

3.1 Aerodynamics

Airfoil: The Boeing HSNLF airfoil is chosen [11] as the baseline geometry for the airfoil design. The airfoil is supposed to provide a large extent of laminar flow and low shock wave losses at transonic Mach numbers. Laminar boundary layer flow of at least up to 50% to 60% of the chord length is expected on the upper surface within the design Reynolds number range of $Re = 15 \cdot 10^6$ to $30 \cdot 10^6$ for $c_l = 0.5$ at $M = 0.78$.

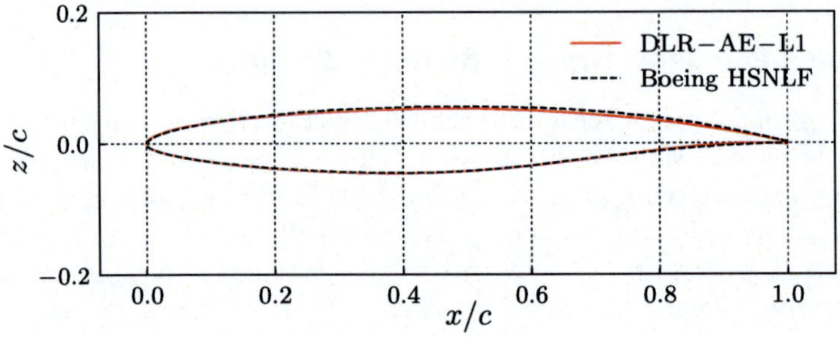

Fig. 1. DLR-AE-L1 and Boeing HSNLF Airfoil: geometry

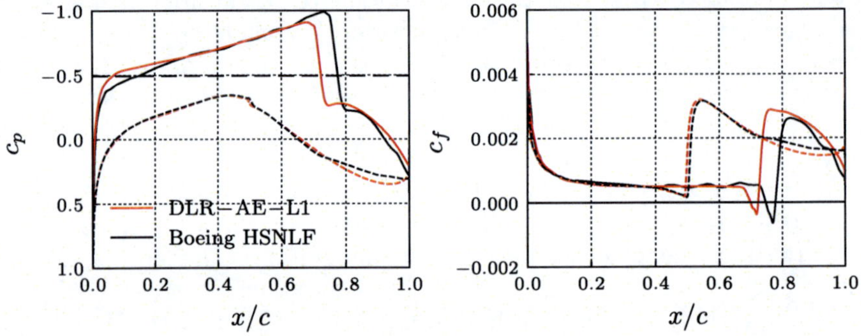

Fig. 2. DLR-AE-L1 and Boeing HSNLF Airfoil: pressure and skin friction coefficient distribution at $c_l = 0.5$, $M = 0.78$, $Re = 15 \cdot 10^6$; full line: upper surface, dashed line: lower surface

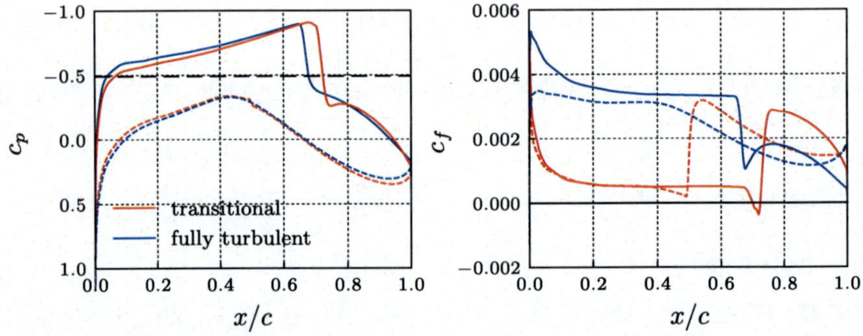

Fig. 3. DLR-AE-L1 Airfoil: pressure and skin friction coefficient distribution at $c_l = 0.5$, $M = 0.78$, $Re = 15 \cdot 10^6$ for the transitional and fully turbulent flow; full line: upper surface, dashed line: lower surface

The Boeing HSNLF is not used directly as it is not possible to recreate the design pressure distribution. In addition, the documented airfoil data do not provide a smooth enough surface for high Reynolds number computations.

The design condition for the Boeing HSNLF of $Re = 15 \cdot 10^6$, $c_l = 0.5$, $M = 0.78$ are kept for the DLR-AE-L1. The airfoil coordinates are modified to move the shock location upstream by reducing the curvature of the upper surface. In addition, the nose geometry is changed to reduce the favorable pressure gradient on the suction side downstream of $x/c \approx 0.1$ to attenuate the growth of crossflow instabilities. The lower aft section is modified to produce more lift by an increased curvature. The modified point data is smoothed using a b-spline interpolation. It is found that the resulting airfoil coordinates are well represented in geometry tools (ModGen), grid generators (ANSYS ICEM, Centaur), and result in a sufficiently smooth pressure distribution at the design Reynolds number.

Figure 1 shows the DLR-AE-L1 airfoil and the Boeing baseline geometry. Figure 2 shows the pressure and skin friction coefficient distribution at $Re = 15 \cdot 10^6$, $c_l = 0.5$, $M = 0.78$. The γ transition model is used for transition prediction. The turbulence level is set to $Tu = 0.05\%$, which represents a very quiet wind tunnel or free flight conditions. The shock on the upper surface is successfully moved upstream and the corresponding boundary layer separation is weakened.

Figure 3 shows the pressure and skin friction coefficient distribution for the DLR-AE-L1 in a transitional and a fully turbulent flow at the same flow conditions as above. The shock location is located more aft for the transitional flow, as the displacement thickness decreases in the partly laminar flow.

Wing: The wing with a semi-wingspan of $b/2 = 1.6\,\text{m}$ and no twist has a leading edge sweep angle of $\Lambda_{LE} = 20°$, a root chord length of $c_{root} = 0.5\,\text{m}$, a tip chord length of $c_{tip} = 0.15\,\text{m}$, and a taper ratio of $\lambda = c_{tip}/c_{root} = 0.3$. For simplicity, the airfoil sections are aligned in flow direction. The geometric parameters give a mean aerodynamic chord of $c_{mac} = 0.35641\,\text{m}$.

The computational grid consists of a structured surface mesh with 350 nodes in chord and 250 in span direction. The prism boundary layer grid consists of 85 layers, for which the maximum y^+ value of the first grid cell is about 0.5 for the investigated flow conditions. The overall grid contains about $17 \cdot 10^6$ nodes. No systematic grid convergence study is performed yet but it is checked that the transition model behaves well within the transition region.

Figure 4 shows the angle of attack, moment, and drag coefficient for a constant lift coefficient of $C_L = 0.5$ with increasing Mach number at $Re_{mac} = 15 \cdot 10^6$. The variation of the angle of attack to maintain the desired lift coefficient and the change in the moment coefficient with Mach number are much more pronounced for the transitional flow. As the drag divergence Mach number is approached, the drag coefficients of the transitional flow remain beneath the drag for the fully turbulent flow. The drag benefit due to lower skin friction is larger than the drag penalty due to a stronger shock and larger boundary

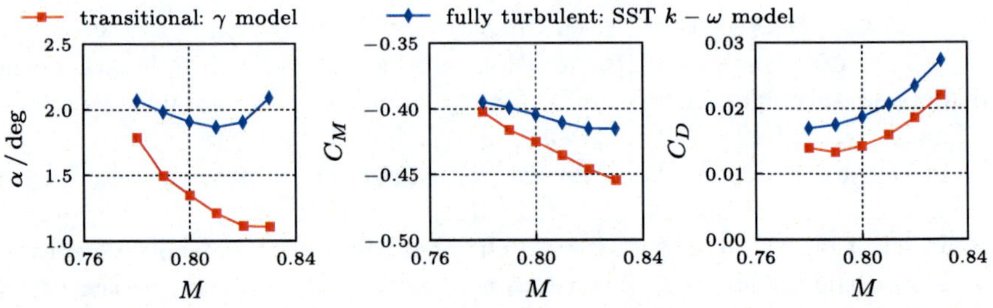

Fig. 4. DLR-AE-L1 Wing: Angle of attack, moment, and drag coefficient for increasing Mach numbers at $C_L = 0.5$, $Re_{mac} = 15 \cdot 10^6$

layer separation. At higher Mach numbers, no steady solution is found for the target lift coefficient, which is most likely caused by shock buffet due to shock-boundary-layer interaction.

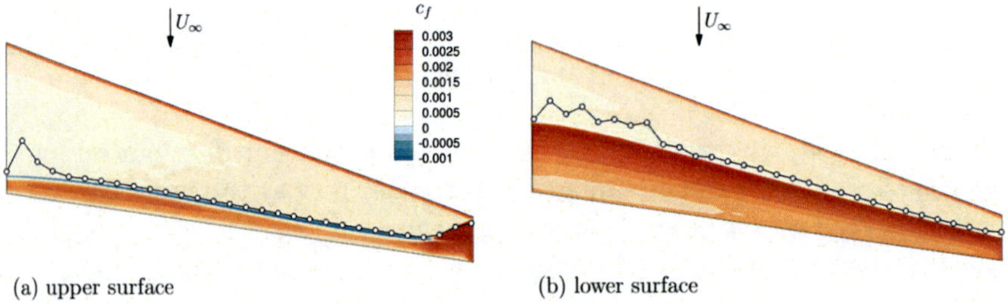

Fig. 5. DLR-AE-L1 Wing: skin friction coefficient distribution computed with the γ transition model and transition line from the stability code LILO

Figure 5 shows the skin friction coefficient distribution computed with the γ model at $C_L = 0.5$, $M = 0.82$, $Re_{mac} = 15 \cdot 10^6$. The transition location is found at the main increase in skin friction coefficient. In addition, the transition location obtained with the TAU transition module [7,8] with the boundary layer code COCO and the stability code LILO for critical incompressible N factors of $N_{TS} = 11.8$ and $N_{CF} = 8.4$ is shown by the line in Fig. 5. The transition locations predicted by both methods match well on both the upper and lower surface of the wing. The stability code predicts a combination of crossflow and Tollmien-Schlichting transition towards the wing root, but the transition location does not shift much upstream. It should be noted that the critical N factors are rather low compared to a turbulence level of $Tu = 0.05\%$. Due to the overall dominance of TS transition, it is assumed that the γ model can be used for transition prediction for these flow conditions.

3.2 Structural Model

For investigating the differences between the flutter behavior in a transitional and a fully turbulent flow, the structural model is tailored to exhibit flutter for the examined range of aerodynamic parameters at realistic wind tunnel stagnation pressures. The structural model representing a wind tunnel model is built in a two-step process. Firstly, the complete volume of the wing is meshed with six-sided and five-sided solid elements, all having the same material properties of a maraging steel. By performing a normal modes analysis in MSC.Nastran (SOL103) with the wing root rigidly clamped, the mode shapes and eigenfrequencies are identified disregarding structural damping.

Secondly, the structural model is artificially weakened by introducing reduction factors for each eigenfrequency to obtain a flutter case. The first eigenfrequency is reduced less to provide a reasonable upper bound on the required simulation time. The thus obtained first ten eigenmodes are used in the CFD-CSM coupling. Table 1 lists the reduction factors and baseline eigenfrequencies together with a description of the associated mode shapes.

Table 1. Structural model: mode shapes and wind-off eigenfrequencies

Mode	Description	f/Hz	Mode	Description	f/Hz
1	1st bending	$0.5 \cdot 17.84$	6	4th bending	$0.25 \cdot 250.32$
2	2nd bending	$0.25 \cdot 60.14$	7	2nd torsion	$0.25 \cdot 351.36$
3	3rd bending	$0.25 \cdot 137.39$	8	5th bending	$0.25 \cdot 398.31$
4	1st in-plane	$0.25 \cdot 161.93$	9	2nd in-plane	$0.25 \cdot 507.85$
5	1st torsion	$0.25 \cdot 193.39$	10	3rd torsion	$0.25 \cdot 527.29$

3.3 Aeroelasticity

For each Mach number, coupled CFD-CSM computations are performed to determine the largest structural deformation at a given freestream pressure. At the Mach number that results in the largest deformation, the freestream pressure is varied to determine the critical pressure, for which flutter occurs with a step size of $\Delta p = 0.2$ bar.

The largest deformations are found for $M = 0.82$. The steady flow condition is shown in Fig. 6 at two span stations in terms of the pressure and skin friction coefficient distribution. For the transitional flow, a shock-induced boundary layer separation covers the whole wing span. The fully turbulent flow only separates in the outer wing section (outboard of $\eta = y/(b/2) = 0.55$) as the shock is weaker and the turbulent boundary layer is less prone to separate.

Figure 7 and 8 show the modal coordinates of the first six eigenmodes (compare Table 1) at the critical pressure. The flutter instability occurs at a lower

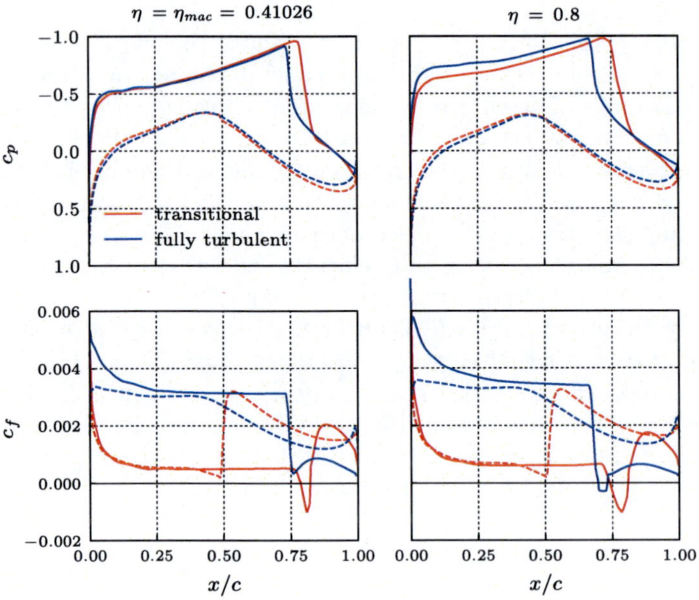

Fig. 6. DLR-AE-L1 Wing: pressure and skin friction coefficient distribution at $M = 0.82$, $C_L = 0.5$, $Re = 15 \cdot 10^6$; full line: upper surface, dashed line: lower surface

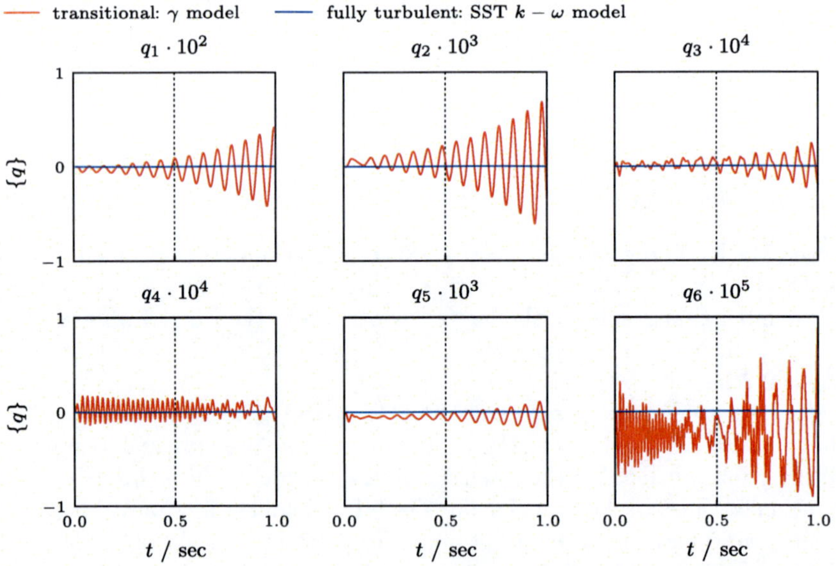

Fig. 7. DLR-AE-L1 Wing: modal coordinate q of the first 6 eigenmodes for $p = 2$ bar

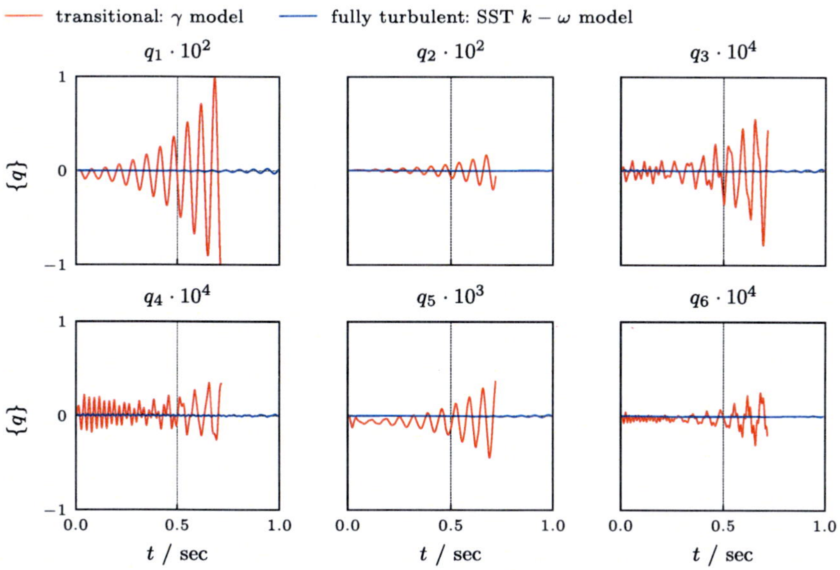

Fig. 8. DLR-AE-L1 Wing: modal coordinate q of the first 6 eigenmodes for $p = 2.2$ bar

pressure of $p = 2$ bar in the transitional flow compared to $p = 2.2$ bar in a fully turbulent flow. Table 2 gives the flutter frequency f_{fl} and the logarithmic decrement $\Lambda = \ln(\widehat{q}_n/\widehat{q}_{n+1})$ defined by two consecutive oscillation maxima $\widehat{q}$ obtained from mode 1. Although the flutter mode is the same for both flow conditions, the transitional flow is more severe in terms of critical pressure and mode excitation. In the case of the transitional flow at $p = 2.2$ bar, the computation is terminated as the grid deformation fails for larger oscillations.

Table 2. Flutter frequency f_{fl} and logarithmic decrement Λ obtained from mode 1

	Transitional		Fully turbulent	
p/bar	f_{fl}/Hz	Λ	f_{fl}/Hz	Λ
2.0	14.306	−0.191		
2.2	14.846	−0.328	14.791	−0.284

To put the resulting modal displacements into perspective, Fig. 9 and 10 show the time history of the lift and moment coefficient. For reference, a quasi-steady lift increase of $\Delta C_L = 0.01$ is found for an angle of attack increase of $\Delta\alpha \approx 0.067°$ for the transitional flow neglecting any unsteady effects.

A time step size of $\mathrm{d}t = 2.62 \cdot 10^{-4}$ s is used for all unsteady computations, for which the inner iterations converge sufficiently well. The spatial and temporal discretization is based on experience for computations with similar flow

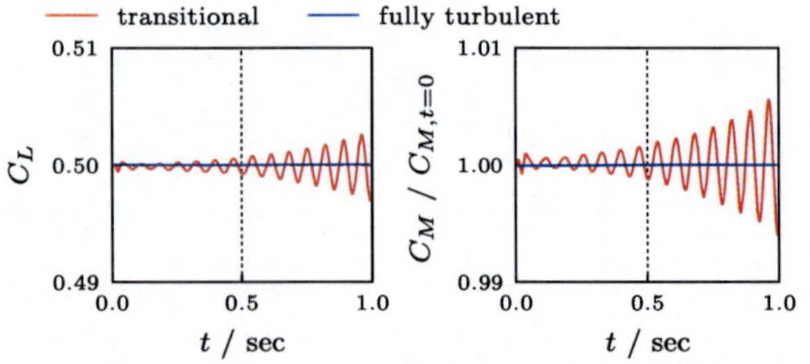

Fig. 9. DLR-AE-L1 Wing: lift and moment coefficient for $p = 2$ bar

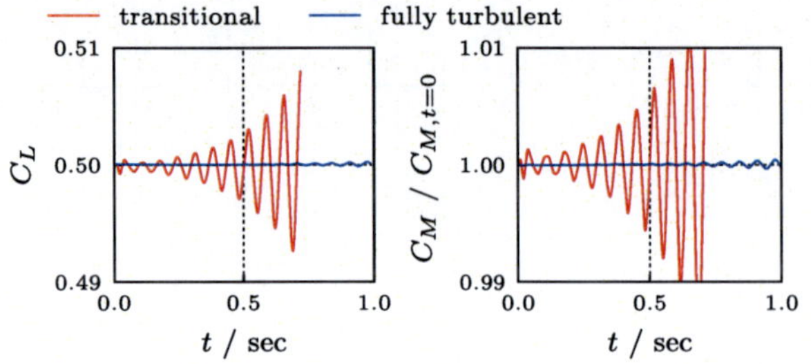

Fig. 10. DLR-AE-L1 Wing: lift and moment coefficient for $p = 2.2$ bar

conditions. However, the grid and time step size need to be put under scrutiny in future studies. In addition, a longer time span needs to be covered to investigate the occurrence of limit cycle oscillations.

4 Conclusion and Outlook

At the moment, there is no proof that a partly laminar boundary layer flow in transonic flow is necessarily more critical in terms of its aeroelastic behavior compared to a fully turbulent flow. However, this research supports the observation that aerodynamic forces in a transitional flow change in a way, that facilitates the occurrence of dynamic instabilities in the transonic flight regime. A more detailed investigation of the character of transitional transonic flutter is required and will be carried out. In addition, a detailed description of the unsteady aerodynamics associated with each structural mode is needed to understand differences between the fully turbulent and transitional flutter behavior. As the numerical computations are only bound by numerical stability and not by any actual structural properties, the occurrence of limit cycle oscillations is likely and the limiting mechanisms need to be described and understood.

More general, two important fields of research emerge for future work: First, the aeroelastic effects in a transitional, unsteady transonic flow need to be understood through wind tunnel experiments and numerical studies. Second, transition modeling approaches need to be placed under scrutiny to develop a sound model representation, that captures boundary layer transition in unsteady flows and complex shock boundary layer interaction.

Acknowledgments. This work was funded within the DLR project VicToria.

References

1. Tichy, L., Mai, H., Fehrs, M., Nitzsche, J., Hebler, A.: Risk analysis for flutter of laminar wings. In: IFASD, IFASD-2017-196, Como (2017)
2. Seitz, A., Hübner, A., Risse, K.: The DLR TuLam project: design of a short and medium range transport aircraft with forward swept NLF wing. CEAS Aeronaut. J. **11**, 449–459 (2020)
3. Lynde, M.N., Campbell, R.L.: Computational design and analysis of a transonic natural laminar flow wing for a wind tunnel model. In: 35th AIAA Applied Aerodynamics Conference, AIAA 2017-3058, Denver (2017)
4. Klimmek, T.: Parameterization of topology and geometry for the multidisciplinary optimization of wing structures. In: Proceedings CEAS 2009 - European Air and Space Conference, Manchester (2009)
5. Schwamborn, D., Gerhold, T., Heinrich, R.: The DLR TAU-code: recent applications in research and industry. In: Proceedings European Conference on Computational Fluid Dynamics ECCOMAS, TU Delft, Delft (2006)
6. Fehrs, M.: Boundary layer transition in external aerodynamics and dynamic aeroelastic stability. Dissertation, TU Braunschweig (2018). ISSN 1434–8454, ISRN DLRFB-2018-11, also NFL-FB 2017–27, Braunschweig (2018)
7. Krumbein, A., Krimmelbein, N., Schrauf, G.: Automatic transition prediction in hybrid flow solver, part 1: methodology and sensitivities. J. Aircr. **46**(4), 1176–1190 (2009)
8. Krumbein, A., Krimmelbein, N., Schrauf, G.: Automatic transition prediction in hybrid flow solver, part 2: practical application. J. Aircr. **46**(4), 1191–1199 (2009)
9. Reimer, L.: The FlowSimulator - a software framework for CFD-related multidisciplinary simulations. In: European NAFEMS Conference: Computational Fluid Dynamics (CFD) - Beyond the Solve, München (2015)
10. Newmark, N.M.: A method of computation for structural dynamics. J. Eng. Mech. ASCE **85**(EM3), 67–94 (1959)
11. Versteegh, T.C.: Natural laminar flow, low wave drag airfoil. European Patent Office. EP0111785A1, status: withdrawn (1983)

Check for updates

Influence of Surface Irregularities on the Expected Boundary-Layer Transition Location on Hybrid Laminar Flow Control Wings

Juan Alberto Franco Sumariva(✉), Alexander Theiss, and Stefan Hein

Institut für Aerodynamik und Strömungstechnik, DLR,
Bunsenstraße 10, 37073 Göttingen, Germany
Juan.Franco@dlr.de

Abstract. Hybrid laminar flow control (HLFC) is a promising concept to reduce the viscous drag on aircraft wings by shifting the expected laminar-turbulent transition (LTT) location downstream. However, the structural joint between the leading edge suction panel and the wing box can diminish this desirable shift in LTT location and thus reduce the efficiency of the HLFC system. This paper introduces a toolchain, which allows to numerically quantify the detrimental effects of two-dimensional surface irregularities for transonic HLFC swept wings at free-flight Reynolds numbers. A backward-facing step and two different gaps are studied, which resemble possible structural joint geometries currently considered within the HLFC-WIN project. The stability analysis performed with the Adaptive Harmonic Linearized Navier-Stokes (AHLNS) approach reveals that none of the three irregularities affects the expected LTT location, suggesting that there might be some potential to relax the surface tolerances initially specified in the HLFC-WIN project using classical criteria.

Keywords: Transition · Surface irregularity · HLFC · AHLNS · PSE

1 Introduction

Hybrid laminar flow control (HLFC) wings have captured the interest of several research projects, in particular the HLFC-WIN project [1], because of the potential of suction for reducing drag on commercial aircraft [2]. This reduction in drag is mainly due to the impact on laminar-turbulent transition (LTT), extending the region along the wings where the flow remains laminar. In such circumstances, the growth of flow instabilities (e.g. Tollmien-Schlichting (TS) waves or crossflow (CF) vortices), might trigger the LTT. Typically, on HLFC wings, suction is applied only in the leading edge region (before the front spar). The flow suction modifies the boundary-layer (BL) profiles in such a way that the growth of the above-mentioned flow instabilities is delayed.

However, the joint between the leading edge suction panel and the wing box may introduce surface irregularities (e.g. backward/forward-facing steps

© The Author(s), under exclusive license to Springer Nature Switzerland AG 2021
A. Dillmann et al. (Eds.): STAB/DGLR Symposium 2020, NNFM 151, pp. 174–184, 2021.
https://doi.org/10.1007/978-3-030-79561-0_17

(BFS/FFS), gaps, etc.) that might enhance the spatial development of those flow instabilities and, therefore, reduce the potential benefit of the suction system [3]. For this reason, it is crucial during the design process to quantify the influence of such surface irregularities in terms of BL instability.

Techniques like Parabolized Stability Equations (PSE) [4] are not suited for BL instability analysis in the vicinity of those surface irregularities: its formulation assumes that streamwise variations of base flow and disturbances quantities are small (with respect to surface-normal variations). On the other hand, approaches like Linearized Navier-Stokes (LNS), Harmonic LNS (HLNS), and Direct Numerical Simulation (DNS) do not impose any assumption regarding the nature of the convective instabilities in streamwise direction. This fact implies that the required numerical resources increase significantly. Alternatively, the Adaptive Harmonic LNS (AHLNS) equations [5] can also handle these large streamwise gradients using a fully-elliptic system of equations (similar to DNS). In contrast, the AHLNS formulation assumes a *wave-like* character of the instabilities (as in PSE), leading to a significant reduction in the number of streamwise grid points required compared with LNS, HLNS, or DNS computations.

Alternatively to the above described numerical approaches, it is very common in the industry to use the criteria defined in the work of Nenni and Gluyas [6], who gave a critical Reynolds number for certain types of surface irregularities, based on incoming flow conditions and geometrical dimensions of the surface irregularity only. The criteria were established for a particular flow condition and airfoil geometry (The X21-wing). Therefore, although the use of the Nenni and Gluyas criteria provides a rapid estimation of the maximum allowable dimensions for a surface irregularity, the range of validity of such estimation is very limited.

The present paper provides a systematic procedure for the study of the effect of 2D surface irregularities (in particular, a BFS and two different gaps) on the expected LTT location for HLFC wings at transonic flow conditions using the recently developed AHLNS approach. To the authors' knowledge, there is no similar work published from numerical computations.

2 Methodology

This section outlines the numerical methods used to obtain the steady laminar base flow and the BL instability characteristics affected by different surface irregularities located at the joint between the leading edge suction panel and the wing box on an HLFC wing.

2.1 Reference Design and Base Flow Computation

The original fully three-dimensional (3D) wing geometry corresponds to a design used in the HLFC-WIN project [1] equipped with a leading edge suction panel that reaches up to about 20% of the chord on the upper side of the wing. The base flow has been obtained numerically using the DLR TAU code, a compressible unstructured and Reynolds-averaged Navier-Stokes code for external flows [10] and was kindly made available by Martin Kruse (DLR).

Above reference flow for the nominally smooth 3D wing geometry with suction had been computed using the following approach. First, a TAU simulation was performed for a wing surface without any prescribed suction rates on the suction panel. However, the impact of the suctioned BL flow was considered within the employed transition prediction toolchain. Based on the TAU-computed pressure distribution without suction, a laminar base flow including suction was generated with the BL code COCO [7] for several line-in-flight cuts. Subsequently, the growth of TS waves and CF vortices was investigated utilizing the local incompressible stability theory within the LILO code [8] and the transition location prescribed in the TAU computations was iteratively updated based on N-factor results using the N_{CF}-N_{TS} transition prediction strategy.

The described procedure (coupling of TAU+COCO+LILO) is, however, not suited to study the impact of surface imperfections on the growth of BL instabilities since no laminar base flows can be generated with the BL code COCO including surface irregularities like steps and gaps. Furthermore, the surface imperfections induce streamwise gradients into the flow, which are not considered in the local stability theory. For this reason, the base flow computation including surface irregularity and suction is completely performed with the TAU code, and the BL stability properties are investigated according to the compressible non-local theory employing the PSE/AHLNS framework, which is briefly outlined in Sect. 2.4.

Moreover, the computational demands for the entire 3D flow field simulation, including the joint between the leading edge suction panel and the wing box, are excessively high. In order to ease the computational effort and thus study various joint geometries and assess their impact on the BL instability characteristics, the flow field of the original 3D wing at a selected line-in-flight cut is approximated by an infinite swept-wing 2.5D flow with similar BL properties.

2.2 Infinite Swept-Wing Approximation

The selected line-in-flight cut is located at around 2/3 of the wingspan because this section has the most conical flow condition compared to other spanwise cuts. To ensure that the BL and instability characteristics in the selected cut are comparable to the original 3D flow field, the same toolchain is utilized as for the reference case without surface irregularity (i.e. BL computation including suction based on the pressure distribution obtained from a TAU simulation without suction, and subsequent incompressible, local instability analysis using LILO together with the N_{CF}-N_{TS} transition prediction strategy).

The reference pressure distribution is approximated in a parameter study in which the freestream Mach number Ma_∞, the angle of attack α, and the sideslip angle γ were varied in the 2.5D computation. In contrast, the airfoil geometry and suction distribution remained unchanged. Moreover, the transition location predicted for the 3D reference configuration in the selected slice ($x_{tr}/c \approx 0.54$) is also prescribed as the transition location in the 2.5D TAU simulation.

The parameter combination that best approximates both the pressure distribution and the spatial evolution of the convective instabilities is $\alpha = 2.6°$, $Ma_\infty = 0.78$, and $\gamma = 14°$ (see Fig. 1).

The depicted N-factor curves in Fig. 1b are envelopes of n-factor curves representing different frequencies at a zero-degree wave angle (TS) or various total wavelengths at zero frequency (CF). The n-factor is a measure of the accumulated growth of a disturbance and is defined as $n(x) = ln(A(x)/A_0)$, where A_0 denotes the maximum amplitude, in the wall-normal direction, of the streamwise velocity component of the corresponding instability mode at the streamwise position where the disturbance starts to grow. The n-factor is set to zero if, during the calculation, the value of A becomes smaller than A_0.

Note that the CF instability is not amplified in the 2.5D flow field. However, in the 3D reference case at the joint location and the subsequent chord section, the TS waves are most amplified and likely to dominate the transition scenario in the selected line-in-flight cut. Therefore, in the present study, the instability analysis solely focuses on the spatial development of TS waves.

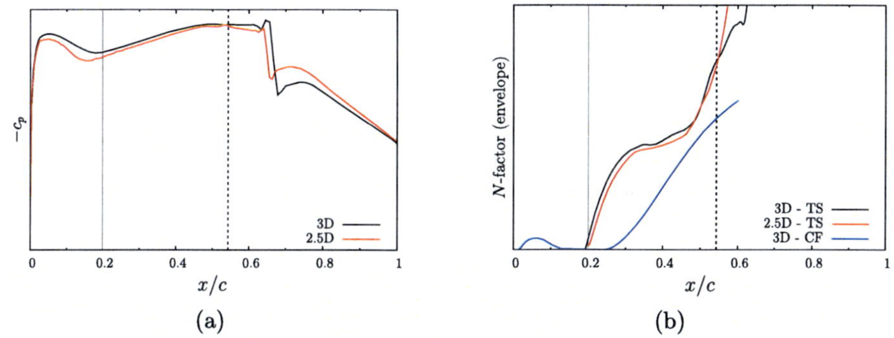

Fig. 1. Comparison of the original 3D flow solution at the selected line-in-flight cut with the approximated 2.5D solution in terms of (a) pressure distribution and (b) instability results. The vertical grey line represents the joint position, while the black dashed line indicates the prescribed LTT location in the base flow TAU computations. The 3D reference data have been provided by Martin Kruse (DLR).

2.3 Surface Irregularity

Once the infinite swept-wing 2.5D flow is defined (see Sect. 2.2), it is feasible to incorporate the joint geometry into the two-dimensional (2D) airfoil shape. DLR's structured grid generator MEGACADS [11] is used for this task. Three examples of irregularities are considered for representing the 2D geometry of the joint between the leading edge suction panel and the wing box. The BFS and the two gaps of different width resemble possible structural joint concepts currently considered within the HLFC-WIN project, and the choice of their geometrical dimensions was motivated by simple criteria similar to the classical Nenni and Gluyas criteria. Their actual geometrical characteristics are given in Table 1.

The geometrical shape is shown in the flow field visualizations in Fig. 2. The position of the surface irregularities is fixed to 20% of the chord for all cases considered.

2.4 Instability Analysis

The Adaptive Harmonic Linearized Navier-Stokes (AHLNS) approach [5], in combination with the Parabolized Stability equations (PSE) [4], offers the most efficient way to compute the effect of surface irregularities on the spatial development of convective instabilities, e.g. TS waves and CF vortices, in terms of n-factor curves. The PSE are applied far from the location of the surface irregularity, where the streamwise variations of the base flow quantities remain small (w.r.t. wall-normal gradients). The AHLNS methodology is used in the vicinity of the irregularity, where the assumption of small variations in streamwise direction of the base flow variables is no longer valid. The efficiency of the AHLNS approach, in comparison with other standard methodologies as HLNS or DNS, lies in the *wave-like* character assumed for describing the convective instabilities, in a similar fashion as PSE. This assumption significantly reduces the number of streamwise grid points required in comparison with DNS computations [5], allowing parametric studies for transonic wings at high Reynolds numbers. An example of the combination of PSE and AHLNS for a parametric study of the effect of a surface irregularity in a compressible flow at a high Reynolds number can be found in the work of Franco et al. [9].

3 Results

3.1 Base Flow Computations

Figure 2 shows the contours of Mach number and two-dimensional streamlines for the three joints chosen. Figure 3 displays the non-dimensional pressure distribution p_{wall} along the airfoil's suction side for all cases studied here. Moreover, the pressure distribution for the clean configuration (label *"No irregularity"*) has been included as reference. The pressure is made non-dimensional by twice the incoming dynamic pressure, i.e. $\rho_\infty \left(U_\infty\right)^2$.

The presence of the backward-facing step introduces an elongated recirculation bubble (see Fig. 2a) with an extension in streamwise direction significantly

Table 1. Surface irregularities (joint geometry) considered in the present work

Name	Type	Depth [mm]	Width [mm]
BFS	Backward-facing step	0.2	–
Gap 1	Gap	0.8	1
Gap 2	Gap	0.8	3

larger than the step height. Regarding the pressure distribution at the wall p_{wall} (Fig. 3), the existence of the BFS is substantially more influential than any of the other two irregularities. Clearly, the pressure distribution for the BFS case deviates from the clean configuration in a much larger extension than the other two gaps considered here. This fact suggests that the BFS might affect the spatial development of the incoming TS waves more significantly than the two gap cases.

On the other hand, it is interesting to notice that the recirculation flow due to the presence of a gap (see Figs. 2b and 2c) does not deflect significantly the streamlines coming upstream of the gap. This fact is confirmed by checking the pressure distribution at the wall p_{wall} (Fig. 3) for both gap cases. The pressure distribution deviates from the clean configuration case only in the close vicinity of the gap location. It is also important to mention the peak in the pressure distribution that appears at the location of each gap. They represent the rapid changes in streamwise direction of the base flow magnitudes provoked by the geometry of the gap and justify the necessity of using the AHLNS approach for the subsequent stability analysis (as it was explained in Sect. 2.4).

3.2 *N*-Factor Curves

A preliminary instability analysis study (not included in the present paper) for the 2.5D clean configuration (i.e. no geometrical joint incorporated into the shape of the airfoil) was done using the PSE code NOLOT [4] for TS instabilities. This study indicated that the most relevant n-factor curves (each one obtained for a particular spanwise wavenumber β and frequency f) were found in the parametric space $\beta \in [200\text{–}400]\ \mathrm{m}^{-1}$ and $f \in [1000\text{–}6000]$ Hz.

The N-factor curve, envelope of all particular n-factor curves computed in the above defined parametric space (β, f), is depicted in Fig. 4 with the label *No irregularity* for the case where the joint does not introduce any surface irregularity. The small increase present for the *No irregularity* case at $x/c \approx 0.20$ is due to the sudden change in boundary conditions for the base flow: from a given suction flow rate at the leading-edge suction panel to a zero-flow rate at the wing box. In the same plot, the N-factor curves for the three joint geometries considered in this study are also included. Clearly, the BFS is the surface irregularity that introduces a larger distortion in the spatial development of the incoming TS waves. This result agrees with the explanation provided for the base flow in Sect. 3.1. However, in the proximity where the transition was fixed for the steady base flow computations (i.e. $x/c \approx 0.54$, see Sect. 2.2), the amplification curves for the BFS and the clean configuration tend to collapse. This result might imply that the BFS considered here does not change the expected transition location significantly (with respect to the expected LTT location of the clean configuration).

For the two gap cases (Gap 1 and Gap 2), their influence on the incoming TS waves is manifestly smaller than for the BFS case. For the Gap 1, the N-factor envelope curve almost overlaps with the amplification curve of the clean configuration along the whole profile. Therefore, it can be concluded that, in terms

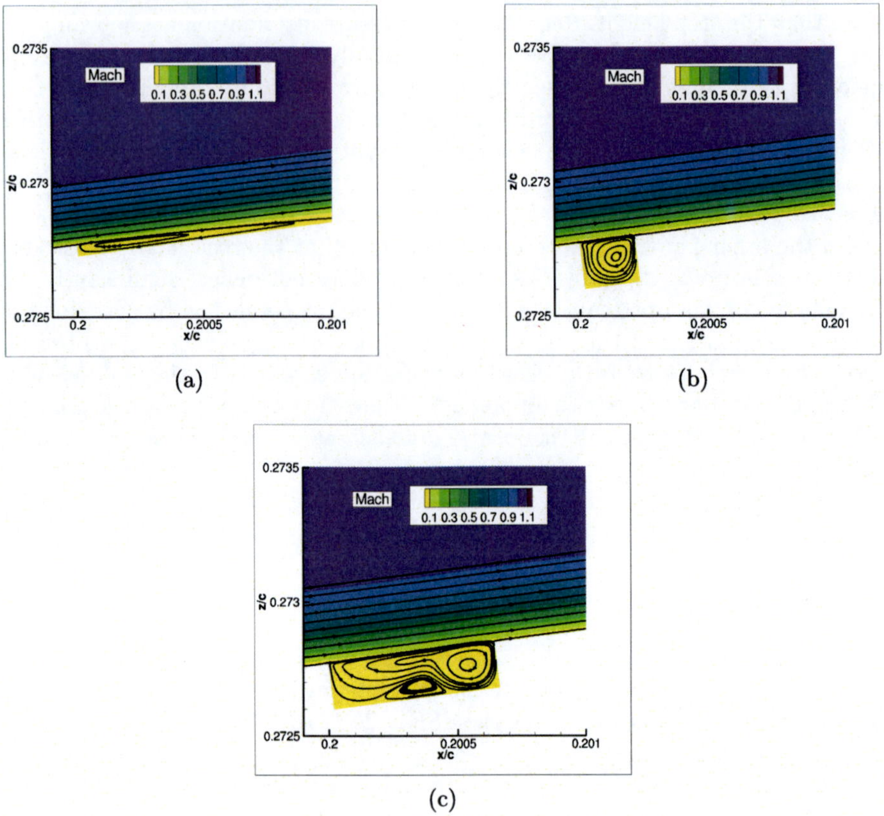

Fig. 2. Steady laminar base flow results. Mach number contours and two-dimensional streamlines for cases (a) BFS; (b) Gap 1; (c) Gap 2.

of stability analysis, the presence of such irregularity is negligible. For the Gap 2, although its width is three times larger than the Gap 1 width, the situation is similar. The N-factor envelope curve deviates from the clean configuration case only in a very small region downstream of the gap location and, after that, both amplification curves coincide. Consequently, it can also be stated that the influence of Gap 2, in terms of expected transition location, is not significant.

Finally, Fig. 5 shows the spatial evolution of an incoming TS wave of frequency $f = 2000$ Hz and spanwise wavenumber $\beta = 200\,\mathrm{m}^{-1}$ passing along the Gap 2 case. The body-attached coordinates are defined as: s represents the arc-length along the airfoil surface (with origin at the leading edge), and η holds for the surface-normal coordinate. For this particular instability, the streamwise wave length is about 6 cm, while the gap width for the surface irregularity is 3 mm

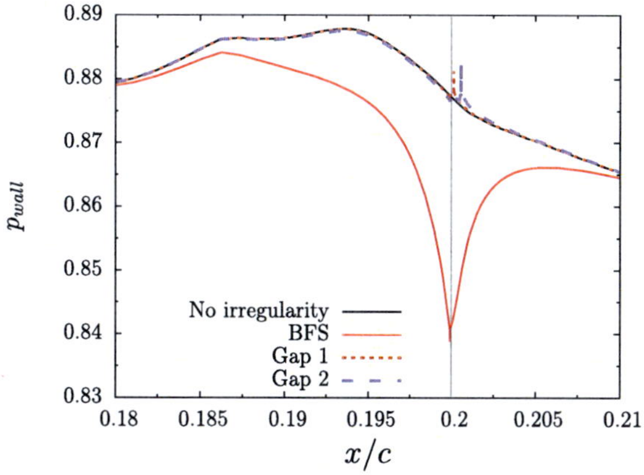

Fig. 3. Steady laminar base flow results. Non-dimensional pressure distribution at the wall p_{wall}. The vertical grey line represents the location of the joint between the leading edge suction panel and the wing box.

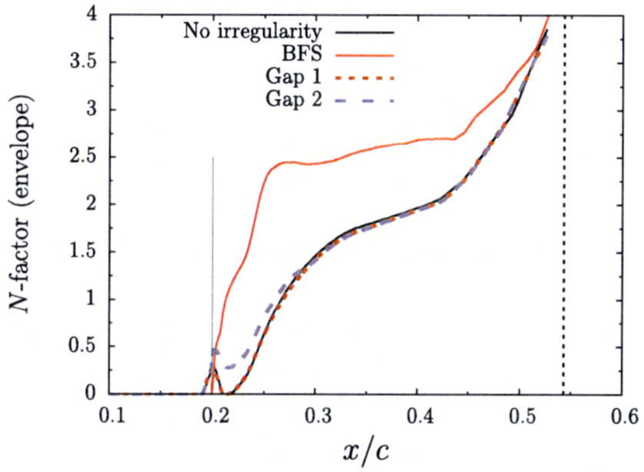

Fig. 4. N-factor envelope curves for all joints considered in the present work. The clean configuration (*No irregularity*) has been included as reference. The vertical grey line represents the location of the joint between the leading edge suction panel and the wing box. The vertical black dashed line indicates the prescribed LTT location in the base flow TAU computations.

(see Table 1). This large difference in streamwise length scales between surface irregularity and instability mode also helps to understand the small influence of the two gap cases considered regarding the expected transition location.

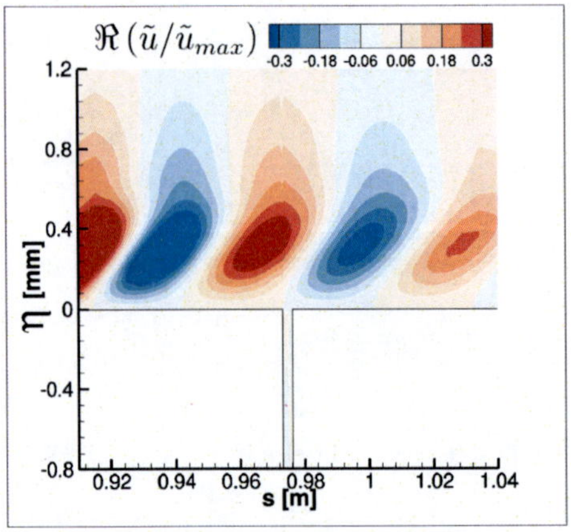

Fig. 5. Real part of the streamwise velocity disturbance $\tilde{u}$ for an incoming TS wave of frequency $f = 2000$ Hz and spanwise wavenumber $\beta = 200\,\mathrm{m}^{-1}$ at an arbitrary instant of time for the Gap 2 case. Body-attached coordinates (s,η) [axes are not to scale].

4 Conclusions

The effect of different two-dimensional surface irregularities on the expected laminar-turbulent transition location on a realistic HLFC wing has been analyzed by means of the AHLNS methodology for the first time. The combined effect of infinite swept-wing configuration, airfoil geometry, transonic flow conditions, and suction represents a qualitative step forward with respect to previous applications of the AHLNS approach, where only zero-pressure gradient and flat-plate configurations were considered. It could be demonstrated that, with the toolchain established within the HLFC-WIN project, it is now possible to take into account the actual laminar flow field and the respective linear stability characteristics when assessing the effect of surface irregularities on laminar-turbulent transition for such configurations, in contrast to the simple criteria of Nenni and Gluyas based on the freestream conditions and a single geometrical dimension only. Thus, after a successful validation against suitable experimental data, the present approach may prove to be much more reliable when used for the specification of surface allowables compared to the simple but much faster classical criteria.

The three surface irregularities studied in the present paper resemble possible structural joint concepts currently considered within the HLFC-WIN project. The choice of their geometrical dimensions was motivated by the maximum tolerances estimated by simple criteria similar to the classical Nenni and Gluyas criteria. For the freestream conditions and suction distribution considered, none of the surface irregularities studied here are expected to have an influence on the transition location. However, the BFS caused a significant rise in the N-factor envelope in the immediate downstream vicinity of the irregularity. Therefore, for flow conditions at which the N-factors at the BFS are higher already, the situation may be different. Hence, further studies at other flow conditions, other dimensions of the surface irregularities, and other chord positions of the joint should be performed in order to check whether the surface requirements for the present HLFC-WIN configuration can indeed be relaxed compared to the rather challenging constraints that were originally imposed using the simple criteria. Moreover, the AHLNS approach is not limited to ideally sharp rectangular geometries, as it was demonstrated in Franco et al. [9], and could also be used e.g. for gap cases with a certain vertical offset between the upstream and downstream edges of the gap. Thus, the present approach can be used to study more realistic joint configurations not covered by the classical criteria for single rectangular FFS/BFS and simple rectangular gaps.

Acknowledgments. The authors would like to thank Martin Kruse (DLR) for providing the complete 3D base flow computation of the HLFC wing and the TAU version for the base flow simulations including suction, and Thomas Haase (DLR) for his careful review of our manuscript and his insightful comments. We also acknowledge valuable contributions from all partners of the HLFC-WIN consortium: Aernnova, DLR, Onera, Sonaca and Airbus. This project has received funding from the Clean Sky 2 Joint Undertaking (JU) under grant agreement No. 945583. The JU receives support from the European Union's Horizon 2020 research and innovation programme and the Clean Sky 2 JU members other than the Union. The results, opinions, conclusions, etc. presented in this work are those of the author(s) only and do not necessarily represent the position of the JU; the JU is not responsible for any use made of the information contained herein.

References

1. HLFC-WIN project. https://www.hlfc-win.eu/
2. Reneaux, J.: Overview on drag reduction technologies for civil transport aircraft. In: ECCOMAS (2004)
3. Methel, J., Forte, M., Vermeersch, O., Casalis, G.: An experimental study on the effects of two-dimensional positive surface defects on the laminar-turbulent transition of a sucked boundary layer. Exp. Fluids **60**, 94 (2019)
4. Hein, S., Bertolotti, F.P., Simen, M., Hanifi, A., Henningson, D.: Linear nonlocal instability analysis - the linear NOLOT code. Internal Report No. DLR IB 223–94 A56 (1995)
5. Franco Sumariva, J.A., Hein, S.: Adaptive harmonic linearized Navier-Stokes equations used for boundary-layer instability analysis in the presence of large streamwise gradients. In: AIAA Paper 2018-1548 (2018)
6. Nenni, J.P., Gluyas, G.L.: Aerodynamic design and analysis of an LFC surface. Astron. Aeronaut. **4**, 52–57 (1966)
7. Schrauf, G.: COCO – a program to compute velocity and temperature profiles for local and nonlocal stability analysis of compressible, conical boundary layers with suction. ZARM Technik report (1998)
8. Schrauf, G.: LILO 2.1 user's guide and tutorial. GSSC Technical report 6, originally issued Sep. 2004, modified for Version 2.1 (2006)
9. Franco Sumariva, J.A., Hein, S., Valero, E.: On the influence of two-dimensional hump roughness on laminar-turbulent transition. Phys. Fluids **32**, 034102 (2020)

10. TAU-Code User Guide. DLR. v.2017.1.0 (2017)
11. Brodersen, O., Ronzheimer, A., Ziegler, R., Kunert, T., Wild, J., Hepperle, M.: Aerodynamic Applications using MegaCads. In: 6th International Conference on Numerical Grid Generation (1998)

Transition Prediction and Analysis of the CRM-NLF Wing with the DLR TAU Code

Sebastian Helm[1(✉)], Michael Fehrs[1], Normann Krimmelbein[2], and Andreas Krumbein[3]

[1] DLR, Institute of Aeroelasticity, 37073 Göttingen, Germany
sebastian.helm@dlr.de
[2] DLR, Institute of Aerodynamics and Flow Technology, 38108 Braunschweig, Germany
[3] DLR, Institute of Aerodynamics and Flow Technology, 37073 Göttingen, Germany
https://www.dlr.de/ae/

Abstract. Promising results are obtained by NASA with the CRM-NLF laminar wing at design conditions. The work presented seeks to develop an understanding of the transition mechanisms involved with the CRM-NLF design under wind tunnel conditions. In addition, the capabilities of DLR tools for RANS-based transition prediction are demonstrated. Particular attention is paid to the effect of turbulent wedges. Numerical results are compared to wind tunnel data made available by NASA.

Keywords: Transonic natural laminar flow · Turbulent wedges

1 Introduction

The CRM-NLF design [1] might be a major step towards the applicability of laminar flow to commercial transport aircraft. It aims at enabling extensive natural laminar flow (NLF) over long range aircraft wings, typically exhibiting leading edge sweep angles beyond 30°. Crossflow (CF) transition is attenuated by an unconventional airfoil curvature at the leading edge [2]. This concept, referred to as CF Attenuated NLF (CATNLF), is considered as a relevant and promising candidate for future laminar transport aircraft design.

The effectiveness of the concept was proven by wind tunnel test in the National Transonic Facility (NTF) [3,4]. Temperature Sensitive Paint (TSP) data reveals laminar areas as well as turbulent wedges. CFD-based numerical simulations are used to understand the transition mechanisms and wind tunnel effects that lead to the experimental data acquired.

The unconventional pressure distribution resulting from the CATNLF design is characterized by an aggressive acceleration of the flow in the vicinity of the leading edge. This is achieved using relatively small leading edge radii. Furthermore, the stabilizing effect of compressibility on Tollmien-Schlichting (TS)

© The Author(s), under exclusive license to Springer Nature Switzerland AG 2021
A. Dillmann et al. (Eds.): STAB/DGLR Symposium 2020, NNFM 151, pp. 185–194, 2021.
https://doi.org/10.1007/978-3-030-79561-0_18

waves is exploited leading to diminished need for acceleration on the wing upper side up to the recompression shock compared to a conventional NLF design. In addition, the CRM-NLF model involves more spanwise variation of the flow characteristics than current NLF wings, e.g. the DLR TuLam wing [5]. It is the objective of this work to develop an understanding of these flow features with regard to future aeroelastic analysis [6].

2 Simulation Set-Up

The CRM-NLF geometry is provided by NASA [7]. A CFD grid of the wing-body half model is generated using the commercial software CENTAUR. The computational grid of the wing-fuselage half-model consists of approximately $29.5 \cdot 10^6$ nodes. The CFD domain is bounded by a semi-sphere with a radius of 100 times the semi-span. The boundary layer grid on the upper and lower wing sides consist of 450 nodes in spanwise direction and 300 nodes in chordwise direction (upper side) with 90 structured layers to resolve the wing boundary layer. The maximum y^+ value is about 0.5 throughout this investigation.

The DLR TAU code is used for the RANS simulations. The SST k-ω turbulence model [8] is applied for fully turbulent computations and for the turbulent part of the transitional flow. Transition prediction is based on the Transition Prediction Module [9] of the TAU code that applies the linear stability code LILO [10] and the e^N method.

On the lower side of the wing a numerical transition tripping is applied close to the leading edge according to the wind tunnel test [11]. For these fixed transition conditions, the turbulence model is activated at a user-defined location. Transition prediction is applied to at least 20 line-of-flight (y=const.) cuts equally distributed on the upper side of the wing. The boundary layer profiles required by LILO are determined by the compressible conical laminar boundary layer code COCO [12]. The default integration strategy (prescribed-frequency and prescribed-propagation-direction for TS waves, prescribed-frequency and prescribed-wavelength for stationary crossflow waves) is used. The transition location is determined where the compressible N-factors exceed the critical value $N_{crit} = 6$ for both CF and TS transition according to Lynde et al. [3].

Numerical results are compared to wind tunnel data acquired in the National Transonic Facility [3,4]. The pressure data provided by NASA contain a correction for wind tunnel effects in order to obtain the best possible comparison to free-flight CFD simulations. The pressure orifices are aligned with the flow. A peniche is used to reduce the influence of the boundary layer at the tunnel wall. Four test cases are considered. The angle-of-attack α, Mach number M and Reynolds number Re_{MAC} based on the mean aerodynamic chord $c_{Ref} = 364.29$ mm (14.342 in.) are given in Table 1. The free-stream turbulence intensity $Tu = 0.24\%$ was determined by Crouch et al. for the NTF [13].

Table 1. Test cases according to ref. [7]

Test case	$\alpha/°$	M	$Re_{MAC}/10^6$	T_{total}/K
1	1.44848	0.856489	14.97197	277.108
2	1.98031	0.856491	14.94591	277.453
3	2.46141	0.856051	14.90909	277.910
4	2.93787	0.855801	14.85308	278.645

3 Results

3.1 Comparison of Numerical and Experimental Data

Extensive laminar flow on the wing upper side (regions of low skin friction on the left side of Fig. 1) is predicted by the DLR TAU code when laminar-turbulent transition is free. The corresponding TSP image acquired in the NTF [7,14] (middle of Fig. 1) shows numerous turbulent wedges typical for the given flow conditions. But apart from that, considerable amounts of brighter areas can be distinguished proving the effectiveness of the CATNLF design. The precise determination of a transition line from the TSP images is an intricate task conducted by Krimmelbein et al. [15]. The present investigation uses an approximated transition line derived from the TSP image, thus, including turbulent wedges. This fixed transition line and the skin friction pattern resulting from the corresponding simulation is shown in Fig. 1 (right).

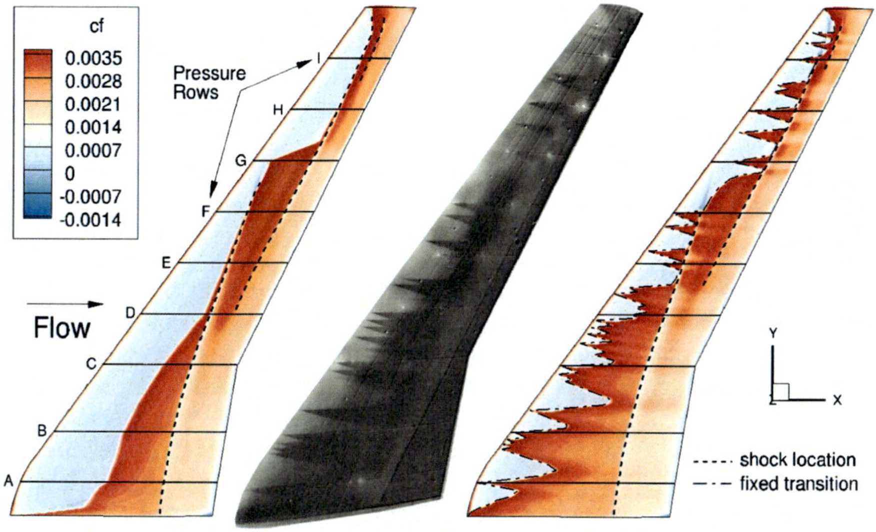

Fig. 1. Skin friction distribution at free transition (left) and fixed transition (right); TSP image [7] (middle); Test case 1 (nominal $\alpha = 1.5°$)

The flow topology at an angle-of-attack $\alpha = 1.5°$ is characterized by a double-shock system on the suction side of the wing. One shock occurs at the inner part of the wing span up to row G and another shock develops further downstream at the outer half of the wing span (dashed lines in Fig. 1). This feature leads to the discontinuity of the transition front near row G, which can be observed in the numerical results as well as in the TSP image. Towards the higher angles-of-attack ($\alpha = 2.5°$ and $3.0°$) the shocks are merging.

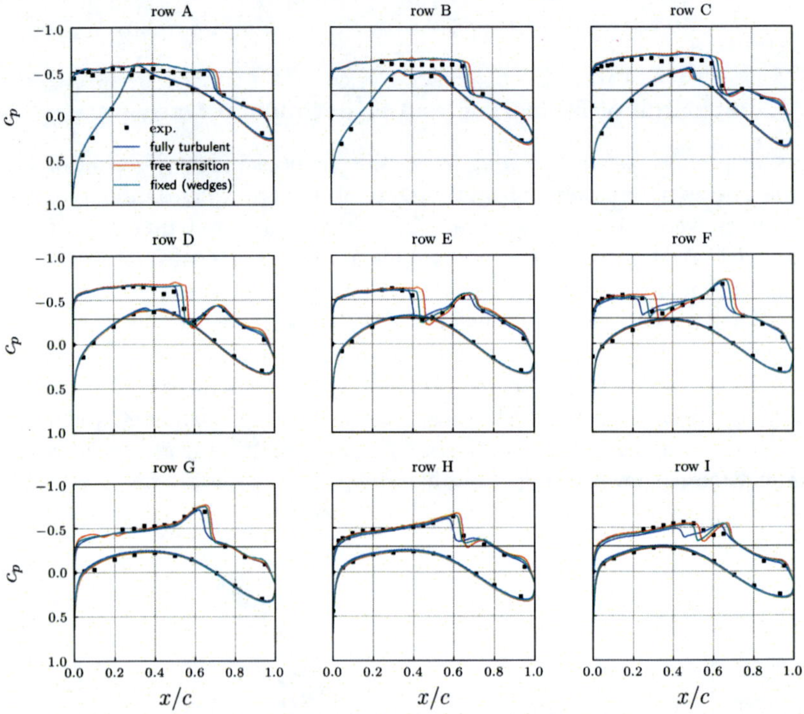

Fig. 2. Pressure distribution at measurement sections, Test case 1 (nominal $\alpha = 1.5°$)

A comparison of experimental and numerical pressure distributions (Fig. 2) shows that the shock is located further downstream in the case of free transition than in the experiment, especially at mid wing. In contrast to that, the fully turbulent simulation agrees better with the pressure measurements, although TSP images suggest a considerable laminar area. This inconsistency can be explained partially by turbulent wedges, which originate in particular at the pressure orifices. The present modeling of turbulent wedges (fixed transition) results in a forward shift of the shock position compared to the clean case (free transition). This is most apparent at the inner shock due to the presence of more turbulent wedges.

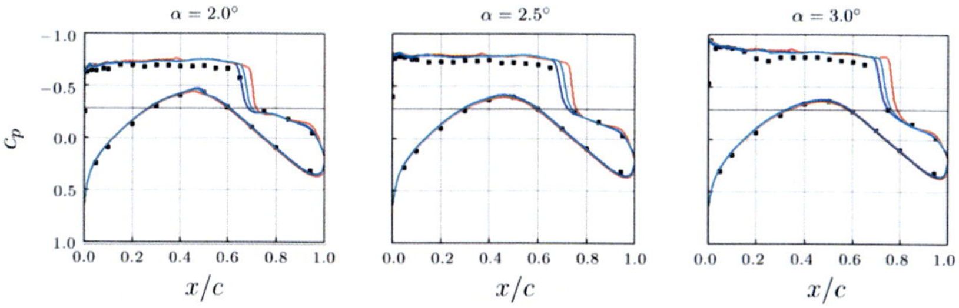

Fig. 3. Pressure distribution at row C, Test cases 2 to 4

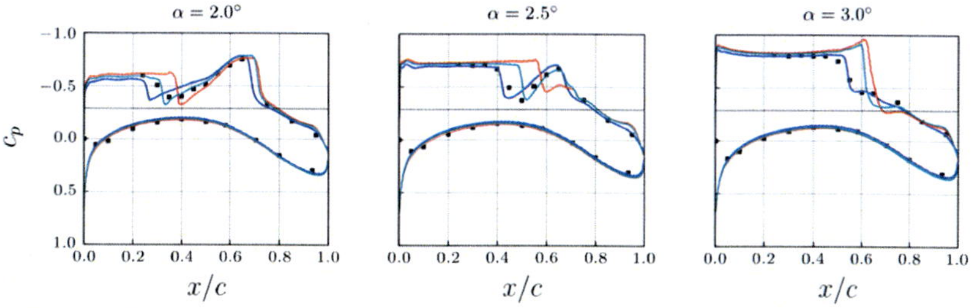

Fig. 4. Pressure distribution at row G, Test cases 2 to 4

The consideration of turbulent wedges leads to an upstream shift of the shocks by 3% of the chord (averaged over all test cases). As such, it is comparable to the actual effect of laminarity on the inner shock position. In other words, the shock at fixed-transition is located 3.5% of the chord downstream compared to the fully turbulent case. It is important to note that the relatively low acceleration (near zero pressure gradient) ahead of the shock leads to a high sensitivity of the shock location compared to experience from conventional laminar wings, e.g. the DLR TuLam wing [5]. Sensitivity studies with artificially larger laminar area up to the shock (not shown) show that the effects of laminarity and turbulent wedges can lead to a shift in the shock position on the order of 5–7% (averaged over the wingspan).

The measurement rows C (Fig. 3) and G (Fig. 4) are selected for more detailed study for the aerodynamic mechanisms at the inner and the outer wing, respectively. At the inner wing transition is moving upstream with the angle-of-attack. This is in accordance with the TSP data.

Unsteady flow effects are reported from the experiment commonly at the outboard wing. Yet, it is unclear if these are caused by inherently unsteady aerodynamics, unsteady inflow conditions or structural dynamics of the model. The resulting variation detected by the pressure orifices is shown in Figures 30 and 31 in reference [4]. Note that the static pressure orifices are not able to measure unsteady pressure fluctuations correctly. Furthermore, an averaging is applied to the pressure data. For comparison of the pressure distributions it is, therefore, important to note that these uncertainties are blurring the experimental pressure data in the vicinity of the shock.

The influence of the turbulence model on the shock position appears to be lower than the influence of laminar-turbulent transition. The Spalart-Allmaras turbulence model predicts the shock location on average about 2% more downstream than the Menter SST k-ω model (determined for test case 2 but not shown in this paper). The former turbulence model is used by Lynde et al. [3] whereas the latter is used throughout the present investigation.

A comparison of aerodynamic coefficients for the wing-fuselage configuration is shown in Fig. 5. It is apparent that there is inconsistency remaining. The numerical results with a transition front from TSP images giving best overall agreement with experimental pressure distributions exhibit an offset with respect to the global coefficients. The good agreement of fully turbulent simulation with the experiment is, therefore, considered as a coincidence.

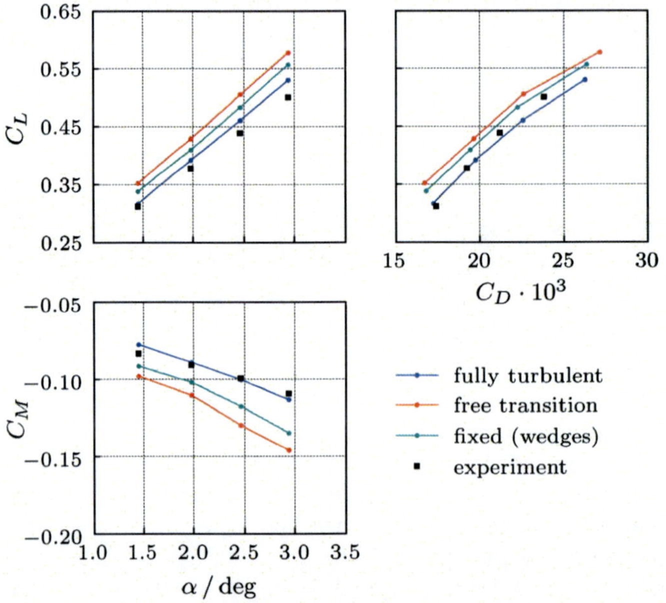

Fig. 5. Aerodynamic coefficients

A major uncertainty lies in the fuselage for which no detailed measurement data is available. The fuselage contributes to the global coefficients with $C_L^{FU} = 0.04$ and $C_M^{FU} = 0.02$. In addition, the juncture flow between fuselage and wing is known to be challenging for turbulence models.

The CFD model was built with the best practices for RANS simulation involving transition prediction. However, no systematic convergence study was performed. Another simplification that might lead to discrepancies is the way transition is realized in the numerical model. The abrupt activation of the turbulence model at transition cannot represent the complex physics involved with the real transition process. Stability analysis is performed on line-of-flight cuts, i.e. neglectable spanwise variation is assumed. This is, however, violated to some extend by the complex 3D base flow (double-shock) and local variations of the effective aerodynamic shape due to turbulent wedges. As pointed out by Streit et al. [16] the parallel flow assumption is another simplification to be revised, in particular for the unconventional leading edge curvature of the CRM-NLF wing.

Deformation of the wind tunnel model is another uncertainty. The CRM-NLF model is built with the objective to match the design geometry when subject to design loads. The varying loads throughout the test cases might lead to static structural deformation, in particular at the relatively high dynamic pressures set during the experiment.

3.2 Influence of Turbulent Wedges

This work does not attempt the detailed modelling of physics involved with turbulent wedges, but rather estimate their overall effect on the pressure distribution. It is well known that the reduced displacement thickness δ_1 of a transitional boundary layer can alter the aerodynamic forces even at high Reynolds numbers [17].

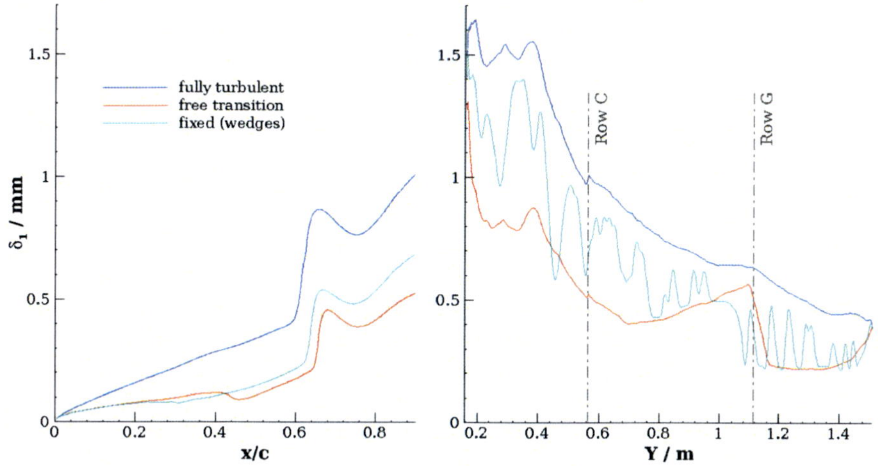

Fig. 6. Displacement thickness at row C (left) and at $x/c = 0.9$ (right)

In comparison to fully turbulent boundary layer flow the viscous decambering is reduced, i.e. the effective airfoil shape is affected. The difference in chordwise displacement thickness is shown in Fig. 6 (left) for row C (close to the mean aerodynamic chord). The displacement thickness is $\delta_1 \approx 1\,\text{mm}$ ($\delta_1/c \approx 0.3\%$)

near the trailing edge in the case of a fully turbulent boundary layer. Free transition is predicted at $x/c = 0.44$ for this section leading to approximately half the displacement effect. Further upstream transition due to each turbulent wedge leads to local increase of δ_1 clearly seen in the spanwise distribution in Fig. 6 (right).

3.3 Stability Analysis

Investigation of the flow quality in the NTF suggests a critical N-factor for TS instabilities in the range of 4 to 8 [13]. NASA uses a critical N-Factor of 6 (compressible linear stability theory) for both, TS and CF instabilities [4]. This strategy is adopted for the present work. The results of the stability analysis presented here are performed based on the test case 1. The fixed-transition case is used as it represents the best overall match with pressure data from the wind tunnel test.

The major transition mechanism is due to TS waves. At the inner wing transition occurs significantly upstream of the recompression shock. At row C the critical N-Factor of 6 is exceeded at x/c = 0.45 (Fig. 7). At midwing (between row D and G) transition occurs due to the pressure rise at the shock. This is also the case for the outer wing in general.

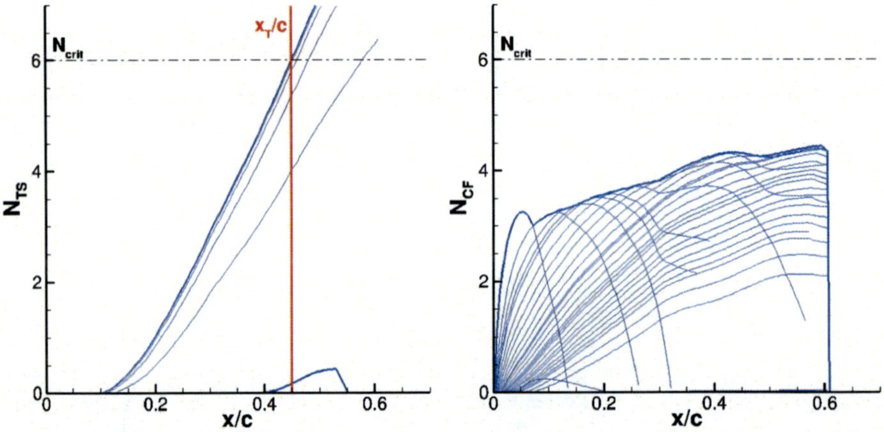

Fig. 7. N-factor growth at row C, Test case 1 (nominal $\alpha = 1.5°$).

However, at the angle-of-attack $\alpha = 1.5°$ CF-instabilities lead to transition slightly ahead of the shock. This can be see at row G (Fig. 8). This midchord CF is limited to a small region at the outer wing and appears only at test case 1 (in accordance with [3]). Besides this, the predicted N-factor curves do not reach the critical value for CF. At higher angles-of-attack ($\alpha = 2.0°$ to $3.0°$) the complete wing is predicted free of CF transition. In that sense, a design optimization using the DLR TAU code in combination with the CATNLF concept could lead to a similar result as the CRM-NLF design.

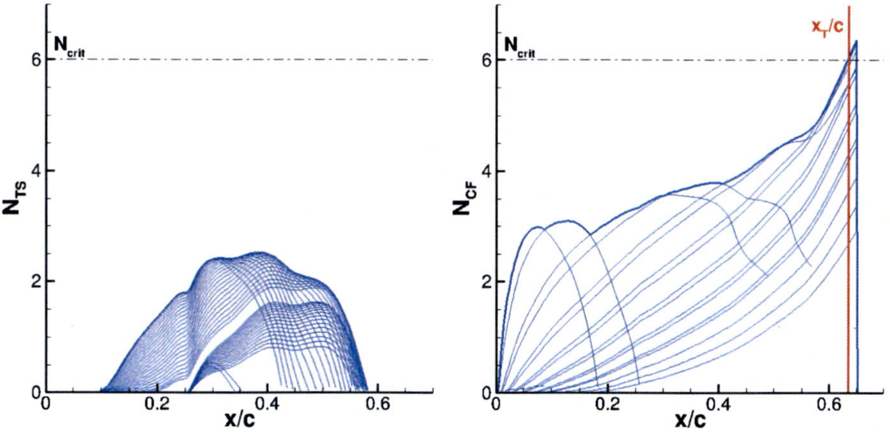

Fig. 8. N-factor growth at row G, Test case 1 (nominal $\alpha = 1.5°$).

4 Conclusion

The CRM-NLF design is analyzed with the DLR TAU code. Good agreement with wind tunnel data is obtained, in particular when turbulent wedges as captured by the TSP technique are taken into account.

The effect of turbulent wedges on the pressure distribution can be significant, i.e. as large as the actual effect of laminarity. Therefore, it should be taken into account for the CFD-based analysis of the wind tunnel test of the CRM-NLF model.

The capabilities of the DLR TAU code to represent the relevant transition mechanisms are demonstrated. The transition prediction agrees well with the results obtained by NASA.

Remaining uncertainties were mentioned. Future work will address static aeroelastic deformation as well as unsteady effects. This work aims at contributing to the understanding of laminar wings with the perspective of aeroelastic analysis.

References

1. Lynde, M.N., Campbell, R.: Computational design and analysis of a transonic natural laminar flow wing for a wind tunnel model. In: 35th AIAA Applied Aerodynamics Conference, AIAA 2017, Denver, p. 3058 (2017)
2. Campbell, R.L., Lynde, M.N.: Natural laminar flow design for wings with moderate sweep. In: 34th AIAA Applied Aerodynamics Conference, Washington, D.C. (2016)
3. Lynde, M.N., et al.: Preliminary results from an experimental assessment of a natural laminar flow design method. In: AIAA SCITECH Forum (2019)
4. Lynde, M.N., Campbell, R.L., Viken, S.A.: Additional findings from the common research model natural laminar flow wind tunnel test. In: AIAA AVIATION Forum (2019)

5. Seitz, A., Hübner, A., Risse, K.: The DLR TuLam project: design of a short and medium range transport aircraft with forward swept NLF wing. CEAS Aeronaut. J. **11**(2), 449–459 (2019). https://doi.org/10.1007/s13272-019-00421-1
6. Tichy, L.: Risk analysis for flutter of laminar wings. In: 17th International Forum on Aeroelasticity and Structural Dynamics, IFASD, Como (2017)
7. Model information on and data from the CRM-NLF wind tunnel test. https://commonresearchmodel.larc.nasa.gov/crm-nlf/. Accessed 28 Sep 2020
8. Menter, F.R.: Two-equation eddy-viscosity turbulence models for engineering applications. AIAA J. **32**(8), 1598–1605 (1994)
9. Krumbein, A., Krimmelbein, N., Schrauf, G.: Automatic transition prediction in hybrid flow solver, part 1: methodology and sensitivities. J. Aircr. **46**(4), 1176–1190 (2009)
10. Schrauf, G.: LILO 2.1 user's guide and tutorial. GSSC Technical report 6, Bremen (2006)
11. General information CRM-NLF wind tunnel test. Version 2. Updated February 10, 2020. https://commonresearchmodel.larc.nasa.gov/wp-content/uploads/sites/7/2020/02/CRM-NLF_Info_v2.pdf. Accessed 28 Sep 2020
12. Schrauf, G.: COCO - a program to compute velocity and temperature profiles for local and nonlocal stability analysis of compressible. Conical Boundary Layers with Suction, ZARM Technical report (1998)
13. Crouch, J., Sutanto, M., Witkowski, D., Watkins, A., Rivers, M., Campbell, R.: Assessment of the national transonic facility for natural laminar flow testing. In: 48th AIAA Aerospace Sciences Meeting including the New Horizons Forum and Aerospace Exposition, p. 1302 (2010)
14. Watkins, N., Goodman, K.Z., Peak, S.: Transition detection at cryogenic temperatures using a carbon-based resistive heating layer coupled with temperature sensitive paint. In: AIAA Scitech Forum, p. 2191 (2019)
15. Krimmelbein, N., Krumbein, A.: Determination of critical N-factors for the CRM-NLF wing. To be Published in New Results in Numerical and Experimental Fluid Mechanics XIII, STAB 2020. Springer (2021)
16. Streit, T.S., Seitz, A., Hein, S., Kunze, P.: NLF potential of laminar transonic long range aircraft. In: AIAA AVIATION Forum, p. 2748 (2020)
17. Fehrs, M.: Boundary layer transition in external aerodynamics and dynamic aeroelastic stability. Dissertation, TU Braunschweig (2018). ISSN 1434–8454, ISRN DLR-FB-2018-11, also NFL-FB 2017–27, Braunschweig

Determination of Critical N-Factors for the CRM-NLF Wing

Normann Krimmelbein[1](✉) and Andreas Krumbein[2]

[1] DLR, Institut für Aerodynamik und Strömungstechnik, Lilienthalplatz 7, 38108 Braunschweig, Germany
normann.krimmelbein@dlr.de

[2] DLR, Institut für Aerodynamik und Strömungstechnik, Bunsenstr. 10, 37073 Göttingen, Germany

Abstract. Temperature sensitive paint images from a wind tunnel campaign for the CRM-NLF geometry are used to extract experimental transition lines to be used to determine critical N-factors based on incompressible and compressible linear stability theory. Based on the e^N-method, transition locations are subsequently predicted for different flow conditions using RANS computations and the results are compared to the experiment.

Keywords: Critical N-factor · Transition prediction · Incompressible linear stability theory · Compressible linear stability theory

1 Introduction

To verify and validate a design method for airfoil shapes that lead to pressure distributions delaying transition on wings with large sweep angles and at high Reynolds number flows (CATNLF, Crossflow Attenuated Natural Laminar Flow [1]) a wind tunnel test campaign in the National Transonic Facility (NTF) at the NASA Langley Research Center has been conducted [2]. The experiments, also aiming at evaluating best practices for laminar flow testing and the general suitability of the National Transonic Facility for laminar flow investigations, where carried out for the flow around a semi-span model of the Common Research Model with Natural Laminar Flow (CRM-NLF). Experimental transition locations have been documented during the tests using temperature sensitive paint (TSP) images by applying different methods, namely a temperature step method and a heating layer approach [3]. The experimental results for the Mach number $M = 0.856$ and the Reynolds number $Re_{MAC} = 15 \times 10^6$ based on the mean aerodynamic chord (MAC) at the angles of attack $\alpha = 1.5°, 2.0°, 2.5°$, and $3.0°$ are used in the present work to extract transition lines for the different flow conditions and to determine critical N-factors for the CRM-NLF wing.

© The Author(s), under exclusive license to Springer Nature Switzerland AG 2021
A. Dillmann et al. (Eds.): STAB/DGLR Symposium 2020, NNFM 151, pp. 195–204, 2021.
https://doi.org/10.1007/978-3-030-79561-0_19

2 Approach

The purpose of this paper is to provide an answer to one of the many questions that arose in the discussions around the CATNLF approach, its suitability for an effective NLF design of a long-range transport aircraft wing with large sweep and the question if it is appropriate to execute the necessary LST analysis in a compressible manner as was done in the underlying approach the CRM-NLF wing design is based on. Compressible LST is favored by NASA [2] and Boeing [5], whereas incompressible LST is the trusted approach at DLR [4] and Airbus. One major point in the discussions which were held mainly on workshops and conferences in the US was the question if the execution of a compressible LST analysis (in contrast to an incompressible analysis) is needed so that an appropriate CATNLF design as achieved for the CRM-NLF wing can be found in the design process. A prerequisite for answering this question is the investigation of how the execution of compressible and incompressible LST analysis might affect the outcome of predicted transition lines for a given configuration given that the correspondingly calibrated critical N-factors are available and consistently applied in the LST analyses. In this respect, the calibration of the critical N-factors is an aspect of major importance in this paper. In contrast to the data from [2, 5] in the present approach also critical Tollmien-Schlichting N-factors from incompressible LST are presented for the NTF, data that have not been published before. In addition, in the present paper the calibration process is described in detail.

To support and extend the existing evaluation of the critical Tollmien-Schlichting and cross flow N-factors for the CRM-NLF at the National Transonic Facility [2, 5], in the present work transition lines have been extracted from TSP images for different angles of attack and stability computations using incompressible and compressible linear stability theory (LST) have been carried out.

The temperature sensitive paint (TSP) images have been obtained during the test campaign in the National Transonic Facility at the NASA Langley Research Center using different transition detection approaches and image-enhancement methods [3]. For transition detection a heating layer approach has been used which has been supported by the results of a more traditional temperature step method. Due to a reduced contrast in the images, a Contrast Limited Adaptive Histogram Equalization technique has been applied to the images of the heating layer approach [3]. Transition lines have been determined by visual inspection of the experimental TSP images from both transition detection approaches. For this, the temperature step and heating layer images have been manually adjusted in terms of, for example, color levels and brightness to identify a clear as possible transition front. The parts of the transition front that have been determined with a certain amount of confidence have been connected by linear interpolation to give a continuous transition line. Interpolation was in particular needed for the unclear transition situation downstream of turbulent wedges or in regions where a lesser contrast of the TSP images prevented a clear identification of the transition front.

The stability computations are based on LST in form of the e^{N-}-method, utilizing the DLR TAU code, DLR's compressible unstructured RANS solver for external flows, and its transition prediction capabilities [4]. Transition prediction in the TAU code is realized through an automated process chain as outlined in detail in [4] consisting of the RANS solver, a procedure for providing the laminar boundary-layer data and a fully-automated

LST code. The boundary-layer data are either provided by a laminar boundary-layer code or extracted directly from the RANS grid in cases where the boundary-layer code cannot provide data, for example, downstream of transonic shocks. The prediction of transition onset locations is realized using the e^N-method and a standard 2-N-factor approach applying critical N-factors for Tollmien-Schlichting (T-S) waves, N_{TS}, and crossflow (CF) vortices, N_{CF}. For the computation of incompressible and compressible amplification rates and N-factors the prescribed-frequency/prescribed-propagation-direction ($\psi = 0°$) integration strategy for the calculation of Tollmien-Schlichting amplifications and the prescribed-frequency (F = 0 Hz)/prescribed-wavelength integration strategy for the calculation of cross-flow amplifications is used. That is, the standard N_{TS}-N_{CF} 2-N-factor approach is applied assuming the highest T-S amplification occur for T-S waves propagating in the direction of the inviscid flow at the boundary-layer edge and stationary CF amplifications as the highest amplified cross-flow vortices. In a first step, critical N-factors have been determined based on boundary-layer code computations along line-in-flight cuts of the wing at different angles of attack. To verify the critical N-factors, in a second step computations with automatic transition prediction have been performed, now with boundary layer profiles for the stability analysis directly extracted from the RANS solution.

A 30×10^6 nodes unstructured computational grid [7] has been used for all computations, with a resolution of the wing surface of approximately 450 grid points in span-wise direction and approximately 300 grid points in streamwise direction. The boundary layer above the wing surface is resolved with 90 structured layers with an initial spacing leading to y^+ values below 0.5 [7]. For the turbulent parts of the flow the negative Spalart-Allmaras (SA-neg) one-equation turbulence model is applied.

3 Approach

The extracted transition lines (solid yellow and orange lines in Figs. 3, 4, 5 and 6), which are based on the basic visual inspection of the TSP images, are extended by linear interpolations (dashed yellow and orange lines in Figs. 3, 4, 5 and 6) at span-wise locations where the actual natural transition front is concealed by turbulent wedges or the contrast in the TSP images complicate a clear identification of the transition front. The wing span of s = 1.344 m extends from y = 0.185 m to y = 1.529 m. As anticipated, both extracted transition lines, from temperature step method (yellow lines in Figs. 3, 4, 5 and 6) and from heating layer method (orange lines in Figs. 3, 4, 5 and 6), are very similar in location and also match those transition locations derived by other authors [2, 6] (magenta lines in Figs. 3, 4, 5 and 6).

Evaluation of the N-factor for Tollmien-Schlichting instabilities at the extracted transition lines reveals an averaged critical Tollmien-Schlichting N-factor of $N_{TS} = 11.7$ for an incompressible stability analysis and a value of $N_{TS} = 4.1$ for a compressible stability analysis (Fig. 1). In addition to the considerable scatter of the predicted N-factors that occurs along the whole wing span and which is very typical for the calibration of critical N-factors based on measurement data there are further uncertainties in the results in the span region with $y > 1$ m (Fig. 1). Here, the evidently clearly identifiable transition fronts often correspond to the shock fronts on the wing, which slightly deviate from the experiment for the computations. This leads to the additional uncertainties in the predicted N-factors (for y > 1 m) when either the shock location in the experiment is located further upstream compared to the CFD computation, resulting in lower values of the N-factor, or the shock location in the experiment is located further downstream compared to the CFD computation, resulting in no values for the N-factor because transition is not detected upstream of the shock. The shock fronts along the wing span can be identified by the grey-shade change from dark to light grey in the left and middle depictions of Figs. 3, 4, 5, and 6 starting at the wing root at about x/c = 0.75. For the angles of attack $\alpha = 1.5°$ and 2.0° in Fig. 3 and Fig. 4 a double-shock system exists with the second shock occurring on the outboard wing along the range from about 75% to about 90% of the wing span.

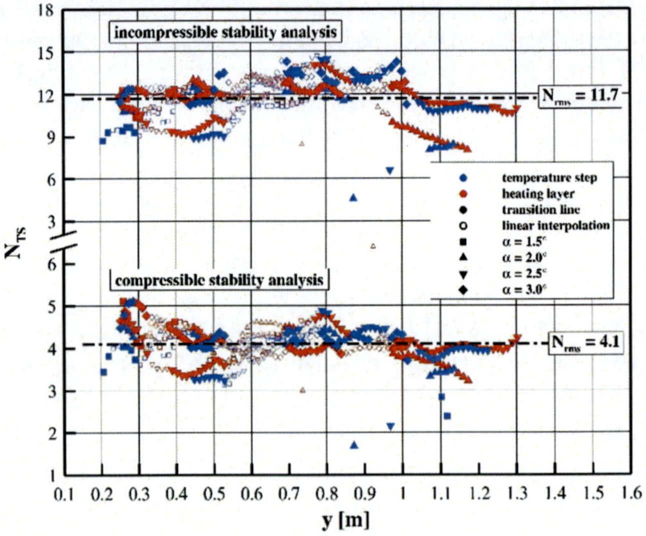

Fig. 1. CRM-NLF wing, Tollmien-Schlichting N-factors at transition along the wing span.

The present circumstance is caused by the application of the boundary-layer code for the computation of the boundary-layer profiles for the stability analysis, which does not provide a solution for separated flows (i.e. a short distance downstream of the strong pressure increase due to the shock). Comparing the compressible stability analysis results, the predicted critical Tollmien-Schlichting N-factors at transition are somewhat lower than used in other stability investigations [2] ($N_{TS} = 4.1$ vs. $N_{TS} = 6.0$) but are still just within the range of critical Tollmien-Schlichting N-factors determined for the flow quality of the National Transonic Facility [5] ($N_{TS} = 4.0 - 8.0$).

Significant cross-flow amplifications that may lead to transition are only present for the lowest angle of attack of $\alpha = 1.5°$. Here, a value of approximately $N_{CF} = 6.0$ is attained for incompressible as well as compressible stability analysis in the mid-span and outboard region of the wing (Fig. 2). However, whenever the cross-flow N factor of $N_{CF} = 6.0$ is reached also the critical Tollmien-Schlichting N-factor is reached (mid-span, Fig. 2) or would be reached a very short distance further downstream due to the existence

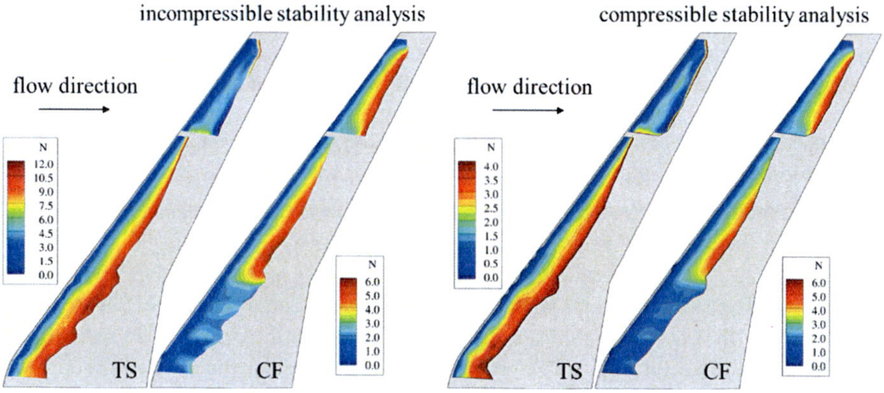

Fig. 2. CRM-NLF wing, incompressible (left) and compressible (right) Tollmien-Schlichting (TS) and cross-flow (CF) N-factors upstream of predicted transition locations, $\alpha = 1.5°$.

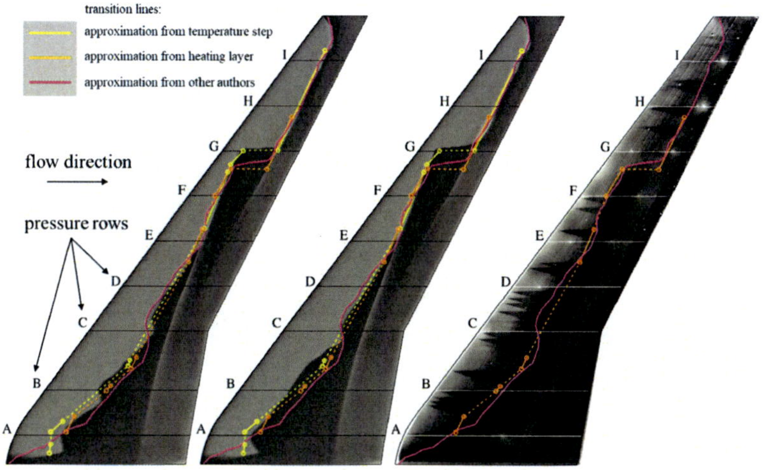

Fig. 3. CRM-NLF wing, comparison of predicted transition fronts, $\alpha = 1.5°$. Left: RANS surface c_f-distribution, incompressible LST. Middle: RANS surface c_f-distribution, compressible LST. Right: TSP image, heating layer method.

of a shock (mid-span and outboard, not shown here, visible for computations where cross-flow transition is not taken into account). As a result, computations with both, Tollmien-Schlichting and cross-flow amplifications considered or computations only taking into account stream-wise instabilities deliver nearly indistinguishable transition lines, leaving it open, if cross-flow instabilities play a role for $\alpha = 1.5°$. For all other angles of attack considered here the cross-flow N-factor never reaches $N_{CF} = 6.0$.

Looking at the predicted transition locations in Figs. 3, 4, 5, and 6 and comparing incompressible (left) with compressible (middle) stability results, it can be seen, that, despite the large difference in critical N-factor, the numerical results are very similar for both methods. Laminar and turbulent regions of the CFD computations are visualized in these figures by the skin friction distribution from the RANS simulations, with light grey areas corresponding to low values of the skin friction coefficient, and hence to laminar flow, and dark grey areas corresponding to higher values of the skin friction coefficient, and hence to turbulent flow. Significant differences between the predicted transition fronts from incompressible and compressible stability analysis can be seen for $\alpha = 2.0°$ (Fig. 4) at the outboard section near the wing tip (around pressure row I) where with incompressible theory the extent of laminar flow is increased compared to compressible theory. Another difference can be seen for $\alpha = 2.5°$ (Fig. 5) at the mid-span region between pressure rows D and F, where a larger extent of laminar flow is achieved for the compressible approach. As already partly visible for $\alpha = 2.5°$ for $\alpha = 3.0°$ (Fig. 6) with incompressible stability analysis transition is predicted upstream of the shock location in the outboard region of the wing (pressure rows F–I), whereas for compressible stability theory transition is predicted at the shock location.

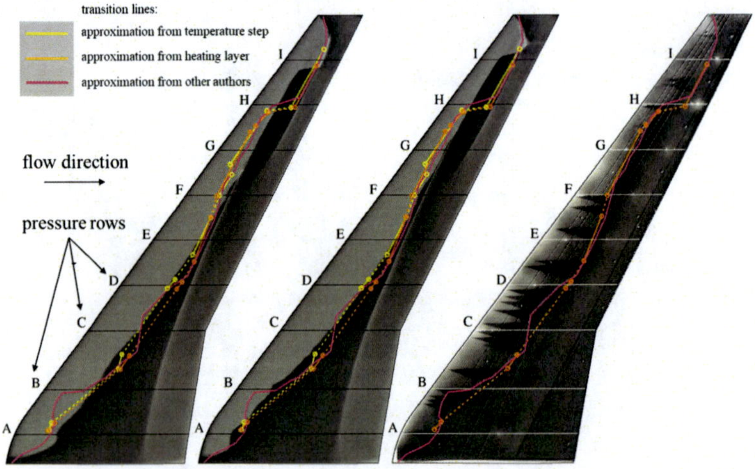

Fig. 4. CRM-NLF wing, comparison of predicted transition fronts, $\alpha = 2.0°$. Left: RANS surface c_f-distribution, incompressible LST. Middle: RANS surface c_f-distribution, compressible LST. Right: TSP image, heating layer method.

Despite the visible minor differences in predicted transition locations among each other, the overall agreement of both prediction methods compared to the experimental data is very good. The most crucial deviation to the experiment can be seen at the outboard area of the wing for $\alpha = 2.0°$ (Fig. 4). Here transition is predicted too far upstream compared to the experiment between pressure rows H and I. Transition has been found to react in this region very sensibly to pressure gradient and shock location so that small deviations in the flow conditions (Mach number, angle of attack) may have a rather large impact on the predicted transition locations. The same is true for the other remarkable deviation between simulation and experiment for $\alpha = 2.5°$ and $\alpha = 3.0°$ (Figs. 5 and 6). Again, at the outboard area of the wing (beyond pressure row F), the predicted transition locations strongly depend on an accurate representation of the actual

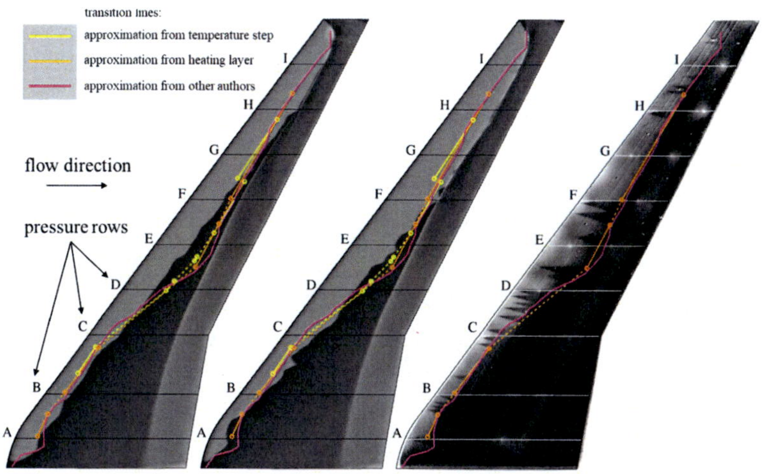

Fig. 5. CRM-NLF wing, comparison of predicted transition fronts, $\alpha = 2.5°$. Left: RANS surface c_f-distribution, incompressible LST. Middle: RANS surface c_f-distribution, compressible LST. Right: TSP image, heating layer method.

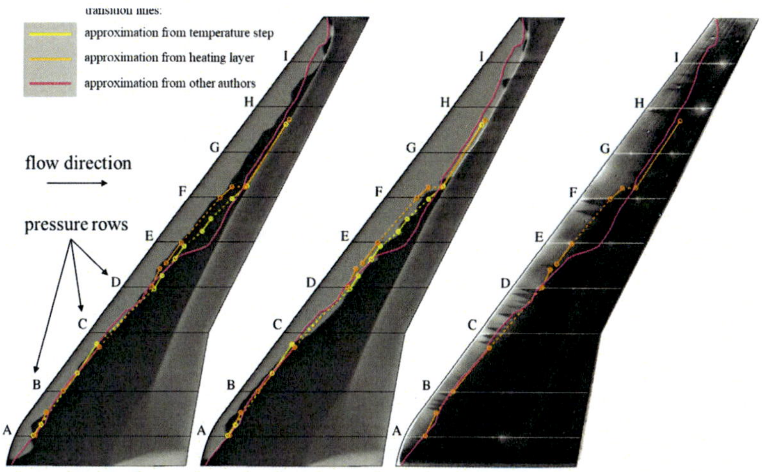

Fig. 6. CRM-NLF wing, comparison of predicted transition fronts, $\alpha = 3.0°$. Left: RANS surface c_f-distribution, incompressible LST. Middle: RANS surface c_f-distribution, compressible LST. Right: TSP image, heating layer method.

flow conditions and small deviations in Mach number or angle of attack may lead to a rather significant impact on the predicted transition lines.

An indication of a somewhat improper representation of the actual flow conditions in the wind tunnel during the numerical simulation can be obtained from the pressure distributions. Comparing numerically predicted pressure distributions for fully turbulent computations and computations with transition prediction with incompressible stability analysis (note: incompressible and compressible stability analysis give very similar transition locations and hence very similar pressure distributions) reveals, that generally the

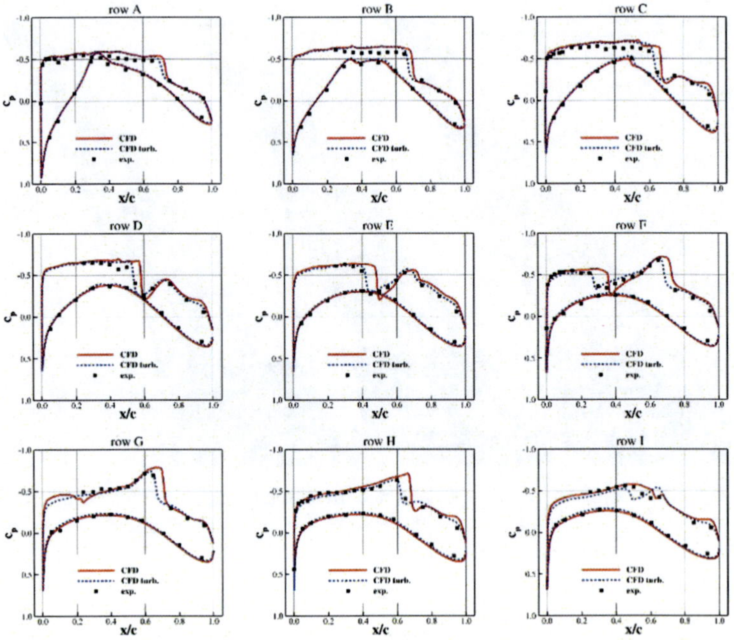

Fig. 7. CRM-NLF wing, pressure distribution at measuring sections, α = 1.5°.

experimental pressure distribution is better matched by the fully turbulent computations for all considered cases (shown here only for α = 1.5° and α = 3.0° in Figs. 7 and 8). For the majority of angles of attack and pressure rows, the shock locations in the transitional computations are located too far downstream compared to the experiment. However, the pressure distributions of the fully turbulent computations are in very good agreement with the experiment. An explanation for this behavior, beyond a possible deviation in flow conditions applied in the computations compared to the experiment, may be seen in the larger amount of turbulent flow in the experiments. For all angles of attack considered here, a large amount of turbulent wedges forms in the experiment (Figs. 3, 4, 5 and 6, right image), leading to a different thickening of the boundary layer compared to laminar flow and thus having a possible impact on the location of the shock and hence on the pressure distribution. Including the turbulent wedges in the numerical computation has already led to an improved prediction of the pressure distribution for transitional flow computations [7].

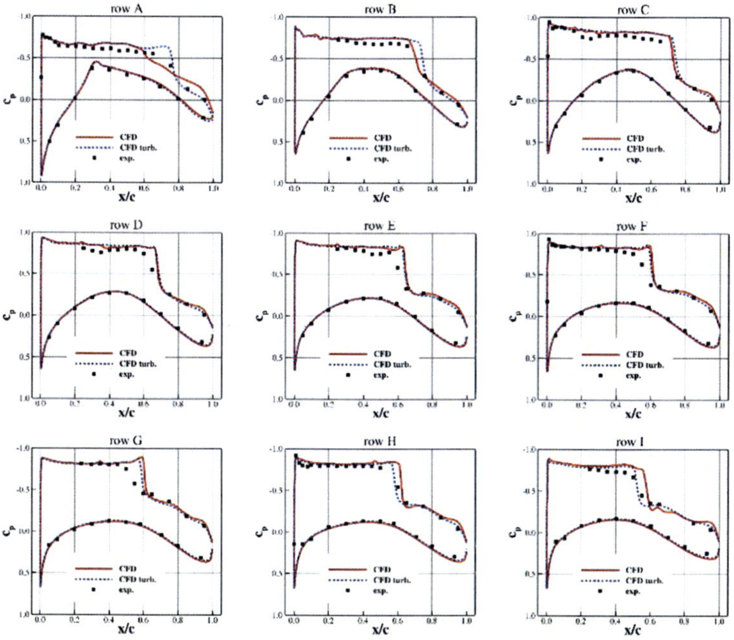

Fig. 8. CRM-NLF wing, pressure distribution at measuring sections, $\alpha = 3.0°$.

4 Conclusion

The determination of critical N-factors for the CRM-NLF wing revealed two different values for Tollmien-Schlichting instabilities, depending on the applied method. For an incompressible stability analysis a critical N-factor of $N_{TS} = 11.7$ was determined, whereas for a compressible stability analysis the critical N-factor is $N_{TS} = 4.1$. The latter is lower than the critical N-factor applied in other investigations using compressible stability analysis but is just within the possible range determined for the facility accommodating the CRM-NLF experiments. The critical N-factor for cross-flow instabilities ($N_{CF} = 6.0$) is independent of applying incompressible or compressible analysis and corresponds to values used in other investigations.

Applying the critical N-factors for transitional computations reveals a good agreement between predicted and experimental transition lines, respectively for incompressible and compressible stability analysis. Slight deviations in the numerical transition lines compared to the experiment may be explained by the sensitivity of the flow to the overall extent of turbulent flow (turbulent wedges in the experiment) and the accurate representation of the actual flow conditions of the experiment in the numerical simulation.

At this stage, the current results confirm the approach that for the prediction of the transition fronts incompressible LST as well as compressible LST can be applied leading to a similarly good agreement between the measured and predicted transition fronts as long as the correspondingly calibrated critical Tollmien-Schlichting N-factors are used. There are arguments in favor of incompressible LST over compressible LST [8] that need further discussion and investigation in the future and for the time being it seems reasonable to follow the established incompressible LST approach further on. On the other hand, it could be shown that in specific cases the application of compressible LST in the design process can lead to airfoil shapes that exhibit a larger laminar length and a lower wave drag [9]. Also in this respect, more work and investigations are imperative in the nearest future in order to achieve more data and knowledge for a better assessible situation.

Acknowledgement. The authors thank M. Fehrs of DLR for the provision of the computational grid and A. N. Watkins of NASA for the provision of heating layer and temperature step floating point TIFF images. The data source of the experimental pressure data is: commonresearchmodel.larc.nasa.gov/crm-nlf/.

References

1. Lynde, M.N., Campbell, R.L.: Computational Design and Analysis of a Transonic Natural Laminar Flow Wing for a Wind Tunnel Model. AIAA-2017-3058 (2017)
2. Lynde, M.N., Campbell, R.L., Viken, S.A.: Additional Findings from the Common Research Model Natural Laminar Flow Wind Tunnel Test. AIAA-2019-3292 (2019)
3. Watkins, A.N., Goodman, K.Z., Peak, S.M.: Transition Detection at Cryogenic Temperatures Using a Carbon-Based Resistive Heating Layer Coupled with Temperature Sensitive Paint. AIAA-2019-2191 (2019)
4. Krumbein, A., Krimmelbein, N., Schrauf, G.: Automatic transition prediction in hybrid flow solver, part 1: Methodology and sensitivities. J. Aircr. **46**, 1176–1190 (2009)
5. Crouch, J., Sutanto, M., Witkowski, D., Watkins, A., Rivers, M., Campbell, R.: Assessment of the National Transonic Facility for Natural Laminar Flow Testing. AIAA-2010-1302 (2010)
6. Helm, S.: Private communication (2020)
7. Helm, S., Fehrs, M., Krimmelbein, N., Krumbein, A.: Transition Prediction and Analysis of the CRM-NLF wing with the DLR TAU-Code. To be published (2021)
8. Schrauf, G., Streit, T.: N-factors and design. Are we expecting too much? In: Presentation Charts, 1st AIAA Transition Modelling and Prediction Workshop, 21–22 January 2021, Virtual Event.
9. Streit, T., Schrauf, G., Hein, S., Seitz, A., Kunze, P.: NLF-Potential für transsonische Langstreckenflugzeuge, Institutskolloquium AS-BS 3. DLR – Institut für Aerodynamik und Strömungstechnik, Braunschweig-Göttingen (Feb. 2021)

Experimental Investigation on Roughness-Induced Transition Under the Influence of Freestream Turbulence

Dominik K. Puckert(✉), Tristan M. Römer, Giacomo Scibelli, and Ulrich Rist

Institute of Aerodynamics and Gas Dynamics, University of Stuttgart, Pfaffenwaldring 21, 70569 Stuttgart, Germany
puckert@iag.uni-stuttgart.de

Abstract. The impact of freestream turbulence (FST) on a roughness-perturbed, laminar boundary layer is investigated in a laminar water channel. Both wire-mesh grid and rising air bubbles are used to induce random disturbances to the flow. The transition Reynolds number depends significantly on the level of random fluctuations. This confirms that the lower transition limit, which is relevant to most engineering applications, depends on external noise and is thus difficult to predict without precise knowledge of related flow parameters. Moreover, visualizations and harmonic forcing with a vibrating wire suggest that the near and far wake may respond differently to freestream turbulence. The near wake can effectively be forced in a region at the height of the roughness, whereas the far wake is less sensitive to this type of forcing.

1 Introduction

Roughness-induced transition is a long-lasting and fundamental problem in fluid mechanics. Its relevance to flight has motivated a great number of workers to predict laminar-to-turbulent transition reliably. It is known that a laminar boundary layer typically causes significantly lower skin friction than a turbulent boundary layer. Designing tolerable roughness sizes on an airfoil is therefore a highly desirable and ambitious means to reduce drag, cost and emissions of future aircraft.

In this context, the determination of the transition Reynolds number has been an important goal in previous investigations [1–3]. However, the spread between most investigations is large and was more recently explained by convective instabilities below the 'critical' Reynolds number associated with a global instability from three-dimensional linear stability theory [4]. An experimental proof of a global instability at the upper limit was provided in [5].

Although the global instability at the upper limit is now better understood, the lower limit remains a puzzle. Resolvent norms and 'inexact' solutions from pseudospectra are believed to be one piece of that puzzle [6]. In [4], such theoretical tools provide an explanation for experimental observations. However, this

© The Author(s), under exclusive license to Springer Nature Switzerland AG 2021
A. Dillmann et al. (Eds.): STAB/DGLR Symposium 2020, NNFM 151, pp. 205–214, 2021.
https://doi.org/10.1007/978-3-030-79561-0_20

method does not yet provide a practical lower transition threshold, which is, to our current understanding, caused by freestream turbulence (FST).

Therefore, an experimental study is needed to extend previous studies with controlled FST. In particular, we aim to clarify the following questions:

- What is the effect of FST on a roughness-perturbed flow?
- How does FST influence the transition Reynolds number?
- Where is the flow receptive to local forcing?
- What is the role of the near and far wake of the roughness?

2 Experimental Set-Up

The experiments in this work have been conducted in the closed-circuit, low-turbulence Laminar Water Channel ('Laminarwasserkanal') of the Institute of Aerodynamics and Gas Dynamics (IAG) at the University of Stuttgart. A comprehensive overview of the facility can be found in [7], whereas a more recent description of methods related to this work are presented in [8].

The set-up of the present investigation is sketched in Fig. 1. The two-dimensional Blasius flow on the flat plate is disturbed by a cylindrical roughness element of height $k = 0.01\,\mathrm{m}$ and diameter[1] $D = k$. The roughness is placed at $x_k = 57k$ from the leading edge and defines the origin of the right-handed coordinate system extending in streamwise (x), wall-normal (y) and spanwise (z) direction. All coordinates in this work are non-dimensionalized by k to comply with nomenclature in similar investigations. Likewise, the Reynolds number is defined as $Re_k = kU_e/\nu$ with the boundary-layer edge velocity U_e and kinematic viscosity ν. Note that other definitions ('roughness Reynolds number' [3]) exist in the literature.

The level of FST in this work is defined as the root mean square of the streamwise velocity:

$$Tu = u'_{\mathrm{rms}}/U_e. \tag{1}$$

Tu is induced via two techniques. The first is based on an array of vertical rods 0.05 m upstream of the leading edge with a spacing of 20 mm and diameter 1.6 mm, yielding a diameter-based Reynolds number sufficiently high for vortex shedding.

The second method is an apparatus that was first presented in [9]. Medical cannulas with an inner diameter of 0.2 mm are mounted on a pressurized tube to form an array with a spacing of 20 mm. When activated, air bubbles emerge from the cannulas and rise in a spiraling movement [10], inducing random disturbances to the flow. This 'bubble generator' is placed near the settling chamber to allow the bubbles to rise to the surface before hitting the flat plate and probe. The turbulence level of this method is further adjusted by an additional wire-mesh as described in [9].

Table 1 summarizes the settings used in this work. 'Reference' is the undisturbed water channel whereas 'FST-1' and 'FST-2' refer to the two techniques described above. Tu and $Re_{k,tr}$ are determined in the next chapter.

[1] The term 'aspect ratio' $\eta = D/k$ is often used in the literature. Here $\eta = 1$.

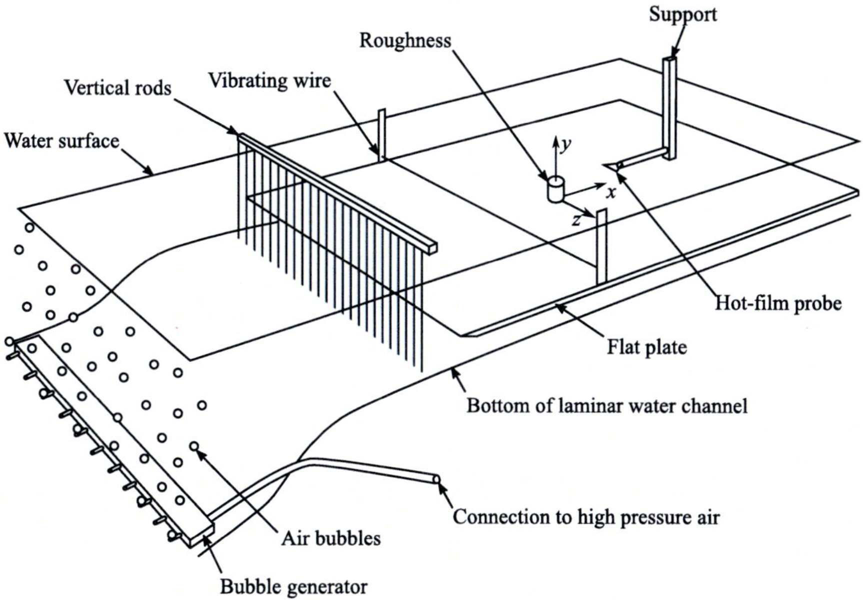

Fig. 1. Test set-up at the laminar water channel.

Table 1. Settings used in this work

	Type	FST generation	Tu $(420 < Re_k < 560)$	$Re_{k,tr}$
Reference	–	No additional FST	0.5%	630
FST-1	Passive	Grid with vertical bars	1.1%	550
FST-2	Active	Rising air bubbles (2 bar)	2.1%	480

A Streamline 90N10 CTA-bridge from the company Dantec, hot-film probe 55R15 and 16-bit A/D converter are used for data acquisition in this work. The probe is calibrated in-situ via a traverse system [8]. Signals are split into mean velocity $\bar{u}$ and fluctuating component u'. No digital filters were applied to include low-frequency disturbances in statistical quantities such as Eq. (1).

3 Results

The qualitative effect of FST can be visualized conveniently in water by using potassium permanganate crystals. These crystals are dense enough to rest on the flat plate and can be put to relevant positions around the roughness. As they dissolve in water, they draw clearly visible magenta streaklines into the flow. For better contrast, a 0.1-mm yellow plastic sheet[2] sticks to the flat plate below the roughness. It is thin enough not to disturb the flow significantly.

[2] www.selbstklebefolien.de.

Figure 2 shows a comparison of the reference case and FST-1 at a Reynolds number $Re_k = 575$. Streaklines that wrap around the roughness indicate the outer edge of a horseshoe vortex, which is a typical feature of this kind of flows [1,2]. Crystals in the near wake of the roughness visualize the footprint of the recirculation zone in the shape of a dark, 90°-tilted heart. In both cases, two thin streaklines extend into the downstream direction. In Fig. 2(a), these streaklines run straight like in a steady, laminar wake. When the flow is disturbed by FST, the streaklines along the centerline of Fig. 2(b) become entangled and mix with the surrounding fluid. During the experiment, this unsteadiness could be well observed by eye in the water channel. The mechanisms causing this difference are investigated in more detail in the following.

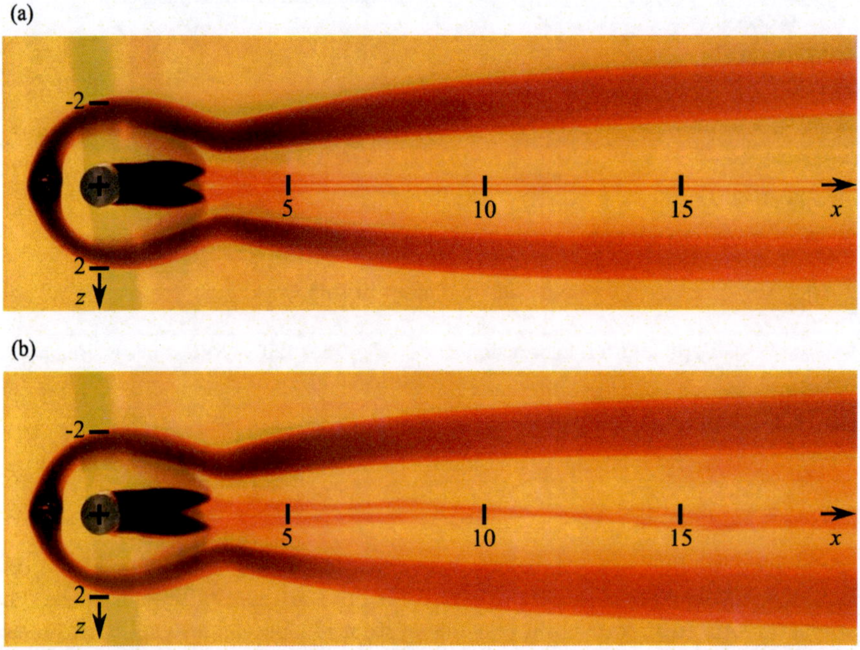

Fig. 2. Visualization of the roughness wake for (a) reference and (b) FST-1.

To enable a qualitative comparison, the FST for the three cases needs to be known. Therefore, the hot-film probe is placed 70 mm above the roughness to measure velocity disturbances for different Reynolds numbers. It can be assumed that the wall-normal distance is large enough to measure disturbances from background noise without noteworthy local impact from the roughness or free water surface, which is 140 mm above the roughness. The duration of one measurement was 600 s with a sampling rate of 100 Hz, providing 60,000 samples for the determination of Tu via Eq. (1). Here, we restrict the parameters to Re_k and Tu, which we consider as the most important ones.

The results of these measurements are plotted in Fig. 3(a). As expected, Tu is higher in both FST-1 ('vertical rods') and FST-2 ('bubble generator + screen') compared to the reference case. The turbulence level of FST-2 also compares well to the results published in [9]. At low Reynolds numbers, large natural fluctuations occur in all three cases. These strong disturbances originate most likely from thermal convection at very low velocities in the settling chamber and can hardly be avoided by conventional design methods. For the Reynolds numbers in this study, this effect is not relevant and therefore not investigated further. More details as well as a non-conventional solution are provided in [11]. The last value of FST-2 seems to be an outlier due to an air bubble sticking to the hot film probe. An average turbulence intensity within a reasonable Re_k-range is provided in Table 1.

In the next experiment, see Fig. 3(b), the probe is placed $x = 100$ downstream of the roughness at $y = 1$ with unchanged acquisition settings. This position allows to determine the intermittency factor γ without being significantly affected by the recirculation zone (compare Fig. 2). The intermittency factor represents the relative amount of time where the signal is turbulent [5]. For instance, fully laminar flow has an intermittency of $\gamma = 0$ whereas a fully turbulent flow yields $\gamma = 1$, or 100%. The criterion for transition is often defined as $\gamma = 0.5$ [12,13]. Figure 3(b) illustrates γ for the three cases. Solid trend lines along the curves as well as a dashed line at 50% are added to determine the intermittency more precisely from the graph. According to the definition above, the transition Reynolds number is $Re_{k,tr} = 630$ for the reference, 550 for FST-1 and 480 for FST-2, respectively. Also note that, according to Fig. 3(a), the turbulence intensity is almost constant in the vicinity of the three respective transition Reynolds numbers, which substantiates the relation between $Re_{k,tr}$ and Tu. This result also answers one of the introductory questions: The transition Reynolds number in this study is reduced by up to 24% due to FST.

To understand the interaction of FST with the roughness flow better, the probe is placed at $(y, z) = (1, 0)$ and traversed in the streamwise direction between $0 < x < 52$ in increments of $\Delta x = 2$. This experiment is repeated for different Reynolds numbers and the root mean square of the streamwise velocity fluctuations of the reference and FST-1 are plotted in Fig. 4. The graphs are similar to [5], where a global instability was detected for different set-ups. For this particular case, the experimental detection of a global instability is not possible because convective instabilities overshadow the expected threshold to global instability. On the other hand, a study on global versus convective instabilities for this set-up was already published in [4]. In this study, it was found that the flow is only marginally stable at $Re_k = 700$, meaning that very small background disturbances can lead to transition without the rise of a global instability. In Fig. 4(a), the disturbance root mean square u'_{rms} rises between $580 < Re_k < 700$ close to the roughness. Interestingly, these disturbances only rise locally in the near wake of the roughness and fade away further downstream. This can be explained with the strong shear layer close to the roughness, which induces instabilities at higher frequencies than the natural instability range of

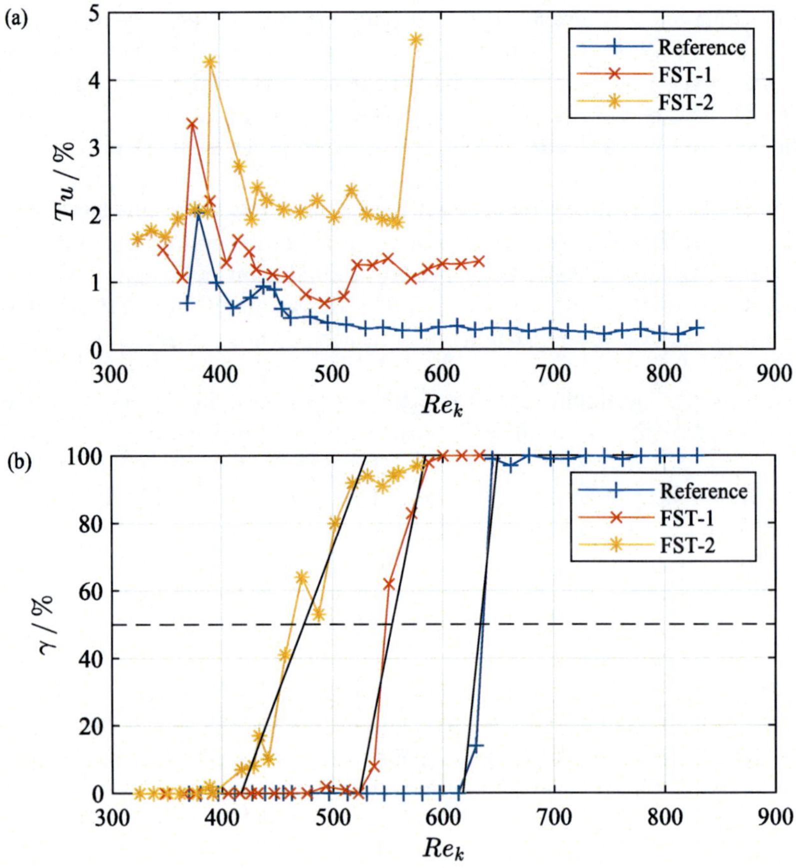

Fig. 3. (a) Turbulence intensity and (b) intermittency function, both versus Reynolds number.

the Blasius flow [8]. When the base flow relaxes downstream of the near wake, these instabilities cannot grow further at their fundamental frequency but may shed subharmonics that survive and even grow further downstream. It is interesting to compare this explanation to the stability diagram in [14].

Figure 4(b) shows similar results, however, the disturbances occur already at lower Reynolds numbers than in Fig. 4(a). This supports the hypothesis of convective disturbances in the presence of FST: higher Tu leads to higher disturbances, thus earlier transition. With a closer look, one can find that the previously determined transition Reynolds number (Table 1) fits well to the end of the domain in Fig. 4(a) but not so well in Fig. 4(b). Furthermore, the red pseudocolor downstream of $x = 20$ between $600 < Re_k < 700$ exhibits a more organized gradient in the reference case compared to FST-1. It seems as if the fate of the flow downstream of the roughness is not yet determined. Also, the FST-disturbed flows have a much larger intermittent range with a flatter slope of γ compared to the reference, see Fig. 3(b). The roughness and its (high-shear) near wake is therefore not the only decision maker whether the flow trips to turbulence or not. The results in this work suggest that the streaky wake further downstream of the roughness plays a large role in FST-induced transition.

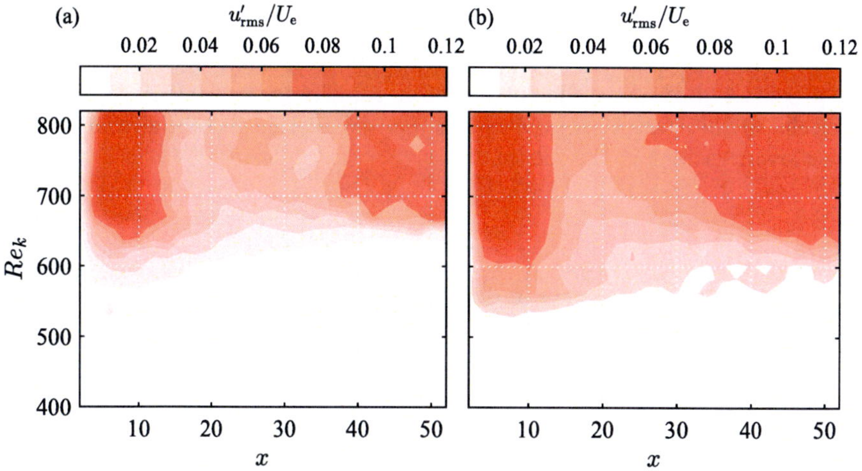

Fig. 4. Disturbances in the roughness wake for (a) reference and (b) FST-1.

In the remaining part, the receptivity of the roughness flow to controlled disturbances is investigated in more detail. In particular, it shall be determined which locations around the roughness are most sensitive to external forcing.

A vibrating wire (see Fig. 1) is therefore placed at different locations in front of the roughness at $Re_k = 550$ between $-9 < x < 7$ with increments of $\Delta x = 1$ and $0.4 < y < 1.4$ with $\Delta y = 0.1$. During this experiment, the probe always remains stationary at $(x, y, z) = (13, 1, 0)$ and records for 60 s for each wire position. The forcing frequency $\omega = 2\pi f k/U_e = 0.8$, with f the physical frequency in Hertz, was determined in preliminary experiments [15] and is in good agreement with the expected vortex shedding at the tip of the roughness [8]. The forcing amplitude is ± 0.25 mm and the remaining details for the vibrating wire can be found in [16].

The result of this measurement, see Fig. 5, reveals locations where localized, harmonic forcing leads to a strong (red pseudocolors) or weak (blue) response $u'_{\rm rms}$ to the forcing. It is interesting to see that a strong response can be triggered by forcing along the roughness height upstream of the roughness. The streamwise distance from the roughness in upstream direction seems to play a minor role along this 'stripe'; any forcing at negative x between $0.8 < y < 1.2$ is effective – except for a likely outlier at $(x, y) = (-2, 0.8)$. In contrast, only little response to forcing can be achieved in downstream direction beyond $x = 6$. For this particular frequency, the amplification seems to be linked to the recirculation zone in the roughness near wake.

Far enough downstream of the roughness, forcing at this frequency has little or even no effect. The chosen frequency of the wire is approximately five times larger than branch II of the neutral stability diagram from linear stability theory and thus not the preferred frequency of the base flow in the far wake. Thus, other types of disturbances, at lower frequency than that of the wire, may become dominant. It is reasonable to assume that spanwise disturbances lead to sinuous meandering of streaks at low frequency. This can visually be confirmed by Fig. 2. It is an important preliminary finding, because sinuous instabilities have previously been very difficult to find in roughness experiments [8]. The role of streaks in the transition process is supposed to be substantial and requires deeper investigations in the future.

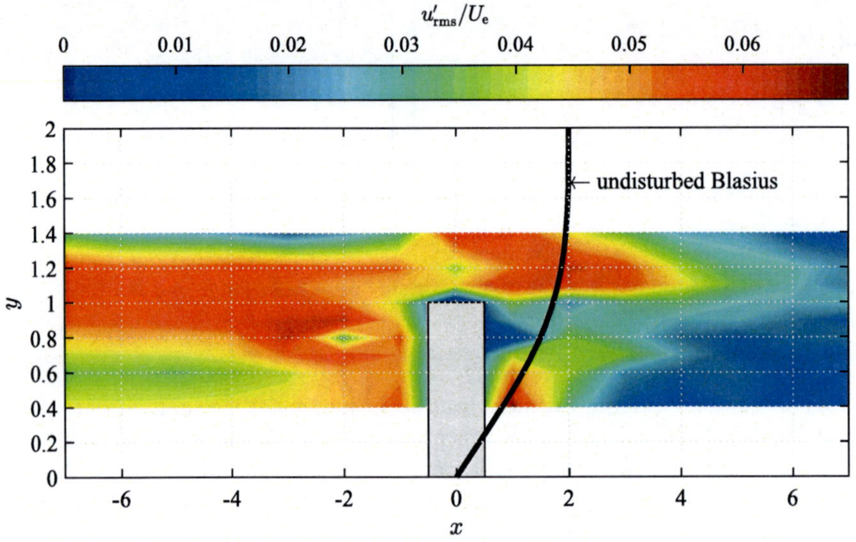

Fig. 5. Excitation map from traversing a vibrating wire. Black line: theoretical Blasius velocity profile for the undisturbed flow without roughness.

4 Conclusions

A laminar Blasius boundary-layer flow containing a cylindrical roughness element was investigated experimentally in a low-disturbance water-channel facility with different levels of controlled freestream turbulence. It was found that laminar-to-turbulent transition in the boundary layer depends on freestream turbulence. Increasing the natural turbulence intensity in the water channel by 120% leads to a reduction of the transition Reynolds number by 21%. Another increase of Tu by 90% reduces $Re_{k,tr}$ again by 13%. This dependency may explain the large spread in the literature [3], where Tu is often not compared. External disturbances have different effects on the near and far wake of the roughness. The near wake was successfully forced with a vibrating wire close to the roughness tip. If the frequency is chosen correctly and the wire placed along a stripe at the height of the roughness, a strong response can be observed. In contrast to the near wake, visualizations reveal that the far wake behaves differently. The meandering of streaks due to FST may play a large role in the final decision between laminar and turbulent state. This hypothesis can be subject to future investigations.

Acknowledgments. The funding of the Deutsche Forschungsgemeinschaft (DFG) under grant number RI 680/39-1 is acknowledged.

References

1. Klanfer, L., Owen, P.R.: The effect of isolated roughness on boundary layer transition. R.A.E. Tech. Memo. 355 (1953)
2. Gregory, N., Walker, W.S.: The effect on transition of isolated surface excrescences in the boundary layer. Aero. Res. Counc. R&M 2779 (1956)
3. von Doenhoff, A.E., Braslow, A.L.: The effect of distributed surface roughness on laminar flow. Bound. Layer Control **2**, 657–681 (1961)
4. Bucci, M.A., et al.: Roughness-induced transition by quasi-resonance of a varicose global mode. J. Fluid Mech. **836**, 167–191 (2018)
5. Puckert, D.K., Rist, U.: Experiments on critical Reynolds number and global instability in roughness-induced laminar-turbulent transition. J. Fluid Mech. **844**, 878–903 (2018)
6. Schmid, P.J.: Nonmodal stability theory. Annu. Rev. Fluid Mech. **39**, 129–162 (2007)
7. Wiegand, T.: Experimentelle Untersuchungen zum laminar-turbulenten Transitionsprozess eines Wellenzugs in einer Plattengrenzschicht. Ph.D. thesis, University of Stuttgart (1996)
8. Puckert, D.K.: Experimental investigation of global instability and critical Reynolds number in roughness-induced laminar-to-turbulent transition. Ph.D. thesis, University of Stuttgart (2019)
9. Siring, M., Puckert, D.K., Rist, U.: Investigations on a mechanism to induce freestream turbulence in a water channel by controlled injection of air bubbles. In: Dillmann, A., Heller, G., Krämer, E., Wagner, C., Tropea, C., Jakirlić, S. (eds.) DGLR 2018. NNFMMD, vol. 142, pp. 265–273. Springer, Cham (2020). https://doi.org/10.1007/978-3-030-25253-3_26
10. Saffman, P.G.: On the rise of small air bubbles in water. J. Fluid Mech. **1**(3), 249–275 (1956)
11. Puckert, D.K., Dieterle, M., Rist, U.: Reduction of freestream turbulence at low velocities. Exp. Fluids **58**(5), 1–5 (2017). https://doi.org/10.1007/s00348-017-2333-y
12. Fransson, J.H.M., Brandt, L., Talamelli, A., Cossu, C.: Experimental study of the stabilization of Tollmien-Schlichting waves by finite amplitude streaks. Phys. Fluids **17**(5), 054110 (2005)
13. Zhang, J., Xu, M., Pollard, A., Mi, J.: Effects of external intermittency and mean shear on the spectral inertial-range exponent in a turbulent square jet. Phys. Rev. E **87**(5), 053009 (2013)

14. Siconolfi, L., Camarri, S., Fransson, J.H.M.: Boundary layer stabilization using free-stream vortices. J. Fluid Mech. **764**, R2 (2015)
15. Scibelli, G.: Experimentelle Untersuchung zum kombinierten Einfluss von Freistromturbulenz und einer Rauigkeit auf die laminare Grenzschicht. Bachelor thesis, University of Stuttgart (2020)
16. Kruse, D.: Implementierung eines Störgenerators für kontrollierte Transitionsexperimente mit anschließender Untersuchung der erzeugten Störung. Bachelor thesis, University of Stuttgart (2016)

A Procedure to Estimate the Size of a Suction Flap for a Passive HLFC System

Geza Schrauf(✉)

DLR, Lilienthalplatz 7, 38108 Braunschweig, Germany
geza.schrauf@dlr.de

Abstract. We describe a simple pre-design method to estimate the size of an air outlet flap for a passive hybrid laminar flow control (HLFC) system to determine the feasibility of that flap. The method is based on an ingenious application of old experimental results. The computational effort is minimal.

Keywords: Hybrid Laminar Flow Control (HLFC) · Passive suction system · Feasibility of passive suction · Size of air outlet · Suction flap

Nomenclature

A_{in}	Inflow cross-section area of the duct to the outlet $[m^2]$.
A_{out}	Outflow cross-section area of the outlet $[m^2]$.
C_{dis}	Discharge coefficient as defined on page 6.
$\dot{m}$	$= \dot{m}_{in} = \dot{m}_{out}$. Mass flow through inflow or outflow cross-section $[kg/s]$.
p_{in}	Static pressure at the inflow cross-section $[Pa]$.
p_{out}	Static pressure at the outflow cross-section $[Pa]$.
p_{ref}	Reference pressure $[Pa]$.
$\bar{v}_{in}$	Mean velocity in inflow cross-section $\bar{v}_{in} = \dot{m}_{in} / (\rho A_{in})$ $[m/s]$.
$\bar{v}_{out}$	Mean velocity in outflow cross-section $\bar{v}_{out} = \dot{m}_{out} / (\rho A_{out})$ $[m/s]$.
v_{ref}	Reference velocity $[m/s]$.
$\dot{V}_{in}$	Volume flow through the inflow cross-section $[m^3/s]$.
$\dot{V}_{out}$	Volume flow through the outflow cross-section $[m^3/s]$.
δ	Flap opening angle $[degree]$.
ρ	Density $[kg/m^3]$.
q_*	Subscript '$*$' denotes a non-dimensional quantity.

1 Introduction

In this paper we present a simple method for estimating the cross-section area, and with that, the size of an outlet that discharges air from an internal flow system into the external flow around an aircraft. The method is an ingenious

© The Author(s), under exclusive license to Springer Nature Switzerland AG 2021
A. Dillmann et al. (Eds.): STAB/DGLR Symposium 2020, NNFM 151, pp. 215–223, 2021.
https://doi.org/10.1007/978-3-030-79561-0_21

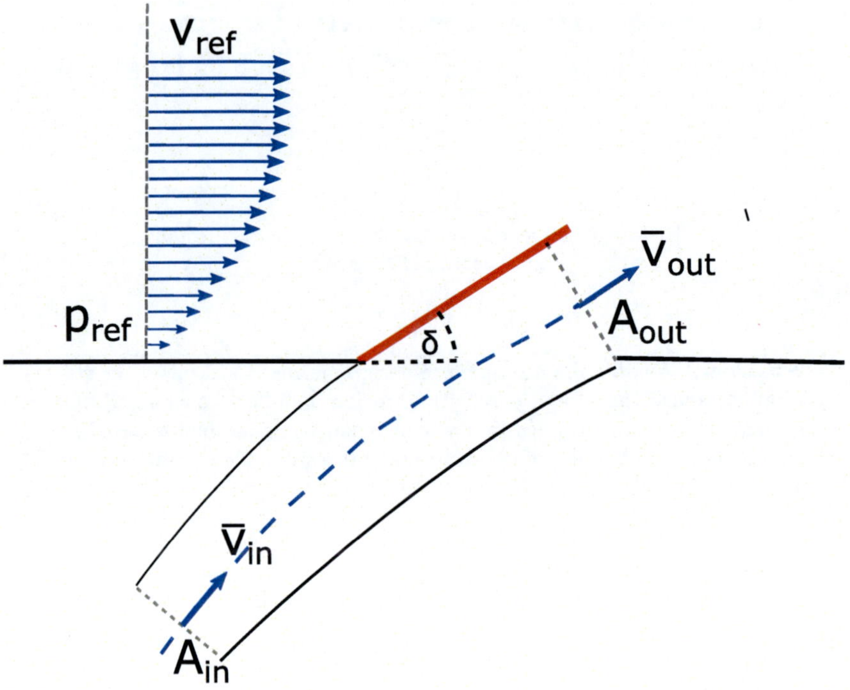

Fig. 1. Sketch of a flap outlet.

application of wind tunnel results from 1938 [1]. It might prove very useful for feasibility and pre-design studies of passive HLFC systems.

We describe the method for a simple flap outlet. However, analogous estimates can be derived for outlets with different shapes, such as circular holes or "Lamellen"-outlets.

2 Review of Rogallo's Results

Figure 1 shows a sketch of a simple flap outlet. The inflow part of the outlet is a tube[1] with cross-section area A_{in}. By opening the flap, we can observe an outflow. The cross-section area A_{out} of an outlet is usually defined as the minimal cross-section area, i.e. for a planar flap, it is measured perpendicular to its surface.

Quantities of the external flow, somewhat upstream of the outlet, are taken as reference quantities for the internal flow. We assume that, on the one hand, this location is close enough to the outlet so that the reference pressure p_{ref} at this location is equal to the surface (or static) pressure at the outlet when the flap is closed. On the other hand, it is assumed that the location is far enough

[1] The shape of the inflow duct is irrelevant, that is the inflow duct can also be a rectangular channel, as, for example, used for the outlets in [1, see p. 5].

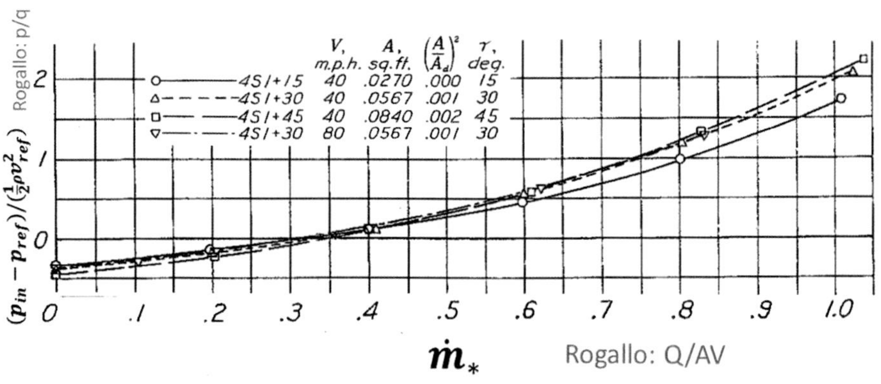

Fig. 2. Part of Figure 43 from [1] with our and Rogallo's notation.

upstream of the outlet so that the reference pressure is not changed when the outlet is opened. This is, of course, an idealization. As reference velocity v_{ref} we take the velocity at the edge of the boundary layer at this location, i.e., somewhat upstream of the flap[2]. Again, we assume that this velocity is not influenced by the opening of the flap.

The input data required for our estimate are the reference pressure p_{ref}, the reference velocity v_{ref}, as well as the inflow static pressure p_{in} and the mass flow $\dot{m}_{in}$ through the inflow cross-section area.

Based on the reference velocity v_{ref}, we define the non-dimensional mass flow through the outlet as

$$\dot{m}_* = \frac{\dot{m}_{out}}{\rho \, A_{out} \, v_{ref}}. \tag{1}$$

For incompressible flow, we can consider the non-dimensional volume flow

$$\dot{V}_* = \frac{\dot{V}_{out}}{A_{out} \, v_{ref}} = \frac{\bar{v}_{out}}{v_{ref}}, \tag{2}$$

where $\bar{v}_{out}$ is the mean velocity through the outflow area.

To make the input pressure p_{in} non-dimensional, Rogallo defines the pressure coefficient $(p_{in} - p_{ref}) \, / \, (1/2 \, \rho \, v_{ref}^2)$ based on the above reference quantities. He then presents the results of his wind tunnel measurements for each specific outlet configuration in the form of

$$\frac{p_{in} - p_{ref}}{1/2 \, \rho \, v_{ref}^2} = f(\dot{V}_*), \tag{3}$$

[2] This definition is similar to the reference velocity we use to describe the influence of forward and backward facing steps on laminar flow. Similar to our case, we use as reference the velocity at the boundary layer edge at the location of the step, however, for a smooth surface without the step.

describing the variation of the pressure coefficient with the non-dimensional volume flow $\dot{V}_*$. In [1] we find plots of such relations for many different types of outlets. Those plots are reproduced in [2]. In Figure 43 of [1], Rogallo presents the results of his wind tunnel measurements for a simple flap with the three opening angles of 15°, 30°, and 45°. In our Fig. 2, we reproduce a part of his figure with our notation for the x- and y-axes. We see that the functions are more or less independent of the opening angle and the flow velocity. From this figure we extract the following numerical values on which we will base our approximation method:

Table 1. Numerical values taken from [1, Fig. 43].

Q/AV	15° 40 $m.p.h.$	30° 40 $m.p.h.$	45° 40 $m.p.h.$	30° 80 $m.p.h.$
	p/q	p/q	p/q	p/q
0.0	−0,32	−0,35	−0,43	−0,36
0.2	−0,16	−0,17	−0,21	−0,18
0.4	0,11	0,12	0,11	0,14
0.6	0,47	0,52	0,55	0,57
0.8	0,98	1,10	1,20	1,20
1.0	1,60	1,80	1,90	

Rogallo's "p/q" is in our notation $(p_{in} - p_{ref})/(1/2\rho v_{ref}^2)$ and his "Q/AV" corresponds to our $\dot{m}_*$. The first line of the table contains the pressure values for zero mass flow. We find that the pressure coefficient is not zero but has an approximate value of −0.4. This means that a flap outlet can produce some underpressure, i.e., that it has a suction capability. This is necessary for a passive HLFC system. Not every outlet has this property, as can be seen, for example, from Figure 25 of [2] for a circular-hole outlet[3].

For zero mass flow, the static pressure p_{out} in the outflow cross-section is equal to the pressure p_{in}, as can be seen from the inviscid Bernoulli equation (6). Using $p_{out} = p_{in}$, we obtain for zero mass flow

$$\frac{p_{out} - p_{ref}}{1/2\,\rho\, v_{ref}^2} \approx -0.4. \tag{4}$$

With this relation, we can estimate the static pressure p_{out} in the outflow cross-section of a flap opening as

$$p_{out} = p_{ref} - 0.4 \cdot 1/2\,\rho\, v_{ref}^2. \tag{5}$$

To derive our estimates, we assume that this relation for the static pressure p_{out} is also valid for non-zero mass flow.

[3] For a circular-hole outlet we obtain from [2, Figure 25] that $p_{in} = p_{ref}$, i.e. a circular hole has no suction power.

Assumption A:

The static pressure p_{out} is independent of the mass (or volume) flow through the outlet and can be approximated by (5).

This assumption seems to be a good first approximation. To support this, we present values obtained from a 3D Navier-Stokes calculation for a simple flap outlet with an aspect ratio[4] of 2.6 for a case with $M_\infty = 0.85$. We see that the pressure p_{out} obtained with the approximation (5) is close to the computed values $p_{out,NS}$, which are obtained as averages over A_{out}. Furthermore, the example shows that the computed pressure $p_{out,NS}$ does not change much when the opening angle δ or the pressure p_{in} in the inflow cross-section are varied.

Table 2. Comparison of computed and estimated values of p_{out}.

δ [*deg*]	p_{in} [*Pa*]	$p_{out,NS}$ [*Pa*]	p_{out} [*Pa*]	Error [%]
7.5°	21879	19715	19329	2.0%
10°	21877	19467	19329	0.7%
10°	20878	19496	19329	0.9%
10°	21877	19497	19329	0.7%
10°	22876	19496	19329	0.9%
10°	23875	19337	19329	0.0%

3 Approximation of the Experimental Results

In this section we show how to use the values of Table 1 to derive the main relation for our approximation method. We assume the flow between A_{in} and A_{out} to be incompressible and inviscid, so that we can apply the Bernoulli equation (6). With this equation we obtain the ideal mean outflow velocity with the help of the inflow pressure p_{in}, the mean inflow velocity $\bar{v}_{in}$, and the static pressure p_{out} as

$$\frac{1}{2}\,\rho\,\bar{v}^2_{out\_ideal} = p_{in} + \frac{1}{2}\,\rho\,\bar{v}^2_{in} - p_{out}\,, \tag{6}$$

or

$$\bar{v}_{out\_ideal} = \sqrt{\frac{2}{\rho}\left[p_{in} + \frac{1}{2}\,\rho\,\bar{v}^2_{in} - p_{out}\right]}\,. \tag{7}$$

The ideal mass flow through the outflow cross-section is

$$\dot{m}_{out\_ideal} = \rho\,\bar{v}_{out\_ideal}\,A_{out}. \tag{8}$$

[4] The aspect ratio of a simple flap is defined as the ratio of spanwise to chordwise extent.

Due to the effect of viscosity (and local flow separation) the real mass flow $\dot{m}_{out}$ through A_{out} will be smaller. Therefore, we define a discharge coefficient[5] as follows:

$$C_{dis} = \frac{\dot{m}_{out}}{\dot{m}_{out\_ideal}} = \frac{\rho\, A_{out}\, \bar{v}_{out}}{\rho\, A_{out}\, \bar{v}_{out\_ideal}}. \tag{9}$$

For incompressible flow through an orifice, the discharge coefficient is reduced to the velocity ratio

$$C_{dis} = \frac{\bar{v}_{out}}{\bar{v}_{out\_ideal}}. \tag{10}$$

With Eq. (7) and $\bar{v}_{in} A_{in} = \bar{v}_{out} A_{out}$, we obtain

$$C_{dis} = \frac{\bar{v}_{out}/v_{ref}}{\bar{v}_{out\_ideal}/v_{ref}} = \frac{\dfrac{\bar{v}_{out}}{v_{ref}}}{\sqrt{\dfrac{p_{in} - p_{out}}{\frac{1}{2}\,\rho\, v_{ref}^2} + \left(\dfrac{\bar{v}_{out}}{v_{ref}}\right)^2 \left(\dfrac{A_{out}}{A_{in}}\right)^2}}. \tag{11}$$

Our p_{in} corresponds to Rogallo's duct pressure p_d and we can write (11) as

$$C_{dis} = \frac{\dfrac{\bar{v}_{out}}{v_{ref}}}{\sqrt{\dfrac{p_d - p_{ref}}{\frac{1}{2}\,\rho\, v_{ref}^2} - \dfrac{p_{out} - p_{ref}}{\frac{1}{2}\,\rho\, v_{ref}^2} + \left(\dfrac{\bar{v}_{out}}{v_{ref}}\right)^2 \left(\dfrac{A_{out}}{A_d}\right)^2}}. \tag{12}$$

According to our main assumption, the static pressure p_{out} in the outlet cross-section can be approximated by (5), so that the second term in the denominator becomes -0.4. The first term $(p_d - p_{ref})(\frac{1}{2}\,\rho\, v_{ref}^2)$ is the "p/q"-value listed in Table 1. This allows us to compute the discharge coefficients directly from the tabulated values for each flap setting.

In the next step we consider the non-dimensional mass flow

$$\dot{m}_* = \frac{\dot{m}_{out}}{\rho\, v_{ref}\, A_{out}} = \frac{\bar{v}_{out}}{v_{ref}} \tag{13}$$

[5] The **discharge coefficient** is a **mass flow ratio** and must be distinguished from the **contraction number** or **contraction ratio** which is defined as an **area ratio**. For incompressible flow, the discharge coefficient becomes a volume flow ratio. When considering the incompressible flow through only one opening, it is then reduced to a ratio of area-averaged velocities. We point out that the discharge coefficient is not well-defined for very small mass flows. For $\dot{m}_{out} \rightarrow 0$, we obtain an expression of the type "0/0".

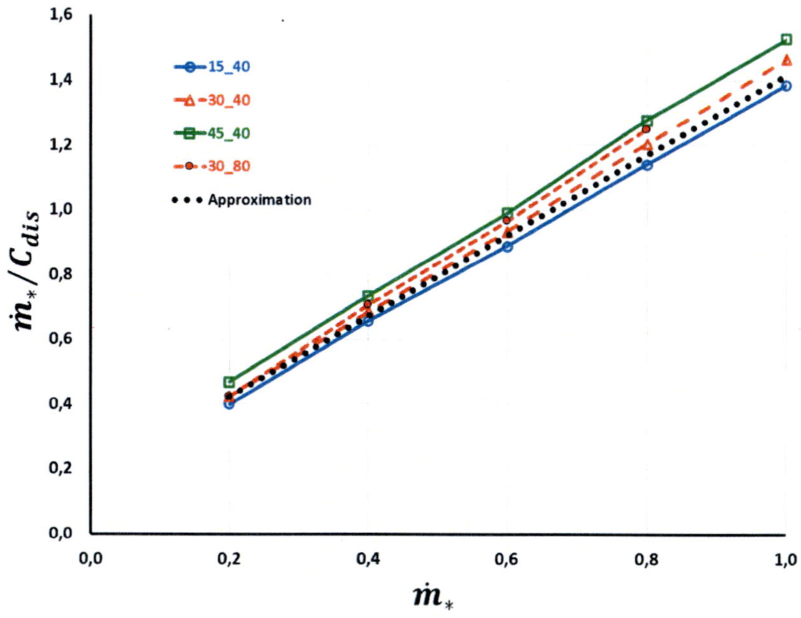

Fig. 3. Plot of the functions f_δ.

and re-plot Fig. 2 in the form[6]

$$\frac{\dot{m}_*}{C_{dis}} = f_\delta(\dot{m}_*). \tag{14}$$

The result is shown in Fig. 3. We see that each flap setting is represented by a nearly linear relation

$$\frac{\dot{m}_*}{C_{dis}} = f_\delta(\dot{m}_*) = a(\delta)\,\dot{m}_* + b(\delta), \tag{15}$$

with coefficients $a(\delta)$ and $b(\delta)$ that can be approximated by

$$a(\delta) = 0.0028\,\delta + 1.18, \tag{16}$$
$$b(\delta) = 0.0016\,\delta + 0.14. \tag{17}$$

In Fig. 3, we include an additional line to illustrate the approximation for a flap angle of 22°.

[6] Initially, we tried to approximate the functions $C_{dis} = \tilde{f}_\delta(\dot{m}_*)$ with polynomials or elliptic curves. However, working with the functions f_δ of the formulation (14) is easier and leads to more accurate estimates, because the functions f_δ are nearly linear.

4 Flap Size Estimation

Now we want to estimate the cross-section area for an outlet flap which is supposed to discharge a certain mass flow $\dot{m}$ for an inlet pressure p_{in}. The mass flow leaving the flap through its outflow cross-section A_{out} is equal to the mass flow entering through A_{in}. Therefore, we have $\bar{v}_{in} = \dot{m}/(\rho\, A_{in})$, and with that, we know the quantities $p_{ref}, v_{ref}, \rho, p_{in}, \bar{v}_{in}$. Our estimation procedure consists of two steps:

Step 1:

With the reference quantities and density ρ we compute p_{out} with formula (5). To generate an outflow, the outlet pressure p_{out} must be smaller than the inlet pressure[7] p_{in}, i.e.

$$p_{out} < p_{in}. \tag{18}$$

If this condition is violated, an HLFC system with the plenum pressure $p_p = p_{in}$ cannot operate in passive mode, and we need to redesign to allow for a higher plenum pressure.

Step 2:

Using formulas (7), (10) and (13) we obtain

$$\frac{\dot{m}_*}{C_{dis}} = \frac{1}{v_{ref}} \sqrt{\frac{2}{\rho}\left[p_{in} + \frac{1}{2}\,\rho\,\bar{v}_{in}^2 \;-\; p_{out}\right]}, \tag{19}$$

and with (15) we get

$$a(\delta)\,\dot{m}_* + b(\delta) = \frac{1}{v_{ref}} \sqrt{\frac{2}{\rho}\left[p_{in} + \frac{1}{2}\,\rho\,\bar{v}_{in}^2 \;-\; p_{out}\right]}. \tag{20}$$

This formula allows us to compute the non-dimensional mass flow $\dot{m}_*$ for a given flap angle δ. This non-dimensional mass flow $\dot{m}_*$ is the main input for our estimations. Because of

$$\dot{m}_* = \frac{\dot{m}_{out}}{\rho\, v_{ref} A_{out}} = \frac{\dot{m}_{in}}{\rho\, v_{ref} A_{out}}, \tag{21}$$

we can either obtain the outflow cross-section area for a given mass flow $\dot{m}_{in}$ as

$$A_{out} = \frac{\dot{m}_{in}}{\rho\, v_{ref}\, \dot{m}_*}, \tag{22}$$

or the mass flow for a given outflow area A_{out} (defined by the opening angle δ) with the help of

$$\dot{m} = \dot{m}_*\, \rho\, v_{ref} A_{out}. \tag{23}$$

---

[7] If Δp is the pressure loss from the plenum to the inlet cross-section, then p_{out} must be smaller than $p_{in} = p_p - \Delta p$.

The area obtained with formula (22) can be used to assess the feasibility of a passive HLFC system: if the area is too large, then the envisaged passive HLFC system with plenum pressure p_p and mass flow $\dot{m}$ is not feasible. Furthermore, with (23), we can estimate the suction capability of a simple flap when installed at a certain location.

The quantity $\dot{m}_\star$ is the basis for our estimation. Therefore, in Table 3, we present values of $\dot{m}_\star$ estimated with (20) and compare them to values $\dot{m}_{\star,NS}$ obtained from a 3D Navier-Stokes calculations.

Table 3. Comparison of computed and estimated values of $\dot{m}_\star$.

$\delta\,[deg]$	$p_{in}\,[Pa]$	$\dot{m}_{\star,NS}\,[-]$	$\dot{m}_\star\,[-]$	Error [%]
7.5°	21879	0.259	0.243	−6.4%
10°	21877	0.266	0.240	−9.9%
10°	20878	0.204	0.157	−23%
10°	21877	0.266	0.240	−9.9%
10°	22876	0.318	0.308	−3.2%
10°	23875	0.367	0.367	0.0%

The computational cases are the same as in Table 2. From this table, we see that we have a reasonable agreement. Only the case $\delta = 10°$ and $p_{in} = 20878 Pa$, i.e. the one with the weakest suction, exhibits a large error. This might be caused by the three-dimensional flow around the flap sides, which becomes dominant when the outflow is weak. In all other cases considered, including flaps with smaller aspect ratios or at different locations, the error in $\dot{m}_\star$ remained below 12%.

5 Conclusions

We present a method to estimate the size of a suction flap for a passive HLFC system. Such an estimate proves useful during the pre-design phase, when we need to find out whether an envisaged flap would be as small as a matchbox or as large as a barn door. If the estimated size is too large for the envisaged location, then we can drop this flap design without performing further calculations with expensive methods such as Navier-Stokes calculation. This saves a lot of time and effort.

References

1. Rogallo, F.M., Gauvain, W.E.: Wind-tunnel investigation of air inlet and air outlet openinge for aicraft. NACA Special Report No. 95 (1938)
2. Rogallo, F.M.: Internal flow systems for aircraft. NACA Report No. 713 (1941)

Receptivity of Swept-Wing Boundary Layers to Surface Roughness and Inhomogeneous Suction

Daniel Simanowitsch[1,2(✉)], Alexander Theiss[1,2], and Stefan Hein[1,2]

[1] Cluster of Excellence SEA – Sustainable and Energy-Efficient Aviation, Technische Universität, Braunschweig, Germany
[2] Institute of Aerodynamics and Flow Technology, German Aerospace Center (DLR), Bunsenstr. 10, 37073 Göttingen, Germany
{daniel.simanowitsch,alexander.theiss,stefan.hein}@dlr.de

Abstract. Suction through porous aircraft wings is a promising concept to reduce its drag, but the modified boundary layer can be more susceptible to external forcing, like surface roughness and inhomogeneities in the suction distribution. This paper analyzes the effects of suction on the linear receptivity of a swept-wing boundary layer to surface roughness and non-uniform suction, focusing on stationary crossflow instabilities. Employing a compressible adjoint approach, receptivity coefficients for a Falkner-Skan-Cooke similarity solution and a transonic swept-wing boundary layer on an A320 vertical tail plane with and without suction are given.

Keywords: Receptivity · Parabolized stability equations · Adjoint

1 Introduction

Aerodynamic drag has a significant influence on the fuel consumption of modern airplanes. Hybrid Laminar Flow Control (HLFC) is a promising concept to reduce the drag of future commercial airplanes. Suction through porous surfaces is used to delay laminar-turbulent transition and thus to reduce the overall skin-friction drag. In three-dimensional boundary layers (BL), suction in the leading-edge region of the transonic swept wings is used to stabilize crossflow (CF) instabilities. The thinner BL can be more susceptible to a given external forcing, however, i.e. the effectiveness with which CF instabilities are generated by environmental disturbances increases. Moreover, suction through porous surfaces typically increases surface roughness and results in a non-uniform suction distribution. Both types of external forcing can trigger stationary CF instabilities. In N-factor based transition prediction with Linear Stability Theory (LST) or with Parabolized Stability Equations (PSE), the processes that lead to the formation of boundary-layer instabilities are not considered at all. Hence, only the stabilizing effect of the idealized spanwise homogeneous suction on the subsequent development of the boundary-layer instabilities is covered. Changes in

© The Author(s), under exclusive license to Springer Nature Switzerland AG 2021
A. Dillmann et al. (Eds.): STAB/DGLR Symposium 2020, NNFM 151, pp. 224–234, 2021.
https://doi.org/10.1007/978-3-030-79561-0_22

the receptivity characteristics of the modified boundary-layer flow and the typically increased external forcing due to suction through porous surfaces, which presumably both contribute to the increase in initial boundary-layer instability amplitudes, therefore have to be compensated by assuming lower CF transition N-factors when predicting laminar-turbulent transition on HLFC wings.

Receptivity coefficients are a local measure quantifying the effectiveness with which a boundary-layer instability mode can be generated by a certain type of external forcing. Only few data exist in literature on receptivity coefficients of realistic transonic swept-wing boundary layers without and with suction. Together with information on the actual surface roughness and suction distribution, such data allow quantifying detrimental effects on the emergence of instabilities and could provide guidelines on how to adjust the corresponding transition N-factors. The negative impact of suction in boundary layers on receptivity has been shown for Tollmien-Schlichting waves in a two-dimensional boundary layer by Crouch [1]. Janke [2] has previously shown that an equivalent forcing model together with a PSE-like method is a suitable tool to predict receptivity of three-dimensional boundary layers to roughness and suction by comparing the results to a Direct Numerical Simulation (DNS). An efficient approach for receptivity modeling, which does not require a new direct solution for each forcing configuration, is to employ the adjoint solution. How adjoint Orr-Sommerfeld and PSE equations predict forced and natural receptivity has been extensively studied by Hill [3,4]. Adjoint solutions have henceforth been employed to e.g. determine sensitivities of BL flows [5], receptivity to roughness, suction, acoustic and free-stream disturbances [6,7]. Schrader [8] employed a DNS solution to determine receptivity coefficients for the self-similar Falkner-Skan-Cooke (FSC) boundary layer, while Piot and Schrader [9] expanded his work and showed that these coefficients are dependent on the forcing shape in spatially evolving flows, when the Fourier component of the shape at the chordwise instability wavenumber is small. Bertolotti [10] suggested to include the streamwise derivative of the shape function as a remedy.

This paper demonstrates the influence of suction on the local boundary-layer receptivity to surface roughness and inhomogeneous suction. We focus on steady crossflow instabilities in swept-wing boundary layers. Compressible PSE or LST are used to quantify the downstream growth of instabilities and adjoint PSE or LST respectively are employed to quantify receptivity. This approach is validated for a Falkner-Skan-Cooke BL and applied to the boundary layer on an A320 vertical tail plane. The flow over this fin is analyzed with and without a generic suction distribution, and the receptivity coefficients for both cases are given.

2 Methodology

In this section, the equations to obtain the linear receptivity coefficients for infinite swept-wing flows are derived. A compressible, non-local formulation for the direct and adjoint problem in curvilinear coordinates is chosen to analyze realistic airfoil flows.

2.1 PSE

To derive the compressible Parabolized Stability Equations (PSE), the total three-dimensional flow is split into the mean flow quantities $Q = \{\bar{\rho}, U, V, W, \bar{T}\}$ and disturbance quantities $q' = \{\rho', u', v', w', T'\}$, which are assumed to be small. In the following, {x, y, z} denote the chordwise, wall-normal and spanwise directions respectively and {U, V, W} the respective velocities. The non-dimensional, compressible Navier-Stokes equations are then formulated in disturbance form by introducing this decomposition and subtracting the equations describing the baseflow (see [6]). The PSE equations are then derived from a Wentzel–Kramers–Brillouin (WKB) ansatz where the disturbance q' is decomposed into a shape function q and a phase function Θ:

$$\begin{aligned} q'(x,y,z,t) &= q(x,y)\,\Theta(x)\,\mathrm{e}^{i\beta z - i\omega t}\,, \\ \Theta(x) &= \mathrm{e}^{i\int_{x_0}^{x}\alpha(x')\mathrm{d}x'}\,. \end{aligned} \tag{1}$$

Hereby $q = \{\rho, u, v, w, T\}$ is assumed to be slowly varying in chordwise direction x and $\hat{q} = q\Theta$. An ambiguity arises due to both shape function and exponential part being functions of x, which is removed with the condition $\int_0^\infty q^H \frac{1}{h_1}\frac{\partial q}{\partial x} h_2 \mathrm{d}y = 0$, ensuring minimal streamwise changes of q across y. The scale factors h_i are necessary for a derivation in an arbitrary orthogonal curvilinear coordinate system. Ansatz 1 is then introduced into the linearised Navier-Stokes equations, a scale separation between slow variations in chordwise direction and faster ones in wall-normal direction is performed, resulting in a parabolic equation system of the form $L_p\, q = 0$, with L_p the linear operator defined as follows

$$L_p = A + B\mathfrak{D}_y + C\mathfrak{D}_{yy} + D\mathfrak{D}_x\,, \tag{2}$$

with

$$\mathfrak{D}_x = \frac{1}{h_1}\frac{\partial}{\partial x}, \quad \mathfrak{D}_y = \frac{1}{h_2}\frac{\partial}{\partial y}, \quad \mathfrak{D}_{yy} = \frac{1}{h_2^2}\frac{\partial^2}{(\partial y)^2}. \tag{3}$$

The operator matrices A, B, C and D can be found in [11]. These equations can be solved marching downstream with boundary conditions

$$T = u = v = w = 0 \quad \text{at} \quad y = 0 \ \text{ and as } \ y \to \infty\,. \tag{4}$$

The equation system of LST can be derived from Eq. (2) by introducing the assumption of locally parallel flow and neglecting non-local terms, i.e. $D = 0$.

2.2 Roughness and Inhomogeneous Suction Boundary Conditions

The surface roughness is not simulated directly, but instead modeled via inhomogeneous boundary conditions [8]. No-slip conditions are projected using a Taylor-series expansion under the assumption that the roughness height is small compared to the boundary-layer thickness, thus truncating it at first order. Therefore, the boundary conditions are:

$$\begin{aligned} \hat{u}(x,0) &= -h_x\frac{\partial U}{\partial y}, \quad \hat{v}(x,0) = -h_x\frac{\partial V}{\partial y}, \\ \hat{w}(x,0) &= -h_x\frac{\partial W}{\partial y}, \quad \hat{T}(x,0) = -h_x\frac{\partial \bar{T}}{\partial y}, \end{aligned} \tag{5}$$

with some spanwise periodic roughness element $h(x,z) = h_x(x)\cdot \mathrm{e}^{i\beta_R z}$, β_R its spanwise wavenumber and $h_x(x)$ the chordwise shape function. A roughness with shape h triggers instabilities with the same spanwise wavenumber β.

Analogously, an inhomogeneous suction distribution $v_s(x,z) = v_b(x)\cdot \mathrm{e}^{i\beta_S z}$ with the spanwise wavenumber $\beta_S = \beta$ is defined on top of an arbitrary spanwise homogeneous distribution. The difference to the homogeneous boundary conditions (BC) in Eq. (4) is:

$$\hat{v}(x,0) = v_b\,. \tag{6}$$

2.3 Adjoint PSE

Knowledge of the initial amplitudes of any disturbance inside the boundary layer is necessary for a precise assessment of the transition location. Receptivity analysis via adjoint equations aims to provide these amplitudes effectively.

The derivation employs a definition for the inner product '$<\cdot,\cdot>$':

$$< a,b >= \int_{X_0}^{X_1}\int_0^\infty a^H b\, h_1 h_2\,\mathrm{d}x\,\mathrm{d}y\,, \tag{7}$$

with $()^H$ being the conjugate transpose. We define an adjoint operator L_p^* and obtain the following Lagrange identity

$$< q^*, L_p\, q >=< L_p^*\, q^*, q > + \int_{X_0}^{X_1}\int_0^\infty \nabla\cdot J(q,q^*)\, h_1 h_2 \mathrm{d}x\,\mathrm{d}y \tag{8}$$

with the adjoint flow vector $q^* = \{\rho^*, u^*, v^*, w^*, T^*\}$ and the bilinear concomitant $J = (J_x, J_y)^T$. Since the adjoint quantities q^* can be chosen arbitrarily, we define them to fulfill $L_p^* q^* = 0$. After partial integration of the left side in Eq. (8), the adjoint operator is derived as:

$$L_p^* = A^* + B^*\mathfrak{D}_y + C^*\mathfrak{D}_{yy} + D^*\mathfrak{D}_x\,. \tag{9}$$

The adjoint operators A^*, B^*, C^* and D^* can be found in Ref. [5]. Again, the equation system of adjoint LST can be derived from Eq. (9) by introducing the assumption of locally parallel flow and neglecting non-local terms, i.e. the D^*-term is zero.

If no external disturbances are introduced inside the domain, the absolute value of the streamwise part of the integrated bilinear concomitant $\int_0^\infty J_x\, h_2\,\mathrm{d}y$

is a constant and therefore poses a natural normalisation value, which is also derived from the integration by parts as:

$$J_x = q^{*H} D q\,, \tag{10}$$

with q^* and q obtained from the unforced problem with boundary conditions Eq. (4).

2.4 Receptivity Coefficients

To calculate amplitudes triggered inside the boundary layers by external disturbances, the Lagrange identity Eq. (8) is expanded:

$$\begin{aligned} 0 = \int_0^\infty [J_x]_{x_0}^{x_1} h_2 \,\mathrm{d}y - \int_{x_0}^{x_1} \Bigg[q^{*H} \left(B - \frac{1}{h_1 h_2} \frac{\partial h_1}{\partial y} C - \mathfrak{D}_y(C) \right) q \\ - \mathfrak{D}_y(q^{*H}) C q + q^{*H} C \mathfrak{D}_y(q) \Bigg]_{y=0}^{\infty} h_1 \,\mathrm{d}x\,. \end{aligned} \tag{11}$$

Equation 11 offers an efficient way to calculate instability amplitudes from the adjoint solution. We define the complex instability amplitude A as the streamwise disturbance velocity at the position y_{max} where its absolute value $|u|$ is maximal, $A(x) = \hat{u}(x, y_{\text{max}})$. To introduce A into Eq. (11), a new variable $\hat{J}$ is defined, $\hat{J} = \int_0^\infty J_x \, h_2 \,\mathrm{d}y \,/ A$. No incoming disturbances are assumed, i.e. x_0 is located upstream of any sources, thus $A(x_0) = 0$. Equation 11 is simplified with the boundary conditions $(u^*, v^*, w^*, T^*) = 0$ at $y = 0$ and as $y \to 0$, and taking into account the entries of matrix C (cf. [5]), Eq. (11) becomes

$$A(x_1) = \frac{1}{\hat{J}(x_1)} \int_{x_0}^{x_1} \left[q^{*H} B q - q^{*H} \frac{1}{h_1 h_2} \frac{\partial h_1}{\partial y} C q - \mathfrak{D}_y(q^{*H}) C q \right]_{y=0} h_1 \,\mathrm{d}x\,. \tag{12}$$

For the two cases considered in this work, surface roughness and inhomogeneous suction, we assume that $\hat{q} = q\Theta$ satisfies the boundary conditions Eq. (5) or Eq. (6) respectively. The receptivity coefficient $c_s(x;\beta)$ for suction is then defined as

$$c_s = \frac{1}{\hat{J}(x)} \left[q^{*H} B \tilde{q} - q^{*H} \frac{1}{h_1 h_2} \frac{\partial h_1}{\partial y} C \tilde{q} - \mathfrak{D}_y(q^{*H}) C \tilde{q} \right]_{y=0}\,. \tag{13}$$

and the receptivity coefficient $c_r(x;\beta)$ for roughness as

$$c_r = \frac{1}{\hat{J}(x)} \left[q^{*H} \frac{1}{h_1 h_2} \frac{\partial h_1}{\partial y} C \tilde{q} + \mathfrak{D}_y(q^{*H}) C \tilde{q} \right]_{y=0}\,, \tag{14}$$

since inserting the boundary conditions Eq. (5) for $q = \Theta^{-1}\hat{q}$ into Eq. (12) results in the B-term vanishing. The receptivity coefficients c_s and c_r are formulated to be independent of the disturbance distribution at the wall (i.e. $v_b(x)$ or $h_x(x)$)

by dividing Eq. (5) and Eq. (6) with the respective disturbance shape (e.g. $\tilde{u} = \hat{u}/(\Theta h_x)$). The amplitude at some position x_1 is then calculated as

$$A(x_1) = \int_{x_0}^{x_1} c_i \, f(x) \, h_1 \, \mathrm{d}x \tag{15}$$

with c_i being c_r or c_s and $f(x)$ being either h_x or v_b for a roughness or suction distribution respectively. Note that c_r and c_s have different units and can be compared in terms of amplitude A if information on $f(x)$ is given.

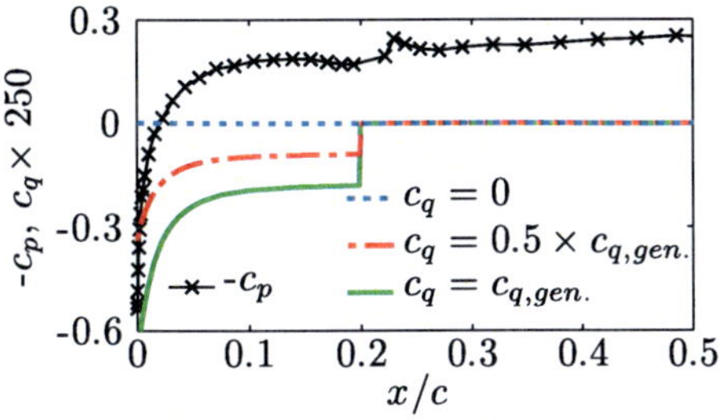

Fig. 1. Pressure and generic suction distributions.

2.5 Baseflow Configuration

The adjoint approach has been employed for the BL on the vertical tail plane (VTP) of an A320 aircraft. The flow data stem from a check-out flight test performed in November 2017 within the AFLoNext project [12]. The suction panel was located on both sides of the VTP, covering the first 20% of the chord length, designed as a double skin structure with a micro-perforated titanium sheet as outer skin supported by an inner sheet with stringers between them [12].

The laminar baseflow for the pressure distribution (Fig. 1) on the starboard side measured at a freestream Mach number of 0.76 and a side slip angle of 0.5° was computed using the boundary-layer code COCO [13]. For the present study, the complicated suction distribution from the actual suction chamber design (see [12]) was replaced by a generic one, $c_{q,\mathrm{gen}}$, resembling the actual distribution in terms of the level of suction and its chordwise evolution. The generic distribution is defined as

$$c_{q,\mathrm{gen}} = \frac{v_b}{Q_\infty} = -0.0007 - 0.002 \cdot (1 + 20\, x/c)^{-3} \quad \text{for} \quad x/c < 0.2 \tag{16}$$

with Q_∞ being the freestream velocity and c the chord length. Three baseflows have been generated by introducing multiples of $c_{q,\mathrm{gen}}$ into COCO: (i) no suction, $c_q = 0$, (ii) half suction, $c_q = 0.5 \times c_{q,\mathrm{gen}}$, and (iii) full suction, $c_q = c_{q,\mathrm{gen}}$.

3 Results

The receptivity coefficients are presented for a FSC similarity solution, verifying our implementation based on the adjoint LST and PSE framework, and for the realistic BL over an A320 vertical tail plane equipped with a suction panel.

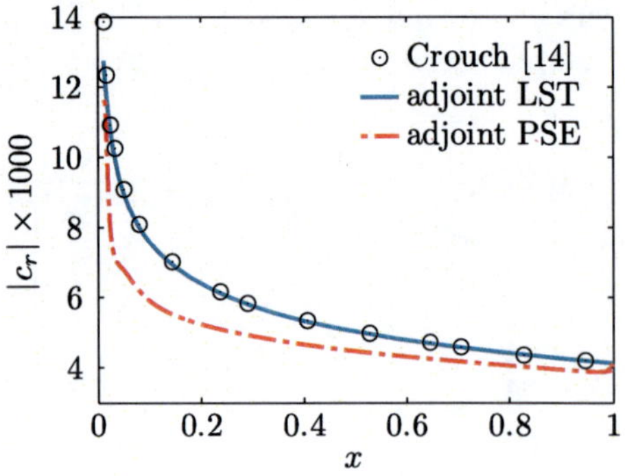

Fig. 2. Roughness receptivity coefficients for a FSC boundary layer.

3.1 FSC Results

The parameters of the FSC test case have been chosen as in [14], i.e. $\beta_H = 0.6$ and $\phi_n(x_n) = 30°$ at $Re(x_n) = 1000$, with the Hartree parameter β_H determining the pressure gradient and $\phi_n(x_n)$ denoting the local sweep angle at the normalization position x_n. We validate the results by following the adjoint method under the assumption of locally parallel flow, similar to [14]. The receptivity coefficients for a steady crossflow instability with a spanwise wavenumber of $\beta = 0.45$ have been obtained utilizing compressible equations according to Eq. (13). The results are shown in Fig. 2 together with those in [14], derived from an incompressible set of equations. The results with parallel flow approximation are in perfect agreement, as compressibility has a negligible influence for this flow. As expected, receptivity is higher towards the leading edge and decreases further downstream, with maximum receptivity at the initial chordwise position. The PSE results show the influence of non-local and non-parallel effects. At larger values of x, the coefficients get similar to those obtained assuming parallel flow, while the discrepancy towards the leading edge is of the order of magnitude of c_r. Hence, local theory over-predicts receptivity to surface roughness.

Note that the depicted adjoint PSE results have also been compared to unpublished receptivity coefficients obtained from a different PSE code within the Telfona project [15]. Those non-local results also show a very good agreement.

3.2 A320 VTP Receptivity Results

Since the flow conditions around the VTP are very similar for the port and starboard side, the following results are limited to the spatial extent of the suction panel of the VTP's starboard side. Figure 3 shows the N-factor results, calculated according to the local stability theory, for stationary crossflow instabilities for the case without suction (Fig. 3(a)) and the case with half of the generic suction rate (Fig. 3(b)). The N-factor is a measure of the accumulated growth of a disturbance and is defined as $N(x) = \ln(A(x)/A_0)$, where A_0 corresponds to the disturbance amplitude at the streamwise location where the instability starts to grow. For the case without suction, boundary-layer transition is expected at $x/c = 0.016$, assuming a typical transition N-factor of 9. In contrast, with only half of the generic suction rate, the overall growth of stationary crossflow instabilities (CFI) is probably insufficient to cause transition up to $x/c = 0.2$. At the full generic suction rate, there is no amplification of stationary crossflow vortices on the suction panel anymore. Taking the surface curvature of the airfoil into account via metric terms in the local instability analysis, the maximum achievable N-factor reduces significantly, and at $c_q = 0.5 \times c_{q,gen.}$, all stationary CFI are also damped. In the following, the receptivity investigations are limited to surface roughness and suction distributions with a spanwise wavenumber of $\beta = 1600 1/\text{m}$. The CFI triggered at this wavenumber reaches one of the highest N-factors at the end of the suction panel (cf. Fig. 3(a)). Figure 4(a) depicts the receptivity of the boundary layer to surface roughness, whereas Fig. 4(b) shows the respective coefficient for non-uniform suction, both computed employing the adjoint LST framework. For all investigated cases, the receptivity coefficient for both roughness and inhomogeneous suction is highest in the leading-edge area and decreases towards the end of the suction panel. Note that the observable kink in the c_r-distribution at $x/c > 0.15$ coincides with the beginning of an adverse pressure gradient region (cf. Fig. 1). In line with the expected trend, the boundary-layer thickness decreases with increasing suction rate and the flow becomes more receptive to disturbances at the wall. Hence, special care should

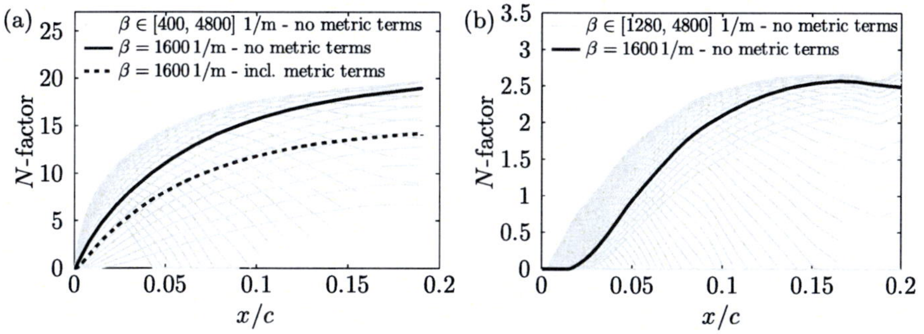

Fig. 3. Streamwise evolution of the N-factor for the stationary crossflow instability for selected spanwise wavenumbers β in case of (a) $c_q = 0$ and (b) $c_q = 0.5 \times c_{q,gen.}$.

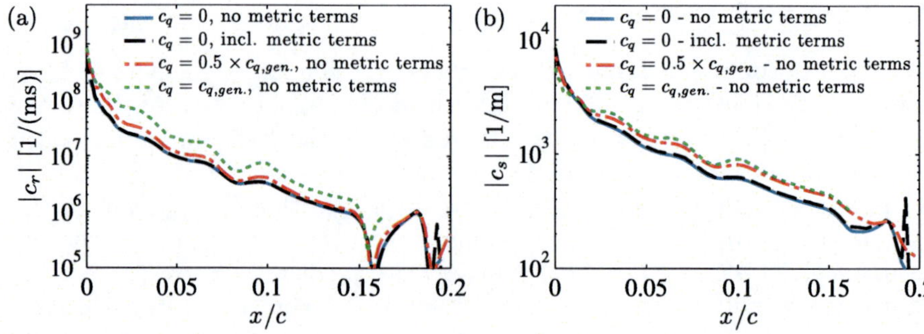

Fig. 4. Impact of the suction rate c_q on the boundary-layer receptivity due to (a) surface roughness and (b) suction with $\beta = 16001/\mathrm{m}$.

be taken when arranging and manufacturing the suction panel to ensure that the desired effect of transition delay is not jeopardized by the increased receptivity efficiency of crossflow-dominated boundary layers to external forcing. The inclusion of the metric terms, which significantly impact the CFI growth rates (cf. Fig. 3(a)), affects the receptivity coefficients only to a small extent, as indicated by the black dashed line for the $c_q = 0$ case in Fig. 4.

Investigations utilizing the adjoint PSE framework are planned to further investigate the effects of nonlocal and non-parallel flow terms on the observed trends.

4 Summary and Conclusions

The impact of varying suction rate on the receptivity of crossflow-dominated boundary layers to surface roughness and non-uniform suction was studied. The corresponding receptivity coefficients were derived from direct and adjoint solutions of the linear local and nonlocal stability equations. The numerical approach was verified by comparing the roughness receptivity coefficients for a Falkner-Skan-Cooke similarity solution with suitable reference data from the literature. Subsequently, the receptivity analysis was extended to a realistic transonic swept-wing boundary-layer flow on the A320 vertical tale plane with a porous suction panel. The amount of suction was systematically varied starting from a generic chordwise suction distribution that resembles the suction distribution of the corresponding flight test data point.

It has been shown that for stationary crossflow vortices, the effects of surface curvature on the receptivity coefficients were negligible, in contrast to their significance for the crossflow disturbance amplification rates. Moreover, as boundary-layer thickness decreases with rising spanwise homogeneous suction rate, the boundary layer can be more susceptible to a given external forcing, like surface roughness and spanwise suction non-uniformities, which both can trigger stationary crossflow vortices. The results from the present local linear receptivity analysis confirm this expected trend and, for the first time, provided quantitative data on the rise of the corresponding linear receptivity coefficients with increasing mean suction rates for a realistic HLFC configuration. The rising receptivity of the boundary layer with suction is one of the two main reasons why typically lower CF transition N-factors have to be assumed in laminar-turbulent transition prediction on HLFC wings. As the same external forcing leads to higher initial amplitudes when receptivity increases, the N-factor at a similiar transition amplitude decreases. The second contribution stems from the increased amount of external forcing due to the presence of the porous suction panel itself. Quantifying the latter effect would require detailed information on the corresponding changes in surface roughness and the suction non-uniformity, however.

Acknowledgments. We would like to acknowledge the funding by the Deutsche Forschungsgemeinschaft (DFG, German Research Foundation) under Germany's Excellence Strategy – EXC 2163/1 – Sustainable and Energy Efficient Aviation – Project-ID 390881007. The authors also express their thanks to Heiko von Geyr (DLR) for providing the A320 flight test data. The flight test activities received funding from the European Community's Seventh Framework Programme FP7/2007-2013, under grant agreement No. 604013, AFLoNext project.

References

1. Crouch, J. D.: Receptivity issues for laminar-flow control. In: IUTAM Symposium on Mechanics of Passive and Active Flow Control. Fluid Mechanics and its Applications, vol. 53, pp. 151–158. Springer (1999)
2. Janke, E.: Receptivity of three-dimensional boundary layers to roughness and suction – Linear and nonlinear effects. In: Wagner, S., Rist, U., Heinemann, H.J., Hilbig, R. (eds.) New Results in Numerical and Experimental Fluid Mechanics III. Notes on Numerical Fluid Mechanics (NNFM), vol. 77, pp. 367–374. Springer (2002). https://doi.org/10.1007/978-3-540-45466-3_43
3. Hill, D.C.: Adjoint systems and their role in the receptivity problem for boundary layers. J. Fluid Mech. **292**, 183–204 (1995)
4. Hill, D.C.: Receptivity in non-parallel boundary layers. In: Proceedings of the 1997 ASME Fluids Engineering Division Summer Meeting, FEDSM97–3108 (1997)
5. Pralits, J.O., Airiau, C., Hanifi, A., Henningson, D.S.: Sensitivity analysis using adjoint parabolized stability equations for compressible flows. Flow Turbulence Combustion **65**, 321–346 (2000)
6. Dobrinsky, A.: Adjoint analysis for receptivity prediction. Ph.D. thesis, Rice University (2002)
7. Tempelmann, D.: Receptivity of crossflow-dominated boundary layers. Ph.D. thesis, Royal Institute of Technology Stockholm (2011)
8. Schrader, L.U.: Receptivity of boundary layers under pressure gradient. Licentiate thesis, Royal Institute of Technology Stockholm (2008)
9. Piot, E., Schrader, L.U.: Receptivity of three-dimensional boundary layers to surface roughness. ERCOFTAC Bull. **74**, 13–19 (2007)
10. Bertolotti, F.P.: Receptivity of three-dimensional boundary-layers to localized wall roughness and suction. Phys. Fluids **12**, 1799–1809 (2000)

11. Hein, S., Bertolotti, F.P., Simen, M., Hanifi, A., Henningson, D.: Linear nonlocal instability analysis - the linear NOLOT code. DLR Internal Report, DLR IB 223–94, A56 (1995)
12. Schrauf, G.H., von Geyr, H.: Simplified hybrid laminar flow control for the A320 fin - Aerodynamic and system design, first results. AIAA Paper 2020–1536 (2020)
13. Schrauf, G. H.: COCO – a program to compute velocity and temperature profiles for local and nonlocal stability analysis of compressible, conical boundary layers with suction. ZARM Technik Report, November 1998
14. Crouch, J.: Receptivity of three-dimensional boundary layers. AIAA Paper 1993–0074 (1993)
15. Tempelmann, D., Hanifi, A., Hein, S., Henningson, D. S.: Validation of receptivity models: comparison with KTH wind tunnel tests. TELFONA (Testing for laminar flow on new aircraft), Deliverable No. D4.2-5 (2009)

DLR Feasibility Study of HLFC Wing Designs for S1MA Wind Tunnel Test

Thomas Streit[1](✉), Heiko Geyr von Schweppenburg[1], David Cruz[2], and Rafael Sanchez[3]

[1] DLR, Institut für Aerodynamik und Strömungstechnik, 38108 Braunschweig, Germany
th.streit@dlr.de
[2] Aernnova Engineering Division, 28034 Madrid, Spain
[3] Aerometallic Components, S.A., 01510 Milano Menor, Alava, Spain

Abstract. This work presents results of a numerical feasibility study in which HLFC wing designs for a wind tunnel model were obtained for a test in the S1MA wind tunnel. The wind tunnel model will be a testbed for HLFC technology in general and specifically a test of the integrated HLFC technology in the ground-based demonstrator of the European project HLFC-WIN. In the feasibility study a wing should be designed which satisfies the following requirements: 1) Inner wing with high values for attachment transition criteria Reynolds number 2) Outer upper wing surface with cruise design pressure distribution corresponding to the HLFC-WIN ground-based demonstrator 3) Obtained surfaces should allow manufacturability of a metallic microperforated leading edge HLFC panel. The results obtained in this work show that a design is possible which satisfies all three requirements. As a result of the requirements and the short aspect ratio planform of the wind tunnel model wing in comparison to the large aspect ratio of the ground-based demonstrator wing, the wing design led to a wing with large 3D properties. Nevertheless, the designed outer wing satisfies an isobar concept and the surface allows the manufacturability of the leading edge suction panel.

Keywords: Laminar boundary layer · HLFC technology · Transonic wing design · Inverse design methods

1 Introduction

Within the Project HLFC-WIN [1] which is part of the European Union Clean Sky 2 Joint Undertaking, DLR designed wings for a proposed HLFC (Hybrid Laminar Flow Control) wing test in the large atmospheric wind tunnel S1MA [2]. HLFC is a laminar flow control technique in which suction is applied in the nose region in order to avoid transition, which usually occurs in the nose due to crossflow instability, whereas downstream of the suction region the laminar boundary is obtained with NLF (Natural Laminar Flow), i.e. appropriately shaping the wing geometry.

© The Author(s), under exclusive license to Springer Nature Switzerland AG 2021
A. Dillmann et al. (Eds.): STAB/DGLR Symposium 2020, NNFM 151, pp. 235–245, 2021.
https://doi.org/10.1007/978-3-030-79561-0_23

In HLFC-WIN, an integrated HLFC wing will be developed by maturing and combining innovative sub-technologies in all disciplines, especially materials and structures. The target is to construct a large ground-based demonstrator in which TRL4 (Technology Readiness Level 4) is proven on integrated level. The ground-based demonstrator has the HLFC wing surface shape of the GBD-DLR-2 geometry. The GBD-DLR-2 geometry is the Airbus long range research geometry XRF1 [3, 4] with an HLFC wing. The wing was designed by DLR and is based on an Airbus HLFC wing design.

In this work a numerical feasibility study was performed by DLR in which HLFC wings were designed for the S1MA-WT (Wind Tunnel). There are several reasons which motivate a test in the S1MA-WT. For typical long range aircrafts, the attachment line transition occurs in the inner wing. Therefore, there is a high interest in obtaining flight relevant results for a test in which very high attachment criteria Reynolds number $\overline{Re}$ [5–8] values are obtained. By placing a small suction panel in the attachment line region, it could be tested if attachment line transition could be suppressed. In addition, the test would be performed at long range cruise condition Mach number, thus providing data which previously only has been obtained at low speed. Furthermore, there is also interest to provide a HLFC-WIN ground-based demonstrator testbed, testing the integrated laminar technology. For that purpose, the outer wing of the HLFC-WIN-S1MA-WT model would have a metal HLFC suction panel similarly constructed as the one of ground-based demonstrator. This would allow to test the HLFC-WIN technical solution for the HLFC suction panel. In addition, the test would include testing the joint between suction panel and wing box, testing influence of steps and gaps, using a wake rake for measuring drag, studying the shock, and influence due to turbulent wedges.

The afore mentioned reasons which motivate a test in the S1MA-WT lead to following requirements for the HLFC-WIN S1MA-WT design: 1) Large $\overline{Re}$ values typical of long range inner wings should be achieved in the inner wing, 2) In the region of the outer wing where the planed HLFC suction panel is placed a wing is designed which on the upper side has the pressure distribution of the corresponding spanwise region of the GBD-DLR2 geometry for cruise condition. This is a challenging task since in contrast to the GBD-DLR-2 geometry, which is a complete cruise configuration for a long range aircraft, the wind tunnel geometry is a short aspect ratio wing. 3) The wing design should allow manufacturability of the metallic microperforated leading edge HLFC panel.

In this work, after the introduction a section is given in which the geometry and flow conditions are described. Then in Sect. 3 design objectives are given. In the paper the numerical methods used for mesh generation, CFD, stability analysis and design are described in Sect. 4. The design strategy and results are presented in Sect. 5. Finally, conclusions are given in Sect. 6.

2 Geometry and Flow Conditions

2.1 GBD-DLR-2 and HLFC-WIN S1MA Geometries

Fig. 1. GBD-DLR-2 geometry

The GBD-DLR-2 geometry is a complete cruise geometry, including wing, body, vertical tail, nacelle and flap track fairing components, see Fig. 1. The HLFC part of the wing is restricted to the upper wing side and extends from a span position close to the nacelle up to the tip. The inner wing has the original turbulent XRF1 geometry.

The HLFC-WIN S1MA wind tunnel model is a short aspect wing with a span of 4.5 m. As initial geometry for the HLFC-WIN S1MA wind tunnel model the GBD-DLR-2 geometry was selected. Specifically, the span between sections $\eta = 0.479$ and $\eta = 0.878$ with η the dimensionless span position η = span/s. For the wind tunnel test, the wing between these sections is scaled to a semispan of s = 4.5 m. Inner and outer wing regions, where requirements 1 and 2 have to be satisfied respectively, are defined by following span regions: inner wing 0.0 m $\leq$ span $\leq$ 1.643 m, outer wing 1.643 m $\leq$ span $\leq$ 4.500 m. Figure 2 compares planforms of the GBD-DLR-2 (in the region of interest) and the HLFC-WIN S1MA model designs. Two wings with different planforms were designed. The first, denoted kink geometry, which has larger leading edge sweep for the inner wing. The second, denoted no kink geometry having identical sweep as the GBD-DLR-2 geometry.

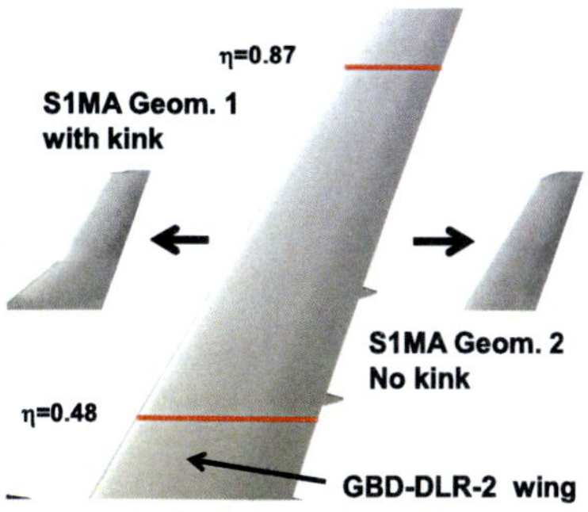

Fig. 2. Comparison of S1MA model geometries with outer HLFC-WIN GBD-DLR-2 geometry.

2.2 Flow Conditions

Design for the HLFC-WIN S1MA model is $M_\infty = 0.83$, $Re_c = 11.13 \cdot 10^6$ for c = 1.00 m. The Mach number corresponds to the design point of the GBD-DLR.2 geometry. At middle span the chord Reynolds number of the HLFC-WIN S1MA wind tunnel model is 0.73 of the corresponding wing section of the GBD-DLR-2 geometry.

3 Design Requirements

The DLR HLFC-WIN S1MA wind tunnel designs are feasibility studies for a wing which must satisfy the following requirements concerning geometry and laminar boundary layer flow properties.

1. Inner wing with high values for attachment transition criteria Reynolds number $\overline{Re}$, with values 700–1000.

2. Outer upper wing surface with cruise design pressure distribution corresponding to the HLFC-WIN ground-based demonstrator wing surface GBD-DLR-2.
3. Manufacturability of metallic microperforated leading edge HLFC panel.

4 Numerical Methods

In this section the different numerical methods used to analyse and design the HLFC-WIN-S1MA-WT geometry are presented.

4.1 CFD Method and Design Method

The analysis and design CFD solutions were obtained using the DLR inverse design solver FLOWer [9, 10]. FLOWer is an Euler and RANS solver for block structured meshes. It has been validated against experimental data and other CFD methods. The FLOWer 3D transonic inverse design solver has been chosen as CFD method for this work since: a) the meshes required here are not too complex (allowing structural mesh generation) and b) FLOWer Inverse is a robust, efficient and accurate CFD method, ideal for a large number of design iterations and extensive off-design analysis. Recently it has been used for the design of laminar wings for transonic flows [11].

4.2 Computational Meshes

Meshes were generated using the DLR in-house mesh generation software [12, 13] for structured wing-body configurations. A CH-topology is used for the wing-body mesh. First wall-normal grid spacings were chosen so that y + was lower than 1. The chosen boundary layer resolution is sufficient for viscous drag determination, but not for stability analysis. Here, stability analysis will be performed with boundary layer calculations based on sectional pressure distributions inputs. In this work analysis of CFD solutions for wing body meshes is performed on meshes with $531 \times 121 \times 81$ points and with 321×65 (chord direction $\times$ span direction) points for the wing surface. In the present work the wind tunnel wall was considered and was approximated with a symmetry wall.

4.3 Stability Analysis

In the design of the GBD-DLR-2 geometry transition positions were determined mostly using local linear stability theory. Local theory allows a robust and efficient stability analysis of 3D flows. Despite more sophisticated stability analyses methods exist, for computational cost intensive design and analysis work, linear local theory remains the preferred stability analysis engineering tool. At Airbus and DLR G. Schrauf's STABTOOL [14, 15] program is used for such work. This program uses the linear local stability theory with the e^N-method. It provides N-factors for Tollmien-Schlichting (TS) N-factors N_{TS} and crossflow (CF) N-factors N_{CF}. In addition, ALI (attachment line instability) is checked using attachment line transition criteria. The design objectives of the HLFC-WIN-S1MA designs did not require transition position determination, but STABTOOL was used to obtain the attachment line criteria.

5 Results

5.1 Design Strategy and Procedure

Attachment line transition is predicted using attachment line criteria [5–8]. One of these criteria is based on the Reynolds number $\overline{Re}$, which in the case of a swept cylinder is given by [16]:

$$\overline{Re} = \sqrt{Q_{\infty}/2\nu \cdot r_{LE} \cdot \sin\varphi_{LE\ eff} \cdot \tan\varphi_{LE\ eff}} \quad (1)$$

With Q_{∞} the free stream velocity, ν the kinematic viscosity, r_{LE} the leading edge radius and φ_{LE_eff} is the effective sweep angle (see Ref [17] for φ_{LE_eff} definition). This equation is exact for a cylinder. For other geometries it is a good approximation for any shape, if a cylinder is considered which approximates the shape [16]. The so-called Poll–Pfenniger attachment criterion says that in order to avoid attachment line transition $\overline{Re} < 250$.

According to Eq. (1), in order to obtain large $\overline{Re}$ values for given Q_{∞} and ν values, following measures can be applied:

a. increase of leading edge radius or absolute airfoil thickness,
b. increase of geometrical leading edge sweep $\varphi_{LE\ eff}$.
c. For a backward swept wing increase of spanwise airfoil thickness variation with span (see Ref. [17]) in order to increase variation between effective leading edge sweep φ_{LE_eff} and geometrical φ_{LE},
d. increase dihedral (for non-zero angle of attack increases leading edge sweep)

The attachment line criteria $\overline{Re}$ is most sensitive to modifications a) and b). Despite the GBD-DLR-2 having dihedral, it was decided that final HLFC-WIN SIMA WT model designs would be without dihedral. In this work the applied modifications to increase $\overline{Re}$ were a combination of measures a), b) and c).

Concerning leading edge sweep for the kink geometry, which was the first geometry studied, the original leading edge sweep was increased from 32° to 52° in the inner wing. Another possibility to increase leading edge is sweeping the wing by rotating it in the wind tunnel. Therefore, a second geometry without kink was designed and it was assumed that this geometry could be rotated in the wind tunnel from 32° to 52°.

Once the inner wing was designed, the 3D transonic DLR inverse design code [3] was used to achieve requirement 2 for the outer wing. Therefore, the design point pressure of the GBD-DLR-2 geometry was used as target pressure distribution for the inverse method. In the target a transition region was defined in which the required outer wing distribution changes from the required outer wing pressure distribution to the inner wing pressure distribution.

Concerning requirement 3) the designs of the inner and outer wing obtained in the previous steps were not able to satisfy it. This was due to the short aspect ratio and design requirements 1) and 2), which led to a 3d geometry with non-acceptable variations of thickness and twist. Therefore, wings were constructed based on a reduced number of defining sections, which however still satisfied requirement 1) and 2). For the final designs for both cases, i.e. the kink and no kink geometry, Aernnova and Aerometallic Components verified that a leading edge metallic microperforated suction panel could be manufactured.

5.2 Inner Wing Attachment Line Criteria Results $\overline{Re}$

Inner wing results for $\overline{Re}$ and effective sweep angle are given in Fig. 3. Note that the requirement for $\overline{Re}$ of values between 700 and 1000 is achieved with the kink geometry, whereas the no kink geometry, even in the case of rotated wing with leading sweep 52° only reaches values of for $\overline{Re} \approx 600$ (see Fig. 3 left side). This is due to the larger absolute leading edge radius of the kink geometry, which in part is due to airfoil sections with larger chord length. As shown in Fig. 3 right side, the effective leading sweep is different from the geometrical leading edge sweep (see Fig. 3 right side). Also, as expected, it decreases in the proximity of the inner wing to the wind tunnel wall. In addition, note that in the test some part of the inner wing will be immersed in the wind tunnel wall boundary layer. An estimate (private communication with Estelle Piot ONERA) gives a wind tunnel wall boundary layer thickness of 120 mm, i.e. it extends to a span region of $\eta < 0.0125$.

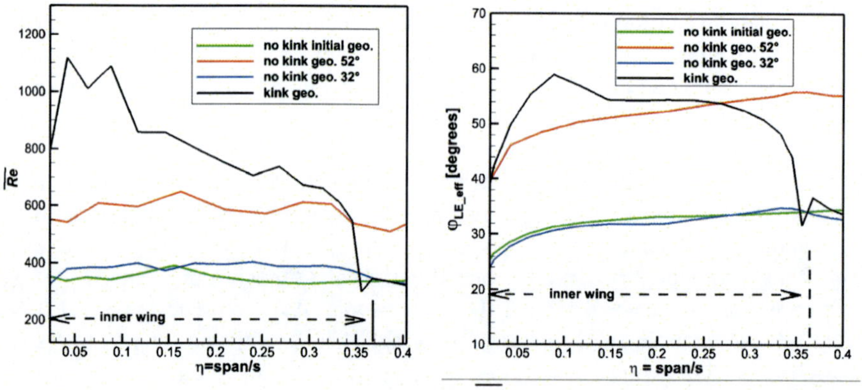

Fig. 3. Inner wing attachment transition criteria $\overline{Re}$ (left side) and φ_{LE_eff} (right side).

5.3 Presure Distributions Results, Especially Outer Wing

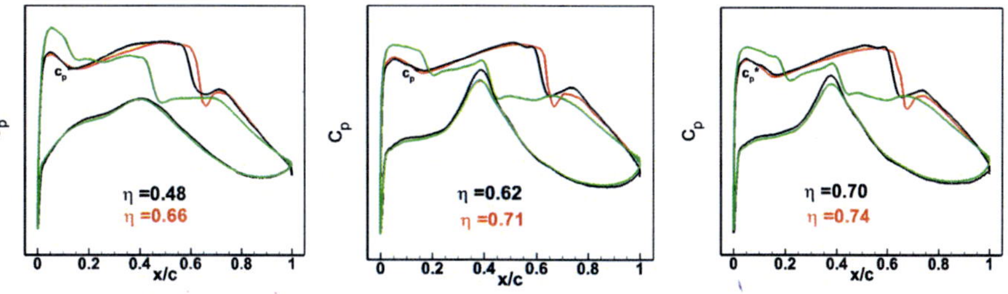

Fig. 4. Results for outer wing sections. Corresponding span position given for GBD-DLR-2 sections in red and for kink geometry in black. Pressure distributions are given for initial geometry (green), the used adapted GBD-DLR-2 target pressure distribution (red) and designed kink geometry (black).

To achieve requirement 2 the 3D DLR inverse design code [3] was applied using as target pressure distribution an adapted GBD-DLR-2 design point pressure distribution. An adapted target pressure distribution is used in order to allow for a smoother transition to the inner wing and tip regions. Figure 4 shows pressure distributions for the initial geometry, the target and the designed kink geometry. Results show that the initial geometry has a large difference to the target pressure distribution, whereas the designed kink geometry is close to prescribed target C_p. For both the kink geometry and the no kink geometry design results are close to the prescribed target HLFC pressure distribution in a span region $0.40 < \eta < 0.80$. Despite the wind tunnel model being a low aspect ratio wing GBD-DLR-2 long range configuration, it was possible to obtain an outer wing with the target pressure distribution. Figure 5 shows isobar contours for the kink geometry indicating that a constant shock position is obtained. For the design point an outer wing was obtained which satisfies requirements regarding pressure distribution in the suction region, N-factor distribution, shock position and isobar concept design. Note that on the upper side the pressure distributions have a typical HLFC characteristic gradient up to the shock in order to have a laminar boundary layer which extends to a position close to the shock. The peak in the leading edge region should not have a larger extent than the leading suction panel, at least in the design point. The suction peak on the lower side of the wing (see Fig. 4 sections $\eta = 62$ or $\eta = 0.72$) was introduced in the GBD-DLR-2 airfoils to increase airfoil thickness. Since the C_p target of the GBD-DLR-2 wing was used for the HLFC-WIN-S1MA-WT designs, this peak is also present in the wind tunnel designs. For the wind tunnel designs it has no relevance. If the loads of the wind tunnel designs were too large, an additional design step could have been done. In such a step a target pressure distribution would have be used which keeps upper pressure distribution constant and modifies the lower side to reduce lift.

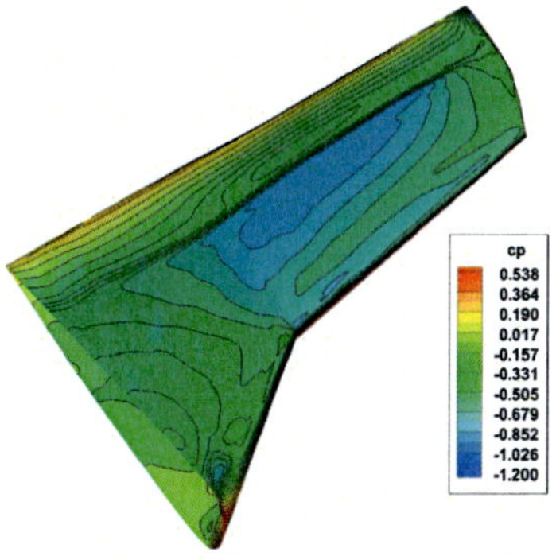

Fig. 5. Upper wing isobar contours for kink geometry design.

The initial geometry without kink had a too strong shock for the root section. Here the initial wing geometry was defined by using the airfoils of the kink geometry design, their chord length scaled to the no kink geometry. The strength of the shock was reduced, by using the inverse design with a new target pressure distribution with reduced the shock for the inner wing sections. Figure 6 shows a comparison between initial and designed pressure distribution for a section close to the root for the no kink geometry.

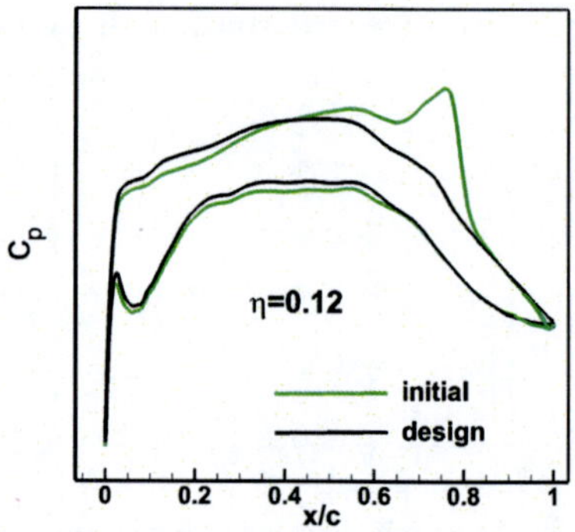

Fig. 6. Pressure distribution no kink geometry inner wing.

As function of angle of attack, the designed wing geometries also develop pressure distributions with characteristics similar to the GBD-DLR-2 wing. Results are given for the geometry with kink and without kink in Fig. 7. Note that in off design conditions the leading edge peak varies. In contrast gradients of the pressure distribution between suction peak and shock remains similar. As expected shock position moves to a more downwind position with increasing angle of attack.

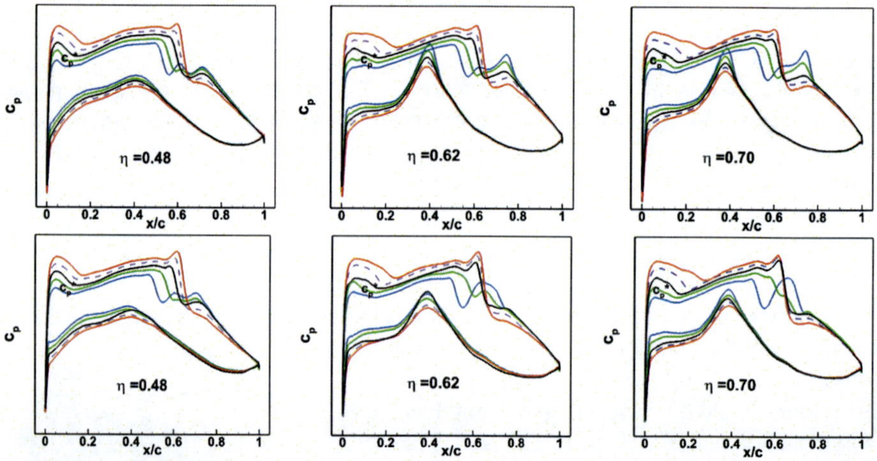

Fig. 7. Pressure distributions for outer wing sections for different angles of attack for designed geometries. Upper side are results for the geometry with kink, lower side for the geometry without kink. Results for angle of attack: $\alpha = 0.99°$ (blue), $\alpha = 1.49°$ (green), design point $\alpha = 1.99°$ (black), $\alpha = 2.49°$ (violet dashed line) and $\alpha = 2.99°$ (red).

For the final designs Aernnova and Aerometallic Components verified that a leading edge metallic microperforated suction panel could be manufactured.

6 Summary

In this work a feasibility study was performed in which wing geometries for a wind tunnel model were designed for tests in the S1MA wind tunnel. Within the European project HLFC-WIN the planned S1MA tests constitute a general HLFC technology testbed. Specifically, the feasibility study should show if a geometry could be designed which could produce large attachment line criteria $\overline{Re}$ values in order to assess open questions concerning long range aircraft inner wing leading edge transition. Furthermore, the designed wings should provide upper wing design point pressure distributions similar to the ones of the wing of the HLFC-WIN project GBD-DLR-2 geometry in order to test integrated HLFC technology. The wing surface of the HLFC-WIN ground base demonstrator has the wing surface of the GBD-DLR-2 geometry which is a long range aircraft cruise configuration with an outer HLFC wing part. A third requirement is that the wing surfaces allows the manufacturability of a metallic microperforated leading edge HLFC panel. In the design the wing was divided into two regions: an inner region leading to large $\overline{Re}$ values and an outer region in which the cruise design point pressure distributions of the GBD-DLR-2 wing are obtained. Two geometries were designed: one with a trapezoidal wing and one in which the inner wing has larger sweep.

For the geometry with larger sweep in the inner wing $\overline{Re}$ values of 700 to 1000 were obtained, thus realizing inner wing long range attachment criteria values. The geometry without larger sweep reached values of $\overline{Re} \approx 600$.

By using the DLR transonic 3D inverse design code, the pressure distribution requirement for the upper side was satisfied for both designs in the outer wing. That this could be achieved was not clear from the beginning since the inner wing has different inner wing pressure distributions in comparison to the inner wing of the complete cruise GBD-DLR-2 geometry. In addition, the wind tunnel model is a short aspect ratio wing in comparison to the large aspect wing of the GBD-DLR-2 geometry. The designs provide for the outer wing a geometry which satisfies the isobar concept with long range aircraft typical HLFC pressure distributions.

For the final designs Aernnova and Aerometallic Components verified that a leading edge metallic microperforated suction panel could be manufactured.

Acknowledgment. This project has received funding from the Clean Sky 2 Joint Undertaking (JU) under grant agreement No 945583. The JU receives support from the European Union's Horizon 2020 research and innovation programme and the Clean Sky 2 JU members other than the Union. The authors would like to thank Thomas Haase and Arne Seitz for their support during this work.

Disclaimer. The results, opinions, conclusions, etc. presented in this work are those of the author(s) only and do not necessarily represent the position of the JU; the JU is not responsible for any use made of the information contained herein.

References

1. HLFC-WIN Homepage: https://www.hlfc-win.eu/
2. SIMA: https://www.onera.fr/sites/default/files/windtunnel/pdf/S1MA-2018.pdf
3. Kroll, N., et al.: DLR project Digital-X: towards virtual aircraft design and flight testing based on high-fidelity methods. CEAS Aeronaut. J. **7**(1), 3–27 (2015). https://doi.org/10.1007/s13272-015-0179-7
4. Pattison, J., Wilson, T., Herring, M.: High fidelity simulation of the folding wing tip for loads alleviation. In: International Forum on Aeroelasticity and Structural Dynamics. St. Petersburg, Russia (2018)
5. Pfenniger, W.: Laminar Flow Control. Laminarization. AGARD-R-654, pp. 3-1-3-75 (1977)
6. Poll, D.I.A.: Some Aspects of the Flow near a Swept Attachment Line with Particular Reference to Boundary Layer Transition. Cranfield Inst. of Techn., CoA Report No. 7805 (1978)
7. Arnal, D., Juillen, J.C., Reneaux, J., Gasparian, G.: Effect of wall suction on leading edge contamination. Aerosp. Sci. Technol. **8**, 505–517 (1997)
8. Hall, P., Malik, R., Poll, D.I.A.: On the stability of the Infinite Swept Attachment Line Boundary Layer, NASA-CR-172300 (1984)
9. Kroll, N., Fassbender, J.K.: MEGAFLOW –Numerical Flow Simulation for Aircraft Design, Notes on Numerical Fluid Mechanics and Multidisciplinary Design (NNFM) vol. 89. Springer Verlag, Closing Presentation DLR Project MEGAFLOW Braunschweig (de), 10–11 January 2002
10. Streit, T., Hoffrogge, C.: DLR transonic inverse design code, extensions and modifications to increase versatility and robustness. Aeronaut. J **121**(1245), 1733–1757 (2017). https://doi.org/10.1017/aer.2017.101
11. Streit, T., Wedler, S., Kruse, M.: DLR natural and hybrid transonic laminar wing design incorporating new methodologies. Aeronaut. J **119**(1221), 1303–1326 (2015)
12. Wichmann, G.: Untersuchung zur Flügel-Rumpf-Interferenz durch Anwendung eines Eulerverfahrens für kompressible Strömung, Dissertation. Institut für Strömungstechnik, Technische Universität Braunschweig, ZLR-Forschungsbericht 93-06 (1993)
13. Streit, T., Wichmann, G., Rohardt, C.-H.: Nachrechnung und Entwurf von transsonischen Tragflügeln für die DO-728/DO-928, Technical Report, DLR IB 129–99/13. Braunschweig, Germany (1999)
14. Schrauf, G.: LILO 2.1 – Users Guide and Tutorial, GSSC Technical Report, July 2006
15. Schrauf, G.: COCO – A Program to Compute Velocity and Temperature Profiles for Local and Nonlocal Stability Analysis of Compressible, Conical Boundary Layers with Suction, ZARM Technik Report, November 1998

16. Forte, M., Piot, E., Vermeersch, O., Casalis, G., François, C.: Research activities of ONERA on laminar airfoils in the framework of JTI Clean Sky SFWA-ITD: transition control. Greener Aviation 2016, Oct 2016, BRUXELLES, Belgium. ffhal-01399238f
17. Streit, T., Wichmann, G., Campbell, R.L., von Knoblauch zu Hatzbach, F.: Implications of conical flow for laminar wing design and analysis. In: AIAA Paper 2011-3808, 29th AIAA Applied Aerodynamics, Conference, 27–30 June 2011, Honolulu, HI, USA

Check for updates

The Effect of 2-D Surface Irregularities on Laminar-Turbulent Transition: A Comparison of Numerical Methodologies

Francesco Tocci[1(✉)], Juan Alberto Franco[1], Stefan Hein[1], Guillaume Chauvat[2], and Ardeshir Hanifi[2]

[1] German Aerospace Center (DLR), Bunsenstrasse 10, 37073 Göttingen, Germany
francesco.tocci@dlr.de
[2] FLOW, Department of Engineering Mechanics, KTH Royal Institute of Technology, SE-100 44 Stockholm, Sweden

Abstract. The applicability of Local Stability Theory (LST), Parabolized Stability Equations (PSE) and the Adaptive Harmonic Linearized Navier-Stokes (AHLNS) approach is investigated in the presence of 2-D surface irregularities through comparison with Direct Numerical Simulations (DNS). Remarkably good agreement between DNS and AHLNS is obtained for the amplification curves of Tollmien-Schlichting (TS) waves in all the cases studied. The LST and PSE results exhibit differences which are discussed in relation to the local distortion of the boundary layer induced by the irregularities.

Keywords: Laminar-turbulent transition · 2-D surface irregularities · DNS · AHLNS · PSE · LST

1 Introduction

Laminar-turbulent transition prediction is of practical interest in aircraft design since transition affects important aerodynamic quantities such as drag and heat transfer. Extended laminar flow on aerodynamic surfaces is an effective way of reducing aircraft drag. One of the major challenges for the implementation of laminar-flow surfaces is the potential for any irregularity to move transition upstream. Under low-disturbance environment, boundary-layer transition results from the growth and breakdown of different flow instabilities. In 2-D flows the scenario is dominated by TS instabilities. Common wing-surface irregularities, such as two-dimensional steps, gaps or waviness can alter the growth characteristics of TS waves and therefore must be taken into account at the design stage [1]. The effect of steps, gaps and humps on the development of TS waves in an incompressible boundary layer has been investigated in [2] using DNS. All the irregularities were found to have an overall destabilizing effect. In the present work,

© The Author(s), under exclusive license to Springer Nature Switzerland AG 2021
A. Dillmann et al. (Eds.): STAB/DGLR Symposium 2020, NNFM 151, pp. 246–256, 2021.
https://doi.org/10.1007/978-3-030-79561-0_24

we compare those results with LST, PSE and AHLNS. In the past, LST and PSE, together with the e^N method, have been successfully applied for transition prediction in cases without surface irregularities where the local streamwise flow gradients were small. However, the validity of the assumptions of such methods becomes questionable with the increased gradients locally induced by the surface irregularities. AHLNS removes some of the inherent limitations present in LST and PSE, and it has been already applied successfully in the presence of humps by Franco et al. [3]. Edelmann et al. [4] used DNS and LST with forward-facing steps introducing, for the latter, a special treatment in the region around the step and Park et al. [5] performed PSE analyses for smooth humps. To the best of the authors' knowledge, there is no published comparison between LST, PSE, AHLNS, and DNS for the development of TS waves encountering different types of surface irregularities.

2 Numerical Methods

In this section we present very brief descriptions of the numerical approaches used to study the linear spatial evolution of convective instabilities in boundary-layer flows. The instability of steady flows to small amplitude perturbations can be analysed using the modal approach. Given an operator that describes the evolution of small perturbations, this approach considers the temporal or spatial development of individual eigenmodes of that operator. The resulting linear stability theory relies on the decomposition of all flow quantities $\boldsymbol{q}$ into a steady basic flow $\overline{\boldsymbol{q}}$ plus an unsteady disturbance flow component $\tilde{\boldsymbol{q}}$ according to

$$\boldsymbol{q}(\boldsymbol{x},t) = \overline{\boldsymbol{q}}(\boldsymbol{x}) + \epsilon\tilde{\boldsymbol{q}}(\boldsymbol{x},t), \qquad \epsilon \ll 1, \tag{1}$$

where $\boldsymbol{x}$ is the space coordinate vector and t is time. Substituting (1) into the Navier-Stokes (NS) equations, subtracting the equations satisfied by the steady flow, and dropping the terms of order higher than ϵ yields the Linearized Navier-Stokes (LNS) equations. The basic laminar state $\overline{\boldsymbol{q}}$, around which the linearization is performed, needs to be calculated for each of the surface irregularities. In [2], the development of TS waves encountering a forward-facing step, backward-facing step, rectangular or smooth hump and gap was studied by means of DNS. That work provides the steady laminar 2-D basic flows for the instability analysis based on LST, PSE and AHLNS, which is the main contribution of this paper. Different assumptions can be made for the form of the basic flow and the disturbances in (1), which then result in the governing equations associated with LST, PSE and AHLNS. Due to space limitations, only the key aspects for each methodology are reported.

Local Stability Theory. In LST the flow is assumed to be parallel thus not accounting for the growth of the boundary layer. In the classical LST, the basic flow is considered homogeneous along two out of the three spatial directions, here streamwise x and spanwise z direction. Therefore, the flow is only dependent on the wall-normal coordinate, and the perturbation has the following form:

$$\begin{aligned} \tilde{\boldsymbol{q}}(x,y,z,t) &= \hat{\boldsymbol{q}}(y)e^{i\Theta(x,z,t)} + \text{c.c.} \\ \Theta(x,z,t) &= \alpha x + \beta z - \omega t, \end{aligned} \tag{2}$$

where ω is the angular frequency, c.c. denotes the complex conjugate, and α, β are the wavenumbers in streamwise and spanwise direction, respectively.

Parabolized Stability Equations. A more general representation of the disturbances is given by:

$$\begin{aligned} \tilde{\boldsymbol{q}}(x,y,z,t) &= \hat{\boldsymbol{q}}(x,y)e^{i\Theta(x,z,t)} + \text{c.c.} \\ \Theta(x,z,t) &= \int_{x_0}^{x} \alpha(x')dx' + \beta z - \omega t. \end{aligned} \tag{3}$$

If a flow varies slowly in one spatial direction, say x, then a scale separation $1/Re$ is introduced between the weak variation in the x direction and the strong variation in the y direction. The normal basic flow velocity component and the streamwise dependence of the basic and disturbance flow scale with $1/Re$, where Re is based on a length scale l^* proportional to the boundary-layer thickness, $l^* = \sqrt{x^* \nu^* / U^*}$ ($*$ denotes dimensional quantities). With the above scaling and ansatz (3) introduced into the LNS equations, considering only terms that scale with powers of Reynolds number up to $1/Re$, we obtain the stability analysis approach based on PSE. The effects of small flow non-parallelism are consistently taken into account and due to the fact that the disturbance characteristics predicted by such a method are influenced by local and upstream flow conditions, this theory is called nonlocal. For a complete explanation about the PSE methodology, the readers are referred to Herbert [6]. Stability calculations were performed with the spatial nonlocal linear stability code NOLOT, which was developed in cooperation between DLR and the Swedish defence research agency FOI and has been extensively validated [7].

Adaptive Harmonic Linearized Navier-Stokes Methodology. The AHLNS equations are obtained in a similar way to the PSE formulation as regards the wave-like character of the disturbances which are divided into an amplitude function and wave function as Eq. (3). However, differently from the PSE approach, the streamwise wavenumber α and the amplitude function $\hat{\boldsymbol{q}}$ are allowed to vary rapidly in streamwise direction. The assumption of slowly varying flow in streamwise direction is removed also for the basic flow, so that all the terms of the LNS equations are kept. A multi-zonal approach is then applied, which exploits PSE relatively far away from the surface irregularity, where the streamwise variations of the basic flow and amplitude function are small, and AHLNS only in the vicinity of the irregularity. In the present paper, results obtained using this multi-zonal approach will be labeled as AHLNS results. For a detailed description of this methodology the readers are referred to Franco et al. [8].

Amplification Curve and Growth Rate. By integrating the spatial growth rate of the perturbations, the amplification curve of their amplitude along the streamwise direction can be obtained. The amplification curves of the DNS results presented in [2] allow a quantitative comparison with LST, PSE and AHLNS through the relation: $\ln (A_u/A_{ref})_{DNS} = \int_{x_{ref}}^{x} \sigma(x')dx'$, where A_u is the maximum amplitude along the wall-normal coordinate ($A_u(x) = \max_y |\hat{u}(x,y)|$) and σ is the spatial disturbance growth rate. In LST, the spatial growth rate is computed from the imaginary part of the complex streamwise wavenumber, $\sigma = -\alpha_i$. In PSE and AHLNS, it is given by $\sigma = -\alpha_i + \Re(\frac{1}{\xi}\frac{\partial \xi}{\partial x})$, where $\hat{u}$ at the wall-normal location where it reaches its maximum value, has been chosen in this paper among the possible choices for ξ. The growth rates of the DNS can also be determined carrying out numerically the differentiation of the amplification curves: $\sigma_{DNS} = d(\ln(A_u/A_{ref}))/dx$.

3 Results

3.1 Laminar Basic Flows

The laminar basic flows used in this work were computed by Tocci et al. [2] through DNS. The surface irregularities are located on a flat plate without pressure gradient at Reynolds number based on the centre position of the irregularity $Re_{x_c} = 4 \cdot 10^5$. The Blasius length scale $l_c^* = x_c^*/\sqrt{Re_{x_c}}$ at this location is chosen for non-dimensionalization. With this scaling, the center of the irregularities is located at $x_c = 632.46$. The non-dimensional height or depth of the geometric features is $h = 0.808$ ($Re_h = 511$) and the non-dimensional width of the humps and gap is $b = 31.62$ ($Re_b = 2 \cdot 10^4$). We refer the reader to [2] for further details on the basic flow characteristics. As example, Fig. 1 shows the steady laminar basic flow for one of the surface imperfections studied, a smooth hump.

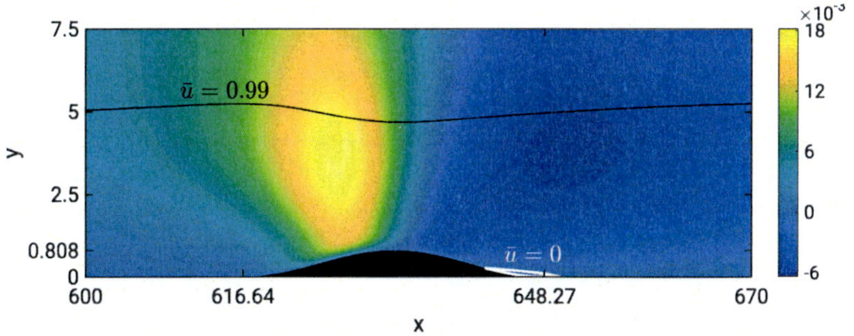

Fig. 1. Coloured contours of the non-dimensional wall-normal velocity component $\overline{v}$ and iso-lines of the non-dimensional streamwise velocity component $\overline{u} = 0.99$ (black) and $\overline{u} = 0$ (white). The axes are not to scale.

3.2 Instability Analysis

Before looking at the surface irregularities, the flat plate solution is considered in Fig. 2. The growth rate obtained from LST differs from the DNS and PSE calculations. This deviation can be explained by nonparallel effects, that is, the influence of growing boundary layer (see also [9]). Moreover, the amplitude curve of LST (Fig. 2, right) shows the integrated error, so that the relative deviation from DNS is higher far downstream. The curves for DNS and PSE of Fig. 2 essentially collapse and therefore show the remarkable agreement of the two numerical approaches in a flat plate boundary-layer flow. The growth rate of the DNS exhibits high values where the disturbances are introduced and thus manifest the generation of disturbances waves. Before considering the overall comparison of the methodologies, we point out some of the key findings concerning the applicability of LST and PSE in the presence of surface irregularities. Since the neutral point for the flat plate ($x = 620$) lies within the irregularities region (Fig. 1), a slighty upstream position ($x_{ref} = 610$) is chosen to normalize the amplitude in order to show the differences in amplification curves far downstream due to the irregularities only.

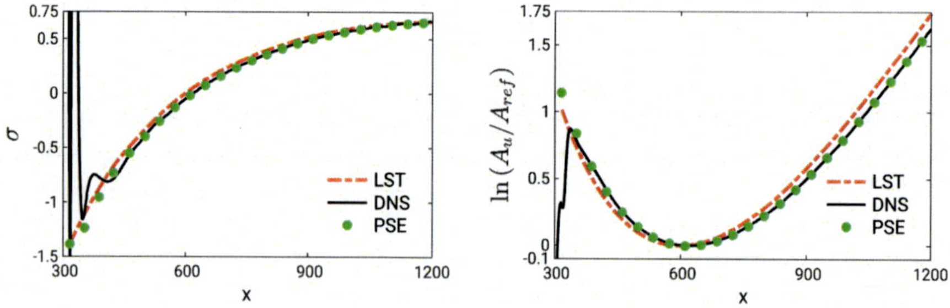

Fig. 2. Dimensional growth rates (left) and natural logarithm of the normalized disturbance amplitude (right) for TS waves at $F = 2\pi f^* \nu^* / U_\infty^{*2} = 49.34 \cdot 10^{-6}$ in incompressible flat plate boundary-layer flow. A_{ref} represents the maximum amplitude of $\hat{u}$ along y at $x_{ref} = 610$.

LST. Particularly critical for the LST calculations is the region around a sharp edge, i.e. rectangular hump or step cases, where the LST computation stops. Basically, the resolution in streamwise direction for the LST can be the same as for the DNS and it is possible to achieve grid independence of the results. However, a small region in the vicinity of the sharp edge has to be omitted (see also [4]). For the forward-facing step, Fig. 3 shows the mesh independence study and the influence of the size of the region which is not considered for the LST computation, labelled with increasing size order as Δx_1, Δx_2 or Δx_3. Similar behaviour can be observed for the other irregularities with sharp edges, while for the smooth irregularities the LST computations could be carried out for all the points in streamwise direction.

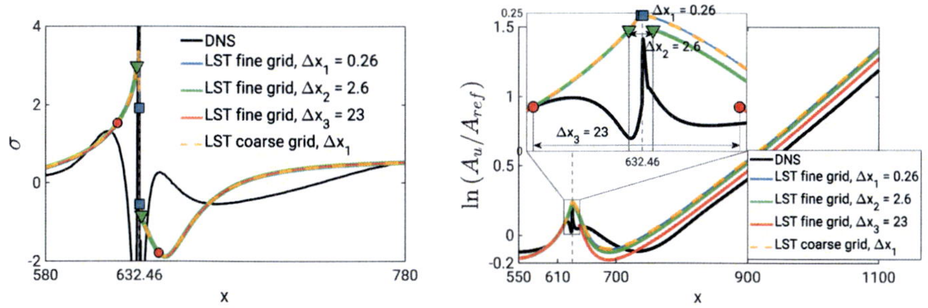

Fig. 3. Dimensional growth rates (left) and natural logarithm of the normalized disturbance amplitude (right) for TS waves at F $= 49.34 \cdot 10^{-6}$ in the presence of a forward-facing step. The dashed grey line shows the location of the step. The markers represent the last and first point of the splitted LST domains across the step: Δx_1, Δx_2 and Δx_3 refer to the increasing skipped region. A_{ref} represents the maximum amplitude of $\hat{u}$ along y at $x_{ref} = 610$.

PSE. The classical nonlocal stability equations, even though parabolized in the sense that second derivatives in the streamwise direction are missing, still contain some residual ellipticity. This leads to rapid oscillations of the solution which can eventually blow up the calculation when the marching step-size is too small. This step-size restriction becomes critical especially in the vicinity of the irregularity where higher resolution is needed. It is possible to march over the irregularity without a breakdown of the calculation only with large enough step-sizes. It can be shown that the residual ellipticity is mostly due to the upstream propagation through the streamwise pressure gradient terms. This issue can be then addressed through modifications of the PSE where the pressure gradient terms are neglected. Even though this technique relaxes the step-size restriction, it turns out that the grid convergence of the results can be obtained only for the smooth irregularities. For the smooth hump case the PSE results show an oscillatory behavior for $\alpha_r \cdot dx < 0.85$ and grid independence is not reached. Dropping the pressure terms allows grid independent results from a step size of $\alpha_r \cdot dx = 0.5$. On the contrary, the above procedure does not help in the presence of the rectangular hump and the forward-facing step. For those cases, it is worth mentioning that the PSE results are considerably affected not only by the step-size restriction, i.e. the number of points, but also by the distribution of the points where the PSE computation is done, especially in the vicinity of the irregularity. This is shown in Fig. 4 where the same number of points is considered but they are shifted such that one point falls right before or after the end of the hump. Figure 5 summarizes the PSE results for the irregularities studied, pointing out the problem of grid dependency in the presence of a sharp corner, i.e. rectangular hump and forward-facing step. We remind the reader that Andersson et al. [10] proposed another stabilization technique that removes the step-size restriction in PSE at the expense of an increasing artificial modification of the governing equations with decreasing step size, however it was not considered in this work.

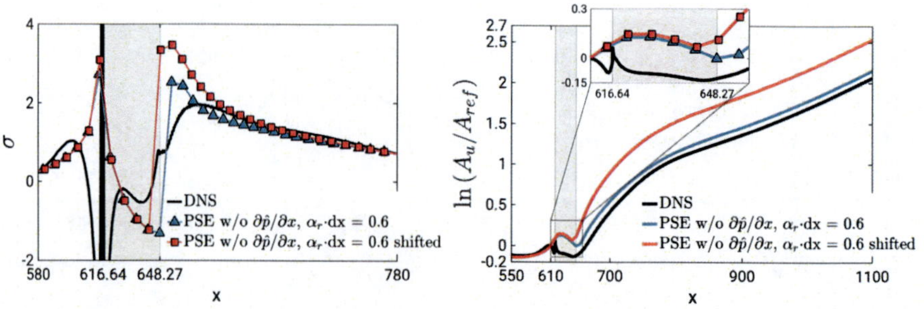

Fig. 4. Dimensional growth rates (left) and natural logarithm of the normalized disturbance amplitude (right) for TS waves at $\mathrm{F} = 49.34 \cdot 10^{-6}$ in the presence of a rectangular hump, indicated by the grey area. The markers represent the point distribution for the PSE computations. A_{ref} represents the maximum amplitude of $\hat{u}$ along y at $x_{ref} = 610$.

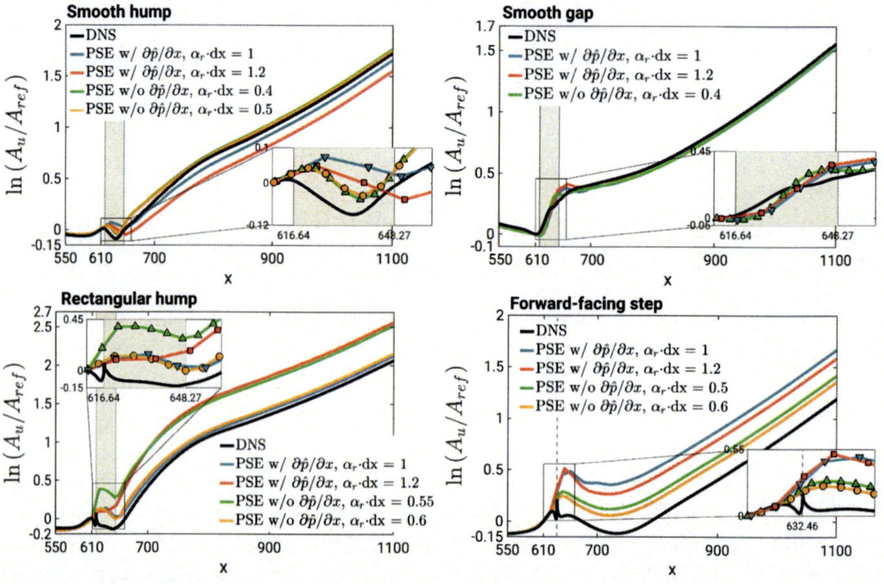

Fig. 5. Natural logarithm of the normalized disturbance amplitude for TS waves at $\mathrm{F} = 49.34 \cdot 10^{-6}$. The grey area indicates the location of the humps and gap, while the dashed grey line shows the location of the step. A_{ref} represents the maximum amplitude of $\hat{u}$ along y at $x_{ref} = 610$.

The comparison between LST, PSE, AHLNS and DNS is shown in Fig. 6 for a rectangular and smooth hump, smooth gap and forward-facing step. A rectangular gap and a backward-facing step were also investigated; however, their instability analyses do not provide any further conclusion which is not already drawn by analysing the other cases, therefore they are omitted here due to

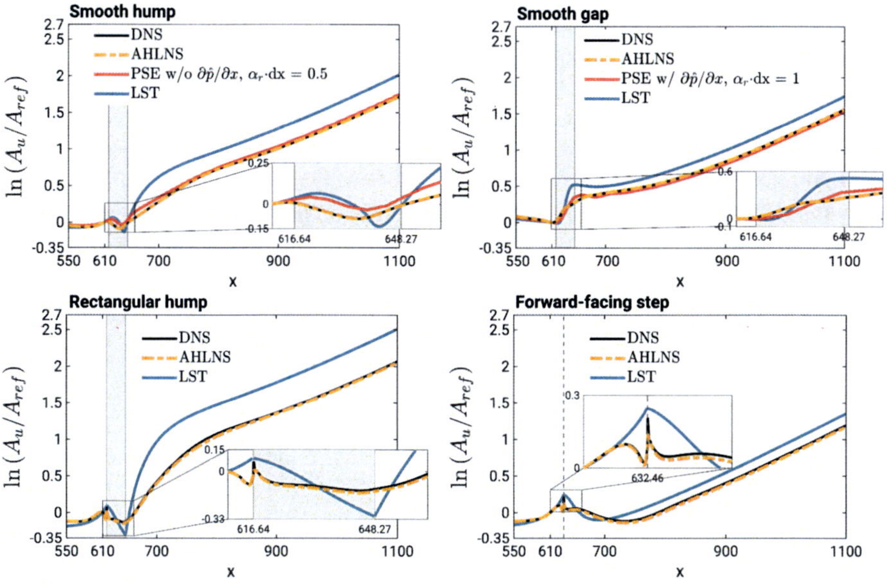

Fig. 6. Natural logarithm of the normalized disturbance amplitude for TS waves at $\mathrm{F} = 49.34 \cdot 10^{-6}$. The grey area indicates the location of the humps and gap, while the dashed grey line shows the location of the step. A_{ref} represents the maximum amplitude of $\hat{u}$ along y at $x_{ref} = 610$.

space limitation. The AHLNS amplification curves are in excellent agreement with those obtained by DNS for all the surface irregularities, demonstrating the validity of this methodology to study the effect of surface irregularities on the spatial development of convective boundary-layer instabilities. PSE is able to predict the development of TS waves over the smooth hump and gap with some deviations in the region of the irregularities which, anyway, do not cause large differences far downstream where the same growth rate of the flat plate is recovered. As mentioned earlier, the smooth hump case required to drop the pressure terms $\partial\hat{p}/\partial x$ in order to reduce the step size and achieve a converged solution. The latter stabilization technique was not necessary for the smooth gap while for the rectangular hump and forward-facing step the results were anyway grid dependent (see Fig. 5) and therefore the PSE curves are not plotted. For all investigated cases, LST is over-predicting the growth in the area of the separation bubble induced by the irregularities. As a result, the amplification level far downstream is higher compared to DNS. For LST, the basic state is approximated as locally parallel, an assumption that does not hold, in particular in the vicinity of the irregularity where the flow is deflected reaching higher value of the wall-normal velocity (Fig. 1). Besides violating the nonparallel assumption, the local distortion of the boundary layer induced by the irregularities might conflict with the PSE requirement of a slowly changing basic flow in the streamwise direction. As expected, the rectangular hump introduces stronger localized streamwise gradients compared to the smooth hump, as shown in Fig. 7 where the derivative $\partial \overline{u}/\partial x$ is plotted. In Fig. 8 the ratio between the maximum along the wall-normal direction of the derivative $\partial \overline{u}/\partial x$ and $\partial \overline{u}/\partial y$ gives an indication of the regions where actually the gradients in the streamwise direction are not negligible compared to that in the wall-normal direction. In the rectangular hump and forward-facing step cases, the peaks associated with high values of this ratio should cast doubt on the applicability of PSE and provide an explanation for the difficulties in obtaining grid-converged solutions (Fig. 4 and Fig. 5).

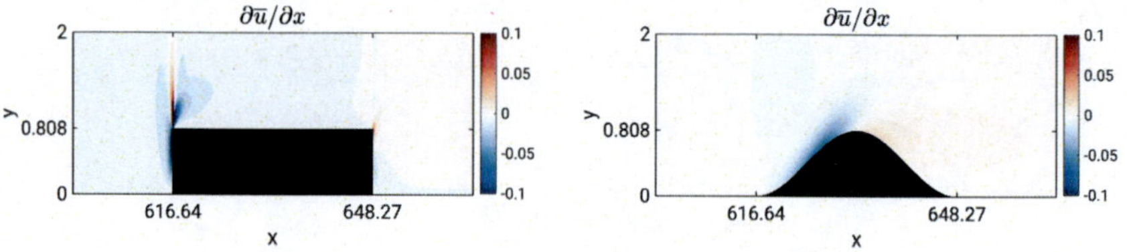

Fig. 7. Coloured contours of the non-dimensional streamwise derivative of the streamwise velocity component for the basic flow $\overline{u}$ (the axes are not to scale). Rectangular hump (left) and smooth hump (right).

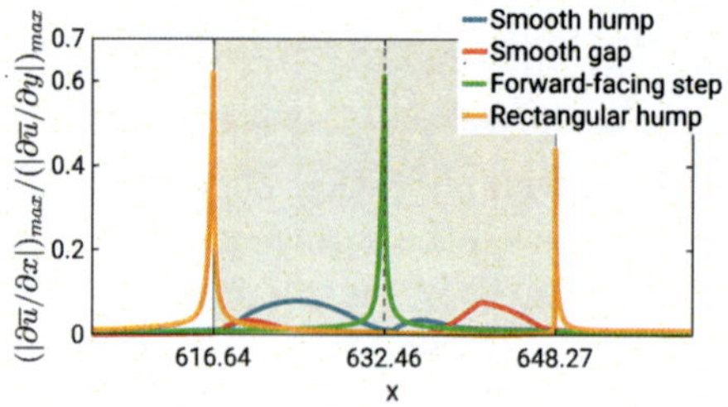

Fig. 8. Ratio of the streamwise and wall-normal derivative of the streamwise velocity component of the basic flow $\overline{u}$. The light grey area indicates the location of the humps and gap, while the dashed dark grey line shows the location of the step.

LST and PSE required a similar computational effort of the order of seconds. The AHLNS computations took around minutes and come at a fraction of the cost of DNS which are orders of magnitude more expensive.

4 Conclusions

The spatial linear instability analysis of TS waves is presented through the results provided by LST, PSE and AHLNS in the presence of different 2-D irregularities on a flat plate in an incompressible flow. Even if limited to a single frequency, comparison with DNS allows to evaluate the accuracy and applicability of the different approaches. The AHLNS approach, given the excellent agreement with DNS for all the cases, proves to be a perfectly suited tool for the boundary-layer instability analysis in the presence of large streamwise gradients. Moreover, with a moderate computational effort compared to DNS, it is well suited for parametric studies. The use of PSE should be avoided in the presence of sharp corners, where the assumption of slowly varying basic flow with respect to x is violated but it could still be valid for smooth irregularities such as the ones studied here. However, we remind the reader that PSE could also provide wrong spatial amplification in the presence of smooth irregularities if they introduce stronger streamwise gradients. In general, PSE requires some attention due to its step-size restriction with respect to the resolution required in the vicinity of the irregularities. In this regard, the smooth gap is the less critical irregularity while the smooth hump case required to neglect the pressure terms $\partial\hat{p}/\partial x$ in order to reduce the step size and achieve grid convergence. Besides these pure numerical considerations, one should always keep in mind that when the characteristic length scale of the perturbation in the streamwise direction, i.e. the TS wavelength, is of the same order as the corresponding characteristic length scale of the streamwise flow variation induced by the surface irregularity, the applicability of PSE is dubious. LST shows differences from the DNS results in all the cases studied. This disagreement is entirely due to the inherent weaknesses of the LST modelling assumptions in the presence of surface irregularities which induce local distortion of the boundary layer.

Acknowledgements. The financial support of SSeMID, Stability and Sensitivity Methods for Industrial Design, funded by European Union's Horizon 2020 research and innovation programme under the Marie Skłodowska-Curie Grant Agreement No. 675008, and from the Clean Sky 2 Joint Undertaking under the European Union's Horizon 2020 research and innovation programme under Grant Agreement No. CS2-AIR-GAM-2014-2015-01 are gratefully acknowledged.

References

1. Nenni, J.P., Gluyas, G.L.: Aerodynamic design and analysis of an LFC surface. Astronaut. Aeronaut. **4**(7), 52–57 (1966)
2. Tocci, F., Chauvat, G., Hein, S., Hanifi, A.: Direct numerical simulations of Tollmien-Schlichting disturbances in the presence of surface irregularities (To appear in Procedia IUTAM, 2020)
3. Franco, J.A., Hein, S., Valero, E.: On the influence of two-dimensional hump roughness on laminar-turbulent transition. Phys. Fluids **32**, 034102 (2020)
4. Edelmann, C., Rist, U.: Impact of forward-facing steps on laminar-turbulent transition in transonic flows. AIAA J. **53**(9), 2504–2511 (2015)
5. Park, D., Oh, S.: Effect of shape of two-dimensional smooth hump on boundary layer instability. Int. J. Aeronaut. Space Sci. (2020)
6. Herbert, T.: Parabolized stability equations, Progress in Transition Modelling, AGARD Report No. 793 (1994)
7. Hein, S., Bertolotti, F.P., Simen, M., Hanifi, A., Henningson, D.: Linear nonlocal instability analysis - the linear NOLOT Code. DLR-IB 223–94 A56 (1995)

8. Franco, J.A., Hein, S.: Adaptive Harmonic Linearized Navier-Stokes equations used for boundary layer instability analysis in the presence of large streamwise gradients. AIAA Paper 2018-1548 (2018)
9. Fasel, H., Konzelmann, U.: Non-parallel stability of a flat-plate boundary layer using the complete Navier-Stokes equations. J. Fluid Mech. **221**, 311–347 (1990)
10. Andersson, P., Henningson, D., Hanifi, A.: On a stabilization procedure for the parabolic stability equations. J. Eng. Math. **33**, 311–332 (1998)

Numerical Study of Dynamic 2D Bumps for Active Gust Load Alleviation

Junaid Ullah(✉), Ferdinand Seel, and Thorsten Lutz

University of Stuttgart, Institute of Aerodynamics and Gas Dynamics, Pfaffenwaldring 21, 70569 Stuttgart, Germany
ullah@iag.uni-stuttgart.de

Abstract. Gust loads on aircraft are critical for structural design. This paper investigates local bump-like modifications in the leading edge region on a generic supercritical airfoil at transonic inflow conditions for the purpose of gust load alleviation. Several parameters are studied to identify a sweet spot for gust load compensation. These include the bump's height, its streamwise position, its streamwise extension and its oscillation frequency. The evaluations of the URANS simulations reveal that the bumps as investigated in this work are not applicable for compensation of gust induced lift. However, dynamic bumps enable a compensation of approx. 20% of the gust induced pitching moment. A comparison of dynamic bumps with an oscillating trailing edge flap (TEF) and leading edge flap (LEF) reveals the superiority of the flaps for gust load mitigation, especially when a load factor of one is targeted throughout the critical gust event. The qualitative effect of dynamic leading edge bumps on the surrounding flow field is found to be similar to the one caused by oscillating LEF.

1 Introduction

Modern aircraft encounter significant gust induced forces during cruise flight, which represent dimensioning loads for structural design [1]. Current approaches to reduce gust loads include mainly passive methods such as additional stiffness, redistribution of mass and restriction of flight speed. Active Gust Load Alleviation (AGLA) by means of moving wing components presents an attractive measure to reduce gust loads. In addition to mitigation of structural loads AGLA systems promise a significant reduction in fuel consumption, passenger comfort, structural weight and endurance. Recent studies on alleviation of maneuver and gust loads at transonic speeds cover mostly dynamic trailing edge flaps (TEF) and ailerons [1–3].

The supersonic region on aircraft wings represents the most distinctive feature at transonic conditions. Thus, a modification of the supersonic region by e.g. precisely controlling the expansion and compression waves inside the sonic region, introducing a new system of shock waves, controlling the size of the sonic region, and shifting the location of the closing vertical shock wave, can significantly contribute to a mitigation of gust induced loads. Local bump-like

© The Author(s), under exclusive license to Springer Nature Switzerland AG 2021
A. Dillmann et al. (Eds.): STAB/DGLR Symposium 2020, NNFM 151, pp. 257–268, 2021.
https://doi.org/10.1007/978-3-030-79561-0_25

modifications inside the supersonic region on the suction side of the wing, represent an attractive measure to introduce disturbances which can affect the flow patterns inside the whole supersonic region. So far bumps are only applied in steady mode in the region of the shock foot. These shock control bumps (SCB) have demonstrated their effectiveness in reducing the wave drag at the point of drag rise and shifting the transonic buffet onset towards higher velocities by spreading the recompression area and delaying shock induced separation through a λ-shock system, respectively [4,5]. 2D bumps in the leading edge area of the wing were introduced by Sobieczky et al. [8]. These bumps generate expansion waves that are reflected at the sonic line as compression waves and thus reduce the strength of the compression shock when appropriately designed. These so-called Expansion Shoulder Bumps (ESB), which exhibit a comparable operation principle as so called "peaky airfoils", have also been combined with their rear counterparts called SCB or recompression shoulder bumps (RSB). Although this approach has been proven to be effective in reducing the strength of the shock, it was not pursued further in other studies. Especially not in terms of its potential for AGLA. Dynamic bumps can be realized as mechanical components via surface modification by an adaptive device [6] or as fluidic components via combined blowing and suction of secondary air [7], both allowing significantly higher actuation frequencies and lower inertia compared to state of the art flaps and spoiler systems [1,6].

In the paper at hand a numerical investigation of parametric dynamic 2D bumps located in the region of sonic expansion is conducted on a supercritical airfoil with the purpose to control the entire supersonic region and thus evaluate their potential for AGLA. Furthermore, a comparison of a dynamic bump to individual and combined actuation by means of a dynamic trailing edge flap (TEF) and a dynamic leading edge flap (LEF) is presented. Current studies of dynamic TEF and LEF by Ullah et al. [9] revealed their high potential regarding alleviation of gust induced lift and pitching moment.

2 Numerical Approach and Flow Conditions

The DLR TAU code [10] is employed for the 2D URANS simulations of the DLR F15 airfoil [13]. The main numerical settings comprise a Spalart-Allmaras Original (SAO) one equation turbulence model, a 2nd order central scheme with scalar artificial dissipation for spatial discretization, and an implicit backward euler time stepping method.

The deflection of the dynamic bump is realized using the mesh deformation capabilities implemented in TAU. Based on the aircraft regulations in the CS25 of the EASA a "$1 - \cos$" vertical gust with a fixed amplitude of $v_z = 10\,\mathrm{m/s}$ is introduced in the simulations by the Disturbance Velocity Approach (DVA) already validated by Heinrich [11] for gust wave length λ_{Gust} to chord length c ratios larger than one. The inflow conditions for the 2D simulations are derived from the design inflow conditions given for the related 3D aircraft configuration $Ma_{3D} = 0.8$, $H = 35{,}000 ft$ and $c_L = 0.5$ [12].

In this first analysis, the bump shape is based on the expansion shoulder bump (ESB) introduced by Sobieczky et al. [8], see Fig. 1. The bump starts at the leading edge $x/c = 0$ and its maximum is placed at 5% chord (c). The ramp exponents of the bump function (n_1 and n_3) are selected in a way to allow for a smooth connection to the airfoil at X_1 and X_3. The crest curvature Z_2" results from a modeling of the bump with a fifth degree polynomial. Figure 2 illustrates a selection of the investigated geometric variations of the bump. The parametric study includes a variation of the bump deflection (Fig. 2 (a)), its streamwise extension (Fig. 2 (b)), its position (Fig. 2 (c)) and its oscillation frequency. As we aim at gust alleviation instead of drag reduction the investigated non-dimensional height of the bumps z/c is increased up to three times the maximum height $(z/c)_{max}$ defined in Fig. 1 and the maximum streamwise extension $(x/c)_{max}$ is increased from $(x/c)_{max} = 15\%$ to $(x/c)_{max} = 25\%$. Furthermore, both, convex and concave bump shapes are realized to assess their impact on gust load mitigation, despite the fact that the latter does not comply with the typical bump shapes known from literature. The selected bump frequency covers a range from a quarter to two times the convective time of the reference gust. The time function of the dynamic bump mirrors a "1 − cos" function similar to the one of the investigated gust shape.

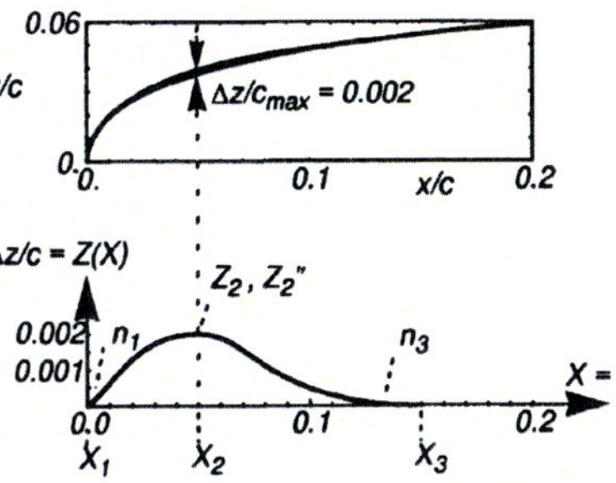

Fig. 1. Geometric parameter for ESB from [8].

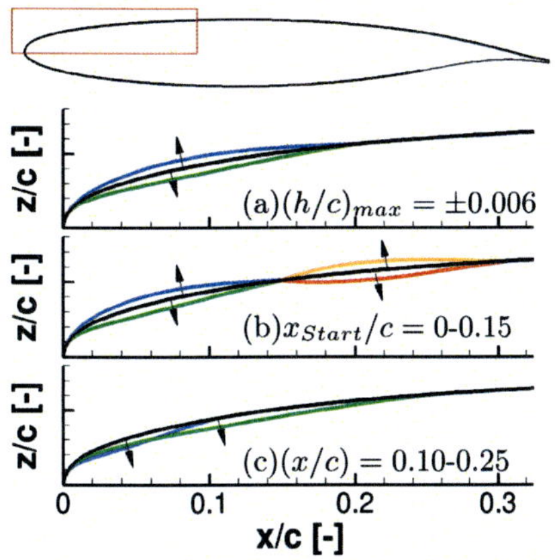

Fig. 2. Selective overview of investigated bump parameters (arrows indicate direction of bump deflection).

3 Prestudy–Critical Load Case

An inevitable first step to estimate the required bump control authority is to conduct computations of gust interacting with the baseline DLR-F15 airfoil. This set of simulations is published by Ullah et al. in [9] and will be briefly reintroduced here to allow for an independent reading of the paper at hand. Figure 3

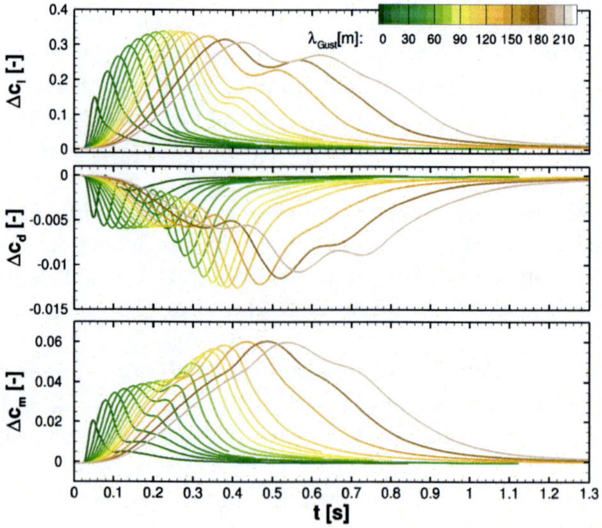

Fig. 3. Gust wavelength study of the DLR-F15 airfoil from [9].

shows the time response of the DLR-F15 airfoil to "1 − cos" vertical gusts of different wave lengths ranging from 18 m to 214 m which is in compliance with CS25 regulations. The force and moment coefficients are plotted relative to their reference baseline state. The time history of the bump deflection from 0–100% is denoted by $\eta(t)$. The time histories show a saturation of the maximum lift coefficient c_l with increasing λ_{Gust}. For $\lambda_{Gust} > 80$ m even a drop in the maximum c_l can be observed which is attributable to shock induced flow separation. The time histories of the moment coefficient c_m are evaluated about the 45% chord point, representing the neutral axis of the wing box of the reference configuration. In general, the airfoil experiences a nose down pitching moment throughout the gust event. Even though the bump is not intended to reduce horizontal loads, the drag coefficient c_d is always monitored to prevent a disproportionate increase in drag. The negative Δc_d during the gust interaction can be ascribed to a rotation of the aerodynamic reference system by the varying gust induced effective incidence angle. The critical load case is selected to be a "1 − cos" gust with $\lambda_{Gust} = 50$ m, as explained in [9].

4 Results

First, the bump effect is evaluated at steady inflow conditions for the parametric setup introduced in Sect. 2. The time response of the force and moment coefficients is depicted in Fig. 4. The gust induced lift and pitching moments have been identified as $\Delta c_{l,Gust,max} = 0.315$ and $\Delta c_{m,Gust,max} = 0.038$ for $\lambda_{Gust} = 50$ m, see Fig. 3. A concave bump deflection creates a negative Δc_l

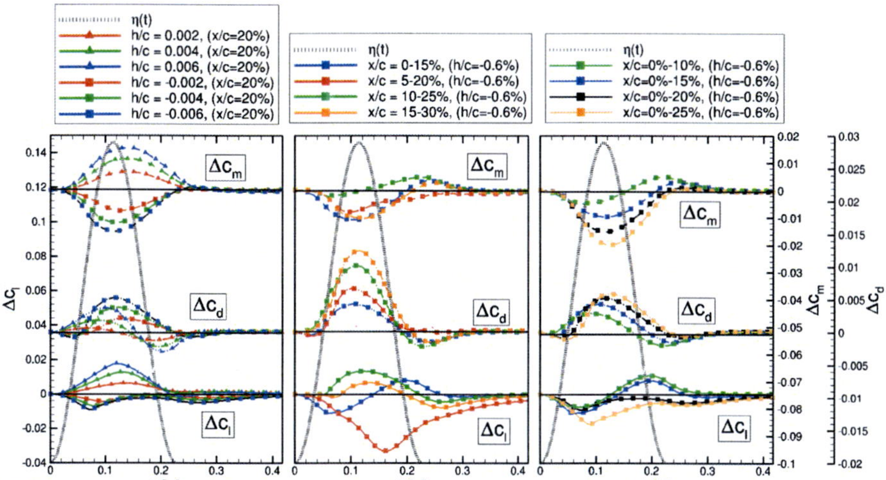

(a) Variation of height (b) Variation of position (c) Variation of length

Fig. 4. Force and moment coefficients for parametric variation of bump parameters.

which counteracts the positive Δc_l impinged by the reference gust. The variation of the bump height does not imply a high control authority over c_l. On the other hand, the concave bumps generate a significant nose down moment, which is inverse to $\Delta c_{m,Gust}$ and on the same order of magnitude. Similar effects can be observed for a variation of the bump location and variation of the streamwise extension of the bump as shown in Fig. 4 (b) and (c), respectively. Only a negative bump deflection of $(h/c)_{max} = -0.006$ is presented in Fig. 4(b) and (c) for reasons of clarity and based on the findings from a variation of the bump height in Fig. 4(a). The most effective bump regarding $\Delta c_{l,Gust}$ mitigation is the bump located at $x/c = 5{-}20\%c$ with $(z/c)_{max} = -0.6\%c$. It indicates only a compensation of approx. 9% of the gust induced Δc_l. The bump ranging from $x/c = 0$ to $x/c = 0.25$ with $(z/c)_{max} = -0.6\%c$ shows the best features in terms of $\Delta c_{m,Gust}$ alleviation. A reduction in $\Delta c_{m,Gust}$ of about 60% can be estimated.

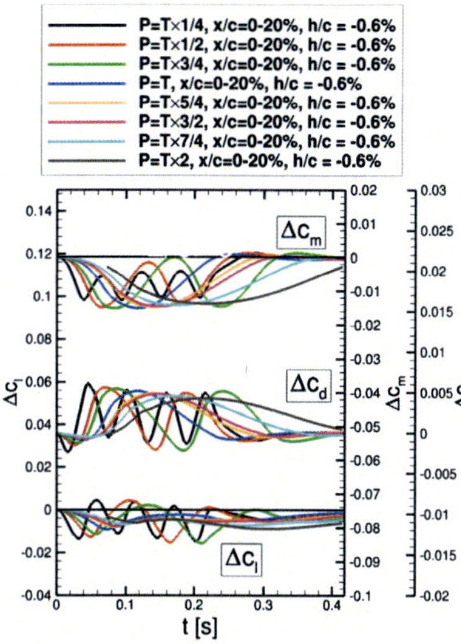

Fig. 5. Force and moment coefficients for variation of bump frequency. P denotes the actuation period and T denotes the reference actuation period derived from the gust interaction time for $\lambda_{Gust} = 50\,\mathrm{m}$.

In addition to a bump geometry variation, the influence of the actuation frequency of the bump is analyzed. As shown in Fig. 5, the amplitudes of the generated forces and moments are hardly affected by the frequency of the bump. A precise load reduction of a specific gust requires an adaptation of the bump frequency to the convective time of the gust. Therefore, the independency of the bump effects from its frequency promotes that the results derived from this study, which considers a "1 − cos" gust with $\lambda_{Gust} = 50\,\text{m}$, are applicable to a variety of gusts with different wave lengths. The predictions above assume a validity of linear superposition of the bump effect at steady inflow conditions and the gust effect on the undeformed airfoil. The evaluations promote a utilization of dynamic bumps for alleviation of $\Delta c_{m,Gust}$. At the same time no significant effect on $\Delta c_{l,Gust}$ is assumed.

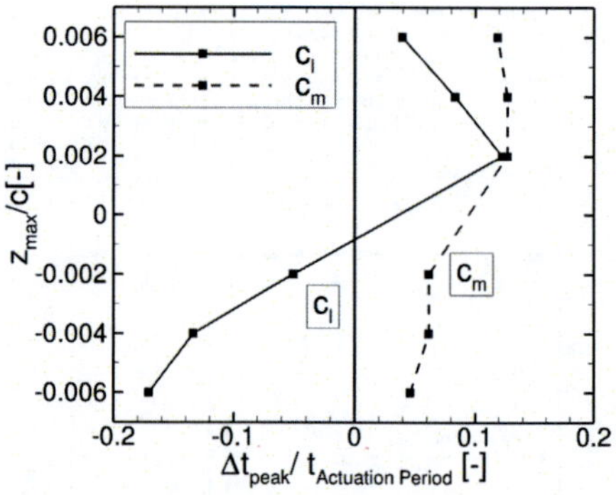

Fig. 6. Phase sh0ift between maximum bump deflection and peak in c_l and c_m time history.

The phase shifts between the dynamic bump deflection indicated by $\eta(t)$ in Fig. 4 and the lift and moment responses are depicted in Fig. 6. The concave bump $(z/c)_{max} < 0$ and the convex bump $(z/c)_{max} > 0$ demonstrate opposite phase shifts between $c_{l,max}$ and the maximum actuation deflection η_{max}. While the magnitude of the phase shift is decreased with increased $(z/c)_{max} < 0$ for the concave bump, this trend is inversed for the convex bump. A phase shift of almost $-0.18 \times t_{Actuation\ Period}$ can be derived for the bump with $(z/c)_{max} = -0.006$. The phase shift for c_m shows rather small variation across the range of the simulated bump deflections. Both bumps, concave and convex, show a positive phase shift for c_m. However, in terms of a realistic operation mode of the bumps a prediction system for the upstream flow field, like Doppler LIDAR sensors are unavoidable and thus it is irrelevant for the feasibility of the bump, if the phase shift has a positive or a negative sign, as long the magnitude of the phase shift is not disproportionate.

The studies conducted in [9] demonstrate a high potential of the LEF for compensation of the nose up moment impinged by the gust. A TEF deflection can be employed individually to fully compensate the $\Delta c_{l,Gust}$ imposed by the considered reference gust [9]. A dynamic concave bump yields similar effects as a downward deflection of the LEF. They mainly affect c_m while c_l is only barely altered. Therefore, the assessment of the bump efficiency includes a comparison with a dynamic LEF at gust inflow conditions. Furthermore, both LEF and bump are simulated in combination with a TEF deflection, because the current results suggest only a reasonable reduction in both, $\Delta c_{l,Gust}$ and $\Delta c_{m,Gust}$, if a combination of LEF and TEF or LEF and bump is utilized. The size of the TEF (25% c) and the LEF (20% c) comply with typical lengths of plain flaps and droop noses, respectively. The time function of the TEF and the LEF deflections is similar to the one of the bump introduced in Sect. 2. In the following analysis a bump ranging from $x/c = 0$–20% with a deflection of $z/c = -6\%$ is selected, as it represents a good compromise between a mechanically feasible surface deformation and impact on c_d and c_m, see Fig. 4. No phase shift between the time of gust interaction and bump deflection is applied, as the investigated concave bump shows only a small phase shift for c_m, see Fig. 6. Figure 7 presents the time histories of the force and moment coefficients which result from an interaction of the reference gust with the airfoil in the following states: undeformed, bump deflection only, upward TEF deflection of 4.5°, combined upward TEF deflection of 4° and downward LEF deflection of 8° and combined TEF deflection of 4.5° and bump deflection.

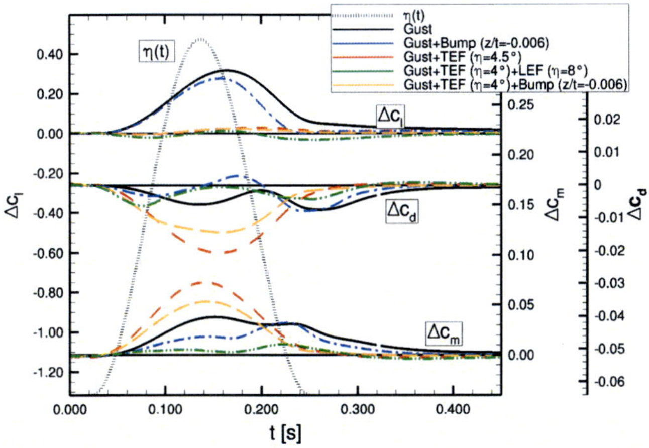

Fig. 7. Time history of force and moment coefficients illustrating effects of Gust, TEF, Bump and LEF. c_l, c_d, and c_m are plotted relative to their cruise reference state.

The phenomena accompanied by the gust interaction with the undeformed airfoil were already discussed above. These curves represent the reference state for all simulations. The bump has a visible effect on the c_l curve which is in the order of magnitude predicted by the bump simulations at steady inflow. The reaction of c_l occurs in a delayed manner, which is also conform with the findings derived from Fig. 6. The gust induced peak in c_m is only reduced by about 21%, which is below the anticipated Δc_m from the steady inflow simulations. However, there is a more substantial decrease in c_m of about 50% when observing the first local maximum. An adaptation of the bump time function which is adjusted towards the time response resulting from the gust airfoil interaction instead of mirroring the "1 − cos" gust shape may lead to some improvements in the overall capacity of $\Delta c_{m,Gust}$ mitigation. The results of the individual TEF deflection from [9] illustrate the capability of the TEF regarding gust induced $\Delta c_{l,Gust}$ compensation. Adding an LEF deflection of $\eta = 8°$ to the TEF deflection suffices for a considerable compensation of Δc_m induced by the gust and the upward deflected TEF. When replacing the LEF with a bump the general effect of Δc_m reduction can be observed as well. However, the bump reaches only an alleviation of 26% of the gust and TEF induced Δc_m. The c_l curve is mostly insensitive towards an LEF and bump deflection when combined with a TEF. The upward directed TEF deflection leads to an enhancement of c_d. The horizontal force induced by the LEF and the bump deflection are directed in streamwise direction. Therefore, a reduction of the c_d magnitude referred to the simulations of the TEF deflection can be observed. Again the LEF has a more pronounced impact on c_d than the bump, allowing a significant reduction of c_d towards the baseline cruise conditions.

Figure 8 shows the characteristic lines in the supersonic area for the baseline case (a), a concave bump (b), and a convex bump (c). The characteristic lines are derived from stationary MSES [14] computations, which solves the Euler equations coupled with an integral boundary layer method. As mentioned in Sect. 1, one purpose of the bumps is to control the depicted compression and expansion waves inside the supersonic area, which should enable a fast and efficient method for gust load alleviation. The location and size of the supersonic area is clearly affected by the bump deflection leading to a considerable influence on the gradients of the characteristic lines. Therefore, the effect of the bumps on the gradients of the expansion and compression waves is indirectly determined by local deceleration and acceleration through curvature modification instead of a precise control of the characteristic lines. This outcome is affirmed by the investigations of the c_p distributions assessed further below.

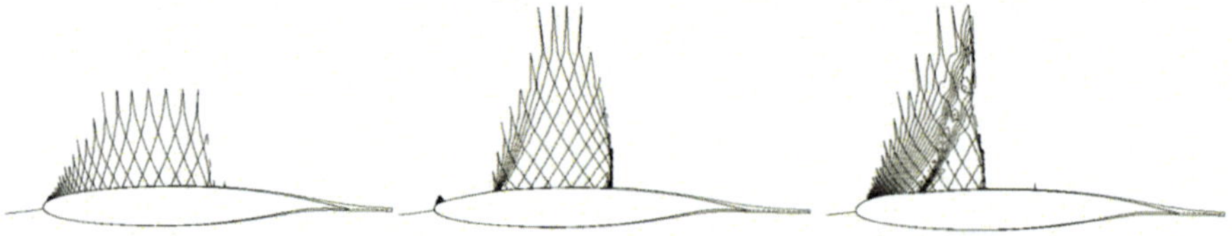

(a) Baseline (b) $z/c = -0.006$ (c) $z/c = +0.006$

Fig. 8. Characteristic patterns within the supersonic region.

Figure 9 shows the numerical schlieren of the baseline solution and the transient dynamic bump solutions at steady inflow conditions. The results with bump contain snapshots at the time step of minimum and maximum bump deflection. Similar flow features as derived from the characteristic lines in Fig. 8 are visible. These are mainly a strong impact of the bumps on the supersonic area in size and position. An inward deflected bump leads to a downstream shift of the supersonic region when neglecting the small supersonic bubble just downstream of the leading edge. The shock location at the time step of minimum bump deflection $z/c = -0.006$ is almost identical to the shock location of the baseline solution. An outward deflected bump creates a supersonic region which

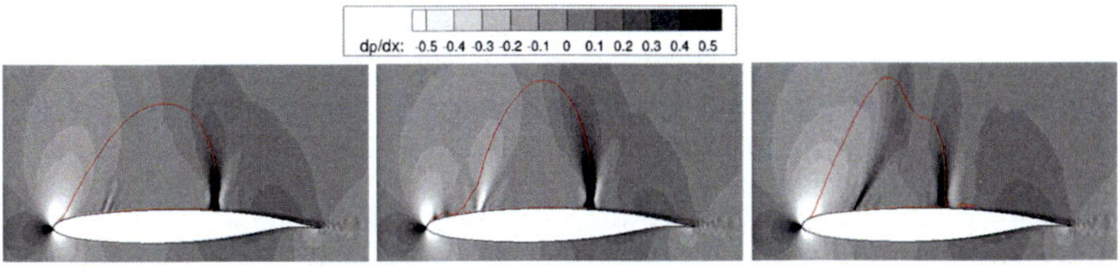

(a) Baseline (b) $z/c = -0.006$ (c) $z/c = +0.006$

Fig. 9. Numerical schlieren, red lines indicate sonic lines.

is divided by an oblique shock, this can also be seen in the image with the characteristic lines (Fig. 8). A discontinuity in the airfoils surface curvature occurs at the downstream end of the bump, despite the smooth connection of the bump to the airfoil at this position, see Fig. 1. This results from an inaccuracy in the mesh deformation procedure. An adaptation of the bump curvature function allowed for a supersonic region without included oblique shock. However, the difference in the force coefficients between the results with and without surface discontinuity is negligible and thus not presented here. The outward deflected bump shows a clear upstream movement of the vertical shock compared to the baseline solution. The shift of the supersonic region and the shock wave affects the transient behavior of c_m. A concave bump leads to a shift of the flow acceleration towards the rear section of the airfoil generating a nose down moment.

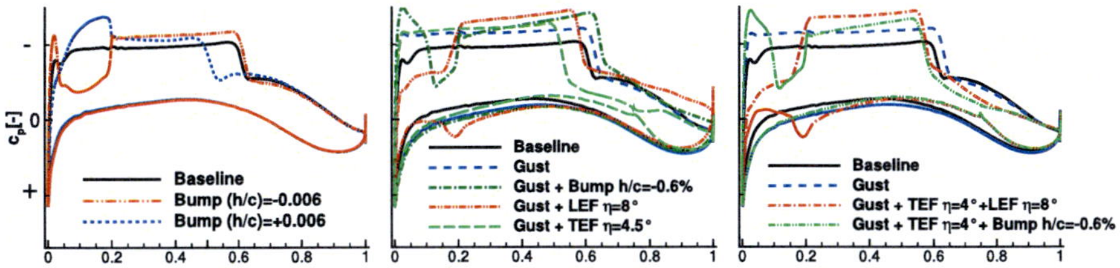

(a) Bump at steady inflow (b) individual actuation with gust (c) combined actuation with gust

Fig. 10. Impact of gust and actuators on the c_p distributions at the time of maximum deflection.

The c_p plots depicted in Fig. 10 confirm the assessments of the characteristic lines and the schlieren. The results at steady inflow conditions in Fig. 10 (a) show the local acceleration and deceleration in the airfoil's leading edge region compared to the baseline case. An increase in the pressure level just downstream of the bump, is visible for both concave and convex bump shape. This is also apparent from the increased height of the supersonic region downstream of the bump as shown in Fig. 9. Compared to the relocation of the shock seen for the convex bump, the streamwise location of the shock is hardly affected for the concave bump. The results in Fig. 10(b)–(c) demonstrate the effects of the upward deflected TEF, the downward deflected LEF and the dynamic concave bump at gust inflow conditions in individual and combined manner. As thoroughly outlined in [12], the effect of a vertical "1 – cos" gust with positive velocity component is similar to an increase in the incidence angle. This is reflected in the c_p distributions of the gust simulations without actuator deflection. Even though there is an interaction between the TEF and the bump and the TEF and the LEF, the effects of the different actuators can be separately assessed. The TEF leads to a constriction of the c_p distribution on the rear section. This leads to a significant loss of c_l as the investigated supercritical airfoil is designed to generate a considerable amount of lift through rear loading [12]. The LEF deflection mainly leads to a deceleration in the nose region, similar to a bump deflection. The deceleration region of the concave bump is somewhat reduced when interacting with the reference gust and the shock is shifted downstream. This shows that a simple superposition of the gust effect and the bump effect at steady inflow conditions in terms of quantitative predictions of their efficiencies should be treated with caution. Therefore, the effect of the bumps on the features of the expansion and compression waves is rather determined by local deceleration and acceleration through curvature modification instead of a precise control of the characteristic lines. All in all, the evaluations clearly suggest that the effect of the investigated concave bump is reached by local flow deceleration and the related relocation of the supersonic region and the shock wave. Thus, the working principle is similar to a downward deflected LEF. However, only a small control authority over c_m can be reached by the bump compared to the LEF.

5 Conclusions

Dynamic nose bumps are investigated on a generic supercritical airfoil by means of URANS simulations with the purpose to alleviate gust loads of a critical gust event. It is found that even the most effective nose bump has only a small impact on the lift coefficient c_l which is one order of magnitude smaller than the gust induced Δc_l. However, approx. 20% of the maximum gust induced nose down pitching moment can be mitigated by a dynamic bump with concave shape. As the aim of future gust alleviation systems is to reach a load factor of one throughout a gust encounter, the investigated bumps are not sufficient in an individual manner, especially when considering their lack of c_l control capacity. A trailing edge flap (TEF) deflection, which is capable of fully compensating gust induced c_l, is employed in combination with a bump and a leading edge flap (LEF). The upward deflected TEF induces an additional nose up moment. The results in this paper demonstrate that a concave bump can only reach approx. 25% of the control authority over gust and TEF-induced pitching moment compared to an LEF. The flow field shows a similar reaction to an LEF deflection and a dynamic bump, which strongly suggests that the effect of the bumps evaluated in the paper is determined by local deceleration and acceleration instead of a control of the supersonic region. Nevertheless, dynamic bumps are still promising actuators for active gust load alleviation (AGLA), especially when considering their potential with regard to agility compared to the rather slow LEF. The results derived from this study represent just a first impression of the capability of bumps for AGLA. A more detailed parameter study, which includes a shape variation of the bump, an adaptation of the bump time function and a transfer to a 3D wing geometry is required to close the gap.

Acknowledgments. This research is funded by the Federal Ministry for Economic Affairs and Energy as part of the LuFo project PoLamin.

References

1. Xu, J., Kroo, I.: Aircraft design with active load alleviation and natural laminar flow. J. Aircr. **51**(5), 1532–1545 (2014)
2. Miller, E.J., et al.: Approach for structurally clearing an adaptive compliant trailing edge flap for flight. In: Society of Flight Test Engineers International Annual Symposium, Lancaster, CA; United States (2015)
3. Prachař, A., Hospodář, P., Vrchota, P.: Gust alleviation of aeroelastic aircraft using CFD simulation. Transp. Res. Procedia **29**, 366–375 (2018)
4. König, B.: Shock Control Bumps on Transonic Transport Aircraft. Dissertation University of Stuttgart, Publisher Dr. Hut (2013)
5. Mayer, R., Zimmermann, D., Wawrzinek, K., Lutz, T., Krämer, E.: Numerical study of three-dimensional shock control bump flank effects on buffet behavior. In: Nagel, Wolfgang E., Kröner, Dietmar H., Resch, Michael M. (eds.) High Performance Computing in Science and Engineering '15, pp. 495–509. Springer, Cham (2016). https://doi.org/10.1007/978-3-319-24633-8_32
6. Geissler, W., Koch, S.: Adaptive airfoil. In: Sobieczky, H. (ed.) IUTAM Symposium Transsonicum IV. Fluid Mechanics and its Applications, vol. 73. Springer, Dordrecht (2003). https://doi.org/10.1007/978-94-010-0017-8_46
7. Hassan, A.A., Osborne, B.A., Schwimley S., Billman, G.: Dynamic bumps for drag reduction at transonic-supersonic speeds. U.S. Patent No. 8,016,245 (2011)
8. Sobieczky, H., Geissler, W., Hannemann, M.: Expansion shoulder bump for wing section viscous/wave drag control. In: Meier, G.E.A., Viswanath, P.R. (eds.) IUTAM Symposium on Mechanics of Passive and Active Flow Control. Fluid Mechanics and its Applications, vol. 53, pp. 29–34. Springer, Dordrecht (1999). https://doi.org/10.1007/978-94-011-4199-4_5
9. Ullah, J., Lutz, T., Klug, L., Radespiel, R., Wild, J.: Active gust load alleviation by combined actuation of trailing edge and leading edge flap at transonic speeds. In: AIAA Scitech 2021 Forum, pp. 19–21, January 2021. https://doi.org/10.2514/6.2021-1831. virtual event
10. Schwamborn, D., Gerhold, T., Heinrich, R.: The DLR TAU-code: recent applications in research and industry. In: ECCOMAS CFD 2006 Conference (2006)
11. Heinrich R.: Simulation of interaction of aircraft and gust using the TAU-code. In: Dillmann, A., Heller, G., Krämer, E., Kreplin, H.P., Nitsche, W., Rist, U. (eds.) New Results in Numerical and Experimental Fluid Mechanics IX. Notes on Numerical Fluid Mechanics and Multidisciplinary Design, vol. 124. Springer, Cham (2014). https://doi.org/10.1007/978-3-319-03158-3_51
12. Klug, L., et al.: Actuator concepts for active gust alleviation on transport aircraft at transonic speeds. In: AIAA Scitech 2020 Forum, Florida (2020)

13. Wild, J.: Mach and Reynolds number dependencies of the stall behavior of high-lift wing sections. J. Aircraft **50**(4), 1202–1216 (2013). AIAA
14. Drela, M.: A User's Guide to MSES 3.05. Massachusetts Institute of Technology (MIT), Cambridge (2007)

High-Agility Configuration

Steady-State Flow Solutions for Delta Wing Configurations at High Angle of Attack Using Implicit Schemes

Arpit Aggarwal(✉), Ralf Hartmann, Stefan Langer, and Tobias Leicht

German Aerospace Center, Member of the Helmholtz Association, Institute of Aerodynamics and Flow Technology, Lilienthalplatz 7, 38108 Braunschweig, Germany
{arpit.aggarwal,ralf.hartmannn,stefan.langer, tobias.leicht}@dlr.de

Abstract. Finding fully converged, steady-state solutions of the compressible Reynolds Averaged Navier-Stokes (RANS) equations for aerodynamic configurations on the border of the flight envelope often poses serious challenges to solution algorithms that have proven robust and successful for configurations at cruise conditions. Examples of such cases are agile configurations at high angles of attack. When trying to compute solutions in these scenarios, one often observes that the solution process breaks down after few iterations or that a steady-state RANS solution, although it may exist, cannot be reached with the employed solution algorithm. While, in general, no clear reason for this behavior can be identified, the complexity of these flows seems to be significantly greater compared to flows around transport aircraft in cruise flight. The flow fields are dominated by the interaction of shock waves with a system of vortices emanating from the leading edges on the upper surface of the wing, leading to massive flow separation. These flow features tend to be inherently unsteady and can be assumed to cause problems in computing a converged solution using an algorithm designed to find steady-state solutions of the RANS equations. To avoid these problems, it is not uncommon to calculate such configurations in an unsteady mode, which often comes at a rather high computational cost. This article demonstrates the necessity for implicit smoothers to approximate fully converged solutions of these challenging simulations. A numerical example is given to confirm that convergence is only possible using an exact derivative together with a suited preconditioner.

1 Introduction, Governing Equations and Discretization

1.1 Introduction

To approximate steady-state solutions of the turbulent compressible Reynolds averaged Navier-Stokes (RANS) equations multistage Runge-Kutta methods are established as smoothing techniques within a Full-Approximation Storage (FAS) multigrid method [1]. In the literature of Computational Fluid Dynamics (CFD) many acceleration and stabilization techniques have been suggested to improve the basic multistage scheme, for example explicit local time stepping [2], point and line implicit methods [3–6] as well as methods exploiting specific approximations to the full derivative or

© The Author(s), under exclusive license to Springer Nature Switzerland AG 2021
A. Dillmann et al. (Eds.): STAB/DGLR Symposium 2020, NNFM 151, pp. 271–281, 2021.
https://doi.org/10.1007/978-3-030-79561-0_26

even using the full derivative combined with an iterative solution method approximating the solution to the linear systems [7]. In general, smoothers having a large amount of implicitness have proven to be more successful compared to explicit or weakly implicit ones. A complete overview of how the various Runge-Kutta smoothers are related can be found in [8].

Although implicit smoothers have proven their potential for many examples, for configurations on the border of the flight envelope one often observes severe difficulties to reliably approximate a steady-state solution. This suggests that further parameters influencing the overall iterative process need to be investigated for increased robustness. For example, each iterative solution process for nonlinear equations starts from an initial guess, typically the free-stream. For complex flow fields, however, the free-stream can be far away from the final solution. Thus, it is not unexpected that simple globalization strategies – even if they work well for smooth flows – are no longer able to find a valid solution path from the initial guess to the final result when applied to flows at the border of the flight envelope.

This article discusses and extends implicit smoothers by globalization (also called regularization) methods to prevent divergence in the beginning of the nonlinear iteration process, in order to overcome the problem of not having a good initial guess. Proposed methods exploit CFL number ramping strategies together with the choice of the underlying time step. Furthermore, a comparative study is performed between these methods and unsteady computations on the same configuration.

1.2 RANS Equations and Discretization

For an open domain $\Omega \subset \mathbb{R}^3$ the three-dimensional RANS equations are considered in conservative variables $\mathbf{W} := (\rho, \rho u_1, \rho u_2, \rho u_3, \rho E)$ written as

$$\frac{\mathrm{d}}{\mathrm{d}t} \int_{\Omega} \mathbf{W}\,\mathrm{d}\mathbf{x} + \int_{\partial\Omega} (\mathbf{f}_c \cdot \mathbf{n} - \mathbf{f}_v \cdot \mathbf{n})\,\mathrm{d}s = 0. \tag{1}$$

Turbulence is modeled using the one-equation turbulence model introduced by Spalart and Allmaras [9, 10]. The additional equation is given by

$$\frac{\mathrm{d}}{\mathrm{d}t} \int_{\Omega} \tilde{\nu}\,\mathrm{d}\mathbf{x} + \int_{\partial\Omega} (\mathbf{f}_{c,turb} \cdot \mathbf{n} - \mathbf{f}_{v,turb} \cdot \mathbf{n})\,\mathrm{d}s(y) = \int_{\Omega} Q\,\mathrm{d}\mathbf{x}. \tag{2}$$

The eddy viscosity μ_t is computed by $\mu_t := \rho\tilde{\nu} f_{v_1}$ where $f_{v_1} := \frac{\chi^3}{\chi^3 + c_{v_1}^3}$, $\chi := \frac{\tilde{\nu}}{\nu_l}$, $\nu_l := \frac{\mu_l}{\rho}$. The definition of the convective terms $\mathbf{f}_c \cdot \mathbf{n}$ and $\mathbf{f}_{c,turb} \cdot \mathbf{n}$, viscous terms $\mathbf{f}_v \cdot \mathbf{n}$ and $\mathbf{f}_{v,turb} \cdot \mathbf{n}$ and the source terms Q can be found for example in [9, 10]. For the results given in this article the recommendations concerning the turbulence model given in [10] are followed.

A finite volume formulation on median dual grids is employed as spatial discretization. Control volumes are defined around the vertices of the primary grids. The inviscid terms are discretized using a central difference scheme with an added matrix valued artificial dissipation [11]. A first order Roe scheme or a second order central difference scheme is applied for the convective part of the turbulence model (2). Gradients required for the viscous terms and for the source terms are computed using a Green-Gauss formulation (see [12]). For a detailed description of the discretization strategy and boundary conditions applied, we refer to [11, 13].

2 Solution Method

2.1 Implicit Smoother

After discretization one obtains the system of ordinary differential equations

$$\frac{d\mathbf{W}}{dt} = -\mathbf{M}^{-1}\mathbf{R}(\mathbf{W}) \qquad \text{where} \quad \mathbf{M} := \operatorname{diag}\left(\operatorname{diag}\left(\operatorname{vol}(\Omega_i)\right)\right). \tag{3}$$

To approximate a steady-state solution of (3), a multistage diagonally implicit Runge-Kutta method is applied to obtain the smoother (see [8])

$$\begin{aligned} \mathbf{W}^{(0)} &:= \mathbf{W}^{n} \\ \mathbf{W}^{(j)} &= \mathbf{W}^{(0)} - \alpha_{j+1,j}\mathbf{P}_j\left(\mathbf{W}^{(j-1)}\right)^{-1}\mathbf{R}\left(\mathbf{W}^{(j-1)}\right), \quad j = 1,\dots,s \\ \mathbf{W}^{n+1} &= \mathbf{W}^{(s)}, \end{aligned} \tag{4}$$

$$\mathbf{P}_j\left(\mathbf{W}^{(j-1)}\right) := (\Delta T)^{-1}\mathbf{M} + \frac{d\mathbf{R}}{d\mathbf{W}}\left[\mathbf{W}^{(j-1)}\right], \quad \Delta T := \operatorname{diag}\left(\operatorname{diag}\left(\Delta t_i\right)\right). \tag{5}$$

To realize (4), the linear equation $\mathbf{P}_j x = \alpha_{j+1,j}\mathbf{R}\left(\mathbf{W}^{(j-1)}\right)$ needs to be solved. A left or right preconditioned Generalized Minimal Residual (GMRes) method [14] is used,

$$\mathbf{Prec}_j^{-1}\mathbf{P}_j\mathbf{x} = \alpha_{j+1,j}\mathbf{Prec}_j^{-1}\mathbf{R}(\mathbf{W}^{(j-1)}), \tag{6}$$

$$\mathbf{P}_j\mathbf{Prec}_j^{-1}\mathbf{u} = \alpha_{j+1,j}\mathbf{R}(\mathbf{W}^{(j-1)}), \quad \mathbf{u} = \mathbf{Prec}_j\mathbf{x}. \tag{7}$$

Here $\mathbf{Prec}_j$ denotes a preconditioner. Then, for each step in the GMRes method, a symmetric Gauss-Seidel method is applied to approximate a solution of

$$\mathbf{Prec}_j\mathbf{w} = \mathbf{P}_j\mathbf{v}_k \tag{8}$$

where $\mathbf{V}_m = (\mathbf{v}_1, \dots, \mathbf{v}_m)$ is the Krylov subspace. The key ingredient is the design of the preconditioner. In the following, two different strategies are proposed.

2.2 Preconditioning Strategies

Strategy 1: The matrix-free approach implements the preconditioner by a lower-upper symmetric Gauss-Seidel (LU-SGS) scheme, that is

$$\mathbf{Prec}_j = (\Delta T)^{-1}\mathbf{M} + \frac{d\mathbf{R}^{1st,scalar}}{d\mathbf{W}}$$

and $\frac{d\mathbf{R}^{1st,scalar}}{d\mathbf{W}}$ is a scalar, compact approximation to $\frac{d\mathbf{R}}{d\mathbf{W}}$ (see [13] for details). To approximately solve the linear systems (8) by a symmetric Gauss-Seidel method the number of forward and backward sweeps can be varied.

Strategy 2: In the second approach the preconditioner is given by

$$\mathbf{Prec}_j = (\Delta T)^{-1}\mathbf{M} + \frac{d\mathbf{R}^{1st}}{d\mathbf{W}}, \tag{9}$$

where $\frac{d\mathbf{R}^{1st}}{d\mathbf{W}}$ is a linearization of a residual based on a first-order discretization using a compact stencil. Then, the linear systems (8) are approximately solved by application of a line-symmetric Gauss-Seidel method.

Remark: For zero steps of the actual GMRes iteration, **Strategy 1** reduces to the LU-SGS method suggested by Yoon and Jameson [15], while **Strategy 2** yields the methods suggested in [7,11].

2.3 Approximation of Jacobian

Within the GMRes method the matrix-vector multiplication of the operator given in (5) applied to a vector is approximated by a finite difference of the residual operator,

$$\frac{d\mathbf{R}}{d\mathbf{W}}(\mathbf{W})\,\mathbf{h} \approx \frac{1}{\varepsilon_2 - \varepsilon_1}\left(\mathbf{R}\left(\mathbf{W} + \varepsilon_2\mathbf{h}\right) - \mathbf{R}\left(\mathbf{W} + \varepsilon_1\mathbf{h}\right)\right).$$

For $\varepsilon_2 = -\varepsilon_1 = \varepsilon > 0$ one obtains a symmetric finite difference, while $\varepsilon_2 = \varepsilon > 0$ and $\varepsilon_1 = 0$ yields a computationally cheaper forward difference. The choice of a suitable ε is not trivial, as cancellation and approximation errors need to be balanced. A possible method can be found in [12, Chapter 6.2.5].

2.4 Choice of Time Step, Stabilization

As a general rule, time step sizes and thus CFL numbers should be chosen smaller (for robustness) in the beginning and larger (for efficiency) in the final phases of the nonlinear solution process. It turns out that for flows at the border of the flight envelope, such as delta wings at high angle of attack, a simple CFL number ramping strategy such as [11]

$$\mathrm{CFL}\,(n) = \min\left\{\mathrm{CFL}_{\mathrm{init}} \cdot f(n), \mathrm{CFL}_{\mathrm{max}}\right\}, \quad f(n) = \begin{cases} 1, & n < 10, \\ \gamma^{n-10}, & n \geq 10, \end{cases} \quad \gamma > 1, \tag{10}$$

sometimes fails and the iteration diverges. Instead, in particular for the initial phase of the algorithm where the initial guess $\mathbf{W}_\infty$ (i.e. the free-stream) is far away from the solution, two different strategies are proposed.

Strategy 1: The matrix of local time steps Δt is replaced by

$$\Delta T^{\mathrm{global}} := \mathrm{diag}\left(\mathrm{diag}\left(\Delta t_{\mathrm{min}}\right)\right), \qquad \Delta t_{\mathrm{min}} = \min\left\{\Delta t_i : i = 1, \ldots N\right\}. \tag{11}$$

Here Δt_i denotes the local time step determined by the spectral radius of local derivatives [8] and N the number of grid points. Then, the Runge-Kutta smoother may be interpreted as a kind of Rosenbrock-method with a global time step Δt_{min}. During the iteration the CFL number might be increased carefully. After a certain number of initial steps the matrix of global time steps can be replaced by local time steps, to speed up the convergence rate.

Strategy 2: The global time step choice (11) is a straightforward idea, but can be computationally expensive, in the sense that it can lead to an overly careful approach, which then requires unreasonably many initial steps. Hence, the idea is to use a CFL ramping strategy, which is not as aggressive as (10), in particular in the starting phase, but which is also not as careful as (11), via including a feedback mechanism to detect emerging divergence. Here, the Switched Evolution Relaxation (SER) [16] technique is proposed to increase the CFL number. Then the mapping $f(n)$ in (10) is given by

$$f(n) = (q(n))^{\alpha}, \qquad q(n) := \|\mathbf{R}(\mathbf{W}_{\infty})\|_2 \,/\, \|\mathbf{R}(\mathbf{W}^n)\|_2, \tag{12}$$

where α is a free parameter. Assuming convergence ($\|\mathbf{R}(\mathbf{W}^n)\|_2 \to 0,\, n \to \infty$) $q(n) \to \infty,\, n \to \infty$ is obtained, and the rate depends on the convergence rate of $(\|\mathbf{R}(\mathbf{W}^n)\|_2)_{n\in\mathbb{N}}$. Hence, the SER technique (12) relates the increase in CFL to the convergence rate of the global normalized residual $q(n)$. As the residual decreases or increases with the iterations, this feedback is looped into the calculation of the increase of CFL. CFL_{max} in (10) is kept at a very high value of about 10^{10} despite the demanding nature of the considered configurations. The parameter α controls damping and smoothing in the initial phase. For large-scale 3D turbulent flows a choice of $\alpha = 2/5$ is used, as in the examples presented here. For simpler cases, as 2D flows, a larger value of α is often possible.

Additionally, to further improve robustness, solver recovery techniques are implemented to monitor, restore and repeat the solution process with a reduced CFL number, for the case that the SER technique has increased the CFL number too large or too quickly.

3 Numerical Results

3.1 Transonic Flow over a Delta-Wing with Attached Sting

As a first example, a transonic flow around a 65° sweep angle delta wing geometry with attached sting from the *Second Vortical Flow Experiment (VFE-2)* [17, 18] is considered with following inflow conditions: Mach number $\mathbf{Ma} = \mathbf{0.869}$, angle of attack $\mathbf{AoA} = \mathbf{24.7^{\circ}}$ and Reynolds number $\mathbf{Re} = \mathbf{59.5 \times 10^6}$. This case is solved on a sequence of three mixed-element meshes including tetrahedra, hexhedra, prisms and pyramids. The number of nodes are about $\mathbf{4.6 \times 10^6}$, $\mathbf{10.0 \times 10^6}$ and $\mathbf{30.9 \times 10^6}$ computed in parallel with 32, 64 and 96 partitions, respectively. A plot of the geometry with mesh topology is given in Fig. 2. Figure 1 hints at the complexity of the solution of the RANS equations for this configuration. A lambda-shock is seen on the upper surface of the wing. A large vortex and flow separation is observed downstream. The interaction of these flow phenomena yields a complex flow topology around the geometry and may lead to a breakdown of weak solution algorithms.

To find a steady state solution, a Full-Multigrid (FMG) method is employed together with **Strategy 1** from Sect. 2.2 using five levels of grid coarsening. The SER technique (12) is used to increase the CFL number from a starting value of 1. The computations are compared with the LU-SGS method (the legacy solution method in the

Table 1. Time taken to achieve convergence.

Grid	LU-SGS	Strat.1
1	22.6 h	23.3 h
2	56.7 h	176.7 h
3	228.7 h	1729 h

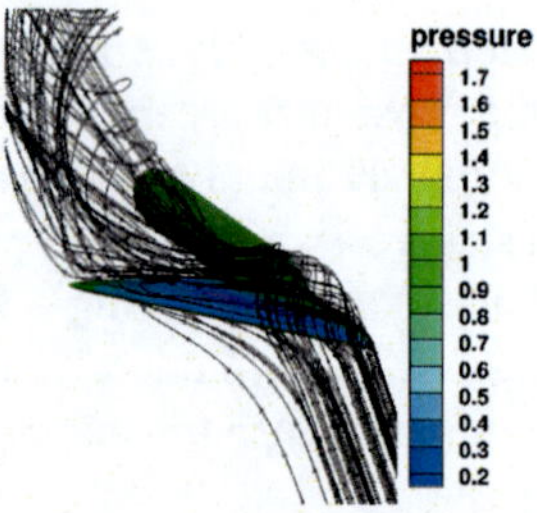

Fig. 1. Pressure distribution with streamlines

Fig. 2. Delta wing with attached sting.

DLR TAU-Code [13]). The latter smoother represents a compromise between acceptable robustness for a large number of test cases, and a small amount of work per iteration. On the other hand, for complex flows and a significant increase in mesh size this method often fails to find steady state solutions. Comparisons of the convergence histories are shown in Fig. 3. The iterations are stopped when the density residual dropped below 10^{-10}, which is reached for both methods. The residual of the turbulence flow equation stagnates for the LU-SGS method on the finer meshes, a hint for a lack of robustness. Although the number of multigrid cycles until convergence is lower for **Strategy 1** from Sect. 2.2, the effort per iteration is significantly greater, leading to an increase in computational time, shown in Table 1. Strictly speaking, such a comparison is not really valid, since no convergence is achieved for the LU-SGS method (regarding the turbulent flow equation). Furthermore, no particular emphasis has been put into optimizing the solver parameters. In particular, on the finer meshes there might be plenty of room for efficiency improvements of **Strategy 1**, for example, considering nonblocking communication and "pipelining" [19] as well as shifting some work from outer Krylov iterations to inner preconditioner iterations, cf. Sect. 3.2.

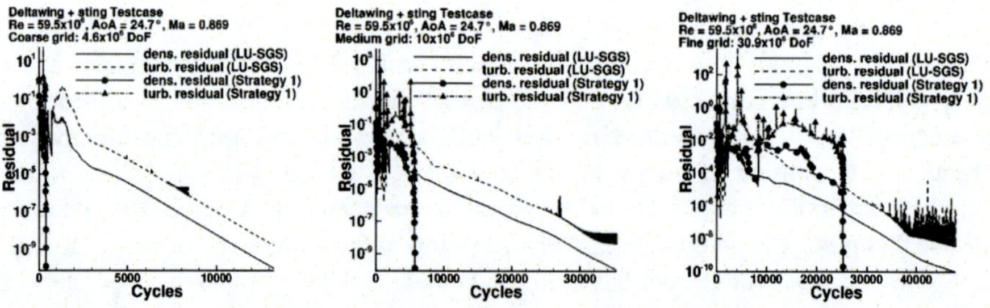

(a) Grid 1, 4.6×10^6 nDoF (b) Grid 2, 10.0×10^6 nDoF (c) Grid 3, 30.9×10^6 nDoF

Fig. 3. Residual convergence for delta-wing with attached sting using **Strategy 1** from Sect. 2.2.

Figure 4 shows a comparison of the computed pressure coefficient C_p and the experimental data [20] over the wing surface at locations $X/c_R = 0.2, 0.6,$ and 0.95 (see [20]). Figure 4(a) shows a reasonable agreement in the two data-sets, except that the computed results predict the shockwave slightly later than the experimental findings. Moving downstream, in Figs. 4(b) and 4(c), the experimental and computational data show a significant offset on the upper surface of the wing. Possible reasons for these deviations are e.g. errors introduced by the turbulence model, missing mesh resolution, and uncertainties in the measurements. An overprediction of vorticity is also a possible known contributing factor. However, the qualitative behavior is roughly similar while both the CFD results are identical and lie on top of each other.

3.2 Low Mach Flow over a Delta-Wing with Load

As a second example, a delta-wing with an attached missile and a deployed slat (Fig. 5) is considered at a low inflow Mach number $\mathbf{Ma} = \mathbf{0.15}$, a high angle of attack $\mathbf{AoA} = \mathbf{20}^\circ$ and the Reynolds number $\mathbf{Re} = \mathbf{3.4} \times \mathbf{10^6}$. The approximation of steady-state solutions for such low speed cases is often not straightforward due to the high complexity of the associated flow fields, which are often characterized by large regions of separation as well as a mixture of local compressibility effects with incompressible flow dominating major parts of the domain as seen in Fig. 5(a) [21]. Thus, solution methods based on simplifications of the smoother (4) can fail.

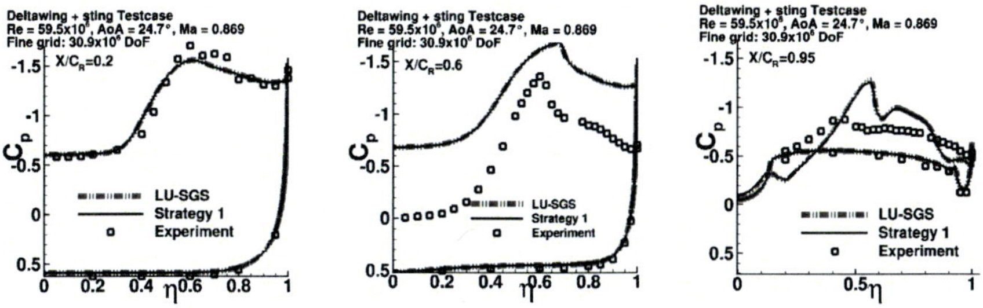

(a) $X/c_R = 0.2$ (b) $X/c_R = 0.6$ (c) $X/c_R = 0.95$

Fig. 4. Comparison of C_p distribution over wing surface between CFD and experimental results.

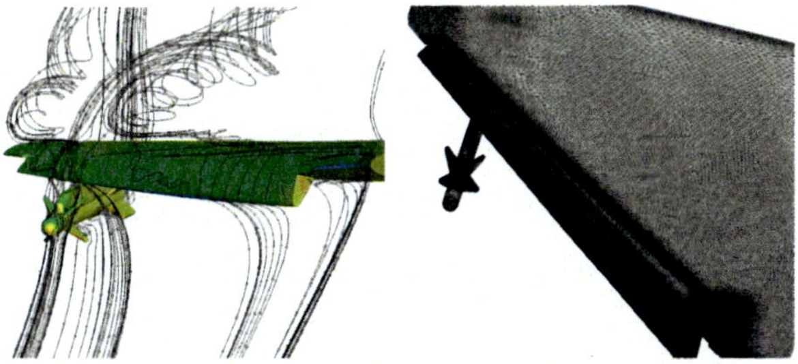

(a) Pressure distribution with streamlines (b) Mesh topology

Fig. 5. Delta wing with load.

To confirm this conjecture, for a coarse mesh with about $\mathbf{5.5 \times 10^6}$ degrees of freedom (nDoF), **Strategy 2** from Sect. 2.2 is applied with a reduced number of GMRes iterations to 0. The convergence history with respect to this setting is plotted in Fig. 6(a). After an initial drop of the residual by about 5 orders of magnitude, the iteration starts to oscillate. Even a reduction of the CFL number from 1000 to 25 does not help. Finally, setting the number of GMRes steps to 200 and thus including an almost exact representation of the residual derivative convergence is enforced. Alternatively, the latter approach can be used directly "from scratch" without major problems, yielding a steady-state solution converged up to machine-accuracy. On the finer mesh having 13.3×10^6 nDoF some work load is shifted into the application of the preconditioner (9) in order to reduce the memory consumption for the GMRes algorithm. The number of Gauss-Seidel sweeps to approximately solve the corresponding linear systems is increased from 5 to 25, while the number of GMRes steps could be reduced to 20. The convergence history is shown in Fig. 6(b). A stable drag and lift coefficient is given for the coarse mesh after about 70, for the fine mesh after about 150 iterations, corresponding to a residual reduction of about 8–9 orders of magnitude, shown in Fig. 6(c). Here, the CFL strategy (10) is used for both, the coarse and the fine mesh. The total computa-

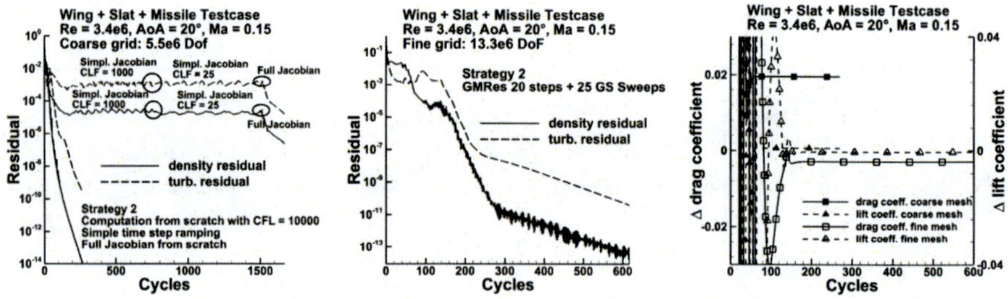

(a) Coarse mesh (b) Fine mesh (c) Delta of drag and lift coefficients

Fig. 6. (a) and (b) Residual convergence for delta-wing with missile; (c) Drag and lift coefficient convergence for delta-wing with missile.

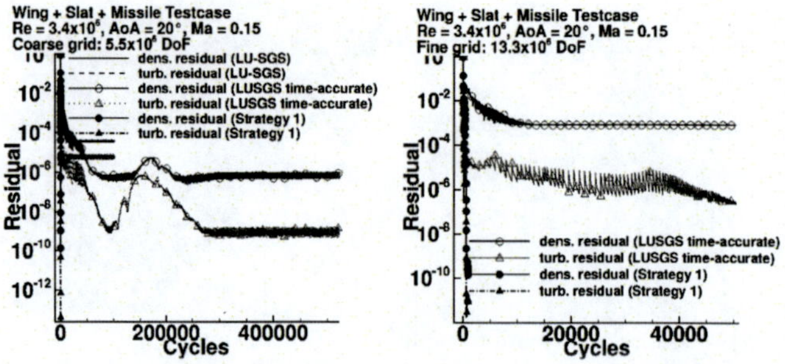

(a) Grid 1, 5.4×10^6 nDoF (b) Grid 2, 13.3×10^6 nDoF

Fig. 7. Residual convergence for delta-wing with missile in highlift configuration using **Strategy 1** from Sect. 2.2.

tional time to achieve a reduction in the density residual of 14 orders of magnitude on the coarse mesh and 128 domains, i.e. about 43000nDOF per domain, is 8h, and on the fine mesh and 384 domains, i.e. about 34000nDoF per domain, about 24h. Consequently, stable values for lift and drag are obtained after about 6 h of computational time on the fine mesh.

For this kind of flows the strongest solution algorithm proposed in this article, that is **Strategy 2** from Sect. 2.2, is necessary to obtain fully converged solutions. In order to support this statement, a further set of comparisons is made using **Strategy 1** from Sect. 2.2. Figure 7(a) shows the convergence histories of the LU-SGS method applied directly as multigrid smoother and as a preconditioner for GMRes (**Strategy 1** from Sect. 2.2) for a steady-state computation and applied as a smoother within an unsteady computation using a BDF2 scheme. For the coarse mesh, the GMRes solver used 200 Krylov vectors with 4 restarts, i. e. , the total number of inner linear iterations is 800 or until a residual reduction of 8 orders from the starting linear residual is achieved. Both the steady-state and time accurate computations stalled after 4 and 6 orders residual reduction respectively. However, the approach including the full Jacobian in the GMRes algorithm reached about 10 orders of reduction for the density residual and about 12 for the turbulence residual. For the fine mesh, the convergence of LU-SGS as a single grid relaxation scheme for time-accurate computations is compared with that of the steady-state GMRes-based solver. The number of Krylov vectors and linear iterations are increased to 300 and 1200, respectively. As seen in Fig. 7(b), the time-accurate solver again stalled after 3 orders of residual reduction while the proposed steady-state solver according to **Strategy 1** converges with 10 orders of reduction in the density residual. The computational time is 194 h on the coarse mesh with 32 partitions and about 231 h on the fine mesh with 64 partitions.

4 Conclusion

This article shows results of recent developments targeted at increasing the robustness and applicability of implicit Runge-Kutta smoothers applied to find approximate solutions of the RANS equations. Two ingredients have been identified as crucial. On the one hand, to overcome problems in the start-up phase, while still allowing strongly implicit schemes with large (pseudo-)time steps for final convergence, several stabilization techniques controlling the time step are proposed. On the other hand, a strong smoother (i.e. having a large amount of implicitness) is required to enforce convergence towards a steady state solution, in particular for configurations at the border of the flight envelope. In particular, for the second example shown in Sect. 3.2, only the method having the largest amount of implicitness had the potential to find a fully converged steady state solution. Incorporating exact derivative information in the linear system to be solved, is an essential first step. Additionally, a suitable preconditioner is required for the GMRes method. For the last example shown in this article, only the preconditioning technique based on a complete Jacobian corresponding to a first order discretization is fully successful, together with a tailored line-implicit iterative linear solver. It should be noted, however, that compared to more simplified methods, these techniques come at the price of increased memory consumption and high computational effort per iteration.

Acknowledgements. Parts of this work have been co-funded by Airbus Defence and Space, Germany. The authors would like to thank Dr. Kolja Elssel for the fruitful discussions, and for providing the second test case, as well as the meshes for both cases.

References

1. Ulrich Trottenberg, C.W.: Oosterlee, and Anton Schuller, Multigrid. Academic Press Inc., Orlando (2001)
2. Jameson, A., Schmidt, W., Turkel, E.: Numerical solutions of the Euler equations by finite volume methods using Runge-Kutta Time-Stepping Schemes. AIAA Paper, AIAA-81-1259 (1981)
3. Mavriplis, D.J.: Multigrid Strategies for Viscous Flow Solvers on Anisotropic Unstructured Meshes. ICASE Report No. 98-6 (1998)
4. Moinier, P., Müller, J.-D., Giles, M.B.: Edge-based multigrid and preconditioning for hybrid grids. In: AIAA 99-3339 (1999)
5. Langer, S.: Investigation and application of point implicit Runge-Kutta methods to inviscid flow problems. Int. J. Num. Methods Fluids **69**(2), 332–352 (2012)
6. Langer, S.: Application of a line implicit method to fully coupled system of equations for turbulent flow problems. Int. J. Comput. Fluid Dyn. **27**(3), 131–150 (2013)
7. Swanson, R.C., Turkel, E., Rossow, C.-C.: Convergence acceleration of Runge-Kutta schemes for solving the Navier-Stokes equations. J. Comput. Phys. **224**(1), 365–388 (2007)
8. Langer, S.: Preconditioned Newton methods to approximate solutions of the Reynolds averaged Navier-Stokes equations. Technical report, Institut für Aerodynamik und Strömungstechnik (2018)
9. Spalart, P.R., Allmaras, S.R.: A One-equation turbulence model for aerodynamic flows. In: AIAA Computational Fluid Dynamics Conference, number 1992–439 in Conference Proceeding Series. AIAA (1992)
10. Allmaras, S.R., Johnson, F.T., Spalart, P.R.: Modifications and clarifications for the implementation of the Spalart-Allmaras turbulence model. In: International Conference on Computational Fluid Dynamics 7, Hawaii, number ICCFD7-1902 in Conference Proceeding Series (2012)
11. Langer, S.: Agglomeration multigrid methods with implicit Runge-Kutta smoothers applied to aerodynamic simulations on unstructured grids. J. Comput. Phys. **277**, 72–100 (2014)
12. Blazek, J.: Computational Fluid Dynamics: Principles and Applications. Elsevier Science Ltd. (2001)
13. Langer, S., Schwöppe, A., Kroll, N.: The DLR flow solver TAU - status and recent algorithmic developments. In: Proceedings of the 52nd Aerospace Sciences Meeting, January 2014, number 2014–0080 in Conference Proceeding Series. AIAA (2014)
14. Saad, Y.: Iterative Methods for Sparse Linear Systems. International Thomson Publishing, Boston (1996)
15. Yoon, S., Jameson, A.: Lower-upper Symmetric-Gauss-Seidel method for the Euler and Navier-Stokes equations. In: 25th AIAA Aerospace Sciences Meeting, AIAA Paper 1987–0600, Reno, NV (1987)
16. Mulder, W., van Leer, B.: Experiments with implicit upwind methods for the Euler equations. J. Comput. Phys. **59**, 232–246 (1985)
17. Hummel, D.: The second international vortex flow experiment (vfe-2). In: Radespiel, R., Rossow, C.-C., Brinkmann, B.W. (eds.) Hermann Schlichting - 100 Years, pp. 105–129. Springer, Heidelberg (2009). https://doi.org/10.1007/978-3-540-70805-6_18
18. Hummel, D., Redeker, G.: A new vortex flow experiment for computer code validation (2001)

19. Yamazaki, I., Hoemmen, M., Luszczek, P., Dongarra,J.: Improving performance of gmres by reducing communication and pipelining global collectives. In: 2017 IEEE International Parallel and Distributed Processing Symposium Workshops (IPDPSW), pp. 1118–1127 (2017)
20. Chu, J., Luckring, J.M.: Experimental surface pressure data obtained on 65 deg delta wing across reynolds number and mach number ranges. vol. 3: Medium-radius leading edge. NASA Technical Memorandum 4645 (1996)
21. Langer, S.: Investigations of a compressible second order finite volume code towards the incompressible limit. Comput. Fluids **149**, 119–137 (2017)

Analysis of Vortex Burst Phenomena on Generic Hybrid Delta Wing Planforms at Subsonic Speeds

Dominik Sedlacek(✉), Florian M. Heckmeier, Artur Usbek, and Christian Breitsamter

Chair of Aerodynamics and Fluid Mechanics, Technical University of Munich, Boltzmannstrasse 15, 85748 Garching, Germany
dominik.sedlacek@tum.de

Abstract. In this study, experimental investigations are performed to analyze the vortex behavior and breakdown characteristics on a generic multiple swept wing fuselage configuration. *Fast Response Aerodynamic Pressure Probe* and *Particle Image Velocimetry* measurements are performed in the low-speed wind tunnel W/T-A facility of the Chair of Aerodynamics and Fluid Mechanics at the Technical University of Munich. The experiments are executed at subsonic speed, symmetric freestream conditions and angles of attack of $\alpha = 16°, 24°$ and $32°$. In a power spectral density analysis of velocity fluctuations helical mode instabilities and shear layer instabilities can be detected.

Keywords: Hybrid delta wings · Fast Response Aerodynamic probe · Vortex breakdown

1 Introduction

High agility aircraft often require an operation at extreme flight conditions. Even at low angles of attack the flow separates at the wing leading-edge or at multiple swept leading-edges and forms a single or multiple vortex systems, respectively. With the separation of the flow from the leading-edge, the shear layer rolls up forming discrete vortical substructures, referred to as *Shear Layer Instabilities* (SLI) [1]. With increasing angle of attack, the fully developed leading-edge vortex breaks down, the cross section of the vortex increases and a transient flow field develops downstream of the breakdown location. For non-slender delta wings the vortex breakdown begins with a less abrupt transition from jet-type to wake-type core flow [2]. In the wake of a spiral type vortex breakdown quasi periodic velocity fluctuations occur corresponding to *Helical Mode Instabilities* (HMI) [3]. With further increase of the angle of attack, the location of vortex bursting is shifted upstream. The vortex breakdown affects the global aerodynamic forces and moments, and therefore, has an impact on typical flight mechanics related instabilities like pitch up and lateral instability at asymmetric freestream conditions[4]. The comprehension of vortex formation, interaction and bursting

© The Author(s), under exclusive license to Springer Nature Switzerland AG 2021
A. Dillmann et al. (Eds.): STAB/DGLR Symposium 2020, NNFM 151, pp. 282–291, 2021.
https://doi.org/10.1007/978-3-030-79561-0_27

is needed. The analysis of the transient flow behavior is vital for investigation of vortex breakdown and the transient flow field downstream of the breakdown location. In the flow field of the burst vortices reverse flow can occur [5].

Previous experiments on transient flow for delta wing configurations were conducted at the Technical University of Munich [6,7]. In most cases a triple wire or X-wire probe was used for hot wire anemometry investigations. In this study *Fast Response Aerodynamic Pressure Probe* (FRAP) measurements are conducted in the low-speed wind tunnel W/T-A of the Chair of Aerodynamics and Fluid Mechanics at the Technical University of Munich (TUM-AER). The probe is chosen for the analysis of the unsteady flow characteristics because of higher robustness and higher reconstructable flow angles compared to hot wire anemometry [8,9]. The flow field is acquired with a FRAP and by *Particle Image Velocimetry* (PIV) and is presented by means of the non-dimensional *Turbulent Kinetic Energy* (TKE). Velocity component fluctuations are investigated in a spectral analysis for particular positions in the vortex field related especially to the vortex breakdown wake field.

2 Configurations

In this study, a generic wing-fuselage configuration with exchangeable wing planforms is considered. The generic configuration is studied in the NATO STO AVT-316 task group. The examined configuration, Fig. 1, is a triple-delta-wing configuration the so-called NA1 W1. The NA1 W1 consists of three wing segments of different leading-edge sweep: The front part with a medium leading-edge sweep of $\varphi_1 = 52.5°$, followed by a wing mid-section with a high leading-edge sweep of $\varphi_2 = 75°$and the rear wing section with similar wing sweep as the front section, $\varphi_1 = \varphi_3$. The model consists of flat plates with a sharp leading-edge. The most important geometrical parameters of the configuration are summarized in Table 1 [4].

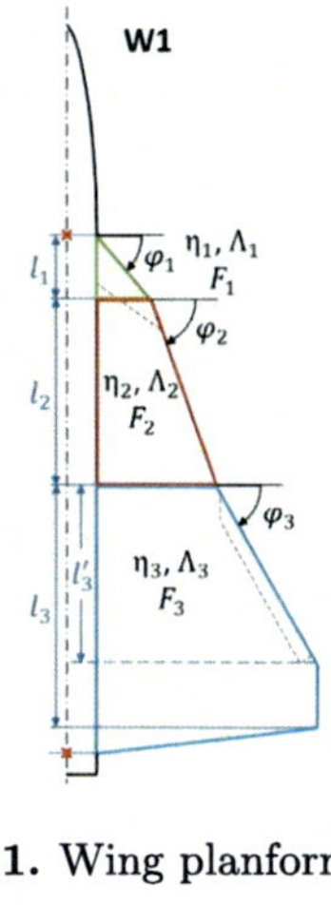

Fig. 1. Wing planform. [5]

Table 1. Geometrical parameters of the NA1 W1 configuration.

		NA1 W1
c_r	$[m]$	0.802
s	$[m]$	0.417
S_{Ref}	$[m^2]$	0.329
λ	$[-]$	0.15
l_μ	$[m]$	0.468
l_1/c_r	$[-]$	0.125
l_2/c_r	$[-]$	0.350
l_3/c_r	$[-]$	0.475
φ_1	$[°]$	52.5
φ_2	$[°]$	75
φ_3	$[°]$	52.5

The measurements are conducted in selected crossflow sections. For FRAP measurements a grid of measurement positions is generated to get the best possible resolution for each section. The measurements are conducted in four different crossflow sections at $x/c_r = [0.125, 0.592, 0.825, 0.950]$. In order to improve the resolution, additional measuring points are added in areas of strong gradients. The spatial resolution varies between the crossflow sections because of the varying window size between $\Delta y/s = \Delta z/s = 0.017$ for $x/c_r = 0.125$ and $\Delta y/s = \Delta z/s = 0.024$ for $x/c_r = 0.950$. Up to 373 measuring points are surveyed per cross section.

3 Experimental Methods

The experiments are conducted in the W/T-A featuring an open test section with cross section dimensions of 1.80 m × 2.40 m (height × width) and a length of 4.80 m. The wind tunnel setup is shown in Fig. 2. The wind tunnel reaches a maximum velocity of $U_{WT} = 65$ m/s and provides a turbulence level of $Tu < 0.4\%$. A three component model support is used to set the angle of attack α, the angle of sideslip β and the roll angle γ. The possible range for the angle of attack is limited to $\alpha = 0° - 40°$.

For Reynolds numbers $Re > 1.0 \cdot 10^6$, it is expected that fully turbulent shear layers develop to constitute the leading-edge vortex systems, shed at the sharp leading-edge contours. PIV measurements are conducted at $U_\infty \approx 50$ m/s and a Reyolds number of $Re = 3.0 \cdot 10^6$ based on the reference length $l_{Re} = 1$ m, while FRAP measurements are carried out at $U_\infty = 30$ m/s, referring to a Reyolds number of $Re = 1.8 \cdot 10^6$ to keep vibrations on the probe support to a minimum. A brief description of the applied measurement techniques is given in the following.

3.1 Fast Response Aerodynamic Probe

To determine spectral quantities of the 3-dimensional velocity field over a hybrid delta wing configuration, intrusive measurements are conducted using a fast-response five-hole probe (FRAP), as depicted in Fig. 2. The additive manufactured FRAP has a hemispheric probe head and the cavities for the placement of the pressure sensors are post-manufactured in a cutting step. Five piezo-resistive pressure sensors (Meggitt Endevco 8507C-2) are placed inside the probe shaft and are operated with the ambient pressure outside of the wind tunnel as the reference pressure of the differential sensor. The probe is connected to NI 9237 data acquisition cards, which transmit the data to a Labview acquisition program. [8]

The probe is calibrated in a calibration wind tunnel for multiple velocity and angle combinations up to 60° cone angle. Furthermore, a temporal calibration of the line-cavity system inside the probe is performed in a frequency test rig for frequencies up to 10 kHz in order to compensate resonance and attenuation effects [9]. For the flow field reconstruction, the acquired pressures are post-processed with the calibration data of both temporal and spatial calibrations. Reconstruction accuracies below 0.2° in both flow angles and 0.1 m/s in the reconstructed velocity can be achieved [10]. The FRAP data are sampled with a sampling frequency of $f_s = 10$ kHz. In order to account for aliasing effects, a low-pass filter below the Nyquist frequency ($f_{lp} \leq f_s/2$) is set in the NI measurement cards. The measurement time is set to $t_s = 7$ s. In case of reversed flow regions or high flow angles, the reconstruction of the flow is equivocal and the data set is discarded.

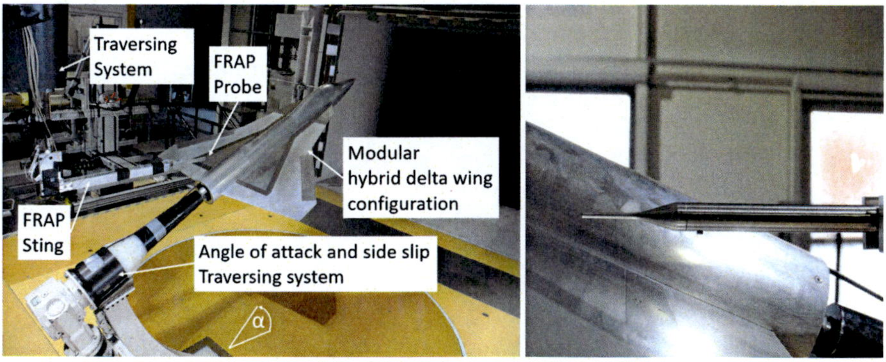

Fig. 2. Left: model and sting support; right: FRAP probe configuration.

3.2 Particle Image Velocimetry

Stereo-PIV measurements are conducted to measure the flow field above the wing in several crossflow sections [5]. For an optimal positioning of the PIV laser sheet, the *Stereo-PIV* system is mounted on a three axis traversing system next to the wind tunnel test section. For an alignment of the cameras and the double pulsed Nd:Yag laser with the angle of attack of the wind tunnel model, the cameras and laser sheet can be rotated around the Y-axis of the traversing system. Two sCMOS cameras with a resolution of 2560×2160 pixel are placed up- and downstream of the measurement plane with an angle of 60°. The sCMOS sensor planes are tilted by Scheimpflugadapters to get a sharp image in the PIV laser sheet, cf. [11]. A spatial resolution of $\Delta d = 1.49 \cdot 10^{-3}$m is provided. Seeding particles with a diameter of $d \approx 1\,\mu$m are infused into the flow. The cameras recorded 400 image pairs per cross section with a sampling frequency of $f_m = 15$ Hz. The presented results are the averaged values determined by 400 samples.

4 Results

Several experiments are carried out in this study in order to investigate the turbulent flow field and the spectral analysis of velocity component fluctuations in the vortex dominated flow field. At this configuration, two vortices develop above the wing: the first in the front region of the wing, the *Inboard Vortex* (IBV), and the second on the third leading-edge region, the *Midboard Vortex* (MBV). Previous studies have shown that at $\alpha = 16°$ the IBV has already a wake type structure at $x/c_r = 0.2$ and the vortex breaks down at $x/c_r \approx 0.475$ [4]. Following, the flow field is presented via the non-dimensional TKE, cf. [6], in selected cross sections.

$$TKE = \frac{u_{rms}^2 + v_{rms}^2 + w_{rms}^2}{2U_\infty^2} \tag{1}$$

4.1 Turbulent Flow Field

In this section, the PIV results for $x/c_r = [0.592, 0.825, 0.950]$ at $\alpha = 16°$ are depicted via the non-dimesional TKE. The axes are normalized with the half span s at the particular cross sections (Fig. 3).

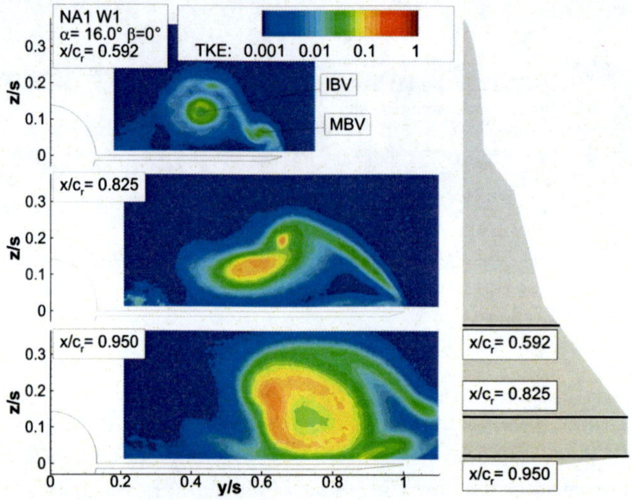

Fig. 3. Distributions of non-dimensional turbulent kinetic energy based on PIV measurements at $\alpha = 16°$ and $Re = 3.0 \cdot 10^6$.

The first cross section shows the flow field downstream of the second wing leading-edge sweep angle change at $x/c_r = 0.592$. Two areas with increased TKE illustrate the development of the named vortices. The IBV is presented by a turbulent vortex core, $TKE_{IBV} = 0.028$, in green and a surrounding less turbulent area in light blue. The MBV spans only over a small area with high TKE values, $TKE_{MBV} = 0.036$. A shear layer between the vortices is visible. This indicates an interaction even in early stages of the MBV. At $x/c_r = 0.825$, the TKE distribution shows a strongly interacting vortex system with recognizable peaks at the particular vortex cores. The IBV is positioned more inboard and closer to the surface. The MBV has moved inboard from the leading-edge and has rotated above the IBV. The absolute values in the vortex cores for the IBV, $TKE_{IBV} = 0.14$, and the MBV, $TKE_{MBV} = 0.15$, are quite similar. The elliptical structure, the reduced non-dimensional axial velocity, $u/U_\infty = 0.41$ and the high TKE values for the IBV indicate an already burst vortex. The similarly high turbulent kinetic energy but still high axial velocity, $u/U_\infty = 1.86$ in the MBV implies a breakdown position more downstream.

In the most downstream located cross section, both vortices have burst, which is revealed by the large area of high turbulent kinetic energy. The highly turbulent flow field represented has increased in area compared to the cross section at $x/c_r = 0.825$ indicating the characteristic annular concentration of maximal turbulence intensities[3]. In the breakdown wake field, a lower turbulent core $TKE = 0.06$ is shown, surrounded by a ring of high turbulent flow, $TKE_{max} = 0.19$.

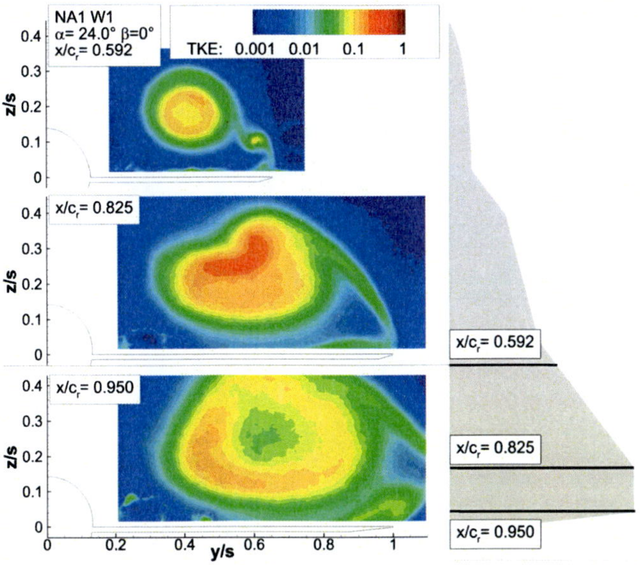

Fig. 4. Distributions of non-dimensional turbulent kinetic energy based on PIV measurements at $\alpha = 24°$ and $Re = 3.0 \cdot 10^6$.

In Fig. 4, the crossflow sections at $\alpha = 24°$ are presented. In the first cross section at $x/c_r = 0.592$, the IBV and the MBV are visible. The IBV has increased in its area and moved slightly upwards. Compared to $\alpha = 16°$, the vortices show a higher TKE as before, $TKE_{max} = 0.13$. In the core of the IBV, a high turbulent ring surrounding a lower turbulent area is present. The vortex has a strongly wake type flow with reversed flow in the vortex core and increased axial velocities surrounding the core.

At $x/c_r = 0.825$, a large and merging vortex wake field is represented. The vortex system is fully burst. Due vortex interaction a high turbulent area, $TKE_{max} = 0.31$, is shown in the upper region of the merging vortex field and the annular concentration of turbulent kinetic energy is disrupted in the region of vortex merging. The high TKE values in the burst IBV area derive mostly from fluctuations in axial direction. However, the turbulences in the region of the burst MBV assemble from fluctuations in all directions. Upstream of the MBV break down location, fluctuations in horizontal and vertical direction dominate. In the third cross section at $x/c_r = 0.950$ an annular TKE concentration is present. Both vortices have already burst and merged. The area with the highest turbulent kinetic energy, $TKE_{max} = 0.14$, has rotated more inboard and is located closer to the wing surface.

The flow field at $\alpha = 32°$ is illustrated in Fig. 5. The first cross section at $x/c_r = 0.592$ shows a fully burst and highly turbulent IBV, $TKE_{IBV,max} = 0.25$. The MBV is already burst and is presented by a turbulent kinetic energy of $TKE_{MBV,max} = 0.38$ with dominating velocity fluctuations in axial and horizontal direction. At $x/c_r = 0.825$ the vortex system is fully burst and is extending over a large area. The annular turbulent concentration is clearly recognizable.

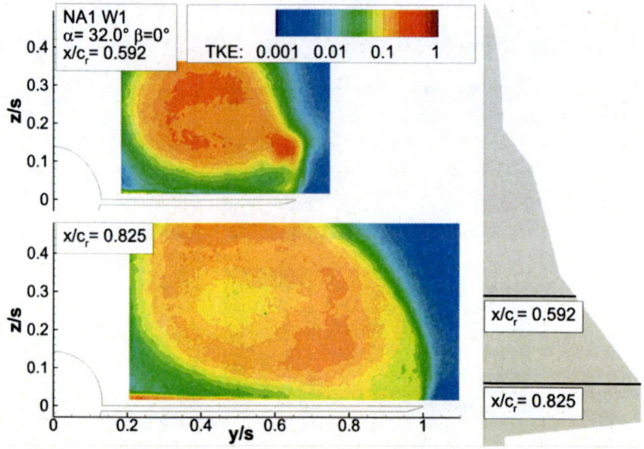

Fig. 5. Distributions of non-dimensional turbulent kinetic energy based on PIV measurements at $\alpha = 32°$ and $Re = 3.0 \cdot 10^6$.

4.2 Spectral Analysis of the Vortex Breakdown

For the analysis of the velocity fluctuations, locations with high turbulent kinetic energy gradient are investigated. In the following, the *Power Spectral Density* (PSD) of the velocity fluctuations determined by transient FRAP measurements is plotted over the reduced frequency k ($k = f \cdot l_\mu / U_\infty$).

Figure 6 presents PSDs at an angle of attack of $\alpha = 16°$. The measurement position in Fig. 6a is located at $x/c_r = 0.592$ in the shear layer between IBV and MBV. At $k \approx 4.5$ increased amplitudes are recognizable for axial and vertical fluctuations. In this region SLI are expected. The frequency range for the SLI is calculated with [12]:

$$k_S = \frac{l_\mu}{x} \cdot \frac{(2000 \div 10000)}{\sqrt{Re_{c_r}}} \tag{2}$$

The expected dominant frequencies for SLI are presented in Table 2. In Fig. 6b, PSD levels occur for higher reduced frequencies, $k \approx 9.04$, in the area of the IBV-MBV merging at $x/c_r = 0.825$.

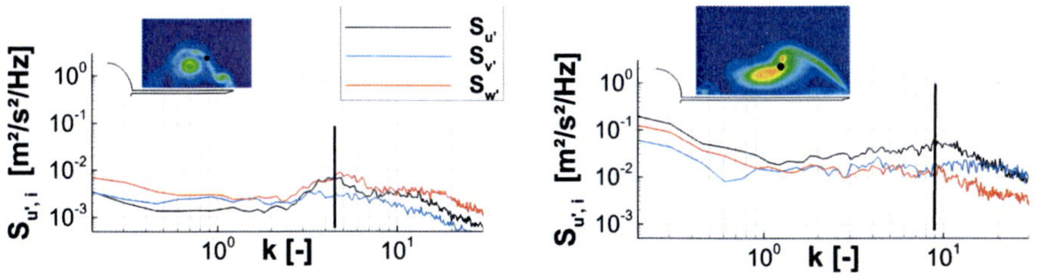

(a) $x/c_r = 0.592$, $y/s = 0.52$ $z/s = 0.15$, (b) $x/c_r = 0.825$, $y/s = 0.66$ $z/s = 0.16$,

Fig. 6. Power spectral densities of the velocity fluctuations at $\alpha = 16°$ and $Re = 1.8 \cdot 10^6$.

In Fig. 7, PSD results for selected positions at an angle of attack of $\alpha = 24°$ are shown. Figure 7a represents the fluctuations at the annular TKE concentration at $x/c_r = 0.592$. A pronounced peak is shown at $k \approx 1.88$. This frequency can be correlated to the SLI as shown in Fig. 2. At $x/c_r = 0.825$, two peaks are visible at $k \approx 0.84$ (solid line) and $k \approx 1.76$ (dashed line). The first peak correlates with a helical mode instability, which is calculated by [3]:

$$k_H = \frac{l_\mu}{x} \cdot \frac{1}{cot(\varphi_{m,eff}) \cdot sin(\alpha)} \cdot (0.28 \pm 0.025) \tag{3}$$

$$\varphi_{m,eff} = \arccos\left(\frac{F_1 \cos\varphi_1 + F_2 \cos\varphi_2 + F_3 \cos\varphi_3}{F_1 + F_2 + F_3}\right) = 59.64° \tag{4}$$

Table 2. Expected frequency ranges for HMI and SLI for different angles of attack and flow positions [3,12].

Angle of attack	Position x/c_r	k_H	k_S
$\alpha = 16°$	0.592	1.6–1.9	1.5–7.3
	0.825	1.1–1.3	1.0–5.3
$\alpha = 24°$	0.592	1.05–1.26	1.5–7.3
	0.825	0.76–0.91	1.0–5.3
$\alpha = 32°$	0.592	0.81–0.97	1.5–7.3
	0.825	0.58–0.69	1.0–5.3

In Fig. 8, the PSDs for $\alpha = 32°$ at $x/c_r = 0.592$, Fig. 8a, and $x/c_r = 0.825$, Fig. 8b, are displayed. The measurement at $x/c_r = 0.592$ is located at the verge of the high turbulent area. The peak at $k \approx 0.97$ indicates a HMI, as expected from Table 2. In Fig. 8b, the measurement position is located at the annular TKE concentration and multiple frequency peaks are visible. The first peak at $k \approx 0.75$ can be again correlated with the HMI. The further peaks at $k = 1.5$ indicate higher harmonics of the HMI.

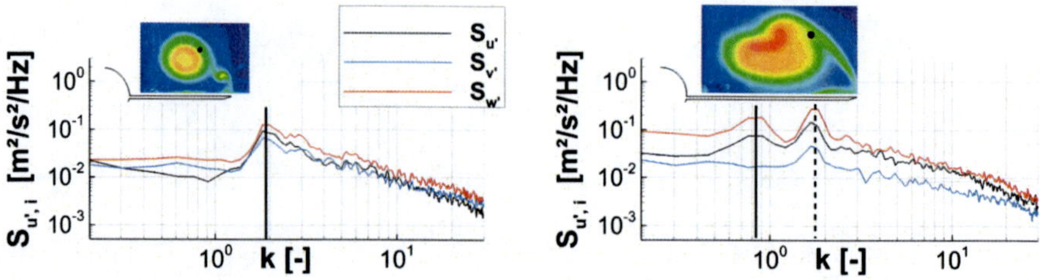

(a) $x/c_r = 0.592$, $y/s = 0.50$ $z/s = 0.23$ (b) $x/c_r = 0.825$, $y/s = 0.78$ $z/s = 0.28$,

Fig. 7. Power spectral densities of the velocity fluctuations at $\alpha = 24°$ and $Re = 1.8 \cdot 10^6$.

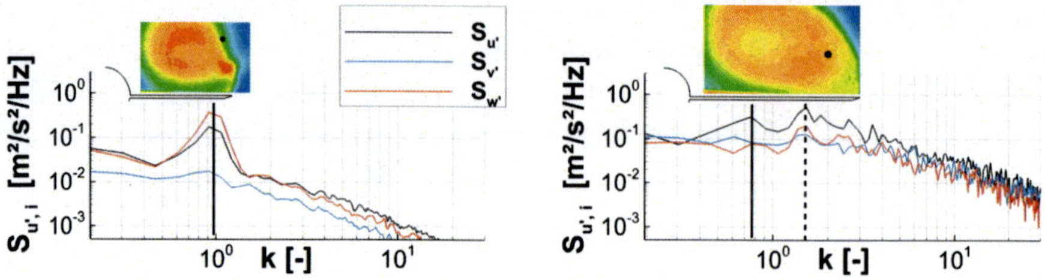

(a) $x/c_r = 0.592$, $y/s = 0.60$ $z/s = 0.28$, (b) $x/c_r = 0.825$, $y/s = 0.83$ $z/s = 0.20$,

Fig. 8. Power spectral densities of the velocity fluctuations at $\alpha = 32°$ and $Re = 1.8 \cdot 10^6$.

5 Conclusion

In this study, experiments with a *Fast Response Aerodynamic Pressure Probe* (FRAP) are conducted to investigate the spectral characteristics of the velocity field fluctuations for a triple delta wing configuration. The flow field pattern is discussed on the basis of *Particle Image Velocimetry* (PIV) data presenting the spatial distribution of the turbulent kinetic energy. The spectral analysis is aimed on the evaluation of instability mechanisms related to vortical flow shear layer and vortex bursting.

The flow field results are presented at $\alpha = [16°, 24°, 32°]$ for three different cross sections. At $\alpha = 16°$, there is a vortex formation of inboard vortex (IBV) and midboard vortex (MBV). At $x/c_r = 0.592$, shear layer instabilities can be detected for a reduced frequency of $k \approx 4.5$, cf. Table 2. The IBV is a wake type vortex even in the front section of the wing. At $x/c_r = 0.825$, the IBV is completely burst and the MBV is merging with the IBV. At $\alpha = 24°$, the IBV burst more upstream and due to the large vortex break down field the MBV is absorbed faster. At $x/c_r = 0.825$, two spectral peaks occur. The low-frequency peak, $k \approx 0.84$, indicates a Helical Mode Instability(HMI), correlating with the calculated frequencies shown in Table 2. The second peak, $k \approx 1.76$, is in the range of the SLI, $k_S \approx 1.0–5.3$. At $\alpha = 32°$, the IBV break down location moves further upstream and the burst vortex field increases in area. At $x/c_r = 0.592$, a frequency, $k = 0.97$, correlating with a HMI is detected. The results indicate that the helical mode instability related to vortex breakdown is a predominant effect while the merging of the vortical structures is accompanied by further shear layer instabilities.

Acknowledgments. The project is funded by the Deutsche Forschungsgemeinschaft (DFG, German Research Foundation) – grant number BR1511/13-1. The fruitful cooperation with Airbus Defense and Space is gratefully acknowledged.

References

1. Kölzsch A.: Active Flow Control of Delta Wing Leading-Edge Vortices. Dissertation, Technical University Munich (2017)
2. Gursul, I., Gordnier, R., Visbal, M.: Unsteady aerodynamics of non-slender delta wings. Progress Aerospace Sci. **41**(7), 515–557 (2005). https://doi.org/10.1016/j.paerosci.2005.09.002
3. Breitsamter, C.: Unsteady flow phenomena associated with leading-edge vortices. Progress Aerospace Sci. **44**, 48–65 (2008). https://doi.org/10.1016/j.paerosci.2007.10.002
4. Pfnür, S., Hövelmann A., Sedlacek D., Breitsamter, C.: Vortex Flow Interaction Phenomena on Multi Swept Delta Wings at Subsonic Speeds. STO-MP-AVT-307, 20 (2019)
5. Pfnür, S., Breitsamter, C.: Leading-edge vortex interactions at a generic multiple swept-wing aircraft configuration. J. Aircraft **56**(6), 2093–2107 (2019). https://doi.org/10.2514/1.C035491
6. Hövelmann, A.: Analysis and control of partly-developed leading-edge vortices. Dissertation, Technical University Munich (2016)
7. Buzica, A., Breitsamter, C.: Turbulent and transitional flow around the avt183 diamond wing. Aerospace Sci. Technol. **92**, 520–535 (2019). https://doi.org/10.1016/j.ast.2019.06.025
8. Heckmeier, F., Iglesias, D., Breitsamter, C.: Unsteady multi-hole probe measurements of the near wake of a circular cylinder at sub-critical reynolds numbers. New Results in Numerical and Experimental Fluid Mechanics XII, pp. 643-652 (2020). https://doi.org/10.1007/978-3-030-25253-3_61
9. Heckmeier, F., Iglesias, D., Kreft, S., Kienitz, S., Breitsamter, C.: Development of unsteady multi-hole pressure probes based on fiber-optic pressure sensors. Engineering Research Express 1, 2, pp. 025023 (2019). https://doi.org/10.1088/2631-8695/ab4f0d
10. Heckmeier, F.M., Breitsamter, C.: Aerodynamic probe calibration using gaussian process regression. Measur. Sci. Technol. **31**, 125301 (2020). https://doi.org/10.1088/1361-6501/aba37d
11. Raffel, M., Willert, C.E., Wereley, S.T., Kompenhans, J.: Particle Image Velocimetry - A Practical Guide. Springer, Cham (2019). https://doi.org/10.1007/978-3-319-68852-7
12. Breitsamter, C.: Turbulente Strömungsstrukturen an Flugzeugkonfigurationen mit Vorderkantenwirbeln. Dissertation, Technical University Munich (1997)

Numerical Modelling of a Wind Tunnel Experiment to Investigate Vortex-Dominated Flow at Medium and High Angles of Attack

Guido Voss(✉)

DLR, Institute of Aeroelasticity, 37073 Göttingen, Germany
guido.voss@dlr.de
https://www.dlr.de/ae/

Abstract. Vortex-dominated flow and their effects on motion-induced air loads in the case of transonic inflow represents a complex and so far not well understood sub-area of aerodynamics. Especially the modeling of such kind of flows with the help of RANS methods represents a major challenge with regard to the choice of initial and boundary conditions. In this article, the results of validation studies are presented using wind tunnel measurement campaigns. The main focus is on the dependence about the angle of attack and the associated characteristics of the vortex-dominated flow.

Keywords: IWEX · Lambda configuration · Vortex dominated flows · Wind tunnel · CFD-simulation · Validation · Reynolds-Stress turbulence model

1 Introduction

In the military sector, unmanned aerial vehicles are gaining increasing influence in order to minimize the risk for human crews in hazardous operational situations and to be able to fly maneuvers that would be too burdensome for pilots in cockpit. Unmanned aircraft, so-called UCAVs (Unmanned Combat Aeronautical Vehicles), are therefore an interesting research branch to investigate the influence of vortex-dominated flows on the aerodynamic and in particular the aeroelastic behavior of such an aircraft configuration.

As part of the DLR-internal projects FaUSST (Fortschrittliche aerodynamische Stabilitäts- und Steuerungsuntersuchungen, advanced aerodynamic stability and control studies) and MEPHISTO, extensive wind tunnel measurement campaigns were carried out to investigate the interaction of the vortex-dominated flow with a forced pitching motion. The results of these measurement campaigns were used to create CFD-based aerodynamic modelling that allows predictions in the future, for example in the area of fluttering of wind tunnel models.

© The Author(s), under exclusive license to Springer Nature Switzerland AG 2021
A. Dillmann et al. (Eds.): STAB/DGLR Symposium 2020, NNFM 151, pp. 292–301, 2021.
https://doi.org/10.1007/978-3-030-79561-0_28

In this paper, the results of unsteady validation calculations are presented, which were considered for two different mean angles of attack with a weak transonic flow ($M = 0.5$) and a pitching motion with small amplitude ($d\alpha = 0.2°$) and high frequency ($f = 12\,\mathrm{Hz}$).

2 Theoretical Considerations

Particularly with high angles of attack and large sweep angle of wings, a complex system of vortices may appear on the upper side of the wing, which influence the lift and moment behavior of the aircraft [1,2]. Especially for military applications, this is used to fly extreme flight maneuvers.

Vortical flow may in some cases assumed as unsteady, i.e. the flow includes time-dependent components. This can manifest itself, for example, in periodic motion of the vortex in space, which in turn can affect the lift behavior of the aircraft. If this time-dependent behavior takes place within narrow limits in numerical investigations, this flow can be represented as steady. Otherwise, the numerical simulation of vortical flow has to be carried out with URANS methods in order to take into account the unsteady components of the flow.

Since there is no general definition of a vortex, different models are used to describe the physical principles of vortex systems, (see [4,7]).

3 Experimental Set-Up

3.1 Transonic Windtunnel Goettingen (TWG)

The Transonic Wind Tunnel Goettingen (TWG) is a wind tunnel of Goettingen type. This means that the air flow circulates in a closed system. Depending on the measuring section used, Mach number ranges with different widths can be obtained. For these investigations, the adaptive measuring section of the TWG was used [7]. Since this unit of the measuring section is not suitable for compensating the pressure fluctuations of unsteady measurements due to the inertia of the hydraulic actuating elements, the adaptive part of the measuring section was preset divergent before the measurements start to take into account the growth of the wall boundary layer and thus to reduce the locking of the wind tunnel through the wall boundary layer. The widening of the adaptive measuring section is about 3 mm per meter. This value was estimated by prior CFD calculations of the empty wind tunnel.

In the region of the pre-chamber the transonic wind tunnel has a cross-section of $A_{pc} = 4 \times 4\,\mathrm{m}^2$. This is reduced via a nozzle to the measuring section to $A_{ms} = 1 \times 1\,\mathrm{m}^2$, whereby the flow is accelerated to the required velocity at the wind tunnel model.

3.2 IWEX Model

For the investigations presented in this paper, the results from three measurement campaigns in the TWG were used. The IWEX model is a half model in a flying wing design made of steel. The IWEX wind tunnel model represents a spin-off of the well-known F17/SACCON configuration [1,3], which has been adapted to the special boundary conditions of the TWG in terms of vortex formation and interaction with wind tunnel walls [5]. The specification of the model is as follows:

- wing span s (from symmetry plane to wing tip): 0.51m
- reference length c (at kink): 0.31132 m
- reference area A (without peniche): 0.16639m^2
- peniche span s_P: 0.03 m
- x-position of rotation axis: 0.3217 m
- y-position of sensor section 1: 0.23 m
- y-position of sensor section 2: 0.355 m
- x-position of sensor section 3: 0.41 m
- x-position of sensor section 4: 0.65 m

The characteristic of the radius of curvature depending on the span is described in more detail in [6,7].

In the area of the symmetry plane, the IWEX model is provided with a peniche to extend the model completely out of the wall boundary layer of the TWG. The span pf the peniche is 0.03 m. In the rear section, this peniche has a special shape to minimize or prevent the formation of the horseshoe vortex and thus detachment on the upper side of the fuselage [5,7,8]. Figure 1 (left) shows the positions of the parts of the IWEX model inside the wind tunnel.

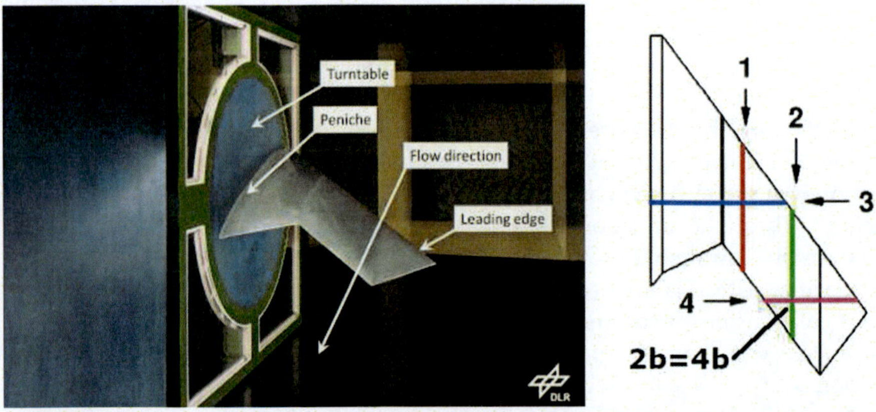

Fig. 1. Position of the IWEX model inside the Transonic Windtunnel Goettingen (photograph, left) and schematic representation of the IWEX wind tunnel model with the position of the unsteady pressure sensors (right)

3.3 Measurement Methodology/Sensors

On the upper and lower side the IWEX model is equipped with four sections of unsteady pressure sensors (kulites) from which the unsteady pressure derivatives can be recorded. The positions of the respective sensor sections are shown schematically in Fig. 1 (right).

4 Numerical Set-Up

4.1 CFD Grid

The computational grid to carry out the numerical investigations was produced with the Centaur grid generation software [9]. This tool can be used to set up unstructured and hybrid computing grids.

The mesh used here contains approximately 7.8 million grid points and approximately 20 million grid elements. The CFD mesh includes the plenum chamber, the nozzle and the measuring section of the TWG. This requires the creation of a new grid depending on the angle of attack of the wind tunnel model. The grid is largely filled up with tetrahedras, the respective boundary layers are built with prisms. In areas of the model in which vortex structures are formed, the model was built with hexahedron blocks. This has proven to be slightly advantageous in terms of computing time in previous studies. Vortex structures also seem to be better resolved. Figure 2 shows the surface grid and the hexahedral block inside the tetraedral mesh of the computing grid. The smooth transition of the flow from the hexahedral mesh to the tetraedral mesh is shown in Fig. 3, where discontinuities in the transition area between the different grid types do not appear. As part of the investigations, an extensive grid convergence study was carried out beforehand. This has shown that the number of grid points and grid elements as well as the grid point density at areas that

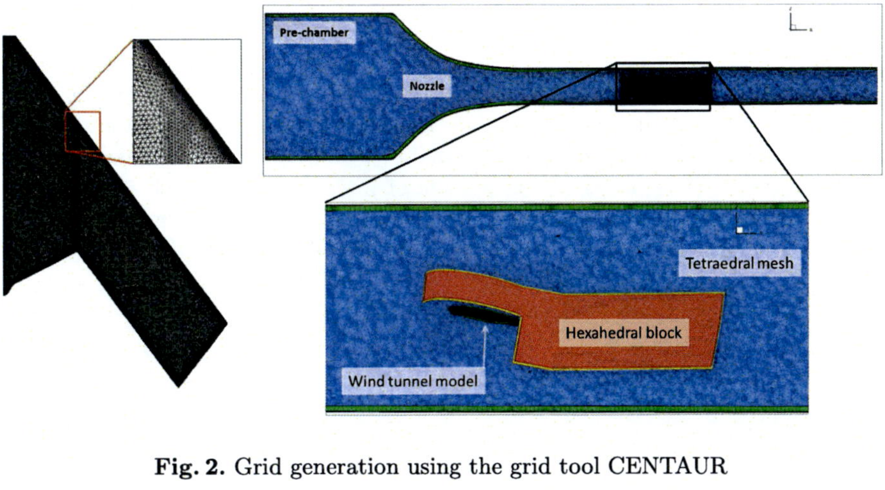

Fig. 2. Grid generation using the grid tool CENTAUR

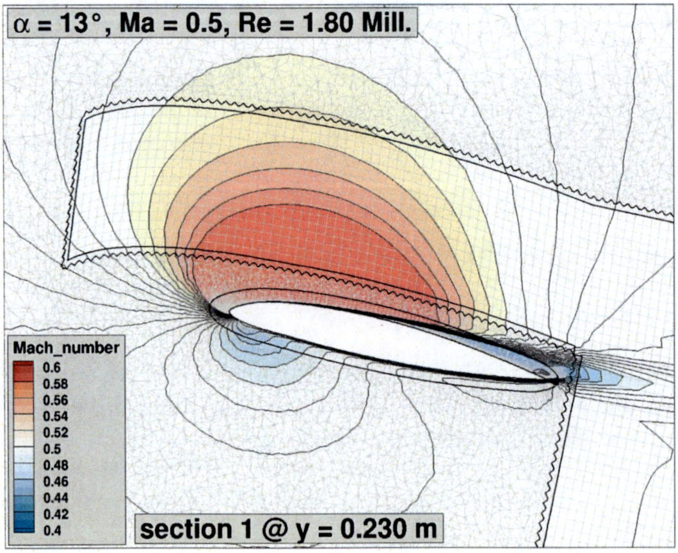

Fig. 3. Mach number distribution in the area of grid type change

are of interest for the investigations are sufficient. The grid points density naturally means a compromise here in order to keep the computing time for unsteady simulations within an acceptable range.

4.2 Flow Solver

The DLR flow solver TAU [13,14] was used for the investigations presented here. It is an unstructured finite volume flow solver that solves the Reynolds-averaged Navier-Stokes equations (RANS). This approach requires the use of appropriate turbulence modeling. For these investigations, the RSM turbulence modeling [11] as well as the one-equation turbulence modeling from Spalart-Allmaras without vortex correction [10] was used.

RSM models additionally take into account anisotropies of the turbulent flow by calculating the components of the Reynolds stress tensor using six additional transport equations. The complex interactions of turbulent flows can thus be better captured. This offers the advantage of higher accuracy of the results, at the cost of longer computing time. In the presented results, a modification of the RSM model is used, the so-called RSM-lnω model. In this case, the natural logarithm is formed from the transport variable ω of the characteristic frequency of the energy-dissipating eddies, which makes the otherwise necessary numerical limitation of ω superfluous [12]. This modification stabilizes the CFD calculations with regard to convergence and can improve the accuracy of the results.

4.3 Boundary Conditions for Wind Tunnel Flow

In this special case, the simulations describe a wind tunnel flow that requires a special structure of the boundary conditions. Wind tunnel flows are conservative in terms of mass and energy conservation if the magnitudes of the mass flow at the inflow and outflow boundaries are identical. For this purpose, special boundary conditions are used and implemented in the TAU code. These are the reservoir-pressure inflow boundary condition and the exit-pressure outflow boundary condition [13].

5 Results

5.1 Steady Calculations

Overall, the steady calculations show good convergence behavior. Cases with a low mean angle of attack in which no vortex system has yet developed behave better in this regard than those in which a vortex system exists, for which the value of the density residual is three orders of magnitude higher at the converged value. In the worst case, the integral coefficients, in particular the drag coefficient C_d, fluctuate at $\alpha = 13°$ in a range of 0.00025, the lift coefficient C_l is in the range of 0.00075. In view of the fact that a sensitive turbulence model was used with a pronounced vortex flow, these are acceptable reference conditions for unsteady calculations. The converged steady calculations are always used as the basis for unsteady CFD simulation calculations.

At the corresponding angle of attack $\alpha = 13°$, the steady c_p distributions at the four pressure sections are shown in Fig. 4 together with experimental data. The red and dark red lines show the results with the RSM-lnω and the standard RSM turbulence model respectively, the blue lines the results computed with the Spalart Allmaras model without vortex correction and the green dots the results of the experimental measurements. The critical c_p value at transonic flows

$$c_{p_{crit}} = \frac{2}{\gamma M^2}\left(\frac{1+\left(\frac{\gamma-1}{2}\right)M^2}{1+\left(\frac{\gamma-1}{2}\right)}\right)^{\frac{\gamma}{\gamma-1}} - 1 \tag{1}$$

of the pressure cuts 1–3 is given in magenta color. In particular for pressure Sects. 1 and 3, and regarding the complex flow conditions, there is a good agreement between the experimentally measured and the numerically calculated values. At the rear area of the pressure Sect. 2, a slight deviation between the experimental and numerical investigations can be observed. This is due to the fact that the main vortex is located here, which cannot be reproduced exactly with the RANS methods with turbulence modeling used here. An even clearer deviation can be found at the pressure Sect. 4. A large part of the main vortex is located in this area. Differences between the standard RSM model and the improved RSM-lnω model can also be observed.

In the area of the front edge of the IWEX wind tunnel model, the pressure coefficient falls below the critical c_p value. In this area there is a small supersonic zone, which causes a weak recompression shock and thus possibly influences the formation of the main vortex system.

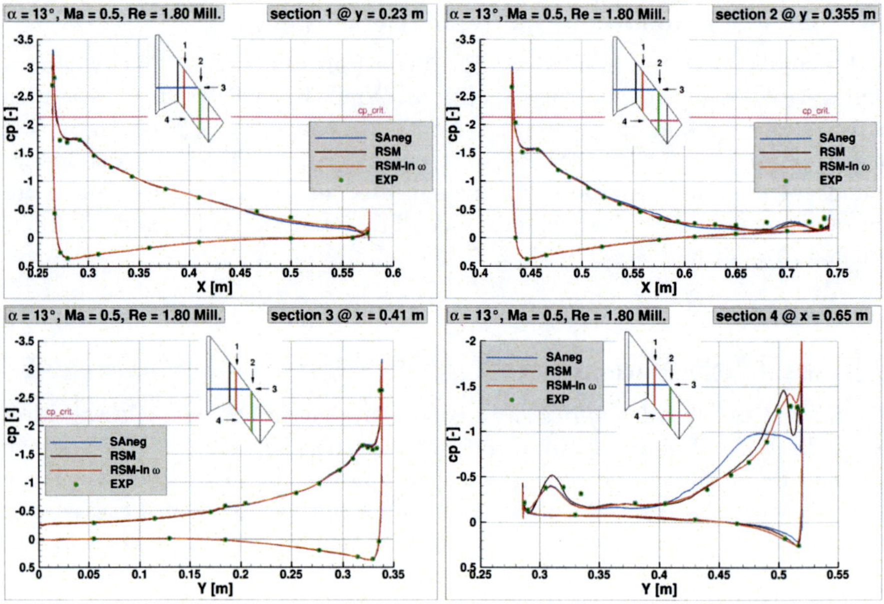

Fig. 4. Comparison of the experimental and numerically determined steady pressure distributions on the four pressure cuts for $\alpha = 13°$

5.2 Unsteady Calculations

The converged steady solutions were used as the restart point for the unsteady calculations. In order to be able to reproduce the framework conditions of the experiments as good as possible, the model was subjected to a monofrequent pitching motion which is applied via grid deformation. The excitation frequency and the excitation amplitude are set according to the values used in the experiment. The excitation frequency is $f = 12\,\text{Hz}$, with a peak to peak value of $d\alpha = 0.21°$ for a mean AoA of $\alpha = 13°$ and $d\alpha = 0.2°$ for a mean AoA of $\alpha = 5°$. The reduced frequency in both cases is about $\omega^\star \approx 0.14$. To ensure the decay of transient behaviour in the unsteady results, a total of four periods of the input signal were simulated. Each period of the computation was resolved with 5000 time steps. With an excitation frequency of $f = 12\,\text{Hz}$, this results in a time step size of $\Delta t = 0.0000167\,\text{s}$. Here, the dual time step scheme for the unsteady calculations was used. This means that an unsteady time step is calculated in pseudo time with a fixed number of inner iterations [15]. In this case, 200 inner iterations were sufficient.

The result presented below is the derivative of the pressure coefficient c_p, which is calculated from the time signal of the system response of the model excited by a pitching oscillation. First, this time signal is converted into the frequency domain using a Fourier transformation (FFT). The result is a complex function that indicates the frequency spectrum of the pressure signal. The transfer function of the time-dependent excited system is obtained by dividing with the Fourier-transformed excitation signal. Since the excitation is a monofrequent sinusoidal signal, only the second component of the frequency signal vector differs from zero once the transient behaviour has disapeared. Thus, only the second component of the vector of the transfer function is used. From this complex value, the magnitude and the phase curve of the pressure derivative are determined for each spatial point on the surface of the wind tunnel model and compared with the experimental values. The curves are shown in Figs. 5 and 6.

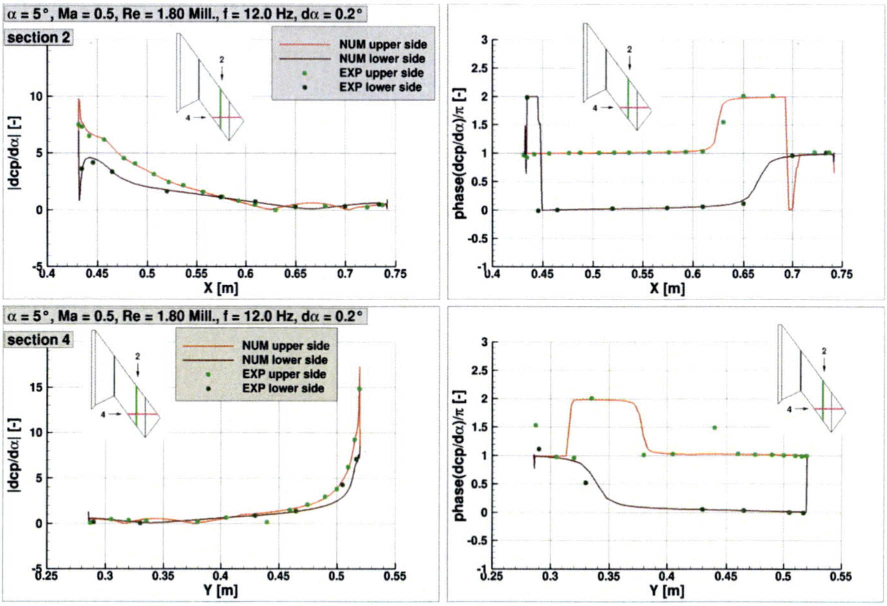

Fig. 5. Magnitude and phase shift of the unsteady c_p derivative at the respective pressure sensor cut for $\alpha = 5°$

Figure 5 shows the unsteady pressure derivative for pressure cuts 2 and 4 for the angle of attack $\alpha = 5°$ calculated with the RSM turbulence model with the included lnω-modification. The magnitude can be seen on the left and the phase curve of the unsteady c_p derivative on the right. No vortex structures have formed in this angle of attack range, so that a good correspondence between experiment and CFD investigation can be observed here.

The comparison of the experimental and numerically determined pressure coefficients at a higher angle of attack is somewhat different (Fig. 6). Especially in the rear outer area of the wing, the quantitative differences are clearly visible due to the distinct vortex topology. A trend can be recognized that the special flow conditions can be well approximated.

In pressure Sect. 2 at point 2b (see Fig. 1, right), a noticeable kink can be seen in the magnitude and the phase progression. This kink is not present in the experimental investigations, it may be a numerical artifact from turbulence modeling. Further investigations are necessary to clarify the reason of this kink.

The differences in pressure cut 4 are even more obvious. The bend in magnitude and phase at point 4b corresponds to point 2b of pressure cut 2. In addition, there is a clear peak at point 4a, which is not visible in the experimental results. In contrast, this peak is not observed in the phase curve.

Overall, even at higher angles of attack, the qualitative trend shows that the flow conditions can be mapped relatively well with the existing CFD methods. The fact that model deformations present in the experiment is not yet included in the numerical investigations. Due to the stiffness of the model associated with high eigenfrequencies, the deformations are expected to be small, but shall be taken into account for future numerical investigations.

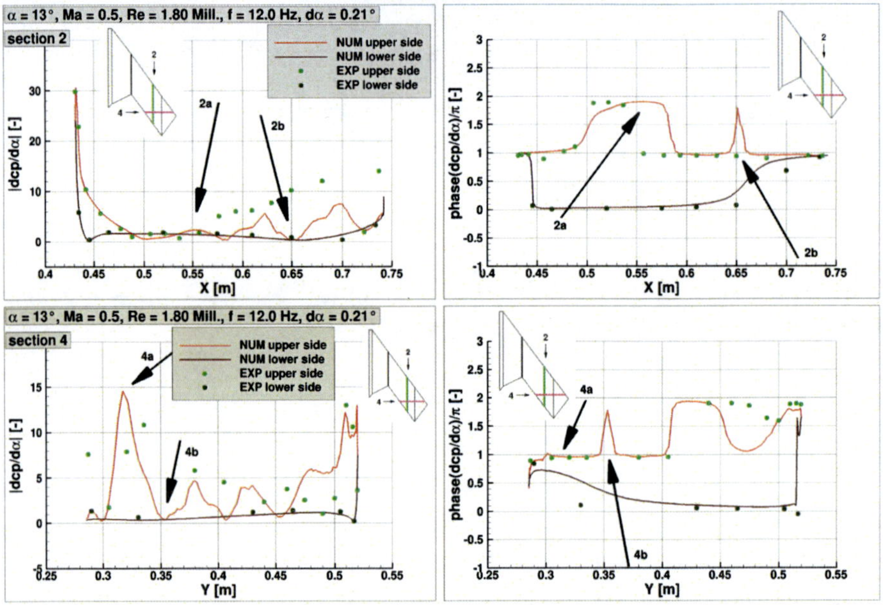

Fig. 6. Magnitude and phase shift of the unsteady c_p derivative at the respective pressure sensor cut for $\alpha = 13°$

6 Conclusion

In this work it has been shown that an appropriate modeling with the current RANS techniques combined with suited RSM turbulence models for vortex dominated flows in the weak transonic regime can be achieved. Particularly in the medium angle of attack range, a very good match can be observed with the experimental unsteady pressure distributions. There are limitations when compared to experimental results, especially at a high angle of attack, involving a clear development of the vortex formation. The special flow conditions make it a complex task to model all details of the flow. The elastic deformations of the IWEX model geometry, which can also have an effect on the pressure distributions, are not yet taken into account in these investigations. In future work, this aspect shall be included in the over-all assessment.

References

1. Schütte, A., Hummel, D., Hitzel, S.M.: Flow physics analyses of a generic unmanned combat aerial vehicle configuration. J. Aircr. **49**(6) (2012)
2. Schütte, A.: Wirbelströmungen an gepfeilten Flügeln mit runden Vorderkanten, Dissertation, Technische Universität Braunschweig. DLR-Forschungsbericht 2015-40 (2015)
3. Voss, G., Schäfer, D., Vidy, C.: Investigation on flutter stability of the DLR-F19/SACCON configuration. Aerosp. Sci. Technol. **93**, 105320 (2019)
4. Haller, G.: An objective definition of a vortex. J. Fluid Mech. **525** (2005)
5. Wiggen, S., Voss, G.: Vortical flow prediction for the design of a wind tunnel experiment with a pitching lambda wing. CEAS Aeronaut. J. **5**(4) (2014)
6. Wiggen, S.: Versuchsbericht IWEX 1. Interner Bericht, DLR-IB 232-2015 C 004 (2015)
7. Wiggen, S.: Bewegungsinduzierte instationäre Luftkräftebei wirbeldominierter Strömung, Dissertation, Technische Universität Braunschweig. DLR-Forschungsbericht 2016-50 (2016)
8. Wiggen, S., Voss, G.: Development of a wind tunnel experiment for vortex dominated flow at a pitching Lambda wing. CEAS Aeronaut. J. **5**(4) (2014)
9. CENTAUR Users Guide. https://www.centaursoft.com
10. Spalart, P.R., Allmaras, S.R.: A one-equation turbulence model for aerodynamic flows. In: 30th Aerospace Sciences Meeting & Exhibit, Reno (1992)
11. Cecora, R.D., Radespiel, R., Eisfeld, B., Probst, A.: Differential Reynolds-stress modeling for aeronautics. AIAA J. **53**(3) (2015)
12. TAU_UserGuide, Tau-Version 2019.1.0
13. Galle, M., Gerhold, T., Evans, J.: Technical documentation of the DLR TAU-Code. Interner Bericht, DLR-IB 233-1997 A 43 (1997)
14. Gerhold, T., Galle, M., Friedrichs, O., Evans, J.: Calculation of complex three-dimensional configurations employing the DLR TAU-Code. AIAA Paper, 19-0167 (1997)
15. Jameson, A.: Time dependent calculations using multigrid, with applications to unsteady flows past airfoils and wings. In: 10th Computational Fluid Dynamics Conference, Honolulu (HI) (1991)

Rotorcraft

Check for updates

Investigation of a Coaxial Propeller Configuration Under Non-axial Inflow Conditions

Michael Cerny(✉), Jan-Arun Faust, and Christian Breitsamter

Technical University of Munich, Boltzmannstr. 15, 85748 Garching, Germany
michael.cerny@tum.de

Abstract. In the last decades, the interest for light-weight air vehicles and unmanned aerial vehicles (UAVs) equipped with small-scale propellers has risen significantly. Such devices are often equipped with counter-rotating coaxial propellers in order to attain higher thrust levels while maintaining a compact package. In response to their operating scope, these propellers are often subject to a wide range of inflow angles. In this work, a coaxial propeller configuration is investigated under different inflow angles ranging from 0° to 180°. Wind tunnel experiments are performed to measure the steady loads and to compare the efficiency against two isolated puller propellers. Furthermore, selected flow fields are captured using smoke visualization. Emphasis is on identifying the interactions of the propeller wakes and the vortex structures. In parallel to the experimental study, a series of URANS simulations are performed in order to study the loads and the flow field in a time-resolved manner.

Keywords: Propeller · Coaxial · Non-axial inflow

1 Introduction

It is the primary motivation for using counter-rotating coaxial propeller configurations to benefit from their capability of producing a higher thrust with the same propeller plane area [1]. Especially, facing non-axial inflow, the specific flow field of coaxial propellers results in increased thrust which is confirmed by other experimental analyses [5]. Furthermore, the arrangement of two counter-rotating propellers shall minimize the swirl in the propeller wake and, hence, reduce the induced drag. Additionally, counter-rotating propellers show specific advantages regarding the stability and maneuverability as well [4].

The existence of these benefits, especially under non-axial inflow conditions, will be analyzed and discussed in this work. Presented are stationary and transient load data as well as flow field analyses, obtained both by smoke visualization as well as by URANS calculations. A detailed description of the theoretical background of the influence of non-axial inflow on propellers in general and particulary on the local inflow characteristics has been presented in [2,3].

© The Author(s), under exclusive license to Springer Nature Switzerland AG 2021
A. Dillmann et al. (Eds.): STAB/DGLR Symposium 2020, NNFM 151, pp. 305–314, 2021.
https://doi.org/10.1007/978-3-030-79561-0_29

2 Experimental Approach

A coaxial configuration consists of two counter-rotating propellers, mounted on a common axis. Here, the puller propeller APC 18x8E and the pusher APC 18x8EP with a diameter of D = 18 in. (457.2 mm) are applied with an axial spacing of $\Delta x = 0.381D_P$. In between them, two independent brushless motors (KDE5215XF-220, electrical power 2 kW) are used to drive the two propellers at the same rate in the opposite rotational direction. Torque balance is not applied.

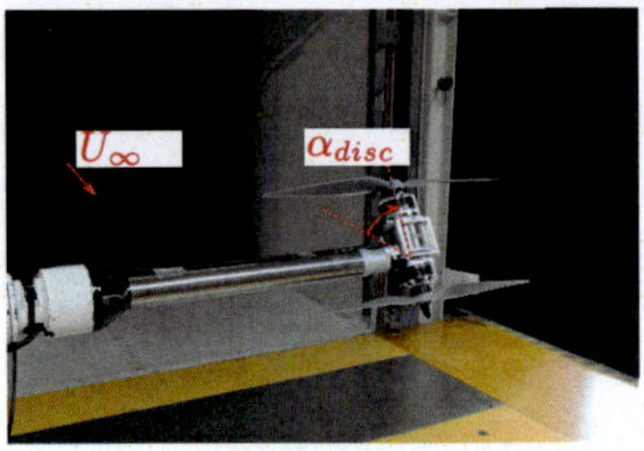

(a)

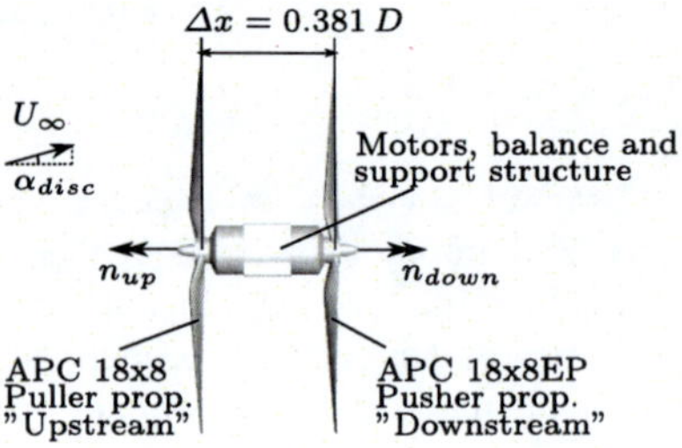

(b)

Fig. 1. **(a)** Picture and **(b)** sketch of the wind tunnel setup.

In the center of the setup, a six-component balance (K6D40 200N/5Nm/CG, ME-Messsysteme GmbH, precision 0.2 N and 0.005 Nm) is mounted to measure the loads of both propellers simultaneously. At the end of a narrow horizontal beam, the model is placed rotatable in the wind tunnel to cover all inflow angles from $\alpha_{disc} = 0°$ to 180° (see Fig. 1).

3 Numerical Approach

The structured computational grid is meshed with the commercial software ANSYS ICEM CFD. It is subdivided into three domains: An outer static hexahedral domain with 19 million cells contains the unmoving model parts like the horizontal beam and the balance. Its edge length is approximately 30 times the propeller diameter. At the inlets, velocity vectors are defined, the outlets are defined as pressure outlets.

At the location of the propellers, cylindrical domains containing the propellers are integrated within the outer domain. For the upstream propeller, 5 million cells are used, the domain of the downstream propeller consists of 11.6 million cells. These domains rotate with the rotational speed of the propeller with sliding mesh conditions between the domains. The temporal resolution is set to one degree per time-step. An impression of the whole mesh at the location of the propeller's cross-section can be seen in Fig. 2c. The other simulation settings are equivalent to the analysis of the isolated propeller which is described in [3]. The SST turbulence model is applied. The thrust, torque, pitching, and yawing moment are derived by a pressure integration over the blades, the hubs, and the motors.

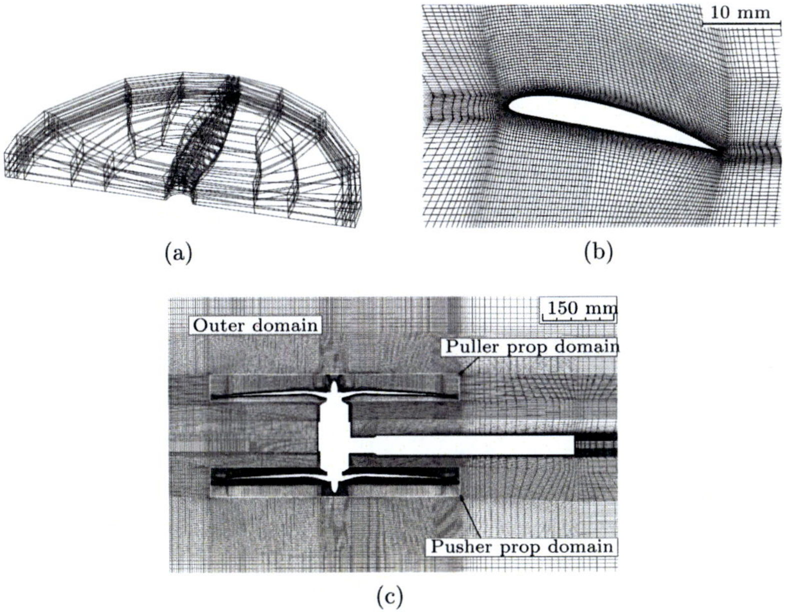

Fig. 2. **(a)** Block structure of the propeller domains, **(b)** mesh around the airfoil at $r/R = 0.7$, and **(c)** mesh of all three domains at $y = 0$.

4 Results

In this section, the results are presented, subdivided into an analysis of the steady and transient forces and a discussion of the occurring flow fields.

4.1 Steady Forces

The steady loads of the coaxial propeller are presented and compared to the data of two isolated puller propellers, derived from [3]. They are plotted over the inflow angle α_{disc} to reveal the influence of a non-axial inflow. The free flow velocity is $U_\infty = 10\,\text{m/s}$, the rotational velocity n is constant giving the advance ratio $J = U_\infty/nD = 0.33$.

In Fig. 3a, the thrust coefficients $c_T = T/(\rho n^2 D^4)$ of both propellers separately as well as the total thrust are plotted, the numerical results being in good agreement with the experiments. The general trend over α_{disc} is similar for all propellers. The thrust increases with a rising inflow angle until it reaches $\alpha_{disc} = 120°$. A detailed comparison of both coaxial propellers reveals that the thrust of the upstream propeller is always higher than the one of the downstream propeller. In general, a downward-oriented interaction occurs which can be divided into two main effects. The swirl of the upstream propeller increases the local inflow angle at the blade sections $\alpha_{section}$ of the downstream propeller.

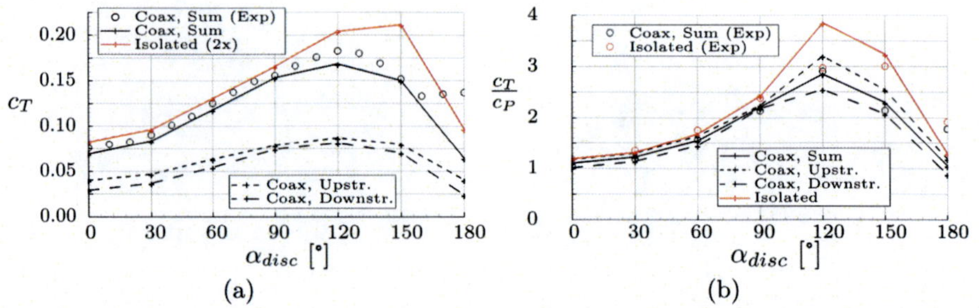

Fig. 3. Normalized coefficients of the **(a)** thrust and **(b)** thrust to power ratio. $J = 0.33$, Experiments and URANS calculations.

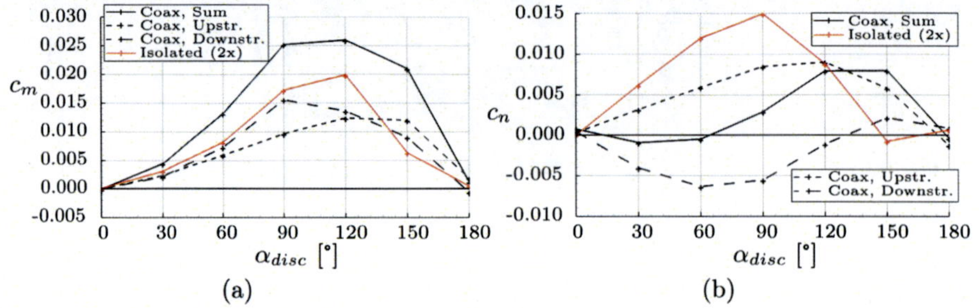

Fig. 4. Normalized coefficients of the **(a)** pitching and **(b)** yawing moment. $J = 0.33$, URANS calculations.

That is typically overcompensated by the increase of the axial velocity at the downstream propeller plane, decreasing the local inflow angles and the downstream's thrust. The pusher propeller shows the same geometry than the puller, especially no pitch-correction is performed, increasing this adverse downstream-oriented interaction. Figure 3b shows, that this interaction affects the ratio of thrust to power coefficient $c_P = P/(\rho n^3 D^5)$ (obtained by electrical current measurements) similarly with the efficiency of the upstream propeller being higher than the one of the downstream propeller. But also for the upstream propeller a reduction in efficiency can be observed in comparison to the isolated propellers revealing the existence of a disadvantageous upstream-oriented interference.

Figure 4a shows the trend of the pitching moment $c_m = M_m/(\rho n^2 D^5)$ over α_{disc}. While it can be neglected for an axial and a reversed inflow, both propellers produce a pitching moment in the same direction for non-axial inflow conditions. Ultimately, the total load outreaches the one of the isolated configuration by around fifty percent. While the downstream propeller reaches its maximum under lateral inflow, the maximum of the upstream propeller occurs at $\alpha_{disc} = 120°$, which flattens the total load peak. One reason for that is the formation of highly asymmetrical inflow conditions for the downstream propeller at lower inflow angles by facing "fresh air" at its windward side and accelerated flow at its leeward side (see Fig. 8a).

The two propellers contribute in opposite direction to the yawing moment coefficient c_n up to inflow angles of $\alpha_{disc} = 120°$ (see Fig. 4b). Hence, except for the inflow angle $\alpha_{disc} = 150°$, the coaxial propeller produces a significantly smaller yawing moment than two isolated propellers wich agrees with the findings of Dunham [6] and is beneficial for the stability characteristics of coaxial propellers.

4.2 Transient Forces

In Fig. 5, the transient thrust over the circumferential propeller position $c_T(\zeta)$ is shown relative to its mean level $\overline{c_T}$ for different inflow angles.

Opposite to the isolated propeller, the coaxial propeller shows significant periodic thrust peaks at $\zeta = 0°$ and $\zeta = 180°$, correlated with the propeller blades passing each other. For the downstream propeller, these fluctuations are significantly higher due to its position in the wake of the upstream propeller and the horizontal support beam. With an increased inflow angle, the thrust peaks flatten and a phase shift of the thrust peaks of up- and downstream propeller can be observed. This leads to the assumption, that with increasing inflow angle, the influence of the blade passings decreases while the influence of the non-axial inflow itself becomes more dominant for each propeller.

The load polars of the upstream propeller show a high similarity with the ones of the isolated propeller for all inflow angles except $\alpha_{disc} = 180°$. The polars of the downstream propeller are phase-shifted and show stronger fluctuations in relation to the isolated propeller. The phase-shift results in a reduction of the total thrust level magnitude and an increase of the frequency of the total thrust. A characteristic feature are the distinct thrust fluctuations of the downstream propeller at very high inflow angles caused by a strongly asymmetrical flow at the propeller plane with a recirculation zone at its windward side and an accelerated flow at its leeward side (see Fig. 8b).

Cameron [1] confirms increasing thrust fluctuations when the lateral inflow magnitude is increased. He performs his measurements without axial inflow, but it underlines the dominance of the non-axial inflow on the resulting load fluctuations while the magnitude of the axial inflow has a less significant effect.

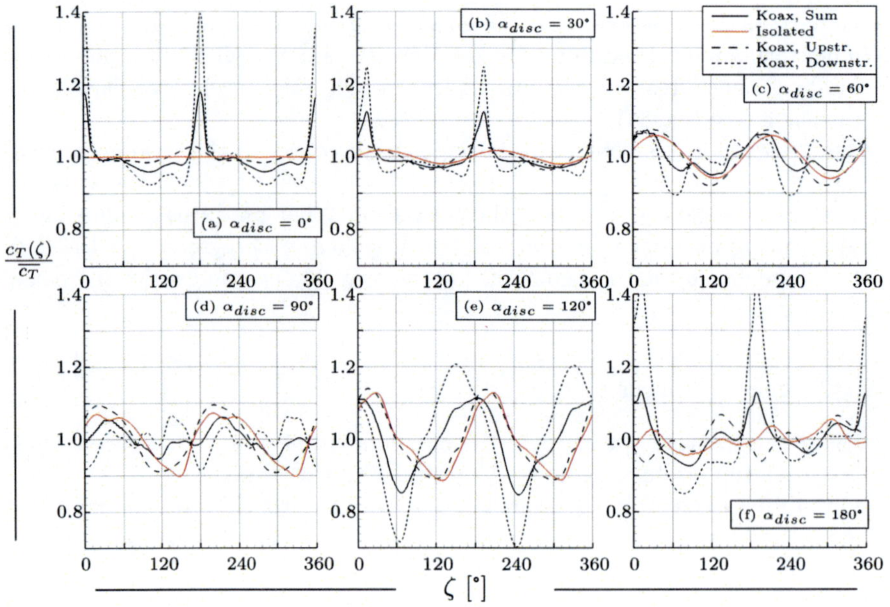

Fig. 5. Transient thrust coefficients in relation to their mean levels over one propeller revolution under different inflow angles. $J = 0.33$, URANS calculations.

4.3 Velocity and Pressure Along Propeller Axis

In addition to the detected force characteristics mentioned above, the progression of the averaged induced axial and circumferential velocity components and the static pressure is analyzed along the propeller axis. In Fig. 6, they are area-averaged over a circular area of radius 1.1 R at five different slice locations A-E and time-averaged over twelve circumferential blade locations. To subtract the inclined inflow and isolate the influence of the propellers, the absolute axial velocity component U_x is normalized writing

$$U_{x,norm} = U_x/U_\infty - \cos\alpha_{disc} \tag{1}$$

while for the circumferential velocity component the induced velocity tangential to the propeller plane is normalized and the magnitude of the component in cicumferential direction at a collocation point P at the location y_P, z_P is used

$$U_{circ,norm} = \left| \frac{\sqrt{U_y^2 + (U_z - U_\infty \cdot \sin\alpha_{disc})^2}}{U_\infty} \cdot \cos\left(\arctan\frac{U_z - U_\infty \cdot \sin\alpha_{disc}}{U_y} + 90^\circ - \arctan\frac{z_P}{y_P} \right) \right| \tag{2}$$

With the absolute lateral and vertical velocity components U_y and U_z. The relation of both components gives the induced swirl angle γ_{swirl}

$$\gamma_{swirl} = \arctan\frac{U_{circ,norm}}{U_{x,norm}} \tag{3}$$

For an axial and minor inclined inflow ($\alpha_{disc} = 0 - 30°$), Fig. 6a shows a continuous increase of the averaged induced axial velocity $Avg\,U_{x,norm}$ along the propeller axis, related to the decreased efficiency of the downstream propeller (see Fig. 3b). On the contrary, the induced circumferential velocity component $Avg\,U_{circ,norm}$ is increased by the upstream propeller and decreased by the downstream one (see Fig. 6c). Simultaneously, the static pressure is increased especially by the downstream propeller, indicating the existence of swirl recovery (see Fig. 6b).

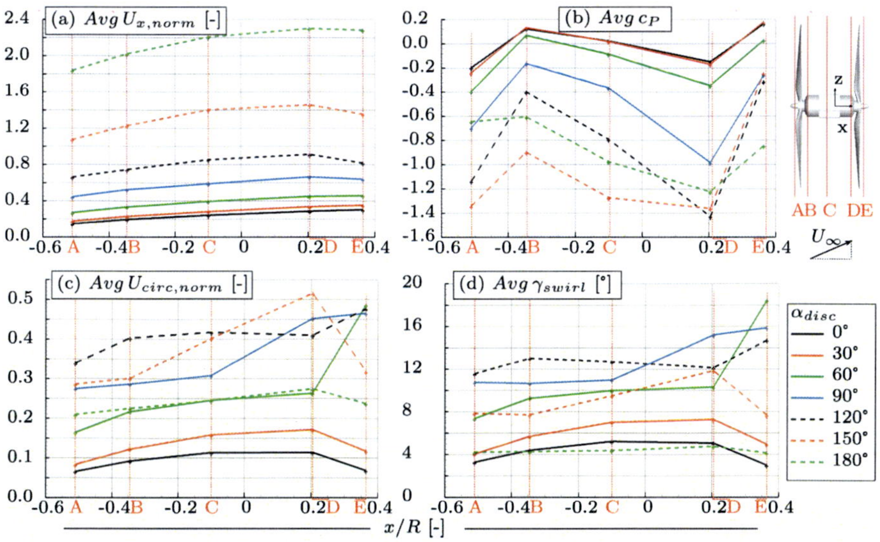

Fig. 6. Progression of the induced velocity and the pressure along the propeller axis for different inflow angles. $J = 0.33$, URANS calculations.

For a higher inflow angle of $\alpha_{disc} = 60°$, the effect of swirl recovery vanishes since the upstream propeller's wake does not influence the downstream propeller plane in a beneficial way anymore but affects the propeller plane indicating the same circumferential velocity direction. Hence, a strong increase of $Avg\,U_{circ,norm}$ and $Avg\,\gamma_{swirl}$ occurs for the downstream propeller. Although this is disadvantageous for the local inflow angles and the produced thrust, the total thrust at $\alpha_{disc} = 60°$ is increased since the reduced absolute axial inflow velocity overcompensates for this effect.

The high total thrust levels between $\alpha_{disc} = 90° - 150°$ (see Fig. 3a) are in agreement with the significant pressure increase over both propeller planes as shown in Fig. 6b. For these inflow angles, the efficiency of the coaxial configuration is significantly lower than the one of the isolated propeller. This can be explained by the disadvantageous superimposing flow acceleration of both propellers which cannot be compensated due the absent swirl recovery.

For reversed inflow, the produced thrust abruptly decreases with the formation of a vortex ring (see Fig. 8c). Concerning that, Fig. 6c shows the normalized circumferential velocity component to be significantly reduced in comparison to $\alpha_{disc} = 150°$ while the induced axial component's level is by far at its maximum, both resulting in a significantly decreased averaged swirl angle γ_{swirl} and hence, decreased local inflow angles $\alpha_{section}$. Additionally, this is accompanied by a significantly reduced pressure increase especially for the upstream propeller (see Fig. 6b).

4.4 Flow Fields

The flow fields of the coaxial configuration are analyzed illustrating the influence of the inflow angle and, in particular, the specific behavior of the two propellers and their mutual interference.

Axial Inflow. A typical application of a coaxial propeller configuration implies an axial inflow. The corresponding flow field is illustrated at the propeller's cross-section by $U_{x,norm}$ (see Fig. 7a) and $U_{circ,norm}$ (see Fig. 7b).

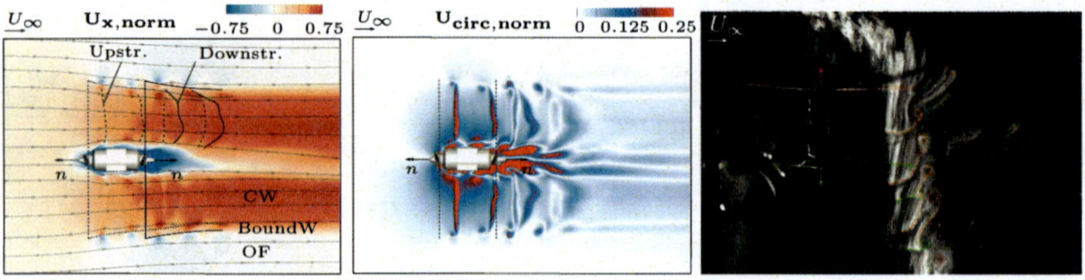

(a) (b) (c)

Fig. 7. Normalized induced **(a)** axial and **(b)** circumferential velocity component. $\alpha_{disc} = 0°$, $J = 0.33$, URANS calculations. **(c)** Smoke visualization in the wind tunnel. $\alpha_{disc} = 90°$, $J = 0.16$.

They show the increased induced axial velocity at the downstream propeller plane, being the main reason for the reduced efficiency of this propeller. The in-plane streamlines show the faster contraction of the upstream's wake which correlates with the higher flow acceleration by the upstream propeller (see Fig. 6a). Taylor showed these phenomena experimentally [8] and other experimental analyses confirm his findings [7].

The contraction of the upstream's wake results in the segmentation of the slipstream into a core wake (CW), a boundary wake (BoundW) where "fresh air" from the side is accelerated by only one propeller and an outer flow (OF). In Figs. 7a, b, one recognizes that the upstream's blade wakes convect rather undeformed downwards. Oppositely, the downstream's blade wakes are bent due to the segmentation of the slipstream with its varying axial velocities. A similar behavior can be seen in Fig. 7c although the smoke visualizes the wake under different operating conditions. Here, the leeward side of the downstream propeller's plane is located in the core wake and its blade wakes (green) convect rather undeformed. Oppositely, the upstream's blade wakes (red) experience a strong deformation due to their widening into the boundary wake with reduced axial velocities.

Non-axial Inflow. The variation of the inflow angle influences the individual local inflow conditions of both propellers as well as the degree of mutual interference due to the occurring lateral deflection of the propeller wakes. Figure 8a illustrates the non-concentrically penetration of the propeller wakes resulting in a strong acceleration of the windward core wake due to a concentrated upstream propeller's slipstream and a less pronounced acceleration of the leeward core wake.

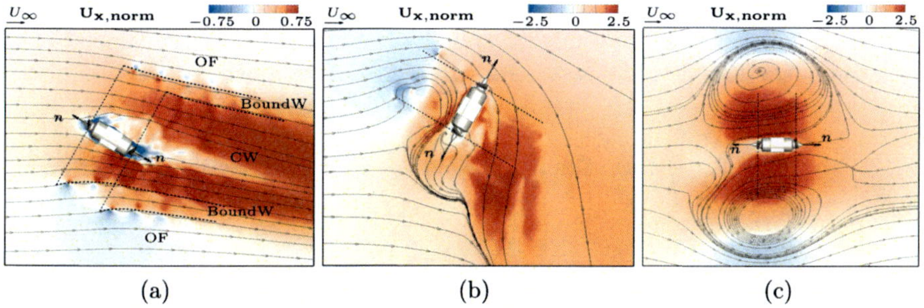

Fig. 8. Normalized induced axial velocity component under **(a)** $\alpha_{disc} = 30°$, **(b)** $\alpha_{disc} = 120°$, and **(c)** $\alpha_{disc} = 180°$. $J = 0.33$, URANS calculations.

Especially, the outward region of the downstream's propeller plane protrudes out of the upstream propeller's slipstream and hence, faces non-accelerated "fresh air" from the side. This results in increased local inflow angles and total thrust (see Fig. 3a). On the contrary, the leeward side of the downstream's propeller plane is located inside the accelerated slipstream of the upstream propeller, decreasing the local thrust. Consequently, the downstream propeller experiences a distinct thrust asymmetry, being the main reason for the dominant pitching moment of the coaxial propeller and outweighing the sum of the pitching moments of two isolated propellers (see Fig. 4a).

At $\alpha_{disc} = 120°$, a vortex structure is formed at the windward side of the downstream propeller, driven by the passing flow above the propeller and the flow reversal by both propellers (see Fig. 8b). The correlated reduction of the axial velocity causes a strong increase of the local inflow angle at the blade sections and a local thrust increase. The flow at the leeward side is accelerated by both propellers, decreasing the lift in this region. The resultant thrust asymmetry explains the maximum of the pitching moment.

A further increase of the inflow angle varies the flow field rapidly. The windward vortex structure is deflected outwards and extended over the whole circumference, forming a vortex ring state. Consequently, the former reduced axial velocity is replaced by a distinct flow acceleration (see Fig. 8c).

5 Conclusion

The steady and transient forces of a coaxial propeller have been analyzed for the whole inflow angle regime. The configuration showed to be less efficient than a combination of two isolated propellers. The coaxial-specific maneuverability could be analyzed by a pitching and yawing moments analysis. The transient loads showed the specific reaction of upstream, downstream, and isolated propeller under non-axial inflow. The analysis showed that although the upstream propeller produces more thrust, especially the impact of the downstream one is responsible for undesired coaxial-specific effects, namely load fluctuations or high pitching moments. Furthermore, the influence of the non-axial inflow on the flow field structures could be analyzed and illustrated using URANS calculations. With them, the found load characteristics could be explained.

Acknowledgements. The authors would like to thank the Bavarian Research Foundation (Bayerische Forschungsstiftung) for funding the HyProp research project under the grant number AZ-1277-17.

References

1. Cameron, C.G.: Comprehensive Aeromechanical Measurements of a Model-Scale, Coaxial. Counter-Rotating Rotor System. Diss. University of Texas, Austin (2016)
2. Cerny, M., Breitsamter, C.: A comparison of isolated and ducted fixed-pitch propellers under non-axial inflow conditions. Aerospace **7**, 112 (2020)
3. Cerny, M., Breitsamter, C.: Investigation of small-scale propellers under non-axial inflow conditions. Aerosp. Sci. Technol. **106**, 106048 (2020)
4. Cheney, M.C.: The ABC Helicopter. AIAA Paper 69-217. Georgia Institute of Technology, Atlanta, Georgia (1969)
5. Coleman, C.P.: A Survey of Theoretical and Experimental Coaxial Rotor Aerodynamic Research, NASA Technical Paper 3675. Ames Research Center Moffett Field, Ca (1997)
6. Dunham, D.M., Gentry, C.L.: Low-speed wind-tunnel tests of single- and counter-rotation propellers. NASA Technical Memorandum TM-87656 (1986)
7. Nagashima, T., Ouchi, H., Sasaki, F.: Optimum performance and load sharing of co-axial rotor in hover. J. Jpn Soc. Aeronaut. Space Sci. **26**(293), 325–333 (1978)
8. Taylor, M.K.: A Balsa-Dust technique for air-flow visualization and its application to flow through model helicopter rotors in static thrust. NACA Technical Note TN-2220, Langley Aeronautical Laboratory Langley Air Force Base, Va (1950)

BOS-Based Three-Dimensional Reconstruction of Rotor Blade Tip Vortex Positions

Nicola Debernardis(✉), Clemens Schwarz, and Johannes N. Braukmann

German Aerospace Center (DLR), Institute of Aerodynamics and Flow Technology, Bunsenstraße 10, 37073 Göttingen, Germany

Abstract. The present work discusses a methodology to reconstruct three-dimensional positions of rotor blade tip vortices based on stereoscopic background-oriented schlieren data. The reconstruction procedure is first applied on a rotor test stand where a comparison with previous PIV measurements is performed to validate the reconstruction scheme. Subsequently, it is adapted for the reconstruction of the vortex system surrounding a model helicopter in free flight during an unsteady takeoff maneuver.

Keywords: Epipolar geometry · 3D reconstruction · BOS · Blade tip vortices

1 Introduction

Over the last years many investigations on rotor blade tip vortices have been carried out in the Institute of Aerodynamics and Flow Technology in Göttingen using optical measurement techniques. In two recent experiments particle image velocimetry (PIV) and the background-oriented schlieren technique (BOS) were combined to investigate the wakes of an isolated rotor on a test stand under cyclic pitch conditions [1] and of a free-flying model helicopter in ground effect [2]. In the present investigation, the BOS data acquired in both experiments are used to reconstruct the corresponding vortex filaments in three-dimensional (3D) space. In addition to its visualization intent, BOS is tested as a method for the extraction of quantitative values. Due to its easier applicability, it constitutes an efficient alternative for the determination of three-dimensional vortex positions in comparison to a volumetric PIV setup. In both cases, the vortex systems were recorded simultaneously from different perspectives using stereo camera systems. By using epipolar geometry it is possible to match discrete vortex points and subsequently triangulate their position in 3D space. This approach has been used in several studies in the past to reconstruct tip vortices of both fixed wing aircraft [4] and helicopters in wind tunnel measurements [5] and in free flight [3,6]. In the current study a modular reconstruction scheme was developed that allowed for the analysis of datasets recorded with different experimental setups

© The Author(s), under exclusive license to Springer Nature Switzerland AG 2021
A. Dillmann et al. (Eds.): STAB/DGLR Symposium 2020, NNFM 151, pp. 315–324, 2021.
https://doi.org/10.1007/978-3-030-79561-0_30

for a multitude of test cases. The methodology was applied to data from the rotor test stand and the model helicopter to analyze the spatial structure of rotor blade tip vortices in different conditions.

2 Rotor Test Stand Experiment (RTG)

2.1 Experimental Setup at the RTG

The experiment was carried out at the rotor test facility in Göttingen (RTG) [1]. The four-bladed rotor had a radius of $R = 0.65$ m, a chord length of $c = 0.072$ m, a DSA-9A airfoil and was arranged horizontally (see Fig. 1), (see [1] for details). The rotor was operated with collective and cyclic pitch settings (reported in [1]) in a wind tunnel which provided a slow axial inflow of 2.5 m/s. The combination of cyclic pitch and an axial inflow represents a generic test case. The cyclic pitch enables to study the resulting aerodynamics in a highly unstable rotor downwash, while the axial inflow ensures known inflow conditions as explained in [1]. With this setup the rotor was operated out of ground effect. A detailed description of the RTG can be found in [7]. The measurement techniques used during the experiment are PIV, to characterize the velocity field, and BOS to acquire volumetric information around the PIV plane.

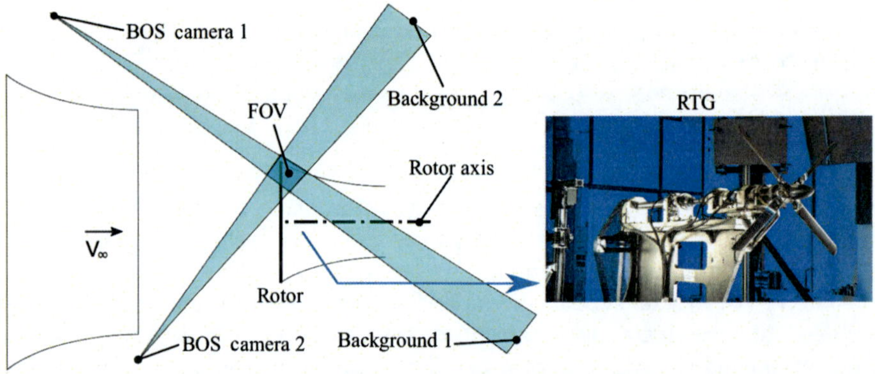

Fig. 1. Sketch of the RTG (top view) with the BOS setup (left) and image of the RTG (right, after [1]). In addition, a sketch of the field of view (FOV) is included.

The BOS cameras were imaging a measurement volume around the PIV plane covering an azimuthal range of 22° (see Fig. 1). A high focal length of $f = 300$ mm was chosen to increase the sensitivity of the setup to density gradients.

2.2 Reconstruction Methodology

Figure 2 depicts sample BOS results recorded instantaneously with both cameras. The BOS result after the correlation is a displacement field of the background pattern. The vortices are visualized by means of the gradient of the displacement field.

The individual vortex filaments appear as bright lines. To detect the vortex filaments in the images, the NeuronJ neurite tracing algorithm designed by Meijering [8] has been used which was applied for the extraction of vortices in [1,12]. The implementation of this tool to the tracing of blade tip vortices was first shown by Overmeyer [10], who successfully traced vortices in shift images of different signal-to-noise ratios (SNR). The points on the vortex filaments, identified with NeuronJ in the image of camera 1 (yellow line), generate a pencil of epipolar lines in the second image. The epipolar lines can be used to establish corresponding points on a vortex filament of camera 1 in the corresponding image of camera 2. This is done by intersecting the epipolar lines from the first camera (orange line in Fig. 2 b) with the vortex filaments in the second camera.

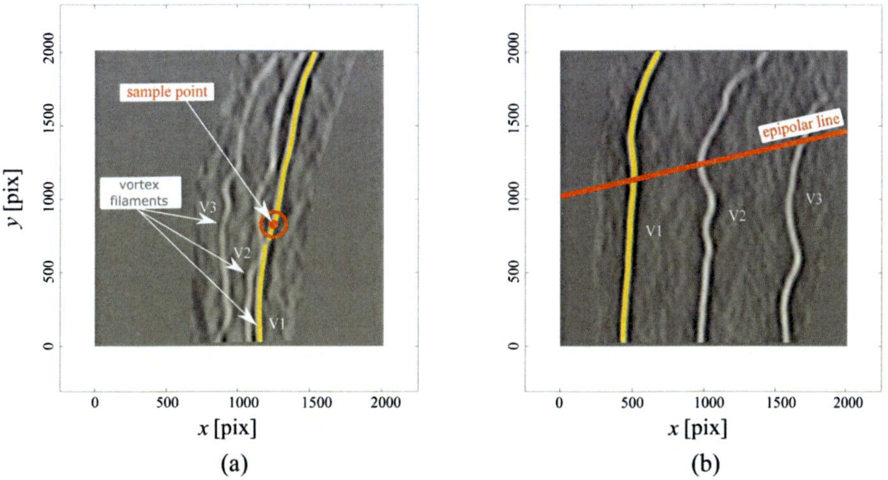

Fig. 2. Sample BOS result from camera 1 (a) and 2 (b) with epipolar line for a point on vortex filament V1 in the corresponding image from camera 2. During the acquisition the blade loading was $\frac{C_T}{\sigma} = 0.157$ and the angular velocity was set to $\Omega = 2\pi \cdot 23.6$ Hz

A fundamental requirement for both the application of epipolar geometry and for the triangulation of points in space is a reliable calibration of the camera system. The calibration was performed using the camera calibration toolbox for MATLAB by Bouguet [11]. A calibration target consisting of a dot pattern was used, the detected markers were used as input for the calibration. The calculated camera matrices serve to derive projection matrices P and P' and subsequently construct the fundamental matrix for the stereo setup. The accuracy in establishing point correspondence was estimated referring to the position of well known targets and resulted in 1.14 pixel (± 0.26 mm).

2.3 Results

The 3D locations of the vortices were triangulated for a number of four test cases. Every selected test case corresponds to a different pitch phase (t/T) at a different azimuthal position, but the same inflow conditions. During each revolution the blades undergo one full pitch cycle. The pitching variation was $\Theta = 24° - 6° \cos(2\pi\ t/T)$. For each test case, a total of 200 images for each camera were taken. A total of four test cases has been analyzed covering wake ages ranging from $\Psi_v = 40°$ to $\Psi_v = 70°$. The first two correspond to wake ages of 40° and 55° relative to blade 1. The two other test cases, on the other hand, contain identical wake ages with respect to blade 3. Since the two sets correspond to different phases during the pitch cycle, the corresponding vortices were created at different pitch angles. The same four cases were measured simultaneously with PIV allowing for a direct comparison between the vortex positions. Figure 3a depicts the resulting vortex locations for a sample data set. The lighter lines represent the individual filaments, whereas the darker lines correspond to the phase average of all the 200 vortex filaments. The averaged vortex locations for all four analyzed test cases are shown in Fig. 3b in a perspective view. The three vortices in the upper right corner were created during the down-stroke motion of the blade, where the blade goes from a pitch phase of $t/T = 0.88$ to $t/T = 0.96$. The other two vortices on the opposite side were created during an up-stroke motion where the blade goes from a pitch phase of $t/T = 0.38$ to $t/T = 0.47$.

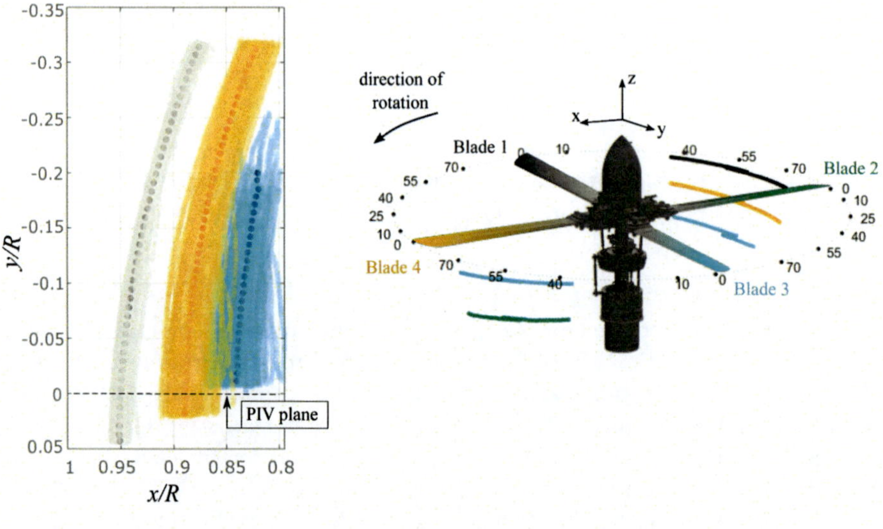

(a) (b)

Fig. 3. Reconstructed vortex positions from 200 images for a sample test at $t/T = 0.96$ and $\Psi_v = 40°$(3a) and 3D representation of all four analyzed test cases (3b). Each vortex is colored based on the blade it originated from.

The reconstructed blade tip positions allow for a determination of the rotor hub position and a new local coordinate system (x, y, z) is defined by moving the origin into the rotor hub center. The vortices are color coded based on the blade they trail from. The vortices with a higher wake age have a greater vertical convection rate that induces larger distances between the vortex filaments.

2.4 PIV Comparison

To validate the reconstruction process, the vortex positions extracted from BOS data are compared to simultaneously recorded PIV measurements. The PIV measurements were taken at the same relative wake ages of 40° and 55° and the same phase (t/T).

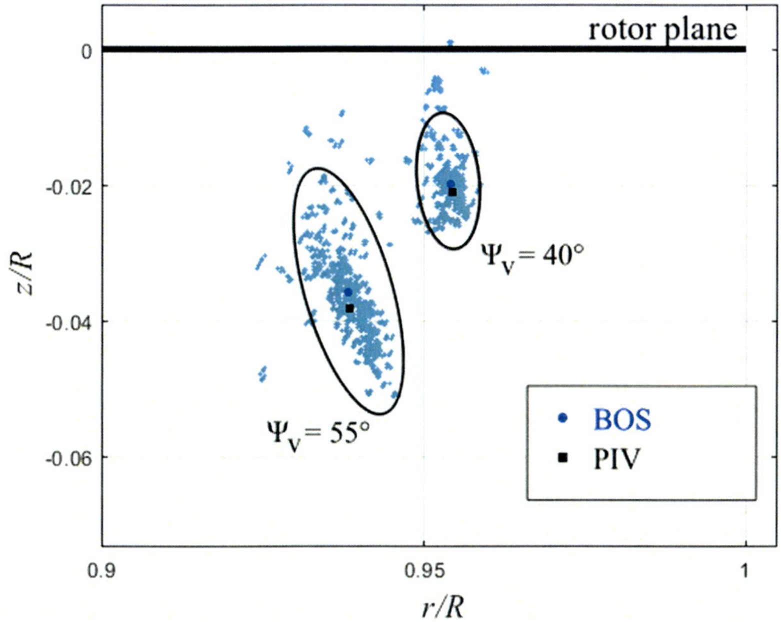

Fig. 4. Instantaneous and mean vortex positions in a radial plane (r, z). The vortex positions refer to phases of $t/T = 0.96$ and $t/T = 0.92$ and wake ages of 40 and 55° respectively.

Figure 4 depicts individual and average vortex positions from BOS (blue) and the average vortex center that was determined based on velocity data from PIV (black) [1]. The vortex positions refer to phases of $t/T = 0.96$ and $t/T = 0.92$ at two different wake ages of 40° and 55° relative to blade 1. In addition, ellipses representing a confidence interval of 95% are included. It can be seen that as the wake age increases the radial vortex position decreases. The cycle-to-cycle scatter in individual vortex positions is inherent to a rotor tip vortex system and was analogously observed in the PIV data [1]. The average locations show a deviation of only 0.09 mm between BOS and PIV. This suggests that stereoscopic BOS can be a valuable tool to gain quantitative information on the spatial structure of a vortex system.

3 Free-Flying Model Helicopter Experiment

3.1 Experimental Setup of the Model Helicopter

In the second experiment, the wake of a free-flying model helicopter in ground effect was investigated. A series of flight tests were performed consisting of both quasi-steady and unsteady maneuvering flight states. A more detailed discussion of the different flight states and the resulting wake structure is given in [2]. The tests were conducted with the DAEDALUS model helicopter from the Institute of Aerodynamics and Gas Dynamics at the University of Stuttgart, shown in Fig. 5. The test rotorcraft is a model helicopter with two untwisted and stiff rotor blades with a radius of $R = 0.7$ m, chord length of $c = 54.5$ mm, and a symmetrical Joukovski 0015 airfoil. In addition to the PIV measurements reported in [2], a multi-camera stereo-BOS system was used to capture the spatial structure of the vortex system surrounding the helicopter. A sketch of the setup is shown in Fig. 5 along with a photograph of the test helicopter and sample BOS results from both perspectives for a takeoff maneuver. Further details on the BOS setup can be found in [9].

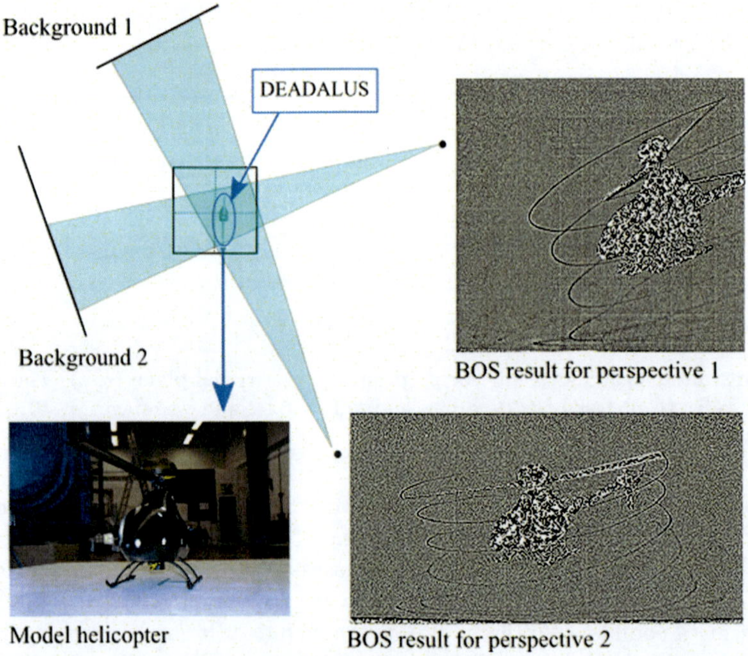

Fig. 5. Top view of the experimental setup along with a photograph of the test helicopter and sample BOS results for both camera perspectives during a takeoff maneuver.

3.2 Reconstruction Methodology

Figure 6 depicts the vortex system for both camera perspectives during a takeoff maneuver. Analogous to the RTG experiment, a reconstruction scheme based on epipolar geometry was applied. A difficulty in identifying corresponding vortex points arises when the vortex axis is nearly parallel to the epipolar line. The green epipolar line in Fig. 6b, for example, intersects the second vortex in two points. In the case of multiple intersection points, false pairing of points on non-corresponding vortices have to be checked manually for their physical validity. While in the case of two discrete and identifiable intersection points this problem can still be solved, the more the epipolar line aligns to the vortex axis, the more the limit in the accuracy is approached. Thus, the reconstruction of adjacent points is limited by the possibility of identifying a univocal intersection point. Another limiting factor is the low SNR in certain regions. This limits especially the reconstruction of older vortices due to their decreasing strength. In the present case only vortices up to a wake age of about 650° could be reconstructed.

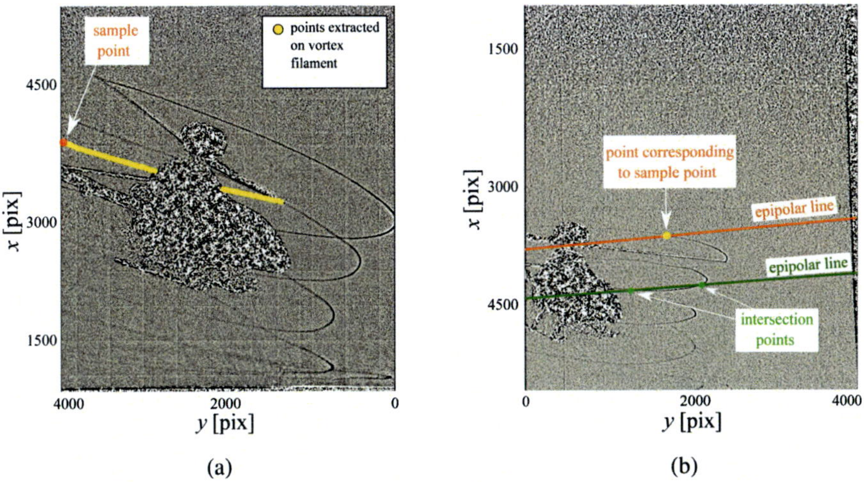

Fig. 6. BOS results from both camera perspectives showing the epipolar intersection with a vortex filament. The individual orange dot on the yellow vortex filament (image 6a), corresponds to the orange epipolar line (6b). The yellow intersection point with the vortex filament corresponds to the orange sample point in image 6a.

3.3 Results

Despite the increased complexity, it was possible to reconstruct significant parts of the vortex system, which are depicted in Fig. 7. For visualization purposes, the whole vortex system is depicted together with a 3D model of the DAEDALUS model helicopter in multiple views. A new local reference system (x, y, z) is defined by moving the origin to the rotor hub center. The vortices are color coded based on the blade they trail from. The model helicopter is added to the plots with its position and orientation determined by triangulating optical markers that were applied to the fuselage (see [2] and [9]). The rotor blade orientation is determined by the reconstructed locations of the blade tips, assuming a stiff rotor. While for the first blade on the advancing side, the blade tips vortex filament match the blade tip where it trails from, for the retreating blade an offset between blade tip and vortex can be detected. This vertical offset can be due to the bending of the blade, which is not accounted for in the 3D model.

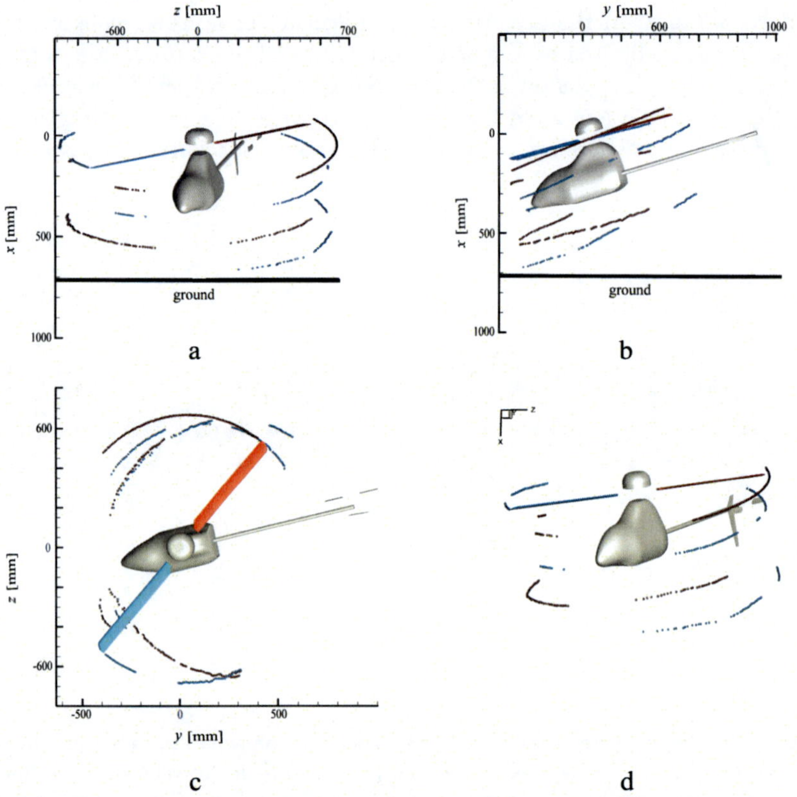

Fig. 7. Different perspectives (a–d) of the model helicopter and reconstructed main rotor vortex system.

The characteristic helical shape of the vortex system is visible. After a first contraction of the wake, a subsequent radial expansion with wake age due to the ground effect occurs. At the time of the image acquisition, the helicopter rotor is about 1 rotor radius R above the ground as was reported in [2]. The evolution of the vortex system's radius versus the wake age is better visible in Fig. 8. After a first contraction, which reaches a value of 0.9 R at the wake age of 340°, the radius of the vortex system expands due to the proximity of the ground starting from a wake age of 470° up to the maximum detectable wake age of 650°. A similar behavior was also observed in data from PIV measurements for the same model helicopter in ground effect [2].

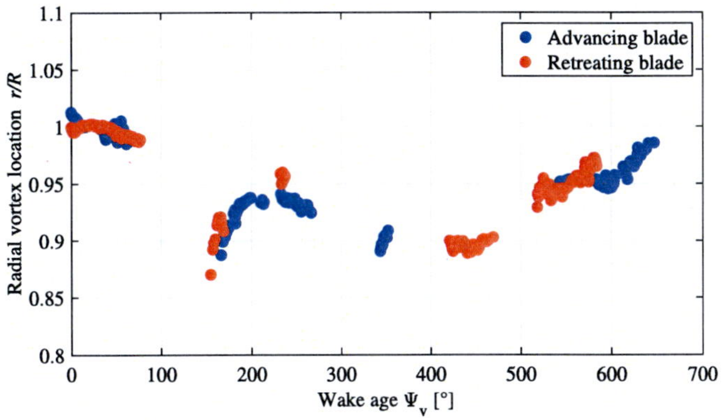

Fig. 8. Radius of the vortex system over wake age.

4 Conclusion

A procedure for 3D vortex reconstruction based on stereoscopic BOS data was presented and applied to two different experiments. In the first part of the study, the vortex system produced by a rotor test stand with cyclic pitch was investigated. Based on a calibrated stereo camera system, fundamental matrices were determined that allowed for the use of epipolar geometry in order to establish point correspondences between the images of two cameras. Individual points on the vortex filaments were matched by calculating epipolar lines for vortex points in the first image and extracting the intersection point with the corresponding vortex filament in the second image. The accuracy of the epipolar based reconstruction was quantified by reconstructing the 3D position of a target whose geometry in 3D space was known with good precision. The comparison with PIV mean vortex positions (see [1]) allowed the validation of the reconstruction process that was further applied to a second experiment. In the second part of the current investigation, the reconstruction scheme was applied to the vortex system of a free-flying model helicopter in ground effect. Despite the low SNR ratio for older vortices and the unfavorable geometric setup with regard to vortex point identification based on epipolar geometry, it was possible to reconstruct significant parts of the vortex system up to a wake age of 650°. Overall, the results showed the feasibility and the value of a 3D reconstruction of vortex filaments from stereo-BOS data. The developed reconstruction scheme allows for a quantitative analysis of large parts of the rotor vortex system and can be applied to an extended database in the future.

References

1. Braukmann, J.N., Wolf, C.C., Goerttler, A., Raffel, M.: Blade tip vortex system of a rotor with cyclic pitch. AIAA J. **58**(2020). https://doi.org/10.2514/1.J058678
2. Schwarz, C., Bauknecht, A., Mailänder, S., Raffel, M.: Wake characterization of a free-flying model helicopter in ground effect. J. Am. Helicopter Soc. **64**(012010) (2019) https://doi.org/10.4050/JAHS.64.012010
3. Klinge, F., Hecklau, M., Raffel, M., Kompenhans, J., Göhmann, U.: Measurement of the position of rotor blade vortices generated by a helicopter in free flight by means of stereoscopic Background Oriented Schlieren Method (BOS). In: 13th International Symposium on Application of Laser Techniques to Fluid Mechanics, Lisbon, Portugal, 26–29 June (2006)
4. Meyn, L.A., Bennett, M.S.: Application of a two camera video imaging system to three-dimensional vortex tracking in the 80-by 120-foot wind tunnel. In: AIAA 11th Applied Aerodynamics Conference, Monterey, CA, USA, 1 January (1993)
5. Schairer, E., Kushner, L.K., Heineck, J.T.: Measurements of tip vortices from a full-scale UH-60A rotor by retro-reflective background oriented Schlieren and stereo photogrammetry. In: American Helicopter Society 69th Annual Forum, Phoenix, AZ, USA, 21–23 May 2013
6. Bauknecht, A., Ewers, B., Wolf, C.C., Leopold, F., Yin, J., Raffel, M.: Three-dimensional reconstruction of helicopter blade–tip vortices using a multi-camera BOS system. Exp. Fluids **56**(1), 1–13 (2014). https://doi.org/10.1007/s00348-014-1866-6
7. Schwermer, T., Richter, K., Raffel, M.: Development of a rotor test facility for the investigation of dynamic stall. In: New Results in Numerical and Experimental Fluid Mechanics X, vol. 132. Springer, Cham (2016) https://doi.org/10.1007/978-3-319-27279-5_58
8. Meijering, E., Jacob, M., Sarria, J.-C.F., Steiner, P.L., Hirling, H., Unser, M.: Design and validation of a tool for neurite tracing and analysis in fluorescence microscopy images. Cytom. Part A J. Int. Soc. Anal. Cytol. **58** (2004). https://doi.org/10.1002/cyto.a.20022
9. Bauknecht, A., Schwarz, C., Raffel, M., Mailänder, S.: Flow measurement techniques for rotor wake characterization on free-flying helicopters in ground effect. In: AIAA Scitech Forum, San Diego, CA, USA (2019). https://doi.org/10.2514/6.2019-2107
10. Overmeyer, A.D.: A rotor tip vortex tracing algorithm for image post-processing. In: American Helicopter Society 71st Annual Forum, Virginia Beach, VA, USA, May 5–7 (2015)
11. Bouguet, J.Y.: Camera Calibration Toolbox for Matlab. http://www.vision.caltech.edu/bouguetj/calib_doc/index.html. Accessed 3 2020
12. Schwarz, C., Bauknecht, A., Wolf, C.C., Coyle, A., Raffel, M.: A full-scale rotor-wake investigation of a free-flying helicopter in ground effect using BOS and PIV. J. Am. Helicopter Soc. **65**(032007) (2020). https://doi.org/10.4050/JAHS.65.032007

Quantification of the Influence of Particle Voids on PIV Measurements via Synthetic-PIV

Vincent Domogalla(✉)

German Aerospace Center (DLR), Göttingen, Germany
vincent.domogalla@dlr.de

Abstract. Particle Image Velocimetry (PIV) is an important technique to investigate complex flow fields but limited to the tracking fidelity of the flow tracers and the uncertainties in the evaluation process. To examine these issues this work implements a physic-based motion simulation for small spherical particles, serving as flow tracers in PIV, with existing tools for image generation. Particular focus set on the behavior in vortical structures and the occurrence of "particle voids" due to centrifugal forces. The results show that a realistic particle distribution ideally can be achieved when applying a statistical distribution of the particle size. Furthermore, they suggest a significant error in the velocity prediction if the region of interest has no sufficient particle seeding.

Keywords: Particle Image Velocimetry · Vortex · Synthetic · Particle force · Particle dynamics · Helicopter · Blade tip · Error · Uncertainty

1 Introduction

Vortex structures are part of the basic phenomena in flow technology and often have a significant influence on the aerodynamic performance of aircraft, e.g. helicopters. When maneuvering, the vortex system of the main rotor can interact with the main rotor itself, the tail, and also the nacelle. The unsteady character of such vortex systems influences aerodynamic performance significantly, as well as acoustics and vibration. For a better understanding the flow field is investigated by computational simulation and wind tunnel experiments, for example by Goerttler et al. [3], Braukmann et al. [1], and Wolf et al. [14] in the rotor test facility of the DLR, German Aerospace Center in Göttingen. In these investigations, vortex development and aging are of particular interest and often examined with optical measurement techniques.

A common technique to measure flow velocities in wind tunnel experiments is Particle Image Velocimetry (PIV). Since the limitations of algorithms utilized for the evaluation of PIV experiments can influence the results significantly, this is a severe source of error and uncertainty. Heavily affected are regions with high rotational velocities and low particle density such as vortex cores. The centrifugal forces diminish the number of tracer particles resulting in poorly resolved data,

© The Author(s), under exclusive license to Springer Nature Switzerland AG 2021
A. Dillmann et al. (Eds.): STAB/DGLR Symposium 2020, NNFM 151, pp. 325–334, 2021.
https://doi.org/10.1007/978-3-030-79561-0_31

named particle voids. An example of those voids is depicted in Fig. 1, showing an exemplary particle image of a PIV experiment from Wolf et al. [14] with voids at several vortex ages Ψ. The red circle marks the vortex position for $\Psi = 325°$ where the void has diminished due to turbulent mixing.

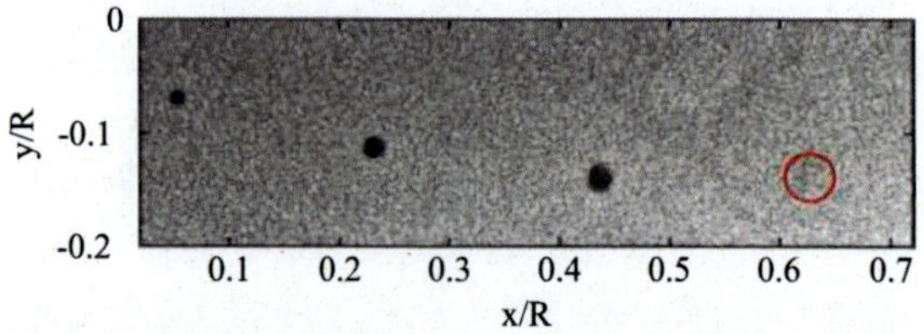

Fig. 1. Particle voids for vortex age from $\Psi = 55°$ to $235°$, $\Psi = 325°$ not visible (red circle), published by Wolf et al. [14]

To validate the evaluation results, the Synthetic Particle Image Velocimetry (SPIV) was developed, not to be confused with the Stereo Particle Image Velocimetry as mentioned by Raffel et al. [8]. In this work, an existing tool was extended with dedicated particle physics to allow for a proper investigation of particle dynamics. The algorithm is based on a software stack previously developed at DLR [4] using the work of Kyle Lynch from Sandia National Laboratories and is implemented in MATLAB. The code of Lynch is used to calculate the optical projection of the particles, represented as voxels, from a three-dimensional volume onto a two-dimensional image plane. It allows the usage of multiple cameras and camera positions and also accounts for optical effects such as the impact of the aperture or the geometric position of the sensor planes relative to the image plane. Thereby, the toolchain is able to generate synthetic particle images that can be evaluated by standard PIV software. Furthermore, the Synthetic Image Generator (SIG) developed by Lecordier et al. [6] is used to generate synthetic calibration images for the subsequent evaluation.

2 Theoretical Background

Blade Tip Vortex

The pressure difference between the upper and lower surface of the blade causes the rotors lift force. Since the rotor blade has a finite length, this pressure difference equalizes at the blade tips. The high-pressure air is no longer blocked from returning into the low-pressure regions above the blade, causing flow around the blade tips and inducing a rotational motion. This flow then propagates into the wake and is known as the blade tip vortex.

Vortices can be quantified by their vorticity ω and circulation Γ. With these characteristics, it is possible to describe the radial velocity profile of a vortex. One of many simplified models for a vortex flow is the Lamb-Oseen vortex. It is derived from the Navier-Stokes equations and can be expressed depending on the radius r and the time t as

$$\omega(r,t) = \frac{\Gamma_0}{4\pi\nu r t}\left(1 - e^{\frac{-r^2}{4\nu t}}\right) \quad (1)$$

with the kinematic viscosity ν. The equation can be rearranged as an expression of the tangential velocity [2,9] to

$$v_{tan}(r) = v_{tan,max}\left(1 + \frac{1}{2\alpha}\right)\frac{r_c}{r}\left[1 - exp\left(-\alpha\left(\frac{r}{r_c}\right)^2\right)\right] . \quad (2)$$

In this case v_{tan} is the tangential velocity, r_c the core radius and $\alpha = 1.25643$ the Lamb constant.

PIV

The particle image velocimetry (PIV) is an often-used technique for the acquisition of flow data from wind-tunnel experiments. The flow field is seeded with small particles that are able to follow even small flow structures. Mostly oil droplets out of Di-Ethyl-Hexyl-Sebacat (DEHS) or vegetable oil, or Helium Filled Soap Bubbles (HFSB) are used for this purpose. To capture the flow field for quantification, the particles are illuminated with a laser, spanning a thin sheet of light. A schematic setup is shown for example in [8]. The evaluation of images, taken at two different points in time, through cross-correlation allows the calculation of the underlying flow field. The cross-correlation splits the particle image into sub-samples and calculates the most probable displacement vector between two time points for each set of sub-samples, which translates together with the time difference between both pictures into a local velocity vector. Sufficient seeding of the flow region is necessary with correlation algorithms requiring several particles (e.g. five to seven) within each sub-sample. If this condition is not met, the results of the cross-correlation can no longer be considered valid.

Particle Mechanics

The movement of rigid particles in a homogeneous flow is explained by the law of momentum conservation. Boussinesq, Basset, and Oseen formulated the first solution for this problem, known as BBO-equation. Maxey and Riley had extended this equation in 1983 [7], and various authors had further modified it. The variant of Sommerfeld [12] for rigid particles and a homogeneous flow states as follows

$$\begin{aligned} m_p\frac{d\mathbf{v_p}}{dt} &= \frac{18\eta_f}{\rho_p d_p^2} m_p\left(\mathbf{u} - \mathbf{v_p}\right) - m_f\frac{D\mathbf{u}}{Dt} + \frac{1}{2}m_f\left(\frac{D\mathbf{u}}{Dt} - \frac{d\mathbf{v_p}}{dt}\right) + (m_p - m_f)\mathbf{g} \\ &\quad +9\sqrt{\frac{\rho_f\eta_f}{\pi}}\frac{m_p}{\rho_p d_p}\int_0^t \frac{\frac{D\mathbf{u}}{D\tau} - \frac{d\mathbf{v_p}}{d\tau}}{\sqrt{t-\tau}}d\tau \ , \end{aligned} \quad (3)$$

where m_p is the particle mass, m_f the mass of the surrounding fluid displaced by the particle, $\mathbf{v_p}$ the particle velocity, and $\mathbf{u}$ the velocity of the fluid. These velocities are depending on the position of the particle x_p at the time t. Furthermore, d_p describes the particle diameter, η_f the dynamic viscosity of the fluid, ρ_f its density, ρ_p the particle density and $\mathbf{g}$ is the gravitational vector. The different terms can be understood as the contribution of different forces, terms on the right side of the equation:

1. Drag force F_D
2. Pressure force F_p
3. Virtual Mass Force F_{VM}
4. Hydro-static lift force F_L
5. Basset force F_B

Further modifications to Eq. 3 and additional forces are also discussed by Sommerfeld [12]. Due to the forces acting on the particle, its response time to a spontaneous acceleration is limited by its mass and inertia, also explaining the reason for particle voids. Since the particle cannot react quickly enough to the change in flow direction within the vortex, it gets carried out of the core. The response time τ_p can be derived from Eq. 3 as

$$\tau_p = \frac{d_p^2 \rho_p}{18 \eta_f} \tag{4}$$

by neglecting every force but the drag force, which dominates in the low Re regime [8].

3 Synthetic Particle Image Velocimetry

The Synthetic Particle Image Velocimetry allows the generation of synthetic particle images from three-dimensional velocity fields, e.g. CFD data. The particle trajectories, caused by different forces acting on the particle, are computed over time. Subsequently, the particle positions for an arbitrary point in time are projected to an image plane and saved as a picture. A set of two pictures with a small time delay, representing the PIV double acquisition, can be analyzed with conventional PIV software to recalculate the velocity field. A comparison of the original data and the recalculated velocity reveals regions of high uncertainty. These are caused by the limitations of the evaluation algorithm in regions of low particle density. In the previous work of Hartje [4], the baseline concept for the SPIV was developed and tested. It is based on the code of Kyle Lynch, which is responsible for the projection of the particles from the three-dimensional computation space onto the two-dimensional image plane and the image calculation itself. The projection is done with the pinhole camera model from Tsai [13]. In this work dedicated particle physics is added to the tool, which was neglected in the previous version.

Modification of the Equation of Motion

To avoid unnecessary computation time and complexity, Eq. 3 was modified by removing negligible forces from the equation, in particular the Basset force, which is only relevant for density rations $\rho_p/\rho_f \approx 1$. Furthermore, Saffmann and Magnus forces, as mentioned by Sommerfeld [12], are negligible because of the small particle diameter. Their influence is estimated by the equations given by Sommerfeld [11]. Other forces such as the Coulomb force or the thermophoretic force are insignificant for the considered case, leaving only the drag F_D, pressure F_P, and virtual mass force F_{VM} as well as gravitational force F_G. To allow a wider range of applications, possible rarefaction effects are considered. The particle drag was modified with the Cunningham correction [12] depending on the Knudsen number Kn. Also, the influence of the particle Reynolds number Re_p had to be accounted for, using the correction of Schiller and Nauman [10]. Combining both terms in the factor K, which is

$$K = \begin{cases} \left(1 + Kn\left[2.514 + 0.8 \cdot exp\left\{\frac{-0.55}{Kn}\right\}\right]\right)^{-1} & 0.1 < Kn < 1000 \ \& \ Re_p < 0.25 \\ 1 & Kn < 0.1 \ \& \ Re_p \leq 1 \\ 1 + 0.15 \cdot Re_p^{0.687} & Kn < 0.1 \ \& \ Re_p > 1\,, \end{cases} \tag{5}$$

gives the final equation of motion as

$$\begin{aligned} \frac{d\mathbf{v_p}}{dt} &= 18\frac{\rho_f}{\rho_p}\frac{K}{d_p^2}\nu_f\left(\mathbf{u} - \mathbf{v_p}\right) + \frac{\rho_f}{\rho_p}\left((\mathbf{u}\cdot\nabla)\mathbf{u} - \mathbf{g}\right) \\ &+\frac{1}{2}\frac{\rho_f}{\rho_p}\left((\mathbf{u}\cdot\nabla)\mathbf{u} - \frac{d\mathbf{v_p}}{dt}\right) + \mathbf{g}\,, \end{aligned} \tag{6}$$

where ν_f is the kinematic viscosity of the fluid and ∇ is the nabla operator.

Implementation

The computation of the equation of motion is implemented following a Lagrangian approach, allowing for continuous tracking of the particles over time. After initializing the flow field, which could either be an artificial case such as a Lamb-Oseen vortex or genuine data from CFD simulation or measurement, a certain number of particles are placed at random positions, ensuring a sufficient particle density after Keane and Adrian [5]. Afterward, the trajectories are computed over time by solving Eq. 6. Transforming the equation from a differential equation of second order to a system of first-order equations allows the usage of the *ode45* differential equation solver of MATLAB with an adaptive timestep to derive the solution. After the simulation is finished, a set of images with the current particle positions is generated for every desired time step. To ensure sufficient seeding, the number of particles within the computed area must be constant. Therefore, the particles exceeding the boundaries of the computed area are replaced with new particles at random positions within the volume. This phenomenon also occurs in real PIV experiments, when tracer particles leaving the light sheet perpendicular to the plane and is known as *out of plane loss*.

Test Case

To verify the results of the particle simulation a generic Lamb-Oseen vortex was used as a reference flow field derived from Eq. 2. The core radius r_c was chosen to 5.4 mm and the max. tangential velocity $v_{tan.max}$ to 29.5 m/s. The vortex parameters are chosen to fit the measurement data from Wolf [14]. Additional parameters are listed in Table 1a. A parameter study was conducted to evaluate the influence of the tracer particle diameter, simulating several unique values and statistical size distributions, as suggested by Raffel et al. [8]. A log-normal distribution with the probability density function (PDF)

$$f_m(d_p) = \frac{1}{\sqrt{2\pi}\, d_p \cdot \sigma_G} \cdot exp\left[-\frac{1}{2}\left(\frac{ln(d_p) - \mu}{\sigma_G}\right)^2\right] \tag{7}$$

is fitted to the data from Raffel et al. [8] with the particle diameter d_p as random variable, giving $\mu = 0.17$ and $\sigma_G = 0.71$. As baseline data, the measurement of the q3-distribution of DEHS particles with a calibrated nozzle at 1 bar was used. Other size distributions (e.g. q1 and q2) were not considered, because the particles are assumed to be a point mass with a constant density. In Fig. 2 a comparison of the curve fit and the measurement data is depicted exemplary for $d_p = 1\,\mu$m. The hatched area marks the particle sizes used for the simulation. Smaller particles were not considered because of missing verification data and the decreasing visibility of the particles, as well as particles larger than 6.5 µm, which are too heavy and would fall to the ground immediately. Important material parameters are listed in Table 1b.

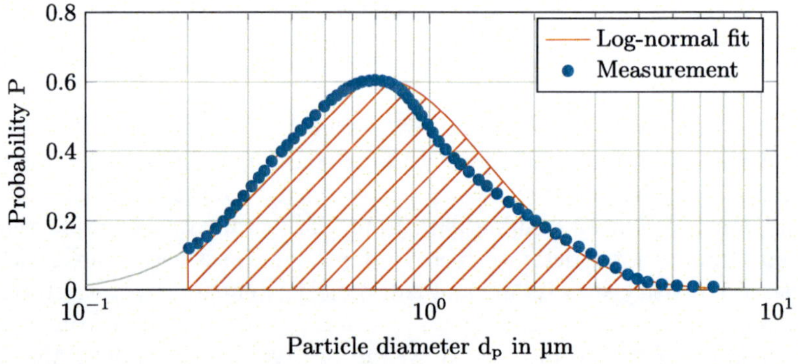

Fig. 2. Comparison of the fitted PDF of the particle diameter with the measurement data from Raffel et al. [8]

Table 1. Overview of the parameters used for simulation and evaluation

(a) PIV parameters				(b) Material parameters			
Simulation Setup				**Fluid: Air**			
Time delay	Δt	25	µs	Temperature	θ_f	20	°C
Sensor plane		16.6×14	mm^2	Pressure	p_f	1013	mbar
Pixel size		6.5	µm	Density	ρ_f	1.205	$kg\,m^{-3}$
Timstep		adaptive	–	Dynamic viscosity	η_f	18×10^{-6}	Pa s
Aperture	f	50	mm	Kinematic viscosity	ν_f	15×10^{-5}	$m^2\,s^{-1}$
Object distance	G	1500	mm	Kolmogorov length	η_K	4	µm
Evaluation Parameters				**Particles: DEHS**			
Windowsize		16	px	Temperature	θ	20	°C
Overlap		75	%	Density	ρ_p	912	$kg\,m^{-3}$

4 Results

Examples of the generated particle images can be found in Fig. 3. The images show the particle void developed for different particle diameters at a vortex age of $\Psi = 55°$ (corresponding to a time of $t = 6.47\,ms$ with a blade frequency of $f = 23.6\,Hz$). From the left to the right, in the first two images (a, b) monodisperse particles with a diameter of 1 µm or 3 µm was chosen, respectively. For the following three pictures (c–e), a non-uniform size distribution was applied as proposed by Raffel et al. [8] and stated in Eq. 7. The corresponding (mean) diameter of the particles d_p ($d_{p,m}$) is given below the images. Monodisperse particles are leading to a rapid increase of the void size, especially for young vortex ages and large particles (3 µm). A non-uniform size distribution results in a much slower and more realistic void development, especially for older vortices. This behavior can be explained by the particle mass and inertia. Increasing the size also increases the mass and inertia, which leads to a slow particle response (see Eq. 4). In the case of the non-uniform size distribution, many small particles remain within the vortex resulting in a much more diffuse void boundary. The analysis of the void size and the following comparison to measurement data is pictured in Fig. 4. The void radius, r_{void}, was determined using the intensity

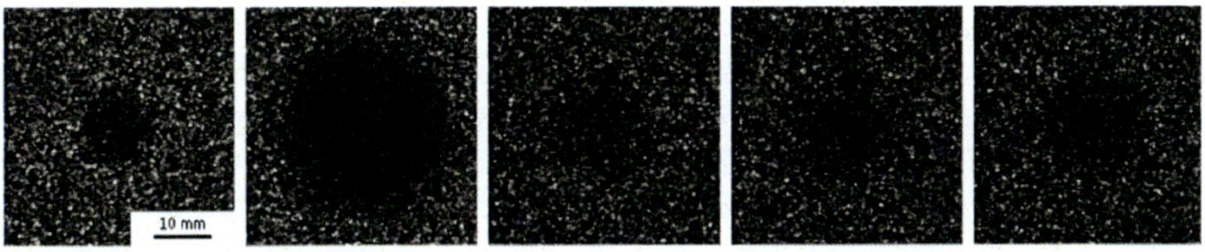

(a) $d_p = 1\mu m$ (b) $d_p = 3\mu m$ (c) $d_{p,m} = 1.18\mu m$ (d) $d_{p,m} = 1.48\mu m$ (e) $d_{p,m} = 1.78\mu m$

Fig. 3. Particle voids for different particles diameter at a vortex age of $\Psi = 55°$ showing the influence of a non-uniform particle size on the void development.

distributions of the SPIV images. Applying a spatial low-pass filter removes individual particles and enables to differentiate between seeded and non-seeded areas using an appropriate threshold. The figure shows that the usage of a statistical particle size distribution in comparison to a monodisperse particle size leads to a better agreement with the measurement data when considering the void size. Nevertheless, the constant particle diameter of 1 µm shows a reasonable replication of the measurement data for a vortex age up to $\Psi = 55°$. For older vortices the influence of smaller particles remaining close to the vortex core increases and the void size grows faster than the measurement data suggests. Since the computation time and memory requirements for smaller particles increases dramatically, the monodisperse approach with a diameter of 1 µm is favorable for younger vortices.

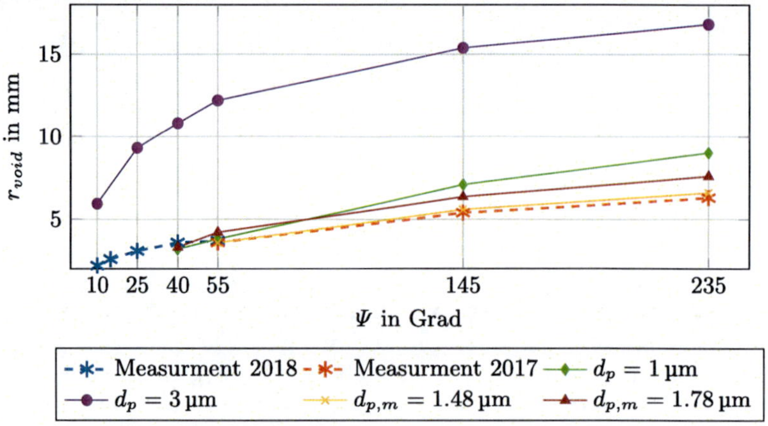

Fig. 4. Development of the void size with measurement data for different particle diameters

In Fig. 5 the velocity distribution for a constant particle diameter of 1 µm (yellow line) and 3 µm (orange line) are compared to the reference velocity distribution (blue line). For the sake of computation time, the monodisperse approach was chosen. The velocities are derived from the generated particle images by the commercial PIV software LaVision DaVis 8.2. The data shows a good agreement between the input reference velocity (blue line) and the SPIV evaluation even within the particle void (purple line). If the particle size exceeds a certain diameter, the velocity profile matches only outside of the particle void (green line). The low particle density within the void leads to a low or even zero signal-to-noise ratio and a high number of outliers. These outliers are detected by the PIV algorithm and replaced with a linear interpolation of the local velocity used as an initial value for subsequent steps in the multi-pass processing. For a particle diameter of 1 µm, this circumstance does not affect the results much because the velocity profile in the void can be considered linear. For larger particles and larger voids, the linear approximation results in a significant error because of the nonlinear velocity profile in the void.

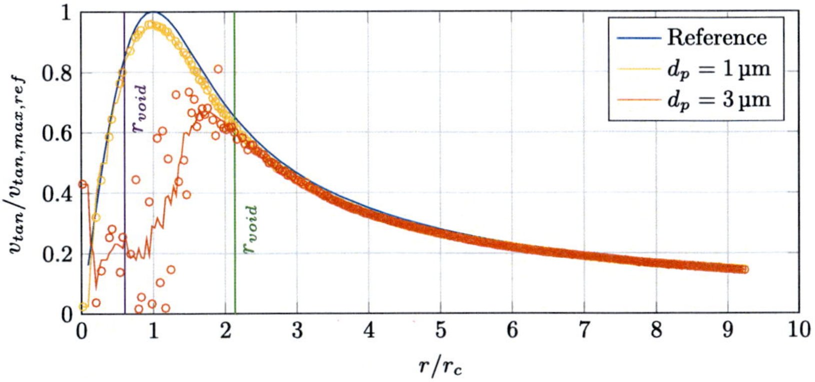

Fig. 5. Relative tangential velocity for a constant particle diameter for a vortex age of $\Psi = 55°$ with $v_{tan,max} = 29,5$ m/s and $r_c = 5,4$ mm, void size 1 µm: purple, 3 µm: green

5 Conclusion

This work describes the development of a tool for the creation of synthetic PIV images and presents results of its application. The tool allows for a physically correct calculation of particle trajectories based on an adapted Maxey-Riley equation (Eq. 6). From the calculated particle positions, synthetic particle images are generated. To verify the results, a generic velocity field of a Lamb-Oseen vortex based on measurement data was used.

The validation of the synthetic PIV images showed an overall good agreement with the measurement data. The variation of the particle size revealed a significant influence of the particle diameter on the results. A non-uniform size distribution with a mean diameter $d_{p,m} = 1.48$ µm was found to match the measurement data best. However, the assumption of a constant diameter of 1 µm showed reasonable results for vortex ages up to $\Psi = 55°$, while saving computational resources. The voids size matches the measurement data (Fig. 4) and the velocity profile in Fig. 5 is reproduced good enough for further investigations. This work also shows that it is not possible to interpolate sparse velocity data within the particle void if the void exceeds the vortex core radius, as seen in Fig. 5.

Acknowledgements. The work presented in this paper is part of my master thesis "Quantitative Erfassung von Wirbeln durch Impfpartikel und bildgebende Messtechniken" (Leibniz University Hannover, 2019), which contains additional aspects and investigations. The author wants to thank Kyle Lynch for providing the source code for the particle projection and image calculation on which this work is based.

References

1. Braukmann, J.N., Wolf, C.C., Goerttler, A., Raffel, M.: Blade tip vortex system of a rotor with cyclic pitch. AIAA J. **58**(7), 2869–2880 (2020). https://doi.org/10.2514/1.J058678
2. Breitsamter, C.: Nachlaufwirbelsysteme großer Transportflugzeuge: Experimentelle Charakterisierung und Beeinflussung. Aerodynamik. Utz, München (2007)
3. Goerttler, A., Braukmann, J.N., Wolf, C.C., Gardner, A.D., Raffel, M.: Blade tip-vortices of a four-bladed rotor with axial inflow. Vertical Flight Society 75th Annual Forum & Technology Display, vol. 75. Vertical Flight Society (2019). https://doi.org/10.4050/JAHS.65.042002
4. Hartje, L.: Untersuchung des Einflusses der Particle Image Velocimetry auf die Charakterisierung von Blattspitzenwirbeln. Master thesis, Leibniz Unisersität Hannover (2018)
5. Keane, R.D., Adrian, R.J.: Optimization of particle image velocimeters. I. Double pulsed systems. Measur. Sci. Technol. **1**(11), 1202 (1990). https://doi.org/10.1088/0957-0233/1/11/013
6. Lecordier, B., Demare, D., Vervisch, L.M.J., Réveillon, J., Trinité, M.: Estimation of the accuracy of PIV treatments for turbulent flow studies by direct numerical simulation of multi-phase flow. Measure. Sci. Technol. **12**(9), 1382 (2001). https://doi.org/10.1088/0957-0233/12/9/302
7. Maxey, M.R.: Equation of motion for a small rigid sphere in a nonuniform flow. Phys. Fluids **26**(4), 883 (1983). https://doi.org/10.1063/1.864230
8. Raffel, M., Willert, C.E., Scarano, F., Kähler, C.J., Wereley, S.T., Kompenhans, J.: Particle Image Velocimetry. Springer, Cham (2018). https://doi.org/10.1007/978-3-319-68852-7
9. Saffman, P.G.: Vortex Dynamics. Cambridge Monographs on Mechanics. Cambridge University Press, West Nyack (2012)
10. Schiller, L., Nauman, A.: Über die grundlegenden Berechnungen bei der Schwerkraftaufbereitung. Zeitschrift des VDI **77**, 318–320 (1933)
11. Sommerfeld, M.: Modellierung und numerische Berechnung von partikelbeladenen turbulenten Strömungen mit Hilfe des Euler-Lagrange-Verfahrens: Zugl.: Erlangen, Nürnberg, Univ., Habil.-Schr., 1996. Berichte aus der Strömungstechnik. Shaker, Aachen (1996)
12. Sommerfeld, M.: Bewegung fester Partikel in Gasen und Flüssigkeiten. In: Stephan, P., Mewes, D., Kind, M., Schaber, K., Wetzel, T., Kabelac, S. (eds.) VDI-Wärmeatlas, Springer Reference Technik. Springer, Berlin (2018). https://doi.org/10.1007/978-3-662-52991-1_88-2
13. Tsai, R.: A versatile camera calibration technique for high-accuracy 3D machine vision metrology using off-the-shelf TV cameras and lenses. IEEE J. Robot. Autom. **3**(4), 323–344 (1987). https://doi.org/10.1109/JRA.1987.1087109
14. Wolf, C.C., Braukmann, J.N., Stauber, W., Schwermer, T., Raffel, M.: The tip-vortex system of a four-bladed rotor in dynamic stall conditions. In: American Helicopter Society 74th Annual Forum (2018)

Development and Analysis of a Coaxial Rotor Test Bench in Axial Flow Conditions

Jan-Arun Faust(✉), Nikolai Herzog, Michael Cerny, and Christian Breitsamter

Technical University of Munich, Boltzmannstr. 15, 85748 Garching, Germany
jan.faust@tum.de

Abstract. Small-scale coaxial-rotor assemblies are commonly utilized in Unmanned Aerial Vehicles to provide a compact design solution for achieving increased thrust potential and flight stability. The presence of two rotors operating in close proximity to each other results in aerodynamic interactions that have a profound effect on the individual rotor performance. A dedicated coaxial rotor test bench was developed for the purpose of testing various setups in a multi-rotor arm configuration. This study investigates the influence of different propeller blade geometries, number of rotor blades, and rotor separation distances on the aerodynamic, electric, and system performance of coaxial rotor configurations in static thrust conditions. Furthermore, the thrust and power measurements for one configuration is presented in axial and reverse flow conditions.

Keywords: Experimental aerodynamics · Coaxial rotor configurations

1 Introduction

Multi-rotor Unmanned Aerial Vehicles (UAVs) capable of Vertical Take-Off and Landing (VTOL) are increasingly deployed for aerial photography, surveillance, reconnaissance, and search operations. To achieve higher lift capabilities, enhance flight stability, and provide redundancy for similar vehicle dimensions, additional rotors can be incorporated into the UAV designs. This can be realized by featuring additional arms (hexa- and octocopters), or by utilizing coaxial assemblies. The emphasis of this study is placed on fixed-pitched coaxial rotor configurations that are commonly featured in multi-rotor UAVs and similarly found in some Urban Air Mobility (UAM) prototypes. Coaxial assemblies usually feature a pair of contra-rotating rotors with a low separation distance that results in a variety of aerodynamic interactions.

The most dominant form of interaction is caused by the trailing wake of each rotor that governs the resultant flow conditions experienced by the other rotor. This generally incurs in a loss of net rotor system aerodynamic efficiency [1], although there have been claims to the contrary, in which performance gains

© The Author(s), under exclusive license to Springer Nature Switzerland AG 2021
A. Dillmann et al. (Eds.): STAB/DGLR Symposium 2020, NNFM 151, pp. 335–344, 2021.
https://doi.org/10.1007/978-3-030-79561-0_32

were observed ranging from 5–11% (as discussed by Ramasamy [2]). The general consensus is that the thrust output of each rotor is reduced in comparison of having two rotors operating in isolation. The resulting influence on the profile and induced power seems to be largely configuration dependent. The conflicting observations could be attributed to variations in blade planforms, Reynolds numbers, and rotor separation distances at which the experiments/simulations were performed.

Ramsamy [2] was able to demonstrate a consistent 9% improvement in hovering efficiency over a single-rotor with equivalent solidity, independent of the blade planform. Performance gains are likely to be associated (in part) with a reduction of swirl losses evident in contra-rotating rotor systems, which was quantified to approximately 4% in hover [3]. A further prominent parameter to consider in coaxial configurations is the rotor separation distance. It impacts the inflow profile of the lower rotor in response to the wake contraction of the upper rotor [4]. Smaller separation distances were found to lead to more equal thrust sharing between the two rotors [2], as both are subjected to more similar (symmetric) inflow conditions [4].

In recent years various coaxial rotor configurations have been studied experimentally, including contra-rotating coaxial rotors [4,5], and co-rotating coaxial rotors [3], in largely static thrust conditions. For the current investigation, a modular and dedicated test bench was developed to study a range of contra-rotating coaxial rotor configurations under arbitrary flow conditions. This paper discusses the results obtained from axial flow and static thrust conditions. The inclined flow measurements are presented separately [6].

2 Experimental Setup

All experiments were performed in Wind Tunnel A at the Technical University of Munich, as shown in Fig. 1(a). The small-scale fixed-pitch rotors were driven by a pair of KDE-5215XF-220 brushless motors. The individual rotor speeds were measured by analysing the three-phase current output of the KDE-XF95AHVC-OTPO electronic speed controller (ESC) using Hobbywing brushless revolutions per minute (RPM) sensors. The motor throttle positions were set using a pair of Arduino Nano PID (proportional–integral-derivative) controllers to achieve the targeted rotor speeds. The developed setup features a six component ME-K6D40 load cell integrated compactly between the electric motors. The load cell has a nominal force of 500 N in thrust direction, 200 N in side forces, 10 Nm in axial torque and 5 Nm in lateral torque with an overall accuracy of 0.5%. Two Cytron-X050 current sensors were installed at the power supply source (constant voltage of 48V) to measure the electrical input power (0.8% measuring accuracy).

The package features exchangeable distance sleeves to alter the rotor separation distance ΔH (see Fig. 1(b)). Sets of three different lengths were manufactured, corresponding to $\Delta H_1 = 0.3 - 0.35D$, $\Delta H_2 = 0.4 - 0.45D$, and $\Delta H_3 = 0.5 - 0.55D$. Three propellers were studied, namely a common model aircraft APC18$\times$8 propeller ($D = 457.2$ mm), two- and three-bladed KDE 18.5$\times$6.3

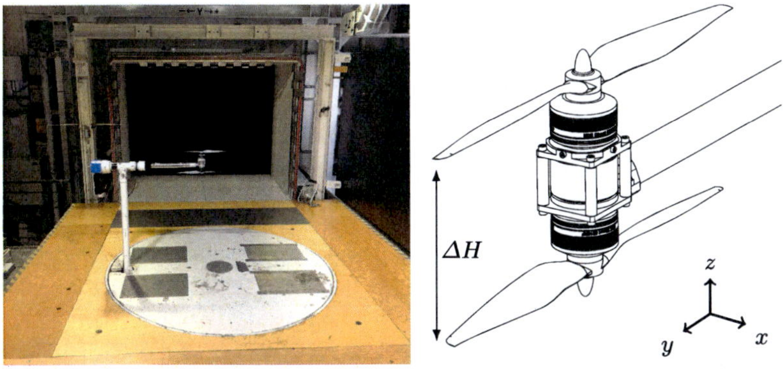

(a) Experimental setup in windtunnel (b) Detailed CAD model assembly

Fig. 1. Experimental coaxial rotor system setup featuring two outrunner brushless motors mounted on a frame with an internal load cell (in upright position).

(D = 469.9 mm) foldable propellers commonly used in UAVs. APC propellers were statically balanced using a propeller balance with corrections made using a thin layer of lacquer. KDE propellers arrived dynamically calibrated.

A steel tube (D = 40 mm) was used to attach the assembly to the floor mount, which contains an integrated stepper motor to enable free rotation about the arm axis to alter the inflow angle. The configuration was aligned with the wind tunnel flow orientation in axial flow and static thrust conditions.

3 Experimental Procedure

3.1 Test Envelope

A series of parameter studies were performed to analyze different rotor configurations, as summarized in Table 1. The primary objective was to quantify the influence of design choices on the configuration performance including different types of propellers, number of rotor blades, and varying rotor separation distances in static thrust conditions. The secondary objective was to gain insights into various characteristic rotor states present in positive inflow and reverse flow conditions, corresponding to fast climb and descent flight. All coaxial rotor measurements presented here were performed at matched rotor speeds.

3.2 Data Processing and Uncertainty Analysis

The load sensor calibration and verification was performed prior to obtaining measurements. Each configuration was tared in order to isolate the rotor influence. The measurements were sampled using a resolution of 3 kHz for a duration of 30 s. Post processing removed the inherent noise using a digital Butterworth filter. Uncertainty analysis was performed for all measurements by considering

Table 1. Summary of test envelope

Data set	Configuration	Rotor(s)	RPM	H/D	V_∞ [m/s]
1	Puller, pusher	APC	2000–7000	N/A	−10, 0, 10
2	Puller, pusher	KDE DUAL	2000–7000	N/A	0
3	Puller, pusher	KDE TRIPLE	2000–6500	N/A	0
4	Coaxial (2+2)*	APC	2000–7000**	0.35–0.55	−10, 0, 10
5	Coaxial (2+2)*	KDE DUAL	2000–7000**	0.3–0.5	0
6	Coaxial (2+3)*	KDE MIXED	2000–6500**	0.3	0
7	Coaxial (3+3)*	KDE TRIPLE	2000–6500**	0.3	0

*Number of rotor blades in coaxial configuration (upper rotor + lower rotor).
**Coaxial rotors were operated at matched rotor speeds (upper rotor = lower rotor).

the uncertainties associated with random errors with repeated measurements (precision uncertainty) and those which are not due to random sources (bias uncertainties). The precision uncertainty was specified as the standard deviation of the measured loads. The bias uncertainty was estimated by linearly scaling the loads with the measuring accuracy.

4 Results and Discussion

To evaluate the system efficiency in static thrust operating conditions, the Figure of Merit FM, was used as an aerodynamic performance metric. It is defined by the ratio between the induced power P_i, and the actual required power P, reading

$$\eta_{aero} = FM = \frac{P_i}{P} = \frac{T^{3/2}/\sqrt{2\rho A}}{P} \tag{1}$$

where T is the rotor thrust, A is the rotor disk area, and ρ is the fluid density. Historically, the Figure of Merit of coaxial rotors utilizes the total thrust and power produced by the coaxial rotor system (i.e. $T_{coax} = T_u + T_l$, $P_{coax} = P_u + P_l$), and the rotor disk area of a single rotor [7]. Using simple momentum theory, Leishman [1] demonstrated the deficiencies of using this definition, which inherently assumes that the induced power is too large for realistic coaxial rotor applications. It was suggested that the coaxial induced power in Eq. 1 can be expressed in terms of the induced power of two isolated rotors $P_{i,isolated}$ (of equivalent thrust), using an induced power interference factor κ_{int}, as given by

$$P_{i,coax} = \kappa_{int} \times 2P_{i,isolated} \tag{2}$$

For two rotors operating in the same plane with equal thrust sharing this value corresponds to a theoretical value of $\kappa_{int} = \sqrt{2}$. However, more realistic values can be derived for two rotors operating in different planes, under the assumption of either equal thrust sharing ($\kappa_{int} = 1.281$) or torque-trimmed rotors ($\kappa_{int} = 1.219$). These lower values provided better agreement with experimental measurements [1,2]. In this analysis the latter value was used, based on the observation that the power load sharing remained similar for a range of configurations. While it may not be entirely representative of the current test case, any definition of FM can be adopted, as long as it is used to compare like-with-like rotors at an equivalent disk loading [1].

Additionally, the electric powertrain driving the rotor also has a characteristic load dependent efficiency which is quantified by

$$\eta_{electric} = \frac{Q\Omega}{UI}, \tag{3}$$

where Ω is the rotor speed, Q is the mechanical torque at the shaft (measured using the load cell), U and I are the (specified) voltage and (measured) input current, respectively. The electrical system efficiency quantifies the losses associated with the cables, ESC, and electric motor. The total system efficiency is the product of both reading

$$\eta_{system} = \eta_{aero} \times \eta_{electric} \tag{4}$$

Single Rotor Configuration Analysis: The analysis begins by investigating the electrical, aerodynamic, and system efficiencies of single rotor configurations. The efficiency of the electrical power train is summarized in Fig. 2(a). The efficiency showed a strong load dependency with a peak of 72–74%. Ideally, small deviations between the different configurations would be expected. The variations observed here are likely associated with the measuring inaccuracies and inconsistent cooling conditions. An analytic fit was made for each rotor configuration to approximate the mechanical power of coaxial configurations as a function of the the electric input power (used to evaluate FM in later section).

The aerodynamic efficiency of the propeller configuration was evaluated using the Figure of Merit, as shown in Fig. 2(b). Both pusher and puller configurations were tested to evaluate the influence of the support arm geometry on the rotor performance. The measured data showed that the APC propeller had a larger thrust and power coefficient in a puller configuration, while the opposite was observed for both KDE propellers. In terms of the aerodynamic efficiency, all pusher configurations demonstrated slightly superior behaviour by approximately 2% under low disk loading ($DL < 100$ N/m^2), which was comparable with the gains found by Theys [5]. All configurations peaked at a value of 72–73%, with no rotor performing significantly better than the other.

Coaxial Rotor Configuration Comparison: Various coaxial configurations were tested to quantify the system sensitivity in response to characteristic aspects, including blade shape, number rotor blades, and axial distance spacing. The influence of the axial spacing distance was tested using two different $(2+2)$ configurations in static thrust conditions. The measured data showed that the axial distance spacing had a significant impact on the aerodynamic performance

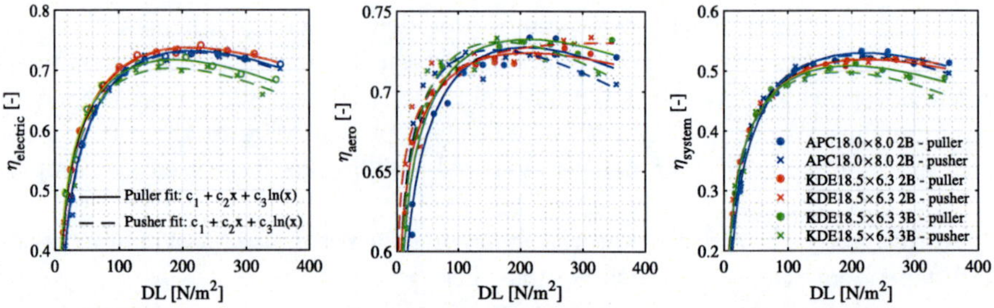

(a) Electric efficiency (b) Aerodynamic efficiency (c) System efficiency

Fig. 2. Efficiency breakdown for the three different single propeller configurations at static thrust.

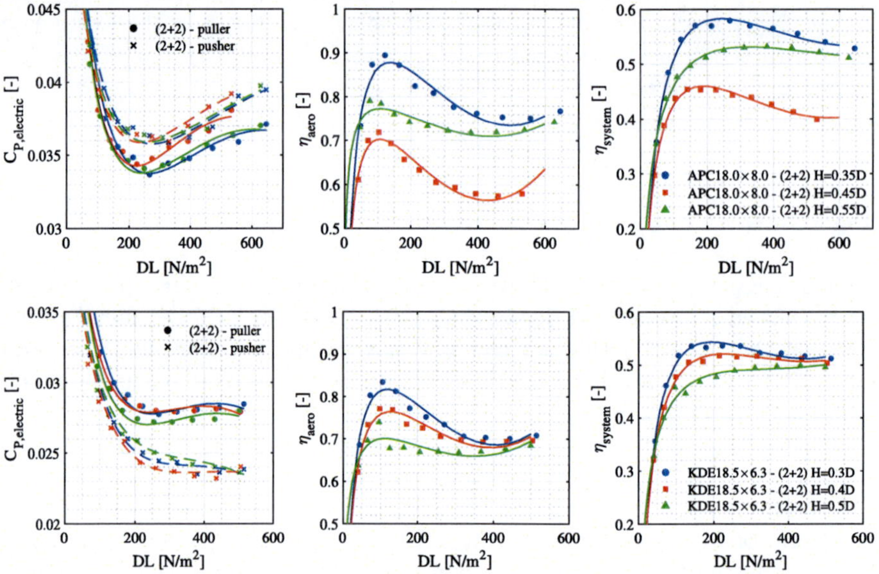

(a) Electric power coefficient (b) Aerodynamic efficiency (c) System efficiency

Fig. 3. Coaxial rotor system power load sharing, coaxial Figure of Merit as the aerodynamic efficiency metric, and the total system efficiency for various axial distance settings using APC and KDE $(2+2)$ configurations.

(see Fig. 3(b)). Interestingly, it was observed that the influence was largely blade dependent. For instance, the KDE $18.5 \times 6.3 - (2+2)$ configuration appeared to show a clear disk loading dependency. An increase in axial spacing was found to decrease the thrust output for a constant rotor speed, while the total required power remained largely unchanged, resulting in a gradual drop in aerodynamic efficiency. At higher disk loading ($DL > 400\,\mathrm{N/m^2}$) all configurations tended to converge to a value of $\eta_{aero} = 70\%$.

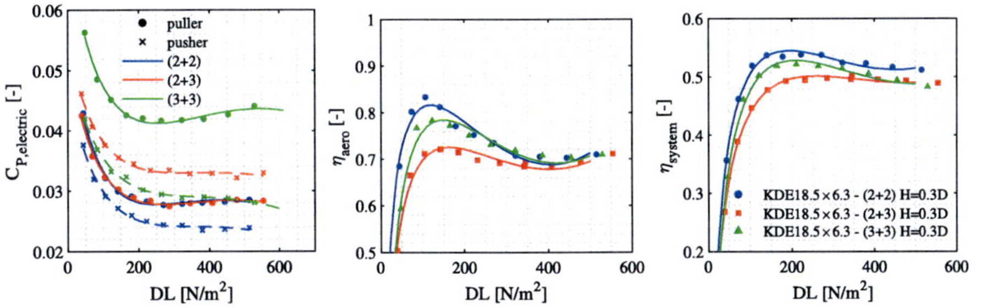

(a) Electric power coefficient (b) Aerodynamic efficiency (c) System efficiency

Fig. 4. Coaxial rotor system power load sharing, coaxial Figure of Merit as the aerodynamic efficiency metric, and the total system efficiency for various blade number configurations using KDE rotors.

On the contrary, the APC $18.0 \times 8.0 - (2+2)$ configuration showed a different sensitivity to the coaxial setup. The changes in the aerodynamic efficiency were significantly larger between different distance settings and showed diverging trend projections. The mid configuration ($H = 0.45D$) was found to generate the least amount of thrust, while requiring the most input power. When analyzing the electric propeller power coefficient ($C_{P,electric} = UI/(\rho n^3 D^5)$) in Fig. 3(a), it can be observed that the power load sharing was also different to the KDE configuration range. In this configuration, the pusher propeller within the setup appeared to be drawing the more power than the puller propeller, which should amount in a higher thrust share. This could suggest as to why the configuration showed a larger dependency on axial distance spacing, with the lower rotor being more susceptible to changes in response to the upper rotor wake contraction and resulting interactions. From both setups it was found that the smallest axial spacing resulted in the most efficient configuration. This could be the result of both rotors experiencing flow conditions that are more similar to each other, resulting in more equal load sharing. This observation appears to be case specific, as Ramasamy [2] found an asymptotic increase in FM with increasing axial spacing of up to $1.5D$.

A second comparison considers different blade number combinations that could be utilized in coaxial assemblies to achieve a range of desired effects. For instance, an increase in the number of rotor blades would enable slower rotor speeds and a reduced Mach tip number. An unmatched rotor configuration (i.e. $2+3$) could potentially also achieve more stable operating conditions by achieving better yaw trim and decreasing the vibrations/noise that coincide with the blade passage frequency. From the current measurement, it was observed that increasing the number of rotor blades did not result in a favorable change in the system efficiency (see Fig. 4).

From all the studied configurations, it was found that the APC$18.0 \times 8.0 - (2+2)$ setup with the smallest axial spacing had the highest peak system efficiency $\eta_{system} = 58\%$. Given the chosen aerodynamic efficiency metric, it performed approximately 6% better than a similarly sized KDE18.5 × 6.3 − (2 + 2) configuration. While it appears that the configuration also performed noticeably better than isolated rotor configurations (15–17% at their respective peaks, see Fig. 2(b) and Fig. 3(b)), a direct comparison is to be cautioned based on the chosen induced power definition. The measured data suggests that the load distribution of coaxial rotors enables each rotor to operate at lower (individual) disk loading and lower Mach tip numbers, resulting in a higher combined hovering efficiency than a single rotor for the same thrust.

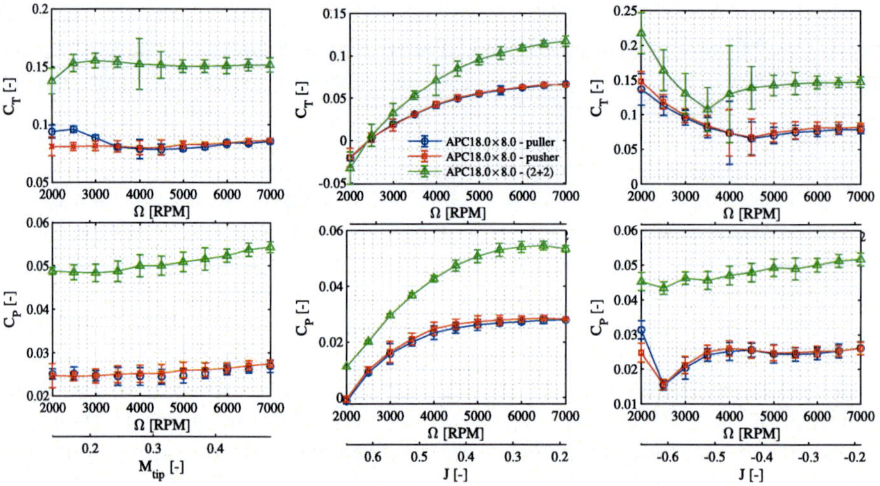

(a) $V_\infty = 0$ m/s (b) $V_\infty = 10$ m/s (c) $V_\infty = -10$ m/s

Fig. 5. APC18.0 × 8.0 single rotor and coaxial rotor (2 + 2) configuration thrust and power coefficient for various axial flow conditions (propeller convention).

Rotor Configurations in Axial Flow Operating Conditions: Various APC18.0 × 8.0 rotor configurations were investigated at different axial flow conditions. The measured thrust ($C_T = T/\rho n^2 D^4$) and power ($C_P = P/\rho n^3 D^5$) propeller coefficients for isolated rotors (puller and pusher) and a coaxial configuration ($\Delta H = 0.35D$) are presented in Fig. 5 for static thrust (hover), positive inflow (fast climb), and reverse flow (fast descent) conditions. A relatively high free-stream velocity was utilized in order to capture a large advance ratio ($J = V_\infty/nD$) range. In static thrust conditions both the thrust and power coefficients remained relatively constant, with a slight linearity related to compressibility effects. Noticeable changes in the thrust and power were observed for the axial flow conditions as the propeller cycles through different operating states. In positive inflow conditions the propeller transitions from windmilling (being powered and driven by the flow) to generating thrust. In reverse flow conditions the dominant rotor-wake interactions lead to unsteady fluctuations in the rotor loads (noted by larger uncertainty regions).

The strength of Blade Vortex Interactions (BVI) are largely dependent on the advance ratio. It governs the position of the tip vortices, as the free-stream velocity displaces the vortices upwards and the rotor induced downwash attempts to do the opposite. Generally various rotor states can be observed in descent flight conditions, namely Propeller Brake, Ring Vortex State, Turbulent Wake State and Windmill State. The thrust coefficient is able to provide an indication of the possible rotor states (see Fig. 5(c)). A clear minimum in both the single and coaxial rotor thrust coefficient can be seen ($J_{single} \approx -0.43$, $J_{coax} \approx -0.52$) which corresponds to the Ring Vortex State. At lower rotor speeds and larger negative advance ratios, the propeller operates in a Turbulent Wake or Windmill State. With increasing rotor speed the rotor thrust coefficient was found to stabilize, as the propeller entered the Brake State, at which point the influence of the free stream velocity diminishes ($J_{single} > -0.35$, $J_{coax} > -0.4$).

The influence of the rotor-rotor interactions were largely dependent on the operating state. The following numbers compare the coefficients related to a coaxial configuration with respect to a single rotor. In static thrust conditions a coaxial rotor generated $1.75 - 1.9\ C_{T,single}$ while requiring $2.0 - 2.1\ C_{P,single}$. In positive inflow conditions the influence of the rotor generated inflow increases, resulting in $1.5 - 1.8\ C_{T,single}$ and $1.8 - 1.9\ C_{P,single}$. In reverse flow conditions the thrust minimum trough was significantly narrower and shifted to a lower rotor RPM range ($1.3\ C_{T,single}$, $1.8\ C_{P,single}$ at 3500 RPM). The presence of a second rotor heavily influenced the occurrence of the Ring Vortex State and it appears that a coaxial configuration was able to produce a more stable thrust output range by entering the Propeller Brake State at lower rotor speeds (up to $2.06\ C_{T,single}$, $1.9\ C_{P,single}$ at 4500 RPM), as it can be seen in Fig. 5(c).

5 Conclusion

In this investigation, a dedicated modular rotor test bench was developed and used to study single and coaxial (contra-rotating) rotor configurations in static thrust and axial inflow conditions. Both the mechanical thrust and torque response of the configurations and the required electric power of the drive system were measured. Based on the results, the following conclusions for the tested propellers can be drawn:

1. Pusher propellers performed more efficiently in hover over a puller configuration with gains up to 2–3% for disk loadings smaller than $100\,\mathrm{N/m^2}$. Smaller differences were observed under high thrust conditions.
2. The influence of the axial spacing between the two rotors was found to have a profound effect on the aerodynamic efficiency (varying as much as 18%). The smallest realizable axial spacing resulted in the optimum configuration.
3. Increasing the blade count for the same blade platform resulted in noticeable decreases in the Figure of Merit (up to 5%).
4. Based on the chosen Figure of Merit definition, coaxial configurations demonstrated superior aerodynamic efficiency over isolated single rotors, with gains up to 17% realized at a disk loading of $150\,\mathrm{N/m^2}$ using APC18.0 $\times$ 8.0 rotors.

5. It was found that coaxial rotors transitions through various states differently to a single rotor setup. The presence of a secondary rotor shifts the Ring Vortex State to a lower RPM range, and thus provides a significantly more stable thrust output in descent flight conditions.

Acknowledgements. The authors would like to thank the Bavarian Research Foundation (Bayerische Forschungsstiftung) for funding the HyProp research project under the grant number AZ-1277-17. Further, we gratefully acknowledge the successful collaboration and support by the project partner Rolls-Royce Deutschland Ltd. & Co. Lastly, the authors would like to thank Luigi Findanno and Wolfgang Lützenburg for their contributions in building the experimental model.

References

1. Leishman, J.G., Ananthan, S.: Aerodynamic optimization of a coaxial proprotor. In: AHS 62th Annual Forum, Pheonix, AZ, USA (2006)
2. Ramasamy, M.: Measurements comparing hover performance of single coaxial, tandem, and tilt-rotor configurations. In: AHS 69th Annual Forum, Phoenix, AZ, USA (2013)
3. Uehara, D., Sirohi, J.: Quantification of swirl recovery in a coaxial rotor system. In: AHS International 73rd Annual Forum & Technology Display, Fort Worth, TX, USA (2017)
4. Shukla, D., Komerath, N.: Drone scale coaxial rotor aerodynamic interactions investigation. J. Fluids Eng. **141**(7), 071106 (2019). https://doi.org/10.1115/1.4042162
5. Theys, B., Dimitriadis, G., Hendrick, P., De Schutter, J.: Influence of propeller configuration on propulsion system efficiency of multi-rotor Unmanned Aerial Vehicles. In: International Conference on Unmanned Aircraft Systems (ICUAS), Arlington, VA, USA (2016)
6. Cerny, M., Faust, J., Breitsamter, C.: Investigation of a coaxial propeller configuration under non-axial inflow conditions. In: STAB Symposium (2020)
7. Coleman, C. P.: A Survey of Theoretical and Experimental Coaxial Rotor Aerodynamic Research. NACA, Moffett Field (1997)

Check for updates

Numerical Investigation of Aerodynamic Fairings on a Cycloidal Rotor

Louis Gagnon(✉)

Institute of Aerodynamics and Gas Dynamics, University of Stuttgart, Stuttgart, Germany
gagnon@iag.uni-stuttgart.de

Abstract. The modifications to the flow caused by fairings placed outside a cycloidal rotor are studied. Six configurations using these passive aerodynamic surfaces are covered. The rotors perform fully developed hover flight with a 2D URANS CFD model. Two principal flow regimes emerge from the six configurations. One regime reproduces the blade vortex shedding typically found in cycloidal rotors which occurs as the blade travels upwards. The other regime does not exhibit it and the blades instead shed two smaller vortices as they reach the bottom of the rotor. This modified flow regime rotates the thrust vector and the wake by roughly 25°. It also significantly reduces both power consumption and drag. A rotor with opened fairings on both sides yields a 6% increase of the thrust to power ratio at equal thrust.

Keywords: Cycloidal rotors · VTOL · Fairings · Thrust · Energy efficiency

1 Introduction

Cycloidal rotors are devices used to produce aerodynamic thrust. Research prototypes were built to propel airships [1], micro-aerial vehicles [2], and drones [3,4]. Attempts were also conducted to build large versions with a high-power density in the hope to provide an efficient means for manned flight in narrow areas. Such attempts are and have been carried by the private sector in countries such as Austria [5], Russia[1], and the United States [6]. One can think of such rotors as being geometrically similar to vertical axis wind turbines and Voith propellers. The blades of such rotors are arranged to have their span axis in the same direction as the axle of the propeller. The thrust they produce can be freely redirected in the 360° circle described by the motion path of their blades. A schematic example is shown in Fig. 1(a).

The motion of a cycloidal rotor blade is challenging to interpret as it combines rotation and oscillation. For this study, a three-blade rotor turning at $\Omega = 18.67\,\text{rad/s}$ is used. The pitch function is a sine of magnitude 25° with no phase and to which a constant offset of 6° is added. Such an offset is common for these

[1] http://arey-group.ru, last verified January 2021.

© The Author(s), under exclusive license to Springer Nature Switzerland AG 2021
A. Dillmann et al. (Eds.): STAB/DGLR Symposium 2020, NNFM 151, pp. 345–354, 2021.
https://doi.org/10.1007/978-3-030-79561-0_33

rotors and causes the local blade pitch angle to be smaller at the top half of the rotor. The blade is a 15 cm-chord NACA0012 airfoil located at radius R = 0.5 m from the center and pivoting around its 25% chord point. This configuration gives a Reynolds number Re = 90,000 and a Mach number Ma = 0.03. The rotor geometry and pitching schedule are unchanged throughout the paper. The motion is highly dynamic, as pointed by the oscillation's reduced frequency $\kappa = 0.15$ of the studied case. This is nearly 6 times that of a helicopter such as the MBB Bo-105. Considerable blade surface velocities result from these blade oscillations. They are plotted in Fig. 1(b) as the blade travels on the rotor path.

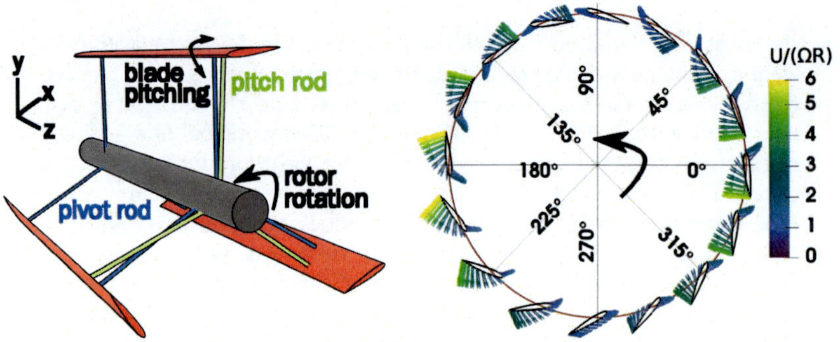

(a) Mechanism. (b) Blade velocities for $\theta = 0° + j22.5°$.

Fig. 1. Cycloidal rotor and vectorial representation of the blade velocities with induced flow. Colored linearly and scaled with the square root.

The arrows present the direction and square root of the magnitude of the combined rotation plus oscillation velocities for various points on the blade's surface. The impinging flow velocity is thus the combination of the inverse of these arrows and the induced flow. The induced velocity nearly doubles as the air travels through the rotor. At the top of the rotor, its maximum is roughly $0.4\Omega R$ which is 10 times smaller than the blade oscillation velocity at its trailing edge. The inflow and wake are shown in Fig. 2(b). It is also known that the thrust generation per blade only occurs over a small portion of the rotation, namely around $270 \leq \theta \leq 300$. This is highlighted in Fig. 2(a) which also shows the reference quadrants. There, the square roots of the averaged blade surface pressures are plotted normal to the airfoil for various segments of a complete rotation. For simplicity, and because the velocities on both suction and pressure sides of the blade are similar, only the arrows pointing outside the blade are plotted. It is thus clear that for the case under study, the oscillation dominates the blade motion. This can be seen by the maximum velocities U at the tail of the blade reaching 6 times the tangential rotor velocity ΩR. Consequently and contrary to a fixed-wing, the relative flow is often more perpendicular to the blades planform than parallel to it. Also, roughly 3% of the rotor moment comes from the wall shear stresses. Viscous forces are thus not inspected, although they are included in the displayed results.

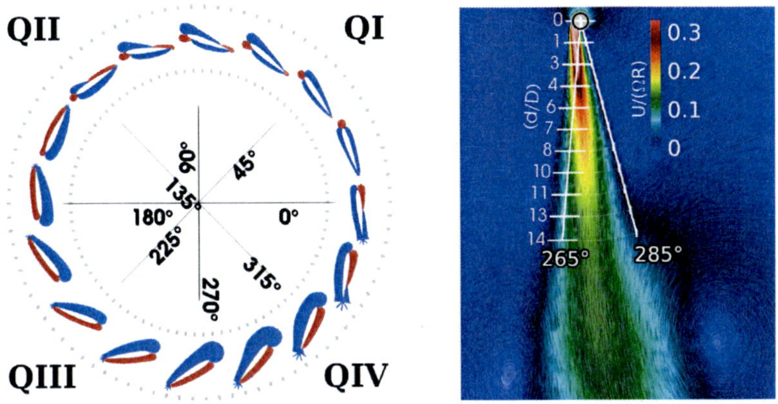

(a) $\sqrt{C_p}$ for $\theta = j22.5°$.

(b) Wake after 150 rotations.

Fig. 2. Naked cycloidal rotor. a) Mean +ve (red) and –ve (blue) surface pressure along a rotation. b) Velocity field averaged over a rotation and graded in distance d/D from the rotor center; video: louisgagnon.com/sv1.

A general idea about the induced flow on the rotor was given by [3], expressed in two dimensions: the flow enters the rotor axially from a roughly 180° wide arc above it and becomes curvilinear while going through the rotor to finally exit vertically downwards. However, it remains unclear what happens in the portion of the rotor cylinder where the induced flow switches from entering the rotor to exiting from it. Computational fluid dynamics (CFD) simulations show that there are strong blade-vortex interactions in that zone. Disturbances are carried down the rotor and influence the second encounter with the blades.

Although the induced velocity and the blade-shed vortices exhibit much lower velocities than the oscillation itself, their influence is nonetheless worth studying. The focus of this paper is to study possible modifications to the flow structure inside the rotor. The aim is to see whether these modifications can improve the energy efficiency and longevity of the rotor and provide means to attach the rotor to an aerial vehicle. To do so, passive aerodynamic fairings are installed on

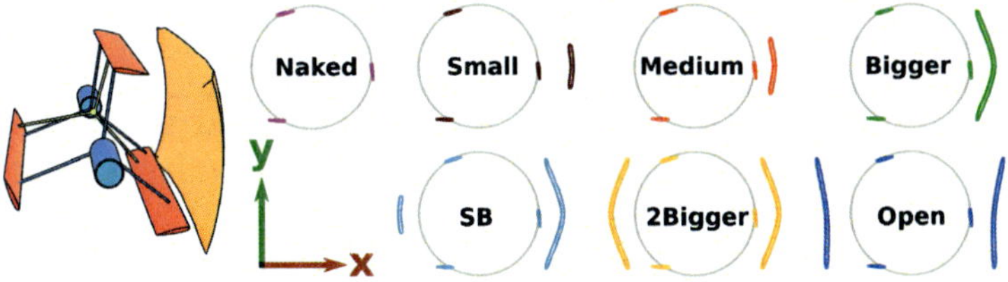

Fig. 3. Faired cycloidal rotor sketch along with the geometrically exact representations of the studied cases. There are 3 fairing sizes of which the larger can be closed or open. The SB case is shorthand for Small and Bigger.

the side of the rotor with the hope that they can redirect or limit the velocity of the flow entering the rotor in the critical zones. A 3D sketch of the concept is drawn in Fig. 3 along with the exact 2D geometries and names of the studied cases. In every case, the fairings extend equally above and below the rotor and are vertically aligned with the rotor. The displayed orientation axes are right-handed and are valid throughout this article whereas the center of the rotor is always located at $(0, 0)$. The work is carried out with a 2D Unsteady Reynolds Averaged Navier Stokes (URANS) CFD simulation with a fully resolved boundary layer flow.

2 Model Setup

The moving parts of the rotor are divided in cylindrical grid zones which each rotate and oscillate as a whole. Between these zones, Arbitrary Mesh Interface (AMI) boundary conditions perform conservative field interpolations [7]. The anisotropic structured mesh of the boundary layer and its transition to an unstructured mesh is generated using the Pointwise meshing software. The unstructured rotor and outer domain meshes are generated by Gmsh [8]. The various case grids have between 500 and 600 thousand cells and consist, where possible, of quadrangles. The y^+ value on the airfoils remains everywhere well below 1. The fully turbulent k-ω-SST model is used. The solution is obtained using the time-marching PIMPLE algorithm without compressibility. As such, inner iterations are performed at each time step using a steady-state approach, namely the SIMPLE algorithm. For these, relaxation is used to enhance stability. These inner solutions are then fed inside the loop of an unsteady algorithm, namely the PISO algorithm. This method, named PIMPLE, allows larger timesteps while ensuring better stability. Time integration is performed using a 2nd order backward-differencing scheme. A variable timestep is used throughout the simulation and is automatically adjusted to maintain the Courant number below 40. This corresponds to a $0.2°$ to $0.33°$ rotor motion per timestep. The outer domain is 60 diameter wide and its boundary is imposed by a freestream condition corrected for dynamic pressure. The fairing walls are slip while the airfoils have imposed velocities which match their motion.

3 Results and Discussion

All the results presented in this section are taken after having let the hovering rotor complete 45 full rotations. This ensures that the starting vortices which are seen at the bottom of Fig. 2(b) are far enough to have little influence on the calculated flow and forces. Verification showed that the average error between the periods is roughly 2%. The flow fields are rendered in Figs. 4 and 5. Pressure thrust and drag at the blade surfaces are given in Fig. 6. For the latter, attention should be paid to the linear distance, along the blade normals, of the pressure areas. These distances are proportional to the square root of the pressure while the drawn areas are unwillingly enlarged by the curvature of the

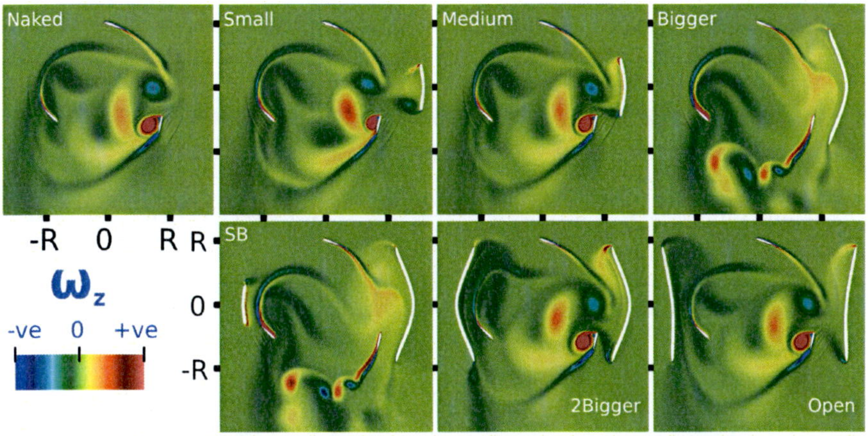

(a) Vorticity around the z-axis for blade at 329°. The following features of the original regime cases listed below are visible: imminent detachment of the LEV on the rightmost blade; both LEV and TEV of the preceding blade being convected towards the center of the rotor so as to follow the typical curvature of cycloidal rotor flow. For the cases of the modified regime, the two pairs of LEV-TEV are visible at the bottommost rotor position and a smooth wake occurs at the position where shedding occured for the original regime. Video: louisgagnon.com/sv2.

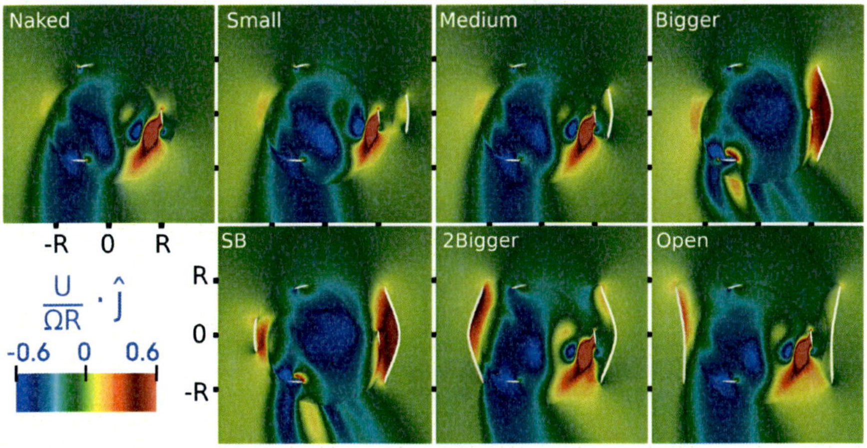

(b) Instant velocity at the start of a rotation shown inside the rotating mesh zone and velocity averaged over 360° shown outside. The jump located at d=0.65R is expected and caused by the rotating interface. The modified regime cases show a wider downflow zone inside the rotor and a fairing-induced upwards flow zone on the right side.

Fig. 4. Vorticity and vertical velocity where the center of the rotor is at 0 and R is the radius. It is seen that the original regime of the Naked case holds for the Small, Medium, 2Bigger and Open cases. The modified regime is seen for Bigger and SB.

airfoil, namely towards the leading edge. In both plots, yellow and green represent desired effects, namely either low drag or high thrust, while red and blue represent undesired effects of their respective pressures. Finally, the values of the comprehensive forces, power, and efficiency find themselves in Table 1 and in the plots of Fig. 7. Most values are self-explanatory and defined in the caption. The thrust varies between the fairing arrangements because both angular velocity and pitching schedule are constant. Consequently, individual power and efficiency values are corrected into P_c and E_c, respectively. This is achieved by using the demonstrated relationships $P \propto \Omega^3$ and $T \propto \Omega^2$. They are used to recalculate the power using the change in angular velocity Ω that makes the case thrust equal to the reference one. The corrected power is thus $P_c = P\xi$ where $\xi = (\frac{T_{ref}}{T})^{3/2}$. Consequently, $E_c = \frac{T_{ref}}{P_c}$.

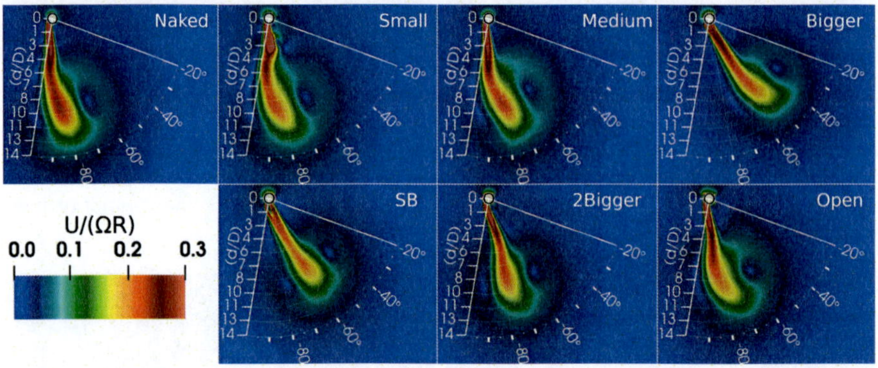

Fig. 5. Average wake velocities over a complete rotation. Scale gives the distance d/D from the center of the rotor in multiples of the rotor diameter. Angles from −100° (260°) to -20° (340°) are indicated.

Table 1. Force and power summary. Mean values over a blade period. Variables are F_y and F_x for vertical and horizontal thrust, respectively; T for the thrust magnitude and angle; T_f for the thrust generated by the fairings in the direction of the total mean thrust of each case; P for rotor power; E for the efficiency as T/P; P_c and E_c as adjusted P and E values to obtain the same thrust as the reference (Naked) case.

Case	F_y(N)	F_x(N)	T(N)	T(°)	T_f(N)	P(W)	E($\frac{N}{kW}$)	P_c(W)	E_c($\frac{N}{kW}$)
Naked	10.10	−1.00	10.15	95.6	0.00	70.27	144	70.27	144
Small	10.53	−0.64	10.55	93.5	0.01	70.35	149	66.39	152
Medium	10.01	−1.08	10.07	96.2	0.13	68.74	146	69.57	145
Bigger	5.23	−3.86	6.50	126.5	5.06	47.14	137	91.99	110
SB	5.17	−3.06	6.01	120.6	4.29	45.15	133	99.17	102
2Bigger	7.16	−2.92	7.73	112.2	−0.67	68.30	113	102.64	98
Open	9.90	−1.80	10.06	100.3	0.53	65.54	153	66.37	152

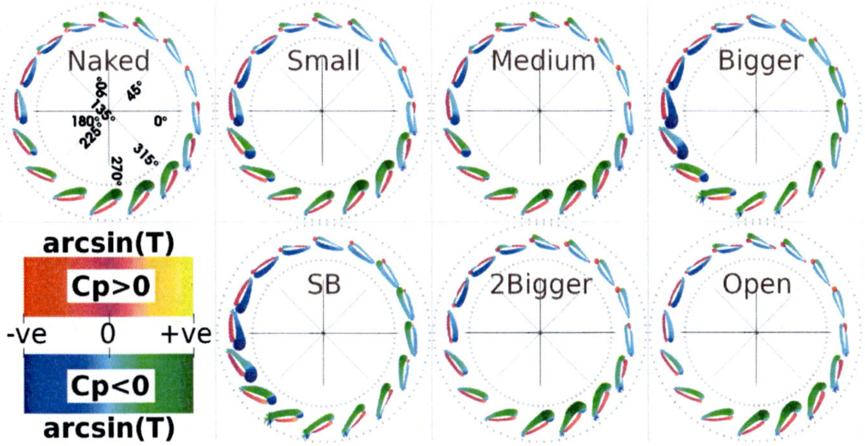

(a) Nondimensionalized pressure-induced thrust T in its mean direction. The wanted positive thrust is shown in yellow and green while the unwanted negative thrust is shown in red and blue. The stronger and longer lasting gains in thrust caused by the original regime's vortex schedding are seen. Sheddings by the new regime occur earlier but do not impede thrust generation until the blade reaches almost the 0° position.

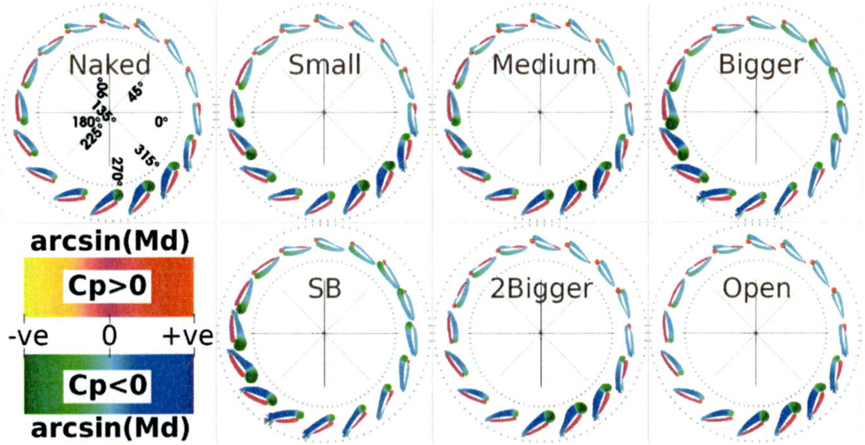

(b) Nondimensionalized pressure-induced angular drag (moment) Md. The wanted negative drag is shown in yellow and green while the unwanted positive drag is shown in red and blue. For both regimes, a stronger drag is seen on both pressure and suction sides of the blade during vortex separation.

Fig. 6. Average coefficients of pressure taken for each 22.5° slice starting a 0°. Lines normal to the blades are scaled by the square root of the pressure coefficient. Underpressures are blue and green while overpressures are yellow and red. The influence on the pressure caused by the large separation zones of both the original regime (Naked, Small, Medium, 2Bigger and Open) and of the modified regime (Bigger and SB) are clearly visible and anticipated by roughly 90° for the modified regime.

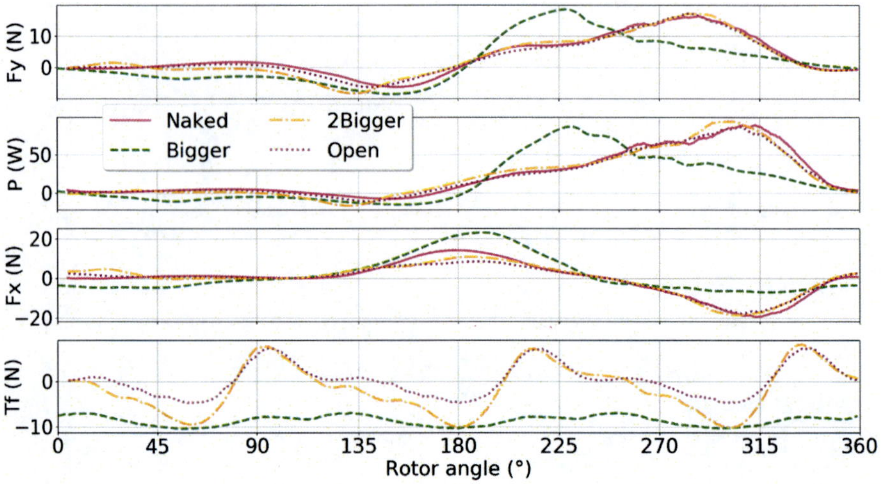

Fig. 7. Selected cases. Instantaneous vertical and horizontal forces along with aerodynamic power consumed by one blade. Thrust generated by the fairings in the direction of their corresponding rotor mean thrust.

Figure 4(a) allows interpretation of the vortex generation and interaction within the rotor. For all cases the blades will at a given time shed a leading edge vortex (LEV) immediately followed by a trailing edge vortex (TEV). For the cases with the large, right side fairing without its left-side counterpart, thus Bigger and SB, the double-shedding occurs twice. The first LEV sheds at roughly 246°, but remains partially attached to then at 296° newly shed. Each of those two LEVs are accompanied by a TEV which closely follows. The resulting 4 vortices are carried out of the rotor by the induced flow and do not interact with other blades. This is not the case for the other cases, where the LEV sheds at roughly 313° and is also accompanied by a TEV. Although there is in these cases, only one pair of vortices being shed, they are sucked towards the bottom-right side of the rotor by the induced flow and interact with the blade which follows. For the Medium case, the two shed vortices are slightly smaller and shed roughly 17° later with respect to the other single-shedding configurations.

The shift of vortex shedding for the Bigger and SB cases also leads to an anticipation of the strong pressure-induced thrust and drag on the airfoil, as shown in Fig. 6. This earlier thrust peak is, however, not purely beneficial. Figure 7 helps to further quantify the influence of the two flow regimes and their different vortex shedding structure given by Figs. 4 and 6. It clearly shows that Bigger will suffer from a drop in thrust due to its early detachment. For the original regime, detachment occurs when the blade is in the process of turning upside down and a drop in lift is thus already expected. A surprising result is that for Bigger and SB, the fairing contributes to 75% of the thrust. This comes from their interaction with the passing blade on the right side.

Finally, there is a small quantity of vorticity being generated by the fairings themselves, and in the Medium case, a shedding which does not share the period of rotation occurs at the lower portion. It, however, does not trigger a difference in the periodic behavior of the fields and forces on the blade passing nearby. The vorticity snapshot also highlights the influence of the large concave fairings of cases Bigger, SB and 2Bigger on the upper inflow. As such, where a large fairing is present, the wake of the blade will not follow a smooth path. This velocity perturbance is also clearly seen on Fig. 4(b) for the same cases. The cancellation of the vortex-shedding on the right side of the rotor for cases Bigger and SB is partly due to the presence of a vortex in the cup formed by the right fairing. The latter vortex submerges the blade traveling upwards first into a right-directed flow which reduces the effect of leading edge pitching. The blade then enters into a flow traveling upwards and thus reducing the velocity of its impinging flow. As hinted above, the flow for these two cases is also modified by displacing the high vorticity zone caused by the shedding of the LEV and TEV. That zone was inside the rotor disk in quadrant I for the reference case and is outside the rotor disk in quadrant IV for cases Bigger and SB.

The cases initially displayed Fig. 3 were elaborated in stepwise phases according to the order of presentation. As such, after the Small case was shown to have little influence on the flow field, Medium was elaborated as being longer and closer to the rotor. The impact remained minimal, and thus Bigger was created, which yield a significant power reduction. From that, the SB and 2Bigger cases were created to counteract the highly rotated thrust and wake. The later achieved its purpose at the cost of the worst calculated efficiency. The last attempt was made with the Open case, which as hoped, returns a high efficiency but does not enact the modified vortex-shedding regime of cases Bigger and SB. It is suggested that the fairings of Small and Medium have little influence on the flow because they are located in a slow velocity field, as shown in Fig. 4(b).

For all cases, Fig. 6 shows that most of the rotor thrust and moments result from the negative relative pressures acting on the blades. The positive C_p values rather only provide little drag and downlift, with the exception of providing lift in the vicinity of shedding the first LEV. Suction pressure instead pulls the nose of the blade in the vicinity of the LEV shedding, as shown in Fig. 6(b). At the same time the suction, however, contributes significantly more drag on the planform of the airfoil. For the standard regime, this suction is responsible for most of the drag produced by the blade.

4 Conclusion

The 2D CFD simulations allowed finding that the flow regime of a cycloidal rotor can be reshaped by aerodynamic fairings. For the 6 cases studied, there were two clearly distinct flow regimes. The flow regime resulting from the cases with the large inward-curved right fairings required less power to run and avoided blade-vortex interactions. However, it also produces significantly less thrust. The Small and Open cases produced the best efficiencies and should be further studied. The results presented lead to believe that a smaller and further away or a more evenly curved Open fairing could improve the efficiency. A configuration may exist where the vortex shedding is suppressed, but was not found by the current study. Finally, open questions for further research are the relationship between trim, thrust vectoring, and the fairings; whether such a fairing could find a use outside of hover; and, the optimization of the fairing shape to impede its separation and improve its impact on the rotor flow. As such, a similar influence on the rotor may also result from modifications to the pitch control function. As for non-hover configurations, research is already undergoing for a dual-wing configuration [9]. And, in the light of the Open case results, attempts to reproduce ducted fans profiles of other VTOL aircraft may yield promising results.

Acknowledgments. The "A Novel and Simple Aircraft Requiring Minimal Power to Hover" project is funded by the Alexander von Humboldt foundation through their Research Fellowships for Postdoctoral Researchers. The author is also grateful the countless advice received from his colleagues at his host institute.

References

1. Ilieva, G., Páscoa, J.C., Dumas, A., Trancossi, M.: A critical review of propulsion concepts for modern airships. Cent. Eur. J. Eng. **2**(2), 189–200 (2012). https://doi.org/10.2478/s13531-011-0070-1
2. Benedict, M., Mattaboni, M., Chopra, I., Masarati, P.: Aeroelastic analysis of a micro-air-vehicle-scale cycloidal rotor. AIAA J. **49**(11), 2430–2443 (2011). https://doi.org/10.2514/1.J050756
3. Yun, C.Y., Park, I.K., Lee, H.Y., Jung, J.S., Hwang, I.S.: Design of a new unmanned aerial vehicle cyclocopter. J. Am. Helicopter Soc. **52**(1), 24–35 (2007). https://doi.org/10.4050/JAHS.52.24
4. Hu, Y., Du, F., Zhang, H.L.: Investigation of unsteady aerodynamics effects in cycloidal rotor using RANS solver. Aeronaut. J. **120**(1228), 956–970 (2016). https://doi.org/10.1017/aer.2016.38
5. Schwaiger, M., Wills, D.: D-Dalus VTOL - efficiency increase in forward flight. Aircr. Eng. Aerosp. Technol. **88**(5), 594–604 (2016). https://doi.org/10.1108/AEAT-04-2015-0104
6. McNabb, M.L.: Development of a cycloidal propulsion computer model and comparison with experiment. Master's thesis, Mississippi State University (2001)
7. Farrell, P.E., Maddison, J.R.: Conservative interpolation between volume meshes by local Galerkin projection. Comput. Meth. Appl. Mech. Eng. **200**(1–4), 89–100 (2011). https://doi.org/10.1016/j.cma.2010.07.015
8. Geuzaine, C., Remacle, J.F.: Gmsh: a 3-D finite element mesh generator with built-in pre- and post-processing facilities. Int. J. Numer. Meth. Eng. **79**(11), 1309–1331 (2009). https://doi.org/10.1002/nme.2579
9. Rodrigues, F.F., Habibnia, M., Pascoa, J.: Novel propulsion system for VTOL aircraft based on cycloidal rotors coupled with wings. In: Volume 1: Fluid Applications and Systems: Fluid Measurement and Instrumentation. American Society of Mechanical Engineers, July 2020. https://doi.org/10.1115/fedsm2020-20292

Application of Fan Boundary Condition for Modelling Helicopter Rotors in Vertical Flight

Anna A. Kostek[1,2,3(✉)], Katarzyna Surmacz[1], Michał Rajek[1], and Tomasz Goetzendorf-Grabowski[2]

[1] Łukasiewicz Research Network - Institute of Aviation, Krakowska Av. 110/114, 02-256 Warsaw, Poland
anna.kostek@dlr.de
[2] Warsaw University of Technology, Warsaw, Poland
[3] Now at German Aerospace Center (DLR), Göttingen, Germany

Abstract. The fan boundary condition in Ansys Fluent was applied to a helicopter rotor in vertical flight using a Robinson R22 geometry. This simplified boundary condition, which is based on a pressure jump over an actuator disk, offers considerable advantages in speed and stability compared to methods using a blade element theory. It was shown that the identification of different rotor working states and preparation of the induced velocity curve is possible by applying the analyzed rotor model. Particular attention was given to the prediction of the vortex ring state - a phenomenon that cannot be described using the momentum theory and which poses a significant threat during helicopter flight. The outcomes were comparable to those of experimental visualizations and simulations performed using the more computationally expensive Virtual Blade Model, thereby proving the viability of the fan boundary condition to model the main rotors.

Keywords: Fan boundary condition · Virtual Blade Model · Computational models for main rotors · Helicopter vertical flight · Robinson R22 · Rotor working states · Vortex ring state

1 Introduction

Vertical flight of a helicopter involves its rotor operating in various flow conditions, most of which can be satisfactorily described by means of the momentum theory. However, some of them, like the vortex ring state, are accompanied by particularly complex aerodynamic phenomena, for which the simplified theory is not applicable. At the same time a thorough flow analysis under such conditions is extremely important because of the risks involved. CFD software, alongside with tunnel experiments, provides a tool to investigate complicated vortex structures. However, the choice of computational methods, including modelling of the main rotor, usually has to be a compromise between numerical costs and

© The Author(s), under exclusive license to Springer Nature Switzerland AG 2021
A. Dillmann et al. (Eds.): STAB/DGLR Symposium 2020, NNFM 151, pp. 355–364, 2021.
https://doi.org/10.1007/978-3-030-79561-0_34

the accuracy level of the results. This means, depending on the purpose of the research, appropriate simplifications should be applied.

Choosing the fan boundary condition available in the Ansys Fluent software as a model of a main rotor significantly reduces preparation and calculation time, however, its application is limited only to cases considering the axial flow through the rotor, possibly with a low lateral velocity. The question is, whether, despite of its simplified assumptions, the fan boundary condition is capable of reproducing the variety of phenomena associated with helicopter vertical flight.

2 Methodology

The fan is one of the boundary conditions of the internal surface type, which introduce step changes in flow parameters. The rotor model in this method is presented as an infinitely thin surface of a pressure jump, which allows for an average effect of blade rotation. In the presented analysis a uniform pressure jump across the disk was assumed (Fig. 1). According to [2] wake swirl can be considered in the calculations after determining tangential and radial velocity distribution along the rotor radius (Fig. 2) as polynomial functions (1) and (2) respectively:

$$u_\theta = \sum_{n=-1}^{N} f_n r^n, \quad (1) \qquad u_r = \sum_{n=-1}^{N} g_n r^n, \quad (2)$$

where f_n, g_n are the function coefficients, r is the radial distance from the hub and $-1 \leq \mathrm{N} \leq 6$.

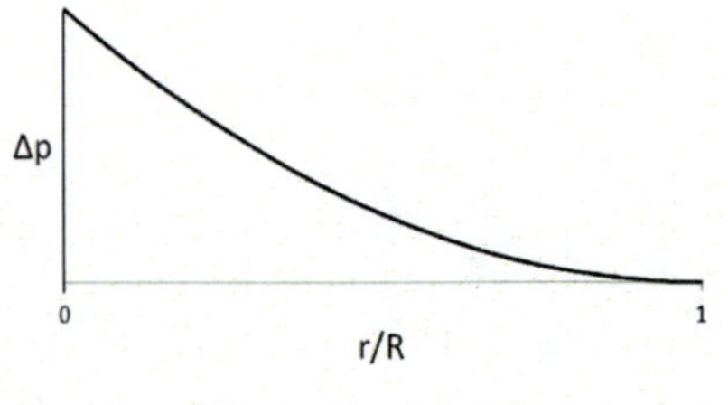

Fig. 1. Uniform pressure jump distribution along the disk radius

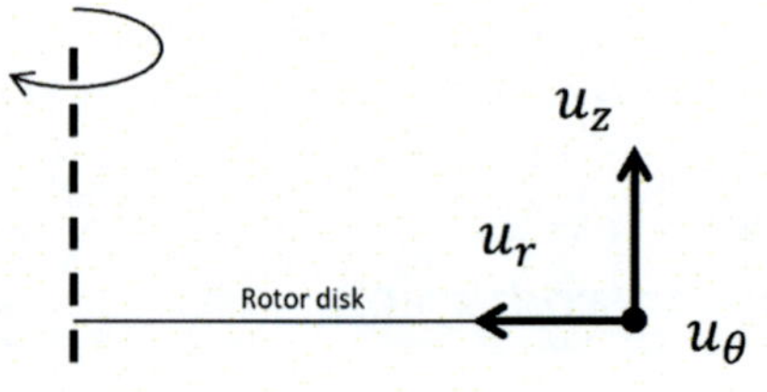

Fig. 2. Velocity components in the applied coordinate system

The Virtual Blade Model (VBM) is a more complex approach based on the blade element theory [13]. In this case the rotor is presented as a single layer of elements (here structural, as shown later on Fig. 5b). The required input includes detailed information on the geometric and mass properties of rotor blades, along with aerodynamic characteristics of profiles, and rotor parameters, e.g. its orientation, rotational speed, collective and cyclic angles. On this basis the time-averaged effects of blades rotation can be simulated using individual cells of the rotor disk as momentum sources, which are then included in Navier-Stokes equations as a vector of external forces.

2.1 Numerical Setup

A model of the Robinson R22 helicopter was reconstructed by means of NX Unigraphics software [12] on the basis of [10] and placed in the middle of a cube-shaped domain with a side length ten times the rotor diameter (Fig. 3). ANSYS Icem software [1] was used to prepare an unstructured tetrahedral grid with layers of prism elements covering the boundary layer around the fuselage (Fig. 4).

Simulations concerned the range of flight velocities V up to 8 m/s for the axial climb and 17 m/s for the axial descent using grids of 1.2 and 3.2 million elements (Mesh I, Mesh II) in a steady mode and additionally with a transient flow for Mesh II. A single-equation Spalart-Allmaras turbulence model was chosen as sufficient for the needs of analysis of large-scale turbulent structures characteristic for vertical flight of the helicopter. The fan boundary condition was set on the main and tail rotor disk surfaces with the constant values of the pressure jump of 107 Pa and 340 Pa respectively, without considering swirl.

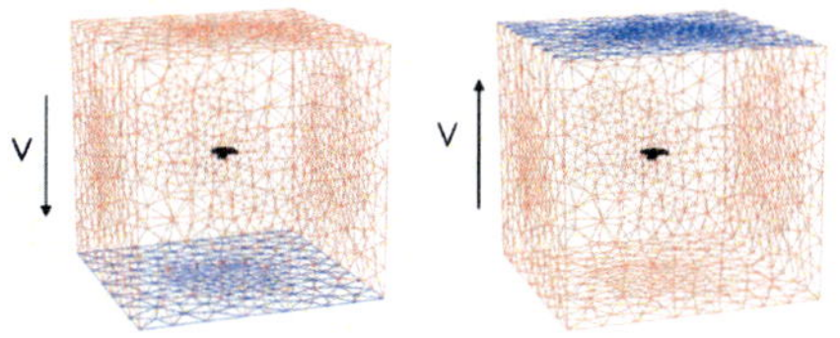

Fig. 3. Computational domain for descent and climb; pink: pressure far-field, blue: pressure outlet

Fig. 4. Grid setup

Wake swirl modelling available in the fan boundary condition was tested for a hover case on the basis of comparison with the VBM results using the same grid setup (Mesh I). The only obvious differences in the grids were related to the rotor model (Fig. 5). In the end, for comparison purposes, the tail rotor operation was not taken into account and a simplified helicopter model with no skids was used.

A function defining the tangential component of the wake swirl velocity in hover was modelled on the basis of two approaches described in [7] - actuator disk vortex theory and momentum theory extended with swirl. The first one assumes for a uniformly loaded rotor that a tangential component of velocity u_θ added to the flow equals the strength of the vorticity Γ:

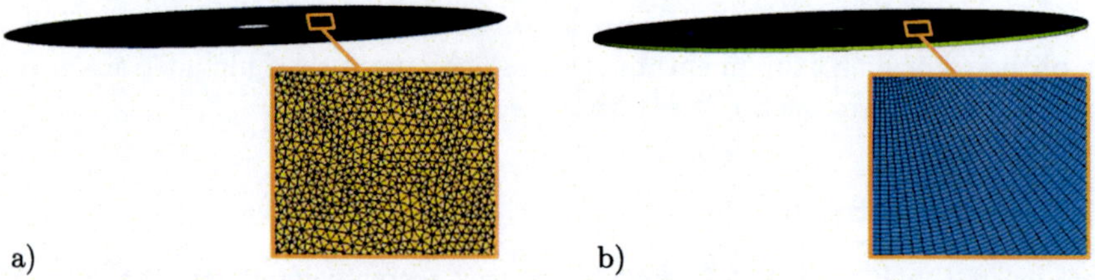

Fig. 5. Main rotor representations; a) fan boundary condition b) Virtual Blade Model

$$u_\theta = \frac{\Gamma}{2\pi r} = \frac{T}{\rho A \omega r}, \tag{3}$$

where T is the thrust force, A the rotor disk area and ω the rotational speed of the rotor. The second equation takes into account the wake swirl arising from induced and profile drag (6), where v_h is the induced velocity in hover and c_l, c_d stand for the lift and drag coefficient respectively.

$$u_\theta = v_h \left[\frac{2 v_h \omega r}{(\omega r)^2 + v_h^2} + 2\frac{c_d}{c_l} \right], \tag{4}$$

Functions $u_{\theta 2}$ and $u_{\theta 3}$, displayed on Fig. 6, are examples of approximations of the tangential velocity distribution according to (3) and (4). The increase in tangential velocity caused by turbulence at the edge of the rotor disk has been considered for the function $u_{\theta 4}$, for which a velocity peak up to the value of v_h was added on the basis of experimental results [3]. For the radial velocity distribution an approximating polynomial function u_{r2} was suggested based on an analytical model [5] (Fig. 7), along with a simple, linear function u_{r1} with the maximum value equal v_h at its end.

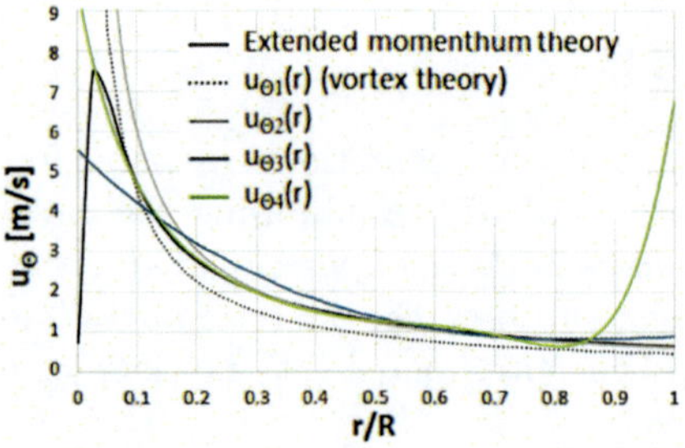

Fig. 6. Functions approximating tangential velocity distribution

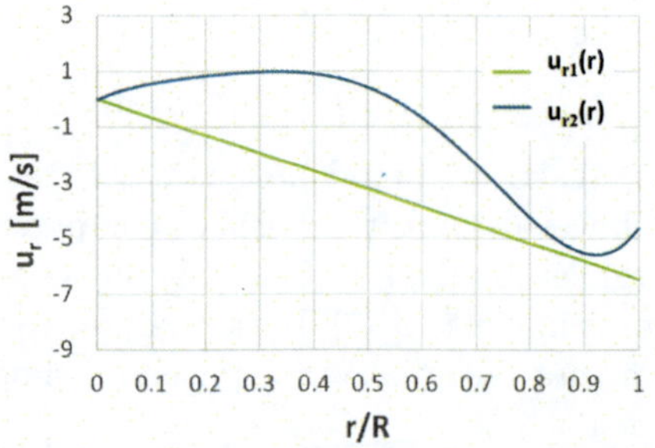

Fig. 7. Functions approximating radial velocity distribution

3 Results

3.1 Identification of Rotor Working States

Figure 8 presents changes in the induced velocity v depending on the vertical flight velocity V obtained from numerical simulations. The effect of these changes is visible on Fig. 9 showing induced power as a function of V according to the momenthum theory (5). The average values of the induced velocity v were calculated from the resulting mass flow rate through the rotor using the equation (6), where ρ is the air density and A is the disk area:

$$\overline{P} = \overline{v} + \overline{V}, \quad (5) \qquad \dot{m} = \rho(v + V)A. \quad (6)$$

Labels $\overline{v}$ and $\overline{V}$ correspond to the values of velocities nondimensionalised by the induced velocity in hover. Similarly, $\overline{P}$ is the power value divided by the power value in hover.

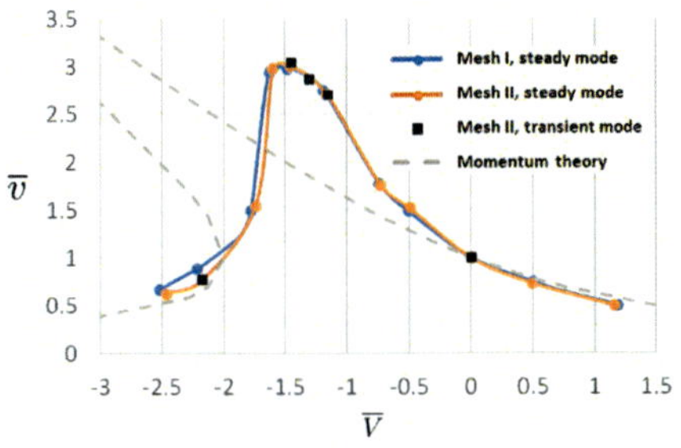

Fig. 8. Induced velocity curve

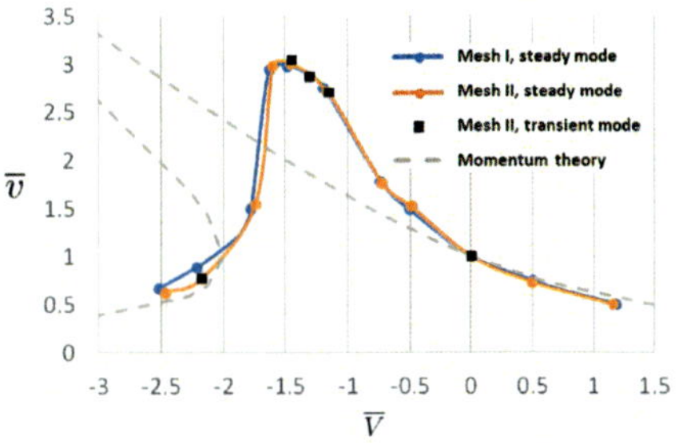

Fig. 9. Power curve

The presented plots show that refinement of the mesh did not influence much the results, so further calculations from Sect. 3.2 were conducted using Mesh I. As expected, for the range of velocities corresponding to the normal working state ($\overline{V} \geq 0$) and windmill brake state ($\overline{V} \leq -2$) obtained results are consistent with the momentum theory [9]. The highest calculated value of the induced velocity is as much as three times higher than the value of the induced velocity in hover and was reached around $\overline{V} = -1.5$. Similar values were also observable during experiments [4, 16]. The resulting loss in induced power causing increased power consumption (Fig. 9) characterizes flight in the vortex ring state. The steep decline in v for higher flight velocities followed by a change to negative power values indicates entering the autorotative flight regime, typical for the turbulent wake state.

The presence of particular flow states in the numerical simulations using the fan boundary condition can be verified closely when compared with the results from experiments [11,14]. Smoke visualizations from the wind tunnel test of a descending helicopter model showed apparent similarities for the corresponding nondimensionalised flight velocity values as presented on Fig. 10.

For low descent speeds, the direction of flow just underneath the rotor is determined by the mass of induced air, resulting in a lack of noticeable circulation on the smoke visualization (Fig. 10a). However, as the descent velocity increases, the coming mass of air moves closer to the rotor disk until it reaches the rotor around $\overline{V} = 1$. From this point on, the increase in flow velocity causes a clear recirculation of air around the rotor, which indicates a fully developed vortex ring state (Fig. 10b). The largest diameter of the vortex with the length of approximately the rotor diameter was observed around the flow velocity $\overline{V} = -1.2$ for both analyzes. The process of the development and vanishing of the vortex ring state was observed in detail by means of PIV (Particle Image Velocimetry) in [14,15]. As a result of the speed increase, the vortex ring starts to shrink and concentrate closer to the rotor disk, while at the same time cores of vortices move gradually over the rotor surface. Similar changes in the vortex ring geometry were obtained from the numerical simulations (Fig. 11).

As the velocity increases further, recirculation weakens and thus the net-flow through the rotor disk is nearly zero ($\overline{V} = -1.7$). Consequently the area of turbulence characteristic for the turbulent wake state appears above the rotor plane (Fig. 10c). For high descent speeds, as the turbulence fades, the flow is again well-defined and does not change significantly with the further increase in velocity. As the flow velocity notably exceeds the induced velocity, smoke reaches far behind the rotor disk. The flow is now directed upwards being fully determined by the mass of coming air (Fig. 10d). The presented visualization reflects the description of the windmill brake state from the momentum theory.

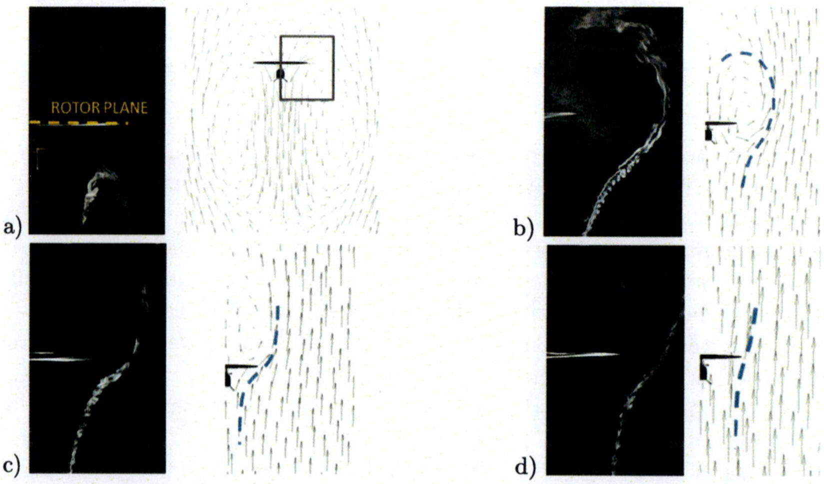

Fig. 10. Comparison between smoke visualizations and numerical analysis of flow around a descending helicopter rotor; a) $\overline{V} = -0.7$ b) $\overline{V} = -1.2$ c) $\overline{V} = -1.7$ d) $\overline{V} = -2.8$

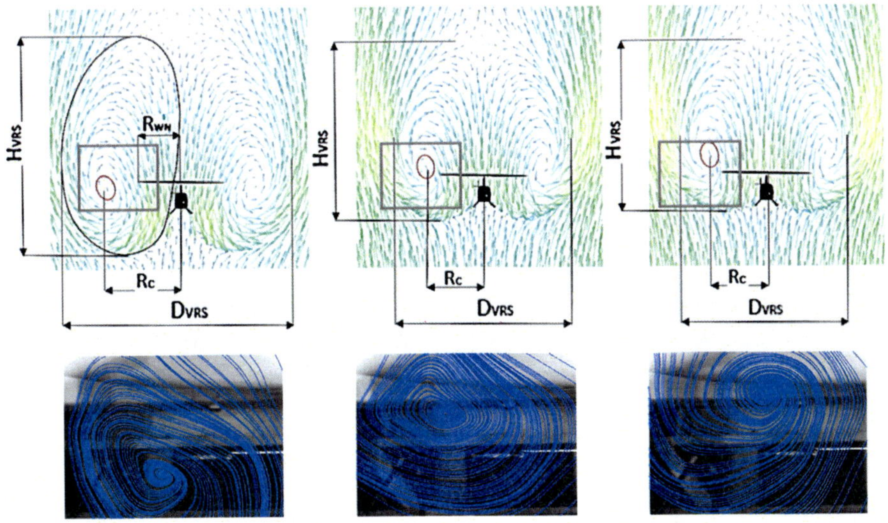

Fig. 11. Change in geometry of vortex ring with increasing descent rate for $\overline{V} = -1.2$, $\overline{V} = -1.3$, $\overline{V} = -1.5$ compared with PIV visualizations, symbols based on [6]

3.2 Wake Swirl in Hover

The obtained distributions of the tangential velocity from approximating functions (Fig. 6) clearly show better agreement with the results produced by means of the VBM than primary settings of the fan (Fig. 12).

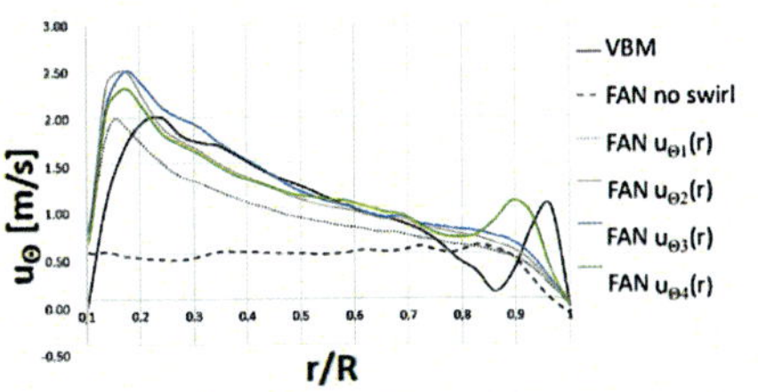

Fig. 12. Tangential velocity distribution obtained from numerical calculations (azimuth 90°)

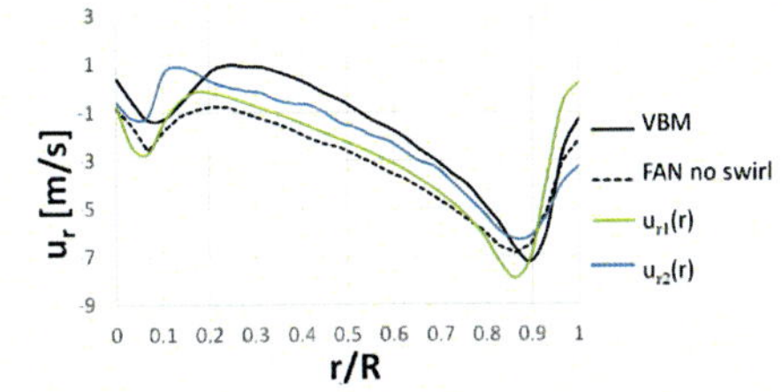

Fig. 13. Radial velocity distribution obtained from numerical calculations (azimuth 90°)

In order to obtain physical results when changing swirl parameters, it is important to define the distribution of tangential and radial velocity components in parallel. After applying function u_{r2} the radial velocity distribution was in better accordance with the VBM results than a primary model without swirl (Fig. 13). However, it also resulted in an incorrect axial velocity distribution, as presented on Fig. 14. More adequate induced velocity values were reached by modelling with a simplified function u_{r1}.

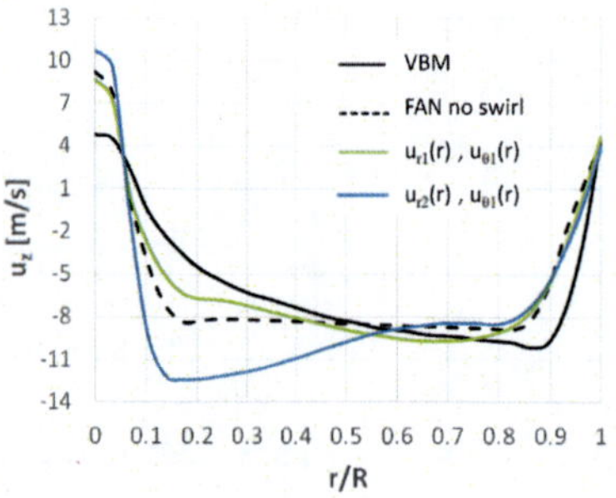

Fig. 14. Axial velocity distribution obtained after adding swirl (azimuth 90°)

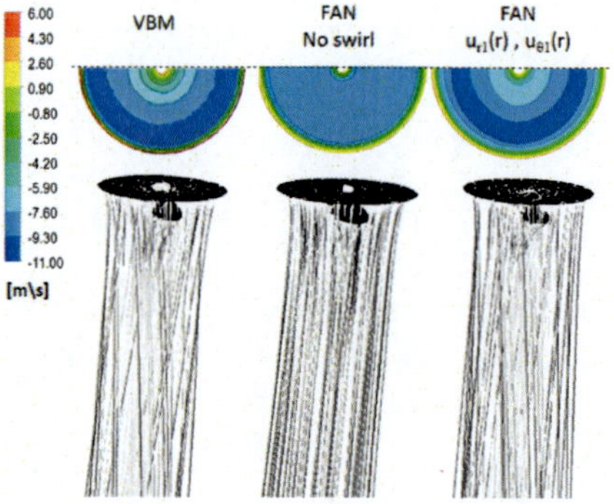

Fig. 15. Comparison of axial velocity contours on the rotor surface and change in wake swirl

4 Discussion and Conclusions

The conducted simulations allow for the identification of particular flow states with the range of their occurrence corresponding to the experimental results. This proves the validity of using the fan boundary condition as a model of a helicopter's main rotor for the purpose of vertical flight simulations. Figure 16 presenting the induced velocity curve additionally includes straight lines, which define theoretical limits of rotor working states during helicopter descent. Presented areas correspond to those obtained from numerical simulations.

Additionally, the application of basic settings of the fan boundary condition enables to observe stages of development of a vortex ring state, which is indescribable on the basis of theoretical approach and thus represents a particularly interesting point of analysis. Results obtained in this way provide the ground for the investigation of threats related to the above phenomenon and the development of escape techniques and prevention methods. However, it should be emphasized that the flow through the rotor in this condition is highly unsteady both in space and time, which means that simulations based on the fan boundary condition offer only a simplified, averaged presentation of the described state. The experiment [14] proved that generated vortex structures periodically expand until they are abruptly detached from the rotor blade. The resulting chaotic nature of the flow leads to fluctuations of the thrust force and high vibrations. As shown on Fig. 8 transient calculations did not significantly change the results compared to less time-consuming stationary analysis. The assumption of a constant value of the pressure jump and consequently a constant thrust force in defining the parameters of the fan boundary condition is a significant limitation for showing non-stationary effects of the rotor flow, regardless of the adopted calculation mode. However, the fan boundary condition settings allow the application of a function describing the pressure jump value depending on the inflow velocity. The example of defining this relation for hover can be found in [8].

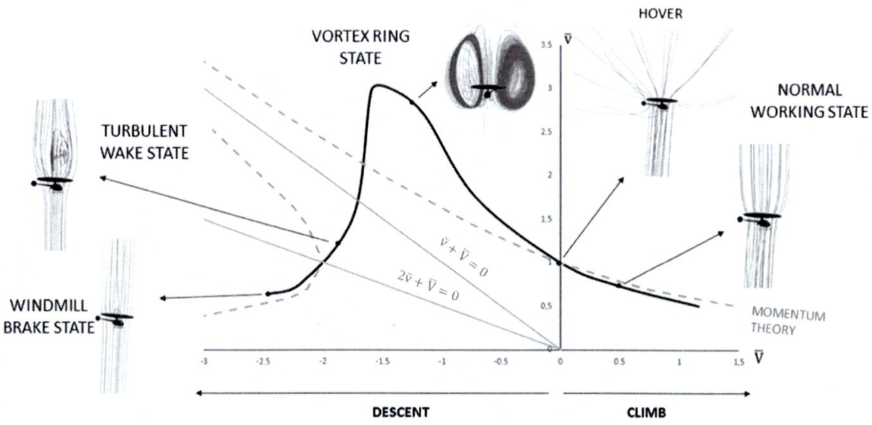

Fig. 16. Identification of main rotor working states based on simulations results

An advantage of the adopted model is its time-effectiveness, when it comes to both numerical costs and the process of preparing the rotor geometry. This is an essential advantage compared to the VBM that requires much input data, high quality grids and calculation times up to 20% longer. The simulation results indicate that the use of additional fan boundary condition settings for wake swirl modelling leads to outputs similar to the VBM model, but does not ensure complete consistency of results. Functions determining the distribution of velocity components along the rotor radius can be defined on the basis of theoretical equations and experimental results. It should be noted, however, that the suggested approximation functions are only applicable for hover, or can be modified for cases within the scope of momentum theory. To derive the above dependencies for the descent in the vortex ring state, if possible, would require much deeper analysis.

When defining the swirl settings, care should be taken with regard to changing associated parameters, as shown on Fig. 14. However, the adjusting of radial and tangential velocity components may allow for an indirect modelling of the axial velocity (Fig. 15). The lack of possibility of azimuthal correction of the introduced functions limits the application of the analyzed model of the main rotor to cases characterized by axisymmetrical flow through the rotor. For this reason, unlike the VBM, the fan boundary condition cannot be used for forward flight simulations. Another natural limitation resulting from adopting the main rotor model as a disk is the inability to simulate effects appearing on individual blades. One of them are changes in the geometry of tip vortices resulting from the increase in axial flight velocity, which were visible on smoke visualizations (Fig. 10). They present an additional feature characteristic for particular flow states, which, however cannot be detected with the use of both the fan boundary condition and the VBM. To summarize, the fan boundary condition adopted as a main rotor model enables to simulate the vertical flight of a helicopter with reduced calculation costs and the obtained results are a good starting point for more precise research.

References

1. ANSYS Inc.: ANSYS ICEM CFD 14.5 User's Manual (2012)
2. ANSYS Inc.: ANSYS Fluent User's Guide (2013)
3. Boatwright, D.: Measurements of velocity components in the wake of a full-scale helicopter rotor in hover. Missisipi State Univeristy (1972)
4. Castles, W., Jr., Gray, R.B.: Empirical relation between induced velocity, thrust, and rate of descent of a helicopter rotor as determined by wind-tunnel tests on four model rotors. NACA-TN-2474 (1951)
5. Conway, J.T.: Exact actuator disk solutions for non-uniform heavy loading and slipstream contraction. J. Fluid Mech. **365**(1), 235–267 (1998)
6. Grzegorczyk, K.: Wykorzystanie metod obliczeniowej mechaniki plynow do modelowania szczegolnego przypadku znizania smiglowca (VRS). Institute of Aviation, Warsaw (2010). [Eng.: Application of CFD methods for modelling a specific case of helicopter descent (VRS)]
7. Johnson, W.: Helicopter Theory. Dover Publications Inc., New York (1980)
8. Kostek, A.A.: Modelowanie wirnikow nosnych za pomoca warunku brzegowego fan na przykladzie smiglowca Robinson R22. M.Sc. thesis, Warsaw University of Technology (2019). [Eng.: Modelling of helicopter main rotors using the fan boundary conditon based on the example of the Robinson R22 helicopter]
9. Leishman, G.: Principles of Helicopter Aerodynamics. Cambridge University Press, Cambridge (2006)
10. Robinson Helicopter Company: R22 Maintenance Manual (2018)
11. Surmacz, K.: Dynamika zjawiska pierscienia wirowego na wirniku nosnym smiglowca. Ph.D. thesis, Air Force Institute of Technology, Warsaw (2015). [Eng.: Dynamics of the vortex ring state on the helicopter main rotor]
12. Siemens Documentation, NX 12 (2017)
13. Stalewski, W., Surmacz, K.: Investigations of the vortex ring state on a helicopter main rotor based on computational methodology using URANS solver. Lukasiewicz Research Network - Institute of Aviation, Warsaw (2019)
14. Stryczniewicz, W., Surmacz, K.: Badania eksperymentalne stanu pierscienia wirowego na wirniku nosnym smiglowca metoda anemometrii obrazowej (PIV). Institute of Aviation, Warsaw (2014). [Eng.: Experimental studies on the vortex ring state on helicopter main rotor using Particle Image Velocimetry (PIV)]
15. Surmacz, K., Ruchała, P., Stryczniewicz, W.: Wind tunnel tests of the development and demise of Vortex Ring State of the rotor. Institute of Aviation, Warsaw (2014)
16. Washizu, K., Azuma, A., Koo, J., Oka, T.: Experiments on a model helicopter rotor operating in the vortex ring state. J. Aircr. **3**(3), 225–230 (1966)

Efficient Aerodynamic and Aeroacustic Optimization of Propeller Sections Using Bayesian Methods

Andreas Kümmel(✉) and Christian Breitsamter

Chair of Aerodynamics and Fluid Mechanics, Technical University of Munich, Boltzmannstr. 15, 85748 Garching, Germany
andreas.kuemmel@tum.de

Abstract. Due to various design variables, like the blade shape and operating conditions, the optimization of propellers is a complex task. The aerodynamic and aeroacoustic design optimization is conducted by using an efficient machine learning optimization approach, the Bayesian optimization. On the one hand, this allows for a fast and efficient optimization process. On the other hand, the knowledge about the optimization problem is extend and the influence of the individual design variables can be identified. Exemplarily, the radial airfoil sections of the propeller blade are optimized regarding the glide ratio and the aeroacoustic footprint.

Keywords: Bayesian optimization · Propeller design · Aeroacoustics

1 Introduction

The upcoming trend of urban air mobility (UAM) creates new points of contact between aviation and society. This results in new challenges, which directly influence the objectives for new aircraft. For example, the societal demand for emission-free, green transportation is higher than ever. As a result, new propulsion systems like electrical propulsion are moving into focus. In the case of UAM, propellers are mostly used as part of the propulsion system [1]. They have to generate the required thrust and eventually lift. Therefore, the aerodynamic efficiency of the propellers significantly determines the performance of the aircraft [2]. Efficient propellers ensure high endurance and cruise range.

Furthermore, the aeroacoustic footprint of a propeller is determining the acceptance of new flight configurations in urban areas. A major contributor to the noise emissions of an UAM are the propellers. Propeller noise is composed by a tonal and a broadband part. The tonal noise is caused by the displacement effect and the thrust generation of the propeller blades. In addition, blade vortex interaction effects (BVI) also contribute to the tonal part. The broadband part is determined by the airfoil self noise [3].

In order to provide low noise emission and aerodynamic efficient propellers, an optimization of the propeller blade shape can be conducted. Propellers pose

© The Author(s), under exclusive license to Springer Nature Switzerland AG 2021
A. Dillmann et al. (Eds.): STAB/DGLR Symposium 2020, NNFM 151, pp. 365–375, 2021.
https://doi.org/10.1007/978-3-030-79561-0_35

a challenge for the optimization process due to the variety of design variables. Within the presented work, only the radial airfoil sections of the propeller blade are considered for demonstration purposes. They have a significant influence on the aerodynamic efficiency of the propeller and the broadband aeroacoustic noise. Due to the high level of complexity and the various design variables an efficient optimization approach is applied. A machine learning based Bayesian optimization algorithm is used, to reduce the calculation time and effort. Other optimization methods are genetic algorithms and gradient based optimization approaches. Bayesian Optimization is most commonly used for the hyperparameter tuning of deep neural networks [4], but gets more popular for other disciplines like robotics or shape optimization [5]. Pure aerodynamic shape optimizations of airfoils using a BO approach are described in [6,7]. The position of an flap is optimized in [6], whereas in [7] the focus is on the design of a natural laminar wing airfoil.

2 Methodology

In the following section, the used methods are presented. The aeroacoustic calculation and the Bayesian optimization approach are described in more detail.

2.1 Parametrization

For the airfoil shape optimization a parameterized geometry model is needed. Therefore, Bézier curves of 4th order for the upper and lower side of the airfoil are used [8]. This allows for high flexibility regarding the airfoil design, while keeping the number of design variables small. Furthermore, through the parameterization with Bézier curves a smooth airfoil contour is ensured. Figure 1 shows the parameterized airfoil model with its design variables and the bounds of each design variable (grey areas). The design vector $\mathbf{Z}$ is composed by the coordinates of the Bézier control points. The x-direction for P_1 and P_4 are fixed, because they define the leading edge of the airfoil.
$\mathbf{Z} = (P_{1,y}, P_{2,x}, P_{2,y}, P_{3,x}, P_{3,y}, P_{4,y}, P_{5,x}, P_{5,y}, P_{6,x}, P_{6,y})$.

2.2 Airfoil-Self Noise

A major contributor to the broadband noise of propellers is the airfoil self noise. The single contributors to airfoil self noise are described in detail in Brooks et al. [9]. For propellers the turbulent boundary layer trailing edge (TBL-TE) noise is important: Different calculation methods are available for the calculation of TBL-TE noise emissions [10]. They can be split in high and low fidelity calculation methods. For the high fidelity calculations, mostly CFD simulations provide the necessary flow data. Whereas, for the low fidelity approaches, empirical formulas or potential flow solvers are used for the calculation of the needed flow data.

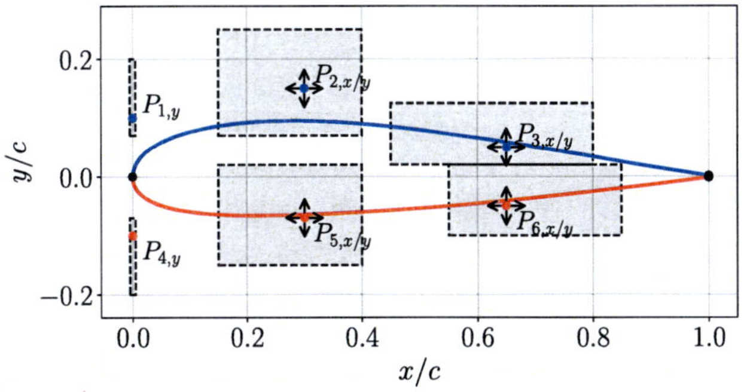

Fig. 1. Parameterized airfoil model with the design variables $P_{i,j}$ and the respective bounds.

Within the low fidelity approaches, the method by Brooks, Pope and Marcolini (BPM) [9], a modified BPM method [10] and the TNO method [11] are most commonly used. In the presented work, the original BPM method is used for the calculation of the TBL-TE noise, because of the fast calculation procedure. The basic equations are described in the following: There are three different mechanisms which contribute to the overall sound pressure level SPL_{TBL-TE}:

$$SPL_{TBL-TE} = 10\log(10^{SPL_\alpha} + 10^{SPL_s} + 10^{SPL_p}) \tag{1}$$

SPL_α is the contribution of the angle of attack α to the emitted noise, Eq. 2. SPL_s and SPL_p are the parts accounting for the turbulent boundary layer on the suction and the pressure side, Eqs. 3 and 4:

$$SPL_\alpha = 10\log(\frac{\delta_p^* M^5\, LD_h}{r_e^2}) + B(\frac{St_s}{St_2}) + K_2 \tag{2}$$

$$SPL_s = 10\log(\frac{\delta_s^* M^5\, LD_h}{r_e^2}) + A(\frac{St_s}{St_1}) + (K_1 - 3) \tag{3}$$

$$SPL_p = 10\log(\frac{\delta_p^* M^5\, LD_h}{r_e^2}) + A(\frac{St_p}{St_1}) + (K_1 - 3) + \Delta K_1 \tag{4}$$

δ_s^* and δ_p^* are the displacement thicknesses of the turbulent boundary layer at the trailing edge on the suction and pressure side, respectively. M is the Mach number of the free stream and L the span of the investigated section. D_h is a directivity term and r_e the observer distance. A and B are spectral shape functions in dependence of a Strouhal number ratio for the TBL-TE and separation noise, respectively. St_1 and St_2 are Strouhal numbers calculated with the free stream Mach number. St_s and St_p are the Strouhal numbers for the displacement thicknesses on the suction and pressure side. The definitions of the different Strouhal numbers St are given in [9]. $K1$, $K2$ and ΔK are experimentally determined constants. Equations 2–4 show a dependency of the emitted noise to the displacement thickness δ^*. Generally, a smaller δ^* reduces the TBL-TE noise.

2.3 Bayesian Optimization

The variety of design variables during a propeller optimization demands for a fast and efficient optimization algorithm to reduce the number of function evaluations and optimization time. Therefore, a machine learning approach based on a Bayesian optimization (BO) algorithm is used. The BO is a global optimization method, best suitable for computationally expensive black box functions. Additionally, no knowledge of derivatives is needed. Therefore, BO is well suited for this optimization problem [12].

The principles of BO are based on the combination of Gaussian processes (GPs) to create a statistical model of the optimization problem and a suitable acquisition function to determine where to evaluate the next sample. The basic steps to create a GP are [13]: First, a probability model, the so called prior, is defined. The prior expresses the probability of the belief. Second, a posterior is calculated by means of the prior and data observations, Eq. 5. And third, the probability model has to be fitted and evaluated. For BO, these three steps are repeated and the probability model is updated for each optimization iteration. The posterior becomes the prior in the next optimization loop.

$$posterior = \frac{likelihood \times prior}{marginal\ likelihood} \tag{5}$$

The prior is composed by a mean function m and a covariance function κ, the so called kernel, see Eq. 6. A detailed description about GPs can be found in [13–15]. General information about machine learning is given in [4].

$$f(x) \sim \mathcal{GP}(m(x), \kappa(x, x')) \tag{6}$$

Within BO, the acquisition function is responsible for determining the location of the data observations. The location of the next sample is chosen to maximize the acquisition function in each optimization iteration. For the presented work, the maximum probability of improvement (MPI) acquisition function is used.

3 Optimization Problem

In a first step, the radial airfoil sections of a propeller blade are purely aerodynamically optimized. The aim is to maximize the glide-ratio ε_{max}. Exemplarily, the radial airfoil section for $r/R = 0.7$ is presented in the following. This section is chosen for the investigation, because at this radius the major part of the propeller thrust is produced and, therefore, it significantly determines the efficiency of the propeller. In a second step, the trailing edge airfoil-self-noise SPL_{TBL-TE} is considered in addition. The objective function is composed by the maximum glide-ratio ε_{max} and the sound pressure level SPL_{TBL-TE}. Additionally, structural constraints must be fulfilled. The different airfoil sections have to withstand the inertial loads and the resulting stress σ evoked by the propeller rotation.

Table 1. Propeller specifications and operating conditions

Propeller radius	0.7	m
Number of blades	3	
Design thrust	300	N
RPM	2000	rev/min
Flight velocity	30	m/s
Chord at $r/R = 0.7$	0.05	m

Thus, the optimization problem for the pure aerodynamic optimzation is given by Eqs. 7 and 8:

$$\text{maximize:} \qquad f(\mathbf{Z}) = \varepsilon_{max} \tag{7}$$
$$\text{subject to:} \qquad g(\mathbf{Z}) \leq \sigma_{max} \tag{8}$$

For the coupled aerodynamic-aeroacoustic optimization the optimization problem is given by Eqs. 9 and 10:

$$\text{minimize:} \qquad f(\mathbf{Z}) = w_1 \cdot \varepsilon_{max,norm} + w_2 \cdot SPL_{TBL-TE,norm} \tag{9}$$
$$\text{subject to:} \qquad g(\mathbf{Z}) \leq \sigma_{max} \tag{10}$$

w_1 and w_2 are used to weigh the individual objective functions. $\varepsilon_{max,norm}$ and $SPL_{TBL-TE,norm}$ are the normalized maximal glide-ratio and the normalized sound pressure level, respectively. The propeller specifications and section properties as well as the operating conditions are shown in Table 1.

GPyOpt is used as the optimizer. *GPyOpt* is a Bayesian optimization framework implemented in Python, which is capable of handling constraints [12]. For the aerodynamic investigations *Xfoil* is used. *Xfoil* is a potential flow solver coupled with an integral boundary layer method suitable for low Reynolds number airfoils [16]. Angle of attack polars in the range of $\alpha = [0° - 10°]$ with a resolution of $\Delta\alpha = 0.5°$ are calculated for each airfoil. The individual objective functions are evaluated at the optimal angle of attack $\alpha_{opt} = \alpha(\varepsilon_{max})$.

4 Results

In the following section, the results for the pure aerodynamic optimization and the coupled aerodynamic-aeroacoustic optimization are presented.

4.1 Pure Aerodynamic Optimization

In Fig. 2a, the convergence of the optimization process is illustrated by the objective function ε_{max}. For the initial 20 candidates, the maximum value is below $\varepsilon_{max} = 60$. Thereafter, an increase of the maximum glide-ratio ε_{max} can be detected. This is a result of the Gaussian process and the created statistical surrogate model. Thereby, the optimal solution can be found in a very efficient manner. Here, the best candidate is found after approximately 50 function evaluations. The design of the best candidate is shown in Fig. 2b. The maximum glide-ratio is $\varepsilon_{max} = 74.24$ at an angle of attack of $\alpha_{opt} = 7.5°$. In comparison, the NACA 0012 airfoil has a maximum glide-ratio of $\varepsilon_{max} = 45.08$ at $\alpha_{opt} = 6.5°$.

Figure 3 illustrates the evaluation of the created statistical surrogate model by slices through the optimal solution. The objective function is shown along the four most significant directions, thus along the coordinates of the Bézier control points with the highest influence on the objective function. The most significant directions are all aligned with the y-direction and directly determine the thickness of the airfoil. The directions in order of the most pronounced influence are the following: $P_{2,y}$,$P_{3,y}$,$P_{1,y}$ and $P_{4,y}$. Figure 3 shows the mean predicted function by the Gaussian process as well as the confidence bound for the prediction. The mean is illustrated by the black line, whereas the blue area defines the confidence bound for the interval of $\pm 1.96\ \sigma_{\mathcal{GP}}$. Additionally, Fig. 3 shows observations for each direction, depicted as red dots. The observations are used to validate the created surrogate model. It can be seen, that the predicted model fits well to the observation data. Especially, near the maximum glide-ratio ε_{max}, the prediction for the mean is very well. For observations further away from the optimum, the mean is marginally underpredicted. All observer points for all directions are within the region of 95% confidence. Furthermore, the limits for the airfoil design regarding the structural constraints can be seen in Fig. 3a and 3b. According to the surrogate model, higher glide-ratios are possible for more slender airfoils. However, the applied structural constraints limit airfoils with even smaller area. The structural constraints are violated for $P_{2,y} < 0.14$ and $P_{3,y} < 0.075$.

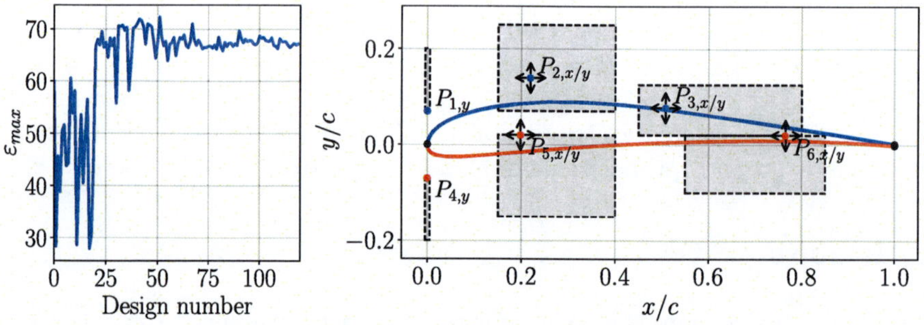

(a) Convergence process (b) Optimized airfoil shape

Fig. 2. Results of the pure aerodynamic optimization, including the convergence of the objective function and the optimized airfoil shape.

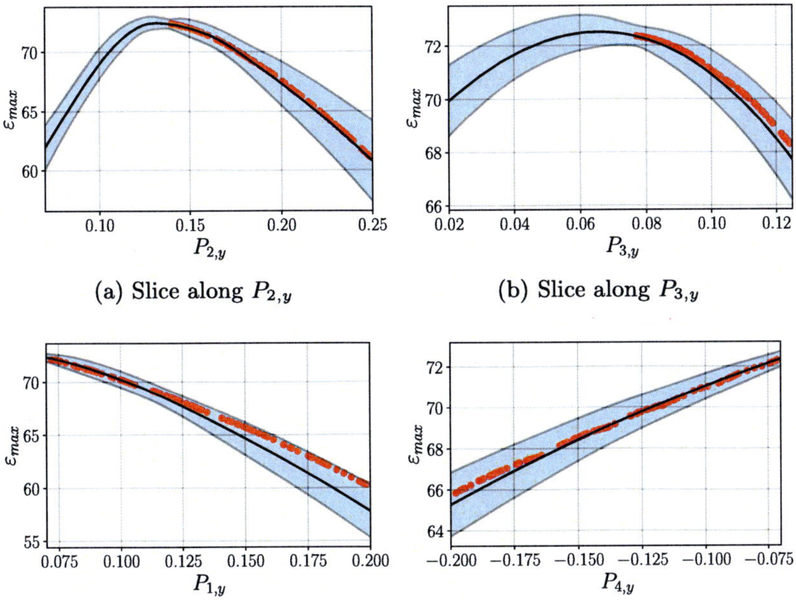

(a) Slice along $P_{2,y}$ (b) Slice along $P_{3,y}$

(c) Slice along $P_{1,y}$ (d) Slice along $P_{4,y}$

Fig. 3. Evaluation of the created surrogate model for the pure aerodynamic optimization, illustrated by slices through the location of the best design. The objective function over the most significant directions is shown (black: mean of predicted objective function, blue: confidence bound and red: observer data for the validation).

4.2 Coupled Aerodynamic and Aeroacoustic Optimization

For the coupled aerodynamic-aeroacoustic optimization, different weights for w_1 and w_2 have been investigated. The aerodynamics is more heavily weighted than the aeroacoustics, to ensure a high glide ratio while reducing the emitted noise. The weights have been set to $w_1 = 0.63$ and $w_2 = 0.37$ for the calculations.

In Fig. 4a, the convergence of the objective function is illustrated. The same trend as for the pure aerodynamic optimization can be seen. Within the initial observations, the returned value of the objective function is well above $f(\mathbf{Z}) = 0.55$. After the first optimization loop, a decrease in the objective function can be detected. The best candidate is evaluated after approximately 80 designs. The design of the best candidate is shown in Fig. 4b. The maximum glide-ratio is $\varepsilon_{max} = 67.0$ at an angle of attack of $\alpha_{opt} = 6.0°$. Compared to the pure aerodynamic optimization ε_{max} decreases by $\Delta\varepsilon_{max} = 9.75\%$. The sound pressure level of the best candidate is $SPL_{TBL-TE} = 72.4$ dB. For the best candidate of the pure aerodynamic optimization the emitted noise is $SPL_{TBL-TE} = 74.2$ dB. Thus, a reduction of $\Delta SPL = 1.8$ dB is achieved. Due to the logarithmic dB-scale, this corresponds to approximately 66.7% of the radiated sound pressure with respect to the purely aerodynamically optimized

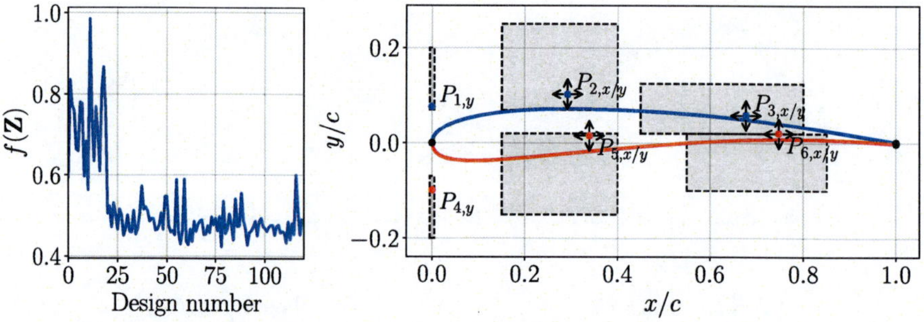

(a) Convergence process (b) Optimized airfoil shape

Fig. 4. Results of the coupled aerodynamic-aeroacoustic optimization, including the convergence of the objective function and the optimized airfoil shape.

airfoil. In comparison, the NACA 0012 airfoil emits $SPL_{TBL-TE} = 75.85$ dB. A summary of the aerodynamic airfoil characteristics is given in Tab. 2.

As already described for the pure aerodynamic optimization, Fig. 5 shows the evaluation of the created statistical surrogate model by slices through the optimal solution. The model is sliced along the four most significant directions, thus along the coordinates of the Bézier control points with the highest influence on the objective function. For the most significant directions, the same behavior can be detected as for the pure aerodynamic optimization. The y-directions of the Bézier control points $P_{3,y}$, $P_{2,y}$, $P_{1,y}$ and $P_{6,y}$ have the highest influence on the objective function. The order of the control points is according to the influence on the objective function. The predicted surrogate model, black line and blue area, fits well to the observations data, red points. The observation data for the coupled aerodynamic-aeroacoustic optimization shows fluctuations, which are a result of the empirical noise model and the step size of $\Delta\alpha = 0.5°$ for the polars. Still, the major part of the observation data is within the region of the 95% confidence bound. Near the optimal solution, the predicted mean value of the objective function is in very good accordance to the observation points. The limits due to the structural constraints can bee seen in Fig. 5a and 5b. No valid designs are created for $P_{3,y} < 0.059$ and $P_{2,y} < 0.1$. The resulting airfoil shape has a bigger nose radius, compared to the purely aerodynamically optimized airfoil. Furthermore, the camber of the aerodynamic-aeroacoustic optimized airfoil is smaller.

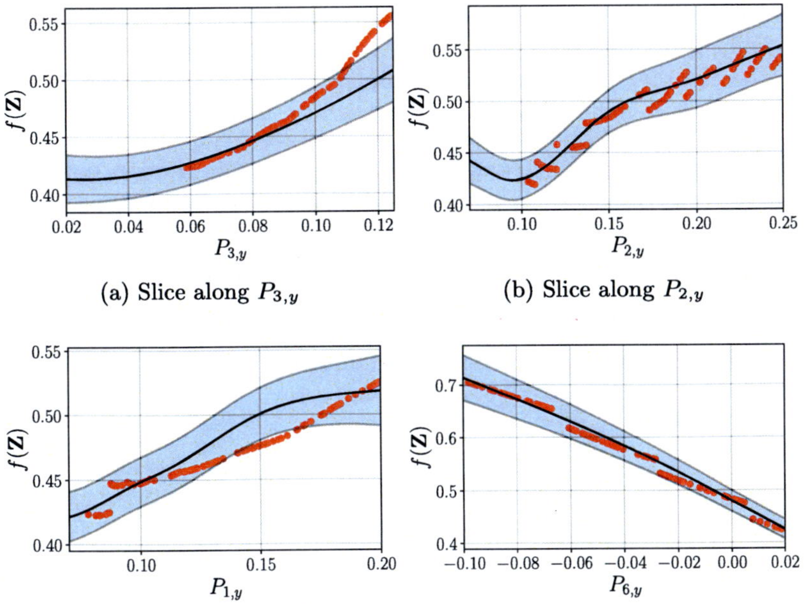

(a) Slice along $P_{3,y}$

(b) Slice along $P_{2,y}$

(c) Slice along $P_{1,y}$

(d) Slice along $P_{6,y}$

Fig. 5. Evaluation of the created surrogate model for the coupled aerodynamic-aeroacoustic optimization, illustrated by slices through the location of the best design. The objective function over the most significant directions is shown (black: mean of predicted objective function, blue: confidence bound and red: observer data for the validation).

Table 2. Aerodynamic characteristics of the optimized airfoils and the reference airfoil NACA 0012.

Airfoil	α_{opt}	ε_{max}	SPL_{TBL-TE}
NACA 0012	6.5°	45.08	75.85
Aerodynamic optimization	7.5°	74.24	74.24
Coupled optimization	6.0°	67.0	72.4

5 Conclusion and Outlook

Within this work, a fast and efficient airfoil optimization by means of a machine learning based Bayesian optimization algorithm is presented. It is possible to optimize airfoils regarding aerodynamic and aeroacoustic characteristics. The main advantage of the applied optimization approach is the efficient calculation of the optimum. Compared to a genetic algorithm the number of function evaluations for the optimization is approximately one order of magnitude lower. Thus, the calculation effort can be reduced and a simultaneous airfoil optimization in connection with a propeller optimization can be performed. This enables a more holistic propeller investigation and thus, a more detailed propeller analysis. Additionally, the Gaussian process delivers reliable results regarding the surrogate model and the prediction of the airfoil performance. This allows to analyze the influence of the different optimization parameters. Thereby, the knowledge and the understanding of the optimization problem can be expanded.

In a next step, the optimization approach will be applied to high fidelity CFD simulations to further improve the results. More accurate observation data will simplify the creation of the surrogate model by the Gaussian process and thus, less function evaluations for the optimization are necessary.

References

1. Silva, C., Johnson, W.R., Solis, E., Patterson, M.D., Antcliff, K.R.: VTOL urban air mobility concept vehicles for technology development. In: Aviation Technology, Integration, and Operations Conference, Atlanta, 25–29 June 2018
2. Hepperle, M.: Electric Flight - Potential and Limitations, Energy Efficient Technologies and Concepts Operations, Lisbon, 22–24 October 2012
3. Vieira, A., Cruz, L., Lau, F., Mortagua, J., and Santos, R.: A new computational framework for UAV quadrotor noise prediction. In: 5th CEAS Air & Space Conference, Delft, 7–11 September 2015
4. Bishop, C.M.: Pattern Recognition and Machine Learning. Springer, Heidelberg (2006)
5. Shahriari, B., Swersky, K., Wang, Z., Adams, R.P., de Freitas, N.: Taking the human out of the loop: a review of Bayesian optimization. Proc. IEEE **104**(1), 148–175 (2016). https://doi.org/10.1109/JPROC.2015.2494218
6. Jeong, S., Murayama, M., Yamamoto, K.: Robust design of transonic natural laminar flow wings under environmental and operational uncertainties. J. Aircraft **42**(2), 413–420 (2005)
7. Sabater, C., Bekemeyer, P., Görtz, S.: Robust design of transonic natural laminar flow wings under environmental and operational uncertainties. In: AIAA Scitech 2021 Forum, Virtual Event, 19–21 January 2021
8. Kümmel, A., Stuhlpfarrer, M., Pölzlbauer, P., Breitsamter, C.: Propeller blade shape optimization with a hybrid BEMT/CFD approach. In: Dillmann, A., Heller, G., Krämer E., Wagner, C., Tropea, C., Jakirlić, S. (eds.) New Results in Numerical and Experimental Fluid Mechanics XII, p. 362–371, Springer, Cham (2019). https://doi.org/10.1007/978-3-030-25253-3_35
9. Brooks, T.F., Pope, D.S., and Marcolini, M.A.: Airfoil self-noise and prediction. In: NASA Reference Publication, NASA-RP-1218, Langley Research Center (1989)
10. Lau, A.S.H., Kim, J.W., Hurault, J., Vronsky, T.: A study on the prediction of aerofoil trailing-edge noise for wind-turbine applications. Wind Energy **20**, 233–252 (2017). https://doi.org/10.1002/we.2003
11. Parchen, R.R.: Progress report draw: a prediction scheme for trailing edge noise based on detailed boundary layer characteristics. Report TNO Institute of Applied Physics. Delft, The Netherlands (1998)
12. The GPyOpt authors: GPyOpt: A Bayesian Optimization framework in Python (2016). http://github.com/SheffieldML/GpyOpt
13. Gelman, A., Carlin, J.B., Stern, H.S., Rubin, D.B.: Bayesian Data Analysis, 3rd ed. Chapmann and Hall/CRC, Boca Raton (2014)

14. Murphy, K.P.: Machine Learning: A Probabilistic Perspective. The MIT Press, Cambridge (2012)
15. Rasmussen, C.E., Williams, C.K.I.: Gaussian Processes for Machine Learning. The MIT Press, Cambridge (2006)
16. Drela, M.: XFOIL: an analysis and design system for low reynolds number airfoils. In: Mueller, T.J. (ed.) Low Reynolds Number Aerodynamics, Lecture Notes in Engineering, vol. 54, pp. 1–12. Springer, Heidelberg (1989). https://doi.org/10.1007/978-3-642-84010-4_1

Check for updates

Approximate Boundary Layer Methods for a Fast Mid-Fidelity Aerodynamics Code for Helicopter Simulations

Philipp Kunze(✉)

German Aerospace Center (DLR), Institute of Aerodynamics and Flow Technology, Lilienthalplatz 7, 38108 Braunschweig, Germany
Philipp.Kunze@dlr.de

Abstract. In order to estimate boundary layer properties and flow separation within the mid-fidelity aerodynamics code UPM, fast, simple, and robust methods are required. Following a review of suitable methods, two approaches were selected and implemented: stripwise analysis using integral boundary layer methods for lifting surfaces and a simplified analysis based on flat-plate analogy for arbitrary non-lifting bodies. The implemented methods were validated for airfoil, wing, fuselage, and rotor simulations. The results show that the implemented methods accomplish the original purpose of estimating friction forces for bodies in attached flows and detecting regions probably dominated by flow separations. However, they neither model detached flows nor predict their impact on body forces and moments.

Keywords: 3D panel code · Boundary layer analysis · Helicopters

1 Introduction

With continuously increasing computing power the use of mid-fidelity methods is becoming feasible in ever earlier phases of helicopter development. One such method is DLR's Unsteady Panel Method (UPM) [19] with free-wake modelling. It allows for capturing many interaction effects which cannot be captured by lower fidelity aerodynamics models commonly used in comprehensive rotorcraft codes. However, the panel method, which is used to model aerodynamic bodies in UPM, does not consider viscous effects. As these effects play an important role in many flight conditions, UPM is extended here by approximate boundary layer (BL) methods able to estimate these effects. The selection and implementation of these methods are sketched and exemplary results are presented in this paper.

As a first step, various suitable analysis methods were compared regarding robustness, computational effort, implementation complexity, and user-friendliness. Finally, two approaches were selected and implemented in UPM. The first one is intended for lifting surfaces and based on stripwise analysis along spanwise/radial segments using laminar and turbulent integral BL methods.

© The Author(s), under exclusive license to Springer Nature Switzerland AG 2021
A. Dillmann et al. (Eds.): STAB/DGLR Symposium 2020, NNFM 151, pp. 376–386, 2021.
https://doi.org/10.1007/978-3-030-79561-0_36

A similar streamline-based approach for arbitrary non-lifting bodies deemed too complex and error-prone. Thus, a simplified BL analysis method based on local flow properties and flat-plate analogy was implemented and combined with simple flow separation criteria in order to predict the occurrence of flow separations on arbitrary bodies.

The newly implemented approximate BL methods were tested and compared for airfoil, wing, rotor, and fuselage test cases (Fig. 1).

2 Computational Method

2.1 Unsteady Panel and Free-Wake Method

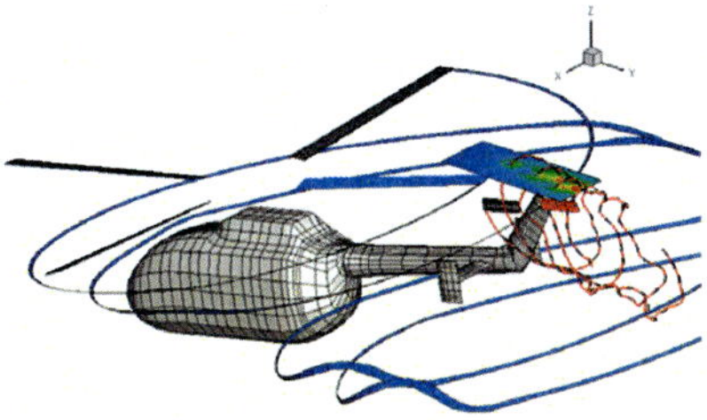

Fig. 1. Examplary UPM simulation [20]

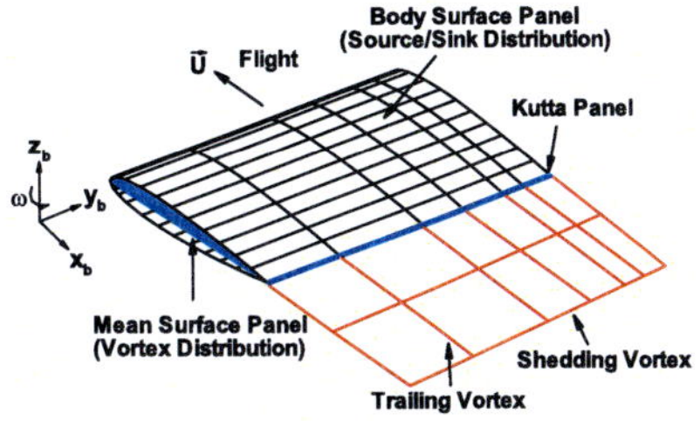

Fig. 2. UPM lifting surface model [20]

UPM uses a velocity based potential flow formulation with Neumann boundary conditions, similar to the classical Hess and Smith [8] panel method. Arbitrary non-lifting bodies are modeled using quadrilateral and triangular panels with constant source/sink distribution. Lifting surfaces consist of quadrilateral panels arranged in a single structured surface patch per wing or rotor blade. In order to account for lift generation, the source panels are superposed by a vortex lattice on the mean surface of the lifting surfaces (Fig. 2). At the trailing edge, a full-span free-wake vortex sheet consisting of constant strength vortex rings is emitted and transported through the flow-field in a force-free way, simulating the unsteady wake rollup. A second-order Adams-Bashforth method is used for temporal integration. The velocities induced by the wake are optionally evaluated using a tip-vortex rollup model. Besides the classical Kutta condition, an iterative pressure Kutta condition is implemented, which enforces equal pressure at the trailing edge panels of the upper and lower sides and thus takes unsteady pressure terms into account when solving the flow around lifting surfaces.

2.2 Approximate Boundary Layer Analysis

Sectional Boundary Layer Analysis. For lifting surfaces, stripwise BL analysis using integral methods is employed to calculate the chordwise and spanwise BL properties. For each segment, the analysis starts with the detection of the stagnation point. Then, the segment is divided into a lower and an upper strip, each starting at the stagnation point and ending at the trailing edge. Next, input data required by the integral BL methods are extracted along both strips. Using this information, an integral BL analysis is performed for each strip. It consists of the following steps:

1. *Laminar Analysis.* The BL analysis starts by employing one of three implemented incompressible integral methods for laminar flow: Schlichting [11], Eppler [4], or Thwaites [15] with Curle's modifications [2]. At the end of each integration step, BL transition criteria are checked. The integration stops if transition is predicted, a condition for laminar flow separation is true or the end of the streamline is reached.
2. *Laminar-Turbulent Transition.* Three transition mechanisms are implemented and may trigger the switch from laminar to turbulent analysis:
   (a) Forced transition at a prescribed location x_{tr}/c relative to chord c.
   (b) Turbulent reattachment after laminar flow separation.
   (c) Transition due to Tollmien-Schlichting (TS) instabilities.
   All empirical criteria for determining TS transition given in [7] were implemented. The Arnal e^n approximation by Würz [18] is used for all calculations in this paper, whith $n_{crit} = 9.0$ if not otherwise stated.
3. *Turbulent Analysis.* If any transition criterion triggers, the BL analysis is continued using one of three implemented integral methods for turbulent flow: Truckenbrodt [16,17], Eppler [4], or Nash/Hicks [9]. It stops if a condition for turbulent flow separation evaluates true or if the end of the strip is reached. If separation is predicted, the BL properties are assumed to be constant until the end of the strip, the skin friction is set to zero.
4. *Map BL Results to Surface.* The integral BL properties determined along the strips in the previous steps are mapped back to the surface.
5. *Calculation of Friction Loads.* Finally, sectional and total forces and moments due to skin friction are calculated by surface integration of the local skin friction determined by the BL analysis.

Laminar, transition, and turbulent methods can be selected by the user and combined arbitrarily. If a fully turbulent analysis is performed, the first two steps are skipped. The current implementation does not model the mutual interaction of BL displacement and potential flow by viscous-inviscid coupling.

Simplified Analysis for Fuselages. For arbitrary non-lifting bodies, a simplified turbulent analysis method was implemented. It is based on local flow properties and flat-plate analogy. In detail, the following steps are performed:

1. *Streamline Arc Length Computation.* An advancing front algorithm starting at stagnation panels and advancing based on a surface flux formulation was developed to calculate the inviscid streamline arc length at each panel collocation point. It works on unstructured triangle and quad surface meshes.
2. *Skin Friction Coefficient Estimation.* The local skin friction coefficient is evaluated using an approximate relation for the turbulent flat plate given by Schlichting [11]: $c_{f,loc} = [2\log_{10}(Re_s) - 0.65]^{-2.3}$, $Re_s < 10^9$
   Here, $Re_s = \frac{Us}{\nu}$ is the local Reynolds number, where s is the local streamline arc length and U is the local potential flow velocity.
3. *Separation Region Estimation.* Two methods are available:
   (a) Simple angle criterion. Flow separation is assumed to occur if the flow is retarded ($\frac{\partial U}{\partial s} > 0$) and the flow redirection angle ρ_{loc} (angle between the local panel normal vector and the flow direction) exceeds a certain critical value ρ_{sep}. As suggested by Ahmed [1] and Seddon [13] for bluff bodies, $\rho_{sep} = 20°$ is used throughout this paper.
   (b) Turbulent Stratford method [14]. It considers c_p history and Reynolds number effects.

   The skin friction coefficient is set to zero in the region of separated flow.
4. *Friction Loads Computation.* The loads on each surface patch and the total loads due to skin friction are computed by integration of the local skin friction in direction of the potential flow.

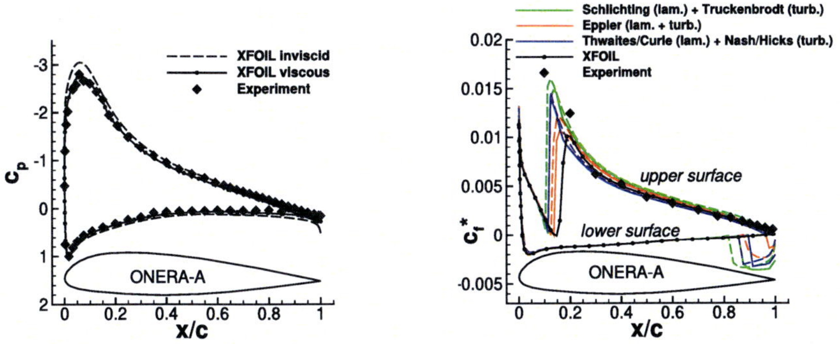

(a) pressure distribution

(b) skin friction coefficient, using inviscid (dashed lines) and viscous (solid lines) XFOIL velocity distribution as input

Fig. 3. ONERA-A airfoil at $M_\infty = 0.15$, $Re = 2.07 \times 10^6$ and $\alpha = 8.1°$

3 Results

3.1 Airfoils

As a first step, the newly implemented integral BL methods were tested and compared for an isolated setup, decoupled from UPM. Therefore, the streamline arc length and velocity distributions from XFOIL [3] calculations were taken as input for the BL module. In order to determine the influence of BL displacement on the solution, each calculation was performed twice, once using the inviscid XFOIL velocity distribution and once using the viscous XFOIL velocity distribution. The results were then compared to BL quantities calculated by XFOIL and measured in experiments. Multiple airfoils were investigated. As an example, Fig. 3 shows results for the ONERA-A airfoil at an angle of attack $\alpha = 8.1°$, Mach number $M_\infty = 0.15$, and Reynolds number $Re = 2.07 \times 10^6$. The airfoil geometry was discretized using 160 points. Figure 3a shows the inviscid and viscous pressure distributions calculated using XFOIL and experimental results [6]. The viscous pressure distribution matches the experimental data well, while the inviscid solution deviates slightly from the experiment, especially in the region of the suction peak. Figure 3b shows surface friction coefficients[1] calculated by the three implemented integral BL methods together with experimental results from [6] and XFOIL results. The sign of the lower surface skin friction was inverted in order to improve clarity. All results agree very well in the laminar region, and also in the turbulent region the differences do not exceed the expected range of accuracy. On the upper side, transition due to laminar separation and turbulent reattachment is predicted by all methods. Schlichting + Truckenbrodt returns the most conservative estimate regarding transition positions and friction forces compared to the experiment and XFOIL. Thwaites/Curl + Nash/Hicks results show the best accordance with XFOIL, especially in the turbulent region. Whether the viscous or inviscid XFOIL results are taken as input for the new BL methods only has a minor effect on the computed skin friction. Nevertheless, the predicted laminar separation locations are slightly shifted downstream when using the viscous velocity distribution as input, i.e. the prediction based on the inviscid velocity distribution is more conservative in terms of total drag. The transition location in the XFOIL results is even further downstream. This is attributed to the fact that XFOIL models laminar separation bubbles with a finite length while the UPM BL module assumes immediate transition after laminar separation. This is also observed in the trailing edge region of the lower side, where all implemented laminar methods predict laminar separation, while XFOIL's laminar model stays attached.

3.2 Wings

After successful validation of the implemented integral BL methods for airfoils using external velocity distributions from XFOIL, the BL module was fully integrated into UPM. For all following simulations, UPM's internal panel geometries and potential flow solutions were used as input for the BL routines. Several wing geometries were simulated using the newly implemented routines. Results of the well-established steady coupled 3D panel and BL code VSAERO [10] were used as reference. As an example, Fig. 4 presents incompressible UPM and VSAERO results for a tapered wing (aspect ratio $\Lambda = 10.67$, taper ratio $\lambda = 0.5$,

[1] All skin friction coefficients c_f^* were calculated using the local onflow velocity U_∞.

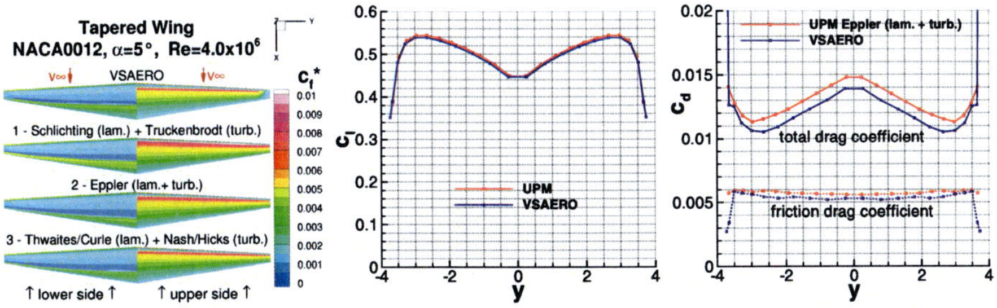

(a) Skin friction coefficient (b) c_l distribution (c) c_d distribution

Fig. 4. Tapered wing at $Re_{root} = 4.0 \times 10^6$, $\alpha = 5°$

NACA0012 airfoil, no twist, $Re_{root} = 4.0 \times 10^6$ and $\alpha = 5°$). The surface of each half wing consisted of 100 (chordwise) × 11 (spanwise) panels. The skin friction distributions of VSAERO and the UPM BL methods (Fig. 4a) show a good qualitative agreement, even though the transition position calculated by VSAERO for the lower side is located slightly upstream of the UPM result and the turbulent c_f predicted by all UPM BL methods is slightly higher. Noticeable differences are observed in the wing tip region. They are attributed to the fact that in VSAERO the wing tips were closed by a planar patch while the wings modeled in UPM had open tips. The method combination of Thwaites/Curle and Nash/Hicks matches the VSAERO result best. This is not surprising, because the VSAERO BL analysis is based on the same methods. The inviscid lift coefficient plot, Fig. 4b, shows a slight lift offset, which is caused by the different wake models (planar wake in VSAERO vs. free-wake in UPM). Figure 4c shows the total drag coefficient and its component due to skin friction calculated by VSAERO and UPM. For UPM, only the results of the Eppler methods are shown. The differences in the computed friction drag are small, relatively even smaller than for the total drag. Again, large deviations in the tip region due to the different wing tip cap and wake modeling are observed.

3.3 Rotors

Finally, the implemented integral BL methods were validated for the 7A rotor in hover. Wind tunnel and whirltower measurements [12] are available for this rotor. Furthermore, CFD results using DLR's RANS codes FLOWer and TAU are available as reference solutions. The 7A model rotor has a radius of 2.11 m and a chord length of 0.14 m. The considered test case had a tip Mach number of $M_{tip} = 0.617$. The surface of each rotor blade was discretized by 98 (chordwise) × 17 (radial) panels. The CFD surface meshes used twice the resolution in both dimensions. Eight rotor revolutions were computed for each data point in UPM.

Figure 5 depicts calculated skin friction distributions for a collective pitch of 5.97°. The upper two solutions shown in Fig. 5a were computed including BL transition prediction, the lower two solutions were computed fully turbulent. The artifacts at the blade root seen in the CFD results are attributed to the setup using a Euler wall boundary at $r/R = 20\%$. Also, the CFD mesh was clamped at the outer blade edge while the UPM blades were not closed at the tip, causing differences in this region. The chordwise skin friction distributions at $r/R = 70\%$ shown in Fig. 5b and 5c give further insight into the results. The transition point predicted by FLOWer on the upper surface is located around 10% upstream of the UPM results. It is probably caused by differences in the pressure distributions and different n_{crit} values, which are determined internally from local turbulence model quantities in FLOWer. On the lower surface, the transition points match very well. The laminar and turbulent skin frictions calculated by FLOWer and UPM have a small offset. Still, the accordance is surprisingly good when considering the many simplifications inherent to the UPM simulations.

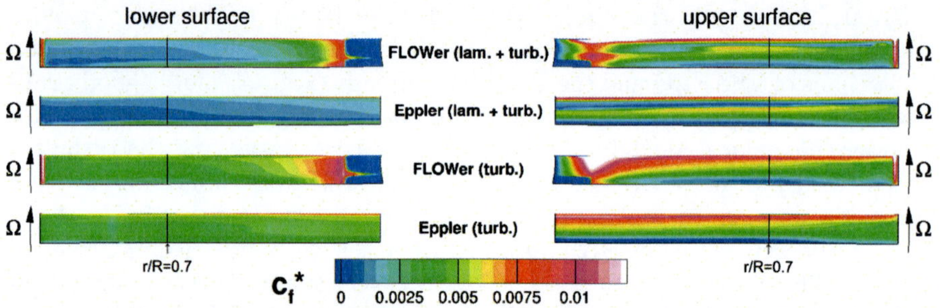

(a) contour plot

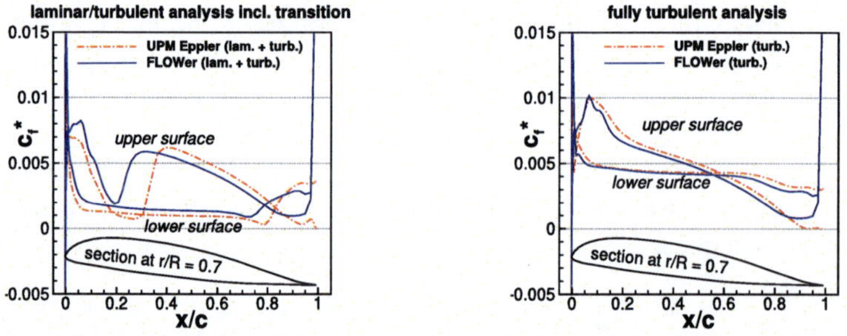

(b) section $r/R = 0.7$ analysis including transition (c) section $r/R = 0.7$ turbulent analysis

Fig. 5. Rotor 7A skin friction coefficient, collective pitch 5.97°

The rotor performance results for four collective pitch angles are compiled in Fig. 6. The thrust coefficients calculated by UPM agree very well with the CFD solutions. For most operating points UPM predicts a marginally lower thrust coefficient. The discrepancy of the two experimental datasets is much larger than the differences observed in the simulated results. Only the inviscid UPM result is plotted because the influence of the friction forces on thrust can be neglected. The power coefficient predicted by UPM when using the BL methods with transition prediction matches the whirltower experiment, especially for the smaller collective pitch angles. But when considering the slight overprediction of thrust for the same whirltower test, a noticeable overprediction of the required power is expected, as it is the case with the FLOWer result. The lower power predicted by UPM could originate from differences in the calculated induced drag or by the missing form drag component in UPM. Nevertheless, the fact that the delta between UPM and FLOWer gets larger for larger pitch angles is an indication that the missing form drag increases at larger collective pitch angles. As expected, the calculated power is shifted upwards when performing fully turbulent simulations.

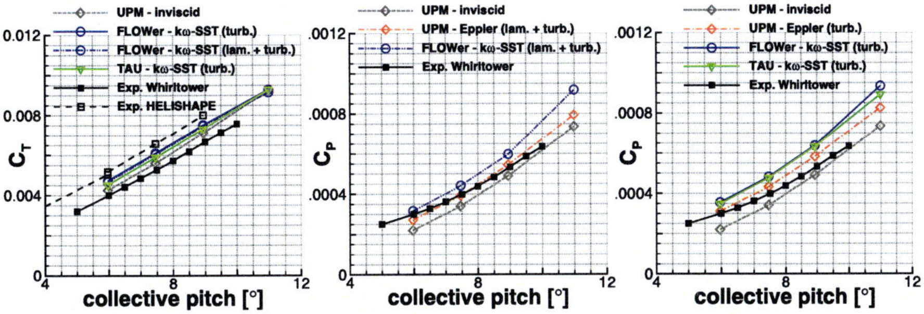

(a) C_T (b) C_P (incl. transition) (c) C_P (fully turbulent)

Fig. 6. Rotor 7A simulated and measured performance coefficients

3.4 Fuselages

The simplified turbulent BL analysis method described in Sect. 2.2 was tested for generic fuselage geometries and compared to VSAERO solutions. Identical surface discretizations were used for both codes. In VSAERO transition was tripped right after the stagnation points in order to obtain turbulent results. VSAERO performs integral BL analyses along surface streamlines seeded by the user. The analysis stops if flow separation is predicted. The results are then mapped back from the streamline points to the surface panels.

Figure 7 depicts surface pressure and friction results for the generic ROBIN helicopter fuselage [5] at $\alpha = 0°$, $M_\infty = 0.1$ and $Re = 1.6 \times 10^6$. The full model consists of 7,854 panels. The inviscid pressure distributions calculated by UPM and VSAERO match very well as seen in Fig. 7a.

The skin friction coefficient is plotted in Fig. 7b. The advancing front algorithm for arc length computation works robustly, even though there are two stagnation points. The calculated skin frictions show qualitatively similar distributions. But the flat plate analogy relation seems to overestimate skin friction

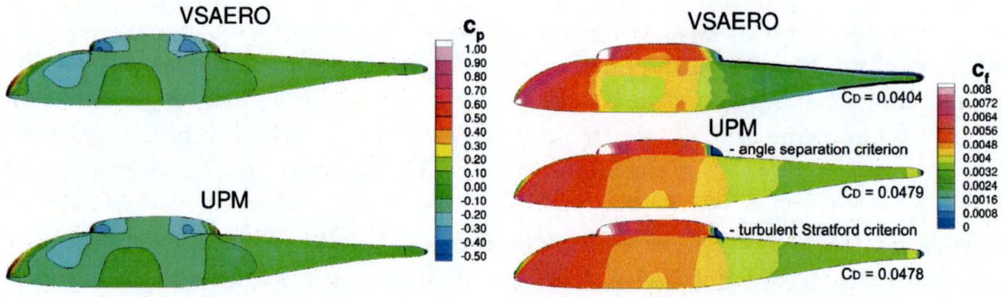

(a) surface pressure (b) skin friction and flow separation line

Fig. 7. ROBIN fuselage at $\alpha=0°$

on a large portion of the surface, returning a conservative value for the total friction drag when compared to VSAERO.

Both, the angle criterion with $\rho_{sep} = 20°$ and the turbulent Stratford criterion were used for the estimation of flow separation regions in UPM. The predicted flow separation lines are highlighted by thick black contour lines in Fig. 7b. The predicted flow separation area at the back of the pylon is larger than the one predicted by VSAERO. But, in contrast to VSAERO, the separation state is not propagated downstream along the streamlines.

4 Conclusion

The UPM code was extended by approximate boundary layer methods in order to estimate the impact of viscous effects. After comparing various analysis methods regarding robustness, computational effort, implementation complexity, and user-friendliness, two approaches were selected and implemented. An approach employing integral methods for the stripwise analysis of lifting surfaces and a simplified approach based on local flow properties and flat-plate analogy for arbitrary non-lifting bodies. The methods were validated for several cases.

The airfoil results showed that, of the implemented integral methods, the Eppler methods offer the best compromise between robustness and accuracy. The Truckenbrodt quadrature consistently overestimated the turbulent skin friction. The Nash/Hicks method delivered accurate results, but had robustness issues in some cases. All methods returned conservative estimates for skin friction, transition, and separation locations when compared to the coupled viscous-inviscid solver XFOIL. These findings were confirmed by 3D wing and rotor results.

The simplified analysis method for non-lifting bodies worked robustly and sufficiently accurate. The predicted fuselage skin frictions agreed qualitatively well with VSAERO results, but the accuracy lagged behind the one of the stripwise analysis for lifting surfaces. The separation criteria gave plausible results. Both implemented separation criteria returned conservative estimates when compared to VSAERO.

One must be aware that the implemented methods do not aim for the accurate quantitative prediction of flows dominated by viscous effects. They rather enable the quantification of uncertainty by detecting regions probably dominated by flow separations.

Acknowledgment. The project this report is based on was funded by the German Federal Ministry for Economic Affairs and Energy under grant number 20H1501B. The responsibility of this publication lies with the author.

References

1. Ahmed, S., Ramm, G., Faltin, G.: Some salient features of the time-averaged ground vehicle wake. Technical report, SAE-Paper 840300, SAE (1984)
2. Curle, N.: A two-parameter method for calculating the two-dimensional incompressible laminar boundary layer. J. Roy. Aeronaut. Soc. **71**(674), 117–123 (1967). https://doi.org/10.1017/S0001924000056189
3. Drela, M.: XFOIL: an analysis and design system for low Reynolds number airfoils, pp. 1–12. Springer, Heidelberg (1989). ISBN 978-3-642-84010-4
4. Eppler, R., Somers, D.M.: A computer program for the design and analysis of low-speed airfoils. Technical report, NASA-TM-80210, NASA, Hampton, VA, United States (1980)
5. Freeman, C., Mineck, R.E.: Fuselage surface pressure measurements of a helicopter wind-tunnel model with a 3.15-meter diameter single rotor. Technical report, NASA-TM-80051, NASA, Hampton, VA, United States (1979)
6. Gleyzes, C., Capbern, P.: Experimental study of two AIRBUS/ONERA airfoils in near stall conditions. Part I: Bound. Layers Aerosp. Sci. Technol. **7**(6), 439–449 (2003)
7. Hepperle, M.: Javafoil user's guide. Technical report, MH AeroTools, Braunschweig (2017)
8. Hess, J., Smith, A.: Calculation of potential flow about arbitrary bodies. Prog. Aerosp. Sci. **8**, 1–138 (1967). https://doi.org/10.1016/0376-0421(67)90003-6
9. Nash, J.F., Hicks, J.G.: An integral method including the effect of upstream history on the turbulent shear stress. In: Kline, S.J., Coles, D.E., Hirst, E.A. (eds.) Computation of turbulent boundary layers - 1968 AFOSR-IFP-Stanford Conference, vol. I, pp. 37–45. Stanford University, Stanford (1969)
10. Nathman, J.: VSAERO user's manual, version 7.02. Technical report, Analytical Methods Inc., Redmond, Washington, USA (2007)
11. Schlichting, H., Krause, E., Oertel, H.J., Gersten, K.: Grenzschicht-Theorie. Springer, Heidelberg (2006). ISBN 9783540230045
12. Schultz, K.J., Splettstoesser, W., Junker, B., Wagner, W., Schoell, E., Mercker, E., Pengel, K., Arnaud, G., Fertis, D.: A parametric windtunnel test on rotorcraft aerodynamics and aeroacoustics (Helishape) - test procedures and representative results. The Aeronaut. J. **101**(1004), 143–154 (1997)
13. Seddon, J.: Aerodynamics of the helicopter rear fuselage upsweep. In: 8th European Rotorcraft Forum, Aix-en-Provence, France (1982)
14. Stratford, B.S.: The prediction of separation of the turbulent boundary layer. J. Fluid Mech. **5**(1), 1–16 (1959). https://doi.org/10.1017/S0022112059000015
15. Thwaites, B.: Approximate calculation of the laminar boundary layer. Aeronaut. Q. **1**(3), 245–280 (1949). https://doi.org/10.1017/S0001925900000184
16. Truckenbrodt, E.: Neuere Erkenntnisse über die Berechnung von Strömungsgrenzschichten mittels einfacher Quadraturformeln. Teil I. Ingenieur-Archiv **43**(1), 9–25 (1973). https://doi.org/10.1007/BF00536575

17. Truckenbrodt, E.: Neuere Erkenntnisse über die Berechnung von Strömungsgrenzschichten mittels einfacher Quadraturformeln. Teil II. Ingenieur-Archiv **43**(2), 136–144 (1974). https://doi.org/10.1007/BF00537146
18. Würz, W.: Hitzdrahtmessungen zum laminar-turbulenten Strömungsumschlag in anliegenden Grenzschichten und Ablöseblasen sowie Vergleich mit der linearen Stabilitätstheorie und empirischen Umschlagskriterien. Ph.D. thesis, Universität Stuttgart (1995)
19. Yin, J., Ahmed, S.: Helicopter main-rotor/tail-rotor interaction. J. Am. Helicopter Soc. **4**, 293–302 (2000). https://doi.org/10.4050/JAHS.45.293
20. Yin, J.: Main rotor and tail rotor blade vortex interaction noise under the influence of the fuselage. In: Proceedings of the 38th European Rotorcraft Forum, Amsterdam, The Netherlands, September 2012

Experimental Measurement of a UAV Rotor's Acoustic Emission

Felix Lößle[(✉)], Anna Kostek, Clemens Schwarz, and Rainer Schmid

German Aerospace Center (DLR), Institute of Aerodynamics and Flow Technology (AS), Bunsenstraße 10, 37073 Göttingen, Germany
felix.loessle@dlr.de

Abstract. Besides aerodynamic performance, most unmanned aerial vehicles (UAVs) are limited by noise emission in their range of application. Especially in urban areas, the acceptance of UAV noise emission by the affected population is required. Therefore, one objective of DLR's project URBAN RESCUE is to investigate the noise emission and noise generation mechanisms of small rotors. In order to study noise emission, a test setup was established in DLR's 1m wind tunnel. For different flight modes, acoustic measurements were conducted using a microphone array. For all test cases the noise emission of the rotor's blade passing frequency (BPF) is higher in front of the rotor (upstream) than behind it. Due to higher relative velocities on the advancing side than on the retreating side, emission is asymmetric with respect to the flow direction. Additionally, a time-resolved background-oriented schlieren (BOS) setup was used to visualize tip vortices and study noise generation mechanisms.

Keywords: Noise emission · Small rotors · Acoustic measurements · Microphone array · Background-oriented schlieren

1 Introduction

Due to the complexity of a UAV rotor's flow dynamics, there is a high variety of tonal as well as broadband sound generation mechanisms. In many respects, noise emission mechanisms of small UAV rotors, helicopter rotors and propellers are similar. While helicopter rotors and propellers have been studied for quite long, UAV rotors have come to the fore in recent years. Therefore, most noise generation theories are based on experiments dealing with helicopter rotors and propellers.

Noise generation mechanisms can be classified into rotational noise, airfoil self-noise and interaction noise. Rotational noise can be subdivided into loading and thickness noise. The first emerges when rotating forces act on the blades' surface, whereas the second is caused by fluid displacement due to blade rotation. Considering the Ffowcs Williams-Hawkings (FWH) equation makes clear that loading noise exhibits a dipole and thickness noise a monopole character. Based on the FWH-equation Gutin [4] and later Hanson [5] derived a theoretical model for propeller rotational noise emission in time and frequency domains. Another approach is given by Succi [11], who derived an approximate solution for the

© The Author(s), under exclusive license to Springer Nature Switzerland AG 2021
A. Dillmann et al. (Eds.): STAB/DGLR Symposium 2020, NNFM 151, pp. 387–396, 2021.
https://doi.org/10.1007/978-3-030-79561-0_37

FWH-equation for a rotating blade element in the time domain. Airfoil self-noise occurs when boundary layer structures and near wakes interact with the airfoil. These noise sources contribute to rotor broadband noise. An overview of airfoil self-noise models is given in Brooks et al. [3]. Regarding interaction noise, blade vortex interaction (BVI) is of primary interest when helicopters are studied. When rotor blades interact parallelly with a tip vortex in forward flight or descent, high pressure fluctuations are induced on the leading edge resulting in strong acoustic pulses, which are rich in higher harmonics of the blade passing frequency (BPF) [14].

In recent years, noise emission of small UAV rotor has been studied experimentally and numerically. Acoustic measurements with an isolated UAV rotor were performed with several microphone positions, for example by Tinney and Valdez [12] and Yang et al. [13]. Acoustic measurements were compared with numerical results by Sinibaldi and Marino [10] and Zawodny et al. [15]. However, noise emission and noise generation mechanisms of small UAV rotors are not as well examined and understood as those of helicopter rotors and propellers. So far a model of noise emission has only been derived for a rotor in hover [12]. In summary, a need for research on the field of small UAV rotor acoustics remains. One objective of DLR's project URBAN RESCUE is to gain a better understanding of noise emission and noise generation mechanisms of small UAV rotors. Furthermore, a comprehensive model for noise emission is going to be derived within the scope of the project. In this paper, a first experimental approach to investigate the properties of an isolated UAV rotor in hover and forward flight is presented. The analysis covers the operation of a UAV rotor at different tilt angles and varying wind tunnel velocities. Microphone measurements are conducted in a plane below the rotor (in the direction of typical observers). Spectra and directional characteristics for different flight modes are compared and the applicability of scaling laws established for helicopter rotors are validated for UAV rotors.

2 Experimental Setup

The test setup was established in DLR's 1m tunnel in Göttingen. The object of investigation was a clockwise rotating "CAM-Carbon Light-Prop" manufactured by aero-naut Modellbau for multicopter applications with a diameter of 40.64 cm (16″) and a pitch of 15.24 cm (6″). The test setup and a 3D-scan of the rotor is shown in Fig. 1. Due to space limitations, the rotor was installed vertically in the wind tunnel test section. In order to allow tilt angle variations from $-90°$ to 25° the rotor and its drive unit were mounted on a rotatable support system. A positive tilt angle represents a rearward-tilt of the rotor's tip path plane. In order to maintain the rotor's position when rotating, the rotor hub was located in the rotation axis of the rotatable support system (illustrated by the red line in Fig. 1 (a)). The wind tunnel's flow velocity was measured with a Prandtl probe in the nozzle outlet. All presented measurements were conducted with the rotor operating at a rotational speed of 4000 rpm (corresponding to a hover tip speed of 84 m/s). Measurements with a load cell showed that with this settings the hovering rotor generates thrust of 13.6 N. The nondimensional rotor advance ratio defined as

$$\mu = \frac{V_\infty}{\Omega R} \tag{1}$$

where V_∞ is the magnitude of free-stream velocity, Ω the rotor's rotational frequency and R the rotor radius is used to characterize the forward motion. The advance ratio varied within a range of 0 and 0.26 (corresponding to wind tunnel velocities between 0 m/s and 22 m/s).

Acoustic measurements were performed in a plane below the rotor plane with a microphone array consisting of two lines of six Brüel & Kjær 1/4″ multi-field microphones (type 4961) each. The array was mounted on the wind tunnel's traverse in a plane parallel to the flow direction at a distance of 1140 mm, which turned out to be sufficient to ensure non-affection of the microphones by the wind tunnel's shear layer. However, the exposure of microphones to the rotor's downwash couldn't be prevented for setups with tunnel velocities lower than 10 m/s. The measurement was carried out with 16 different array positions in order to cover an area of 1524 mm × 1375 mm with increments of 100 mm in flow direction (x-axis) and 125 mm in the perpendicular direction (y-axis), as illustrated in Fig. 2. Each microphone's signal was calibrated with a Brüel & Kjær pistonphone (type 4231) before and after the experiment. The wall and floor behind the microphone array was covered with acoustic foam to reduce reflections. The acoustic signals were sampled for 20 s at 50 kHz. The power spectral density (PSD) was calculated with Welch's method (20 non-overlapping Hann-windowed data segments of 50000 samples each).

In order to visualize the blade tip vortices of the rotor a time-resolved BOS system operated at 1.33 kHz was used. A Phantom Veo 640L camera with a sensor consisting of 2560 × 1600 pixels was used and equipped with a 180 mm focal length lens at an aperture setting of f/22. The camera was placed at a distance of 3.26 m to the rotor hub and 6.31 m to the background. The background consisted of retroreflective foil printed with randomly placed dots and was illuminated by a high-power light-emitting diode (LED). The fundamentals of this measurement technique have previously been summarized by Raffel [9] and are not given here.

Geometric Deviations of Directional Characteristics

In order to get the rotor's directional characteristics, the acquired sound pressure level (SPL) of each microphone position was normalized to a distance of one meter relative to the rotor hub. Linear interpolation was carried out between grid points.

Directional characteristics calculated from measurement data should be carefully interpreted: First, acoustic waves are refracted passing the wind tunnel's shear layer. Second, there is atmospheric absorption when the pressure signal propagates. However, both effects are negligible due to low wind tunnel Mach numbers (up to 0.064) and small distances between shear layer and microphones [1].

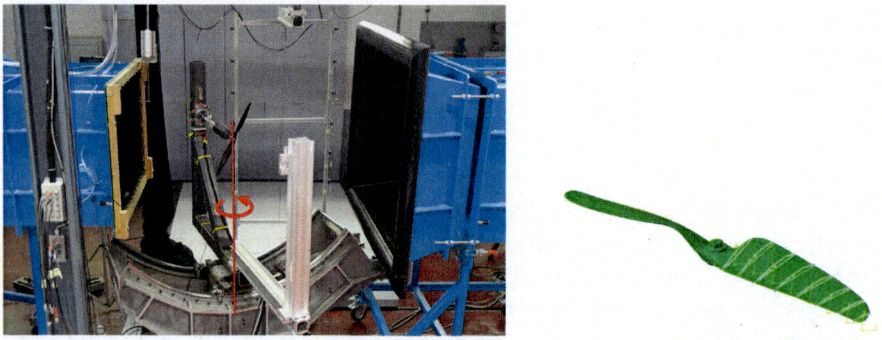

(a) Positioning of the rotor under investigation mounted on the rotatable support system in the wind tunnel; microphone array in the background

(b) Laser 3D-scanned geometry of the rotor of investigation

Fig. 1. Experimental setup of the acoustic rotor measurement in DLR's 1m tunnel

Third, the array's distance to the rotor is not sufficient to allow measurements in the geometric far field (i.e. measuring distance is not much greater than the source region's extent). Hence the rotor's spatial extent causes two distorting effects on the noise hemisphere. On the one hand, if the effective acoustic source is not positioned in the rotor hub (as assumed for the hemisphere projection), the discrepancy between assumed and effective distance between source and microphones leads to SPL deviations. On the other hand, the rotor's spatial extent results in a deviation of the true emission angle. Both effects depend on the microphone position and can be estimated by geometric considerations. The maxima of deviations are expected if the acoustic source was located in the blade tips and are illustrated in Fig. 3. While the variation of SPL increases with elevation and reaches a maximum of 0.75 dB, the deviation in emission angles peaks at elevations between 10° and 20° and doesn't exceed an absolute value of 10°.

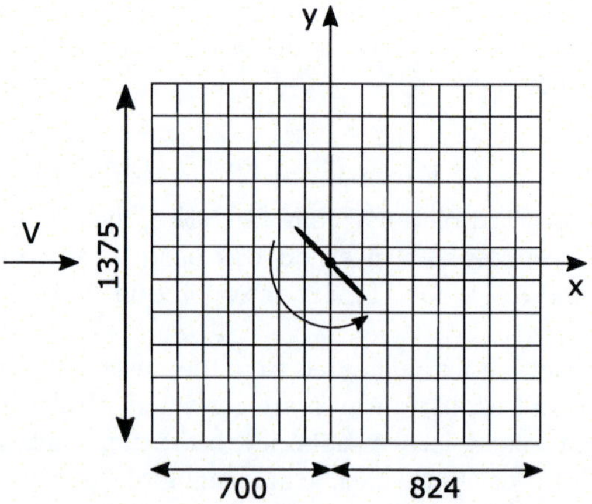

Fig. 2. Grid of microphone positions (view on the rotor's suction side)

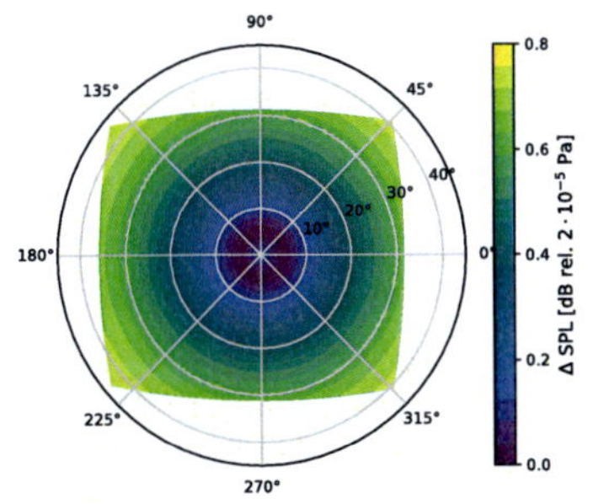

(a) Maximum SPL deviation

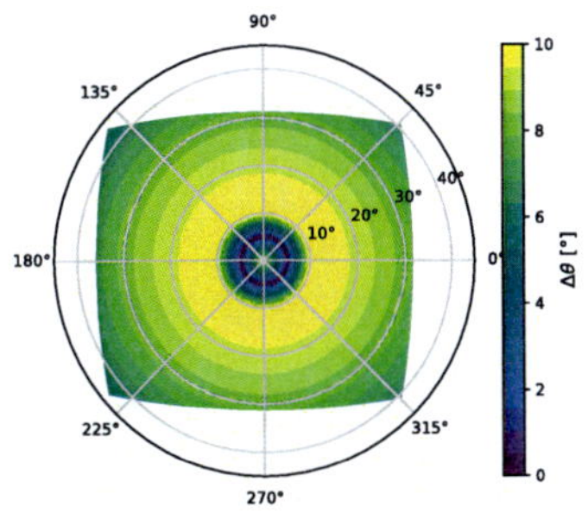

(b) Maximum angular distortion

Fig. 3. Deviations due to the source region's extent, if measurements take place in the rotor's geometric near field

3 Results and Discussion

Results of Acoustic Measurements for a Single Observer

In this section, the spectra of a single microphone located in front of the rotor are considered ($x = -700\,\text{mm}$, $y = 62.5\,\text{mm}$). An overview about spectra at advance ratios μ between 0.05 and 0.26 and tilt angles α between $-10°$ and $10°$ is given in Fig. 4. Two spectra at an advance ratio $\mu = 0.19$ are presented in Fig. 5 in more detail. All spectra show tones at the BPF and its harmonics as well as broadband noise. In the following, the characteristics of these three types of noise are discussed separately.

First of all, the BPF is considered. The spectra show that the SPL of the BPF exceeds the level of higher harmonics by more than 10 dB and therefore determines the overall sound pressure level (OASPL). Figure 6 shows the influence of advance ratio μ and tilt angle α on the SPL of the first harmonic. (The SPL of the first harmonic was obtained by integrating the PSD in the frequency range of BPF $\pm 10\,\text{Hz}$.) The dependence of SPL on advance ratio μ is theoretically predicted: Due to small blade volume, loading noise is expected to be the dominant noise generation mechanism [6]. Since loading noise has dipole character, sound pressure follows a sixth-power law scaling with Mach number [7], which has been experimentally confirmed by Aravamudan et al. [2]. Moreover, scaling of forces with Mach number must be taken into account for rigid rotors. Considering this effect, Lowson and Ollerhead found an eighth power law for rotor rotational noise [7]. Based on this theory, noise is mainly generated at fast moving blade areas, in particular at the blade tips on the advancing side. Using the tip Mach number of the advancing side, the blue and red line in Fig. 6 represent the sixth and eighth power law. The eighth power law captures the rise of the first harmonic's SPL at advance ratios higher than 0.1 well, but underpredicts the increase for smaller advance ratios.

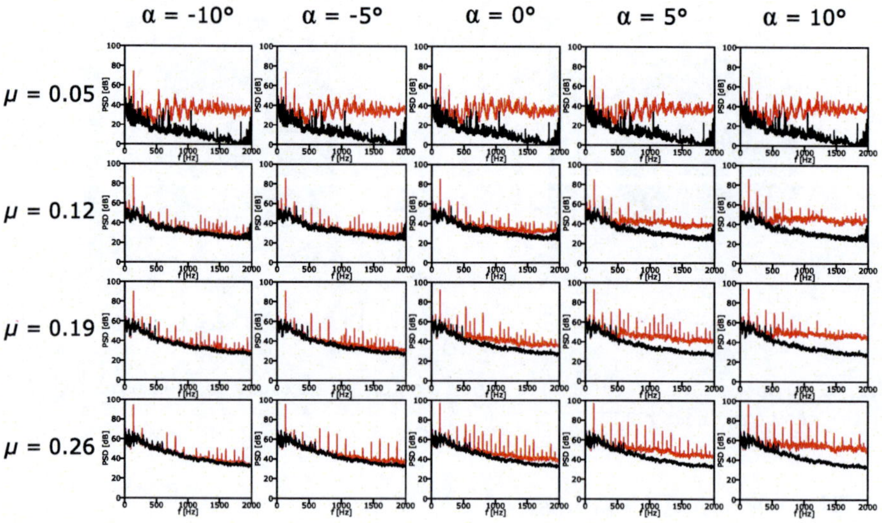

Fig. 4. Spectra of rotor noise at several advance ratios μ and tilt angles α at position $x = -700\,\text{mm}$, $y = 62.5\,\text{mm}$, the acoustic signal of the rotor is indicated by the red lines, background noise by the black lines

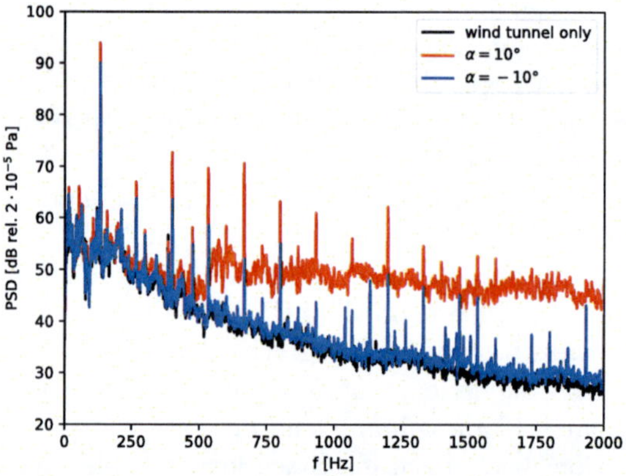

Fig. 5. Measured PSD at position $x = -700\,\text{mm}$, $y = 62.5\,\text{mm}$, the rotor was operated at an advance ratio $\mu = 0.19$

Finally, the characteristics of higher harmonics and broadband noise are addressed. The SPL of the higher harmonics grows with increasing advance ratio and with increasing tilt angle. The level of broadband noise emission is dependent on the rotor's tilt angle. At negative angles, the rotor's broadband noise is drowned out by the wind tunnel and therefore can't be studied. However, at tilt angles of 0° and higher, the rotor's broadband noise is sufficiently loud to be distinguishable from the environmental noise for frequencies greater than 500 Hz.

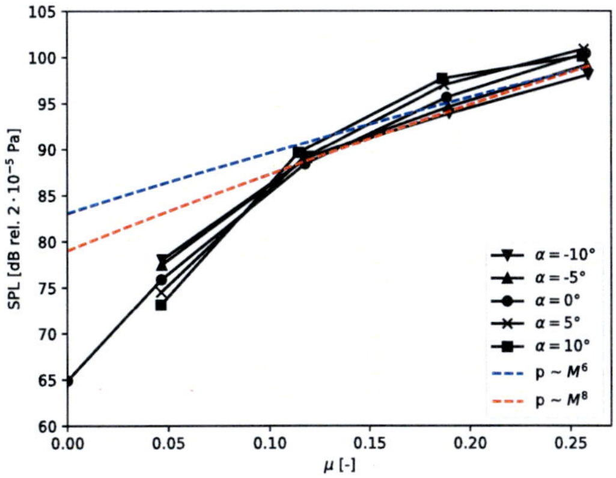

Fig. 6. SPL of the rotor's first harmonic at several advance ratios μ and tilt angles α at position $x = -700\,\mathrm{mm}$, $y = 62.5\,\mathrm{mm}$

Two aerodynamic effects explain the dependence of the raise of higher harmonics and broadband noise on the advance ratio and tilt angle mentioned before. On the one hand, flow separation occurs in an intensified way at higher advance ratios and tilt angles. Flow separation mainly influences broadband noise emission [3]. On the other hand, the interaction of rotor blades with tip vortices constitutes an additional noise mechanism. Measurements performed on a model helicopter showed that BVI results in the increase of higher harmonics [8]. In order to discover whether it comes to an interaction of rotor blades and tip vortices, tip vortices were visualized with background-oriented schlieren (BOS). The results at a tilt angle of 0° and three different advance ratios are illustrated in Fig. 7. In hover and slow forward flight tip vortices are uniformly convected downstream. On the contrary, the convection is found unsteady at advance ratios $\mu = 0.12$ and higher. Due to flat angles of convection the vortex system remains in the vicinity of the rotor for longer, increasing the probability of interaction with rotor blades.

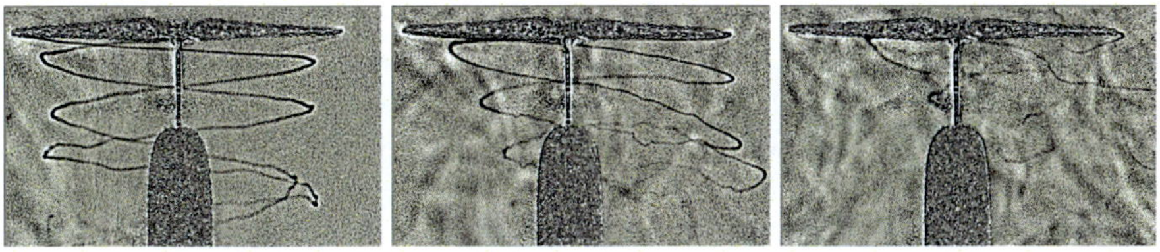

(a) $\mu = 0$ (b) $\mu = 0.05$ (c) $\mu = 0.12$

Fig. 7. Visualization of tip vortices at three different advance ratios μ using BOS

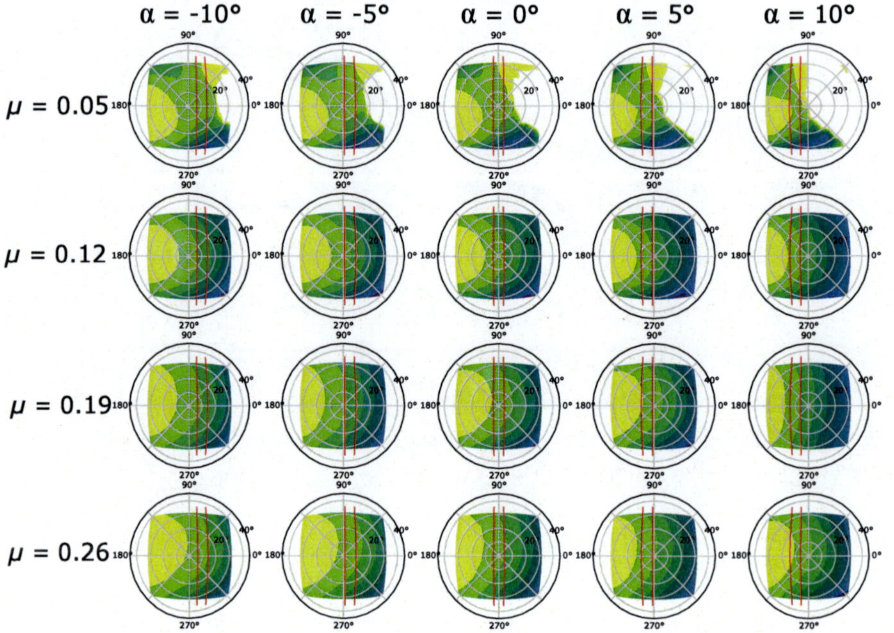

Fig. 8. Measured noise emission spheres of the rotor's first harmonic at several advance ratios μ and tilt angles α, elevation is in radial and azimuth in circumferential direction plotted

Directional Characteristics

As mentioned before, the first harmonic determines the rotor's OASPL in all test cases. Thus, the directional characteristic of only the rotor's first harmonic is considered in the following. In Fig. 8, SPL spheres at advance ratios μ between 0.05 and 0.26 and tilt angles α between $-10°$ and $10°$ are given. One test case SPL sphere is illustrated in more detail in Fig. 9. At an advance ratio $\mu = 0.05$, the rotor's downwash influences the microphone array, which leads to areas in the rotor's rear where the microphones are saturated. These areas are masked in Fig. 8 (blank areas). The color themes of the individual hemispheres are independent from one another and correspond to the minimum and maximum within the individual test cases. The vertical red lines represent the contour of the rotor support structure projected onto the hemisphere. Apparently, the SPL sphere of the rotor's first harmonic is not noticeably influenced by the suspension's shadowing due to its small width relative to wave length. For all test cases, the noise emission of the rotor is higher in front of the rotor (upstream i.e. flight direction) than behind it. Moreover, emission is asymmetric in relation to the inflow direction. At an advance ratio $\mu = 0.05$, noise emission is slightly stronger on the rotor's retreating side than on its advancing side. However, with increasing advance ratio noise emission shifts from the retreating side to the advancing side. At an advance ratio $\mu = 0.12$, the directional characteristic is nearly

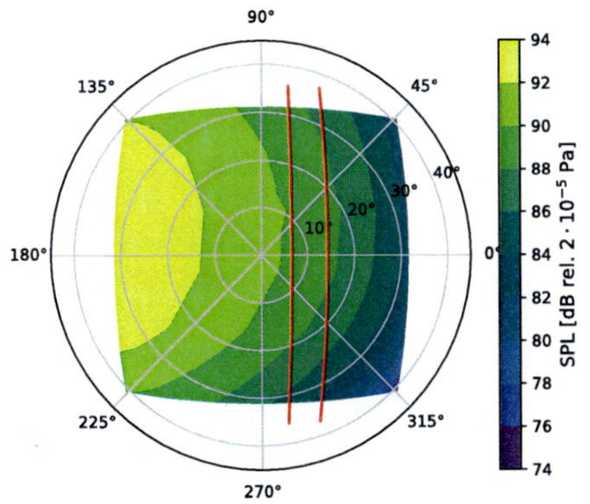

Fig. 9. Measured noise emission sphere of the rotor's first harmonic, the rotor was operated at an advance ratio of $\mu = 0.19$ and a tilt angle of $\alpha = -10°$, elevation is in radial and azimuth in circumferential direction plotted

symmetric. Higher advance ratios result in emission levels which are higher on the advancing side than on the retreating side. As mentioned above, the spanwise asymmetry in the emission directivity can be traced back to relative velocities which are higher on the advancing than on the retreating rotor side. Future work will focus on a better understanding of emission effects causing the directivity.

4 Conclusion

An acoustic test setup for small rotors was established in DLR's 1m tunnel, using a microphone array for measurement. The results show that the rotor's first harmonic determines the OASPL for all test cases and depends on the advance ratio. At advance ratios higher than 0.1 a correlation was found between measurement and Lowson and Ollerhead's eighth power scaling law (in reference to the tip Mach number of the advancing side). For higher harmonics and broadband noise an increase of SPL with growing advance ratio and rotor tilt angle can be observed. We consider the onset of flow separation and interactions of rotor blades with tip vortices to be responsible. The visualization of tip vortices using BOS for a tilt angle of 0° indicates that interaction occures at an advance ratio of 0.12 and higher. The directivity of the first harmonic is asymmetric. At low advance ratios the noise emission is slightly stronger on the rotor's retreating than on its advancing side. With increasing advance ratio the higher noise emissions shift from the retreating side to the advancing side.

For further experiments it's advisible to use an aeroacoustic wind tunnel with anechoic character, if available. In order to exclude the possibility of a significant impact of the room's acoustic properties on the directional characteristics, measurements will also be conducted with a counterrotating rotor. In order to derive a comprehensive model for noise emission, experiments combining acoustic and load measurements are planned. Furthermore, additional experiments with rotors of varying size, number of blades and pitch angles expands the model' scope of application. Future work will also focus on assessing the sound emission effects in more detail, including localizing their exact sources and characterizing the underlying flow effects using beamforming.

References

1. Amiet, R.K.: Refraction of sound by a shear layer. J. Sound Vib. **58**(4), 467–482 (1978)
2. Aravamudan, K.S., Lee, A., Harris, W.L.: A simplified Mach number scaling law for helicopter rotor noise. J. Sound Vib. **57**(4), 555–570 (1978)
3. Brooks, T.F., Pope, D.S., Marcolini, M.A.: Airfoil Self-noise and Prediction. NASA Reference Publication 1218 (1989)
4. Gutin, L.: On the sound field of a rotating propeller. Translated in NACA TM 1195 (1936)
5. Hanson, D.B.: Helicoidal surface theory for harmonic noise of propellers in the far field. AIAA J. **18**(10), 1213–1220 (1980)
6. Hanson, D.B.: Influence of propeller design parameters on far-fild harmonic noise in forward flight. AIAA J. **18**(11), 1313–1319 (1980)
7. Lowson, M.V., Ollerhead, J.B.: Studies of helicopter rotor noise. USAAVLABS Technical report 68-60 (1969)
8. Martin, R.M., Hardin, J.C.: Spectral characteristics of rotor blade/vortex interaction noise. J. Aircr. **25**(1) (1988)
9. Raffel, M.: Background-oriented schlieren (BOS) techniques. Exp. Fluids **56**(3), 1–17 (2015). https://doi.org/10.1007/s00348-015-1927-5
10. Sinibaldi, G., Marino, L.: Experimental analysis on the noise of propellers for small UAV. Appl. Acoust. **74**(1), 79–88 (2013)
11. Succi, G.P.: Design of quiet efficient propellers. In: SAE Business Aircraft Meeting, Society of Automotive Engineers Paper SAE79-0584, pp. 2039–2052 (1979)
12. Tinney, C.E., Valdez, J.: Acoustic scaling for small rotors in hover. In: AHS International 75th Annual Forum, Philadelphia (2019)
13. Yang, Y., Liu, Y., Li, Y., Arcondoulis, E., Wang, Y., Li, W., Huang, B.: Aerodynamic and aeroacoustic characteristics of a multicopter propeller during forward flight. In: AIAA Propulsion and Energy Forum, Cincinnati (2018)
14. Yu, Y.H.: Rotor blade-vortex interaction noise. Prog. Aerosp. Sci. **36**, 97–115 (2000)
15. Zawodny, N.S., Boyd, D.D., Burley, C.L.: Acoustic characterization and prediction of representative, small-scale rotary-wing unmanned aircraft system component. In: American Helicopter Society 72nd Annual Forum, West Palm Beach (2016)

Aerodynamic Analysis and Optimization of a Coaxial Helicopter Fuselage

Lukas Rottmann(✉)

Deutsches Zentrum für Luft- und Raumfahrt e.V. Braunschweig, Lilienthalplatz 7, 38106 Braunschweig, Germany
lukas.rottmann@dlr.de

Abstract. The flight dynamics and flight envelope of coaxial ultralight helicopter CoAX 2D were investigated. For this purpose, series of flight tests was carried out. This paper presents the computation of the fuselage aerodynamics of the CoAX 2D with the DLR flow solver TAU. For cruise flight, fuselage polars with different angles of attack and yaw angles without rotor downwash were calculated. Furthermore, different airspeeds, climb and sink rates were simulated in combination with rotor downwash. To simulate the rotor downwash from the coaxial rotors, the Actuator Disc approach is used. The lift of the horizontal stabilizer is compared between flight test and simulations for validation purposes. The lift was determined in the flight test by deflection of the tail boom. Finally, concepts for the optimization of the fuselage were developed. The landing gear was selected as a possible optimization potential. The changes from an aerodynamic point of view related to the shape of the skid supports. The aerodynamics of the new landing gear were compared with the old configuration and a reduction of drag could be shown.

Keywords: Coaxial ultralight helicopter · Actuator Disc · CoAX 2D · Helicopter fuselage aerodynamics · TAU

1 Introduction

The main difference between a classic helicopter configuration with main and tail rotor to a coaxial helicopter is, no tail rotor is required. The rotor torque is balanced by the arrangement of counter-rotating main rotors. This saves the power required for the tail rotor. The omission of the tail rotor allows a compact and very light construction which makes the coaxial helicopter also interesting for urban operations [1]. From the pilot's point of view, the coaxial helicopter offers advantages in terms of better handling qualities. For example, only little yaw control is required when the collective pitch of the main rotor changes [2].

The company edm-aerotec in Geisleden, Germany took advantage of these benefits to develop the CoAX 2D, and had a successful maiden flight in 2015 [3]. In the context of the Lufo project CURoT (Coaxial Ultralight Rotorcraft Technology), the CoAX 2D was used in a cooperation between edm-aerotec, the Technical University Munich

© The Author(s), under exclusive license to Springer Nature Switzerland AG 2021
A. Dillmann et al. (Eds.): STAB/DGLR Symposium 2020, NNFM 151, pp. 397–406, 2021.
https://doi.org/10.1007/978-3-030-79561-0_38

and DLR to investigate aeromechanical and aerodynamic phenomena that occur especially for coaxial helicopter configurations. For this purpose, edm-aerotec performed measurement flights with the CoAX 2D. Sensors were available to determine the GPS position, acceleration and roll, pitch and yaw angles. A nose boom measured the inflow condition. In addition to the measurements taken to determine the flight condition, the pilot's control forces and the control input were also recorded. For the analysis of the aerodynamic forces at the stabilizers, the structural deflection and the torsion moments at the tail boom were measured by means of strain gauges. Two rotating measurement units for the upper and lower main rotor blades were mounted to record the flap and lead-lag moments as well as the teeter angle of the individual blades [4].

To analyze the flight tests, an analysis model using the comprehensive code Camrad II was developed by the Technical University of Munich and validated with the flight measurement data [5]. Based on these results, the DLR Institute of Aerodynamics and Flow Technology contributes its competences in the field of fuselage aerodynamics to this Lufo project and investigated the fuselage cell of the CoAX 2D by using the flow solver TAU developed by DLR. Further optimization potential is proposed.

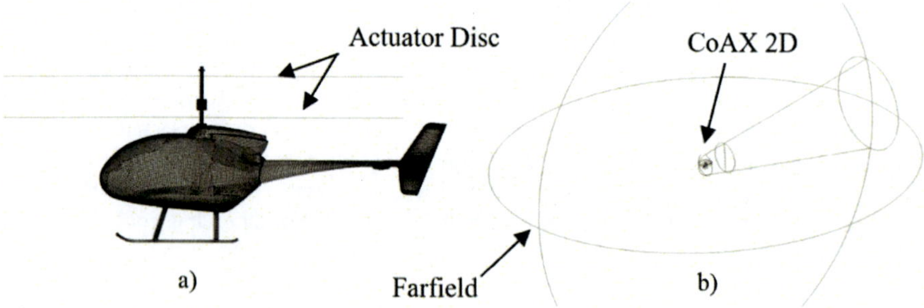

Fig. 1. a) Side view from Fuselage surface mesh, b) Mesh topology

2 Methodology and Numerical Setup

The CoAX 2D coaxial helicopter is an ultra-light helicopter with a maximum take-off mass of 450 kg and was developed by edm-aerotec in Geisleden [6]. The main rotor blades of the upper and lower main rotor plane are identical and are rectangular rotor blades with a twist. The rotation direction of the lower rotor is counterclockwise while the upper rotor rotates clockwise. The fuselage of the CoAX 2D is symmetrical. The fuselage cell is tilted by 4° to the rotorshaft. The landing skids and rotor shaft are perpendicular to each other, to see in Fig. 1a.

The DLR flow solver TAU was used to analyses the CoAX 2D fuselage. The TAU-code solves the compressible Reynolds-averaged Navier-Stokes equations on hybrid unstructured meshes around complex geometries [7]. For all simulations the second order central scheme was used for spatial discretization. Matrix dissipation was used for the central dissipation scheme. The temporal discretization was performed using implicit

Backward Euler. As turbulence model the two-equation turbulence model Menter SST was used. Furthermore, all surfaces were simulated fully turbulent. No multi-grid procedure was used to accelerate the calculations. For the simulation of the rotor downwash, the Actuator Disc approach in the TAU code was chosen [8]. A convergence of at least six orders of magnitude is ensured for all simulations.

The analysis of the CoAX 2D fuselage cell divided into three steps. In the first step, fuselage polars are simulated by varying the angle of attack and yaw angle at constant airspeed and without rotor downwash. The second part includes the rotor downwash modeled using the Actuator Disc. Two Actuator Discs with different input data were used for the upper and lower rotor planes of the coaxial rotor. Each input corresponds to a trimmed flight condition. In addition to different flight speeds, climb and descent speeds are also considered. The input for the Actuator Disc comes from the comrehensive rotor code Camrad II. The model of the CoAX 2D in Camrad II was realized by the Technical University of Munich. Camrad II is a comrehensive rotor code for the simulation of rotorcraft. Multi-body-dynamics, nonlinear finite element, structural dynamics and rotorcraft aerodynamics as well as trim procedures are considered [9]. The model considers the fuselage and horizontal stabilizer as well as the rotor blades of the coaxial rotor and was validated using flight test data. More information can be found in [5]. The input data for the Actuator Disc consisted of axial as well as tangential forces. Radial forces are not processed by the Actuator Disc. Finally concepts for optimizing the fuselage are then developed on the basis of the previous results.

For the first and second part a hybrid mesh with 7.8 mio. points were created. The underlying CAD geometry of CoAX 2D has been simplified. The rotor head is replaced by a simplified geometry. In addition, all ducts within the fuselage are neglected and all gaps were closed. All attachments such as antennas were also omitted. Also, the mounting of the landing gear to the fuselage is simplified. The fuselage surface was discretized in a combination of structured and unstructured way, to see in Fig. 1a. The boundary layer was resolved with 50 points. The first cell height was designed for a $y+ < 1$. The farfield has a spacing of about 40 rotor diameters and is discretized in an unstructured way. Ninety percent of all points are around the fuselage and in the wake of the Actuator Discs. The Mesh topology can be seen in Fig. 1b. For the optimization of the landing gear new meshes were created. The mesh topology described above was retained. The number of mesh points slightly differs.

3 Results

3.1 Fuselage Polar

For the determination of the fuselage polar without influence of the rotor downwash, a polar angle of attack of 18° to −18° is simulated for a cruising velocity of 33.3 m/s for the yaw angles of 0°, 2° and 4°. Due to the symmetry of the fuselage, no negative yaw angles are simulated. Figure 2 shows the drag coefficient over the angle of attack of the fuselage for different yaw angles. Positive angles of attack denote a pitch up, negative angles denote a pitch down. The CoAX 2D has its drag minimum at minus eight degrees angle of attack even at different yaw angles. The cell is aligned with the flow at minus four degrees. With increasing and decreasing angle of attack, the drag of the fuselage also increases.

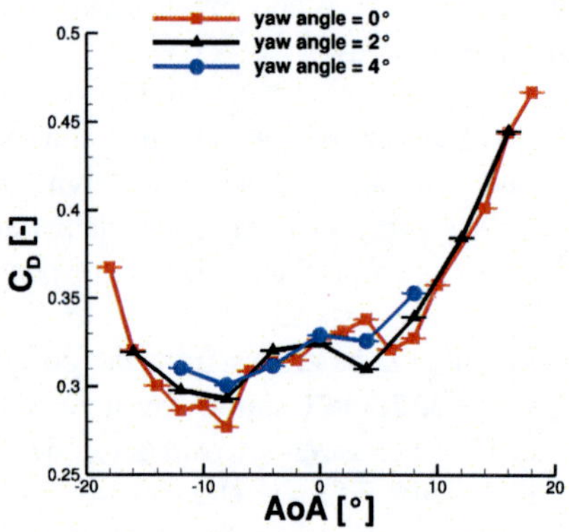

Fig. 2. Drag coefficient over angle of attack

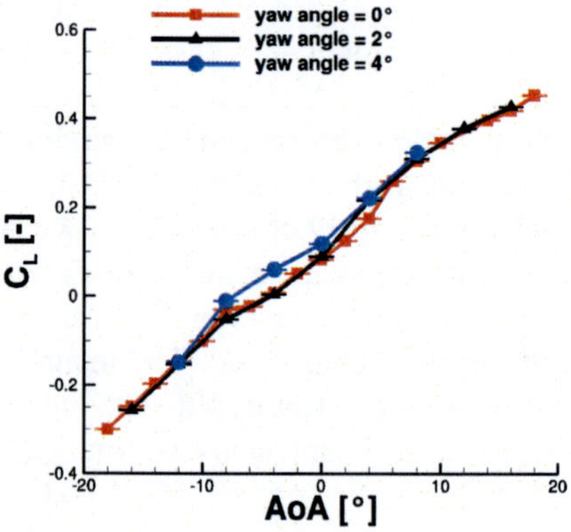

Fig. 3. Lift coefficient over angle of attack

The lift coefficient of the CoAX 2D is sketched in Fig. 3. The simulation shows a linear behavior with respect to the lift coefficient even for different yaw angles, no collapse of the lift coefficient is detected in range ±18°.

With regard to the pitching moment, Fig. 4 shows that the CoAX 2D has a consistently slightly negative pitch torque coefficient in the range between ±5°.

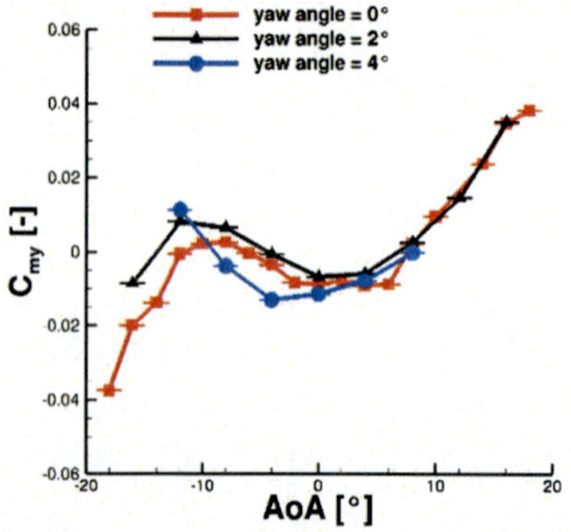

Fig. 4. Pitching moment coefficient over angle of attack

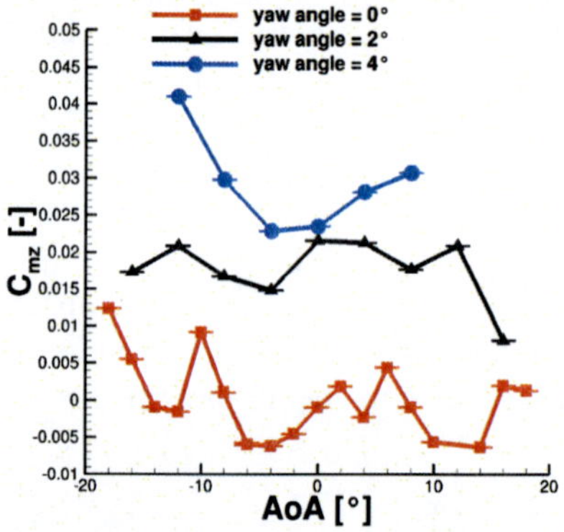

Fig. 5. Yawing moment coefficient over angle of attack

A negative pitching moment leads to a nose down motion. The moment reference point is located on the rotor axis of the CoAX 2D. With decreasing angle of attack, a negative pitch moment is increasingly detected and with increasing angle of attack a positive pitch moment can be observed. The course shows that the fuselage is stable between −10° and −5° Whereas at large negative and positive angles of attack an aerodynamic instability can be recognized. As the yaw angle increases, there is also a raise in the yaw moment, as seen in Fig. 5.

3.2 Flight Polar

To investigate the flow of the CoAX 2D in more detail, the angle of attack and airspeed were varied in this section. To increase the accuracy of the simulation, the rotor downwash is modeled with two Actuator Discs for the upper and lower rotor planes of the coaxial rotor. The input data for the Actuator Disc as well as the angle of attack of the fuselage cell are provided by the Camrad II model of the CoAX 2D. Each input corresponds to a trimmed flight condition. The flight conditions calculated in Camrad II are based on the flight tests performed by edm-aerotec [4, 5]. In addition to various forward flight speeds, climb and descent flights are also calculated.

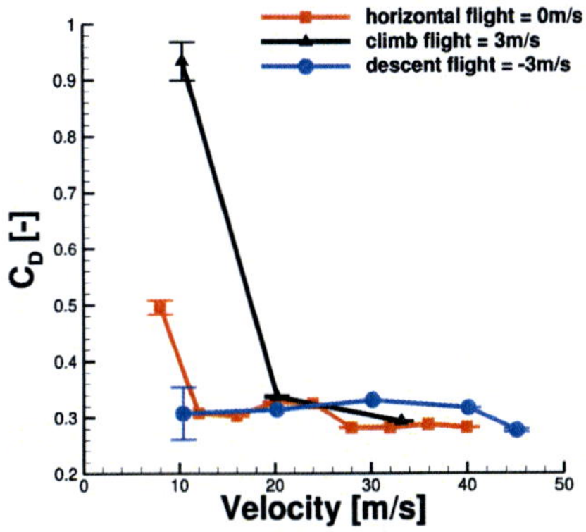

Fig. 6. Drag coefficient over velocity

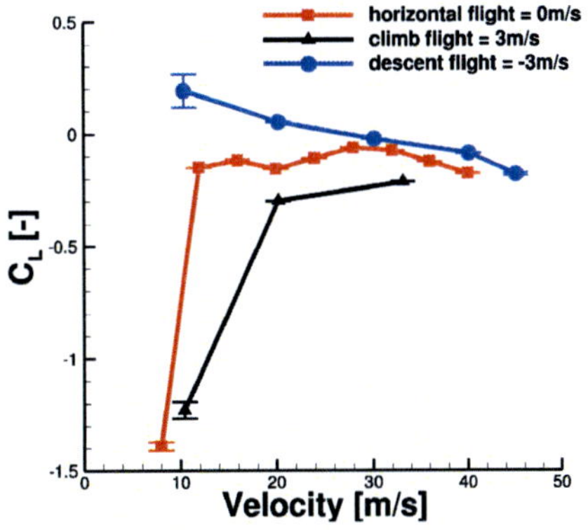

Fig. 7. Lift coefficient over velocity

Figure 6 shows the drag coefficient over speed. It is shown that the coefficient of drag moves in a range from 20 m/s to 45 m/s also for different climb and sink rates in a similar range. Figure 7 shows the lift coefficient over velocity. Positive lift coefficients counteract the weight of the helicopter. In the range between 20 m/s and 45 m/s, no major changes in the lift coefficient are detected too. Whereas the lift coefficient changes between descent flight, horizontal flight and climb flight. This shows that the fuselage generates additional lift during descent and thus relieves the main rotors. A comparison of the flight polars and fuselage polars for a speed of 32 m/s shows that both, the lift coefficient and the drag coefficient, are reduced by the rotor downwash.

Further, it is clarified why the CoAX 2D shows for horizontal flight and climb flight an increase of the drag coefficient on the one hand and a decrease of the lift coefficient on the other hand for speeds below < 20 m/s for climb flight and below < 12 m/s in horizontal flight. Figures 8–11 show the dimensionless eddy viscosity for different speeds in the symmetry plane. The eddy viscosity is an artificial viscosity added to the RANS equations by the turbulence model and may be used as an indication for regions of separately flows. Figure 8 shows for 8 m/s in horizontal flight a different flow separation area behind the fuselage than for 12 m/s in Fig. 9. If the speed continues to increase, the flow separation area remains similar to 12 m/s (Fig. 9). A flow separation at the horizontal stabilizer for 8 m/s (Fig. 8) created by rotor downwash explains the decrease in the lift coefficient and shows that the horizontal stabilizer operates at a high negative angle of attack. Similar behavior can be observed during climb flight.

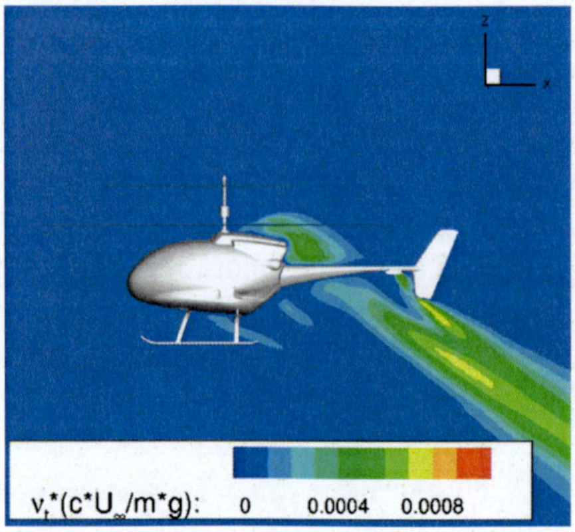

Fig. 8. Horizontal flight, velocity = 8 m/s

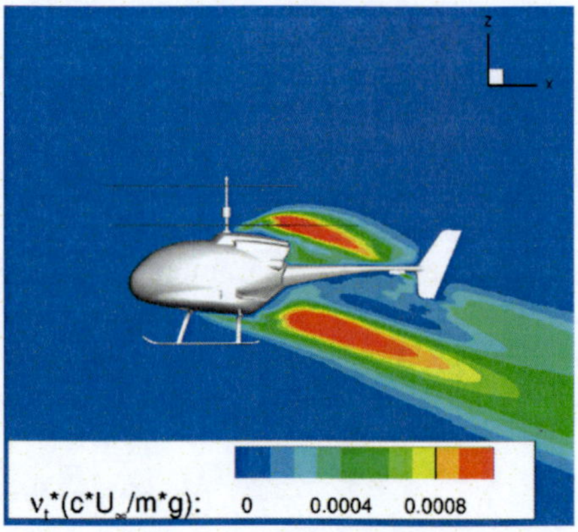

Fig. 9. Horizontal flight, velocity = 12 m/s

In order to find an explanation for the difference in drag and lift coefficient between climb and descent flight, the dimensionless eddy viscosity is also shown in Fig. 10 for a climb speed of 3 m/s and Fig. 11 for descend speed of −3 m/s with a horizontal speed of 10 m/s. It is seen that especially the climb has a much larger and different separation area than the descent. There is also a similarity between the flow separation area for the descent flight in Fig. 11 and the horizontal flight in Fig. 9. This is also shown in similar drag coefficients (Fig. 6). During climb the local eddy viscosity is smaller, but a flow separation of the horizontal stabilizer created by rotor downwash is observed and shows that the horizontal stabilizer operates at a high negative angle of attack.

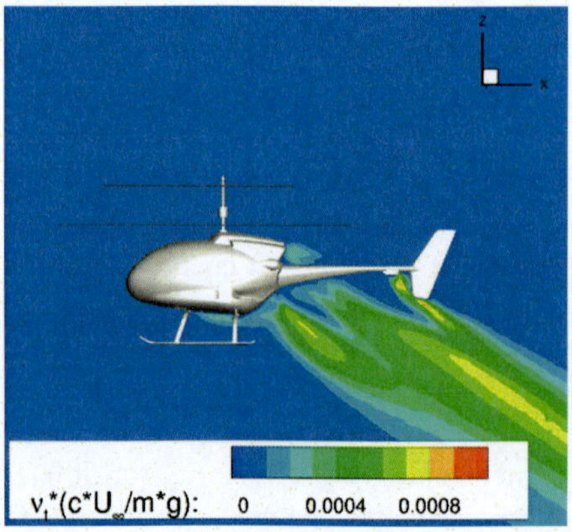

Fig. 10. Climb flight, velocity = 10 m/s

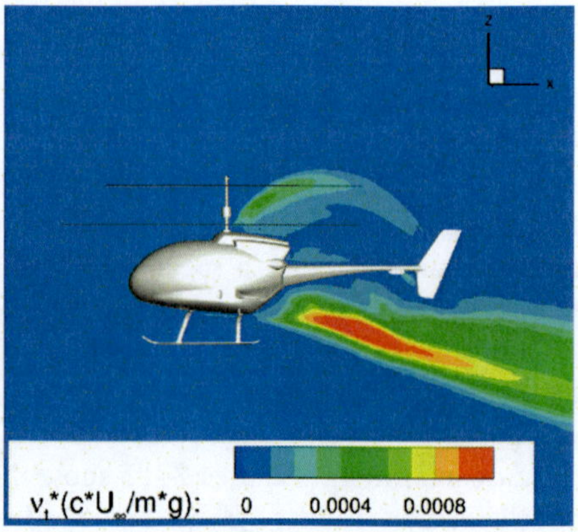

Fig. 11. Descent flight, velocity = 10 m/s

This is noticed in the lift coefficient Fig. 7 and in the pitching moment coefficient in Fig. 12, where the pitching moment coefficient has its highest value both for horizontal flight at 8 m/s and for climb flight at 10 m/s. For descent flight there is no flow separation of the horizontal stabilizer at 10 m/s speed. The higher the speed, the smaller the influence of the main rotors and the pitching moment coefficient decreases. A comparison of the yawing moments in Fig. 13 shows an increase from descent to climb.

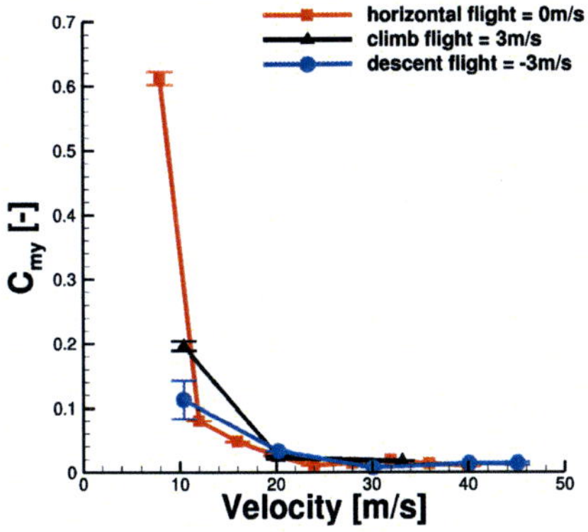

Fig. 12. Pitching moment over velocity

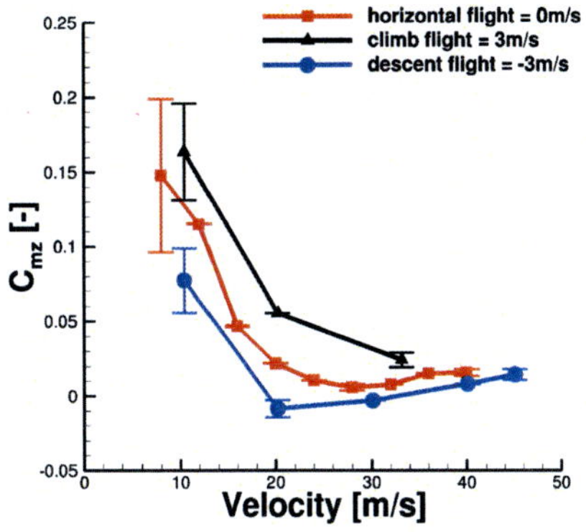

Fig. 13. Yawing moment over velocity

To validate the simulations, the results of the flight test are compared with the results of the flight polar curve. Figure 14 shows the lift of the elevator for the experiment as well as for the Camrad II simulation and the TAU results. A positive lift value indicates a down force at the horizontal stabilizer of the CoAX 2D and has the same direction as the weight force at the helicopter. It shows a good agreement between the simulations and the experimental data.

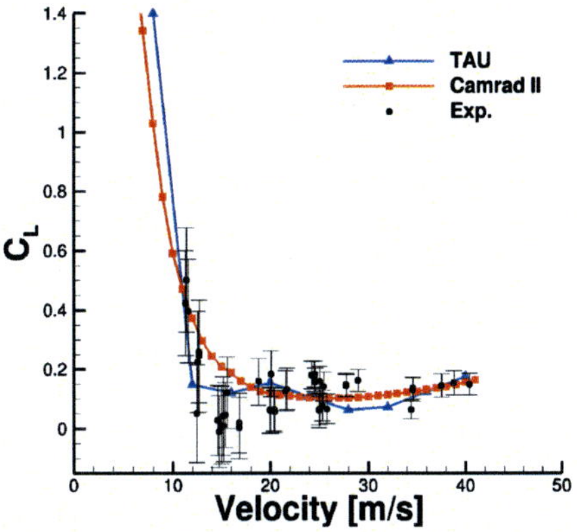

Fig. 14. Comparison the force at the horizontal stabilizer between measurement data and simulations

3.3 Fuselage Optimization

In consultation with the company edm-aerotec, an optimization of the landing gear was decided upon, because in the airworthiness requirement LTF-ULM 2019 [10] the maximum take-off weight was raised from 450 kg to 600 kg and the current landing gear had to be adapted to the higher weight, thus giving room for aerodynamic optimization. The current landing gear consists of two skid supports which are mounted in the front and rear area of the fuselage and which connect the skids with the fuselage. The skid supports consist of a rectangular profile with rounded corners, which in the case of the CoAX 2D are mounted underneath the fuselage. The skids are made with a circular profile and are bent upwards at the beginning and end. In order to evaluate possible modifications of the landing gear and to determine the potential of a landing gear optimization, a fuselage simulation without landing gear is carried out. The data point is selected with a speed of 32 m/s in horizontal flight from the flight polar and recalculated on a new mesh without landing gear. Care is taken to ensure that the topology of the new and the original mesh are similar in many areas as possible. Due to the missing landing gear in the new mesh, it has 0.9 mio fewer mesh points than the original mesh.

Figure 15 shows the dimensionless wall shear stress coefficient. Red areas are negative wall shear stress values and are synonymous with separated flow. The blue areas are positive wall shear stress values and are synonymous with attached flow. By omitting the landing gear, the drag coefficient is reduced by 23.4% for the overall configuration. This is partly due to the reduction of the flow separation in the lower fuselage area. In order to exploit as much of this potential, the drag of the landing gear and the interference drag would have to be reduced. It is possible to reduce the interference drag by minimizing the separation between the skid support and the fuselage. This is achieved by moving the connection between the fuselage and skid supports into the fuselage interior. In this type of assembly, the skids protrude sideways from the fuselage and do not disturb the flow below the fuselage compared to the previous configuration. A comparable skid support arrangement was successfully demonstrated on the Bluecopter Demonstrator (modified EC135) [11]. Due to the extensive structural changes and the associated redesign of the CoAX 2D, this type of optimization was not pursued any further. In order to find an optimization, which can be realized with reasonable effort on the CoAX 2D, a change of the skid support profile as well as the distribution of the skid support profiles are changed. Based on the chassis of the Bluecopter, a double ellipse is selected as the profile. The ellipse is adapted so that the height and thus the structural design of the previous rectangular profile is retained.

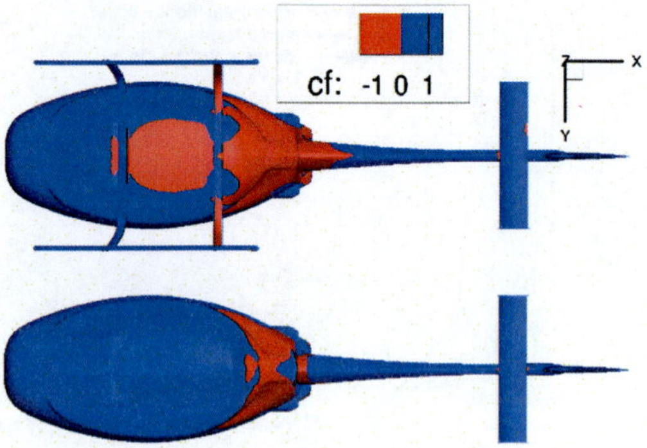

Fig. 15. Skin friction coefficient between with and without landing gear

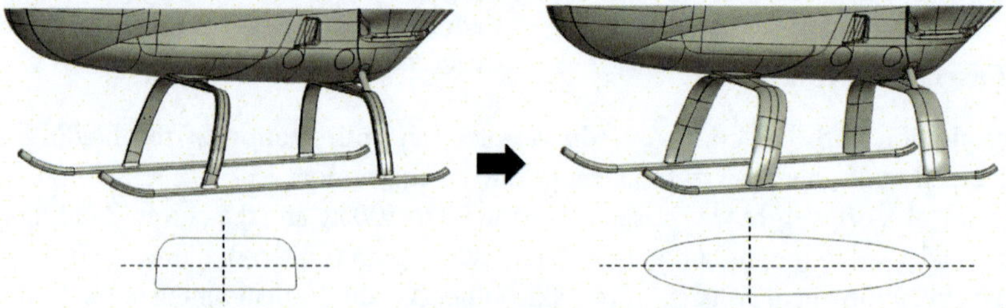

Fig. 16. Comparison between new and old geometry

Figure 16 shows the old skid support profile on the left and the new double elliptical profile on the right. In order to minimize the flow separation area between the skid beam and the fuselage behind the front skid beam, the front skid beam angle is changed.

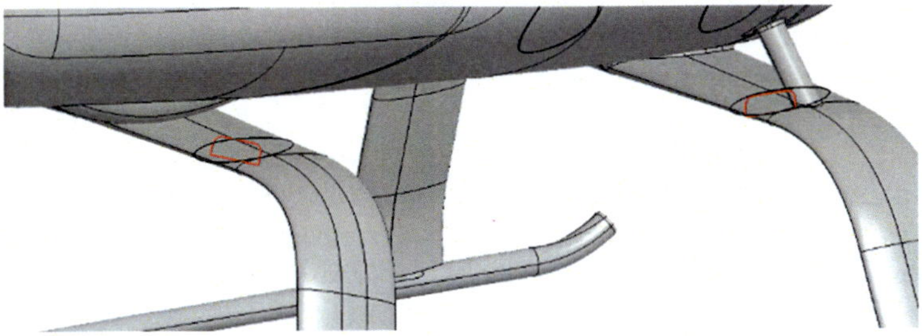

Fig. 17. Comparison between the old and new distribution

Figure 17 shows the reduction of the angle of attack in the front area of the skid support in comparison to the old profile (red). The distribution and the angle of attack of the old profile are taken over from the rear skid support. For the new landing gear, a new calculation mesh is also created and the data point from the flight polar with 32 m/s flight speed in horizontal flight was selected for comparison. Due to the modification of the landing gear, the new computing mesh has 0.3 mio mesh points more than the previous mesh. As seen in Fig. 18, these modifications led to a reduction of the total drag coefficient of 8.29%. On the one hand, this is due to the minimization of the flow separation in the area between the skid support and the airframe. On the other hand, the separation area behind the skid support is reduced as shown in Fig. 19.

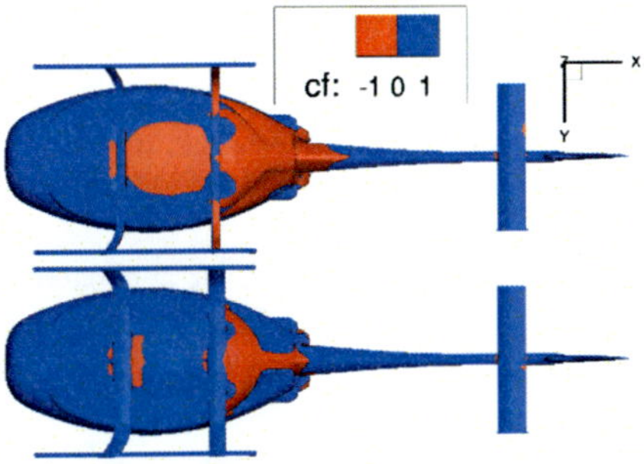

Fig. 18. Skin friction coefficient between old (above) and new landing gear (below)

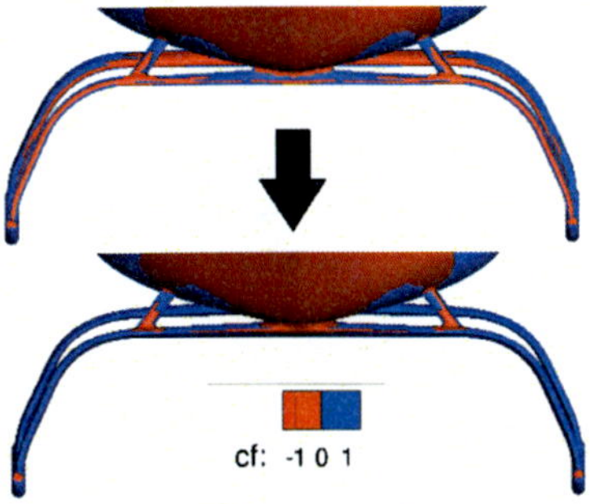

Fig. 19. Skin friction coefficient between old (above) and new landing gear (below), rear view

4 Conclusions

In this paper, the fuselage aerodynamics of the CoAX 2D ultra-light helicopter are examined. In order to get a comprehensive picture, the fuselage aerodynamics is evaluated. For this purpose, a variation of the angle of attack and the yaw angle at constant speed is simulated. The simulation is carried out without rotor downwash. In order to consider the influence of the rotor downwind, in the second step the downwash is modelled by means of Actuator Disc. In addition to the angle of attack, the speed is also varied. A comparison of the flight polars and fuselage polars shows that both, the lift coefficient and the drag coefficient, are reduced by the rotor downwash. Comparisons between the simulations and the flight data showed a good agreement. Based on these results, an optimization of the landing gear is carried out. It is shown that the total drag can be minimized by changing the landing gear around 8%.

Acknowledgments. The author of this paper would like to thank Roland Feil from the Technical University of Munich for providing the input data for the Actuator Disc. Further thanks go to the company edm aerotec. The Projekt CuRot was funded by the German Federal Ministry for Economic Af-fairs and Energy (BMWi) based on a resolution of the German Bundestag.

References

1. Wayne, J. (NASA Ames Research Center), Elmore, J.F., Keen, E.B., Gallaher, A.T., Nunez, G.F. (US Army Aviation Development Directorate - AFDD): Coaxial Compound Helicopter for Confined Urban Operations
2. Mikheyev, S.: 50 Years Experience, Status, Prospects. In: 33rd European Rotorcraft Forum, Plenary Meeting, Kazan, Russia, September 11–14, 2007
3. www.edm-aerotec.de
4. Feil, R., Rinker, M., Hajek, M.: Flight testing of a coaxial ultralight rotorcraft. In: American Helicopter Society 73rd Annual Forum, American Helicopter Soc., Fairfax, VA, pp. 1–14 (2017)
5. Feil, R., Eble, D., Hajek, M.: Comprehensive analysis of a coaxial ultralight rotorcraft and validation with full-scale flight-test data. J. Am. Helicopter Soc. **63**, 042004 (2018)
6. Fliegermagazin: Die UL-Helis kommen 05.2015
7. Schwamborn, D., Gerhold, T., Heinrich, R.: The DLR Tau-code: recent applications in research and industry. In: European Conference on Computational Fluid Dynamics (2006)
8. Raichle, A.: Flusskonservative Diskretisierung des Wirkscheibenmodells als Unstetigkeitsfläche, Dissertation, DLR-Forschungsbericht, DLR-FB-2017-62, 225 S (2017)
9. Johnson, W.: Technology drivers in the development of CAMRAD II. In: American Helicopter Society Aeromechanics Specialists' Conference, San Francisco, CA, January 19–21, 1994
10. Bauvorschriften für Ultraleichte Tragschrauber (einmotorig) BUT vom 15.01.2019
11. Bebesel, M., et al.: Bluecopter Demonstrator – an Approach to Eco-efficient Helicopter Design. In: 41st European Rotorcraft Forum (2015)

Effect of Vortex-Wake Interaction on Vortex-Rotor Interference and on Rotor Trim

Berend G. van der Wall(✉)

German Aerospace Center (DLR), Institute of Flight Systems (FT), Lilienthalplatz 7, 38108 Braunschweig, Germany
berend.vanderwall@dlr.de

Abstract. The flight of a helicopter within the wake of a preceding fixed-wing aircraft, for example during air-to-air refueling, is accompanied by vortex-rotor interactions. The full mutuality of vortex and rotor wake interactions requires a free-wake solution that is applied at different advance ratios in this paper and compared to existing simplified models. The results indicate that the simplified models are valid for higher advance ratios whereas very small advance ratios require the use of a free-wake approach.

1 Introduction

Helicopters can encounter wing tip vortices or those generated by deflected flaps from preceding fixed-wing aircraft, which becomes important during air-to-air refueling of helicopters: when approaching to the refueling position, while refueling, and when departing from it. Then the heavy tanker aircraft flies close to its minimum speed with deflected flaps, while the rotorcraft flies close to its maximum speed. Figure 1(a) shows this type of interaction and the fundamental kind of the resulting disturbance within the rotor disk is illustrated in Fig. 1 (b).

Aerodynamic interference is mentioned in most textbooks as an important subject that drives the design of components of a helicopter [1, 2]. Its analytic treatment is often very complicated and left to computational fluids dynamics (CFD) codes, for example the fuselage-rotor, rotor-tail rotor, rotor-empennage and other interaction problems. A good overview of the multitude and mutuality of interactions amongst helicopter components and with the outer environment is given in [3]. The problem of an external vortex interacting with the helicopter main rotor and its wake has been investigated since the 1970s regarding the helicopter flight into a wide body aircraft wake with respect to flight dynamic response and handling qualities of the rotorcraft [4–6]. It was concluded that the heavier (larger) the helicopter, the less impact the aircraft wake has on the helicopter. Recently helicopter flight within offshore wind parks was the subject of a GARTEUR research activity [7]. The work encompassed a survey of experimental and numerical studies about the wind turbine wake characteristics in the near and far field behind them; computer simulations to assess stability and handling qualities of a helicopter flying across wind turbine wakes; and piloted simulations using ground-based simulators.

© The Author(s), under exclusive license to Springer Nature Switzerland AG 2021
A. Dillmann et al. (Eds.): STAB/DGLR Symposium 2020, NNFM 151, pp. 407–417, 2021.
https://doi.org/10.1007/978-3-030-79561-0_39

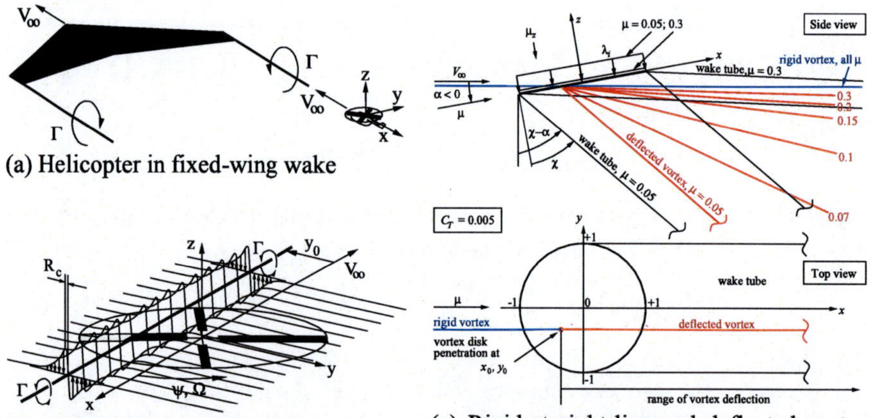

(a) Helicopter in fixed-wing wake

(b) Helicopter rotor with a line vortex

(c) Rigid straight-line and deflected vortex models for different advance ratios

Fig. 1. Illustration of vortex – rotor interaction.

In the past, most numerical investigations considered the vortex-rotor interaction by assuming the vortex as space-fixed (rigid straight-line vortex model) and only considering its induced velocities at the rotor blades; while vortex-wake mutual interactions were ignored. This article introduces a unidirectional interaction by deflecting the vortex when penetrating the rotor disk (deflected vortex model) and the full mutuality of interaction by application of a free-wake method to both the external vortex and the rotor wake. The strongest simplification is assuming a rigid straight-line vortex without any vortex-wake interaction at all as illustrated in Fig. 1 (c) by the blue line. This is the most conservative approach, leading to the strongest vortex impact on rotor trim. For a vortex lying planar in the rotor disk an analytical solution was derived for the rotor controls needed for disturbance rejection [8], or the perturbations developing in rotor blade flapping [9], or the thrust and power disturbances that occur when no controls are applied [10]. A vortex with its axis perpendicular to the rotor plane was investigated in [11].

A simplified unidirectional interaction where the vortex is deflected into the rotor wake slipstream, but without impact on the slipstream itself, was recently investigated in [12]. This is exemplified in Fig. 1 (c) for a wide range of advance ratios μ for a constant thrust coefficient C_T by the red lines, indicating the deflected part of the vortex after reaching the rotor disk. Because of diminishing induced inflow ratio λ_i at high speed the wake skew angle χ and with it the vortex deflection becomes very small and the vortex remains close to the rotor disk. In [13] CFD was applied to the problem for a few cases and results were compared to rigid vortex assumptions. The full mutuality of interactions between vortex and rotor wake showed a significant reduction of vortex-rotor interaction impact on trim.

An intermediate level of fidelity can be obtained by a free-wake simulation where both the vortex and the rotor wake interact with each other. This is the approach used in this paper and the obtained results will be compared to the other methods. All the simplified approaches – including free-wake methods – are typical for comprehensive rotor codes, while CFD codes inherently include all interactional effects, but at a computational effort some orders of magnitude larger.

2 Technical Approach

Rigid Straight-Line Vortex Model. A Vatistas' type of vortex swirl model as given in Eq. (1) is used, which computes the normal vortex-induced velocity at any rotor blade element (x, y) or radial and azimuth position (r, ψ) for a vortex axis lateral position y_0 with respect to the rotor center. Γ_V is the dimensional and λ_{V0} the nondimensional vortex circulation strength, referenced to the interacting rotor's tip speed ΩR and radius R. λ_{V0} is a linear measure for the impact on the trim of the encountering rotor, expressed by the perturbations of the collective and cyclic control angles $\Delta\Theta_0$, $\Delta\Theta_S$, $\Delta\Theta_C$ needed to mitigate the vortex effect on trim. A nonlinear measure is the nondimensional vortex core radius $r_c = R_c/R$. The wing vortex core radius R_c will be large compared to the rotor blade chord length c [10]. r_c is kept almost constant for all computations.

$$\lambda_V = \frac{\Gamma_V}{2\pi\Omega R^2}\frac{y - y_0}{(y - y_0)^2 + r_c^2} = \lambda_{V0}\frac{r\sin\psi - y_0}{(r\sin\psi - y_0)^2 + r_c^2} \tag{1}$$

Deflected Vortex Model. This model [12] makes use of momentum theory as outlined in [1, 2] to compute the wake skew angle χ as a function of the rotor thrust coefficient C_T, the advance ratio μ and the inclination angle α of the rotor's tip path plane. The induced inflow ratio λ_i needs to be solved from the quartic equation given in Eq. (2), where λ_h is the induced inflow ratio in hover. From the point where the vortex reaches the tip path plane at (x_0, y_0), see Fig. 1 (c), the vortex gets deflected by the wake skew angle χ. For the majority of computations $x_0 = 0$ is used throughout in the paper because this allows for the widest range of vortex lateral positions inside the rotor disk $-1 \leq y_0 \leq 1$, see Fig. 1 (c), top view, where for a range of advance ratios the deflection angles are indicated by the red lines.

$$\begin{aligned} &0 = \lambda_i^4 + 2\mu_z\lambda_i^3 + \mu^2\lambda_i^2 - \lambda_h^4; \qquad \mu = V_\infty\cos\alpha/(\Omega R); \qquad \mu_z = -\mu\tan\alpha \\ &\chi = \arctan\left[\mu/(\lambda_i + \mu_z)\right]; \qquad \lambda_h = \sqrt{C_T/2} \end{aligned} \tag{2}$$

Free-Wake Solution. The free-wake code applied here is based on [16], but the classical vortex lattice method has been replaced by continuous bilinear vorticity distributions for the trailed and shed vorticity in the wake elements, who are combined with a "thickness" representing a core radius. This avoids adaptive grid refinement when evaluation points get close to wake panels. A grid coarsening for points far away from the wake panels is applied to accelerate the computation. The external vortex is generated by a fixed wing located one radius upstream of the rotor. Its circulation strength is computed by the angle of attack and lift curve slope of the wing. In order to have just one tip vortex interacting with the rotor, the wing is oriented vertically with its upper tip just above the rotor hub center and its lower tip a wing span below. This way the concentrated vorticity in the vicinity of the wing tip enters the rotor located downstream of the wing and mixes with the rotor wake, while the other wing tip vortex remains sufficiently away below the rotor wake. The only purpose of this wing is to act as a vortex generator.

Rotor Simulation Tools. DLR's high resolution comprehensive rotor simulation code S4 [15] was applied to the Bo105 main rotor and HOST – the Helicopter Overall Simulation Tool of Airbus Helicopters [14] – was applied to a Bo105 helicopter fixed in space and subjected to an aircraft wing tip vortex [10]. S4 computes the dynamic response of elastic rotor blades with flap, lag and torsion degrees of freedom. The induced velocities caused by the rotor thrust are represented by the Mangler/Squire model and the resulting induced inflow distributions depend on C_T, μ, α. Unsteady aerodynamic modeling of blade element forces and pitching moment are incorporated. For the free-wake solution DLR's VAST (Versatile Aeromechanic Simulation Tool) is used with an articulated rigid blade model of the Bo105 main rotor. The model includes flapping and lead-lag motion with equivalent hinge offset and springs to match rotating natural frequencies of first flap and lead-lag modes. No torsion is taken into account. Linear incompressible aerodynamics are applied to the rotor blade elements and a free-wake is used for both the rotor wake and the wake of the fixed wing located upstream of the rotor. VAST is under development and no publications about it are available.

Rotor, Vortex and Wing. The Bo105 main rotor blades with $R = 4.91$ m are discretized into 10 blade elements in VAST and 20 within S4 with progressively smaller elements towards the blade tip in order to better resolve the increasing dynamic pressure along span. The external vortex' core radii, referenced to R, are $r_c = 0.081$ in VAST and 0.1 in HOST and S4, thus the results using VAST with its free-wake may lead to larger control angles than those of S4 with larger core radius and finer blade resolution. A lift curve slope of $C_{l\alpha} = 6.9$ is used in VAST and a quadratic drag coefficient polar is employed. While the analytical results are based on $C_{l\alpha} = 2\pi$, the S4 uses airfoil polars including compressibility. An azimuthal resolution of 3.6 deg is used in VAST, while the S4 code was run with 2 deg resolution. The rotor shaft angle of attack is prescribed. The wing has a span of 5 m and an aspect ratio of 5, discretized into 10 equally spaced blade elements. The core radius in the free-wake code was set to $R_c = 0.4$ m.

Rotor Trim and External Vortex Strength. First, a rotor trim without an external vortex and thus with undisturbed air at the rotor is performed for a range of advance ratios from $\mu = 0$ (hover) to 0.4 representing high speed flight. The rotor thrust is prescribed to 24200 N ($C_T = 0.0055$) and the hub moments to 0 Nm. These are the "Baseline" cases. The majority of VAST computations is limited to four selected advance ratios $\mu = 0.1, 0.2, 0.3, 0.4$ and the vortex or wing is inserted into the flow with its lateral position relative to the rotor hub center varied in the range $-2 \leq y_0 \leq 2$ at 11 positions. At any of these a retrim of the rotor was performed and the difference in control angles relative to the baseline trim is evaluated. One computation with VAST is performed at $\mu = 0.05$ with $y_0 = 0$, but obtaining a converged trim solution at this low advance ratio takes much more time than at higher μ. This is due to the current trim algorithm in VAST when using free-wake, which will be changed in future. In hover no simulation was performed with VAST because with this setup the wing would be standing in still air thus would not generate a vortex at all, leaving the rotor in undisturbed air. The wing circulation strength depends on μ. In the analytical model and in S4 the external vortex is modeled as space-fixed and this can be done at any advance ratio from zero to high speed and the vortex parameters are prescribed. Due to the differences in modelling the results obtained with the different methods can be compared only qualitatively.

3 Results

The rigid straight-line vortex model is used by the analytic model, S4 and HOST, the deflected vortex model by S4 and the free-wake model in VAST includes the full mutuality of vortex-wake interaction. To eliminate the linear influence of λ_{V0} on the results the controls (in rad) required to retrim the rotor are divided by it: $\Delta\Theta/\lambda_{V0}$ [8–12].

Very Small Advance Ratio ($\mu = 0.01$). This is very close to hover with a shaft angle of $\alpha_S = -0.01°$ and the slipstream angle χ is very small. The semi-empirical vortex deflection model can be applied, but it is still close enough to hover to be compared with hovering results using the rigid straight vortex assumption. Results from the analytical model, HOST, S4 with rigid straight-line and deflected vortex and are given in Fig. 2.

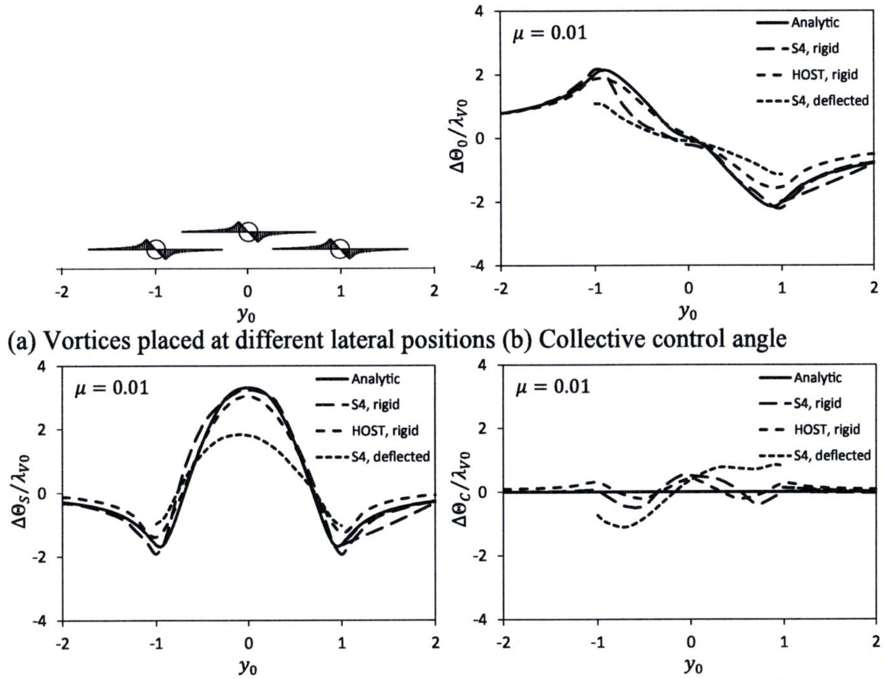

(a) Vortices placed at different lateral positions (b) Collective control angle

(c) Longitudinal cyclic control angle (d) Lateral cyclic control angle

Fig. 2. Vortex influence on rotor controls, $r_c = 0.1, x_0 = 0$.

The collective control curves in Fig. 2 (b) for the analytical model and S4 using the rigid straight-line vortex are very close, while HOST predicts a little less magnitude, which is caused primarily by a coarser blade element resolution [10]. The vortex deflection model requires only about half of the control magnitude, because the vortex is straight and almost parallel to the rotor disk only in its front half while from the center on the vortex is steeply inclined downwards and therefore far away from the rear half of the rotor. Thus, its influence is mainly confined to only half of the disk and only half of the control magnitude is needed, compared to a rigid vortex [12]. The same observations can be made for the longitudinal control angles given in Fig. 2 (c). While the analytical model, S4 and HOST with rigid vortex assumptions generate results that are very close to each other, the deflected vortex model requires only about half of the control magnitude. The results obtained for the lateral control angle differ as shown in Fig. 2 (d). The analytical model requires no control input because of the longitudinal symmetry of induced velocities, the assumption of a central hinge with rigid blades, only flapping degree of freedom, and linear steady aerodynamics. S4 and HOST make use of elastic blades in flap, lag and torsion; therefore some coupling effects result in a minor amount of lateral control angle. The deflected vortex model requires more control angle input than the rigid straight-line vortex model because of the fore-aft asymmetry of vortex-induced velocities: a deflected vortex placed at $y_0 = -1$ induces downwash in the front half of the disk as sketched in Fig. 2 (a), but not in the rear half of the disk due to its deflection at $x_0 = 0$. This longitudinal asymmetry of downwash requires a negative lateral cyclic control to compensate the loss of lift in the front side. In reverse, a vortex position at $y_0 = +1$ induces upwash in the front half of the disk, but not so in the rear, and a positive lateral cyclic control angle is required to compensate this.

Variation of Advance Ratio. Results for advance ratios from $\mu = 0$ to 0.4 are obtained here only with VAST and free-wake, Fig. 3. For the hovering case with $\mu = 0$ no computation was performed with VAST, but obtained by physical consideration. At

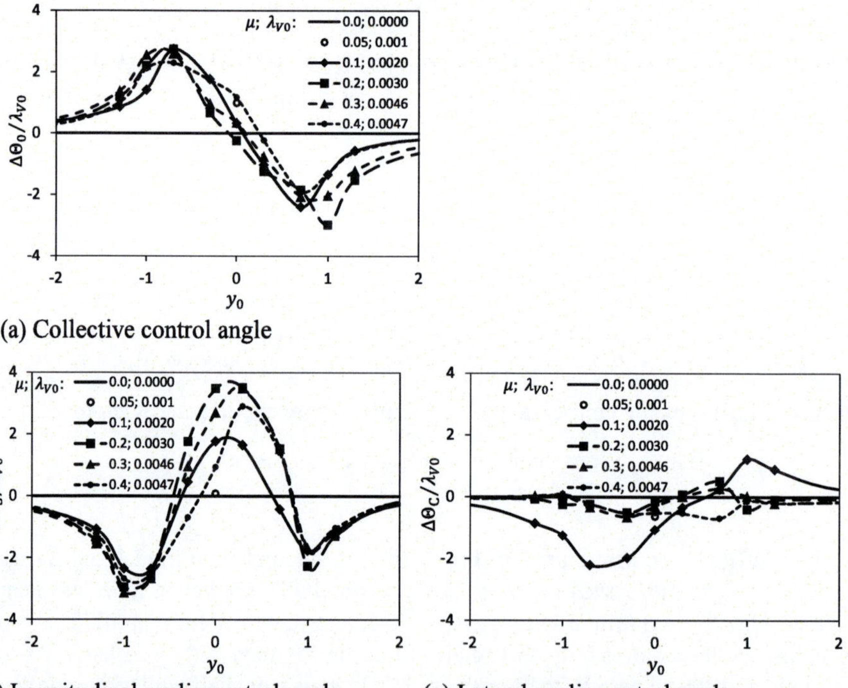

(a) Collective control angle

(b) Longitudinal cyclic control angle (c) Lateral cyclic control angle

Fig. 3. Influence of the advance ratio on rotor controls, free-wake results.

$\mu = 0.05$ the convergence of trim was very time consuming (due to the current trim algorithm in VAST when using free-wake that will be changed in future), therefore only one case with $y_0 = 0$ was computed (circle symbol).

The collective control angle shown in Fig. 3 (a) appears rather similar for every advance ratio. The lateral symmetry at small advance ratios gets progressively deformed with increasing advance ratio due to the lateral asymmetry of dynamic pressure acting on the rotor blades. The peak values become smaller with increasing advance ratio. More variation with advance ratio is found in the cyclic control angles, Fig. 3 (b) and (c). The longitudinal cyclic control angle shows less magnitude at the small advance ratio of $\mu = 0.1$, largest values for $\mu = 0.2$ and 0.3, and smaller again at $\mu = 0.4$, but with increasing asymmetry due to increasing difference in dynamic pressure on advancing and retreating side. Interesting is the larger amount of lateral control angle required at $\mu = 0.1$, which in trend was predicted by S4 with the deflected vortex model, but for an even smaller advance ratio, see Fig. 2 (d).

The hypothesis that the controls should be zero in the hovering case is supported especially by the results for the longitudinal control angle, Fig. 3 (b), that has significantly smaller magnitudes for $\mu = 0.1$, compared to $\mu = 0.2$, especially visible at the vortex center position when $y_0 = 0$. A further reduction of μ to 0.05 (circle symbol) reduces the longitudinal cyclic control even more towards zero as assumed in hover. The lateral cyclic control angle shown in Fig. 3 (c) also has a large reduction when the advance ratio is reduced from $\mu = 0.1$ to 0.05.

In order to provide a feeling for the absolute values of control angles required to mitigate vortex effects, lets assume a vortex strength of $\Gamma_V = 67$ m^2/s, leading to $\lambda_{V0} = 0.01$ with Bo105 rotor radius and tip speed, using Eq. (1). Then, maximum values for vortex influence on controls are in the order of $\Delta\Theta/\lambda_{V0} \approx 4$, thus $\Delta\Theta \approx 0.04$ rad, or 2.3 deg. For larger vortex strength this becomes larger proportionally. At constant vortex strength Γ_V and constant rotor tip speed ΩR the magnitude of $\Delta\Theta$ is proportional to $1/R$, thus $\Delta\Theta$ is larger for smaller helicopters and smaller for larger helicopters than the Bo105.

4 Visualization of Fixed-Wing and Rotor Wakes

Small Advance Ratio ($\mu = 0.05$). Due to the small freestream velocity the wing circulation also is very small in this case: $\lambda_{V0} = 0.001$. The rotor wake developing without the wing and its vortex is shown in Fig. 4 (a), denoted as "Baseline", and in (b) with the wing wake. This side view indicates hardly anything has changed for the rotor wake, but the wing vortex, due to upwash induced by the rotor, is convected upwards before reaching the rotor disk. When reaching the rotor it is sucked into the rotor slipstream away from its formerly horizontal path of convection. Therefore, the simple assumption of the deflected vortex model is essentially valid, provided that the location of deflection is properly computed.

Moderate Advance Ratio ($\mu = 0.1$). The free-stream velocity still is relatively small and the wing vortex strength not large; in this case $\lambda_{V0} = 0.002$. As before, the isolated development of the rotor wake is shown in Fig. 5 (a). Compared to Fig. 4 the stronger

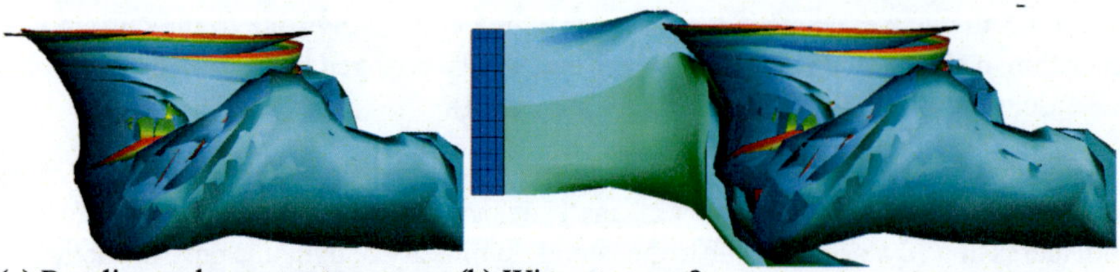

(a) Baseline wake, no vortex (b) Wing at $y\ = 0$

Fig. 4. Vortex-wake interaction, $\mu = 0.05$, side view.

wing circulation results in a stronger deformation of the rotor wake. It can also be seen that the wing's vortex (the blue colored part of its wake) does not enter the rotor disk in its center, but rather more shifted to the right by about 10% of the rotor radius, although the wing is placed at $y_0 = 0$. The side view is given in Fig. 5 (c) for the combination of both where the wing is located at $y_0 = 0$. Compared to the small advance ratio of $\mu = 0.05$ shown in Fig. 4 the rotor wake skew angle is already much larger and agrees with the momentum theory result given in Fig. 1 (c). The wing's wake is sucked into the rotor slipstream right behind the front of the rotor disk, and the entire wing wake is convected almost parallel to the rotor's slipstream. The bottom of the rotor wake is slightly lifted up relative to the Baseline wake, better seen in Fig. 5 (b), which is due to the sense of rotation of the wing vortex, inducing upwash on the rotor's retreating side and downwash on the other, as indicated by the red arrows.

(a) Baseline wake, no vortex (b) Wing at $y_0 = 0$

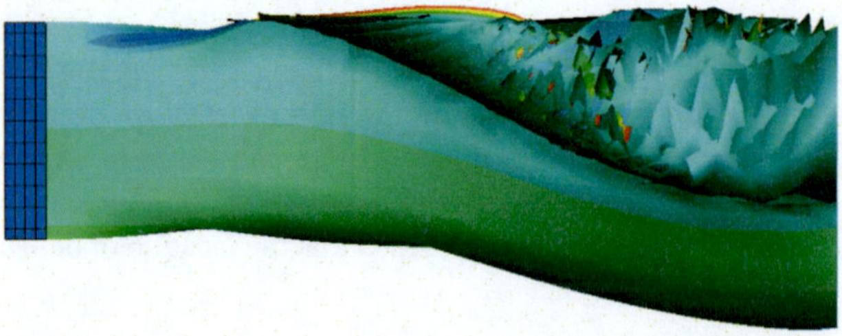

(c) Side view

Fig. 5. Vortex-wake interaction, $\mu = 0.1$.

High Advance Ratio ($\mu = 0.3$). Now the wing's vortex circulation is $\lambda_{V0} = 0.0046$, the rotor wake becomes flatter and due to the high freestream velocity both wakes are quickly convected downstream. The combination of both wakes for a central wing position at $y_0 = 0$ is given in Fig. 6 (b). Compared to the smaller advance ratios shown before, the wing's wake is now more strongly deforming the rotor's wake in the area where it is immersed into it. The wing's wake is deformed by the rotor wake to a much lesser degree. The side view on the wing and rotor wake is given in Fig. 6 (c), where the rotor wake appears more as a system of rolled-up blade tip vortices well separated from each other, Fig. 6 (b), which remains as such when the wing's wake is affecting the rotor. The wake skew angle is very shallow in agreement with momentum theory, compare with Fig. 1 (c).

(a) Baseline wake, no vortex

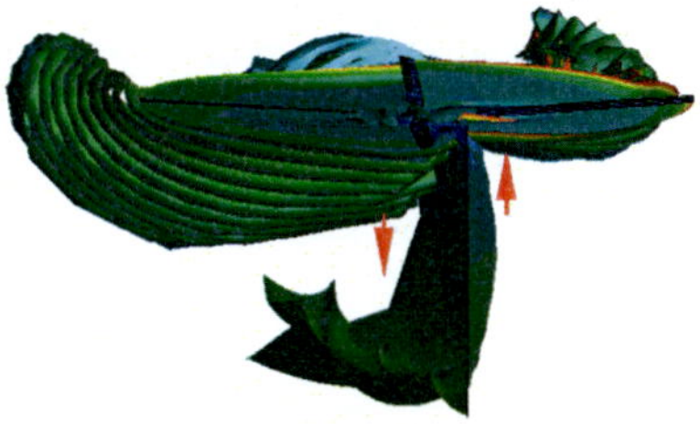

(b) Wing at $y_0 = 0$

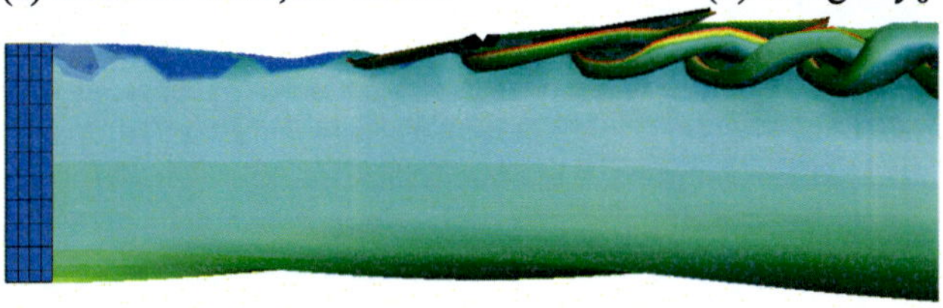

(c) Wing at $y_0 = 0$

Fig. 6. Vortex-wake interaction, $\mu = 0.3$.

5 Conclusions

Three different modeling approaches of increasing complexity are applied to the problem of an external vortex interacting with a rotor and its wake and the results compared to each other, namely

- a rigid straight-line vortex model without any vortex-wake mutual interaction
- a deflected vortex model representing a unidirectional interaction from the rotor wake on the otherwise straight-line vortex
- a free-wake solution including all mutuality of the vortex acting on the wake and vice versa

The main findings are summarized below.

1. For advance ratios $\mu \geq 0.2$ the simple methods using a rigid straight-line or deflected vortex are appropriate, because the vortex-wake mutual interactions are rather small within the rotor disk and develop mainly behind it, due to the large convection speed.
2. For smaller advance ratios around $\mu \approx 0.1$ the rigid straight-line vortex assumption is inadequate due to the strong wake deflection relative to the flight direction. The deflected vortex model generates reasonable results here, but the full mutuality of vortex-wake interactions provides better results, albeit requiring much higher computational effort.
3. For very small advance ratios down to hover also the deflected vortex model becomes questionable, because any vortex will be blown away by the rotor wake and the steady-state condition will be clean air. This can only be predicted by the free-wake approach, but at much higher computational effort compared to the comprehensive codes or the analytical method.

References

1. Johnson, W.: Rotorcraft Aeromechanics. Cambridge University Press, New York (2013)
2. Leishman, J.G.: Principles of Helicopter Aerodynamics. Cambridge University Press, New York (2000)
3. Sheridan, P.F., Smith, R.P.: Interactional aerodynamics – a new challenge to helicopter technology. J. Am. Helicopter Soc. **25**(1), 3–21 (1980). https://doi.org/10.4050/JAHS.25.1.3
4. Mantay, W.R., Holbrook, G.T., Campbell, R.L., Tamaine, R.L.: Helicopter response of an airplane's trailing vortex. J. Aircr. **14**(4), 357–363 (1977). https://doi.org/10.2514/3.58784
5. Azuma, A., Saito, S., Kawachi, K.: Response of a helicopter penetrating the tip vortices of a large airplane. VERTICA **11**(1), 65–76 (1987)
6. Turner, G.P., Padfield, G.D., Harris, M.: Encounters with aircraft vortex wakes: the impact on helicopter handling qualities. J. Aircr. **39**(5), 839–849 (2002). https://doi.org/10.2514/2.3004
7. Bakker, R., et al.: Wind turbine wakes and helicopter operations (overview of the GARTEUR HC-AG23 activities). In: 44th European Rotorcraft Forum, Delft, Netherlands, 18–21 September 2018
8. van der Wall, B.G., van der Wall, L.B.: Analytical estimate of rotor controls required for a straight vortex disturbance rejection. J. Am. Helicopter Soc. **62**(1), 015001-1–4 (2017). https://doi.org/10.4050/JAHS.62.015001
9. van der Wall, B.G.: Analytical estimate of rotor blade flapping caused by a straight vortex disturbance. J. Am. Helicopter Soc. **62**(4), 045001-1–6 (2018). https://doi.org/10.4050/JAHS.62.045001
10. van der Wall, B.G., Lehmann, P.H.: About the impact of wind turbine wake vortices on helicopter trim and rotor blade motion. In: 43rd European Rotorcraft Forum, Milano, Italy, 12–15 September 2017
11. van der Wall, B.G.: Estimate of rotor controls and blade flapping caused by an orthogonal vortex disturbance. J. Am. Helicopter Soc. **63**(2), 025000-1–6 (2018). https://doi.org/10.4050/JAHS.63.025001
12. van der Wall, B.G.: Effect of vortex deflection on vortex-rotor interaction. In: 45th European Rotorcraft Forum, Warsaw, Poland, 17–20 September 2019

13. Whitehouse, G.R., Brown, R.E.: Modelling a helicopter rotor's response to wake encounters. Aeronaut. J. **108**(1079), 15–26 (2004). https://doi.org/10.1017/S0001924000004954
14. Benoit, B., Dequin, A.-M., Kampa, K., von Grünhagen, W., Basset, P.-M., Gimonet, B.: HOST, a general helicopter simulation tool for Germany and France. In: 56th Annual Forum of the American Helicopter Society, Virginia Beach, VA, 2–4 May 2000
15. van der Wall, B.G., Lim, J.W., Smith, M., Jung, S.N., Bailly, J., Baeder, J.D., Boyd, D.D.: The HART II international workshop: an assessment of the state-of-the-art in comprehensive code prediction. CEAS Aeronaut. J. **4**(3), 223–252 (2013). https://doi.org/10.1007/s13272-013-0077-9
16. van der Wall, B.G., Roth, M.: Free-wake analysis on massively parallel computers and validation with HART test data. In: 53rd Annual Forum of the American Helicopter Society, Virginia Beach, VA, 29 April–1 May 1997

Check for updates

Comparisons of Different Spatial Schemes and Limiters for Helicopter Flows

Gunther Wilke(✉)

German Aerospace Center (DLR), Institute of Aerodynamics and Flow Technology, Lilienthalplatz 7, 38106 Braunschweig, Germany
gunther.wilke@dlr.de

Abstract. The requirements for a computational fluid dynamics solver to simulate the vortex-dominated helicopter rotor flows correctly are high. In recent years, many higher order methods and strategies have evolved to meet these requirements. This paper investigates a 3rd and 4th order accurate upwind scheme and compares this with a hybrid approach, where a 2nd order solution is generated near the rotor blades and a 4th order solution is computed in the wake. The investigated test cases are an inviscid vortex, an inviscid NACA0012 computation at zero-lift, two model rotors in hover and one model rotor in descent flight. From the exploration of different combinations of numerical schemes for the inviscid flux computation, it is seen that higher order methods are necessary to capture the trends of helicopter flows correctly. A modification to the upwind scheme is proposed to further decrease numerical dissipation and increase tip-vortex preservation. The outcome is that the 4th order accurate upwind scheme with the added modification prevails over the existing strategy.

Keywords: Higher order schemes · HARTII · SLAU2 · JST · Pade · MUSCL

1 Introduction

There is much ongoing research on what are the best practices for resolving the helicopter flow efficiently. The popular HART-II rotor test [1] has been subject of many code validations studies [2]. The conclusion is that a high spatial resolution as well as temporal resolution are necessary to resolve the blade tip vortices and the associated effects. Recent works by Lim et al. [3] demonstrate grid adaption to improve the prediction capabilities, while Jain et al. [4] depart from the classical 2nd order spatially accurate solution and employ a 5th order accurate scheme to enhance the simulation. Kowarsch et al. [5] also underline the necessity to use a higher order scheme to simulate a H-145 helicopter in descent flight. Öhrle et al. [6] combine grid adaptation with a 6th order method to further improve vortex preservation.

DLR's current approach to model helicopter flows with high precision is to use a mix of the spatially 2nd order central Jameson-Schmid-Turkel (JST) [7]

© The Author(s), under exclusive license to Springer Nature Switzerland AG 2021
A. Dillmann et al. (Eds.): STAB/DGLR Symposium 2020, NNFM 151, pp. 418–427, 2021.
https://doi.org/10.1007/978-3-030-79561-0_40

and the 4th order implicit Pade [8] scheme [9]. With the goal to further enhance vortex preservations, the SLAU2 scheme by Kitamura and Shima [10] has been implemented into DLR's legacy CFD code FLOWer with a 4th order reconstruction (FMCT) for structured meshes according to Yamamoto et al. [11], similar to JAXA's helicopter flow solver rFlow [12]. To further decrease the effective numerical dissipation, the minmod limiters of the FMCT are replaced by van Albada type limiters.

This paper explores the advantages of purely using the 2nd order central scheme, compared to the either 3rd or 4th order SLAU2 scheme and combining them with a 4th order implicit Pade scheme. To evaluate the efficiency of the various possible approaches, an inviscid vortex, an inviscid NACA0012 at zero angle-of-attack, the 7A/7AD rotors in hover [13], and the HART II rotor in descent flight are investigated.

2 Methodology

This section covers the individual methods used. The core flow solver is FLOWer [14], a block-structured solver. For unsteady simulations dual-time stepping is employed, for steady simulations local time stepping. The pseudo-time is advanced with a 5-stage Runge-Kutta scheme, which is further accelerated with implicit residual smoothing and the multi-grid technique. The SA-R turbulence model [15] is used for the viscous test cases. The helicopter simulations are coupled with the comprehensive code HOST [16] to take into account correct trim conditions as well as elasticity.

The Jameson-Schmidt-Turkel (JST) scheme is a central scheme of 2nd order and implemented into the finite-volume environment of FLOWer including matrix dissipation [7]. Through a mixture of first and third differences used in the dissipation terms, the scheme enables shock capturing, while also avoiding black-and-white ordering. The dissipation is also scaled with the perpendicular eigenvalues allowing it to handle highly stretched cells, which are often used in boundary layers. The dissipation constants for this scheme are chosen to be $k_2 = 1/2$ and $k_4 = 1/64$. The matrix dissipation is blended with scalar dissipation by 50% each for robustness.

The Pade scheme was originally developed by Lele [8]. The shortcomings of this scheme are that it does not capture shocks and grid quality requirements are high, thus it is impractical to be used for the blade meshes. However, previous research by the author [9] showed it is very well suited to be used in the Cartesian background grids, where it conserves the blade tip vortices well, while the 2nd order JST scheme is employed for the near-body meshes of the blades and fuselage. The dissipation coefficient α is set to 0.49.

The second version of the Simple Low-dissipation AUSM (Advection Upstream Splitting Method) by Kitamura and Shima [10] has recently been implemented into FLOWer in its rotational variant. The SLAU2 scheme is a upwind scheme and may be therefore understood to have a scalar dissipation, yet it yields robust results for a large range of Mach numbers and thus is well suited for helicopter flows from this perspective. No dissipation coefficient must be chosen, yet the required slope reconstruction employs limiters having a similar effect and introduces parameters to be chosen.

The fourth-order accurate compact MUSCL TVD (FCMT) scheme for structured meshes has been developed by Yamamoto et al. [11]. Like rFlow, the reconstruction of this scheme has been implemented into FLOWer. Since some modifications to the original version are made for FLOWer, the original version is stated first and briefly explained:

$$q^L_{l+1/2} = q_l + 1/6(\Delta^*\bar{q}_{l-1/2} + 2\Delta^*\tilde{q}_{l+1/2}) \tag{1}$$

$$q^R_{l+1/2} = q_{l+1} - 1/6(2\Delta^*\bar{q}_{l+1/2} + \Delta^*\tilde{q}_{l+3/2}) \tag{2}$$

$$\Delta^*\bar{q}_{l-1/2} = limiter2(\Delta^* q_{l-1/2}, \Delta^* q_{l+1/2}) \tag{3}$$

$$\Delta^*\tilde{q}_{l+1/2} = limiter2(\Delta^* q_{l+1/2}, \Delta^* q_{l-1/2}) \tag{4}$$

$$\Delta^* q_{l+1/2} = \Delta q_{l+1/2} - 1/6\Delta^3\hat{q}_{l+1/2} \tag{5}$$

$$\Delta^3\hat{q}_{l+1/2} = \Delta\hat{q}_{l+3/2} - 2\Delta\hat{q}_{l+1/2} + \Delta\hat{q}_{l+1/2} \tag{6}$$

$$\Delta\hat{q}_{l-1/2} = limiter3(\Delta q_{l-1/2}, \Delta q_{l+1/2}, \Delta q_{l+3/2}) \tag{7}$$

$$\Delta\hat{q}_{l+1/2} = limiter3(\Delta q_{l+1/2}, \Delta q_{l+3/2}, \Delta q_{l-1/2}) \tag{8}$$

$$\Delta\hat{q}_{l+3/2} = limiter3(\Delta q_{l+3/2}, \Delta q_{l-1/2}, \Delta q_{l+1/2}) \tag{9}$$

$$\Delta q_{l+1/2} = q_{l+1} - q_l \tag{10}$$

with $q^{L/R}_{l+1/2}$ the primitive variables to be reconstruction on either left or right side of the face $l + 1/2$. The computation of $\Delta^*\bar{q}_{l+1/2}$ and $\Delta^*\tilde{q}_{l+3/2}$ are analog to $\Delta^*\bar{q}_{l-1/2}$ and $\Delta^*\tilde{q}_{l+1/2}$, while $\Delta^* q_{l-1/2}$ and $\Delta^* q_{l+3/2}$ are analog to $\Delta^* q_{l+1/2}$. If $\Delta^* q_{l\pm1/2} = \Delta q_{l\pm1/2}$, then the reconstruction reduces to 3rd order. The constants b_1 and b_2 are set to the recommended values of 4 and 2. In the original paper, the *limiter* functions are based on the minmod limiter and read:

$$minmod2(x, y) = S_x \cdot max[0, min(|x|, b_1 \cdot S_x \cdot y)] \tag{11}$$

$$minmod3(x, y, z) = S_x \cdot max[0, min(|x|, b_2 \cdot S_x \cdot y, b_2 \cdot S_x \cdot z)] \tag{12}$$

$$\text{with } S_x = sign(x) \tag{13}$$

This limiter is relatively robust, but also fairly dissipative. Sweby [17] investigated many different forms of upwind limiters. A less dissipative limiter is the van Albada limiter. Thus, a modification of the original reconstruction is proposed, where the limiter2 and limiter3 function is replaced by an albada2 and an albada3 function:

$$albada2(x, y) = x \cdot (\hat{x^2} + x \hat{\cdot} y)/(\hat{x^2} + \hat{y^2}) \tag{14}$$

$$\Delta^*\bar{q}_{l-1/2} = albada3(\Delta q_{l-1/2}, \Delta q_{l,avg}) \tag{15}$$

$$\Delta q_{l,avg} = albada3(\Delta q_{l-1/2}, \Delta q_{l+1/2}) \tag{16}$$

$$albada3(x, y) = (\hat{x^2} \cdot y + \hat{y^2} \cdot x)/(\hat{x^2} + \hat{y^2}) \tag{17}$$

with $\hat{x^2} = x^2 + \epsilon$, $\hat{y^2} = y^2 + \epsilon$, $x \hat{\cdot} y = x \cdot y + \epsilon$ and ϵ as a safeguard to avoid division by zero. van Albada et al. [18] claim an insensitive behavior to the original limiter with respect to ϵ. For the inviscid 2D cases 10^{-4} and for the viscous 3D cases 10^{-7} is used as they allowed for good and robust convergence. $\Delta\phi_{l,avg}$ is the van Albada averaged value of the initial differences for the left and right state and is used as the limiting value for the 4th order terms. Note that the albada2 function is only inspired by the original van Albada limiter, but has been modified to work for the 3rd order terms as it would otherwise degrade the solution to 2nd order, if $\Delta^* q_{l\pm1/2} = \Delta q_{l\pm1/2}$.

Additional robustness is employed by a pressure switch similar to the JST scheme [19], which is applied to the FMCT reconstructed values.

3 Ehrenfried Vortex

The first test case is the inviscid Ehrenfried vortex [20], specifically designed for numerical investigations.

Cartesian meshes with 201^2, 101^2, 51^2, and 25^2 grid points are used. The vortex is placed inside these meshes without convection and either 2000, 1000, 500 or 250 global time steps with dimensionless time steps ($t = \gamma L/a$) of 0.0025, 0.005, 0.01 and 0.02 to reach at the final time of 5. For each of these grids, the analytical solution, as well as the numerical solution is generated. The JST scheme with matrix dissipation, a central scheme with 6th order Pade filtering (Pade-Filter), the 4th order Pade scheme with 6th order filtering (Pade), and the SLAU2 scheme with 3rd and 4th order reconstruction and either the minmod (MM) or van Alabada (vA) limiter are investigated.

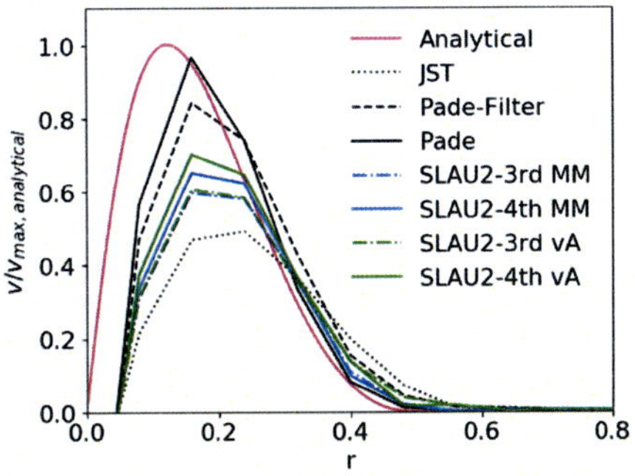

Fig. 1. Velocity profile of Ehrenfried vortex on coarsest grid (25^2) at time T = 5.

The velocity profiles of the coarsest mesh solution with 25^2 are shown in Fig. 1. It is seen that none of the methods is able to accurately resolve the vortex yet, however it is evident that the Pade scheme and Pade-Filter scheme are closest to resolving the vortex followed by the SLAU2 scheme with 4th order reconstruction and van Albada limiters. The 2nd order JST scheme only captures half of the vortex peak velocity. For increasing grid densities, the vortex preservation improves and on the 201^2 mesh all solutions match the analytical expression.

While this plot gives a good qualitative assessment of the schemes, Fig. 2(a) and Fig. 2(b) show the grid convergence with respect to the discrete error integral for each scheme as well as the cost associated to generate the individual solutions. The Pade and Pade-Filter scheme achieve the best performance and can be considered grid converged already on the 51^2 grid. The quality of each scheme is interpolated by fitting the following polynomial $a + b/\sqrt{t} + c/t + d/t^2$ where t is the computational time spent on a single-core CPU. If an error of 1e–3 is to

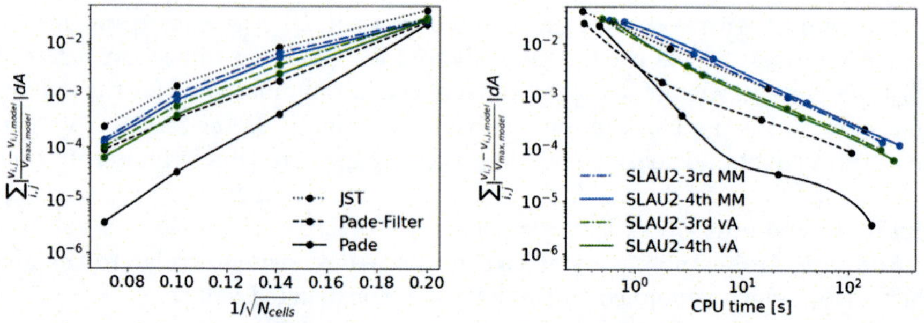

(a) Grid convergence (b) Time to solution

Fig. 2. Effect of grid points and resources on vortex preservation at time T = 5.

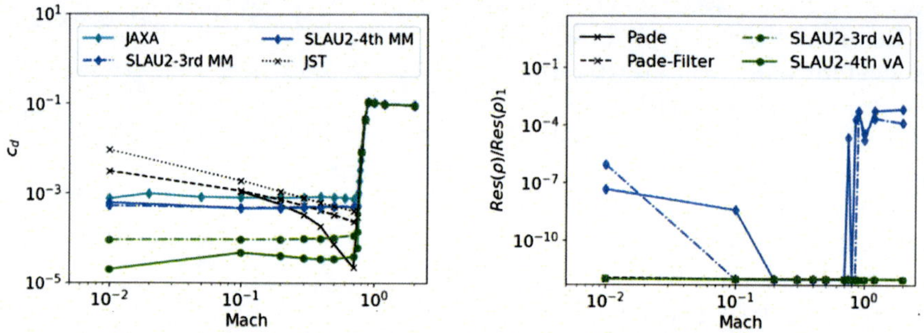

(a) Drag coefficient (b) Density residual

Fig. 3. Final drag coefficient and density residual of the NACA0012 test case.

be achieved, then the Pade scheme is the most efficient one with 1.9 s followed by the Pade-Filter scheme with 3.4 s and SLAU2 4th order van Albada limited scheme with 12.1 s. On the other side of the spectrum is the JST and SLAU2 4th order minmod limited scheme with 27.7 and 30.8 s, respectively.

4 Inviscid NACA0012

The NACA0012 airfoil is computed in inviscid flow at zero angle of attack. The test case is performed in a similar fashion and intent as to the rFlow paper by Tanabe et al. [12]: to study the effect of numerical dissipation onto the drag coefficient and the robustness of the individual schemes at various Mach numbers. The samples range from $M = 0.01$ to $M = 2.0$. Note that for subsonic flow, the compressibility effects vanish and thus the drag coefficient should go to zero for these inviscid cases. The same mesh Tanabe et al. [12] used is applied here to be able to compare results. It is of O-type and has (chordwise $\times$ normal) = (401×101) grids points.

The drag coefficients computed are visualized in Fig. 3(a) along with the achieved respective residual in Fig. 3(b) for the various schemes after a maximum of 200,000 iterations. The Pade and Pade-Filter schemes diverge for points M > 0.75, which is not surprising as they do not have shock capturing properties. Additionally, the Pade scheme diverged for the lowest Mach number (0.01). For the SLAU2 scheme with the minmod limiter, convergence could only be reached for subsonic Mach numbers and at very low Mach numbers difficulties start to arise again. These results, while looking plausible at first sight, should be considered critically as non-converged solutions are ambiguous.

The SLAU2 scheme with the van Albada limiter allows for the lowest drag coefficients for the subsonic Mach numbers, while the minmod limited solution is a magnitude higher and the JAXA is solution close to it. In contrast to the JST, Pade and Pade filtered schemes, the drag coefficients remain rather stable throughout the subsonic Mach numbers, while the JST, Pade and Pade filter schemes drop below the minmod limited solutions for Mach number greater Mach 0.4, 0.4 and 0.2 respectively. At Mach 0.7 the Pade scheme drops even below the van Albada limited solutions. The supersonic Mach number results are very close to each other for all schemes with the exclusion of the Pade based schemes for which no solution exists.

5 7A/7AD Rotors in Hover

The 7A and 7AD rotor in hover are simulated. The 7A rotor features a rectangular planform while the 7AD rotor has a tapered tip with a slight anhedral. Both rotors have a radius of 2.1 m. Data points 61 and 95 of the Helishape [13] wind tunnel campaign are selected for this case. In hover, the rotor blades constantly operate in the downwash and the vortices of the previous blades, therefore making it a very vortex dominant flight condition.

A periodic mesh is utilized with three grid sizes investigated. The finest mesh consists of 4 million mesh points and has a farfield distance of 6 radii above and below the rotor. The second mesh is generated by leaving out every other grid point in each index direction of the finest mesh reducing the number of points by a factor of 8. The same is done for the third mesh and thus only satisfies a $y+ < 2$. Dual-time stepping is applied with an equivalent time step to a rotation of $\Delta\psi = 0.5^o$ and a total of 16 revolutions is computed. The convergence control is done by requiring at least 10 inner iterations during each time step and a maximum of 100. The stopping criterion is a drop of four orders of magnitude of the initial density residual. The observed average number of inner iterations per time step is between 10.5 and 13.0 among the schemes. Only at the beginning of each coupling step the 100 inner iterations are used up.

The JST scheme and the SLAU2 scheme with a 3rd and 4th order reconstruction are tested. The Pade scheme is not applicable here due to the presence of shock waves and the grid quality. Again, for the slope reconstruction the minmod as well as the van Albada limiter are investigated.

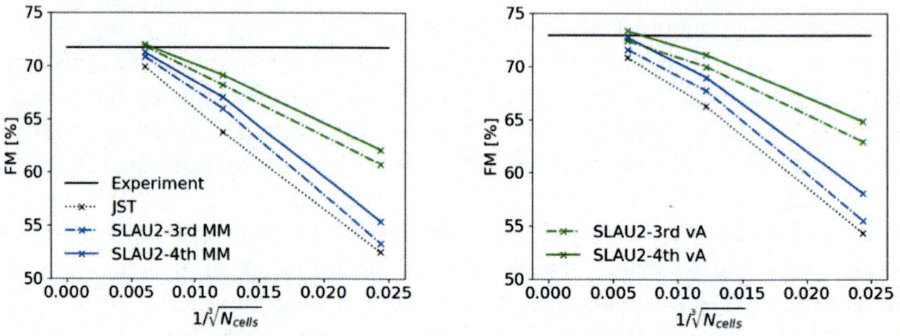

(a) 7A

(b) 7AD

Fig. 4. Figure of Merit for different grid sizes and rotors in hover.

As a metric for comparison, the resulting Figure of Merit of the coupled simulation is shown along with the experimental results of each rotor in Fig. 4. In both cases, increasing the number of grid points brings the obtained Figure of Merit closer to the experimental value, which is then recovered well on the finest mesh. The general trend is that the van Albada limited solutions are superior to the minmod limited solutions, even if the theoretical order is the same.

6 HART-II Rotor in Descent Flight

The HART-II baseline test case is simulated to study the effect of using higher order methods. It is also more computationally expensive than the hover test case, thus only two grid sizes are investigated. The coarse mesh is again derived from the finer mesh by leaving out every other grid point. The fine mesh has a total of 29 million grid cells and the coarser mesh 3.6 million grid cells. The same time stepping strategy as in the previous case is applied and the average number of inner iterations has increased and ranges from 13 to 24 inner iterations.

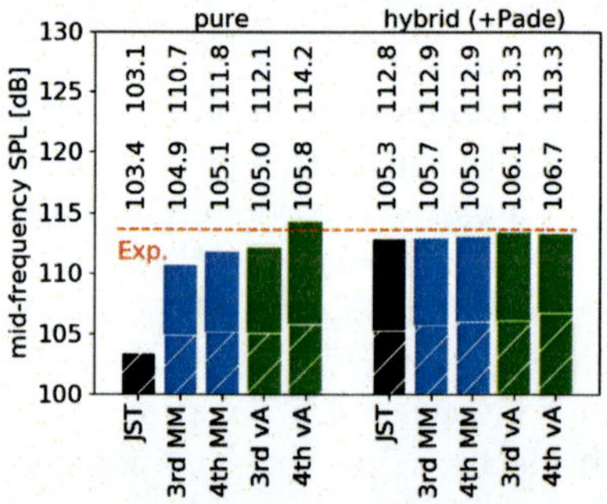

Fig. 6. Mid frequency peaks on the advancing side for the HARTII rotor. Dashed lines/bottom numbers are for the coarse grid, top numbers for the fine grid.

Besides the two grid densities, the investigation also covers the application of the Pade scheme in the background grid. Therefore, computations using the so far applied schemes are denoted 'pure' versus computations featuring the Pade scheme in the background grid are marked 'hybrid' simulations.

The metric looked at for comparison is the peak of the mid-frequency sound pressure level computed on the locations of the experimental noise carpet. The

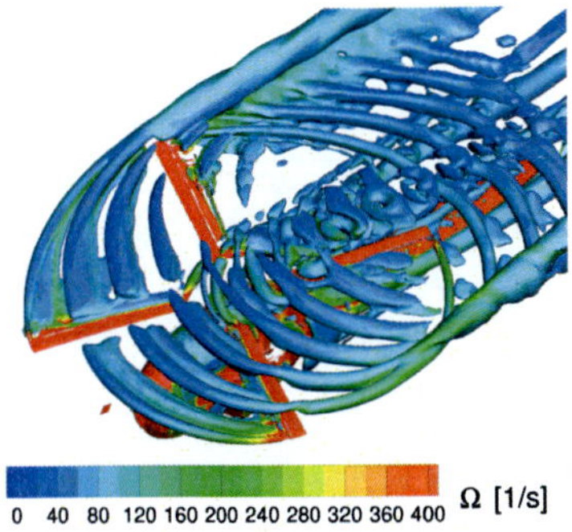

(a) JST

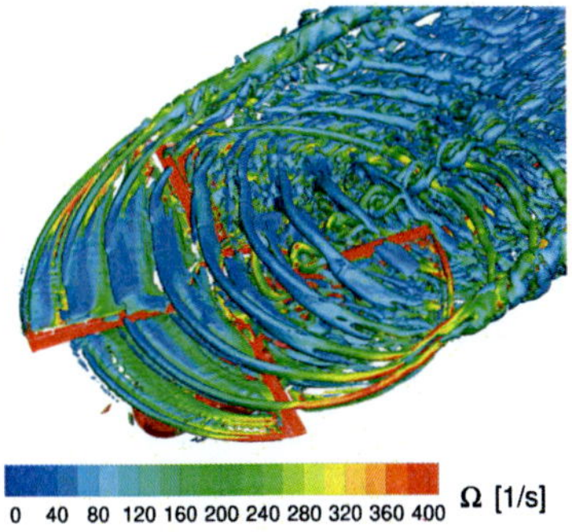

(b) SLAU2-4th vA

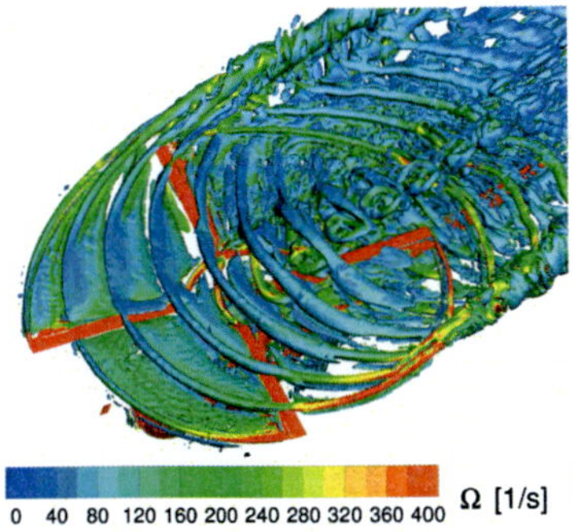

(c) JST+Pade

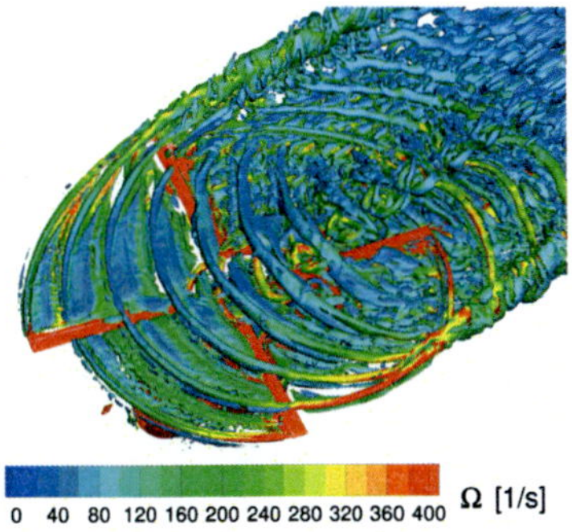

(d) SLAU2-4th vA+Pade

Fig. 5. Vortex fields visualized using λ_2-criterion ($\lambda_2 = -0.1$ [1/s]) on fine grid.

results are listed in Fig. 6. The effect of using a different scheme is little for the hybrid schemes since the vortex convection is accomplished by the Pade scheme. On the coarse mesh, the noise prediction slightly improves by using a higher order scheme and the van Albada limiter delivers slightly better results on top of that. However, on the fine mesh, the results are even closer to each other than for the coarse mesh. Since finding the trim solution is also an iterative process, the convergence of trim control angles to a residual angle of 0.01° degree may therefore not be sufficient anymore to strictly separate the effects of the trim convergence with the effect of the order as is the case for the 4th order minmod limited solution and the 3rd order van Albada limited solution.

A visual demonstration of the vortex preservation is given in Fig. 5 for a selection of the investigated schemes.

7 Summary

Within the cooperation between JAXA, ONERA and DLR concerning rotor optimization, the SLAU2 scheme from Kitamura and Shima [10] as well as a 4th order MUSCL reconstruction for structured meshes according to Yamamoto et al. [11] have been implemented into DLR's legacy flow solver FLOWer in a similar fashion as to JAXA's rFlow flow solver [12]. The effectiveness of this approach has been compared with current best practices by DLR, namely the usage of the JST scheme with Matrix dissipation [7] near the rotor blades and the Pade scheme by Lele [8] for the convection of the vortices and wake. An additional change to the MUSCL scheme has been made by exchanging the so far used minmod limiter with a van Albada limiter to reduce numerical dissipation further.

A range of test cases has been investigated: an inviscid vortex, an inviscid NACA0012 airfoil without lift, two hovering rotors and a rotor in descent flight. The conclusion is that the application of the higher order is generally beneficial over the 2nd order JST scheme. Additionally, employing the van Albada limiter brought further improvements over the minmod limited solutions. In the descent flight condition, much of the vortex preservation is already achieved by using the Pade scheme in the background mesh. Thus, for robustness, the pure application of the SLAU2 scheme with the 4th order van Albada limiter is recommended, while for efficiency the Pade scheme may alternatively be employed in the background mesh.

Aknowledgements. The author would like to thank the helicopters group at JAXA for their support and shared insight. Namely the rFlow developers Yasutada Tanabe and Hideaki Sugawara, but also Masahiko Sugiura for helping me with living in Japan during my exchange and enabling this research.

References

1. van der Wall, B.G.: A comprehensive rotary-wing data base for code validation: the HART II international workshop. Aeronaut. J. **115**(1164), 91–102 (2011)
2. Smith, M.J., Lim, J.W., van der Wall, B.G., Baeder, J.D., Biedron, R.T., Boyd, D.D. Jr., Jayaraman, B., Jung, S.N., Min, B.-Y.: An assessment of CFD/CSD prediction state-of-the-art using the HART II international workshop data. In: 68th Annual Forum of the American Helicopter Society, Ft. Worth, TX (2012)
3. Lim, J.W., Wissink, A., Jayaraman, B., Dimanlig, A.: Helios adaptive mesh refinement for HART II rotor wake simulations. In: Annual Forum 68 of the American Helicopters Society, May 2012
4. Jain, R.K., Lim, J.W., Jayaraman, B.: Modular multisolver approach for efficient high-fidelity simulation of the HART II rotor. J. Am. Helicopter Soc. (2015)
5. Kowarsch, U., Öhrle, C., Keßler, M., Krämer, E.: Aeroacoustic simulation of a complete H145 helicopter in descent flight. J. Am. Helicopter Soc. **61**(4), 042001–1 – 042001–13 (2016)
6. Öhrle, C., Schäferlein, U., Keßler, M., Krämer, E.: Higher-order simulations of a compound helicopter using adaptive mesh refinement. In: AHS International 74th Annual Forum & Technology Display (2018)

7. Swanson, R.C., Radespiel, R., Turkel, E.: On some numerical dissipation schemes. J. Comput. Phys. **147**(2), 518–544 (1998)
8. Lele, S.K.: Compact finite difference schemes with spectral-like resolution. J. Comput. Phys. **103**(1), 16–42 (1992)
9. Wilke, G.: Efficient aero-acoustic simulation of the HART II rotor with the compact Pade scheme. In: 42nd European Rotorcraft Forum (2016)
10. Kitamura, K., Shima, E.: Towards shock-stable and accurate hypersonic heating computations: a new pressure flux for AUSM-family schemes. J. Comput. Phys. **245**, 62–83 (2013)
11. Yamamoto, S., Kano, S., Daiguji, H.: An efficient CFD approach for simulating unsteady hypersonic shock-shock interference flows. Comput. Fluids **27**(5), 571–580 (1998)
12. Tanabe, Y., Saito, S.: Significance of all-speed scheme in application to rotorcraft CFD simulations. In: The 3rd International Basic Research Conference on Rotorcraft Technology, Nanjing, China (2009)
13. Schultz, K.J., et al.: A parametric wind tunnel test on rotorcraft aerodynamics and aeroacoustics (HELISHAPE). Test procedures and representative results. In: 22nd European Rotorcraft Forum (1996)
14. Raddatz, J., Fassbender, J.K.: Block structured Navier-Stokes Solver FLOWer. In: Kroll, N., Fassbender, J.K. (eds.) MEGAFLOW - Numerical Flow Simulation for Aircraft Design, pp. 27–44. Springer, Heidelberg (2005)
15. Dacles-Mariani, J., Kwak, D., Zilliac, G.: On numerical errors and turbulence modeling in tip vortex flow prediction. Int. J. Numer. Meth. Fluids **30**(1), 65–82 (1999)
16. Benoit, B., Dequin, A.-M., Kampa, K., von Grünhagen, W., Basset, P.-M., Gimonet, B.: HOST, a general helicopter simulation tool for Germany and France. In: 56th Annual Forum of the American Helicopters Society, Virginia Beach, VA, May 2000
17. Sweby, P.K.: High resolution schemes using flux limiters for hyperbolic conservation laws. SIAM J. Numer. Anal. **21**(5), 995–1011 (1984)
18. van Albada, G.D., van Leer, B., Roberts, W.W., Jr.: A comparative study of computational methods in cosmic gas dynamics. Astron. Astrophys. **108**(1), 76–84 (1982)
19. Radespiel, R., Kroll, N.: Accurate flux vector splitting for shocks and shear layers. J. Comput. Phys. **121**(1), 66–78 (1995)
20. Ehrenfried, K., Meier, G., Obermeier, F.: Sound produced by vortex-airfoil interaction. In: 17th European Rotorcraft Forum (1991)

Check for updates

Experimental and Numerical Study on Helicopter Acoustic Scattering

Jianping Yin(✉) and Karl-Stéphane Rossignol

DLR/Institut für Aerodynamik und Strömungstechnik, Lilienthalplatz 7, 38108 Brunswick, Germany
jianping.yin@dlr.de

Abstract. The scattering of noise generated by helicopter rotors has been recognized as having a significant influence on both the noise spectra and the noise directivity. This paper presents the research by DLR within the collaborative effort in the GARTEUR Action Group HC/AG-24. Shielding experiments were conducted in DLR's Acoustic Wind tunnel in Braunschweig (AWB) using point and rotor sources. For the numerical simulation of the various test cases, DLR's boundary element method was applied. The acoustic scattering predictions are compared to both test data and analytic solution.

Keywords: Noise scattering · Rotor acoustics · Boundary element method

1 Introduction

With the rising concern for environmental issues and increasingly stringent noise regulations, the noise emission has gained importance in the helicopter's design process. Extensive work, both theoretical and experimental, helped to deepen the understanding of the noise generating mechanisms. Even though the scattering of rotor generated noise by the fuselage has been recognized as having a significant influence on the noise spectra and directivity generated, the research effort towards acoustic installation effects, especially the scattering of tail rotor noise, has not been studied extensively [1–3]. To further understand helicopter noise propagation the GARTEUR Action Group HC/AG-24 [4, 5] was established. The objectives of this AG were (1) to develop and validate numerical prediction methods and (2) to generate a unique noise scattering database through wind tunnel tests using generic configurations. In the current paper, the results from DLR including numerical simulations, wind tunnel tests as well as numerical and test comparisons are addressed. This paper is organized as follows: the experimental approach used in the acoustic scattering test, including noise source, wind tunnel model, acoustic instrumentation and data reduction will be first described; the methodologies applied in the numerical simulations by DLR on the acoustic scattering will be then presented; the acoustic scattering predictions from a generical fuselage model will be analyzed and compared with the test results.

© The Author(s), under exclusive license to Springer Nature Switzerland AG 2021
A. Dillmann et al. (Eds.): STAB/DGLR Symposium 2020, NNFM 151, pp. 428–437, 2021.
https://doi.org/10.1007/978-3-030-79561-0_41

2 Description of the Test Setup for the Fuselage Scattering

The GARTEUR AG24 shielding experiments were conducted in DLR's small-scale low-speed Acoustic Wind tunnel in Braunschweig (AWB) [6]. There were three test campaigns carried. The first two involved generic sphere configurations of various dimensions and materials and two point-sources [5]. The third one involved acoustic wave scattering for a generical helicopter model using two point-noise sources and one rotor source. Figure 1 shows the test set up for the third acoustic scattering measurements. The helicopter model consists of an assembly of generic forms such as an ellipsoid fuselage and cylindrical tail boom. A generic "tail rotor" noise source from Polimi [7] is used (e.g. Figure 1 bottom). Acoustic measurements were conducted using an in-flow microphone (e.g. Figure 1 top). The inflow acoustic measurement plane was set at 400 mm below the axis of the fuselage.

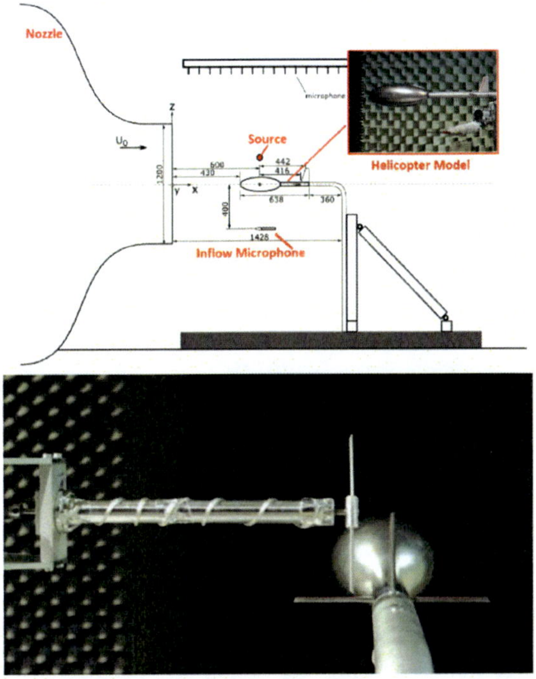

Fig. 1. Scattering experiments on a generic helicopter (Top: test set-up with the fuselage, Bottom: two-blade tail rotor, looking upstream).

2.1 DLR Laser Point Source

By focusing a high energy laser beam to a point, it is possible to initiate the formation of a small plasma which rapidly expands [8, 9], thus forming a pressure wave about its boundary which propagates through the surrounding medium, as shown in Fig. 2. (top), where the combination of one concave lens and two convex lenses is used to ensure plasma formation at about 500 mm from last convex lens.

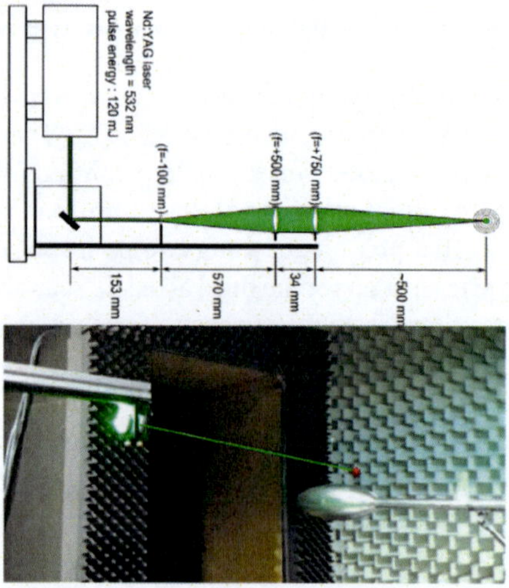

Fig. 2. Optical setup for the laser sound source (top), test set-up for the point source with the fuselage (bottom).

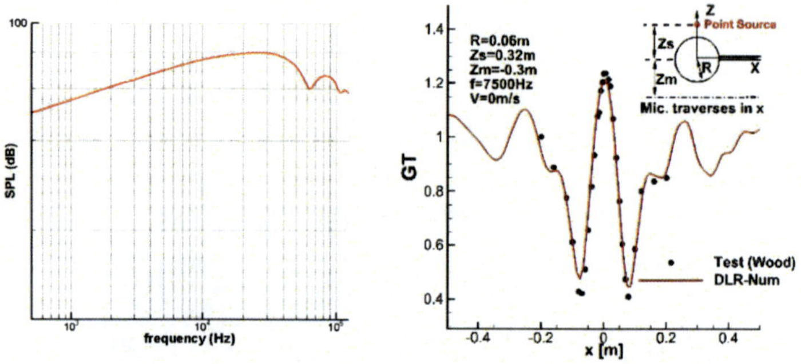

Fig. 3. Left: typical spectrum from DLR laser pulse source (in dB), right: comparison of noise shielding factor for a sphere (at x = 0) with support strut with simulation.

The DLR laser sound source has the advantage of being non-intrusive for both mean flow and scattered acoustic field. Because of its small size and uniform directivity, it can be represented as a point monopole source. It is characterized by a short time impulse waveform with a total duration of approx. 0.1 ms. Its acoustic spectrum is broadband as shown in Fig. 3 left with a useful frequency range extending from approx. 3 kHz to 100 kHz. During the acoustic data reduction, the proper length of the temporal window is important to exclude background reflections and extract a clean signal. An analysis window of 0.8 ms was found to be sufficient. Very good comparisons of a sphere scattering test case with the numerical simulations using a monopole [5] demonstrated the very good quality of the results produced using DLR laser sound source technique, see Fig. 3 right. The comparison of the shielding factor GT for the test and numerical results is conducted. GT is defined as the ratio of total sound pressure and incident sound pressure. The advantage of using GT to evaluate the scattering effect is that no corrections on signal amplitude are required.

2.2 Polimi Tail Rotor Source

A simplified double-blade tail rotor model was designed for this test activity [7]. The rotor has a 0.075 m radius and consists of flat plates blades with 10 mm chord and no twist. The blades are mounted to the hub with fixed pitch angles of 10° as shown in Fig. 4. Pitch setting can be reversed so that the rotor can operate either as a tractor or pusher propeller. The rotor is driven by a brush-less low-voltage electrical motor with an electronic controller. The main design criterion is that the generic rotor noise source should resemble the main characteristics of the tail rotor with clear harmonic components of the rotor noise. Figure 4 right shows the sound pressure level spectrum taken from the microphones located in the rotor rotational plane. A combination of rotor, motor and broadband noise is observed. In addition, an increment of the broadband noise is observed in the high frequency area. In the acoustic signal processing of the rotor noise, averaging is required in order to increase SN (Signal-to-Noise) ratio, and reduce background noise as well as random noise. To derive the final spectrum for each microphone position, two averaging techniques were tested, time averaged and spectrum averaged. For the time averaging procedure, a 1/rev signal from a mounted Hall-effect sensor is used, so that the time signal for a fixed number of revolutions is first overlapped and then averaged. The spectrum in Fig. 4 right is derived from the time averaged signal and fits very well with DLR numerical simulations as shown by the empty red dots. The numerical simulations are computed using DLR unsteady free wake 3-D panel code UPM [10, 11] and the Ffowcs Williams-Hawkings (FW-H) [12] code APSIM, including only monopole and dipole sources. The deviation upwards of 4 kHz is due to high motor noise which is not simulated.

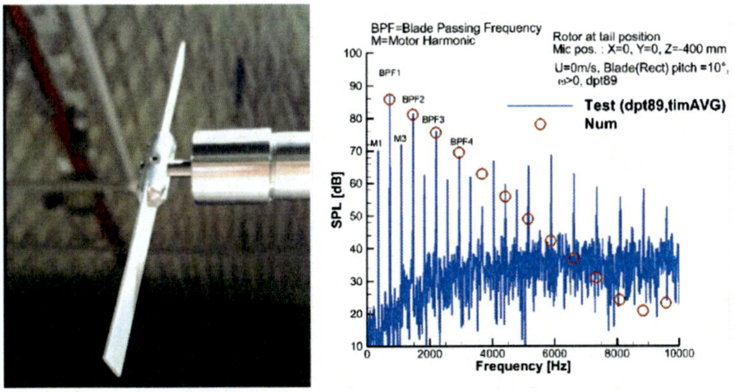

Fig. 4. Left: the tail rotor model, right: comparison of measured SPL spectrum with simulation.

3 Description of DLR Acoustic Scattering Prediction Methodology for the Helicopter Purpose

The noise field, received at an observer can be decomposed into incident or direct and scattered components. The incident acoustic field can be determined through a combined aerodynamic-aeroacoustics analysis. For very high frequency, the scattering field is usually solved by a ray tracing method. For low and middle frequency, BEM (Boundary Element Method) or finite element method (FEM) based on Helmholtz boundary value problem can be applied. In addition, the scattering formulation based on the FW-H equation or computational aeroacoustics method can also be applied. In this section the DLR method for solving the scattering filed is described. The method for obtaining the incident or the direct acoustic field for helicopter rotors will be presented afterwards. A detailed description on the application for the rotor noise shielding effects by the helicopter fuselage in forward flight can be found in [13].

3.1 DLR Acoustic Scattering Prediction

DLR is using the Fast Multipole Boundary Element Method (FMBEM) code [3] which solves the exterior Helmholtz problem for the scattered pressure field. It is a BEM method which employs the Fast Multipole Method (FMM) for triangulated surfaces. FMBEM uses an iterative solver from PETSc library as well as OpenMP/MPI parallelization for the fast evaluation of matrix-vector products so that no storage of the matrix is required. In addition, the Burton-Miller approach [14] is used to guarantee the uniqueness of the solution. Based on the assumption of low Mach number potential flow, a Taylor transformation of the convected wave equation into the Helmholtz equation is used to include mean flow effects [15]. A post-processing is used to deal with the mean flow effects.

3.2 The Rotor Noise Incident Field

For considering the scattering effects using BEM, the incident field is required. When the incident field is created by distributed, rotating and moving sources, such as the rotor noise source, the BEM must be run for each individual incident field generated by each discretized source position and then the intensities (spectra) are summed. Therefore, the method of the Kirchhoff surface source or wave-packet noise source proposed in [16] was used. In this method, the noise source is first propagated to a simplified and fixed source surface which encloses the original complex source. On the fixed source surface, the acoustic pressure and its normal derivatives are calculated, which are required for the further processing using the Kirchhoff formulation in BEM. Therefore, the fixed source surface is also called the Kirchhoff surface source. The Kirchhoff source surface has the form of a closed cylinder, as shown in Fig. 5 right. In general, the unsteady blade surface pressure was computed using DLR's unsteady free wake 3-D panel code UPM. For the field incident to the Kirchhoff surface the DLR FW-H code APSIM was applied, using UPM unsteady blade pressures.

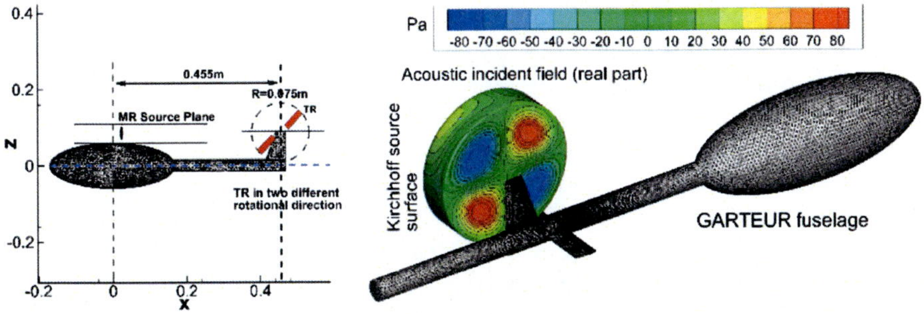

Fig. 5. Left: geometric location of tail rotor source relative to GARTEUR fuselage; right: a Kirchhoff surface around the rotor for further simulation of the acoustic scattering.

4 Results and Discussion

4.1 The Fuselage Scattering from the Laser Point Source

The laser point source is modelled as a point monopole in DLR's Boundary Element Methods simulations. Two source positions, marked as 2 and 3 are chosen for the comparison, as shown in Fig. 6 left, where the P2 and P3 are located within the main rotor source plane as demonstrated in Fig. 5 left. The selected positions of the point source representing main rotor noise source position. Position P3 and P4 represent possible positions occurring for main rotor blade vortex interaction (BVI) noise. P2 position is used to verify the measurement on maximum possible scattering effect.

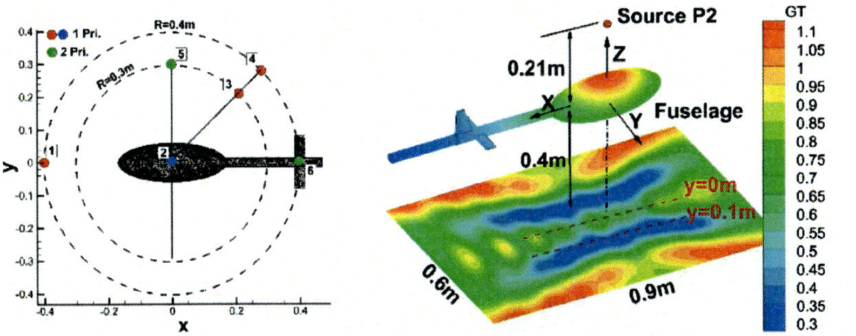

Fig. 6. Left: geometric location of the point source relative to the fuselage (top view); right: the microphone plane below the fuselage for the noise scattering simulations and measurements.

A contour plots of the simulated shielding factor GT on a receiving plane (microphones) at Zm = −0.4 m for the source P2 (Zs = 0.21 m) is demonstrated in Fig. 6 right. The deviation of GT from the value 1 indicates the effect of the fuselage scattering on the noise source. Figure 7 shows the comparison of the contour plot of the shielding factor GT at the frequency 2500 Hz and 5000 Hz respectively. The rectangular dashed line in the numerical simulations indicates the size of the measurement area. Similar shielding characteristics can be observed in both the simulation (right) and the test (left).

In general, at 2500 Hz, the shielding factor GT is always less than one (complete shadow region) in the measurement area. The strongest attenuation region is located on the side of the fuselage. The "silent zone" directly below the fuselage, where no incident wave can be propagated, is entirely determined by the diffracted waves. The maximum sound attenuation (blue areas) increases with an increase in frequency, while a slight shift towards y = 0 can be observed. The diffracted waves cause the enhancement (GT > 1) on the edge of measurement area for 5000 Hz. There are very good correlations in terms of the general shielding characteristics and values with the numerical simulations. Since the carpet plot used in the numerical simulation is obtained with much more microphone positions than the experimental ones, the contour lines are relatively smooth compared to the test. Slightly asymmetrical patterns were observed for all the frequencies in the test. The comparison of the shielding factor from the test with the numerical simulations at two linear sets of positions in y direction is given in Fig. 8. The comparisons show very good correlations in terms of the contour amplitude and pattern. The results for the source point P3 (Fig. 9), show an increased sound level in the area directly ahead of the fuselage, indicating the enhancement of the noise level instead of the cancelation. In general, a cancelation of the sound waves (GT < 1 area) is observed, but is relatively weak in comparing with the source position P2 as the source P3 is relatively far from the fuselage.

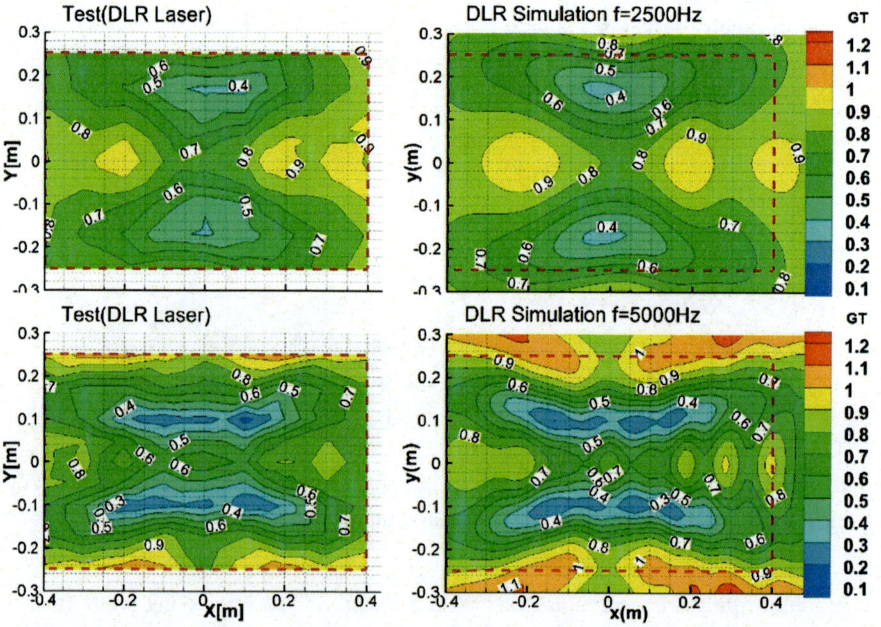

Fig. 7. Contour plots of the shielding factor for a point source at P2, with V0 = 0 m/s, measurement (left), simulation (right).

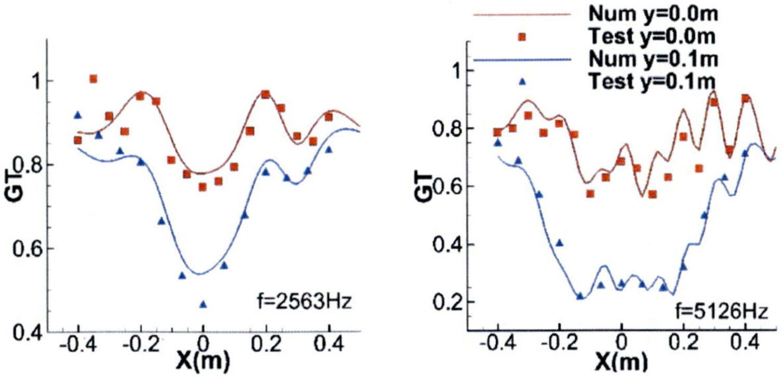

Fig. 8. Comparison of measured shielding factors with simulations at y = 0 m and y = 0.1 m (as shown in Fig. 6 right) for two different frequencies, 2563 and 5126 Hz.

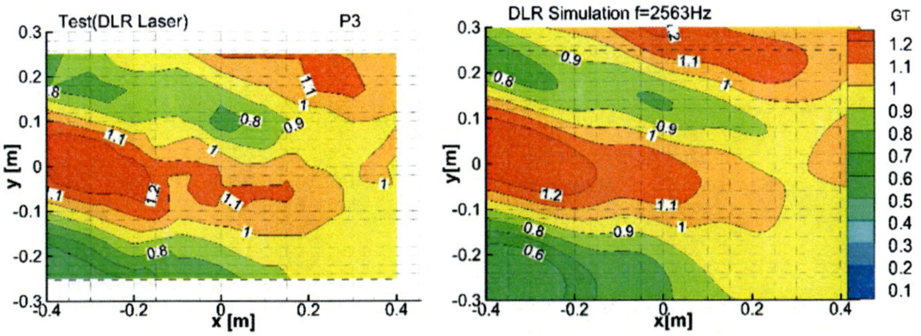

Fig. 9. Contour plots of the shielding factor for a point source at P3, with V0 = 0 m/s, measurement (left), simulation (right).

4.2 Rotor Noise Scattering by the Fuselage

Rotor noise scattering by the fuselage was evaluated at several microphone positions and two different rotor rotational directions. The results for a tail rotor position, as shown in Fig. 5 left, are chosen for the comparison. Figure 10 shows the measured (left) and the simulated contour (right) plots of the shielding factor GT for the tail rotor noise scattering from the fuselage at three different tail rotor blade passing frequencies (BPF), BPF1(737 Hz), BPF2(1474 Hz) and BPF3(2211 Hz), respectively. For BPF1, the shadow region (pointed by the arrow) on the starboard side of the fuselage is much larger in the measurements. In addition, a hot spot appears in the measurements around x = −0.2 m, which is not present in the simulation. The contours for the simulation and the test better fit for BPF2 and BPF3 for the position of the wave enhancement (marked by red dashed-dot line) and the shadow region (pointed by the arrow). However, the prediction accuracy is not as good as for the point source cases. There are several reasons which may be responsible for the discrepancy in the comparisons, such as the differences in the prediction of strength and directivity of the tail rotor noise incident field or the influence of the experimental setup.

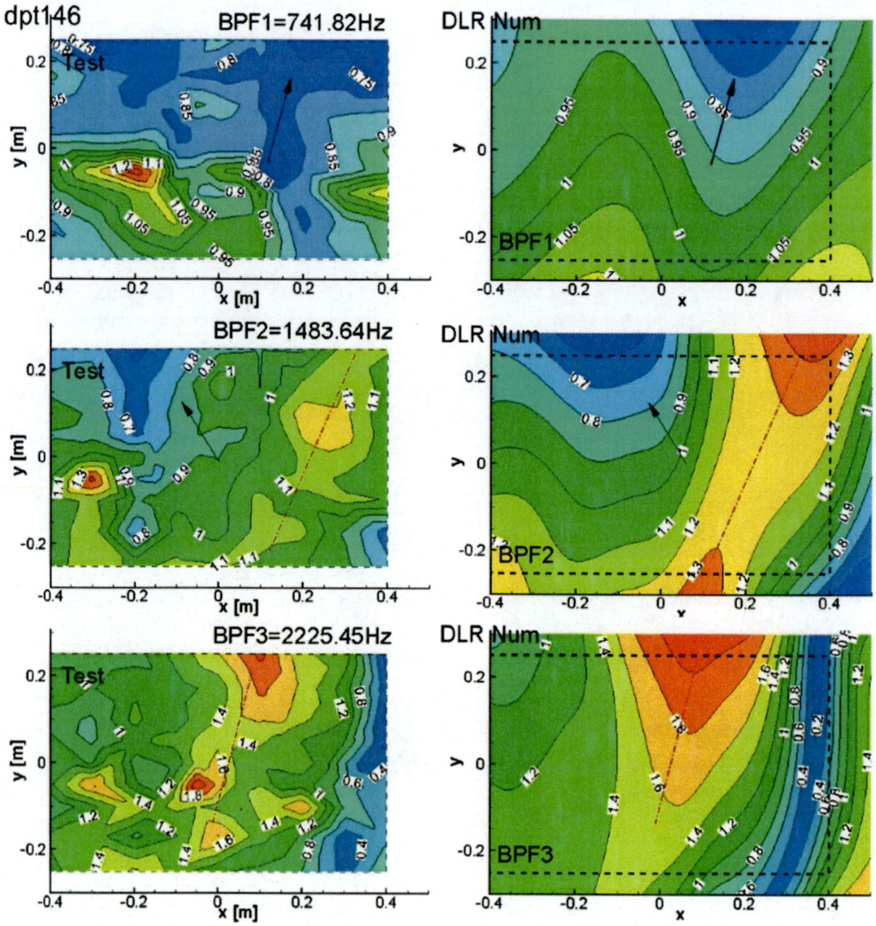

Fig. 10. Contour plot of the shielding factor for fuselage-rotor noise scattering at BPF1, BPF2 and BPF3, V = 0 m/s, right: simulations, left: experiment.

5 Conclusion Remarks

In this paper, experimental and numerical investigations of the shielding characteristics of a generic fuselage are presented. The test results are derived from the noise source generated from either a point or a rotor source.

The following concluding remarks can be drawn:

The noise shielding results using DLR laser-based non-intrusive sound source provide clear and consistent trends for all considered cases. DLR's laser point noise source can be represented in the numerical simulation as a point monopole source.

The numerical simulation results for fuselage scattering using the point source indicate that the BEM method has a high level of accuracy.

A more realistic scattering situation was experimentally studied for a generic tail rotor. Numerical simulations showed an overall good correlation with experimental data, but the prediction accuracy was lower than for the point source.

References

1. Lee, S., Erwin, J.P., Brentner, K.: Acoustic scattering of rotorcraft noise. In: 64th AHS Forum of the American Helicopter Society, 29 Apr– 1 May, Montréal, Canada (2008)
2. Casalino, D., Barbarino, M., Visingardi, A.: Simulation of helicopter community noise in complex urban geometry. AIAA J. **49**(8), 1614–1624 (2011)
3. Lummer, M., Richter, C., Proeber, C., Delfs, J.: Validation of a model for open rotor noise predictions and calculation of shielding effects using a fast BEM. In: 19th AIAA/CEAS Aeroacoustics Conference, 27–29, Mai, Berlin, Germany (2013)
4. Yin, J., et al.: Helicopter fuselage scattering effects for exterior/interior noise reduction. Terms of Reference for the GARTEUR Action Group HC/AG-24, April 2015
5. Yin, J., et al.: GARTEUR activities on acoustical methods and experiments for studying on acoustic scattering. CEAS Aeronaut. J. **10**(2), 531–551 (2018). https://doi.org/10.1007/s13272-018-0333-0
6. Pott-Pollenske, M., Delfs, J.: Enhanced capabilities of the aeroacoustic wind tunnel Braunschweig. In: 14th AIAA/CEAS Aeroacoustics Conference, No. 2008–2910, May, Vancouver, British Columbia, Canada (2008)
7. Yin, J., Zanotti, A., Rossignol, K-S., Gibertini, G., Vigevano, L.: Design of a generic rotor noise source for helicopter fuselage scattering tests. In: 44rd European Rotorcraft Forum, 18–21 September, Delft, The Netherlands (2018)
8. Rossignol, K.-S., Delfs, J., Boden, F.: On the relevance of convection effects for a laser-generated sound source. In: 21th AIAA/CEAS Aeroacoustics Conference, AIAA, 2015–3146 June, Dallas, USA (2015)
9. Rossignol, K.-S., Delfs, J.: Analysis of the noise shielding characteristics of a NACA0012 2D wing. In: 22nd AIAA/CEAS Aeroacoustics Conference, Lyon, France (2016)
10. Yin, J., Ahmed, S.R.: Helicopter main-rotor/tail-rotor interaction. J. Am. Helicopter Soc. **45**(4), 293–302 (2000)
11. Yin, J., van der Wall, B.G., Wilke, G.: Investigation of a simplified aerodynamic modelling technique for noise predictions using FW–H propagation. CEAS Aeronaut. J. **7**(4), 551–566 (2016). https://doi.org/10.1007/s13272-016-0208-1
12. Ffowcs Williams, J.E., Hawkings, D.L.: Sound generation by turbulence and surfaces in arbitrary motion. Philos. Trans. R. Soc. **264**(1151) (1969)
13. J. Yin.: Investigation of rotor noise shielding effects by the helicopter fuselage in forward flight. J. Aircr. **56**(4) (2019). American Institute of Aeronautics and Astronautics (AIAA)
14. Burton, A.J., Miller, G.F.: The application of integral equation methods to the numerical solution of some exterior boundary-value problems. Proc. R. Soc. Lond. A **323**, 201–210 (1971)
15. Taylor, K.: Acoustic generation by vibrating bodies in homentropic potential flow at low Mach number. J. Sound Vib. **65**(1), 125–136 (1979)
16. Agarwal, A., Dowling, A.: The calculation of acoustic shielding of engine noise by the silent aircraft airframe. In: 11th AIAA/CEAS Aeroacoustics Conference (26th AIAA Aeroacoustics Conference), 23–25 May, Monterey, California (2005)

Technical Flows

Bi-stability Detection in the Flow Around a Sphere by Means of Experiments and Lattice Boltzmann Simulations

Christian Bauer[1,2(✉)], Max Müller[1,2], Klaus Ehrenfried[1], and Claus Wagner[1,2]

[1] Institute of Aerodynamics and Flow Technology, German Aerospace Center, Göttingen, Germany
christian.bauer@dlr.de

[2] Institute of Thermodynamics and Fluid Mechanics, Technische Universität Ilmenau, Ilmenau, Germany

Abstract. With the aim to detect bi-stability in the highly separated flow around a sphere mounted on a cross-stream rod, wind tunnel measurements and numerical simulations with a Lattice Boltzmann method (LBM) are performed and compared. In both, LBM simulation and experiment, a bi-stable flow behaviour is detected at a Reynolds number of $\mathrm{Re} = 10^5$ based on the sphere diameter, the inflow velocity and the kinematic viscosity, whereas such behaviour is absent at Reynolds numbers as high as $\mathrm{Re} = 3 \times 10^5$. The bi-stable behaviour detected in the time series of the aerodynamic side force acting on the sphere for $\mathrm{Re} = 10^5$ is reflected by a bi-modal probability density function. For $\mathrm{Re} = 3 \times 10^5$, on the contrary, the probability density function is nearly Gaussian. Further, in agreement with the experiment, two counter-rotating vortex tubes originating from the intersection of the leeward and cross-stream rod-ward side of the sphere are generated.

1 Introduction

Unsteady forces acting on ground vehicles are currently studied in experiments with transient inflow conditions, e.g. on cars in full scale [1], or on simplified car models in wind tunnels [2]. Additionally, those effects are investigated with detached-eddy simulation (DES) studies, e.g. by Krajnović [3]. Moreover, there are unsteady forces induced by wake instabilities that occur in separated flows. With regard to ground vehicles, these wake instabilities lead to bi-stable forces. In wind tunnel experiments, this bi-stability is investigated using simplified car models, for example by Grandemange et al. [4] and Pavia et al. [5]. Gentile et al. [6] show that the wake instabilities are sensitive to even small misalignments. In terms of numerical simulations, such instabilities are investigated using

© The Author(s), under exclusive license to Springer Nature Switzerland AG 2021
A. Dillmann et al. (Eds.): STAB/DGLR Symposium 2020, NNFM 151, pp. 441–450, 2021.
https://doi.org/10.1007/978-3-030-79561-0_42

direct numerical simulations (DNS) at low Reynolds numbers, as done by Bury and Jardin [7], with the assumption that the results are valid at higher Reynolds numbers as well. Additionally, a DES study performed by Constantinescu and Squires [8] suggests that numerical methods which are not resolving all flow scales might also be able to predict the dynamics of a separated wake. In the present case a Lattice Boltzmann method (LBM) simulation of the flow around a sphere mounted on a cross-stream rod (CSR) is performed. The corresponding experimental setup (described in detail in Müller et al. [9]) consists of a sphere mounted on an internal six-component balance attached to a CSR. The wind tunnel measurement performed by Müller et al. [9] provides information on the dynamical behaviour of a separated flow over a sphere at a CSR, which leads to bi-stable side forces at certain Reynolds numbers. In the following, the results obtained from the latter measurement are compared to the predictions by the LBM simulation in order to determine, whether experiment and simulation agree in detecting wake-induced bi-stable forces at similar Reynolds numbers.

2 Experimental Setup

The experimental setup, see Fig. 1, was initially designed to induce transient sinusoidal side winds with a four-wing gust generator called *wing system*. This wing system was installed in the test section of the experimental setup. It was located upstream of the test model at a distance of ten times the chord length of one wing (300 mm). However, as this paper focuses on the inherent unsteady aerodynamics of the wake of a sphere, the flaps remained at the same position at 0° incidence and therefore do not impose any crosswinds. The model itself was a sphere with a diameter of 100 mm mounted on a 6-component balance attached to a CSR with a diameter of 25 mm. The clearance between the sphere and the floor of the test section was 295 mm. The forces acting on the sphere were measured with a sampling rate of 1 kHz and each measurement lasted 100 s. The Reynolds number,

$$Re = \frac{U_\infty d}{\nu}, \tag{1}$$

with U_∞ being the inlet velocity, d the sphere diameter and ν the kinematic viscosity, was set in the wind tunnel control and kept constant by a margin of 0.5% of its set value. Results obtained for the Reynolds numbers 10^5, 2×10^5 and 3×10^5 were recorded. Additionally, oil-film visualisations were conducted in order to highlight the flow structure.

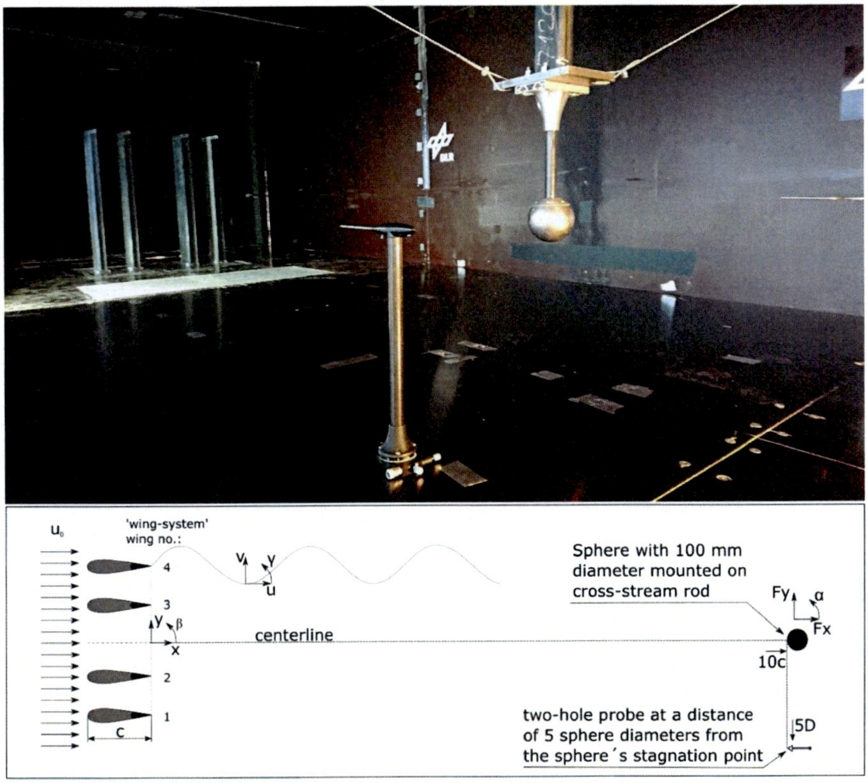

Fig. 1. Setup for the experimental investigation of the flow around a sphere mounted on a CSR with wing system. Top: picture of the test stand in the crosswind test facility (Seitenwindkanal Göttingen, SWG). Bottom: technical sketch of the test stand.

3 Numerical Methodology

As it has proven its ability to capture the dynamics of instationary, separated flows at Reynolds numbers as high as 3×10^5, the LBM was chosen to solve the governing equations via 3DS SIMULIA PowerFLOW®. In contrast to traditional methods of computational fluid mechanics, not the Navier-Stokes but the Lattice Boltzmann equations were solved in PowerFLOW®. The latter describe the dynamics of a particle velocity distribution function on a mesoscopic scale. Macroscopic quantities were then deducted via integration of mesoscopic variables. The discrete equations solved with the LBM read

$$f_i(\mathbf{x} + \mathbf{e}_i \Delta x, t + \Delta t) = f_i(\mathbf{x}, t) + \Omega_i(f(\mathbf{x}, t)), \quad i = 0, 1, \ldots, M, \tag{2}$$

where $\mathbf{x}$ is the location in space, t is the location in time, f_i is the particle velocity distribution function along the ith direction and Ω_i is the collision operator representing the rate of change of f_i due to collision [10]. Moreover, the collision term is approximated by the Bhatnagar, Gross, and Krook model using a linear relaxation [11].

Similar to the Navier-Stokes equation, the LBM can be used for the DNS of fluid turbulence, when the lattice Δx resolves the smallest scales. Due to the high Reynolds numbers considered here, only large scales could be resolved, which corresponds to a large-eddy simulation in combination with subgrid-scale modelling and wall modelling. PowerFLOW® provides a two-equation model based on the turbulent kinetic energy k and dissipation ε. These two variables were used to derive turbulence quantities of the subgrid-scales, such as turbulent length scales $l = k^{3/2}/\varepsilon$, time scales $\tau = k/\varepsilon$ and turbulent viscosity $\nu_t = C_\mu k^2/\varepsilon$.

Complementing the experiments of Müller et al. [9], LBM simulations for the three Reynolds numbers were performed as presented in Table 1. The computational flow domain with dimensions $L_x \times L_y \times L_z = 45.4d \times 12.2d \times 8.3d$ consisted of seven different grid refinement levels with a refinement towards the sphere and the CSR. As boundary conditions a constant velocity U_∞ and turbulence intensity of $0.01U_\infty$ was given at the inlet, whereas a constant static pressure was set at the outlet. All other boundaries of the computational domain were modelled as "frictionless walls", i.e. impermeability and free-slip boundary conditions. The sphere and the CSR were both modelled as smooth solid walls. The flow domain was initialised with the modelled fluid air—having a kinematic viscosity $\nu = 1.568 \times 10^{-5}\,\mathrm{m}^2/\mathrm{s}$, a density $\rho = 1.17\,\mathrm{kg/m}^3$ and the speed of sound $c = 347.2\,\mathrm{m/s}$—at velocity U_∞ and turbulence intensity of $0.01U_\infty$.

Table 1. Simulation cases. $\mathrm{Re} = U_\infty d/\nu$ is the Reynolds number based on the sphere diameter d, the inlet velocity U_∞ and the kinematic viscosity ν. Δx_{min} is the minimum grid spacing. n is the number of time steps. Δt is the simulated physical time.

Case	Re	# voxels	$\Delta x_{min}/d$	n	$\Delta t U_\infty/d$
SR100	10^5	97.8×10^6	0.005	2.57×10^6	1944
SR200	2×10^5	97.8×10^6	0.005	2.58×10^6	1678
SR300	3×10^5	97.8×10^6	0.005	2.00×10^6	1940

4 Comparison of the Predicted and Measured Results

Time series of the force coefficients obtained from LBM simulations are depicted in Fig. 2. As indicated in Fig. 2a for $\mathrm{Re} = 10^5$, small scales in the drag force coefficient (C_x), as well as in the lift force coefficient (C_z), fluctuate around one constant average value, whereas the side force coefficient appears to be fluctuating with respect to two different values. The latter is also denoted as bi-stable behaviour. With increasing Reynolds number, this behaviour vanishes (Fig. 2b, c). For $\mathrm{Re} = 10^5$, instantaneous snapshots of the flow at each stable state are displayed in Fig. 3. Here, the wake flow behaviour at $t = 9.75\,\mathrm{s}$ (top) and at

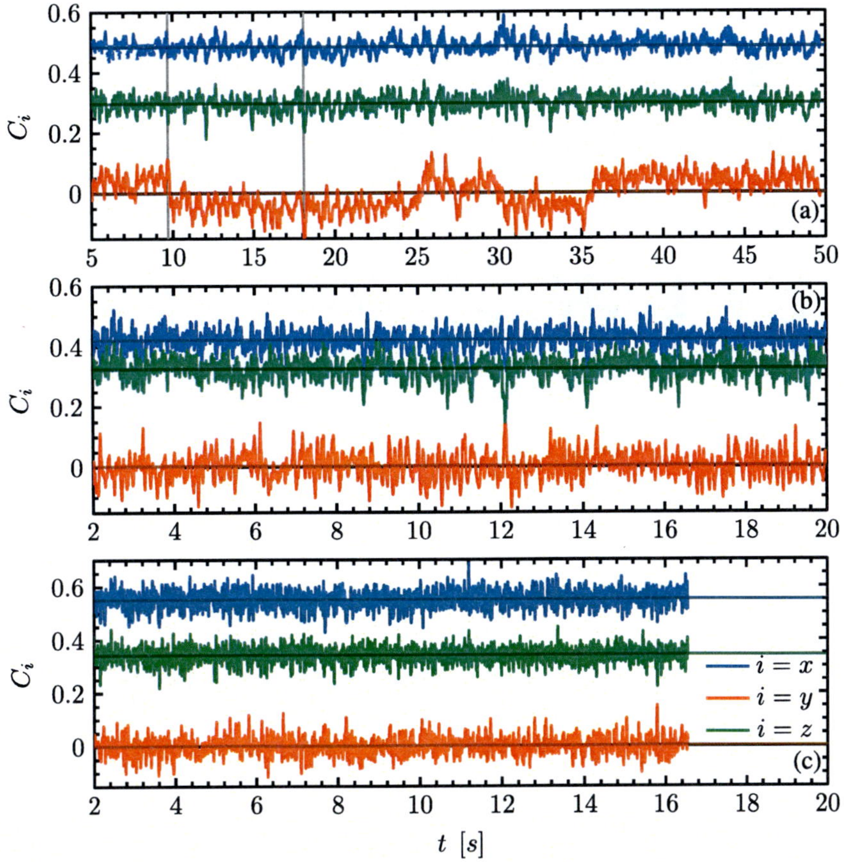

Fig. 2. Force coefficients predicted using LBM: —, C_x; —, C_y; —, C_z. (a) Re $= 10^5$; (b) Re $= 2 \times 10^5$; (c) Re $= 3 \times 10^5$. Dark shaded lines indicate averaged values. Grey lines indicate times of snapshots in Fig. 3.

$t = 18.14$ s (bottom) is visualised via iso-surfaces of the λ_2-criterion introduced by Jeong and Hussain [12]. As indicated by the grey auxiliary lines, the wake region is slightly skewed with respect to the y-axis, resulting in the side force reflected by the side force coefficient in Fig. 2a at corresponding times. The sign of the side force coefficient is determined by the direction in which the wake is skewed. Figure 4 presents the side and drag force coefficients obtained from the experiment. Note that the experimentally obtained side force coefficient time series have been filtered with a 10 Hz low-pass filter, a procedure further described in Müller et al. [9]. The observed bi-stability in the predicted side force coefficient at Re $= 10^5$ matches the experimental data shown in Fig. 4a. Although the absolute values of the side force coefficient are higher in the experiment, the expected bi-stable behaviour can be observed in both data sets. Moreover, the time range in which the flow remains in one of the two stable states

Fig. 3. Instantaneous flow field realisations showing iso-surfaces of the λ_2-criterion ($\lambda_2 = -10^5/s^2$), coloured by the velocity magnitude ranging from 0 (blue) to U_∞ (red) for $\mathrm{Re} = 10^5$. View from top in negative z-direction. Top: Flow field at $t = 9.75$ s. Bottom: Flow field at $t = 18.14$ s. Grey lines indicate the sphere centreline and its horizontal boundaries.

is similar in the experimental and the numerical results for $\mathrm{Re} = 10^5$ (compare Figs. 2a and 4a). For $\mathrm{Re} = 2 \times 10^5$, the experimental side force coefficient shows a bi-stable behaviour with a strong tendency to one side (Fig. 4b), which is not predicted by the LBM simulations (Fig. 2b). For $\mathrm{Re} = 3 \times 10^5$, the side force coefficient shows no sign of bi-stability, neither in the experiment (Fig. 4c), nor in the LBM simulation (Fig. 2c). Further, the drag force coefficients obtained in the LBM simulation and the experiment are compared in Fig. 5, where they are plotted against the Reynolds number.

While the drag force coefficient deviates as much as 20% for $\mathrm{Re} = 2 \times 10^5$, the coefficients for the lowest and the highest Reynolds number agree better with deviations of ten percent and less than one percent for $\mathrm{Re} = 3 \times 10^5$ and $\mathrm{Re} = 10^5$, respectively. Due to the above-discussed deviations between the results obtained for $\mathrm{Re} = 2 \times 10^5$, only the lowest ($\mathrm{Re} = 10^5$) and the highest Reynolds number ($\mathrm{Re} = 3 \times 10^5$) are considered in the following analysis. Consequently, probability density functions (PDFs) are presented in Fig. 6 for the last mentioned Reynolds numbers.

The PDFs confirm the observations for the above-described time series statistically. Although the PDFs of the experiments and the LBM simulations do not collapse for $\mathrm{Re} = 10^5$ (Fig. 6a), both PDFs show a clear bi-stable behaviour. At $\mathrm{Re} = 3 \times 10^5$, the flow tends to become Gaussian, which is reflected in both PDFs, i.e. of the LBM simulation and the experiment. At this Reynolds number ($\mathrm{Re} = 3 \times 10^5$) no bi-stability is observed and the PDFs collapse nicely.

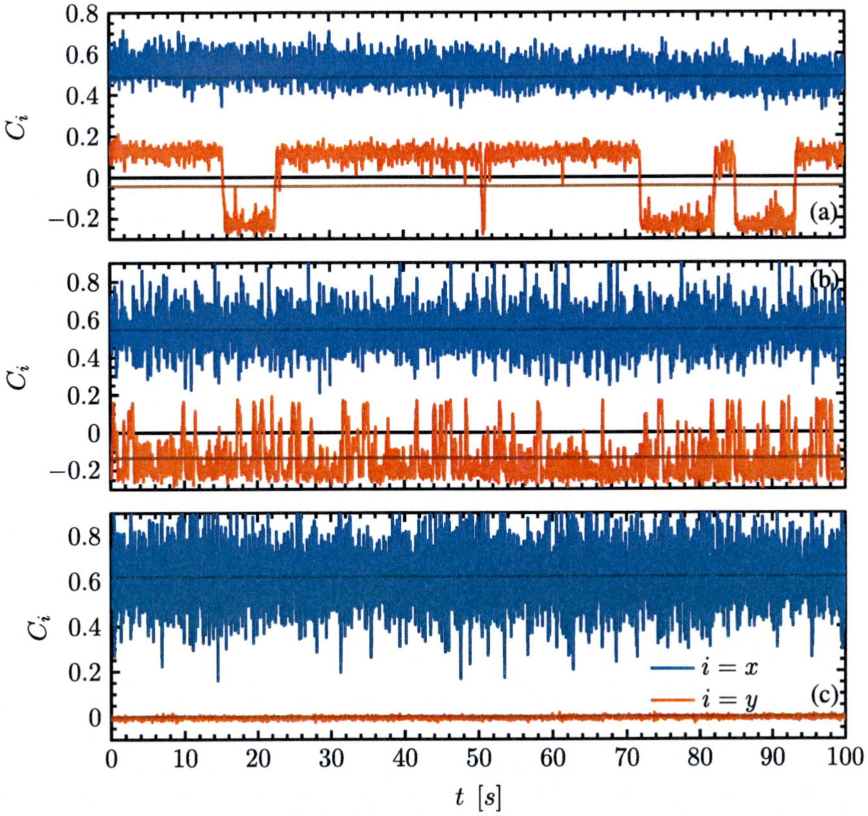

Fig. 4. Force coefficients from experiments: —, C_x; —, C_y. (a) Re = 10^5; (b) Re = 2×10^5; (c) Re = 3×10^5. Dark shaded lines indicate averaged values.

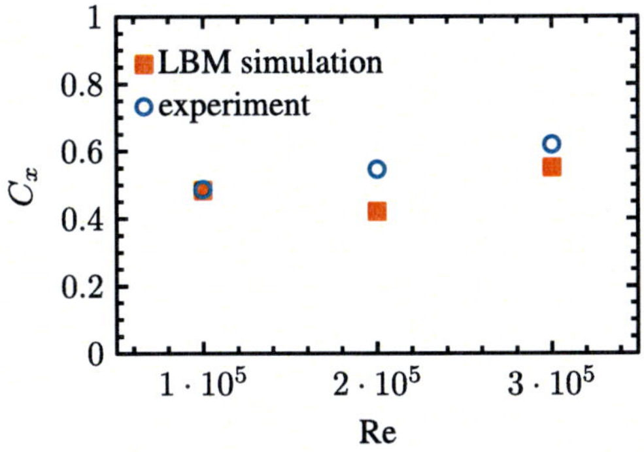

Fig. 5. Drag force coefficient C_x versus Reynolds number Re predicted in LBM simulations and measured in wind tunnel experiments.

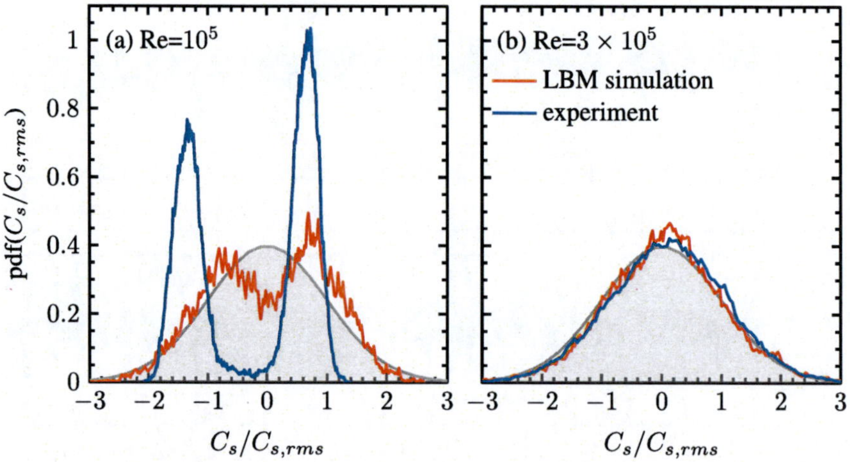

Fig. 6. Probability density functions of the side force coefficient. LBM simulation (—) and experiment (—) for $\mathrm{Re} = 10^5$ (a) and $\mathrm{Re} = 3 \times 10^5$ (b). Grey shaded areas indicate a Gaussian distribution $f(C_s/C_{s,rms}) = 1/(2\pi)^2 \cdot \exp\left[-0.5(C_s/C_{s,rms})^2\right]$.

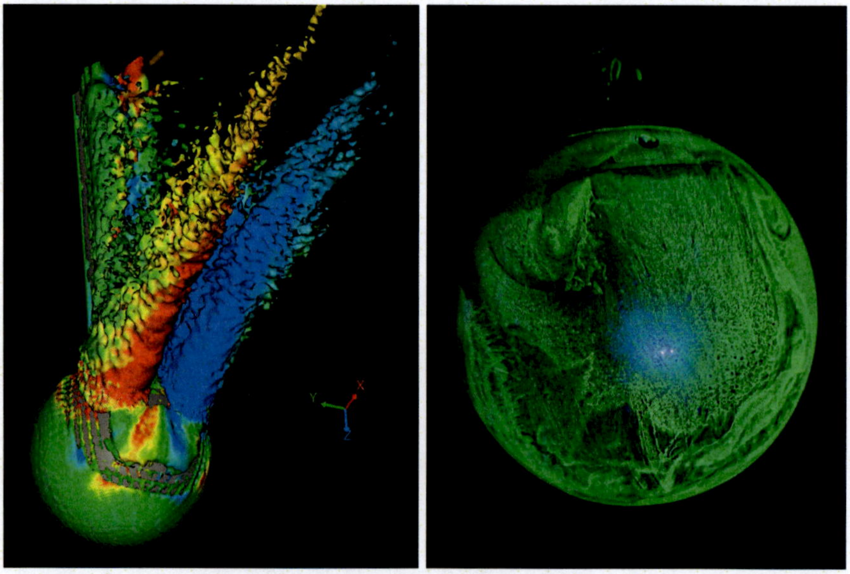

Fig. 7. Left: λ_2-iso-surface of the mean flow structure ($\lambda_2 = -7000/\mathrm{s}^2$) obtained from averaging over 500 snapshots at $\mathrm{Re} = 3 \times 10^5$. Pseudo-colours indicate the x-vorticity component ranging from $-300/\mathrm{s}$ (blue) to $300/\mathrm{s}$ (red). Right: Oil-film visualisation of the leeward side of a sphere at CSR at $\mathrm{Re} = 3 \times 10^5$ (yz-plane).

Finally, the average flow structure of the flow around a sphere mounted on a CSR at $\mathrm{Re} = 3 \times 10^5$ is considered. In the following, the average flow structure is visualised by means of the λ_2-criterion applied to an average flow field in the LBM simulation and by means of an oil film in the experiment (Fig. 7). Both visualisations (Fig. 7a and Fig. 7b) reflect a dual-vortex structure. The λ_2-iso-surfaces in Fig. 7a depict two vortex tubes originating from the intersection of the leeward and the CSR-ward side of the sphere. The colouring of the vortex tubes by the x-vorticity indicates their rotational direction. Looking along the positive x-direction, the red-coloured vortex tube rotates counter-clockwise, whereas the blue-coloured one rotates clockwise. In addition, the footprint of these vortical structures can be seen in the oil-film visualisation shown in Fig. 7b. Thus, the flow fields predicted in the LBM simulation and measured in the experiment agree well for $\mathrm{Re} = 3 \times 10^5$, not only in one-point statistics and instantaneous time series, they also both uncover the mean flow structure.

5 Conclusion

The phenomenon of bi-stability in the flow around a sphere mounted on a CSR has been studied by means of experiments and LBM simulations. For $\mathrm{Re} = 10^5$, both methods show bi-stable side forces, involving similar time scales. For $\mathrm{Re} = 2 \times 10^5$, on the contrary, the experiment still reflects bi-stability, whereas the LBM simulation does not. This also leads to a lower drag coefficient in the LBM simulation, as bi-stability is known to be a drag contributor [5]. For $\mathrm{Re} = 3 \times 10^5$, neither the LBM simulation nor the experiment show any bi-stability. Regarding the average structure of the wake flow for $\mathrm{Re} = 3 \times 10^5$, an oil-film visualisation agrees well with the vortical structures found in the flow fields predicted by the LBM.

Further work needs to be contributed to the observed differences at $\mathrm{Re} = 2 \times 10^5$, as it is not clear, why the LBM is not able to resolve the bi-stability. The quantitative differences between the results predicted by the LBM and measured in the experiment for $\mathrm{Re} = 10^5$ might have multiple reasons: First, the inflow conditions are different. In the experiment, for example, the gust generator acts as an obstacle in the approaching flow. This was not considered in the LBM simulation, which was performed for an inflow with a constant velocity modulated by artificial turbulence. Second, small geometrical imperfections of the sphere used in the experiment might have affected the wake, as the latter is known to be sensitive to small changes. In the LBM simulation, on the other hand, the sphere is modelled as a smooth surface. Third, the LBM simulation requires turbulence modelling, as it is not possible to resolve all scales for the Reynolds number regime of interest. Nevertheless, it was shown that the LBM is able to predict the wake instability for $\mathrm{Re} = 10^5$, and match the turbulent wake for $\mathrm{Re} = 3 \times 10^5$.

Acknowledgements. Provided computing resources on DLR's CARA cluster and proof-reading by Annika Köhne are gratefully acknowledged.

References

1. Schroeck, D., Krantz, W., Widdecke, N., Wiedemann, J.: SAE Int. J. Passenger Cars Mech. Syst. **4**(1), 108 (2011). https://doi.org/10.4271/2011-01-0154
2. Ferrand, V.: J. Fluids Eng. **136**(1), 011101 (2014). https://doi.org/10.1115/1.4025466
3. Krajnović, S.: Int. J. Railway **1**(3), 99 (2008)
4. Grandemange, M., Gohlke, M., Cadot, O.: J. Fluid Mech. **722**, 51 (2013). https://doi.org/10.1017/jfm.2013.83
5. Pavia, G., Passmore, M., Sardu, C.: Evolution of the bi-stable wake of a square-back automotive shape. Exp. Fluids **59**(1), 1–20 (2017). https://doi.org/10.1007/s00348-017-2473-0
6. Gentile, V., van Oudheusden, B.W., Schrijer, F.F.J., Scarano, F.: J. Fluid Mech. **813**, R3 (2017). https://doi.org/10.1017/jfm.2017.4
7. Bury, Y., Jardin, T.: Phys. Lett. A **376**(45), 3219 (2012). https://doi.org/10.1016/j.physleta.2012.09.011
8. Constantinescu, G., Squires, K.: Phys. Fluids **16**(5), 1449 (2004). https://doi.org/10.1063/1.1688325
9. Müller, M., Ehrenfried, K., Bell, J., Wagner, C.: Analysis of model mount configurations with regard to force measurements with transient inflow. In: Dillmann, A., Heller, G., Krämer, E., Wagner, C., Tropea, C., Jakirlić, S. (eds.) DGLR 2018. NNFMMD, vol. 142, pp. 684–694. Springer, Cham (2020). https://doi.org/10.1007/978-3-030-25253-3_65
10. Chen, S., Doolen, G.D.: Annu. Rev. Fluid Mech. **30**(1), 329 (1998). https://doi.org/10.1146/annurev.fluid.30.1.329
11. Bhatnagar, P.L., Gross, E.P., Krook, M.: Phys. Rev. **94**(3), 511 (1954). https://doi.org/10.1103/PhysRev.94.511
12. Jeong, J., Hussain, F.: J. Fluid Mech. **285**, 69 (1995). https://doi.org/10.1017/S0022112095000462

Convolutional Neural Networks for Approximation of Blood Flow in Artificial Lungs

Clemens Birkenmaier[1,2](✉) and Lars Krenkel[1,2]

[1] Ostbayerische Technische Hochschule Regensburg, Biofluid Mechanics, 93053 Regensburg, Germany
clemens.birkenmaier@oth-r.de

[2] Regensburg Center of Biomedical Engineering, OTH and University Regensburg, 93053 Regensburg, Germany

Abstract. Blood flow in channels of varying diameters $< 500\,\mu m$ exhibits strong non-linear effects. Multiphase finite volume approaches are feasible, but still computationally costly. Here, the feasibility of applying convolutional neural networks for blood flow prediction in artificial lungs is investigated. Training targets are precomputed using an Eulerian two-phase approach. To match with experimental data, the interphase drag and lift, as well as intraphase shear-thinning are adapted. A recursively branching regression network and convolution/deconvolution networks with plain skip connections and densely connected skips are investigated. A priori knowledge is incorporated in the loss functional to prevent the network from learning non-physical solutions. Inference from neural networks is approximately six orders of magnitude faster than the classical finite volume approach. Even if resulting in comparably coarse flow fields, the neural network predictions can be used as close to convergence initial solutions greatly accelerating classical flow computations.

Keywords: Deep learning fluid mechanics · Multiphase blood flow

1 Introduction

Coagulation and bleeding are severe but still frequent complications in lung replacement therapies. The flow mechanics inside artificial lungs is of great interest, since observed coagulation phenomena are believed to be mediated by prevalent flow [11]. The geometry which blood encounters between the hollow-fibre membranes in artificial lungs, as illustrated in Fig. 1, might be perceived as confined channels with ever-changing widths that range from approximately $75\,\mu m$ to $375\,\mu m$. Blood flowing through such confined channels exhibits strong non-linear effects. This breakdown of the Newtonian assumption is due to the actual particulate nature of blood and the complex properties of these particles. Blood is a heterogeneous suspension of corpuscles, such as red blood cells (RBCs) and

© The Author(s), under exclusive license to Springer Nature Switzerland AG 2021
A. Dillmann et al. (Eds.): STAB/DGLR Symposium 2020, NNFM 151, pp. 451–460, 2021.
https://doi.org/10.1007/978-3-030-79561-0_43

platelets, in a protein containing fluid. RBCs are highly deformable, oval biconcave disc-like corpuscles, additionally prone to biological processes. At length scales from 75 μm to 750 μm, blood can simplified be perceived as a mixture of particulated RBCs (diameter 7.5 μm) suspended in a continuous Newtonian fluid phase.

Computing blood flows in complex geometries at aforementioned length scales is a non-trivial task. Multi-phase finite volume approaches using an Euler-Euler formulation are feasible, but still computationally costly. Especially with a potential application in patient specific studies, which requires a huge number of individual parameters that are costly to determine.

Neural network (NN) as an alternative method for universal approximation of differential equations have proven to be computationally efficient and still sufficiently accurate compared to established methods such as the finite volume method (FVM) [7,12]. In laminar, steady, single-phase flows at low Reynolds and Mach numbers, convolutional neural networks (CNNs) perform well in predicting velocity and pressure fields or aerodynamic coefficients [1,8,14]. The general advantage of employing deep learning (DL) models, is their inference efficiency, outperforming classical computational methods by far. Another benefit might be that there is no need for exact modelling of effects, as they are learned automatically, given adequate training data. FVM are based on spatial discretisation and resulting flow fields are represented only on discrete and distributed grid locations. Since a Cartesian, equidistant grid basically is a raster graphics, image regression techniques can be applied to predict phase velocity fields as well as particle and pressure distributions in confined geometries from simple mass flow boundary conditions.

Here, the feasibility of employing NNs for predicting internal non-Newtonian multiphase blood flows in wavy geometries is demonstrated. Different NN architectures and design features are analysed and compared, in terms of prediction accuracy.

2 Methods

Three different NN architectures with various design parameters are tested. For this, training data capturing the significant non-Newtonian effects of blood flowing through small and non-straight channels is required. As target metric, a root mean squared error, normalized by the data range and subsequently averaged over all flow variables, is used.

2.1 Training Data

Following the actual geometry in artificial lungs, a simplified two-dimensional (2D) channel with varying width is used for computing parametric training data. The geometry layout is illustrated in Fig. 1b. The changing width is constructed from circular segments, stitched together with a tangential continuity constraint. The dimensions of the wavy geometry are normalized to be fully described with

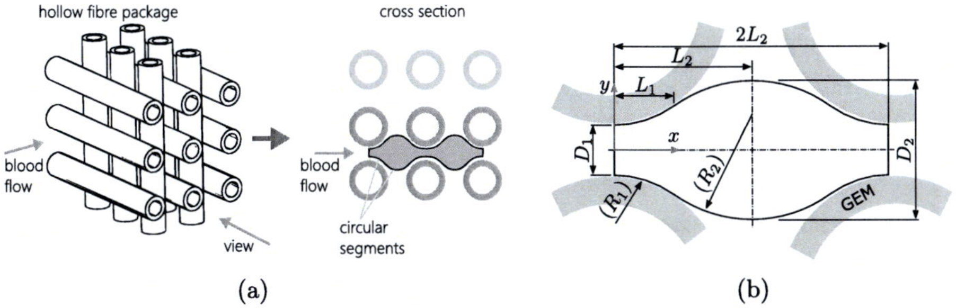

Fig. 1. 2D geometry. **a** The geometry comprising circle segments is derived from the tubular gas exchange membranes (GEMs) in hollow fibre membrane lungs. **b** The non-straight confined fluid domain is parametrised, depending only on the particle diameter D_p. Circular segments (radii R_1, R_2) are stitched together with tangential continuity

only four dimensionless parameters, based on the particle diameter. The geometry is defined by the set $F = D_1/D_p$, $D = D_2/D_1$, $L = L_2/D_1$, $L = L_2/D_1$, $\lambda = L_1/L_2$, where D_p is the particle diameter, F a factor for the throat width, D_1, D_2, L_1 and L_2 are the dimensions according Fig. 1b. Here, a particle diameter of 7.5 µm and a throat factor of $F = 10$ is used. The geometry is meshed in Ansys Meshing with triangles without inflation layers and a constant element size of $0.5D_p$.

The flow fields used as training targets are computed using an Eulerian two-phase model. This mixture theory approach models the fluid phase (blood plasma) and the particulated phase (RBCs) as interacting and interpenetrating continua, exchanging momentum in the form of lift and drag, in the present case. The interphase drag can be described by Stoke's drag for a single spherical particle with a correction for RBCs not being spherical and many interacting particles being present. The latter is implemented by means of a generalized Rourke & Ernestine type hindrance function $H = (1-\phi/\hat{\phi})^{r_e}$, where the parameters are defined to match with experimental results as $\hat{\phi} = 0.85$ and $r_e = -2$ [13]. Additionally, the non-sphericity of RBCs is accounted for by means of an empirical correction factor $C_S = (C_{S,\infty} + (1-C_{S,\infty})/(1+(c_1 Re_p)^{c_2})^{(\hat{\phi}-\phi)}$, where $C_{S,\infty} = 10$ is the high Reynolds correction, $c_1 = 2.5$ and $c_2 = 4.5$ are parameters determining the transition and are found empirically to fit the experimental results. The single particle lift is described by a pure Saffman lift. To take account of the lift breakdown due to the nearby presence of other particles, a generalizing function of the same form as the drag hindrance function is applied, but with a positive exponent $r_e = +2$. Additionally to this inter-phase exchange, there are intra-phasic effects, such as reversible aggregation of RBCs, for example. The shear dependent aggregation and disaggregation of RBCs is modelled as a shear-thinning viscosity of the particulated phase with a correction for phase fraction, which aims on collisional dissipation. Here, a Carreau-type model $\eta(\dot{\gamma}) = \eta_\infty + (\eta_0 - \eta_\infty)(1+(\lambda_p\dot{\gamma})^2)^{0.5(n_p-1)}$ with phase fraction correction is used, since it is applied only to the RBC phase. Therefore, the infinite and zero shear limit is $\eta_\infty = 0.05\,\text{Pa s}$ and $\eta_0 = 0.5\,\text{Pa s}$, receptively, while $\lambda_p = 3.313$ and $n_p = 0.3568$ [5,6].

The pressure based steady and laminar solver in Fluent 19.5 is employed, solving each volumetric phase fraction independently. This improves convergence as it initially allows more deviation from continuity. Additional stability is incorporated by using an implicit under relaxation scheme that allows different relaxations factors for each equation to be set and adapt during computation automatically. As outlet boundary condition (BC) a constant pressure, which is shared by both phases in Euler-Euler models, is used. The mass flow rate for each individual phase is used as inlet BC, which is derived from the inlet Reynolds number. Conversion is done by scaling with mixture viscosity and phase fraction ϕ. In the DL approach, the Reynolds number is used directly as BC.

The entire precomputed data set comprises 5,000 cases with equally gridded parameters spanning $Re \in [0.01, 10]$, $\phi \in [0.1, 0.5]$, $D \in [2, 4]$, $L \in [2, 6]$, $\lambda \in [0.2, 0.5]$. For training the NNs the set is split randomly without considering underlying statistics into a training, validation and test set at ratio of 0.75, 0.15 and 0.1, respectively.

2.2 Neural Network Designs

Considering a non-linear function in the general form of $\hat{Y} = \hat{f}(X, r)$, NNs in general can be applied to approximate the true function $\hat{f}$ in the form $Y \approx f(X, r)$. A suitable representation of $\hat{f} \rightarrow f$ that is capable of approximating $\hat{Y} \approx Y$ from the input X needs to be found. As approximating function f, NNs can be used, given a sufficient degree of freedom or model capacity r and correspondingly sufficient training data to circumvent overfitting [2,12].

A basic design choice is to use convolutional layers as they are especially suited for working on grid-structured inputs with local context [2]. For gridded data, working on different length scales is beneficial as it combines maximum locality, i.e. where in the domain, with maximum context, i.e. what are the neighbours [10]. The necessary spatial down-sampling while expanding the feature space represents the increasing context and is loosely related to the idea of information being preserved, just like energy in any physical system [4]. Regressing a flow field, which has a certain spatial extent and a certain number of variables from this down-sampled feature rich abstract space, requires a subsequent spatial up-sampling with feature reduction in order to match the final spatial dimensions and number of predicted field variables. Here, the upsampling deconvolution is a transposed convolution with learnable interpolation. The repetitive up-sampling can be lossy, as information about locality is lost since neighbour relations are split. This local correspondence information then needs to be re-trained or simply passed through the network using paths without down-sampling [10]. Therefore, a way to preserve this locality while still incorporating context information is the use of skip connection [9]. Figure 2 illustrates the tested CNN designs. In order to reduce overfitting and enhance model robustness, each convolution is followed by a drop out layer at a rate of 0.5. The entire DL workflow is automated

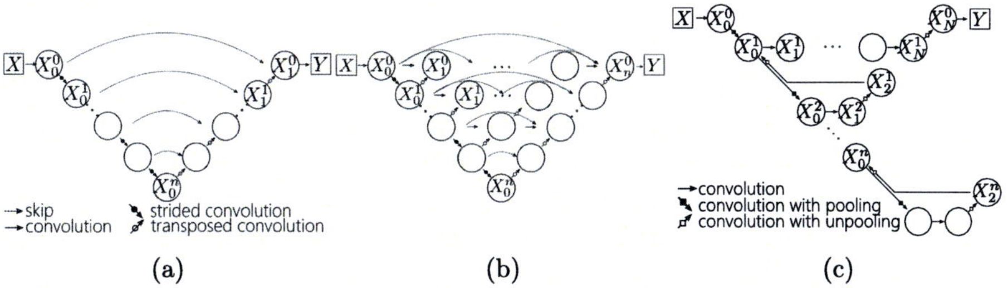

Fig. 2. Neural network architectures. **a** CDRN with plain skip connections. **b** CDRN with densely connected skip connections. **c** RBDN, comprising recursive branches

in Matlab R2019b. The basic CNN design idea with feature expanding down-sampling, or encoding, and subsequent feature reducing up-sampling, or decoding, is referred to as convolution-deconvolution regression network (CDRN). The basic structure of skip connections in CDRNs is a well performing approach in medical image segmentation [9]. Also the idea of incorporating additional convolution paths in the skip connections, i.e. dense skips, stems from medical image segmentation [15]. The concatenation of information originating from different length scales can be done in a straight and horizontal manner, as in CDRNs. An alternative way is to use recursive branches in a vertical arrangement. In Recursively Branched Deconvolutional Networks (RBDNs) also a series of convolution layers act as non-linear transformation. Down-sampling branches with recursive up-sampling incorporate and provide information on different length scales early on in the non-linear transformation path.

2.3 Loss Manipulation

If the regression task comprises physical processes, there might be governing laws that are known. This a priori knowledge, such as the conservation of mass in flow mechanics, can be incorporated in the training process. Furthermore, matter can not cross walls, i.e. impermeable domain boundaries. This can be incorporated in the learning by adding an extra penalty to the loss functional, subsequently forming a steeper gradient on non physical solutions. Deviations from mass being conserved are implemented as additional penalty in the mean squared error (MSE) loss functional. Impermeable walls are translated into the assumption that fluid or particle velocities outside the actual fluid domain are not allowed and are therefore penalized as well.

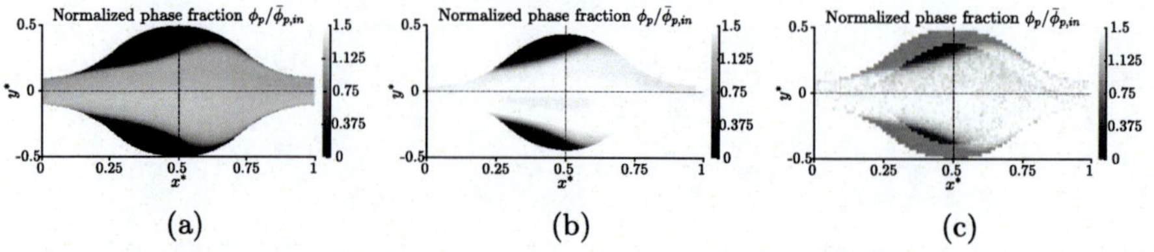

Fig. 3. Resulting fields of normalized phase fraction ($Re = 10.4$, $\phi = 0.2$, $D = 4.2857$, $L = 4.4285$, $\lambda = 0.3823$). **a** experimental. **b** FVM computations. **c** NN prediction

2.4 Hyper Parameter Search

In order to find good performing parameter combinations, first parameters that actually impact the prediction performance are identified by a coarse random search and subsequent Pearson correlation among best performing parameter sets. The three most important parameters for each CNN design then are optimized with a Bayesian search. This approach makes use of the exploitations vs. exploration trade off and hence chooses more promising parameter combinations according to a surrogate model it updates on runtime. The tested design parameters are encoder depth (CDRN) or recursive branches (RBDN) $n \in [1..7]$, transformations (RBDN) $N \in [1..7]$, convolution kernel size $k \in [2..16]$, initial feature map size $f \in [4..64]$, activation leak slope $l \in [0, 0.1]$, L_2 regularisation $\lambda_R \in [10^{-10}, 10^{-1}]$. Training is done using an ADAM optimizer, with a mini batch size of 30.

3 Results

Figure 3 shows fields of normalized phase fractions for one case, from experiments (Fig. 3a), FVM computations (Fig. 3b) and predicted using the best performing densely connected CDRN (Fig. 3c). The FVM multiphase model captures the relevant behaviour of the RBCs and phase fraction fields match well, except for the wall regions near detachment and reattachment areas. Additionally, there is an asymmetric area in the computed phase fraction fields comprising higher particle densities, that does not connect to experimental results. The CNN prediction is coarser in spatial resolution compared to experimental and FVM results, and it bears some noise. Besides potential upscaling noise on a pixel sized scale, also an upscaling alias on the field scale is obvious. However, the CNN prediction still captures the essence of the respective flow pattern.

The mean total wall clock duration of FVM computations on seven AMD EPYC 7451 cores was 4.04×10^3 s, while mean inference duration from CNNs was 5.68×10^{-3} s on a NVIDIA Titan Xp.

For each CNN a best performing hyperparameter set is identified. Exemplary, Fig. 4 shows some results of the fine Bayesian search by means of a normalized root mean squared error (NRMSE). In general, a deeper network, i.e. more convolution/deconvolution stages, are beneficial, but more impact on the mean error

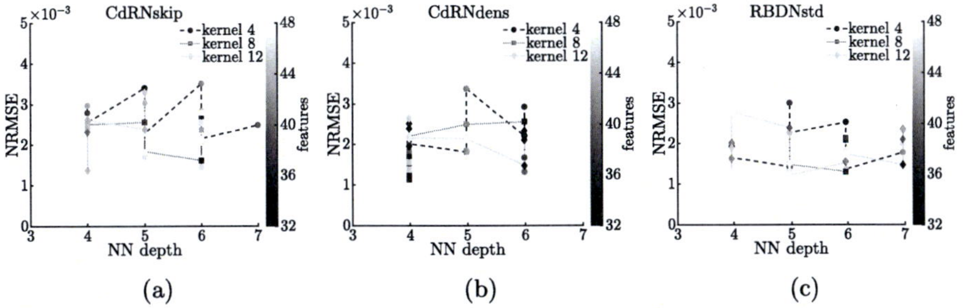

Fig. 4. Bayesian hyperparameter optimization. **a** CDRN with plain skip connections. **b** CDRN with densely connected skip connections. **c** RBDN, comprising recursive branches

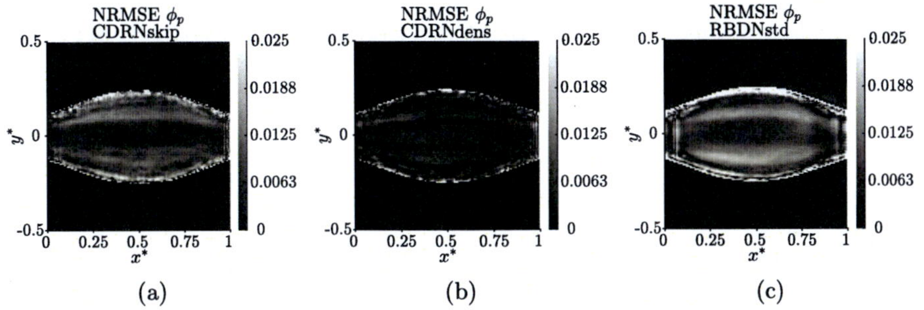

Fig. 5. NRMSE fields of phase fraction. **a** CDRN with plain skip connections. **b** CDRN with densely connected skip connections. **c** RBDN, comprising recursive branches

is due to number of features. Additionally, the convolution kernel size is also more important compared to network depth. All three CNN designs are comparably accurate, by means of the NRMSE.

Figure 5 shows fields of normalized deviations from the precomputed target phase fraction fields. The mean error is taken over 20 previously unseen cases that have the same geometry but differ in inlet Reynolds number and phase fraction. The boundary region where relative RBC fractions change sharply exhibit high errors with all CNNs. The center area in contrast, where most likely RBCs are present in every case, even with different fractions, the error is low. In the CDRN with skips in Fig. 4a additionally quadratic patterns are visible. On a slightly coarser scale, this patterns are also present in the RBDN, shown in Fig. 4c. The prediction accuracy of the densely connected CDRN in terms of phase fraction fields is approximately four times better, compared to the RBDN and the CDRN with plain skip connections.

The patterns in the NRMSE fields in Fig. 5 are also present in the highest activation maps, which are shown in Fig. 6. Activations of the inner convolution paths in the densely connected CDRN in Fig. 6b are shown below the white line, while the main downsampling/upsampling path is shown above. In case of the

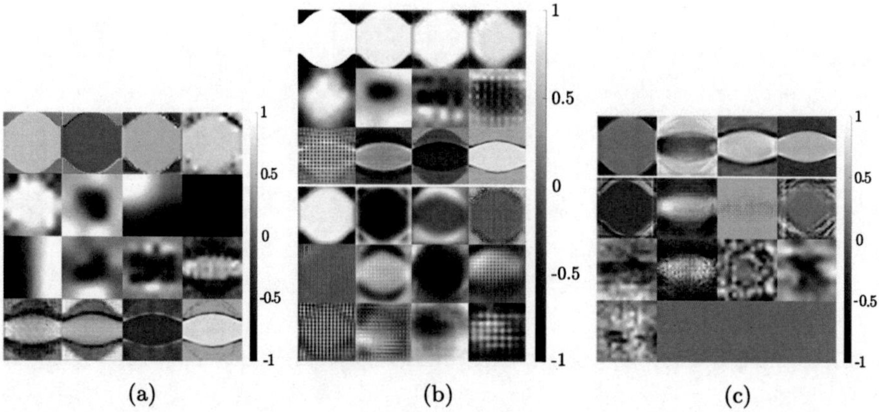

Fig. 6. Highest mean activations. **a** CDRN with plain skip connections (inner dense convolutions below white line). **b** CDRN with densely connected skip connections. **c** RBDN, comprising recursive branches (recursive branch convolutions below white line)

RBDN in Fig. 6c, the non-linear transformations are presented above the white line, while the convolutions in the recursive branches are shown below. In general, the downsampling in CDRN, i.e. the first half of non-linear transformations in RBDN encodes strongly the channel geometry, while the upsampling or the second half of non linear transfigurations, more activates on the phase fraction. In all cases the phase fraction is the most activated feature, since no velocity feature map appears in any case. The gridded pattern, which is also prominent in the NRMSE fields, only appears in the deconvolutional upsampling paths, regardless the CNN design.

4 Discussion

This study demonstrates the feasibility of applying NN as approximating model for blood flows in artificial lungs. Even the largest NN models with $> 10^8$ learnables infer flow fields approximately six orders of magnitude faster compared to FVM. Taking only inference durations into account in this comparison might not be quite valid. The training takes a relevant amount of computing effort and the actual inference might only be perceived as interpolation between the elaborately trained cases. If the training and search for hyperparameters is also considered in comparing computation times, it depends on the number of inferred solutions when break even is reached.

If the accuracy and resolution of CNN predictions is not sufficient for a particular application, it can be easily used to create near convergence initial solutions for other, more accurate and also more trusted methods, which then ideally only have to correct the upscaling interpolation. This upsampling interpolation in CNNs needs special attention since the required additional information needs to be encoded in the NN. The highest activations and also occurring error are due to phase fraction features. This might lead to an underrepresentation of details in velocity fields.

Since the parameter space is large and not smooth, it is hard to draw general conclusions or design rules for NN parameters. However, it seems that larger kernels can be beneficial. Also more features increase the performance, while deeper networks might not improve predictions in general, even if no significant overfitting occurred and thus, the model capacity might not fully exploited yet. Additionally, there is no superior NN design.

However, the fact that there is no clear optimum (even if only a local one) in hyper parameters or NN design might point towards another issue. If the noise in the training targets is too strong according to the bias-variance tradeoff in order to allow for any one-to-one correspondence, there is no valid learning goal. Obviously, if a specific input relates to more than one output, there is no simple rule to learn. Therefore, the training error can not become smaller than a certain level, regardless of the hyperparameters. The exemplary phase fraction field computed using FVM, shown in Fig. 3b, might indicate a significant noise level. What actually are bearable noise levels, strongly depends on the problem at hand and the employed regularization measures.

Another possible reason for not finding a clear optimum might be the difference in spatial resolutions, i.e. more exactly, the differences in interpolation schemes. The particle density is quite homogeneous, except for the transition from RBCs being present or not, which is quite sharp. Similar to the channel wall, this contour needs to be mapped in the coarse DL grid. Therefore, if the contour is not properly aligned, even if the values are correctly predicted, the error becomes large. Instead of correcting every value by its own through the MSE back propagation, a region of interest (ROI) align approach could allow for simply learn the shift of the inherently correct field to form a better prediction [3].

5 Conclusion

Computing multiphase blood flows is costly. Approximating the respecting flow fields using CNNs is not only possible but also by far more efficient. Different encoder-decoder and recursive branching NN designs are tested by performing a two-staged hyperparameter search using precomputed FVM results. Besides the feasibility and inference speed, the results indicate that the used data set is to small in order to allow for generalized design rules for CNNs. A key point of model agnostic DL is, that no knowledge on governing relations is required. In contrast, these relations are encoded automatically within the weights of the NN. Analysing those using tools commonly summarized as explainable artificial intelligence (xAI), might give insights into the underlying physical relationships and could be used for model discovery or parameter identification from experimental data, for example.

References

1. Afshar, Y., Bhatnagar, S., Pan, S., Duraisamy, K., Kaushik, S.: Prediction of aerodynamic flow fields using convolutional neural networks. arXiv:1905.13166v1 (2019)
2. Aggarwal, C.: Neural Networks and Deep Learning. Springer, Basel (2018). https://doi.org/10.1007/978-3-319-94463-0
3. Dai, J., He, K., Sun, J.: Instance-aware semantic segmentation via multi-task network cascades. arXiv:1512.04412v1 (2015)
4. Ghiasi, G., Fowlkes, C.C.: Laplacian pyramid reconstruction and refinement for semantic segmentation. In: Leibe, B., Matas, J., Sebe, N., Welling, M. (eds.) ECCV 2016. LNCS, vol. 9907, pp. 519–534. Springer, Cham (2016). https://doi.org/10.1007/978-3-319-46487-9_32
5. Karimi, S., Dabagh, M., Vasava, P., Dadvar, M., Dabir, B., Jalali, P.: Effect of rheological models on the hemodynamics within human aorta: CFD study on CT image-based geometry. J. Nonnewton. Fluid Mech. **207**, 42–52 (2014). https://doi.org/10.1016/j.jnnfm.2014.03.007
6. Pries, A., Neuhaus, D., Gaehtgens, P.: Blood viscosity in tube flow: dependence on diameter and hematocrit. Am. J. Physiol. **263**(6), H1770–H1778 (1992). https://doi.org/10.1152/ajpheart.1992.263.6.H1770
7. Raissi, M.: Deep hidden physics models: deep learning of nonlinear partial differential equations. J. Mach. Learn. Res. **19**(25), 1–24 (2018)
8. Ribeiro, M.D., Rehman, A., Ahmed, S., Dengel, A.: DeepCFD: efficient steady-state laminar flow approximation with deep convolutional neural networks. arXiv:2004.08826v2 (2020)
9. Ronneberger, O., Fischer, P., Brox, T.: U-net: convolutional networks for biomedical image segmentation. arXiv:1505.04597v1 (2015)
10. Santhanam, V., Morariu, V.I., Davis, L.S.: Generalized deep image to image regression. In: IEEE Conference on Computer Vision and Pattern Recognition (CVPR) 2017, pp. 5395–5405 (2017). https://doi.org/10.1109/CVPR.2017.573
11. Steiger, T., et al.: Accumulations of von Willebrand factor within ECMO oxygenators: potential indicator of coagulation abnormalities in critically ill patients? Artif. Organs **43**, 1065–1076 (2019). https://doi.org/10.1111/aor.13513
12. Thuerey, N., Weissenow, K., Mehrotra, H., Mainali, N., Prantl, L., Hu, X.: Well, how accurate is it? A study of deep learning methods for Reynolds-averaged Navier-Stokes simulations. arXiv:1810.08217 (2018)
13. Wu, W.T., Aubry, N., Massoudi, M.: On the coefficients of the interaction forces in a two-phase flow of a fluid infused with particles. Int. J. Non-Linear Mech. **59**, 76–82 (2014). https://doi.org/10.1016/j.ijnonlinmec.2013.11.006
14. Zhang, Y., Sung, W.J., Mavris, D.: Application of convolutional neural network to predict airfoil lift coefficient. arXiv:1712.10082v2 (2017)
15. Zhou, Z., Siddiquee, M.M.R., Tajbakhsh, N., Liang, J.: UNet++: A nested U-net architecture for medical image segmentation. arXiv:1807.10165 (2018)

Modelling of Heat Transfer for Droplet Condensation in Mixed Convective Duct Flow

Christian Brückner[1](✉), Andreas Westhoff[1], and Claus Wagner[1,2]

[1] German Aerospace Center (DLR), Institute of Aerodynamics and Flow Technology, Bunsenstr. 10, 37073 Göttingen, Germany
christian.brueckner@dlr.de

[2] Technische Universität Ilmenau, Institute of Thermodynamics and Fluid Mechanics, Helmholtzring 1, 98693 Ilmenau, Germany

Abstract. Temperature and humidity measurements are conducted in mixed convective humid-air duct flow with condensation. The latent and total heat transfer during the experiment are determined through the thermal balance for inlet temperatures from 27.5 °C to 35.5 °C, relative humidities from 30% to 55% and at four Reynolds numbers (2000–8000). The experimental results are compared with a heat transfer model from the literature. Adjusted in terms of the geometry and surface properties, the model shows partial agreement for the cases with forced convection but has to be further adjusted regarding the influence of thermal convection.

Keywords: Condensation · Duct flow · Heat transfer

1 Condensation

The widespread use of condensation heat transfer in the energy sector and in air conditioning systems has led to an increased interest in the development of prediction models. For more than 100 years, formulations of models have been investigated by many researchers in order to describe heat transfer in relation to thermal conditions, geometry, material properties and flow conditions [7].

Especially the presence of non-condensable gas (ncg) fractions complicates or even impedes the predictability of the heat transfer as it leads to local accumulation of these gas fractions at the condensate interface. This in turn results in the formation of a diffusion layer and to additional resistance for the latent

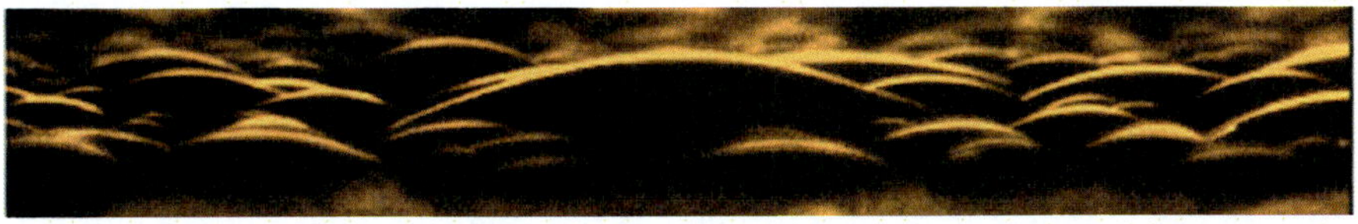

Fig. 1. Condensate droplets on the cooled wall

© The Author(s), under exclusive license to Springer Nature Switzerland AG 2021
A. Dillmann et al. (Eds.): STAB/DGLR Symposium 2020, NNFM 151, pp. 461–471, 2021.
https://doi.org/10.1007/978-3-030-79561-0_44

Nomenclature

Dimensionless Numbers

Gr	Grashof	$\frac{g\,\beta(T_{in}-T_w)d_h^3}{\nu_{air}^2}$
Nu	Nusselt	$\frac{h d_h}{\lambda_{air}}$
Pr	Prandtl	$\frac{\eta c_p}{\lambda_{air}}$
Re	Reynolds	$\frac{U d_h}{\nu}$
Ri	Richardson	$\frac{Gr}{Re^2}$
Sc	Schmidt	$\frac{\nu}{D}$

Symbols

A	area
B_i	mass transfer driving force
C	correction factor
c_p	heat capacity
C_D	drag coefficient
D	diffusion coefficient
d_h	hydraulic diameter
F	force
g	gravitational acceleration
H	(duct) height
h	heat transfer coefficient
h_v	heat of evaporation
L	duct length
$\dot{m}$	mass flow
$\dot{q}$	heat flux
r	radius
T	temperature
u	bulk velocity
$\dot{V}$	volume flow
W	duct width
w	mass fraction
β	thermal expansion coefficient
γ	surface tension
η	dynamic viscosity
λ_{air}	thermal conductivity
ν	kinematic viscosity
θ	contact angle
ζ	contact line radius

Subscripts

a	ambient
air	airflow
D	drag
d	droplet
dp	dewpoint
g	gravity
i	condensate interface
in	inlet
lat	latent heat transfer
nc	non-condensable fraction
vo	outlet
ow	opposite wall
sen	sensible heat
s	surface tension
test	test section
t	total
water	water fraction
w	cooled wall
x,0	begin of test section
x,L	end of test section
x,m	mean of test section

heat transfer. In case of very high fractions of ncg, i.e. humid air, the dominant factor changes from a local depletion of vapour in the diffusion layer to convective mass transfer in the humid air, which is determined by the flow conditions. Through manipulation of the condensate substrate surface towards hydrophobic behaviour and for small rates of condensation on other surfaces, droplet condensation is observed. A sample result of droplets on a condensate substrate is shown in Fig. 1. The size of such droplets is too small to have a direct effect on the wall-bounded flow. However, the increased surface area compared to a plain wall or condensate film adds to the latent heat transfer [2].

Danilo et al. analysed the condensation heat transfer in turbulent convective humid air flow [3]. In their horizontal experiment condensate mass is increased linearly until the critical droplet size is reached. When agglomeration of the droplets starts and a continuous film forms, the global heat transfer stagnates with a reduced amount of convective heat transfer. The latent heat transfer depends on the humidity level and dominates the global heat transfer at high relative humidities. For very small droplets, no impact on the convective heat transfer was observed. Additionally, different humidity levels had no significant impact on the convective heat transfer at similar drop sizes and distributions.

1.1 Model Despcription

Eimann et al. [5] used the diffusion layer model of Peterson et al. [8] and Caruso et al. [2] and developed the model *further* based on their own condensation experiment. The experiment consists of a horizontally oriented rectangular duct with a cross section of 12 mm × 32 mm in which a vertical cooled plate of 9 mm × 6 mm is installed. The cooled plate has the additional function of a heat flux sensor in order to directly record the total heat flux resulting from the latent and sensible heat. The heat flux is recorded in a Reynolds number range $3000 < Re < 22000$ for 36 °C to 55 °C for 36 % to 55% relative humidity.

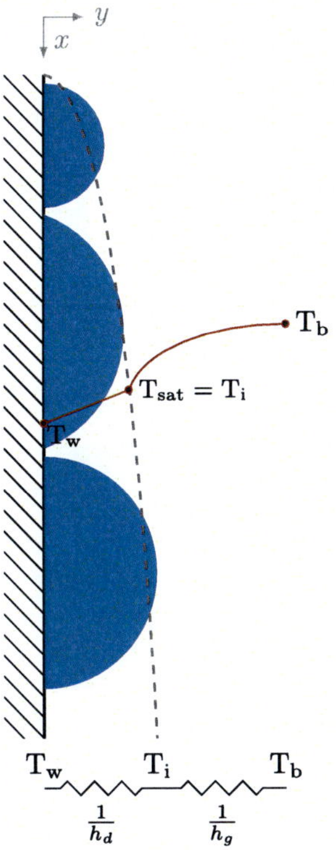

Fig. 2. Scheme of the droplet condensation

The fundamental approach to the heat transfer for dropwise condensation is considered to be a cascade of thermal resistances, the reciprocal of which are the separate heat transfer coefficients. In Fig. 2 the temperature profile between the bulk temperature T_b, the droplet interface temperature T_i and the wall temperature T_w as well as the separate thermal resistances for the droplet $1/h_d$ and the gas-liquid interface $1/h_g$ are illustrated schematically. The sum of the thermal resistances gives the total heat transfer coefficient h_t

$$h_t = \left[\frac{1}{h_d} + \frac{1}{h_g}\right]^{-1} . \tag{1}$$

The individual heat transfer coefficients h_g and h_d are determined with the help of the experimentally obtained total heat flux $\dot{q}_t$ and the temperature of the condensate interface T_i. The partial heat fluxes are determined by the local thermal balance

$$\dot{q}_t = h_t(T_b - T_w) = h_d(T_i - T_w) = h_g(T_b - T_i). \tag{2}$$

Eimann et al. [4] used thermographic imaging to determine the interface temperature for a given droplet size and derived a temporal and spatially averaged formulation for h_d that they extended for the parameter range of $2000 < Re < 21400$

$$h_d = \frac{0.347 Re_b}{1 + 1.75\exp(-330.5 B_i)} \tag{3}$$

$$\text{with} \quad B_i = \ln\left(\frac{w_{nc,i}}{w_{nc,b}}\right).$$

Here, B_i denotes the driving force for the mass transfer with w_{nc} being the mass fraction of non-condensable gas fraction at the interface and in the bulk, respectively. To calculate the mass fraction at the interface, saturation at the interface is assumed and the Gibbs-Dalton law is applied as formulated by Caruso and Di Vitale Maio [2].

The heat transfer coefficient at the gas-liquid interface h_g is derived from the corresponding Nusselt number Nu_g, which in turn is the sum of sensible and latent Nusselt numbers

$$Nu_g = C(Nu_{sen} + Nu_{lat}). \tag{4}$$

Both Nusselt numbers are determined using $Nu - Re$ relations from the literature for duct flows. The correction factor C is introduced to compensate for the influence of the droplets. It is correlated with the critical droplet radius r_{max} for each case and defined as a linear function for $0.2 < r_{max} < 1.61$ mm [5].

Additionally, an iterative procedure is introduced to predict the total heat transfer coefficient based on a measured Reynolds number Re, bulk temperature T_b, wall temperature T_w and non-condensable mass fraction in the bulk $w_{nc,b}$.

The objective of this work is to test the model approach of Eimann et al. against the experimentally determined heat fluxes of a vertical duct flow with condensation on a cooled wall. The comparison of experimental and model data is discussed in terms of the transfer to a larger condensation surface as well as regarding the influence of thermal convection on the heat transfer.

2 Experimental Setup

The experiments are performed in a vertically oriented, differentially cooled duct with a rectangular cross section. The cross section of the duct is 50×533 mm and the cooled surface of the test section has a length of 2.05 m [1]. The temperatures of the duct are controlled with a water-cooled aluminium sheet on one side and a four-zone electrically heated aluminium sheet to achieve isothermal conditions on either side. A sketch of the duct setup is shown in Fig. 3. Upstream of the test section, the inlet section ensures a well-defined, homogenous inflow and downstream of the test section, an outlet section ensures an undisturbed flow in the test section. The aspect ratio of 10.66:1 is chosen to minimize the influence of the side walls, thermal or through the boundary layer. The temperature of the heated wall T_{ow} is equal to the inlet temperature T_{in} in order to minimize the heat flux at this side of the duct.

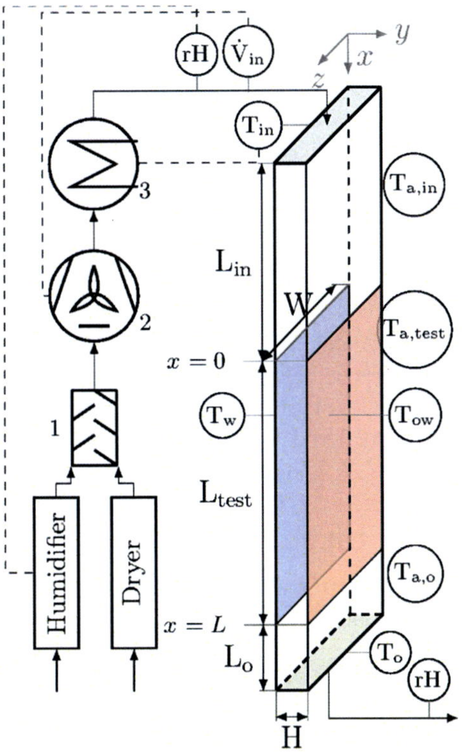

Fig. 3. Process flow diagram of the vertical duct with isothermally cooled wall (T_w, blue) and isothermally heated opposite wall (T_{ow}, red). (1) static mixer; (2) radial fan; (3) heat exchanger

The measurements are conducted for $2000 < Re < 8000$, an inlet temperature range of 27.5 °C to 35.5 °C and a relative humidity range of 30 % to 50% at the inlet. In total, 35 data points are collected. The temperature difference between cooled wall T_w and inlet temperature T_{in} was set to 26.5 K in all experiments to keep Gr in the order of $2.4 \times 10^6 . Further$, Pr and Sc are at near constant levels at 0.75 and 0.65, respectively. With a variation of Re at constant Gr, the Ri yields 0.62 to 0.033.

2.1 Thermal Balance

To determine the heat flux of the experiment, the observed temperatures T_{in}, T_o, T_w, the dew point temperatures $T_{dp,in}$, $T_{dp,out}$ and the air mass flow $\dot{m}_{air}$ are time-averaged and the thermal balance is

$$\dot{q}_{air} = \dot{q}_t + \dot{q}_{lat} + \dot{q}_a, \tag{5}$$

with the contributing total heat flux from fluid to the wall

$$\dot{q}_t = h_t \mathrm{L}_{\text{test}} \mathrm{W} (T_{x,m} - T_w), \tag{6}$$

heat flux due to the change in heat capacity of the air

$$\dot{q}_{air} = \dot{m}_{air}(c_{p,in}T_{x,0} - c_{p,o}T_{x,L}), \tag{7}$$

the latent heat flux, directly calculated from the condensate mass

$$\dot{q}_{lat} = \dot{m}_c h_v, \tag{8}$$

and the ambient heat flux through the side walls

$$\dot{q}_a = h_{a,t} 2\mathrm{HL}_{\mathrm{test}}(T_{x,m} - T_{a,t}). \tag{9}$$

The thermal balance is solved analytically with the idealisation of a spatially averaged heat transfer coefficient h_t for the total heat transfer from the fluid to the wall. The heat transfer coefficient h_t and the mean temperature $T_{x,m}$ are then determined by iteration to match the observed outlet temperature [10]. All material properties are considered at the determined mean temperature $T_{x,m}$ (Fig. 3).

2.2 Adjustment of the Droplet Force Balance

Since the air flow in this experiment is in vertical direction, the contribution to the force balance for the droplets has to be adjusted compared to the horizontal flow in the Eimann model.

$$F_g + F_s + F_D = 0 \tag{10}$$

The gravitational force F_g and the drag force F_d act in the same direction, while the surface tension force F_s keeps the droplet on the wall. All three forces depend on the droplet radius r [9]:

$$F_g = \varrho_{water}\, V(r)\, g \tag{11}$$

$$F_s = -\gamma \int_0^{2\pi} \zeta(r) \cos\theta \cos\phi \, d\phi \tag{12}$$

$$F_D = \frac{1}{2} C_D\, \varrho_{air}\, u^2\, A_{proj}(r) \tag{13}$$

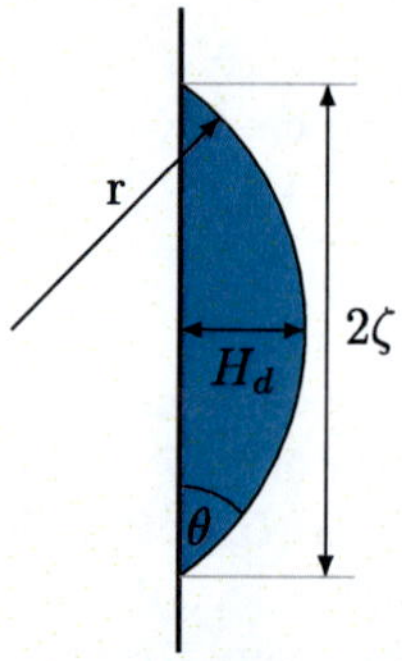

Fig. 4. Vertical projection of the droplet geometry with $\theta < 90°$

Due to contact angles significantly deviating from 90°, the volume $V(r)$ is calculated for a spherical cap rather than a half sphere

$$V = \frac{\pi}{3} r^3 (2 + \cos\theta)(1 - \cos\theta)^2. \tag{14}$$

The vertical projection area $A_{proj}(r)$ of a droplet can be simplyfied to a circle segment

$$A_{proj} = \frac{r^2}{2}\left(2\bar{\theta} - \sin(2\bar{\theta})\right). \tag{15}$$

The mean contact angle $\bar{\theta}$ has a limited validity and can only be used for small deformations due to gravity.

3 Results

The first step of the model approach was to identify the force balance of the droplet in order to determine the critical droplet radius r_{max}. In the experiment, an otherwise untreated anodized aluminium plate was used as condensation surface. The contact angle θ was measured to be $63.3° \pm 15°$ at 18 positions of the surface in a horizontal position. Three sample droplets are shown in Fig. 5 revealing the large spread of contact angles at different positions of the condensation surface. The high variance comes with the large area as well as the long-term use of the aluminium plate and is seen as a more realistic, commonplace condensation surface.

The maximum radii for the different Re as calculated via the force balance and determined experimentally are shown in Fig. 6 for all cases. The force balance gives a similar radius of 3.6 mm for all cases. In the experiment, the maximum radius observed was smaller with values from 2 mm to 2.8 mm. Since the contact angle is significantly smaller than 90° for single droplets and further distorted through coalescence, this difference in radius does not translate into a major difference of the droplet forces. It does, however, significantly impact Eimanns correction factor C, which is only defined up to 1.61 mm and will be discussed further with the total heat transfer.

The experiment provides a direct method of calculating the total condensate mass using the thermal balance. The comparison of the measured latent heat and the projected latent heat by the correlation taken from the VDI Heat Atlas [6] is shown in Fig. 7 in case of an inlet temperature of $T_{in} = 31.5\,°\text{C}$. For the lower Re, the predicted latent heat transfer is too small (Fig. 7a and 7b). While the correlation does consider the transition from laminar to turbulent flow, it does not consider the additional influence of thermal convection introduced to the flow by the cooled wall.

Since the temperature difference, and therefore the Grashof number, is constant for all cases, the influence of the thermal convection is larger at lower Re, where the forced convection is not yet dominant ($0.13 < Ri < 0.62$). In this case, the thermal convection leads to an increased latent heat transfer, probably caused by the accelerated air close to the wall. Due to continuity, the acceleration in the boundary layer leads to a deceleration in the bulk flow and thus to a longer dwell time of the humid air. To compensate for the thermal convection, a modified latent Nusselt number was formed using Ri

$$Nu_{lat}^{*} = (1 + C_{Ri}\,Ri)Nu_{lat} \quad \text{with}\, C_{Ri} = 3.012. \tag{16}$$

The correction factor $C_{Ri} = 3.012$ fits the experiment data best.

Fig. 5. Various contact angle samples on horizontal surface

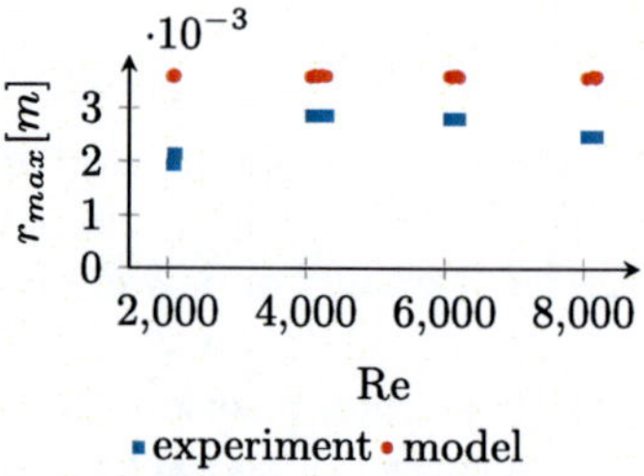

Fig. 6. Critical droplet radius r_{max} over the Re for all cases

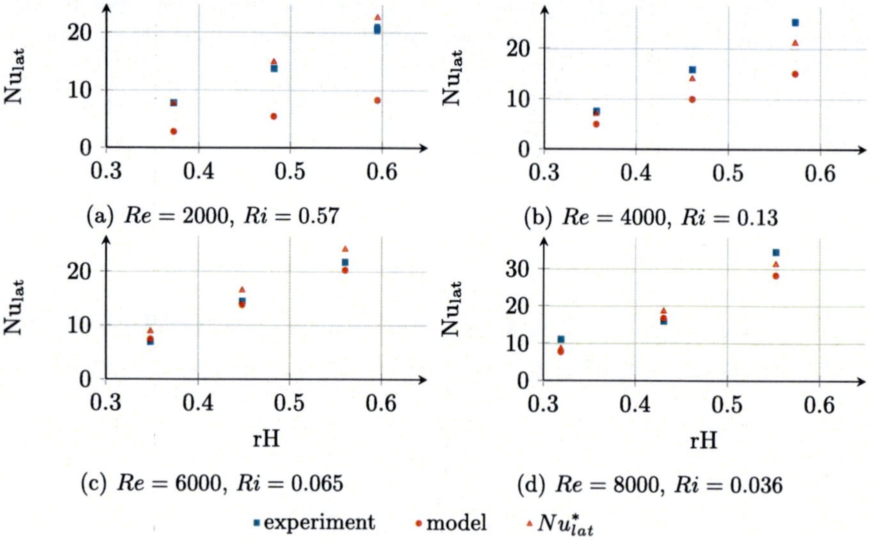

Fig. 7. Nusselt number for latent heat transfer over relative humidity for Re at varying velocities at $T_{in} = 31.5\,°\text{C}$. The relative uncertainty of Nu is at approx. 3.5% (still inside the square mark)

For the highest Re, the influence of thermal convection is very small ($Ri = 0.036$) and thus, the prediction fits reasonably well with the measured latent heat transfer (Fig. 7c and 7d). In all four cases, the prediction is more accurate for the lower relative humidity. This might be an indicator, that even for these comparably low relative humidities, the fog formation has a relevant influence that has to be taken into account.

Figure 8 shows the Nusselt number for the total heat flux from the humid air to the cooled wall for the experiment and the model in two configurations. The first uses the critical radius observed in the experiment, and the second uses the droplet height H_d instead. For the observed contact angles, the droplet

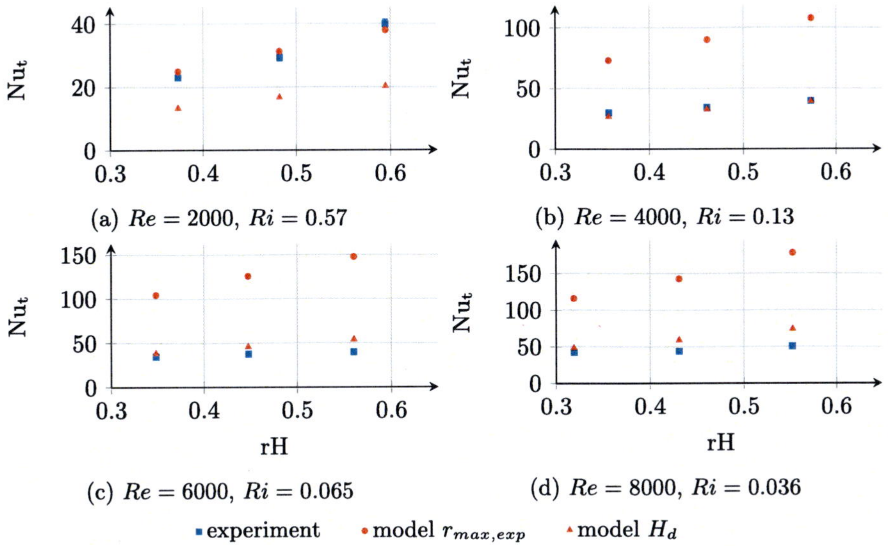

Fig. 8. Nusselt number for total heat transfer over relative humidity for *Re* at varying velocities at $T_{in} = 31.5\,^{\circ}\mathrm{C}$

height and radius diverge significantly compared to the idealised 90° originally used in the model. Since the correction factor is intended for the deviation from a homogenous film, the droplet height appears to be a reasonable replacement. For *Re* 2000 in Fig. 8a, the model prediction using the critical radius shows reasonable agreement. For the other cases *Re* 4000, *Re* 6000 and *Re* 8000 (Fig. 8b, 8c and 8d), the predicted heat transfer amounts to two or three times the value of the experiment. Using the model with the droplet height results in much better agreement in these cases, even though the model still overpredicts the Nusselt number at *Re* 6000 and *Re* 8000. In combination with the information about the latent heat transfer, the deviation at *Re* 6000 and *Re* 8000 shows that the predicted sensible heat is too high, especially for the higher relative humidities. The *Re* 2000 case adjusted to the missing latent heat would still not agree better with the observed droplet radius, but again the influence of thermal convection is not accounted for by the model.

Based on these results, the droplet height is a more suitable substitute for the critical radius in cases with contact angles θ smaller than 90°

$$H_{d,max} = r_{max}\,(1 - \cos(\bar{\theta})). \tag{17}$$

4 Conclusion

Condensation heat transfer in a mixed convective duct flow has been studied experimentally and the results have been compared to a prediction model for dropwise condensation based on the diffusion layer model. The measurements show an increase of the latent and sensible heat transfer with increasing relative humidity and humid air velocity. The prediction model was adjusted to the differing experimental geometry and flow direction as well as to the droplet characteristics. The comparison of the latent heat revealed a significant influence of the thermal convection and a correction based on the Richardson number Ri is introduced.

The maximum droplet radius as the core parameter in the prediction model was replaced by the corresponding droplet height in order to adjust the model to be used with contact angles smaller than 90°. While this still leads to an overprediction of the sensible heat transfer, especially for higher relative humidities, it shows reasonable agreement with the experimental data. The remaining overprediction is attributed to the depletion of water vapour over the condensation surface. This effect has to be further analysed in the future in order to find an equivalent droplet height that factors in the effect of both, surface and condensation rate.

Acknowledgement. The authors like to thank Annika Köhne and Konstantin Niehaus for their contribution to this paper.

References

1. Brückner, C., Bahavar, P., Westhoff, A., Wagner, C.: Reynolds number dependency of the heat and mass transfer in mixed convective duct flow with condensation at a cooled wall. In: Dillmann, A., Heller, G., Krämer, E., Wagner, C., Tropea, C., Jakirlić, S. (eds.) DGLR 2018. NNFMMD, vol. 142, pp. 523–532. Springer, Cham (2020). https://doi.org/10.1007/978-3-030-25253-3_50. ISBN 978-3-030-25253-3
2. Caruso, G., Di Maio, D.V.: Heat and mass transfer analogy applied to condensation in the presence of noncondensable gases inside inclined tubes. Int. J. Heat Mass Transfer **68**(1–3), 401–414 (2014). https://doi.org/10.1016/j.ijheatmasstransfer.2013.09.049. ISSN 00179310
3. Danilo, S., Dominique, C., Frédéric, P.: Experimental dropwise condensation of unsaturated humid air - influence of humidity level on latent and convective heat transfer for fully developed turbulent flow. Int. J. Heat Mass Transfer **102**, 846–855 (2016). https://doi.org/10.1016/j.ijheatmasstransfer.2016.06.001. ISSN 00179310
4. Eimann, F., Zheng, S., Philipp, C., Fieback, T., Gross, U.: Convective dropwise condensation out of humid air inside a horizontal channel - experimental investigation of the condensate heat transfer resistance. Int. J. Heat Mass Transfer **127**, 448–464 (2018). https://doi.org/10.1016/j.ijheatmasstransfer.2018.08.015. ISSN 00179310
5. Eimann, F., Zheng, S., Philipp, C., Omranpoor, A.H., Gross, U.: Dropwise condensation of humid air - experimental investigation and modelling of the convective heat transfer. Int. J. Heat Mass Transfer **154**, 119734 (2020). https://doi.org/10.1016/j.ijheatmasstransfer.2020.119734. ISSN 00179310
6. Gnielinski, V., et al.: VDI Heat Atlas. VDI Gesellschaft Verfahrenstechnik und Chemieingenieurwesen, 2nd edn (2010). https://doi.org/10.1007/978-3-540-77877-6_94. ISBN 9783540778769
7. Huang, J., Zhang, J., Wang, L.: Review of vapor condensation heat and mass transfer in the presence of non-condensable gas. Appl. Thermal Engi. **89**, 469–484 (2015). https://doi.org/10.1016/j.applthermaleng.2015.06.040. ISSN 13594311

8. Peterson, P.F., Schrock, V.E., Kageyama, T.: Diffusion layer theory for turbulent wapor condensation with noncondensable gases. J. Heat Transfer **115**(4), 998–1003 (1993). https://doi.org/10.1115/1.2911397. ISSN 15288943
9. Sommers, A.D., Ying, J., Eid, K.F.: Predicting the onset of condensate droplet departure from a vertical surface due to air flow-applications to topographically-modified, micro-grooved surfaces. Exp. Thermal Fluid Sci. **40**, 38–49 (2012). https://doi.org/10.1016/j.expthermflusci.2012.01.031. ISSN 08941777
10. Westhoff, A.: Experimental study of heat and mass transfer in convective flows of moist air with droplet condensation as a function of surface roughness and wetting properties. In: Proceedings of the 5th International Conference on Experimental Fluid Mechanics, pp. 692–697 (2018). https://doi.org/10.18726/2018_2. ISBN 978-3-943207-32-3

Check for updates

Experimental Simulation of the Human Respiration

Andreas Kohl, Pascal Lange(✉), and Daniel Schmeling

German Aerospace Center (DLR), Institute of Aerodynamics and Flow Technologies, Bunsenstraße 10, 37073 Göttingen, Germany
{andreas.kohl,pascal.lange,daniel.schmeling}@dlr.de

Abstract. In this paper, we present a new mobile respiration simulation system (RSS), which can be connected to existing thermal manikins. With the objective to simulate the human respiration process as realistic as possible, the system was validated on the basis of literature data and results obtained from human subject tests. The RSS reproduces realistic respiration cycles characterized by a sine wave of a typical normal breathing flow rate. The provided flow rates as well as the breathing frequency – representing the time of inhalation and exhalation – were verified by literature values. Since the new system additionally allows to enrich the exhaled air with carbon dioxide (CO_2), experimental studies addressing the indoor air quality are also feasible. Here, the amount of CO_2 emitted by the RSS corresponds to the average amount of CO_2 exhaled by test persons. In addition, the flow characteristics occurring in a human nose are simulated using a self-developed facial mask, in combination with the new system. The result is a breathing thermal manikin based on a mobile respiration simulation system, which can easily be connected to heated passenger models. Accordingly, the system can be installed at any seat within a passenger compartment. This offers the advantage of individually defining the location of the manikin, which can effortlessly be adapted during a measurement campaign. Therefore, the system especially suitable for studies addressing the performance of ventilation systems in passenger compartments and indoor environments.

Keywords: Indoor air quality · Breathing thermal manikin · CO_2 distribution · Passenger comfort · Ventilation efficiency · Airborne spreading

1 Introduction

For many decades, thermal manikins (TMs) have been widely used in research activities addressing the thermal passenger comfort in cabins, such as trains or airplanes, as well as for experimental studies regarding the airflow in indoor environments, e.g. in office buildings [1]. The manikins simulate the geometrical blockage and the sensible heat release of humans. TMs provide repeatable and comparable experimental setups with well-known and precise boundary conditions required for the implementation in computational fluid dynamics (CFD)

© The Author(s), under exclusive license to Springer Nature Switzerland AG 2021
A. Dillmann et al. (Eds.): STAB/DGLR Symposium 2020, NNFM 151, pp. 472–482, 2021.
https://doi.org/10.1007/978-3-030-79561-0_45

simulations. Additionally, so-called "breathing thermal manikins" (BTM) are used to simulate the human respiration, i.e. the forced convective airflow caused by in- and exhalation. The combination of the room ventilation, the thermal convection induced by the human heat release and the forced respiration flow determine the airflow around the human body [2]. The present report describes the development and testing of a human respiration simulation system (RSS), which can be attached to existing TMs currently used at the DLR in Göttingen. In addition to the simulation of the airflow through inhalation and exhalation, the system can also emit the same amount of the respiration waste gas CO_2 as a real human would do. Here, CO_2 is an important indicator for the indoor air quality affecting comfort, health and productivity of the passengers [3]. Modern passenger trains take this issue into account by adjusting the volume flow rate of the fresh air based on interior CO_2 concentrations.

The rapidly developing situation of the corona pandemic in 2020 also leads to the fact that studies regarding air quality, air pollution level and airborne transmission of infectious agents become more and more relevant. Recent studies have proven that tracer gas is a suitable surrogate for exhaled aerosols (smallest liquid droplets) [4]. For this reason, during tracer gas measurements the RSS can be used as a source to dose gas while simulating the respiration process. Consequently, the study of these transport processes forms the basis for developing new ventilation scenarios, focusing on a high contaminant removal efficiency.

In the following sections, the developed RSS is introduced and first test measurements are presented in order to validate the new system. The latter comprise measurements to analyze the performance of the system with regard to typical human breathing profiles and CO_2 exhalation rates. In a second step, the latest results of the application of the RSS in a generic train laboratory are shown.

2 Human Breathing Characteristics

This chapter includes the description of the basic parameters and characteristics of the human respiration process given in the literature. The newly developed RSS is based on the findings regarding the respiration cycle of a human as well as on the breathing flow caused by human respiration. The mechanical parts, the electrics as well as the measurement and control system have been designed to simulate the human respiration process as realistically as possible.

The human respiration process is driven by the lifting and lowering of the chest and the diaphragm resulting in a compression and decompression of the lungs. Therefore, the air inside the lungs is exchanged with the environmental air through the nose and mouth. Inside the lungs, oxygen (O_2) contained in the inhaled air is exchanged with CO_2 in the blood which is produced during cellular respiration in the human metabolism. The average breathing frequency (BF) of humans is between 10 and 15 breaths per minute for normal tidal breathing without physical exertion [5]. The volume that is exchanged during tidal breathing corresponds to the tidal volume of about 0,6 l. These values of the volume and BF are implemented as standard breathing setting of the RSS. Further, during normal breathing, a human emits between 250 and 300 ml of CO_2 per minute [5] which corresponds to about 4% of the tidal volume. The RSS also facilitates the provision of the mentioned CO_2 quantities. Moreover, breathing during physical or mental exertion can also be simulated by the RSS by modifying the settings (e.g. raise of BF, tidal volume and emitted CO_2 quantity).

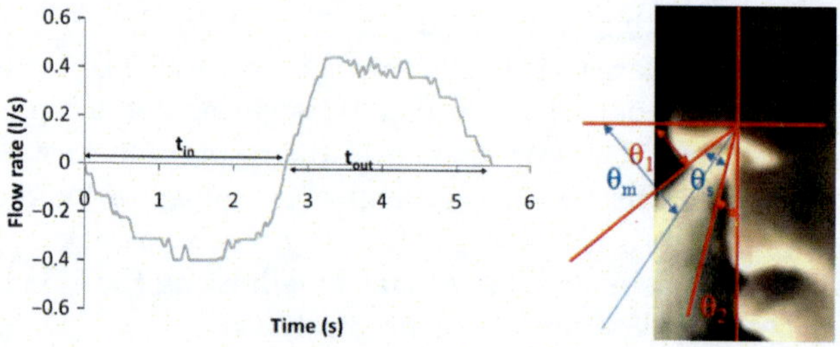

Fig. 1. Flow rate of humans during tidal breathing [6] (left). Side view of the visualized exhalation jets through the nose [6] (right).

The work of Gupta *et al.* [6] analyzes the breathing flow rate of 25 test persons. The typical volume flow of the respiration process is sinusoidal as shown in Fig. 1 (left) and reaches peak flow rates of around $\pm$ 0.4 l/s for an average human and may vary depending on the physiological state of the individual person [6]. In addition to the simulation of the tidal breathing, the system is designed to measure the generated flow rate and to analyze the inhaled air in terms of CO_2 level, temperature and humidity.

Figure 1 (right) shows the human airflow caused by exhalation through the nose according to [6]. The findings regarding the mean flow angles of spreading (Θ_s) and from the side (Θ_m) are given in Eq. (1).

$$\Theta_s = 23° \pm 14° \quad \Theta_m = 60° \pm 6° \tag{1}$$

Moreover, an average nose opening for male subjects during breathing was determined in [6] and amounts to $0,71 \pm 0,23\,\text{cm}^2$. These criteria serve as a guideline for designing a male face mask, which can be connected to the RSS in order to reproduce the corresponding flow angles through the nose.

3 System Layout and Implementation

3.1 Mechanical Implementation

The mechanical implementation is shown as a technical principle in Fig. 2 on the left. The stepper motor is connected via clutch to a spindle, which rotates in both directions. The stepper motor operates in closed-loop control for better acceleration and to avoid step loss as well as resonance frequencies. The rotation is translated into a linear movement by a screw nut and a linear stage. This linear movement y compresses and decompresses the two bellow cylinders, which simulate the human lungs and provide the breathing volume flow $\dot{V}$. In one bellow cylinder the inhaled air is analyzed and in the second cylinder CO_2 is dosed by a solenoid valve (see Fig. 2, right). Detailed information regarding the system components and the design process are given in [7].

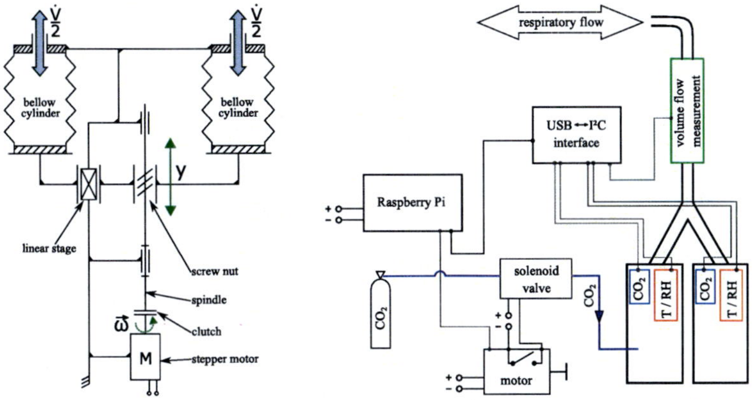

Fig. 2. Technical principle of the RSS [7] (left). Sketch of the control system including the sensor positions, represented by green, blue and red rectangles (right).

3.2 Sensors and Control System

The control system is based on a small single board computer ("Raspberry Pi"), which is responsible for controlling the sensors and sending trigger signals to the motor (see Fig. 2, right). The used sensors comprise a bidirectional volume flow sensor (accuracy: ±1.5%), several CO_2 detectors (accuracy: ±30 ppm + 3%) as well as temperature (accuracy: ±0.1 °C) and humidity probes (accuracy: ±1.5%). The motor features its own system to ensure the closed-loop control in order to control the movement as well as the opening of the solenoid valve, as indicated in Fig. 2, right. The amount of CO_2 emitted in the system depends on the opening time of the valve (τ_V) and is precisely adjustable in steps of 0,1 ml (corresponds to an opening time of 1 ms). The CO_2 is stored in a pressure bottle connected to the system via tubing under a predefined pressure. For detailed information regarding the electrical setup of the RSS see [7].

3.3 Development of a Realistic Human Breathing Flow via Facial Mask

As shown in Fig. 3 (left), the face mask was constructed using a three-dimensional scan of a male face [8] and manufactured using a 3D printing process. The mask is connectable to the RSS using standardized hose parts. Subsequently, the mask is attached to the manikin's face. So far only the respiration through the nose has been simulated. However, a mask supporting breathing through the mouth has recently been developed. The resulting exhalation flow angles are depicted in Fig. 3 (right) and will be discussed in the following section.

4 System Validation and Application

4.1 Validation of Human Breathing Characteristics

Before the RSS can be used for experimental simulation of the human breathing process, both the volume flow characteristics and the amount of exhausted CO_2

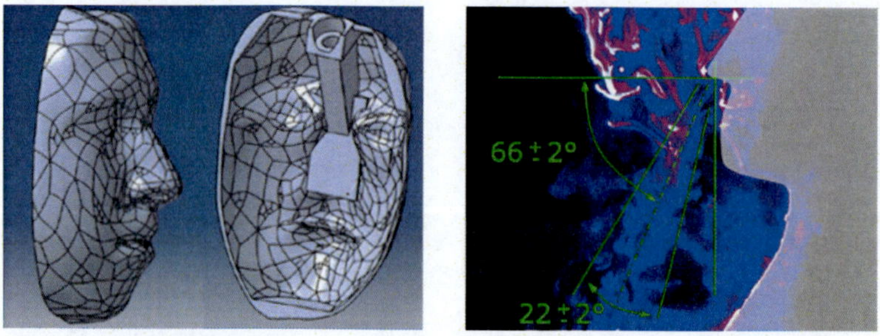

Fig. 3. 3D model of a human facial mask (left). Analysis of the resulting flow pattern through the facial mask (right).

must be calibrated based on the real human metabolism. These characteristics are adjusted by controlling the motor control curve and the opening time of the magnetic valve, respectively. In a first step, the volume flow rate provided by the RSS is compared to the data given in [6], see Fig. 4 (left). For a system's BF of 11 min^{-1} (represented by red triangles), the flow rate of a breathing cycle matches the literature data [6] (grey circles) with the following mean deviations of 1.5% for the tidal volume, 0.5% for the peak values and 0.4% for the frequency.

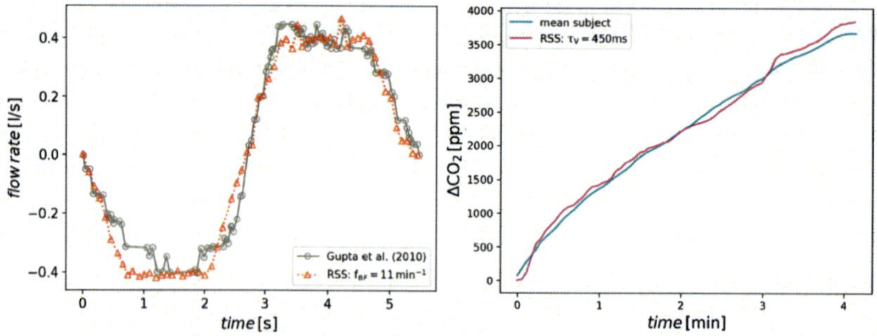

Fig. 4. Validation tests of provided breathing flow rate (left). Results of the validation regarding emitted CO_2 amount (right).

In a second step, the emitted amount of CO_2 is validated by measurements with test persons. For this purpose, the subjects breathed into an enclosed environment (volume $\approx 230\,\text{l}$) over a period of four minutes. The resulting enrichment of CO_2 (cyan colored line), up to a maximum value of 3675 ppm, is shown in Fig. 4 (right) and serves as a benchmark for the RSS. By modifying the opening time of the solenoid valve (see Fig. 2, right), the emitted amount of CO_2 provided by the system can be varied. As shown in Fig. 4 (right), an opening time of $\tau_V = 450\,\text{ms}$ (magenta colored line) results in a CO_2 exhaust rate, which matches the mean subject values with a mean accuracy of 1.8%. The last step comprises the validation of the flow characteristics through the human facial mask. In order to do so, a flow visualization with laser light and smoke is conducted. The resulting flow pattern including the flow angles provided by the generic facial mask is analyzed using a graphical editing software. As illustrated in Fig. 3 (right), the determined angles agree with the values given in Eq. (1) within measurement accuracy. In conclusion, it should be noted, that the human breathing process is highly individual and furthermore strongly depend on the level of activity. For the calibration of the RSS we used mean values from the literature, hence, the RSS represents a "standardized average human".

4.2 Generic Train Laboratory

The application tests are carried out in a generic train laboratory, a full-scale mock-up with the realistic geometry of the DLR's next generation train (NGT), see Fig. 5 (left). The lower deck is suitable for experimental studies addressing air conditioning, thermal passenger comfort, energy efficiency and indoor air quality. The mock-up is equipped with 24 real train seats from a regional train (see Fig. 5, right), which are separated in six seat rows. TMs are used to simulate the sensible heat impact and the obstruction of passengers.

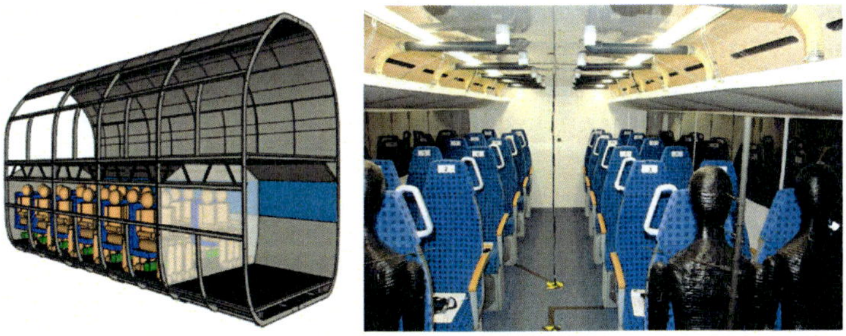

Fig. 5. CAD model of the generic train laboratory [9] (left). Cabin partly equipped with TMs and with installed hatrack-integrated air supply inlets (right).

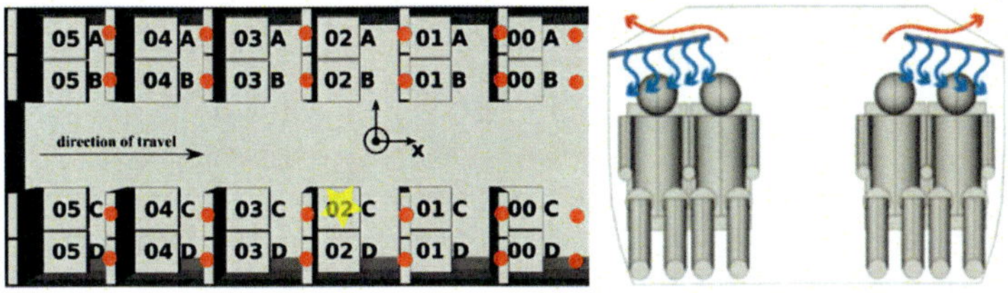

Fig. 6. Seating plan with CO_2 sensor positions (red dots) and CO_2 source location (yellow star), (left). Illustration of HLMV [10] (right).

A comprehensive sensor system comprising local temperature, humidity and CO_2 probes is installed. As indicated in Fig. 6 (left), the respective CO_2 sensors are attached to the backrest of the seats. The generic train cabin in combination with an air conditioning system facilitates experimental studies of different ventilation scenarios. As shown in Fig. 5 (right), a novel ventilation concept, called hatrack-integrated low-momentum ventilation (HLMV), was installed during the presented studies. It is characterized by a low-momentum air supply above the passenger heads through planar, large-surface outlets, which are integrated in the lower part of the overhead luggage racks (see Fig. 6, right) [10]. After circulation, the air leaves the compartment through slit-shape outlets in the lateral parts of the ceiling.

4.3 Spatial CO_2 Distribution in the Generic Train Laboratory

As a sample application, studies with the RSS are compared to investigations using a human subject as a CO_2 source in a generic train cabin. In both cases, the source is placed on seat position 02C (see, Fig. 6, left). During the tests, the ventilation scenario described in Sect. 4.2 is installed. The temperature of the inflowing air is set to 17.5 °C and maintained with a maximum deviation of less than 2%. The total volume flow rate amounts to 225 ± 5 l/s, which is about twice the amount of fresh air for 24 passengers in accordance with EN13129 [11]. As long-distance trains are typically operated with a recirc fraction of 50%, the volume flow rate in the generic mock-up is similar to the total volume flow rate in real train compartments. During the investigations, the passengers are simulated using TMs, each operating at a constant heat release of 75 W.

Novel Breathing Thermal Manikin as CO_2 Source: For the application test with the new RSS, the latter is connected to a facial mask positioned on a TM's face (see Fig. 7, left).

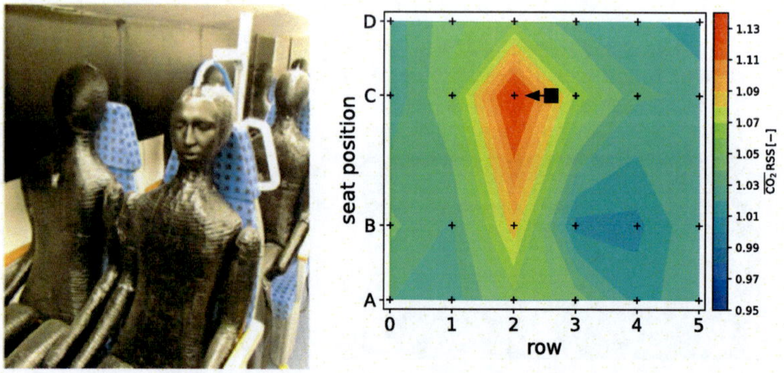

Fig. 7. RSS as a CO_2 source connected to a facial mask at the TM's face (left). Resulting CO_2 distribution within the generic train laboratory caused by the system (right).

After thermal conditions within the cabin have reached a state of equilibrium, the RSS is operated for 45 min. The parameters of the RSS (i.e. BF and opening time of the valve) are set as described in Sect. 4.1. The resulting CO_2 distribution – averaged over the last 25 min of emission – is shown in Fig. 7 (right). Here, the data is normalized using the mean background concentration of 444 ppm. The black square symbolizes the position of the CO_2 source, while the arrow represents the breathing direction. The supporting measurement positions are marked as black crosses. The resulting distribution shows a significant CO_2 accumulation in the immediate vicinity of the source. The maximum values of 1.13 are found at position 02C. The propagation of the gas is limited to the row in which the source is located, since increased values occur for position 02B, 02C and 02D. The concentration at the seats in front and behind the source increases only slightly. The values at the remaining seat positions are about 1.0 corresponding to background concentration.

Human Subject as CO_2 Source: Subsequently, a human subject is used as CO_2 source. After steady-state conditions are achieved in the lab, the test person sits at position 02C (see Fig. 8, left) for 45 min. The subject was instructed to sit as still as possible and breathe calmly during the test phase. The concentration distribution – averaged over the last 25 min of breathing – is normalized with the background concentration of 460 ppm and depicted in Fig. 8 (right). The black square and the corresponding arrow represent the exact position of the source and the direction of breathing, respectively. The black crosses illustrate the location of the sensors. A maximum normalized concentration of 1.1 occurred at position 02C. Generally higher values are measured for the adjoining seats, such as 03C and 03D. This means that the source clearly induces a significant concentration increase at the adjacent seats behind the source. Considering the seat rows two and three, the released CO_2 remains localized in the area of the lab near the source (seat position C and D) since no noticeable rise in concentration occurred at seat positions 02B/01B and 02A/01A. In this case, values of about 1.0 (representing background concentration) are determined at the remaining seat positions.

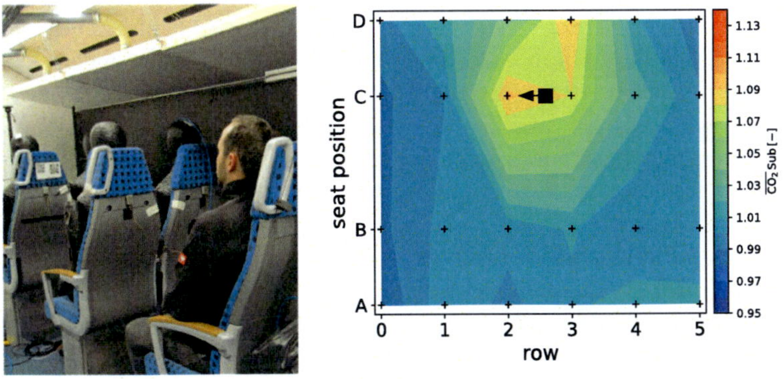

Fig. 8. Human subject as a CO_2 source at seat 02C (left). Resulting CO_2 distribution within the generic train laboratory caused by the test person (right).

Comparison of Mechanical and Human CO_2 Source: To compare the two different sources, the resulting CO_2 distributions in the generic train cabin are taken into account. In both cases, the maximum concentration values occurs directly in front of the source, at seat position 02C. Considering the concentration already existing in the air (background concentration), both types of sources cause the same mean concentration increase (at least for this particular seat position). Therefore, the adjustments to the respiration system, based on the validation tests (see Sect. 4.1), seem to be appropriate to simulate the human breathing characteristics. However, the resulting CO_2 propagation in the cabin shows significant differences for the two investigated sources. Using the RSS as a CO_2 source results in a localized dispersion near the source. Mainly, the emitted gas disseminates in the seat row, where the source is located. Using a human as a CO_2 source also leads to a localized increase in concentration, however, in this case, the source clearly induces accumulations in the adjacent seats, right behind the source location. Further, the gas accumulation seems to remain on the side of the cabin, where the source is located (seat position C and D). In contrast, a propagation in the seat row across the aisle (to position A and B) cannot be observed. A reason for the differences in the results of both investigations could be differently developed flow patterns in the vicinity of the source. The body surface of the TM is smaller compared to a male human. Thus, the TM might emit a lower heat flux density than the test person. Further, the human subject was wearing a jacket, which represents a thermal insulation for the heat release. These facts influence the buoyancy-driven flows and consequently result in differences for the developed flow structures near the source. Furthermore, when using a test person as a CO_2 source, even slight changes of the seating position or small changes of the discharge direction (due to movements of the head) also affect the resulting concentration distribution within the compartment.

5 Conclusion

We introduced a new respiration simulation system, designed and set up at the German Aerospace Center in Göttingen. The system is attachable to a heat-releasing thermal manikin in order to simulate a breathing passenger, especially for research activities addressing the air quality in passenger compartments or indoor environments. To simulate a realistic breathing process, the breathing flow rate was analyzed and compared with literature data, which are based on results of human subject tests. A system's breathing frequency of $11\,\text{min}^{-1}$ leads to a similar respiration curve with a mean deviation of 1.5% and 0.5% for the tidal volume and peak values, respectively. The system also facilitates the dosage of respiration gases, in particular CO_2, similar to the emissions of real persons. Hereto, validation tests with human subjects were conducted to provide realistic CO_2 quantities. The system provide the same CO_2 exhaust rate with a mean deviation of less than 2%. Furthermore, since the gas emission as well as the breathing flow rate of the system are variable, the simulation of different breathing cycles (i.e. during physical exertion) is feasible. Moreover, connecting a facial mask to the respiration system simulates realistic outflow characteristics through the nose.

In a second step, the performance of the respiration simulation system was verified in a generic train cabin for an alternative ventilation system – hatrack-integrated low-momentum ventilation. Additionally, investigations with a human subject as a CO_2 source were conducted. The resulting spatial CO_2 distribution reveal similar concentration peaks directly in front of the source. However, the distribution paths of the gas are slightly different. While the CO_2 provided by the system remains localized in the seat row where the source is located, the test person induces a concentration increase for the adjacent seats behind the source. The differences in the gas propagation are not only attributed to the system's settings but rather to the different body shape and heat emission of the thermal manikin in contrast to the human subject. Furthermore, the discharge direction during exhalation affects the resulting propagation of CO_2 in the passenger compartment. The human subjects influence the discharge direction due to minor changes in sitting posture and head movements, whereas the breathing thermal manikin sits perfectly still. In spite of these explained smaller deviations in the near field, the new respiration simulation system was successfully taken into operation. The validation tests confirmed the realistic breathing frequency, tidal volume, amplitude values and CO_2 enrichment of the system compared to real humans.

Acknowledgement. The authors acknowledge Annika Köhne for proofreading this manuscript.

References

1. Holmer, I.: Thermal manikin history and applications. Eur. J. Appl. Physiol. **92**(6), 614–618 (2004)
2. Ivanov, M., Mijorski, S.: CFD modelling of flow interaction in the breathing zone of a virtual thermal manikin. Energy Procedia **112**, 240–251 (2017)
3. Dorgan, B., Dorgan, E., Kanarek, M., Willman, A.: Healthy and productivity benefits of improved indoor air quality. ASHRAE Trans. **104**, 658 (1998)
4. Ai, Z., Mak, C., Gao, N., Niu, J.: Tracer gas is a suitable surrogate of exhaled droplet nuclei for studying airborne transmission in the built environment. Build. Simul. **13**, 489–496 (2020)
5. Oczenski, W., Andel, H., Werba, A.: Atmen, Atemhilfen: Atemphysiologie und Beatmungstechnik, 9th edn. Georg Thieme Verlag, Stuttgart (2012)
6. Gupta, J., Lin, C.-H., Chen, Q.: Characterizing exhaled airflow from breathing and talking. Indoor Air **20**(1), 31–39 (2010)
7. Kohl, A.: Entwicklung eines Systems zur experimentellen Simulation der Atmung von Passagieren. Master-thesis, TU Ilmenau (2020)
8. GrabCAD Homepage. http://grabcad.com/library/male-face. Accessed 18 June 2020
9. Schmeling, D., Bosbach, J.: On the influence of sensible heat release on displacement ventilation in a train compartment. Build. Environ. **125**, 248–260 (2017)

10. Schmeling, D., Volkmann, A.: On the experimental investigation of novel low-momentum ventilation concepts for cooling operation in a train compartment. Build. Environ. **182**, 107116 (2020)
11. EN 13129: Railways Application - Air conditioning for mainline rolling stock - Comfort parameters and type tests. German Version EN13129:2016 (2020)

Characterization of a Mixed Convection Cell Designed for Phase Transition Studies in Moist Air

Konstantin A. Niehaus[1,2](✉), Andreas Westhoff[1], and Claus Wagner[1,2]

[1] German Aerospace Center (DLR), Institute of Aerodynamics and Flow Technology, Göttingen, Germany
Konstantin.Niehaus@dlr.de

[2] Institute of Thermodynamics and Fluid Mechanics, Technische Universität Ilmenau, Ilmenau, Germany

Abstract. A detailed description of an experimental set-up designed for upcoming investigations of latent and sensible heat transfer in a cuboid sample with air in- and outlets is given. The container is heated at the rear and cooled at the transparent front wall. Temperature measurements reveal that both sides exhibit a mean temperature deviation below 2% relative to the temperature difference between the mean plate temperature and the ambient temperature. This is a suitable temperature distribution for such measurements. Tomographic particle image velocimetry covering the entire volume of the mixed convection cell exhibits a large-scale circulation due to forced convection with a buoyancy-induced flow close to the temperature controlled surfaces. Forced convection origins from a flow between the inlet and the outlet channel with a mean deviation of 1% from the mean velocity and a maximum absolute deviation of 0.04 m/s. Measurements were performed for Reynolds numbers ranging from $300 < \mathrm{Re} < 2000$ and Grashof numbers $\mathrm{Gr} < 1.2 \times 10^8$. A representative flow field obtained at $\mathrm{Re} = 620$ and $\mathrm{Gr} = 1.1 \times 10^8$ is presented as an example.

1 Introduction

The heat transfer in mixed convective air flows with phase transition is a phenomenon which occurs in nature and a plethora of technical applications. Condensate forming droplets leads to an increased heat transfer rate in comparison with film-wise condensation and therefore is advantageous in heat exchangers [13]. In contrast, condensation has negative effects in many other applications. For instance, fogging on the windshield or headlights of motor vehicles leads to restrictions when driving and a reduction of optical transparency influences road safety. In addition, a considerable amount of thermal energy is required for defogging, which leads to a range reduction when it comes to electric driving [16]. The design and modification of these components in terms of optimizing dehumidification or preventing misting is often based on experience and intuition [11]. Despite the enormous progress in numerical methods in recent years, there is still

© The Author(s), under exclusive license to Springer Nature Switzerland AG 2021
A. Dillmann et al. (Eds.): STAB/DGLR Symposium 2020, NNFM 151, pp. 483–493, 2021.
https://doi.org/10.1007/978-3-030-79561-0_46

a lack of reliable and applicable models for the numerical simulation of condensation in general, and in particular of droplet condensation on surfaces [10]. Due to the complexity of the physical processes that determine the heat transport and consequently the fogging on surfaces, reliable simulations are expensive and time-consuming, and therefore not suitable for industrial applications. Thus, the overall goal of this study is to develop a model, which allows the prediction of the condensation behavior on surfaces. In such a configuration the mass transport of vapor due to phase transition, the resulting latent and the sensible heat transfer are determined by the physical processes of convection, diffusion, the boundary conditions and the material properties of the surfaces. In a configuration where phase transition occurs, even smallest changes in the boundary conditions lead to significant changes in the vapor mass transfer. A major challenge of this study is the design and construction of an experimental set-up with the appropriate measurement technology. The focus lies on the characterization of the boundary conditions to ensure reproducibility for future experiments at high measurement accuracies.

In the present paper we introduce our set-up in detail and discuss the results of the flow and temperature characterization. The long-term goal is to control the effects of thermal and forced convection independently to map all physical processes that determine the fogging of headlamp windshields. In principle, the set-up represents a generic replica of a vehicle headlamp. The air flow inside automotive headlights is the result of the superposition of forced and thermal convection. In this case, the forced convection is characterized by the Reynolds number $Re = {}^{Ul}/_{\nu}$ and originates from ventilation holes at the rear wall of the mixed convection cell. In contrast thermal convection is the result of the heat emission from the internal light sources on the headlight's back and is characterized by the Grashof number $Gr = l^3 g \Delta T^{\beta}/_{\nu^2}$. Here, β, g and ν are the thermal expansion coefficient, gravitational acceleration and the kinematic viscosity of the working fluid air as a function of temperature and humidity. The experimental set-up consists of a rectangular cavity that is isothermally heated at the rear and cooled at the front. To compensate for pressure changes caused by temperature fluctuations, car headlights are designed as open systems to exchange air with the engine compartment through air inlets and outlets. The set-up simulates this using two venting channels. For the present configuration the characteristic velocity U is equal to the mean inlet velocity and l is a characteristic system length, which is the cell height in Gr and the inlet channel height for Re. Further, $\Delta T = \langle T_h \rangle - \langle T_c \rangle$ is the temperature difference between the mean temperature of a heating and a cooling plate. If T_c falls below the dew point T_{dp}, vapor mass flow $\dot{M}_v$ occurs due to phase transition. Fogging of surfaces thus depends on the flow velocity U, the air temperature T_{in}, the surface temperature T_c, the heat emission of the lamps $\dot{Q}_l$ as well as on the temperature of the light source T_h and the dew point temperature T_{dp}. We designed an experimental set-up which controls these parameters with high accuracy.

One of the first experimental studies on the issue of condensation inside automotive headlights was published by Hoines and Jiao in 1998 [2]. They studied the influence of the inlet positions on condensation and internal vapor density distribution. Subsequently, numerical and experimental studies on fogging in consumer headlights by Siozawa et al. [14], Liu et al. [9] and Kim et al. [5] reveal a typical temperature difference between front lense and light source of $\mathcal{O}(10\,\mathrm{K})$ for semiconducting light sources and $\mathcal{O}(10^2\,\mathrm{K})$ for incandescent bulbs. Additionally, Kim et al. [5] studied the flow by means of stereoscopic particle image velocimetry (PIV). Their observed maximum velocities of $\mathcal{O}(10^{-1}\,\mathrm{m/s})$ match the results of Guzej and Zachar [1] and Zenin [17]. Further, Kim et al. [5] revealed three-dimensional flow structures and mixed convection. A set-up similar to the set-up presented in this paper was realized by Langebach [8, p. 35], who investigated the heat transport towards the light source to study the influence of venting positions and a sample geometry on bulb temperature. When comparing studies on this topic, it is evident that there is no canonical choice for the length scale within a headlight. So far, the following options for the lenght scale exist: path length of the flow between the vents, the devices width or height (i.e. Guzej [1]) or the inlet width (i.e. Langebach [8]). This results in maximum Reynolds numbers of $\mathcal{O}(10^5)$ and maximum Grashof numbers of $\mathcal{O}(10^9)$. Langebach [8] reported Reynolds numbers of $10^3 - 10^4$ and Grashof numbers of $\mathcal{O}(10^6)$ using the inlet width. Guzej et al. [1] investigated a case with $Re = 9.3\times10^3$ using the cell height in combination with the mean velocity above the front wall ($\mathcal{O}(10^{-2}\,\mathrm{m/s})$). In this work we employ the definition by Langebach [8].

2 Experimental Set-Up

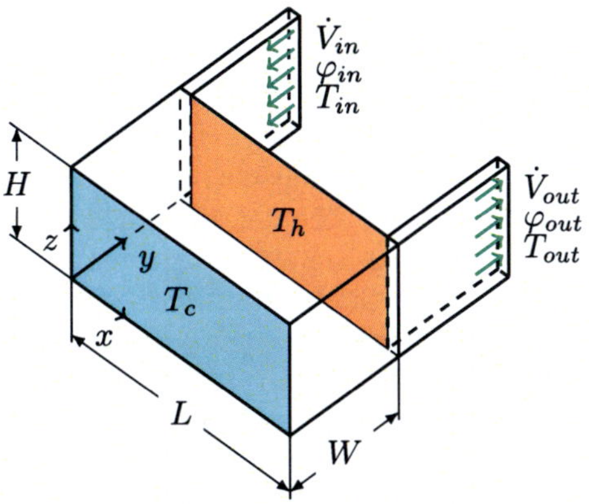

Fig. 1. Schematic set-up of the condensation cell: The cooling plate is colored blue (temperature T_c) and the heating plate is colored red (temperature T_h). Inlet and outlet conditions are characterized by means of humidity φ, temperature T_{in} and volume flow $\dot{V}$.

The set-up consists of a cuboid container with air in- and outlets. The aspect ratios of the cavity are $\Gamma_{yz} = W/H = 1$ and $\Gamma_{xz} = L/H = 2$, where $L = 500\,\mathrm{mm}$ denotes the length and $H = 250\,\mathrm{mm}$ the height of the sample which is equal to

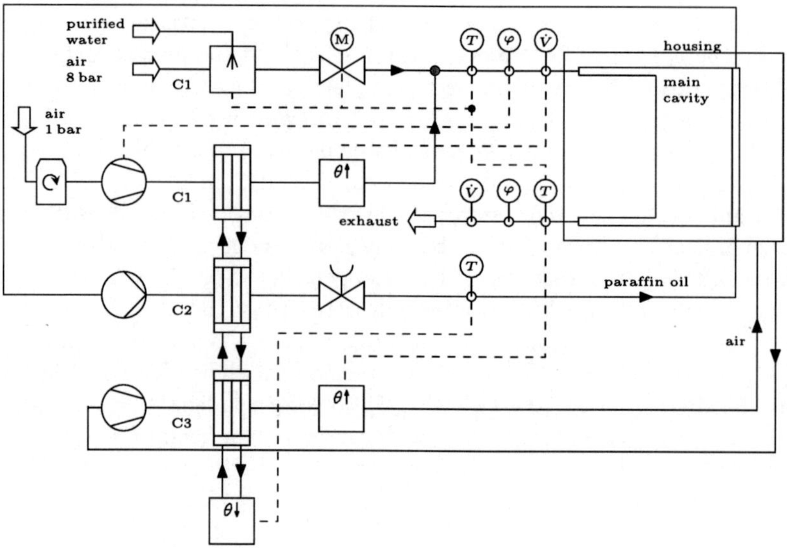

Fig. 2. Three control circuits for inlet temperature, humidity and volume flow of the inlet (C1), the cooling plate temperature (C2) and the ambient temperature of the condensation cell (C3). T, φ and $\dot{V}$ indicate temperature, humidity or volume flow measurements, respectively. A red dot marks the junction of humidifier and the volume flow section of C1.

its width W. Figure 1 depicts a schematic sketch of the sample. Forced convection originates from an inlet on the left and an outlet on the right side of the sample. Both extend over the entire cell height and are $w = 25\,\text{mm}$ wide. The flow direction is indicated by turquoise arrows in Fig. 1. The red-colored area represents a 10 mm copper plate located between the vents with an electrical heating mat mounted on the back. The heating device extends over the entire cell height and has a length of 430 mm. 10 mm polyvinyl chloride is used as an insulation between the vents and the heating plate. To provide a homogeneous temperature distribution, the heating plate is covered by 60 mm styrodur with a thermal conductivity of $\lambda_D = 0.034\,\text{W/m K}$. Fifteen resistance temperature devices (RTD) with an accuracy of approximately 0.1 K are distributed evenly over the entire heat plate. They are used to measure the mean temperature $\langle T_h \rangle$ and the spatial temperature distribution T_h. The probes are mounted in holes of 3 mm depth filled with a heat-conducting paste. The blue area in Fig. 1 indicates the condensing surface, which is composed of two parallel, 6 mm thick glass plates with a 20 mm gap. In between flows a countercurrent with respect to the air flow direction of paraffin oil at temperature $\langle T_c \rangle$. With the objective to obtain isothermal boundary conditions on the cooled surface, the device is equipped with eight volume-flow-controlled openings equally distributed over the height on both sides with a mixing section mounted in front of the inlets. All other walls of the sample are made of 10 mm thick glass plates. The sample itself is placed in an air-filled, temperature controlled cuboid box a with height of 1 m, a length of 1.2 m and a width of 1 m. The ambient temperature T_∞ in the box is set to the mean sample temperature $(\langle T_{\text{in}}\rangle - \langle T_{\text{out}}\rangle)/2$ to obtain adiabatic boundary conditions. Furthermore, the configuration provides optical accessibility for the measurement of velocity fields.

Two flow meters are used to record the volume flow $\dot{V}$ at the in- and outlet of the sample with an accuracy of $\sigma_{\dot{V}} = 1.5\dot{V} + 0.05\,\text{l/min}$. The mean inlet velocity $\langle U\rangle$ is calculated using $\dot{V}/A$ with A being the inlet cross section. Before entering the cell, air passes a diffuser similar to the one studied by Köthe et al. [6] and an array of stainless steel meshes with size-to-wire thickness ratios of 311 µm/112 µm, 165 µm/80 µm and 95 µm/46 µm. These are followed by a honey comb flow straightener positioned 10 cm downstream of the inlet opening with a cell width of 3.2 mm and a wall thickness of 25 µm. The temperature of the air entering or leaving the sample is measured using eight RTD probes with an accuracy of 30 mK. The probes are evenly distributed over the cell's height. Figure 2 shows an overview of the set-up used to control the boundary conditions. The scheme is divided into three control cycles: C1 to C3. The cavity is shown on the top right inside the temperature-controlled housing. Before air enters (indexed with in) or leaves the cell (indexed with out), the vents' volume flow, mean temperature and mean humidity $\langle\varphi\rangle$ are measured in control cycle C1. The volume flow is generated by a radial fan. Additionally, the air system is equipped with an air dryer. Since the amplitude of the sensor noise scales with its absolute value, τ depends on the Reynolds number and is chosen such that the standard error of the mean is below 4 l/min. This is equivalent to a mean velocity deviation of 0.01 m/s. A circulation chiller with a 5 kW cooling output at 20 °C is used to set the front plate temperature. 16 RTDs with an approximate accuracy of 0.1 K are used to measure the mean cooling plate temperature.

Fig. 3. Photograph of the housing with a low momentum ceiling ventilation (white) and the test section inside with the transparent cooling plate and its tubing.

The chiller is used to set T_c, which is the lowest temperature in the system. Consequently, C1 and C3 are equipped with electrical heaters to set T_{in} and T_∞ independently. The dry and temperature-controlled air mixes with humidified air at the joint indicated with a red dot.

Since the results presented here are obtained in a dry environment with no phase transitions, further details on the control accuracies and measurement uncertainties of the humidification system will be given in a consecutive work. Inside the housing, the temperature is measured using RTDs in combination with SHT85 sensors. Figure 3 provides an overview of the housing with the test section inside. The housing is equipped with a low-momentum ceiling ventilation system in order to provide a homogeneous spatial temperature distribution at a low flow velocity. The black box

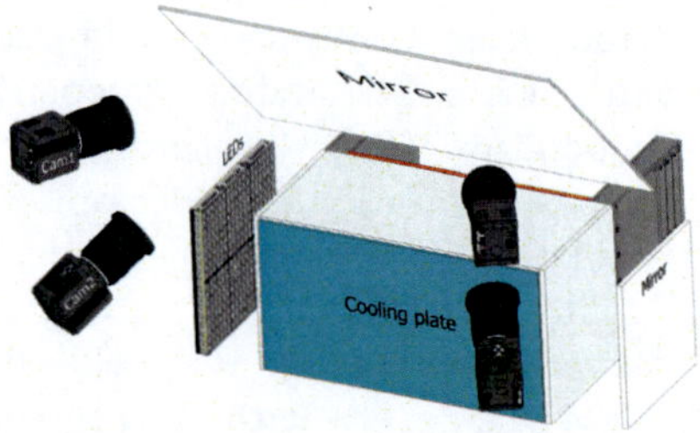

Fig. 4. Tomographic PIV set-up: Four cameras imaging the cell via a mirror mounted above the cell. A LED array illuminates the cell from the left side. The light is reflected back into the cell via a second mirror on the right.

between the set-up's supports is filled with approx 0.7 t of sand to minimize vibrations. Tomographic PIV is used to determine the internal velocity distribution $U(\mathbf{x})$. The PIV measurement set-up is depicted in Fig. 4. Four cameras imaging the main cavity via a surface mirror mounted on top of the cell. LEDs are mounted on the left wall close to the inlet section. Seeding particles are introduced close to the opening of the inlet channel. The measurement volume covers the entire cell. It is lit using two multi-LED illuminators positioned at a distance of 15 cm from the sidewall. On the opposite side, a mirror is mounted to increase the light intensity in the measurement volume. For a detailed description of the light source, see Stasicki et al. [15]. As seeding particles, buoyancy-neutral helium-filled soap bubbles (HFSB) are injected into the flow at the beginning of the inlet to provide a homogeneous particle distribution at the sample inlet. The mean diameter of the HFSB is 0.3 mm.

By injecting the HFSB, an additional volume flow of maximum $\dot{V}/10$ is generated. However, the total volume flow remains constant. The time series of images are recorded by four PCO Edge 5.5 (double shutter) in Scheimpflug condition together with $f = 21$ mm (Distagon T* 2.8, Carl Zeiss), whose aperture is set to $f/11$. Volumetric self-calibration, reconstruction and correlations are obtained using an in-house code that is described in detail in Kühn et al. [7]. The final interrogation volume used for the cross-correlation step in the multi-grid process includes $32 \times 32 \times 32$ voxels with a 50% overlap. To characterize the internal convective flow, a wide parameter range is studied. The overview of the parameter combination of forced (Re) and thermal convection (Gr) is given in Fig. 5. The graphic depicts the set of tuples (Gr and Re), with bars indicating 1σ-uncertainties.

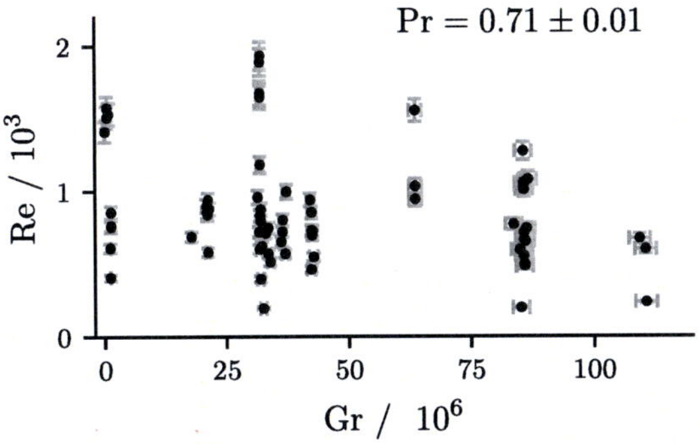

Fig. 5. Parameter space covered in the PIV measurement campaign. A dot indicates a measurement run with corresponding parameters with bars indicating the uncertainty.

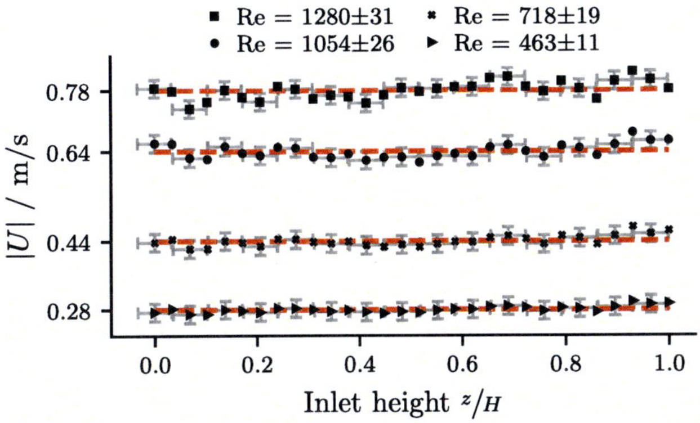

Fig. 6. Inlet velocity profile ($x = 12.5\,\text{mm}$) of the inlet channel in z-direction for different Re with characteristic length $w = 25\,\text{mm}$. Dashed red lines indicate the average velocity.

3 Results

At $x = 12.5\,\text{mm}$ the vertical inlet velocity distribution was measured using an omnidirectional draught probe with an uncertainty of $\pm(0.02\,U + 0.02\,\text{m/s})$. The probe was positioned at 30 sites along the inlet height. The measurement duration was 10 s with a sampling frequency of 10 Hz. Figure 6 shows the time-averaged velocities and the above defined uncertainties as a function of Re. Uncertainties of the mean are displayed only for every other value for the sake of visibility.

The numbers along the y-axis and the dashed line presents the mean velocities. For all Re and measurement positions the relative deviation from the mean lies below 1% i.e. below 0.04 m/s. This is below the deviations of 4% reported by Kühn [4, p. 88], who investigated forced and mixed convection by means of tomographic PIV. In addition, the homogeneity of the temperature distribution

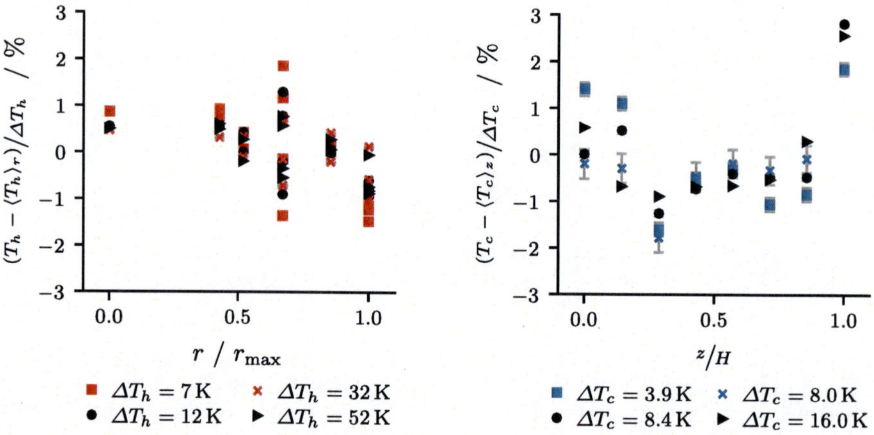

(a) Temperature deviation $T_h - \langle T_h \rangle_r$ normalized by the temperature difference $\Delta T_h = |\langle T_h \rangle_r - T_\infty|$ in percent as a function of the distance r from sensor to plate center for different ΔT_h.

(b) Temperature deviation $T_c - \langle T_c \rangle_z$ normalized by the temperature difference $\Delta T_c = |\langle T_c \rangle_h - T_\infty|$ in percent as a function of the heigth for different ΔT_c.

Fig. 7. Temperature distributions for heating and cooling plate measured using RTD temperature probes at various mean surface temperatures.

of the heating plate was measured. Figure 7a shows the spatial temperature distribution of the heating plate, with the radial distance from the sensor position to the plate center normalized with $r_{\max} = 280\,\mathrm{mm}$ on the ordinate. Different markers correspond to varying temperature differences ΔT_h. The measurement case with the smallest temperature difference shows the highest relative temperature deviations. The radial distribution of deviations from the mean heating plate temperature (Fig. 7a) exhibits a maximum relative deviation below 2% everywhere. The maximum of the absolute deviation occurs at $\Delta T_h = 16\,\mathrm{K}$ with a temperature difference of 0.4 K for $|T_\infty - \langle T \rangle_{t,z}|$. This deviation is of the same magnitude as the uncertainties reported by Schmeling et al. [12] ($\Delta T = 21.2\,\mathrm{K}$, $\sigma = 2.2\%$). The distribution of temperatures inside the cooling plate is shown in Fig. 7b. Measurements were conducted for various temperature differences $\Delta T_c = |T_\infty - \langle T_c \rangle_z|$ between the mean housing temperature and the mean cooling plate temperature. This difference represents the normalized temperature variations over the height of the cooling plate. The temperature distribution in Fig. 7b exhibits relative values similar to those of the heating plate, while the maximum absolute deviation is 0.4 K at a temperature difference of 16 K with a relative deviation of 2.6% at $z = H$. This may be caused by an air cavity in the vicinity of the sensor, the missing insulation during the measurement or a malconfigured volume flow in this area and will be adjusted in future measurements. The deviations are still below those reported by Kästner et al. [3] who also employed a transparent temperature-controlled surface (heated) in a convection experiment. At a temperature difference of 16 K they report a mean deviation of 1.7%, whereas the deviation for the set-up in this study is 0.7%. Moreover, except the value at $z = H$, all deviations are smaller than those reported by Schmeling et al. [12].

In addition to the inflow and boundary conditions, the resulting flow patterns in the sample are of utmost importance since the structures have a strong impact on the heat and thus, on the vapor mass transfer. Examples of mean velocity fields plane through the cell center parallel to the xy-plane and the yz-plane are given in Fig. 8 with $\mathrm{Gr} = 1.1 \times 10^8$ and $\mathrm{Re} = 620$. The in-plane flow is indicated by vectors, while the out-of-plane component is color-coded. The time-averaged vector fields are calculated by means of 630 instantaneous vector fields with a recording frequency of 25 Hz. Due to reflections areas close to in- and the outlet are masked. The flow originates at the inlet $(x, y) = (0, W)$ and is directed towards the cooling plate. At the front wall the flow is redirected parallel to it and detaches at $x/L = 7/8$. Between the cooling plate and the outlet the flow separates into one part leaving the cell through the outlet and another part returning to the inlet parallel to the heating plate. The center line of this large scale circulation can be seen in yz-plane on the left of Fig. 8 with $u = 0$. It clearly shows an asymmetric shape along the z-axis and the 3D flow structure. The presence of thermal convection can be seen in both velocity fields and is indicated by $u_z < 0$ close to the cooling plate and $u_z > 0$ at the heating plate. The influence of Gr and Re has to be characterized in an in-depth analysis of all measurement cases summarized in Fig. 5. However, the present mean velocity fields give a general understanding of the flow structure and demonstrate, that a tomographic PIV measurement of the entire cell has been implemented successfully.

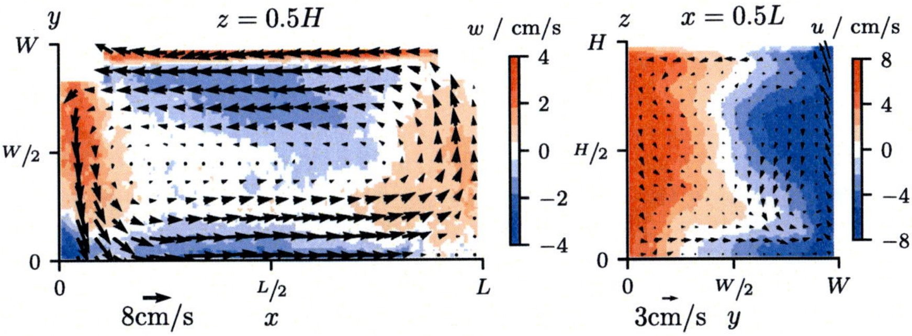

Fig. 8. Time-averaged (26 s, 630 frames) velocity field at $z = 0.5H$ parallel to the xy-plane (left) and at $x = 0.5L$ parallel to the yz-plane (right) with $\mathrm{Re} = 620$ and $\mathrm{Gr} = 1.1 \times 10^8$. Vectors indicate the in-plane and colors the out-of-plane flow.

4 Conclusion

In the present paper we give a detailed description of an experimental set-up designed to investigate latent and sensible heat transfer in a cuboid sample with air in- and outlets. The container is heated at the rear and cooled at the front. Both sides exhibit a measured mean temperature deviation below 2% relative to the temperature difference between the mean plate temperature and the ambient temperature. Forced convection origins from a flow between an inlet and an outlet channel. The homogeneity of the inlet velocity profile is measured at different Reynolds numbers with a mean deviation of 1% from the mean velocity and a maximum absolute deviation of 0.04 m/s. Measurements were performed for Reynolds numbers ranging from $300 < \mathrm{Re} < 2000$ and Grashof numbers $\mathrm{Gr} < 1.2 \times 10^8$. The accuracies of the boundary conditions are compared to other experiments focusing on the investigation of convective air flows. The goal to achieve boundary conditions with uncertainties equal to or below those of experiments investigating similar flow conditions is reached. Hence, the set-up is suited for reproducible measurement campaigns. Furthermore, the optical accessibility allows for tomographic PIV or particle tracking velocimetry measurements covering the entire sample. A first result of this campaign exhibits a large-scale circulation due to forced convection with a buoyancy- induced flow close to the temperature controlled surfaces.

Acknowledgment. The authors like to thank Annika Köhne for proof-reading this work.

References

1. Guzej, M., Zachar, M.: CFD simulation of defogging effectivity in automotive headlamp. Energies **12**(13), 2609 (2019)
2. Hoines, L., Jiao, J.: Environment leading to condensation in automotive lamps. SAE Trans. 682–685 (1998)
3. Kästner, C., Resagk, C., Westphalen, J., Junghähnel, M., Cierpka, C., Schumacher, J.: Assessment of horizontal velocity fields in square thermal convection cells with large aspect ratio. Exp. Fluids **59**(11), 1–13 (2018)
4. Kühn, M.: Untersuchung großskaliger Strömungsstrukturen in erzwungener und gemischter Konvektion mit der tomografischen Particle Image Velocimetry. PhD thesis, TU Ilmenau, August 2011
5. Kim, D., et al.: Velocity field measurement on natural convection inside an automotive headlamp using time-resolved stereoscopic particle image velocimetry. Int. J. Heat Fluid Flow **77**, 19–30 (2019)
6. Köthe, T., Herzog, S., Wagner, C.: Shape optimization of aircraft cabin ventilation components using adjoint CFD. In: Engineering Optimization 2014, pp. 675–680. CRC Press, September 2014
7. Kühn, M., Ehrenfried, K., Bosbach, J., Wagner, C.: Large-scale tomographic PIV in forced and mixed convection using a parallel SMART version. Exp. Fluids **53**, 91–103 (2012)
8. Langebach, J.: Beiträge zur thermischen Auslegung von Automobilscheinwerfern. Univ.-Verlag Ilmenau (2008)

9. Liu, H., Jiang, L.F., Hu, C.G., Chen, X.L., Lin, J.D.: Method of lamp fogging discrimination. In: Applied Mechanics and Materials, vol. 248, pp. 179–184. Trans Tech Publications (2013)
10. Oxyzoglou, I., Tejero, A.: Prediction of condensation forming in automotive headlights using CFD. Technical report, University of Thessaly (2018)
11. Poorman, T., Bielecki, J., Chang, M., Belsare, S., El-Khatib, F.: Automotive lighting thermal performance prediction methods. In: SAE Technical Paper Series. SAE International (2001)
12. Schmeling, D., Bosbach, J., Wagner, C.: Simultaneous measurement of temperature and velocity fields in convective air flows. Meas. Sci. Technol. **25**(3), 035302 (2014)
13. Schmidt, E., Schurig, W., Sellschopp, W.: Versuche über die Kondensation von Wasserdampf in Film- und Tropfenform. Forsch. Ingenieurwes. **1**, 53–63 (1930)
14. Shiozawa, T., et al.: Thermal air flow analysis of an automotive headlamp: the PIV measurement and the CFD calculation for a mass production model. JSAE Rev. **22**(2), 245–252 (2001)
15. Stasicki, B., Schröder, A., Boden, F., Ludwikowski, K.: High-power LED light sources for optical measurement systems operated in continuous and overdriven pulsed modes. In: Lehmann, P., Osten, W., Gonçalves, A.A. (eds.) Optical Measurement Systems for Industrial Inspection X. SPIE, June 2017
16. Westhoff, A.: Experimentelle Untersuchung des Einflusses der O-ber-fläch-en-be-schaf-fen-heit von Scheiben auf die Kondensatbildung. Technical report, FAT, December 2017
17. Zenin, S.: Numerische und experimentelle Untersuchungen zum Wärmetransport in einem Automobilscheinwerfer. Univ.-Verlag Ilmenau (2007)

Check for updates

Study on the Influence of Turbulence on Thermal Comfort for Draft Air

Daniel Schiepel(✉) and Andreas Westhoff

German Aerospace Center (DLR), Institute of Aerodynamics and Flow Technology, Göttingen, Germany
Daniel.Schiepel@dlr.de
http://www.dlr.de

Abstract. Today you can find various standards to assess thermal comfort. However, it lacks on standards for the assessment of thermal comfort in car cabins, particularly in case of high momentum air-flow or draft. Thermal comfort is often perceived differently in cars compared to e.g. in buildings or coaches. With the objective to overcome this problem, an experimental study of the impact of the draft on thermal comfort is carried out. In this paper, we present results of an investigation focusing on the influence of turbulence intensity on thermal comfort. Draft is simulated in a generic car cabin in the following parameter ranges: temperature $T = 17\,°\mathrm{C}$–$29\,°\mathrm{C}$, velocity $U = 0.25\,\mathrm{m/s}$–$2.5\,\mathrm{m/s}$ and turbulence intensity $I_{LT} = 16\%$ and $I_{HT} = 32\%$. The thermal comfort is determined by means of a thermal manikin and infrared thermography to evaluate the thresholds of thermal comfort based on the German Industrial Norms [1,2].

Keywords: Thermal comfort · Equivalent temperature · Cold draft sensibility

1 Introduction

According to a recent study (2019) people spend a considerable amount of their time in vehicles (51 minutes per day) [3]. Hence, for the manufacturers thermal comfort in cars is a critical argument in terms of the competitive edge. The thermal comfort is generally influenced by temperature, radiation, air flow properties of the air-conditioning system and individual sensation. The most frequently used standards to characterize thermal comfort are based on the concept of equivalent temperature (T_{eq}) as defined in the German Industrial Norm [1] originating from the evaluation of thermal comfort for the low-momentum ventilation of coaches and buildings. However, the ventilation in cars is typically realized by means of high-momentum air flow through air outlets located in the area of dashboard. Moreover, the temperature sensation for cars and buildings is different due to e.g. the activity level of the passenger and the influence of the environment. In the present investigation we address the following questions: What are the thresholds of thermal comfort? And what are the impact quantities in case of high-momentum air flow which determine the thermal comfort?

© The Author(s), under exclusive license to Springer Nature Switzerland AG 2021
A. Dillmann et al. (Eds.): STAB/DGLR Symposium 2020, NNFM 151, pp. 494–503, 2021.
https://doi.org/10.1007/978-3-030-79561-0_47

Our aim is to prove whether the existing standards are still valid in case of draft in cars.

The study consists of two parts: In the first part the objective thermal comfort is determined and assessed at different temperature levels based on the concept of equivalent temperature [1]. In addition, in part two a subject study is performed in order to validate the subjective human sensation. Finally, the results of the objective and subjective study are compared to determine whether the common standards are valid with regard to draft. In the present paper, the results of the experimental study focusing on the impact of turbulence intensity, temperature and air velocity on thermal comfort will be presented.

2 Experimental Setup

The experiments are performed in a climate chamber in order to provide well-defined temperature and humidity conditions. The test environment for the thermal comfort measurements is a so-called Single-Person Cabin (SPC). Figure 1 (A) shows a photograph of the SPC with the following dimensions: width = 1000 mm, height = 1200 mm and length = 1500 mm. The SPC has a generic car-like shape with closed sidewalls, floor and ceiling. It is equipped with a transparent windshield, while the front and the back are open in order to provide similar and homogeneous temperature conditions between the environment and the insides of the SPC. This also prevents the accumulation of humid air or CO_2 during the subject trails. At the position of the dashboard, there are two air inlets (Fig. 1 (B)), which ensure defined inflow conditions in terms of velocity, turbulence intensity and temperature.

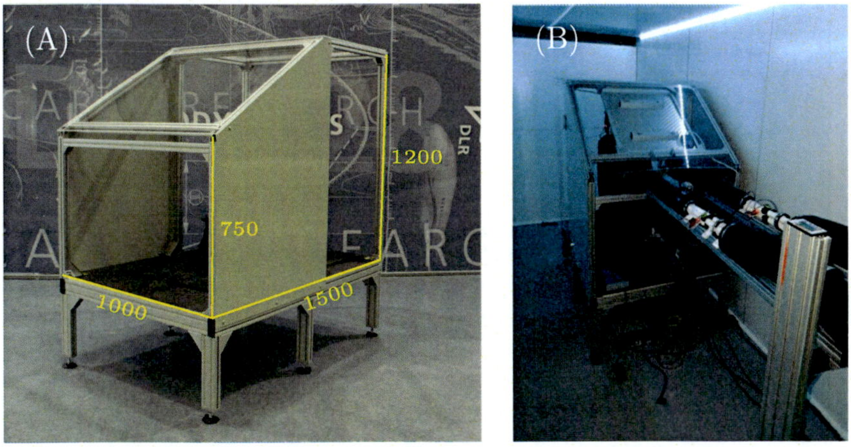

Fig. 1. (A) Photograph of the SPC with the corresponding dimensions in mm. (B) SPC in the climate chamber. The air supply with the two air inlets and the volume flow and temperature measurement system are shown.

The objective of this study is to investigate how turbulence influences the thermal comfort in a car. Therefore, two different turbulence intensities are generated by installing fractal grids in front of the air outlets. The turbulence intensity varies depending on the grid spacing. Grids with denser structures and more obstruction yield higher turbulence intensities compare to grids with larger and less dense structures [4]. For the present study two different grids were 3D printed in a metal printer to generate for stable and precise structures. In Fig. 2 (A) the grid with the larger structures (lower turbulence, in the following termed LT case) is shown. In Fig. 2 (B) the fully mounted device, which produces the higher turbulence (HT case) with an additional punch-hole metal mesh covering the smaller grid further increasing the turbulence, is shown. For protection purposes a standard fan grid is installed as a last layer at the outlet. At the bottom of the image the leveling screws used to adjust the air outlet position are displayed. A stable horizontal alignment is achieved using the metal brackets at both sides of the grid.

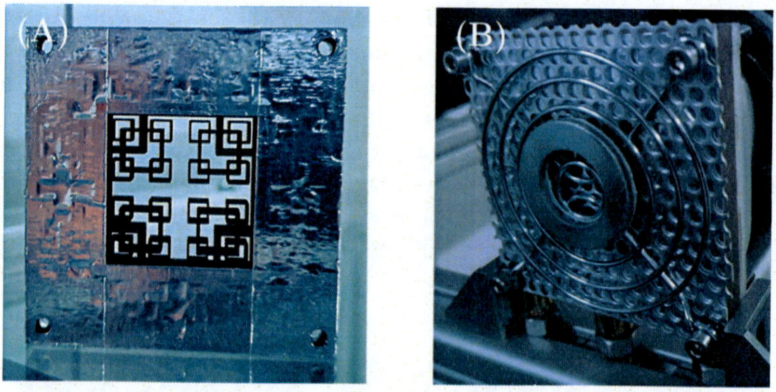

Fig. 2. Turbulence-generating grids. (A) Grid with larger structures (lower turbulence). (B) Fully mounted higher-turbulence version with a punch-hole metal and a standard fan grid.

For the evaluation of the equivalent temperature (T_{eq}), the SPC is equipped with a thermal manikin (TM), see Fig. 3 (A). The manikin's core consists of a melanin foam filling and heating wires covered by a black aluminum surface. This allows for the simulation of the sensible heat emission of a person, typically between 60 and 120 W. The air outlets are aligned so that the center of the air jets from the air outlets hits the left/right chest. This position is indicated by the cross-points of the green lines in Fig. 3 (A). With the described configuration it is possible to investigate the impact of draft on thermal comfort with regard to the upper body as the TM has the function of a measurement tool to determine the comfort based on DIN14505. For the investigation of cold draft scenarios, the comfort rating according to a summer case is investigated. Details on the procedure and method using the TM as a measurement tool are described [5].

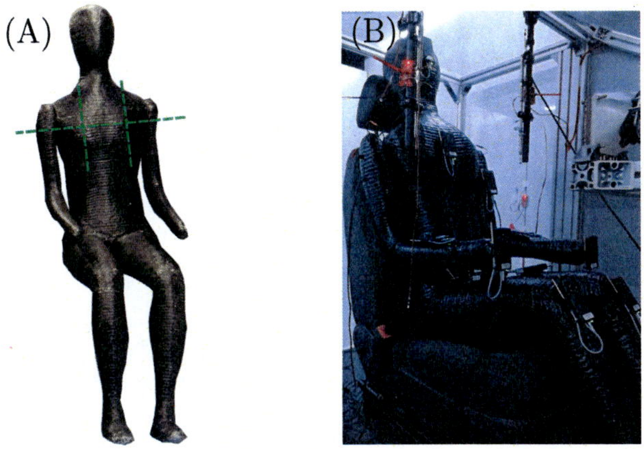

Fig. 3. (A) TM with the impact points of the incoming air indicated by the cross-points of the green lines. (B) Front view of the TM on the seat with the inflow devices and the hot-wire velocity measurement probes.

For the sake of comparability it is necessary to define a reference velocity and the corresponding turbulence intensity. In our experiment the reference velocity for assessing the draft is the velocity at the position of the green line's cross-section in Fig. 3 (A). Therefore, the mean velocity and the turbulence are determined by means of hot-wire anemometry at these positions without the TM being present on the seat (60 cm from the outlet). Since the hot-wires have to be removed during the comfort measurements, a calibration of reference velocity to volume flow is performed. The volume flow for each inlet is determined using three Sensirion SFM3000 mass flow meters. Figure 3 (B) shows the configuration of the velocity measurements. Two Dantec ComfortSense probes, mounted on the ceiling, are adjusted before the TM is removed from the SPC for the calibration of the air flow volume and the corresponding velocity.

The mean velocity U and the turbulence intensity $I = \sigma_U/U$ with the standard deviation σ_U at the chest position of the TM are chosen to characterize the velocity inflow condition for both outlet configurations. Based on the velocity and the volume flow measurements, a functional relation between both quantities is calculated. The velocity as a function of the volume flow is illustrated in Fig. 4 (A). A rather linear dependence is revealed for the LT inflow device. Overall, when comparing the results of the HT device with the LT device, smaller velocities for the different volume flows and an exponential increase in the velocity for HT are apparent. In addition, the turbulence intensity is presented in Fig. 4 (B) as a function of U. Both devices start at low turbulence intensities close to zero. Generally, increasing U leads to an increase in I. The LT case reaches a plateau at $U = 0.15\,\text{m/s}$, while the HT case reaches its fully turbulent state, represented by the plateau-region, at $U = 0.50\,\text{m/s}$. As a result, excluding low velocities ($U < 0.50\,\text{m/s}$), the mean turbulence intensity for the plateau region of the configuration LT is $I_{LT} = 16\,\%$ and for HT it is $I_{HT} = 32\,\%$.

Based on this configuration, it is concluded that an air flow with well-defined conditions regarding temperature, velocity and the turbulence intensity is provided.

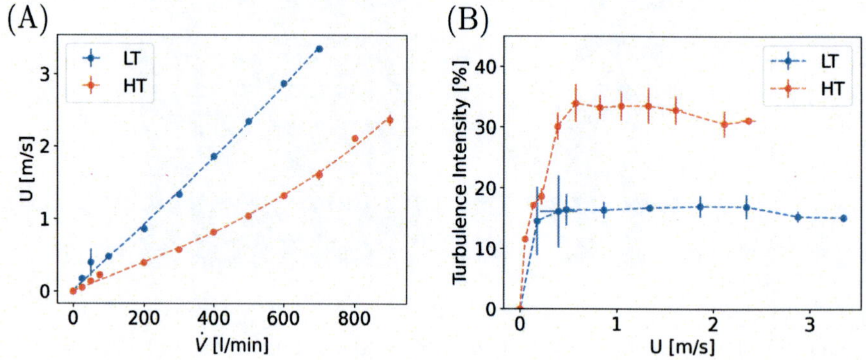

Fig. 4. (A) Reference velocity U as a function of inlet volume flow $\dot{V}$. Blue denotes the LT and red the HT device. (B) Turbulence intensity I as a function of U for LT (blue) and HT (red).

3 Results

With the objective to determine the impact of turbulence on the thermal comfort for different draft air scenarios, measurements are performed for two different turbulence intensities, $I_{LT} = 16\,\%$ and $I_{HT} = 32\,\%$. Additionally, the dependency of the thermal comfort on the reference velocities in the range of $U = 0.25$ m/s–2.50 m/s and air temperatures of $T = 17$, 20, 23, 26 and 29 °C is investigated. To simulate the buoyancy flow caused by the heat release of a human body the TM operates with a constant heating power of 75 W for all measurements. In the following an exemplary discussion of the results for $T = 23$ °C is presented - followed by a discussion of the critical body parts for all temperature levels.

Figure 5 depicts the results determined using infrared (IR) thermography for $U = 1.25$ m/s. (A), (B) and (C) illustrate the surface temperature (T_s), T_{eq} and the thermal comfort assessment in accordance with DIN14505 for a summer case. On the left the results for the LT case are shown and the results for the HT case are presented on the right. A comparison of T_s (Fig. 5 (A)) reveals a significant impact of the turbulence intensity. In case of HT the mean temperature of the chest is more than 2 K lower in relation to LT. Here, the higher turbulence results in an increased heat transfer on the TM's surface and thus in lower values for T_s. Moreover, the higher turbulence also leads to lower surface temperatures for the other body parts in comparison with the lower turbulence flow cases. Here, temperatures up to $1 - 2$ K lower are observed in case of HT. Consequently, significantly higher values for T_{eq} for almost all body parts are obtained in case of LT compared to HT (Fig. 5 (B)). The corresponding comfort rating in accordance with DIN14505 is visualized in Fig. 5 (C). Here, the cold ratings (1) are colored in blue while the warm ratings (5) are colored red with the neutral rating in between given in green (3). The different turbulence intensities result in an objective comfort rating which is rather cold in the chest region for LT and significantly too cold for the HT case. While the upper arms reveal a rather cold

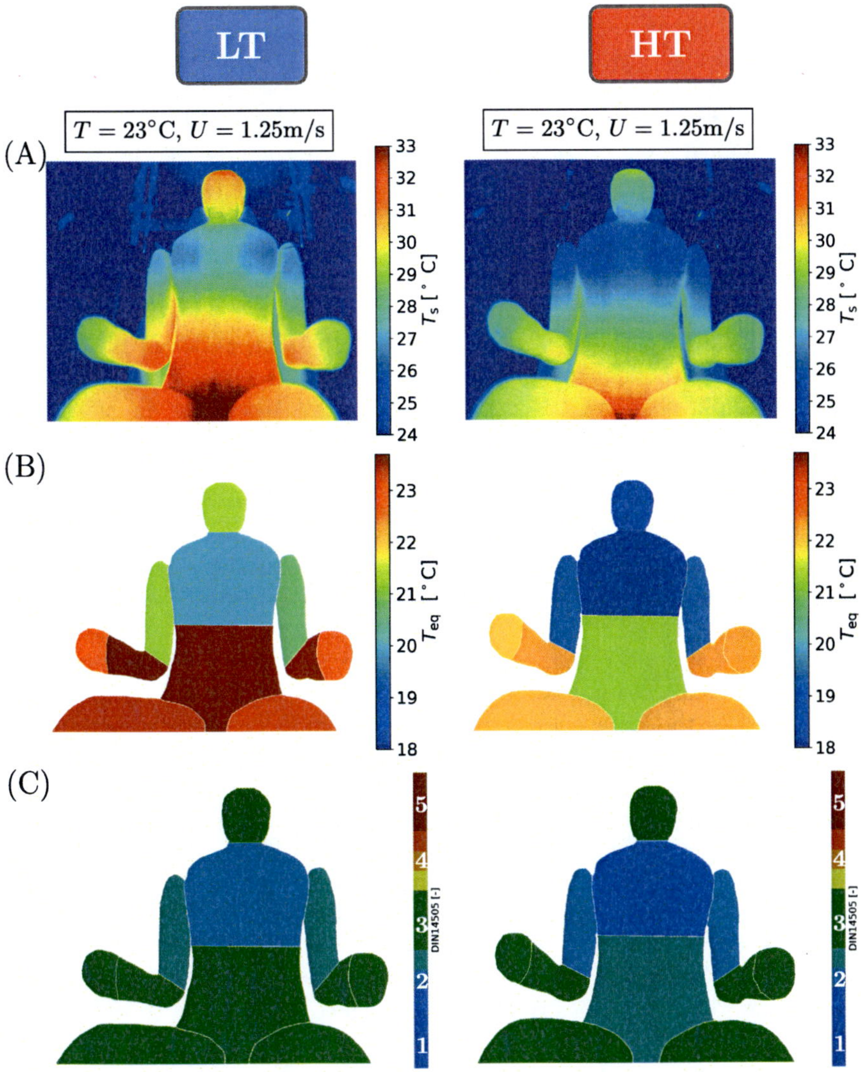

Fig. 5. For $U = 1.25\,\mathrm{m/s}$ (A), (B) and (C) present T_S, T_{eq} and the thermal comfort assessment in accordance with DIN14505 for a summer case. On the left the results for the LT case are shown and the results for the HT case are presented on the right.

rating for LT a cold rating is found in case of HT. Furthermore, for the lower chest the comfort rating changes from neutral to rather cold for HT.

An overview of the equivalent temperature and the comfort rating as per DIN14505 for the chest and the upper arms is given in Fig. 6. It depicts T_{eq} as a function of the reference velocity for $T = 23\,°C$. In addition, the comfort limits are displayed by dashed lines. Values below the light blue line indicate the comfort range of cold but still neutral, and the area between the green and the light blue lines indicates neutral comfort. The data for LT and HT are shown as dark blue and red crosses with a corresponding second-order polynomial trend line. In Fig. 6 (A) the results for the head region are presented. For the lowest velocities $U = 0.5\,m/s$, the equivalent temperature is $T_{eq} = 23.5\,°C$ in case of LT. With increasing U, a decrease of T_{eq} is found with it converging at $T_{eq} \approx 20\,°C$ for $U \geq 2.0\,m/s$. A similar relation can be observed for HT where T_{eq} is 2 K lower for the whole velocity range. Figure 6 (B) shows the results for the region of the right hand. Here, the impact of increasing U on T_{eq} is much smaller compared to the chest. This is an effect of the two air streams not directly affecting the hand region. In case of LT the temperature decrease is just 0.25 K for the full velocity range, while for HT it is four times higher with a value of 1.0 K. The more significant impact in case of HT is the result of the stronger mixing caused by the higher turbulence.

For the upper chest region (Fig. 6 (C)), the area where the air jets hit the surface of the TM directly, the behavior is similar to the behavior in the head region. Here, the temperature change for $0.5\,m/s < U < 1.75\,m/s$ is approximately $\Delta T_{eq} = 1.5\,K$. In addition, similar to the chest region, no significant changes for T_{eq} with increasing U are observed for higher velocities. The reason for this phenomenon is the fact that even with increasing velocity the heat cannot be dissipated fast enough and thus the heat transport from the TM to the airflow is limited. Further, the values for T_{eq} are also lower for the whole velocity range for the HT case. Regarding the comfort, both cases indicate that the comfort level is cold but still neutral. Figure 6 (D) shows the corresponding investigation for the lower chest. Even though this region is not directly hit by the air coming from the outlets, it is significantly affected. T_{eq} behaves similarly for HT and LT: T_{eq} is 24 °C for $U = 0.5\,m/s$ and decreases nearly linearly to approximately 20 °C, with HT decreasing slightly faster and becoming cold but still neutral already at $U = 1.0\,m/s$ as opposed to LT, where the values fall below the limit for $U = 1.75\,m/s$. However, in contrast to the upper chest, a convergence of T_{eq} with increasing U, in particular for LT, is not found in the investigated velocity range.

Figure 6 (E) and (F) present the results for the upper left and right arm. The behavior for both regions is quite similar: In case of LT and $U = 0.5\,m/s$ is $T_{eq} \approx 23\,°C$. The increasing flow velocity leads to a decrease in T_{eq} limited at a convergence of 19 °C for $U \geq, 2.25\,m/s$. In case of HT a 1.5 K lower equivalent temperature is found compared to LT. In addition, for HT the lower T_{eq} limit is already observed for $U \geq 2.0\,m/s$. For both upper arm regions, the LT case is classified as neutral up to 0.75–1.0 m/s, with a rather cold rating for $U >$

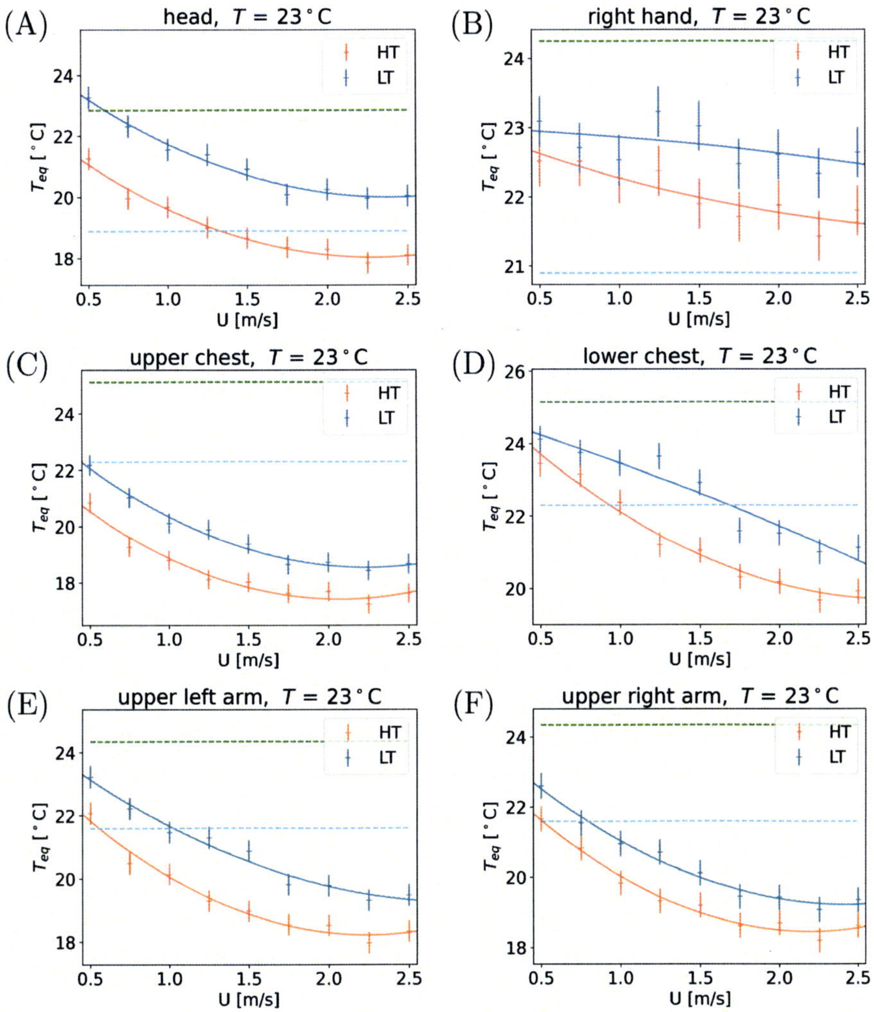

Fig. 6. For an inlet and ambient temperature of 23 °C T_{eq} is presented as a function of the reference velocity $U = 0.5 - -2.5\,\mathrm{m/s}$ with the data points reflected as crosses and a second-order polynomial fit shown as colored lines. The LT case is shown in dark blue and the HT case in red. The neutral comfort rating (3) is indicated as a dashed green line and the rather cold rating (2) as a dashed light blue line. Different body regions are presented in the subplots (A) to (F).

1.0 m/s. For HT the comfort is rated as neutral for $U = 0.5\,\mathrm{m/s}$ and as rather cold for $U \geq 0.75\,\mathrm{m/s}$. Once again, the finding that the higher turbulence leads to lower surface temperature also applies to the upper arms.

Since the HT cases revealed a higher sensitivity in terms of thermal comfort, the three most influenced body regions - head, upper and lower chest - are further investigated with respect to the inflow temperature in the range $T = 20$–$29\,°\mathrm{C}$.

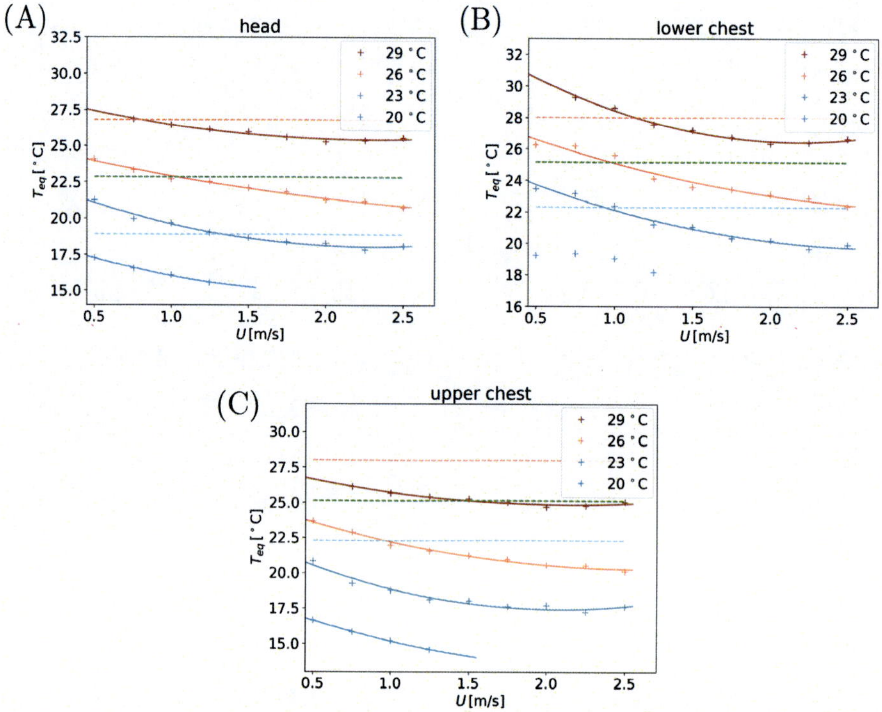

Fig. 7. For the HT case T_{eq} is presented as a function of the reference velocity $U = 0.5$–$2.5\,\mathrm{m/s}$ with the data points reflected as crosses and a second-order polynomial fit shown as coloured lines. The results for three different body regions are presented in the subplots (A) to (C) with four different temperatures each.

In Fig. 7 T_{eq} is presented as a function of $0.5 \geq U \geq 2.5\,\mathrm{m/s}$ with the data points marked as crosses and a second-order polynomial fit is depicted as colored lines. The thresholds for the thermal comfort are indicated by the dashed lines. Values below the light blue line indicate the comfort range of cold but still comfortable, whereas values above the dashed red line indicated warm but still comfortable ratings. The region in between is the neutral comfort zone with the dashed green line indicating its center. In Fig. 7 (A) the results for the head region are shown. A decreasing equivalent temperature is also found for the temperature level $T = 29\,°\mathrm{C}$. Here the change is $\Delta T_{\mathrm{eq}} = 2.5\,\mathrm{K}$. The equivalent temperature passes from the rather warm comfort rating to the neutral comfort classification for $U = 1.0\,\mathrm{m/s}$. For $T = 26\,°\mathrm{C}$ a similar behavior is observed with T_{eq} starting at $24.5\,°\mathrm{C}$. For this scenario the comfort rating remains neutral for the entire investigated velocity span. Starting at $T_{\mathrm{eq}} = 21.5\,°\mathrm{C}$ for $T = 23\,°\mathrm{C}$ at $U = 0.5\,\mathrm{m/s}$, the neutral comfort rating zone is left at $U = 1.5\,\mathrm{m/s}$. For the lowest temperature level, $T = 20\,°\mathrm{C}$, and for the lowest velocity the ratings are already rather cold. Hence, additional measurements for $U > 1.25\,\mathrm{m/s}$ were not performed. Overall, a decrease in the temperature level by $9\,\mathrm{K}$ leads to a $10\,\mathrm{K}$

lower T_{eq}. The results for the lower and upper chest show a similar behavior of the temperature as for the head region. While the results for the lower chest are similar to the results for the head in terms of comfort rating, the upper chest reveals some deviations. The latter tends to a generate a slightly colder comfort rating. The comfort rating is neutral even for the lowest values for U at $T = 29\,°\text{C}$, while in case of the lower chest the comfort is rated as rather warm for $U \leq 1.0\,\text{m/s}$. The cases $T = 20\,°\text{C}$ and $23\,°\text{C}$ are always rated as rather cold. Thus, for a certain temperature the comfort rating can be influenced using a higher or lower turbulence generating grid to adapt the comfort sensation.

4 Conclusion

Thermal comfort in a generic single-person car cabin was studied using a thermal manikin (TM) and infrared thermometry as measurement tools. In the experiments the following parameters and their combinations were investigated: two turbulence intensities, $I_{LT} = 16\%$ and $I_{HT} = 32\%$, reference velocities, $U = 0.25$–$2.5\,\text{m/s}$, and identical inlet and ambient temperatures, $T = 17$–$29\,°\text{C}$. For the investigated parameters the higher turbulence leads to an increased heat transfer between the TM's surface and the air flow compared to the cases with the lower turbulence. This results in decreasing equivalent temperatures and thus a colder objective temperature sensation based on DIN14505. Moreover, for the upper chest region it is observed that changing the inlet air temperature by 1 K leads to an almost identical change of 1 K in T_{eq} - with less significance for regions further away from the impact region of the air. In the next step the results of the subject study will be compared with the results of the objective comfort parameters in order to verify whether EN/ISO14505 is valid for draft or not.

Acknowledgment. This research has been funded by the Research Association of Automotive Technology (FAT).

The authors like to thank Annika Köhne for proof-reading this work.

References

1. ISO: Ergonomics of the thermal environment - Evaluation of thermal environments in vehicles - Part 2: Determination of equivalent temperature. Beuth Verlag, Berlin (2006)
2. ISO: Ergonomics of the thermal environment - Evaluation of thermal environments in vehicles - Part 3: Evaluation of thermal comfort using human subjects. Beuth Verlag, Berlin (2006)
3. Kim, W., Anorve, V., Trefft, B.:American Driving Survey, 2014–2017 (2019)
4. Discetti, S., Ziskin, I.B., Astarita, T., Adrian, R.J., Prestridge,K.P.: PIV measurements of anisotropy and inhomogeneity in decaying fractal generated turbulence. Fluid Dyn. Res. **45**(6), 061401 (2013)
5. Lange, P., Westhoff, A., Schmeling, D., Bosbach, J.: Cost-effective human comfort manikin with realistic thermal load for studies of convection-driven ventilation systems. In: Kosonen, R., Ahola, M., Narvanne, J. (eds.) Roomvent & Ventilation 2018, SIY Indoor Air Information Oy, pp. 367–372, Juni 2018

Check for updates

On the Simulation of a Heavy Vehicle Wake in OpenFOAM with Real-World Data

Keith A. Weinman[1(✉)], Henning Wilhelmi[1,2], James R. Bell[1], Daniela Heine[1], and Claus Wagner[1,2]

[1] DLR Göttingen, Institute of Aerodynamics and Flow Technology, Bunsenstr.10, 37073 Göttingen, Germany
keith.weinman@dlr.de

[2] Institute of Thermodynamics and Fluid Mechanics, Technical University of Ilmenau, Helmholtzring 1, 98693 Ilmenau, Germany

Abstract. Two computational approaches are taken to characterize the drag of a car driven behind a heavy vehicle under real conditions. The on-road approach uses velocity measurements, obtained from an array of static five-hole probes, to construct on-flow boundary conditions replicating the atmospheric dynamics encountered during an on-road measurement. The wind tunnel approach uses an oscillating flap system to control flow time and length scales upstream of a wind tunnel model. The amplitude and frequency of the flap motion is calibrated to reproduce length and time scales at on-road conditions under respective Reynolds and Strouhal number scaling. These approaches are evaluated against experimental measurements using Computational Fluid Dynamics (CFD) and demonstrates that they reproduce the aerodynamic drag and a significant part of the measured onflow condition for the car.

Keywords: DDES · Boundary reconstruction · Heavy vehicle wake · Real-world data · On-road tests · Oscillating flaps · RANS · Radial basis function · Vehicle drag prediction · Wind tunnel tests

1 Introduction

The paper is concerned with the prediction of drag, under real world conditions, of a compact car (length: 4.2 m, height: 1.4 m, width: 2m) following a medium box truck (length: 8 m, height: 3 m, width: 2.24 m) at a fixed distance of 50 m and at a constant speed of 27.776 m/s [1,2]. The configuration conforms to the definition of a two-vehicle platoon. The energy-saving potential of platoon configurations is discussed in several publications. For example, [3] demonstrates that numerical methods provide industrially relevant drag estimates for platoon configurations under both wind tunnel and real world conditions. The paper investigates how well CFD assesses real world unsteady flow effects on drag prediction for a two vehicle platoon configuration. An important difference between the on-road approach discussed in this paper and other published work is that the

© The Author(s), under exclusive license to Springer Nature Switzerland AG 2021
A. Dillmann et al. (Eds.): STAB/DGLR Symposium 2020, NNFM 151, pp. 504–513, 2021.
https://doi.org/10.1007/978-3-030-79561-0_48

influence of the leading vehicle on the trailing vehicle is modeled using an inflow boundary condition. By comparison the wind-tunnel approach generates onflow conditions characteristic of the on-road conditions for a wind tunnel model by directly perturbing the internal flow upstream of the model.

2 Methodology

In the on-road approach velocity measurements are sampled 250 Hz using a 2D array of eleven five-hole probes (located 1 m in front of the car). Lateral distances between the probe elements vary from 0.49 m to 0.98 m. Nine probes are located at a vertical height z = 0.5 m above ground with a further two being located along the array's symmetry axis at z = 1.1 m and z = 1.7 m. Velocity and pressure measurements are taken in the wake of the truck (data set W) and also with only the test vehicle on the road (data set A). Section 2.1 describes the construction of an inflow boundary condition, emmulating real test conditions using the data sets A and W. This boundary condition is used for Delayed Detached Eddy Simulations (DDES) within the framework of OpenFOAM [4]. For the wind tunnel approach experiments, using a 1:4 scale wind tunnel model, are conducted in the cross-wind facility Seitenwindkanal Göttingen (SWG) [5]. Oscillating flaps are used to perturb the oncoming flow. Flap frequency and amplitude are chosen to recreate scaled turbulent length and time scales deduced from on-road measurements. A 1:4 scaled version of the on-road measurement array is located 0.25 m in front of the wind tunnel model. Further details are found in [6] and [2].

2.1 Simulation of the On-Road Incoming Flow

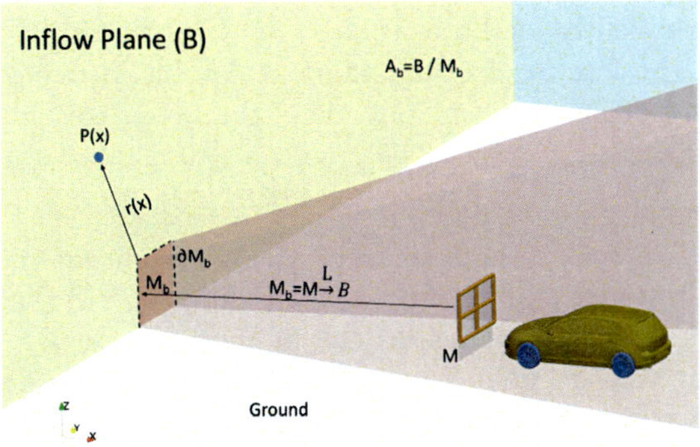

Fig. 1. The velocity field on M is reconstructed from probe measurements and is then mapped linearly (L) to the projection of M on the inflow plane B, M_b. The velocity on M_b is representative of the wake or the unsteady atmospheric boundary layer (ABL). Velocities of any point $P(x)$ on A_b, the relative complement of B on M_b, are computed by summing the atmospheric boundary layer velocity estimates at $P(x)$ with a weighted value ($W_f(x)$) of the velocity at the nearest point on ∂M_b (boundary of M_b). The weighting function $W_f(x)$ is a function of $r(x)$, the distance between $P(x)$ and the nearest point on ∂M_b.

The W and A data set correspond to several minutes of velocity measurements at a constant interval of 0.004 s between data points for each probe (both sets of data are assumed to start at t = 0). Figure 1 illustrates the car, the probe array, and a plane M extending from the ground to the top of the probe array with a width slightly larger than that of the probe array. The truck can be imagined to be positioned directly to the left of the inflow plane. The inflow boundary B is divided into two parts. The first part M_b, given by the projection of M onto B, is representative of either the resolved wake or the resolved ABL. The second part A_b, given by the relative complement of M_b on B such that $x \subset B \| x \not\subset M_b$, is representative of the ABL averaged across the span of the inflow plane. A no-slip boundary condition is defined on the edge on M lying along the ground with zero-gradient boundary conditions being defined on the remaining sides of M. A radial basis function method [7] is used to reconstruct the velocity field on M from either of the measured datasets A or W at each time step. The reconstructed velocity field on M is linearly translated (L) onto M_b. Two interpolation meshes are created on M and on B: the interpolation mesh on B is used by OpenFOAM tools to create the computational inflow boundary data at each time step. Spacing between nodes on the interpolation meshes is determined by the RBF shape parameter. A least-squares fit for a neutral ABL profile at each time step is generated with the W data sets using the friction velocity $u^s(t)$ as the design variable. Other log layer parameters are estimated from the literature. The velocity fields on the wake inflow M_b can be considered as resolved spatially filtered turbulent velocity data (the filtering is due to the influence of the reconstruction and measurement processes) while the velocities on A_b are time-varying mean velocities. The ABL velocity estimates are not conservative but provide a crude estimate of the spanwise averaged ABL conditions when the road tests were undertaken. Inflow turbulence on A_b is modeled directly via the turbulence model. Inflow turbulence intensities can be rescaled independently for both M_b and A_b. The weighting function $W_f(x)$ returns unity on the interior and boundary of M_b but decays to 0 as distance from the boundary of M_b increases outside of the wake. For this work $W_f(x)$ is defined as

$$W_f(x) = 1 - A \tanh(Br(x)), \tag{1}$$

where A and B are scalar coefficients of O(1). Weighting removes potential numerical difficulties and mimics the upstream diffusion of velocity between the wake and ABL. Turbulence intensities can be modified on M_b by rescaling the velocity fields The velocity field, now fully defined on B, is written as an OpenFOAM input file and stored on disk for each time step. Internal OpenFOAM tools linearly interpolate these files (in space and time) onto the computational mesh boundary during the solution process.

2.2 Simulation of Incoming Flow for the Wind Tunnel Model

Figure 2a illustrates the 1:4 scale model of the test vehicle inside the SWG facility. The CFD geometry includes a moving belt to simulate relative ground motion as well as tire rotation. The geometry of the flapped system is shown in Fig. 2b. The flap motion parameters are set with an oscillation frequency 10 Hz and an oscillation amplitude of 10 degrees. Sliding mesh algorithms are used to simulate flap motion. A 2D measurement rake at 1:4 scale is placed in the same relative position to the model as for the on-road measurements.

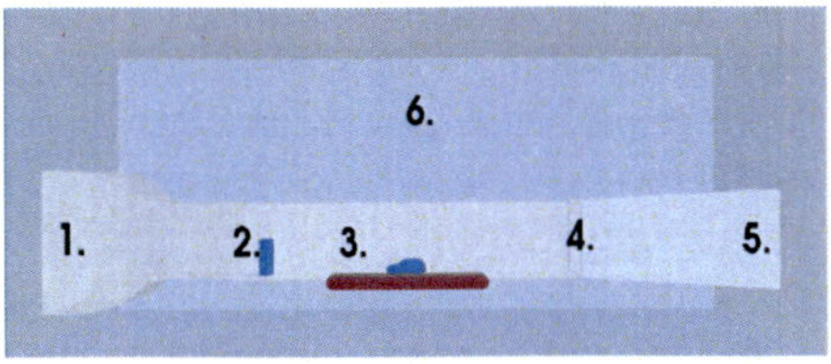

(a) View of SWG facility

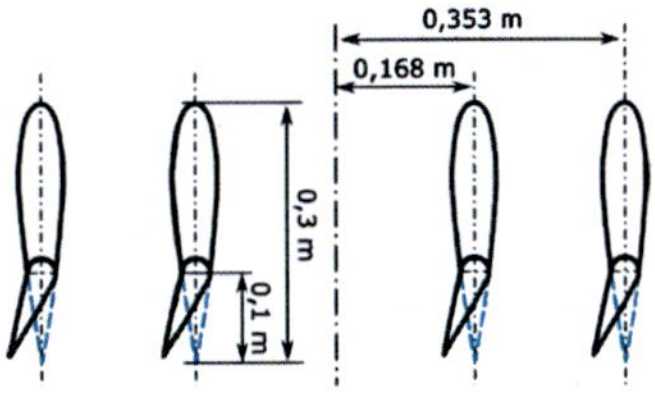

(b) Oscillating flap arrangement.

Fig. 2. Model mounted in SWT. SWT components: 1. Wind tunnel nozzle inflow, 2. oscillating flap system, 3. moving band, 4. pressure equalization gap, 5. WT diffuser outlet, 6. Plenum

2.3 Computational Set-Up

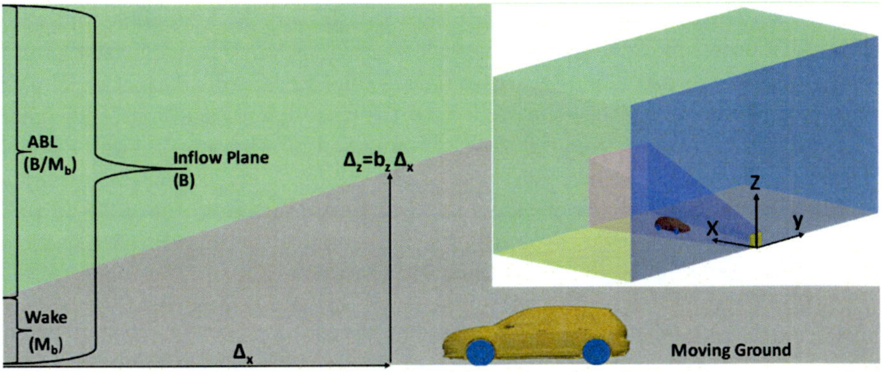

Fig. 3. The left figure illustrates the computational domain in the x-z plane. It shows the definition of the refinement region growth rate parameter b_z and the location of the wake portion of the inflow boundary plane M_b. The right figure illustrates the car within the computational domain. The purple region is a refinement zone to minimize the numerical dissipation influence on the wake velocity convected in from the boundary. Dimensions of the computational domain are given by $\Delta_x = 100\,\text{m}$, $\Delta_y = 100\,\text{m}$, $\Delta_z = 40\,\text{m}$. The car mid-axle position is located at x = 0 m and y = 0 m. Standard OpenFOAM inflow and moving ground boundary conditions are used, together with a farfield boundary at the outflow and other computational domain faces.

Figure 3 illustrates the computational domain. The interface between the wake region and the ABL is treated as a planar mixing layer as flow travels downstream, and the width of the wake region ($b_y = \Delta y_w/\Delta x_w, b_z = \Delta z_w/\Delta x$) grown in the downstream direction according to published measurements of mixing-layer growth. This interface bounds the refinement zone (set during mesh generation) where cell volumes are refined to reduce the on-flow numerical dissipation losses. The distance between the car and the onflow boundary can be varied only by generation of a new mesh. Distances of 50 m and 30 m are examined in this work. The current work uses the Engys OpenFOAM release 2.3.1 [8]. Baseline turbulence models are the Menter k-ω SST [9] and Spalart-Allmaras [10] turbulence models. The Delayed Detached Eddy versions of these models are used for the hybrid RANS-LES calculations. The filter width is chosen as 2Δ, where Δ is the cube root of the cell volume. A segregated pressure-based incompressible solver (pisoFoam) was used for all unsteady calculations. Steady calculations were performed using the simplecFoam solver. For all calculations an initial velocity field is generated using a potential flow method (potentialFoam). Time steps are chosen so that the Courant number is less than one for all control volumes. Second-order central differences are used for viscous fluxes, while a hybrid central-upwind scheme is used for inviscid fluxes. Best results were obtained using the local Courant number as the switching parameter between inviscid schemes. The linear system is solved using an Algebraic Multigrid method for the pressure correction equation, and with a stabilized Krylov Subspace method for the velocity and turbulence variables.

The CAD model used for the on-road calculations is taken from real geometry data provided by the manufacturers and is rescaled for the wind tunnel work. For this work OpenFOAM meshing tools return satisfactory high Reynolds number meshes. The meshing guidelines used follow previous work [11]. Surface length scales of 2 mm are used for the car and tires. The surface discretization length scale varies from 2 mm near the vehicle to the base length scale of 50 cm near the freestream boundaries. Eight levels of refinement are used in the wall normal direction of all no-slip surfaces. Writing L^n as the mesh length scale at a refinement level n, then $L^{n+1} = L^n/2$. Mean $y+$ values are in the ranges of 20–40 wall units. On the basis of several mesh refinement studies the final meshes used in this work are 180 million points for the on-road calculation and 64 million points for the wind tunnel calculation. A refinement level of 4 is used for the resolved wake region. Refinement is limited by available computational resources. For the on-road calculations the CFL number is limited to O(1), leading to times steps of the order 10^{-7} s. Identical CFD practices are used for the wind tunnel calculations.

Drag is not measured during the on-road tests and the wind tunnel measurements performed by FKFS in Stuttgart [6] provide validation data for drag estimation. The DDES variants of the base turbulence models return estimates of drag with differences of less than 3 drag counts against the experiment whereby one drag count is defined as 0.001 Cd. The averaged onflow velocity is used in the normalization of surface pressure forces. Studies showed that Reynolds-averaged Navier Stokes (RANS) simulations returned errors of over 30 drag counts due to inadequate prediction of boundary layer separation (and reattachment) and vortex breakup effects. On this basis the DDES approach was adopted for turbulence modeling despite a significant increase in computational costs (see Sect. 5). Boundary layer resolving meshes require further computational overhead and were found to be very difficult to create with the standard OpenFOAM meshing library. Therefore wall functions are used to model near-wall turbulent stresses. Note that moving ground and tire rotation effects are modeled using standard OpenFOAM source term treatments.

3 Assessment of Incoming Velocity Fields

A qualitative assessment of the incoming flow generated by the on-road approach, comparing instantaneous streamwise u velocity components in time, is shown in Fig. 4. The velocity is underestimated in comparison with the experiment. Finer scale structures seen in the measured data are not present in the DDES results: the velocity field reconstruction filters out any length scales at the order of the inter-probe distance and below, and numerical dissipation effects are cumulative during a fluid particle's lifetime within the computational domain. Figure 4 also illustrates the frequency at which the maximum spectral energy is found for the spanwise v-component of velocity. Experiment and CFD are qualitatively in agreement but the spectral energy magnitudes are underpredicted at all probe positions with a significant one decade difference at z = 1.1 m. This height corresponds to a position where wake/ABL interface effects are expected and suggests that either additional time is required for higher-order velocity statistics to develop and/or that the computational modeling in this region is not fully adequate. Mesh resolution requirements need to be further increased in this region. Approximately 85% of the total control volumes are computed in DES mode but the DES length scales are too coarse for the proper development of the mixing layer interface.

In comparison to the differences observed in Fig. 4, the improvement in comparisons between the measured and computed onflow velocities for the wind tunnel approach is shown in Fig. 5. In this approach onflow components are precisely specified by single Fourier modes. The onflow boundary condition, given by the test section bulk velocity corrected for continuity at the nozzle inlet, is also well understood. The onflow velocity is dominated by the sinusoidal motion at the flap forcing frequency.

For both approaches the starting onflow disturbances are allowed to travel three car lengths past the vehicle before averaging is started. The averaging duration is more than 8 convective time units. Table 1 reflects that the on-road approach underpredicts the u component by up to 25%. This is due to the high-pass filter effect of the reconstruction process as well as the accumulation of low-order dissipation effects over the lifetime of the fluid in the computational domain. At a turbulence intensity of 6%, the mean u components for the reconstructed inflow boundary at $\Delta x = 30\,\mathrm{m}$ (reconstructed from data collected at a

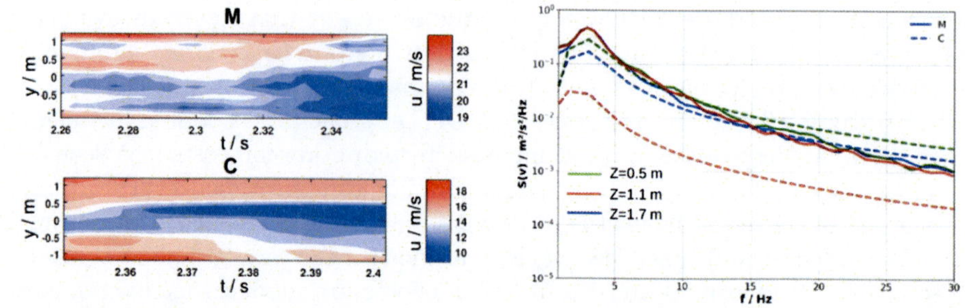

Fig. 4. The left figure shows a comparison of the measured (M) and computed (C) u-velocity component at z = 0.5 m above ground. The right figure shows a comparison of the computed and measured v-component power spectral densities for the three probes located on the array axis of symmetry.

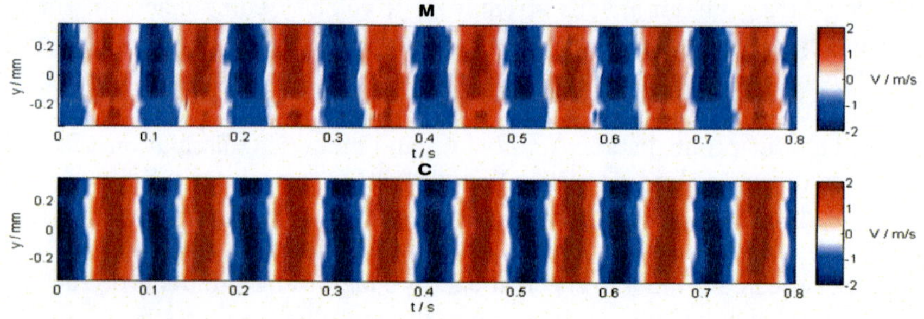

Fig. 5. Development of the v velocity component in the wind tunnel at z = 0.2 m above the test section floor. M: measured, C: CFD.

Table 1. Onflow statistics at the probe positions. M: Measured, B: CFD with block inflow profile. C: CFD with on-road approach. ∇ is the percentage difference against experiment.

	Mean						Standard deviation					
	M	B	C	C	C	C	M	B	C	C	C	C
On-road approach												
Tu_x %	6	6	2	6	6	10	6	6	2	6	6	10
Δx (m)	50	50	50	50	30	50	50	50	50	50	30	50
u (m/s)	24.5	22.1	18.7	18.8	19.3	18.1	2.8	0.46	1.05	1.53	0.9	1.73
v (m/s)	0.3	0.04	−1.9	−1.73	0.0	−2.7	1.8	0.23	0.50	1.77	0.5	1.37
w (m/s)	1.6	−0.4	−0.5	−0.5	0.77	0.5	1.0	0.27	0.28	1.13	0.8	0.61
Wind tunnel approach												
	M (m/s)		C (m/s)		∇%		M (m/s)		C (m/s)		∇%	
u	30.2		29.8		1.3		0.5		2.4		6.3	
v	0.0		−0.2		0.6		0.5		0.5		0.0	
w	0.0		0.2		0.6		0.3		0.4		0.3	

vehicle separation of 30 m) is comparable to that obtained for the reconstructed inflow boundary at $\Delta x = 50$ m. This is a significant observation and implies that if the onflow boundary is to be moved further downstream in order to reduce computational costs, then the inflow boundary velocity must be appropriately rescaled. However, if Δx is of the order of the wake core length, the flow impacting on the vehicle is highly correlated spatially and is not representative of on-road conditions at the vehicle separation distance. Differences between the velocity components v and w (magnified by the smaller magnitudes of these components) are due to slight differences in the atmospheric conditions. Removal of the mean signal from the measured data should be considered in future to avoid this uncertainty. Although the reconstruction method acts as a low-pass filter, the standard deviations of velocity for $\Delta x = 50$ m and a turbulence intensity of 6% agree qualitatively with the measured data. Most spectral energy is concentrated in the lower frequency modes (see Fig. 4) suggesting that further high-frequency filtering would be possible. The first and second statistical moments for the drag obtained with the wind tunnel approach agree more closely with measurements.

4 Assessment of Predicted Drag

Table 2. This table compares numerical estimates of the difference of the mean drag coefficient against drag at reference conditions. Standard deviations are given by δ. For both approaches the drag force has been normalized by the mean local onflow velocity. Tu_x is the inflow boundary turbulence intensity.

Δ_x (m)	$Tu_x\%$	ΔC_d	δ_{C_d}
On-road approach: Reference [6] $C_d = 0.289$ at Δx $= 50$ m, $Tu_x = 6\%$			
50	2	29	0.13
50	6	0	0.13
30	6	6	0.03
50	10	−19	014
Wind-tunnel approach: Reference from experiment: $C_d = 0.291$			
–	6	8	0.02

Results obtained using the on-road approach in Table 2 suggests a functional dependence of aerodynamic drag on the inflow turbulence intensity. However, the data does not identify a trend. From the preliminary work noted in Sect. 2.3 the changes in drag between simulation techniques (RANS and DDES) are of the order of 20–30 drag counts, but were less than 10 counts for two different DDES models. Changes due to variations in the inflow turbulence intensity appear to be as significant as the change across simulation techniques, and this might support the notion that the onflow turbulence intensity is a parameter that should be stated when cross referencing vehicle drag estimations from different sources. The CFD and experimental determinations of drag differ by less than 10 drag counts for the wind tunnel approach. There may be slight differences between the CFD and wind tunnel models which contribute to these differences. Another potential source of error is the prediction of the model blocking effect as well as the boundary layer growth on the wall of the tunnel. Care was taken to ensure that these are computed as accurately as possible, but slight differences may still accumulate and contribute to the 8 count difference observed. The standard deviation of the drag returned using the on-road approach is significantly higher for $\Delta x = 50\,\text{m}$ in comparison to $\Delta x = 30\,\text{m}$, however the variation in the mean drag is less than 10 counts. Correspondence between both CFD and experiment across both approaches for the mean drag is less than 10 drag counts.

5 Computational Resources

Resources of the Simulation Center of Aerodynamic Research in Transportation [12] are used for the work presented in the paper. Requirements for DDES simulations using reconstructed inflow boundary conditions scale at about 14 days for 1000 CPU's on a 120 million node mesh for one convective time unit. By comparison the RANS calculations, although significantly less precise, require approximately 48 h for convergence on 256 CPUs. DDES calculations performed for the wind tunnel require about 50% of the computational resources required for the reconstructed inflow simulations. This is due to the considerable smaller spatial resolution requirements.

6 Conclusions

The paper investigates the effect of changing on-flow conditions on the aerodynamic drag of a compact car which is driven behind a box truck under normal road conditions. Two CFD drag prediction approaches are assessed against both on-road and wind-tunnel experiments The on-road approach uses velocity measurements taken during on-road tests to reconstruct on-flow boundary conditions representative of those experienced during the on-road test. The wind-tunnel approach uses an oscillating flap system to control the on-flow turbulent length and time scales upstream of a wind tunnel model so that on-road conditions are approximately matched.

Both approaches create on-flow conditions which are consistent with real world observations. Both approaches return useful mean pressure estimations on the vehicle surface. Mean drag estimates are accurate to within 10 drag counts of the experimental data for both approaches using DDES. However significant resources are needed for both approaches when predicting second order statistical moments of the oncoming velocity and surface pressure. The wind tunnel approach uses well understood modeling and demonstrates the best agreement with measured oncoming flow dynamics, while the on-road studies suggests a dependence of mean drag on inflow turbulence levels that may be significant when comparing different drag data sets. Both of the approaches studied are complimentary and will be useful in investigations of unsteady vehicle aerodynamics.

However, some work remains in optimizing the on-road approach, particularly with regard to the computational cost of the technique and improving the energy content of the resolved flow.

Acknowledgement. The German Aerospace Center (DLR) would like to acknowledge the FAT Working Group 6 "Aerodynamics" for the realization of the Project and for excellent cooperation during the project period. The authors would also like to acknowledge Ms Annika Köhne for her assistance in the preparation of this paper.

References

1. Wilhelmi, et al.: Aerodynamic characterization of a compact car overtaking a heavy vehicle. In: New Results in Numerical and Experimental Fluid Mechanics XII. Notes on Numerical Fluid Mechanics and Multidisciplinary Design. Springer, Cham (2020). ISBN 978-3-030-25252-6, no. 794804
2. Wilhelmi, H., et al.: Simulation of transient on-road conditions in a closed test section wind tunnel using a wing system with active flaps. SAE Int. J. Passeng. Cars Mech. Syst. (0688), 1–13 (2020). https://doi.org/10.4271/2020-01-0688, ISSN 1946-3995
3. Davila, A., del Pozo, E., Aramburu, E., Freixas, A.: Environmental benefits of vehicle platooning. SAE Technical paper 2013–26-0142 (2013). https://doi.org/10.4271/2013-26-0142
4. Weller, H.G., Tabor, G., Jasak, H., Fureby, C.: A tensorial approach to computational continuum mechanics using object-oriented techniques. Comput. Phys. **12**(6), 620-631 (1998)
5. Haff, J., Richard, U., Kowalski, T., Loose, S., Wagner, C.: Wind tunnel experiments with a high-speed train model subject to cross-wind conditions. In: Pombo, J. (ed.) Proceedings on the First International Conference on Railway Technology: Research, Development and Maintenance, vol. 24. Civil-Comp Press, Stirlingshire (2012)
6. Wilhelmi, et al.: Aerodynamic characterization of a compact car driving behind a heavy vehicle. Notes on Numerical Fluid Mechanics and Multidisciplinary Design. Springer, STAB (2020)
7. Majdisova, Z., Skala, V.: Radial basis function approximations: comparison and applications. Appl. Math. Model. **51**, 728–743 (2017)
8. HELYX 3.2.1 - HELYX-Core User's Guide (2020)
9. Menter, F.R., Kuntz, M., Langtry, R.: Ten years of industrial experience with the SST turbulence model. Turbul. Heat Mass Transfer **4**(4), 625–632 (2003)
10. Spalart, P.R., Deck, S., Shur, M.L., Squires, K.D., et al.: A new version of detached-eddy simulation, resistant to ambiguous grid densities. Theoret. Comput. Fluid Dyn. 20(3), 181195 (2006). https://doi.org/10.1007/s00162-006-0015-0
11. Weinman, K.A., Fragner, M., Deiterding, R., Heine, D., et al.: Assessment of the mesh refinement influence on the computed flow-fields about a model train in comparison with wind tunnel measurements. J. Wind Eng. Ind. Aerodyn. **179**, 102117 (2018). https://doi.org/10.1016/j.jweia.2018.05.005
12. http://scart.dlr.de/site/index.pdp (2020)

Aerodynamic Characterisation of a Compact Car Driving Behind a Heavy Vehicle

Henning Wilhelmi[1,2](✉), Christoph Jessing[3], James R. Bell[1], Daniela Heine[1], Joachim Wiedemann[4], Andreas Wagner[4], and Claus Wagner[1,2]

[1] Institute of Aerodynamics and Flow Technology, German Aerospace Center Göttingen, Bunsenstr. 10, 37073 Göttingen, Germany
henning.wilhelmi@dlr.de

[2] Institute of Thermodynamics and Fluid Mechanics, Technische Universität Ilmenau, Am Helmholtzring 1, 98684 Ilmenau, Germany

[3] Institute of Automotive Engineering, University of Stuttgart, Pfaffenwaldring 12, 70569 Stuttgart, Germany

[4] Research Institute of Automotive Engineering and Vehicle Engines Stuttgart, Pfaffenwaldring 12, 70569 Stuttgart, Germany

Abstract. The transient incoming flow a compact car experiences whilst driving 10 m – 100 m behind a heavy vehicle on a runway has been characterised. The incoming flow was measured using a 2D array of 11 five-hole probes mounted 1 m in front of an operational, full-scale compact car. Additionally, 188 surface pressure taps were used to measure the effect of the incoming flow conditions on the compact car. The experiments were performed under ideal conditions on a 2.9 km long runway in Faßberg near the DLR in Trauen, Germany.

The experiments showed that the heavy vehicle and the distance between the two vehicles have a significant impact on the flow characteristics: with decreasing distances between the vehicles the fluctuations increase from ±0.5 m/s to ±6 m/s and exhibit coherent dominant frequencies at 2.5 Hz in the spanwise direction that are attributed to the vortex shedding in the wake of the heavy vehicle. The turbulence intensities increase from 1% to 25% while the length scales decrease from 25 m to 0.25 m. The change in the incoming flow has a demonstrable influence on the surface pressure of the compact car. The pressure coefficient fluctuates by up to ±0.6 and $c_{p,\,rms}$ increases from 5% to 30%. The effect of the heavy vehicle on the pressure distribution indicates that the global forces acting on the compact car are influenced.

Keywords: Automotive aerodynamics · On-road experiments · Traffic · Driving behind · Platooning

1 Introduction

The aerodynamic conditions a vehicle experiences on-road are diverse and dependent on factors such as naturally occurring winds, other vehicles and road-side obstacles

© The Author(s), under exclusive license to Springer Nature Switzerland AG 2021
A. Dillmann et al. (Eds.): STAB/DGLR Symposium 2020, NNFM 151, pp. 514–524, 2021.
https://doi.org/10.1007/978-3-030-79561-0_49

(Lindener et al. [1], Wordley et al. [2] and McAuliffe et al. [3]). Wind tunnel experiments and numerical flow simulations for aerodynamic assessment and optimisation are typically conducted under idealised conditions. These flow conditions are not necessarily representative of the diverse incoming flows experienced while driving in traffic on the road. Recent, advanced investigations by Schröck et al. [4] and Sims-Williams [5] that have included the simulation of these conditions indicated that they can have an effect on the forces and moments of automotive vehicles. This is why the understanding of on-road conditions and their effects on unsteady automotive vehicle aerodynamics needs to be improved.

On-road experiments measuring the incoming flow which a real-world vehicle experiences have been conducted by Lindener et al. [1], Wordley et al. [2] and McAuliffe et al. [3]. They observed turbulent intensities of up to 16%, length scales between 1 m and 10 m and anisotropy characteristics depending on the environmental conditions (road-side obstacles and traffic density). These on-road conditions are significantly different from the idealised conditions (low, isotropic turbulence intensities of less than 1%) typically simulated in wind-tunnel experiments or numerical simulations. These investigations have characterised the average incoming flow conditions, which can subsequently be replicated in experiments and simulations to assess the average aerodynamic characteristics of vehicles. However, this average is based on a number of different driving manoeuvres: isolated driving, crosswinds and interaction with other vehicles, each having different characteristics and overall levels of impact on a vehicle.

Driving behind another vehicle is a very common high-speed driving scenario, particularly on the German Autobahn. The authors have previously observed that the mean incoming flow velocity can be reduced by up to 33% when driving behind another vehicle (Jessing et al. [6]). Further, the average flow characteristics when driving 50 m behind a heavy vehicle in isolated conditions are representative of the average characteristics of general driving on the German Autobahn. In experiments conducted by Wilhelmi et al. [7] and Bell et al. [8] on overtaking and exposure to gusts, unique incoming-flow characteristics were identified that clearly affect the vehicles surface pressure - indicating an influence on the vehicle's global forces. Zabat et al. [9] and *Davila et al.* [10] have specifically identified significant effects of driving closely behind other vehicles in the form of platooning – in this case a benefit: drag reductions predicted in the range of 45%–80%. This is why the understanding of the transient flow and its effects on a vehicle driving behind another vehicle needs to be improved.

In this work, measurements of the incoming flow and surface pressures measurements of a full-scale compact car on a test track are analysed for driving behind a heavy vehicle and single vehicle driving manoeuvre. These results characterise and provide insight into the cause of the transient incoming flow characteristics and their effect on the vehicle. Further, they can be used as input and validation data in aerodynamic assessment and optimisation experiments as well as simulations attempting to replicate the real-world conditions (see Wilhelmi et al. [11] and Weinman et al. [12]). This research is part of a collaborative project between the *German Aerospace Center (DLR)* and the *Institute of Automotive Engineering – University of Stuttgart (IFS),* that is supported by the *Research Association of Automotive Technology (FAT).* Ideally, such work leads to more accurate experiments and simulations, enabling aerodynamic optimisation of vehicles, improving efficiency, performance and safety.

Fig. 1. Test vehicles (left: heavy vehicle, right: compact car) on the runway in Faßberg

2 Experimental Methodology

2.1 Test Site

The test track was an airfield at the German army base in Faßberg next to the *DLR* in Trauen. The runway is 2.9 km long and 25 m wide. This test-track was selected since it provides an isolated environment without traffic – apart from the test vehicles. The surroundings are characterised by areas of low grass for at least 250 m on each side of the runway followed by tree lines (Fig. 1). The weather conditions were recorded during the experiments. The presented experiments were conducted under low wind conditions of less than 3 m/s.

2.2 Test Vehicles

The compact car was a full-scale, operational Golf VII (2015) provided by *Volkswagen*. The interior was modified to accommodate the measurement and data acquisition systems on board. The heavy vehicle was a *Mercedes-Benz* Sprinter (2017) with a box body. The compact car's velocity over ground (v_c) was measured with a *Kistler* Correvit sensor at 250 Hz. During the manoeuvre the compact car was behind a heavy vehicle in the same lane at distances of $\Delta x = 10$ m, 20 m, 30 m, 50 m and 100 m between the car's front and the heavy vehicle's back. An example for a distance of 10 m between the vehicles is shown in Fig. 1. The distances during the run along the runway were kept constant with the help of markings boxes in the front window. These markings were placed before the runs in a static setup at the desired distance. They outlined the silhouette of the heavy vehicle on the compact car's window from the driver's perspective. During each run the driver then kept the view of the leading heavy vehicle within the outlines to allow for positioning of the two vehicles relative to each other.

2.3 Measurements

The two vehicles were accelerated to a constant velocity. The experiments were carried out at $v_c = 27.8$ m/s. Based on the compact car's length of 4.24 m a Reynolds number of 7.3×10^6 results. The distance was kept constant for each driving-behind manoeuvre.

The measurement time was approximately 100 s and the sampling frequency was 250 Hz. Each run was repeated 4 times in each direction of the runway to avoid directional influences. Additional 4 runs in each direction with a single vehicle (SV) – the compact car in atmospheric free-stream conditions – were carried out.

Incoming Flow Measurement A 2D array of 11 *Vectoflow* five-hole dynamic-pressure probes measured the incoming flow. The array was mounted on struts 1 m in front of the compact car, stabilised by steel cables (Fig. 2a). The array consisted of a 3 × 3 grid with a horizontal spacing of 0.98 m – which results in a marginally wider array than the width of the vehicle – and a vertical spacing of 0.6 m, beginning 0.5 m above the ground (Fig. 2b). The bottom row consisted of two additional five-hole probes, reducing the horizontal distance between the probes to 0.49 m. The horizontal centre of the array was aligned with the horizontal geometrical centre of the car.

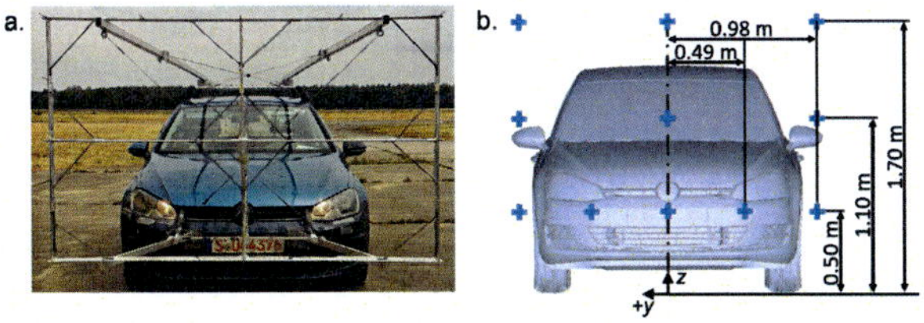

Fig. 2. a: 2D array of five-hole probes in front of the full-scale compact car for the experiments on the test-track. b: Positions of the 11 five-hole dynamic pressure probes (blue dots) in front of the car.

The five-hole probes had 3 mm tip diameters and protruded 100 mm from 12 mm thick aerofoil-shaped beams of the array. The aerofoils housed the pressure tubes. The probes have a calibrated cone of acceptance of ±50°. The pressure taps were connected to a 64-channel *PSI* pressure transducer module by 1.5 m long silicone tubes with an inner diameter of 1.4 mm. The amplitudes and phase distortions due to resonances in the tubing were calculated and corrected based on a theoretical approach using the systems geometries. The pressure transducers had a precision of ±2 Pa at a range of ±5000 Pa. Thus, the flow velocity v_A was calculated with a precision of ±0.6 m/s.

Surface Pressure Measurement. The transient surface pressure was measured at 188 pressure taps over the cars surface (Fig. 3). The measurements were conducted in additional runs without the array to avoid an influence of the attachment's wake on the surface pressures. The pressure taps were distributed over the surface of the car as shown in Fig. 3. The same 1.5 m long silicone tubes as for the five-hole proves with an inner diameter of 1.4 mm were used and the same corrections were applied as for the array. The pressure was measured by 4 *PSI* systems with 32 channels and 1 system with 64 channels. The pressure p at each tap was converted to a normalised pressure coefficient:

$$c_p = (p - p_\infty)/\left(0.5 \cdot \rho v_c^2\right) \quad (1)$$

The atmospheric pressure p_∞ was measured in a reservoir inside the car. The density was calculated by means of the temperature. The compact car's speed over ground (v_c) was selected as reference velocity.

Fig. 3. Distribution of the surface pressure taps. All taps are presented as black dots with individually selected taps highlighted by colours in a. and b. The taps on the front and rear bumper are highlighted in c.

3 Incoming Flow

3.1 Turbulence Intensities, Length Scales and Isotropy

The three components (I_u, I_v, I_w) of turbulence intensity and length scales (L_u, L_v, L_w) – measured by each probe during each run – are presented in Fig. 4. Operating as a single vehicle (SV – designated by black points in Fig. 4), the compact car experiences incoming flow with a low turbulence intensity of ~2% and large length scales of up to 25 m. In contrast, as the compact car drives behind the heavy vehicle, the turbulence characteristics of the incoming flow are influenced significantly: As the distance between the two vehicles decreases from 100 m to 10 m in the driving-behind manoeuvres (designated by green to red respectively in Fig. 4), the turbulence intensity increases from $I_{u,v} = 3\%$ to 25%, while $L_{u,v}$ decreases from 15 m to 0.25 m. The mean turbulence characteristics measured in this specific driving-behind manoeuvre ($I_{u,v} = 4\% - 16\%$, $L_{u,v} = 1 - 10$ m) are within the range measured by Lindener et al. [1], Wordley et al. [2] and McAuliffe et al. [3] in their general on-road measurements.

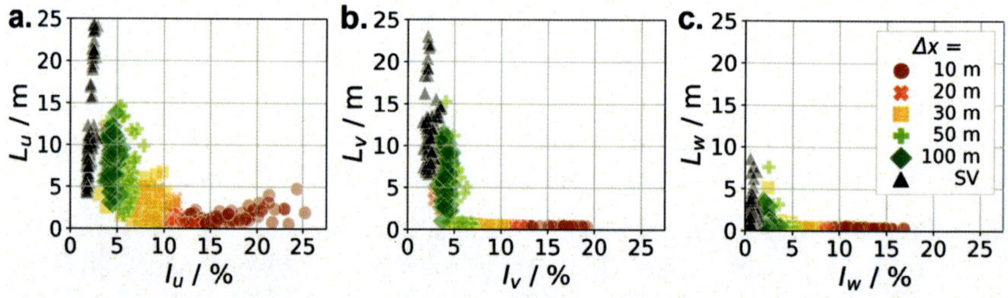

Fig. 4. Turbulence intensities I_u, I_v, I_w and length scales L_u, L_v, L_w, (in a, b, c respectively) at all probes for all runs of the driving-behind and the single-vehicle manoeuvre.

The relative magnitudes of each turbulence intensity component at each distance in Fig. 4a–c are relatively constant. The ratio for a single vehicle was I_u: I_v: I_w = 1:1:0.4. With the compact car driving 100 m behind the heavy vehicle, the ratio changes to 1:1:0.5. The ratio remains unchanged for I_u and I_v, while I_w increases slightly. I_w continues increasing to a ratio of 1:1:0.8 at 10 m with decreasing distances between the vehicles. The results show that the compact car experiences anisotropic turbulence. Wordley et al. [2] and McAuliffe et al. [3] observed similar ratios of 1:1:0.6 and 1:1:0.25.

Except for the close distances of $\Delta x = 10$ m and 20 m, the ratios for driving behind a heavy vehicle are between their results.

3.2 Velocity Flow Field

The transient flow fields provide further insight into the cause of the significant difference in turbulence characteristics between the compact car driving as a single vehicle, and driving behind the heavy vehicle. Contour plots of the spanwise velocity components (v') are presented from the 5 probes at $z = 0.5$ m over the time t in Fig. 5.

The single vehicle (Fig. 5a) experiences minor fluctuations of $v' = 0.5$ m/s. When driving behind the heavy vehicle (Fig. 5b–f), the fluctuations are consistent across the width of the compact car and appear periodic. The fluctuations increase from $v' =$ 1.5 m/s to 6 m/s with decreasing vehicle distances from $\Delta x = 100$ m to 10 m (Fig. 5b–f). These fluctuations, and the aforementioned turbulence characteristics, arise from the compact car experiencing the wake of the heavy vehicle. The signature of the heavy vehicle's transient wake is visible in the incoming flow through coherent, periodically re-occurring structures that are clearly visible at distances from $\Delta x = 100$ m to 20 m (Fig. 5b–e). The flow topology at $\Delta x = 10$ m (Fig. 5f) is not as coherent or visibly periodic. At this very close distance, the compact car seems to be within the wake region of the heavy vehicle that is highly turbulent and dominated by transient, large-scale separation.

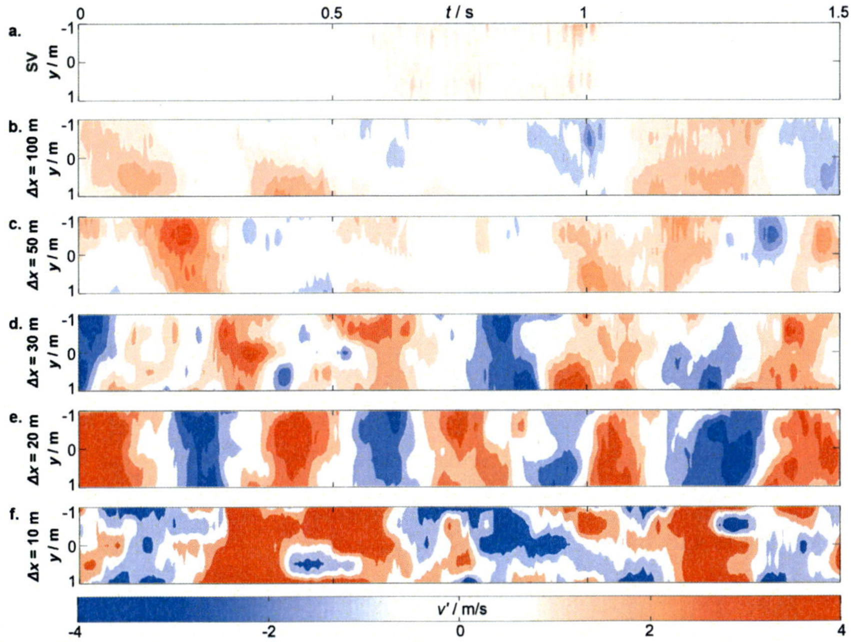

Fig. 5. Contour plot of the flow field v' over the width of the car in $z = 0.5$ m and over time for a single vehicle (a) and when driving behind a heavy vehicle at different vehicle distances (b–f)

3.3 Frequency Spectra of the Velocity Fluctuations

The frequency spectra S of the velocity fluctuations u' and v' measured by the centre probe ($z = 0.5$ m) are presented in Fig. 6 The values measured by this centre probe are representative of the other probes in the array. To calculate the spectra, Welch's method with a *Hanning*-window of 4 s length and a 50% overlap was applied on the 100 s long signals. The mean spectrum over all runs at the same distance was then calculated and is presented in Fig. 6.

The velocity spectrum for the single vehicle (black lines) shows a broad distribution, with reducing energies at higher frequencies f for both components. During the driving-behind manoeuvres, higher energies are visible, with increasing energies as the vehicle distances decrease ($x = 100$ m $-$ 10 m indicated as green to red lines). For frequencies below 3 Hz, plateaus exist in $S(u')$ before dropping off at higher frequencies (Fig. 6a). Interestingly, a clear maximum of ~2.5 Hz is visible in $S(v')$ for the driving-behind manoeuvre at distances of $\Delta x = 50 - 10$ m (light green to dark red lines in Fig. 6. This confirms the periodic structures observed in the flow field shown in Fig. 5b–f. This periodicity can be attributed to horizontal von Kármán-type vortex shedding originating at the heavy vehicle's trailing side edges; based on the vehicle's width of 2.21 m. The frequency of 2.5 Hz corresponds to a Strouhal number of ~0.2.

The results described in Sect. 1 indicate that the incoming flow characteristics (turbulence intensity, length scales, anisotropy, periodicity and flow field coherence) a vehicle experiences when driving behind a heavy vehicle are significantly different to driving alone as a single vehicle in the real world – even more so when compared to the idealised conditions modelled in typical aerodynamic investigations. The amplitudes of the spectra increase with increasing fluctuations in the flow when the distance between the vehicles is reduced. These effects can have a noticeable influence on the aerodynamic characteristics of a vehicle (Schröck et al. [4] and Sims-Williams [5]), This is investigated in the following section, by analysing the transient surface pressure.

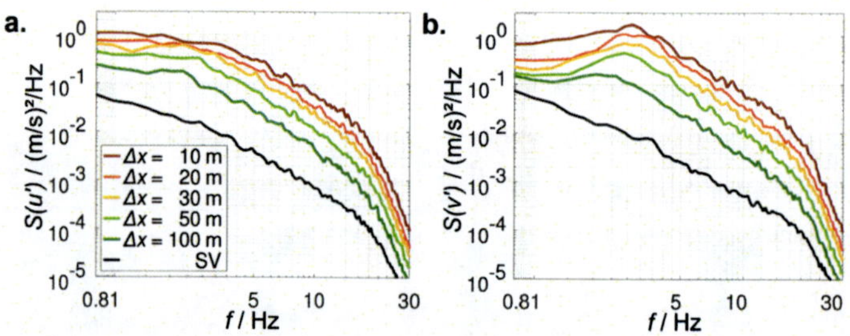

Fig. 6. Spectra of the velocity fluctuations u' (a) and v' (b) for the SV (black) and the driving-behind manoeuvre at the centre probe (0.5 m above ground)

4 Surface Pressure on the Compact Car

4.1 Time-Averaged Pressure

The time-averaged c_p-values (solid lines and circles) and their fluctuations $c_{p,rms}$ (dashed lines, hollow circles) across the front and rear bumper are presented in Fig. 7. The positions of the pressure taps on the compact car are shown in Fig. 3c. The single vehicle (black line) shows increasing pressure coefficients of up to 0.9 towards the centre at the front bumper (Fig. 7a) and values of −0.4 at the centre of the rear bumper (Fig. 7b). The pressure fluctuations remain low at the front with 4.8%–7% and 3%–8% at the back of the single vehicle. Behind the heavy vehicle, the front pressure decreases compared to the single vehicle. The centre's pressure coefficients for the compact car following the heavy vehicle are reduced from 0.86 down to 0.5 with decreasing Δx (red to green lines). The pressure fluctuation increases to 19% with the Δx decreasing. At the rear bumper the pressure of the single vehicle remains at $c_p = -0.4$ over almost the entire vehicle's width, while the fluctuations remain at 3% in the centre and increase to 8% towards the sides. This is the position where the detachment of the flow at the rear of the compact car occurs. The rear pressure and the pressure fluctuations with the heavy vehicle in front do not deviate significantly to the single-vehicle manoeuvre. The pressure reduction occurring with decreasing vehicle distances at the front – while the pressures at the back remain constant – indicates a reduced drag behind the heavy vehicle.

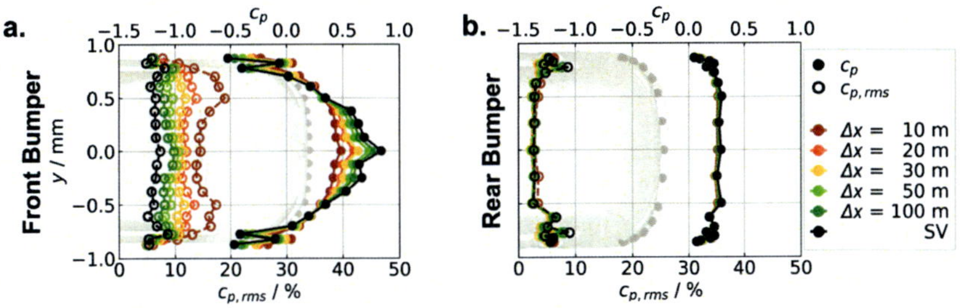

Fig. 7. Pressure distribution (solid circles) and the pressure fluctuation (hollow circles) at the front (a.) and rear (b.) bumper for driving-behind and single-vehicle manoeuvre.

4.2 Transient Pressure

The fluctuating surface pressure c_p' at different locations is presented in Fig. 8. For the single-vehicle manoeuvre, the pressure at the different positions shows only minor fluctuations of $c_p' = \pm 0.1$ over the front and centreline (top and middle row of Fig. 8a). For the driving-behind manoeuvre (Fig. 8b–d), larger fluctuations occur at the vehicle's front, with up to $c_p' = \pm 0.6$ at $\Delta x = 10$ m (blue lines, top row in Fig. 8). The fluctuations increase with decreasing distances between the vehicles and a periodicity visible at $\Delta x = 10$ m as well as at 20 m (Fig. 8-d). The pressure at the front bumper also exhibits an inverse relationship between the (oscillating) windward and leeward side, which is probably due to yawed incoming flow. This effect is more clearly visualised in Fig. 9, where the pressure across the entire front bumper is presented as a contour plot. Coherent oscillations between low and high pressure are visible between the left and right side of the vehicle. These can be clearly identified at $\Delta x = 20$ m (Fig. 9c), but also appear at $\Delta x = 50$ m and 10 m, even though weaker and more chaotic respectively.

(Fig. 9b, d). Due to the shared scaling of the pressure fluctuation, the low fluctuations at the single vehicle are barely visible at all in Fig. 9a but are shown for the sake of completeness. These pressure characteristics are directly attributable to the coherent fluctuations in the incoming flow (Fig. 5). Along the centreline of the vehicle, the pressure fluctuations decrease from the front to the rear (middle row, blue-to-red lines in Fig. 9). At the rear (bottom) the pressure fluctuations are not visibly influenced by the driving manoeuvre, with consistent fluctuations of $c_p{}' = \pm 0.4$ (red lines). These rear fluctuations are expected to be self-induced, dominated by large-scale separation in the wake of the compact car.

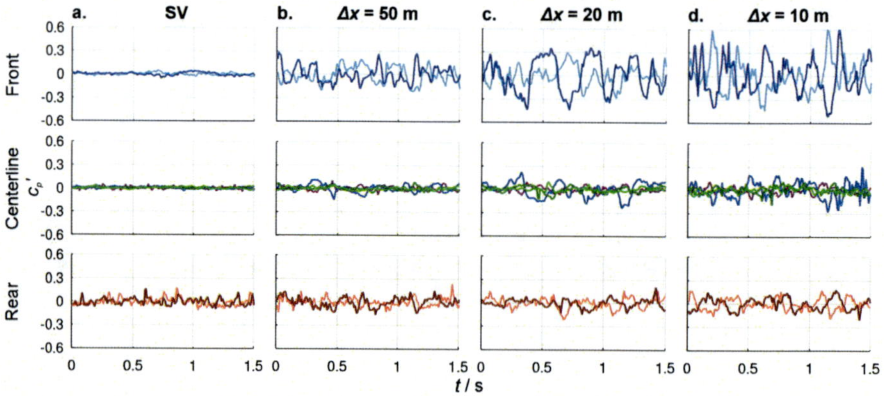

Fig. 8. Pressure fluctuations at pressure taps at the front bumper (top), along the roof's centre line (middle) and the rear bumper (bottom) for the single vehicle (a) and driving behind manoeuvre at 50 m, 20 m and 10 m (b–d). The line colours correspond to the diagram in Fig. 3.

The pressure results provide insight into the transient aerodynamic forces a compact car would experience in the real-world driving-behind and single-vehicle manoeuvre; the different incoming flow characteristics translate into different transient surface pressure characteristics. The signature of the transient velocity is directly observable in the transient surface pressure, specifically at the vehicle's front, but also in the time-averaged pressure distributions. These results indicate that the different incoming conditions that arise from driving behind a heavy vehicle would indeed affect the global aerodynamic forces and moments of the vehicle, that which could not be directly measured in the experiment.

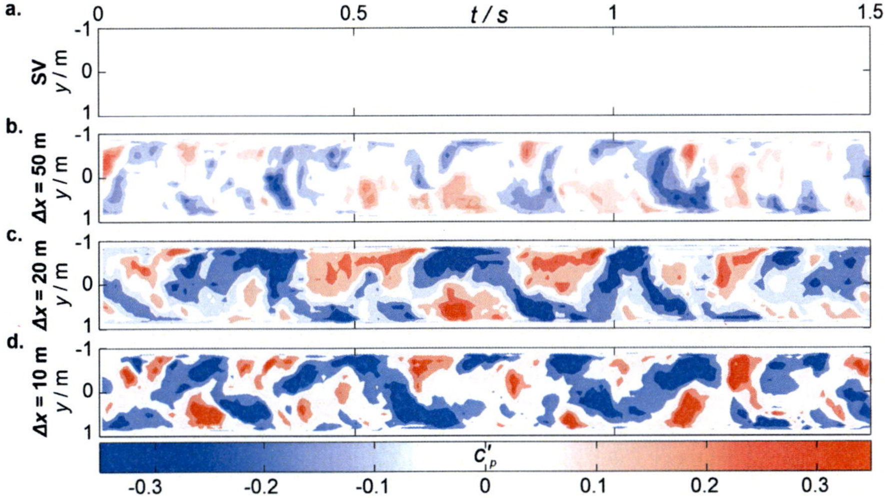

Fig. 9. Transient surface pressure at the front bumper of the compact car for the single-vehicle (a.), driving-behind manoeuvre at 50 m (b.), 20 m (c.) and 10 m (d).

5 Conclusions

The transient incoming flow and surface pressure of a full-scale compact-car on a test track have been characterised in terms of driving 10 to 100 m behind a heavy-vehicle and in a single-vehicle manoeuvre. An array of 11 five-hole probes and 188 surface pressure taps with a temporal resolution of 250 Hz identified a compact car driving behind a heavy vehicle experiences a significant change in incoming flow conditions, compared to driving as a single vehicle (and by extension, the idealised conditions typically used in aerodynamic assessment and optimisation of vehicles). With decreasing distances between the vehicles, the turbulence intensities increase from 2 % to 25 % and the integral length scales decrease from 25 m to 0.25 m for all velocity components. In all tested configurations the compact car experiences anisotropic turbulence ratios of 1:1:0.5 – 0.8. These characteristics are attributed to a transient flow field that exhibits horizontal velocity fluctuations of up to ±6 m/s that are coherent over the width of the compact car with a dominant frequency of ~2.5 Hz attributable to vortex shedding in the wake of the heavy vehicle.

The incoming flow has a demonstrable influence on the surface pressure of the compact car, indicating that the vehicle's global aerodynamic forces and moments would also be affected. The fluctuating incoming flow leads to inversely correlated pressure fluctuations between the (oscillating) leeward and windward side at the front of the vehicle. Pressure fluctuates up to $c_p' = \pm 0.6$ at the front of the compact car and is reduced towards the rear. The time-averaged pressure is also affected, particularly at the rear of the compact car, where pressure decreases with decreasing vehicle distance.

The incoming flow and pressure data can be used as input as well as for validation in future experiments and simulations that include these real-world conditions. These methods can subsequently be used to perform advanced aerodynamic assessment and optimisation of automotive vehicles in realistic driving manoeuvres; ideally leading to the design and development of safer, more efficient automotive vehicles in the future.

Acknowledgements. The authors acknowledge the *FAT* working group 6 for initiating and financing the project as well as their input. *The German Army* and *DLR Trauen* are acknowledged for providing the space for the experimental campaign and their local support. Furthermore, the authors thank Annika Köhne for her support in proofreading the paper.

References

1. Lindener, N., Miehling, H., Cogotti, A., et al.: Aeroacoustic Measurements in Turbulent Flow on the Road and in the Wind Tunnel. SAE Technical Paper 2007–01–1551 (2007). https://doi.org/10.4271/2007-01-1551
2. Wordley, S., Saunders, J.W.: On-road turbulence. SAE Int. J. Passeng. Cars – Mech. Syst. **1**, 341–360 (2009). https://doi.org/10.4271/2008-01-0475
3. McAuliffe, B.R., Belluz, L., Belzile, M.: Measurement of the on-road turbulence environment experienced by heavy duty vehicles. SAE Int J Commer Veh **7**, 685–702 (2014). https://doi.org/10.4271/2014-01-2451
4. Schroeck, D., Krantz, W., Widdecke, N., et al.: Unsteady aerodynamic properties of a vehicle model and their effect on driver and vehicle under side wind conditions. SAE Int. J. Passeng. Cars – Mech. Syst. **4**, 108–119 (2011). https://doi.org/10.4271/2011-01-0154
5. Sims-Williams, D.: Cross winds and transients: reality, simulation and effects. SAE Int. J. Passeng. Cars – Mech. Syst. **4**, 172–183 (2011). https://doi.org/10.4271/2011-01-0172
6. Jessing, C., Wilhelmi, H., Wittmeier, F., et al.: Investigation of transient aerodynamic effects on public roads in comparison to individual driving situations on a test site. ASE Int. J. Adv. Curr. Prac. Mobility **2**, 2585–2595 (2020). https://doi.org/10.4271/2020-01-0670
7. Wilhelmi, H., Jessing, C., Bell, J., Heine, D., Wagner, C., Wiedemann, J.: Aerodynamic characterisation of a compact car overtaking a heavy vehicle. In: Dillmann, A., Heller, G., Krämer, E., Wagner, C., Tropea, C., Jakirlić, S. (eds.) DGLR 2018. NNFMMD, vol. 142, pp. 794–804. Springer, Cham (2020). https://doi.org/10.1007/978-3-030-25253-3_75
8. Bell, J.R., Wilhelmi, H., Heine, D., et al.: Aerodynamic characterisation of a full-scale compact car exposed to transient crosswind. SAE Int. J. Passeng. Cars—Mech. Syst. **14** (2020). https://doi.org/10.4271/06-14-01-0001
9. Zabat, M., Stabile, N., Farascaroli, S., et al.: The Aerodynamic Performance of Platoons: A Final Report (1995)
10. Davila, A., Aramburu, E., Freixas, A.: Making the Best Out of Aerodynamics: Platoons. SAE Technical Paper 2013–01–0767 (2013). https://doi.org/10.4271/2013-01-0767
11. Wilhelmi, H., Jessing, C., Bell, J.R., et al.: Simulation of transient on-road conditions in a closed test section wind tunnel using a wing system with active flaps. . SAE Int. J. Adv. Curr. Prac. Mobility **2**, 2604–2616 (2020). https://doi.org/10.4271/2020-01-0688
12. Weinman, K., Wilhelmi, H., Bell, J.R., et al.: On the simulation of a heavy vehicle's wake in OpenFOAM with Real-World Data. In: Dillmann, A., Heller, G., Krämer, E., Wagner, C. (eds.) STAB/DGLR Symposium 2020. NNFM, vol. 151, pp. 504–513 (2022). https://doi.org/10.1007/978-3-030-79561-0_48

Aeroelasticity and Structural Dynamics

High-Fidelity Aeroelastic Loads Calculation for a Transport Aircraft Configuration Including Pitch and Roll Maneuvers

Johan Moritz Feldwisch(✉) and Matthias Schulze

German Aerospace Center (DLR), Institute of Aeroelasticity,
Bunsenstr. 10, 37073 Göttingen, Germany
Johan.feldwisch@dlr.de
https://www.dlr.de/ae/

Abstract. It is assumed, that the presence of nonlinear effects in transonic flow influences the aerodynamic load distribution and hence must be considered in the aeroelastic loads analysis of maneuvers. In this study a steady pull-up and two quasi-static rolling maneuvers are examined using the Reynolds-Averaged-Navier-Stokes (RANS) equations and compared to the vortex lattice method (VLM). In contrast to the RANS equations, the VLM is widely used for fast analysis of many load cases but does not include transonic effects. The pitch and roll rates are accounted for by the flow solver. A fully elastic transport aircraft configuration is analyzed. Elastic deformations are calculated by the modal approach and are loosely coupled with the aerodynamic model.

The resulting loads are compared to maneuver analysis based on the VLM. For all maneuvers, the aerodynamic loads based on the RANS equations yield higher loads at the wing's root by 1% to 2% and require larger negative balancing forces on the tail plane.

Keywords: Loads · FSI · CFD · Coupled · CSM · RANS · XRF1

1 Introduction

For aircraft design, aerodynamic and inertia loads for multiple load cases need to be considered according to CS25 [1]. An extensive loads analysis requires those loads for several maneuvers, multiple mass configurations, flight speeds, altitudes and aircraft configurations (clean, flaps out etc.). An already reduced analysis like in [2] considers 1062 load cases. Those loads analyses require simplifications and robust methods, as a vast set of flight operation points has to be investigated.

Depending on the geometry (complexity, movables settings) and the flight point different physical effects may become dominant in the flow field and hence, influence the aerodynamic force distribution. Although the VLM provides a robust tool to determine aerodynamic loads at any flight point, it needs

© The Author(s), under exclusive license to Springer Nature Switzerland AG 2021
A. Dillmann et al. (Eds.): STAB/DGLR Symposium 2020, NNFM 151, pp. 527–536, 2021.
https://doi.org/10.1007/978-3-030-79561-0_50

some corrections for geometry, flow-physics and control surfaces. Boundary layer effects and compression shocks are not captured by the VLM. Since the simplifications of the VLM lead to inaccurate results, the trend is to progress towards high fidelity RANS simulation as the computational methods improve and the computational performance still increases.

In previous work of Voß [3], a comprehensive loads analysis for pull-up, push-down and rolling maneuvers of a geometrically simple flying wing configuration showed increasing differences between the VLM and the Euler solution at Mach numbers greater than 0.8. However, the Euler could not be used for trimming all points, as a recompression shock system reduced the aileron effectiveness too strongly.

In [4] multidisciplinary design studies have been performed using the RANS equations for performance evaluation and load computation for pull/push maneuvers. However, only the elasticity of the wing is accounted for and no pitch rates are included.

To progress the capabilities of CFD based loads analysis, the in-house loads framework UltraFLoads is developed. A first application is provided by this study which focuses on the calculation of quasi-steady aerodynamic loads for the fully elastic aircraft in pull and rolling maneuver conditions at transonic flow. The results are compared to loads based on the VLM for a trimmed solution. For future studies UltraFLoads will be used to analyze the limits of CFD for loads analysis and how the limits could be shifted towards the full flight envelope.

2 Loads Analysis

2.1 Steady Maneuvers

For the steady maneuvers of this paper, the following conditions hold. The path angle $\gamma(°)$ and the path azimuth are zero, so the path coordinate system (suffix p) and the geodesic coordinate system (suffix g) coincide. Also, the bank angle and the yaw angle are zero. For steady maneuvers, the translational acceleration $\dot{\mathbf{v}}_b(\mathrm{ms}^{-2})$ and the rotational acceleration $\dot{\boldsymbol{\Omega}}_b(°\mathrm{s}^{-2})$ in the body coordinate system (suffix b) are zero, or at least constant for quasi-static maneuvers. The transformation T_{jl} is used to rotate quantities from frame l to j.

For flight mechanics, the aircraft is condensed to a point (center of gravity) with six degrees of freedom (rigid body acceleration) and a mass matrix M. The aircraft is exposed to external forces $\mathbf{F}_b(\mathrm{N})$ and moments $\mathbf{M}_b(\mathrm{N\,m})$, which are balanced by the inertia forces and moments.

$$\mathsf{M}\begin{Bmatrix}\dot{\mathbf{v}}\\ \dot{\boldsymbol{\Omega}}\end{Bmatrix}_b = \begin{Bmatrix}\mathbf{F}\\ \mathbf{M}\end{Bmatrix}_b^{\text{applied}} + \begin{Bmatrix}\mathbf{F}\\ \mathbf{0}\end{Bmatrix}_b^{\text{centrifugal}} + \begin{Bmatrix}\mathbf{F}\\ \mathbf{0}\end{Bmatrix}_b^{\text{gravity}} . \tag{1}$$

As can be seen from Eq. 1, the body acceleration depends on the applied loads, centrifugal forces and gravitational forces. However, those loads depend on multiple parameters like the position of the aircraft in space, the thrust setting, the control surface and control plane deflections. The general trim problem can be formulated to find the trim parameters such that the difference between body acceleration and the target acceleration is zero. More detailed information on a general solution to the trim problem can be found in [5].

The load factor $\mathbf{n}_b$ is defined in the body coordinate system and relates the body acceleration to the gravitational acceleration. The airframe has to withstand load factors in the vertical direction from −1. up to 2.5. Thus it is one important parameter for the maneuver definition.

$$\mathbf{n}_b = -\frac{1}{g}\left(\dot{\mathbf{v}}_b + \mathsf{T}_{bp}(\boldsymbol{\Omega}_p \times \mathbf{v}_p)\right) + \mathsf{T}_{bg}\begin{Bmatrix} 0 \\ 0 \\ 1 \end{Bmatrix} \tag{2}$$

Pull/push maneuvers describe the curved flight path of the aircraft in the x_g-z_g-plane, which can be approximated by a circular motion. The path angle rate in Eq. 3 can then be approximated for small pitch angles by the vertical load factor n_z and the true air speed. Further information is provided for example by Lomax [6].

$$\dot{\gamma} \approx (n_z - 1)\frac{g}{V_{TAS}}. \tag{3}$$

Rolling maneuvers can be split into three different steady maneuvers which should mimic the initialization of the roll, the steady roll and the roll reversal. For the initial roll, the roll rate $p = 0°\mathrm{s}^{-1}$ and a target roll acceleration should be reached ($\dot{p} \neq 0°\mathrm{s}^{-2}$). For the steady roll, the roll rate $p \neq 0°\mathrm{s}^{-1}$ and the roll acceleration is zero. The last phase is then at a constant roll rate and an opposite constant roll acceleration. While the steady roll is the only real steady maneuver, the other two are rather quasi-steady maneuvers. Here, the control inputs are set to yield an applied loads distribution, which would result into some moment imbalance, causing rotational acceleration. Thus, it is assumed that the forces build up instantaneously. Additionally, the roll maneuvers are defined for different load factors. Those load factors are derived from the circular motion as for the pull/push maneuver. The pitch rate from Eq. 3 is used.

2.2 Inertia Loads

The mass related loads are known, as the maneuvers are defined by their accelerations. Once the pitch angle is determined, the inertia loads can be computed. Since the inertia loads do not depend on the aerodynamic model, they are not further included in the post processing of this study.

2.3 Elastic Deformation

If the elastic deformation of the aircraft significantly changes the aerodynamic load distribution, it needs to be considered in the analysis [1]. In this work the elastic deformation is accounted for by the modal approach for the free-free aircraft. Instead of the solution of the structural degrees of freedom, the generalized coordinates $\mathbf{q}_{\text{elastic}}(\text{m})$ of each mode are determined. As only the first 20 elastic modes ϕ_{elastic} are considered, the resulting structural displacements $u(\text{m})$ are approximate.

$$\mathbf{u} \approx \phi_{\text{elastic}} \mathbf{q}_{\text{elastic}}. \tag{4}$$

Using the generalized stiffness matrix Λ, the modal equation for the steady case is then:

$$\mathbf{q}_{\text{elastic}} = \Lambda^{-1} \phi_{\text{elastic}}^{T} \{\mathbf{F}^{T} \mathbf{M}^{T}\}^{T}. \tag{5}$$

3 Models and Software

The XRF1 is used as a reference aircraft, which is an Airbus provided industrial standard multi-disciplinary research test case representing a typical configuration for a long range wide body aircraft. The geometry and CFD meshes are provided by Airbus, the structural model and the mass models are based on the XRF1-DLR [2].

3.1 Aerodynamic Models

On the one hand there is the simplified VLM model solving the potential equations and on the other hand there is the high-fidelity CFD model solving the RANS equations. The VLM model is generated by the aeroelastic structural design process cpacs-MONA [2] developed at the DLR Institute of Aeroelasticity. The planar VLM model is geometrically corrected and therefore accounts for camber and twist. The aerodynamic forces based on the VLM are computed by MSC.Nastran solution sequence 144 [7]. Both aerodynamic models are available for the full aircraft model configuration shown in Fig. 1. While the VLM model includes the engine outer geometry approximation as lifting surfaces, they are not included for the CFD simulation. Thrust is not accounted for in both models, which results into larger (negative) balancing forces at the HTP. Thus the aerodynamic forces on the wing will be higher compared to a simulation with engine, as the balancing forces need to be accounted for.

The hybrid mesh for the CFD analysis is provided by AIRBUS. It consists of 10.1×10^6 nodes, 4.4×10^6 volume cells and 9.2×10^6 boundary layer cells. Although no extensive mesh convergence study was done, a mesh with 40 % more nodes did not show a different distribution of the bending moment or the shear force. Thus, for loads analysis the mesh density appears to be appropriate. The Spalart-Allmaras turbulence model is selected. The RANS equations are solved using the DLR flow solver TAU [8]. The velocity conditions are not defined at the far-field of the mesh, but on the surface mesh directly. Hence, translatory and rotatory velocities are set by the external motion module, enabling pitch, roll and yaw motion.

The VLM model and the hybrid meshes have definitions for the ailerons and the horizontal tail plane, which can be rotated. For the CFD grids, the control-surface deflections are applied by mesh deformation using a blended approach. This approach introduces a modeling error, as no gaps or steps between the wing

Fig. 1. Pressure distribution for a roll maneuver solved by VLM and RANS.

and the aileron occur. However, this approach is simple, poses less challenges on the mesh generation and is less complicated than handling Chimera meshes. The Chimera approach will be included to model spoiler deflections in future loads studies. Rotating the HTP of the CFD mesh by small angles caused problems, as the trailing edge moves out of the rear fuselage. A Chimera approach could be used, but the aerodynamics of the HTP is not of the main focus of this study. Instead of deflecting the HTP, it was decided to use an elliptical force distribution applied to the HTP structure. A scalar factor s_{HTP} can be used to scale the artificial force distribution and balance the pitching moment of the aircraft. Thus, a higher load on the wing due to the compensation of the downward force on the HTP is captured.

3.2 Structural Models

The structural models and mass models are also generated by cpacs-MONA. Further information about the model can be found in [2]. Instead of the full finite element model, a reduced model for the degrees of freedom of the loads reference axis is used (Fig. 2). An additional interface model is needed for the spatial coupling between the CFD surface and the structural model, shown in Fig. 2. The degrees of freedom of those nodes are rigidly linked to the degrees of freedom of the loads reference axis. For the solution of the modal approach, the modes of the interface model are used.

The actual coupling is component based, so forces and displacements are only transferred between each aircraft component separately. The force interpolation is achieved by a nearest neighbor search. Due to the high density of the interface model, lever arms to the surface mesh are considered to be small and hence moments are neglected. Displacements calculated by the modal approach are then interpolated on the CFD surface mesh using a thin-plate-spline.

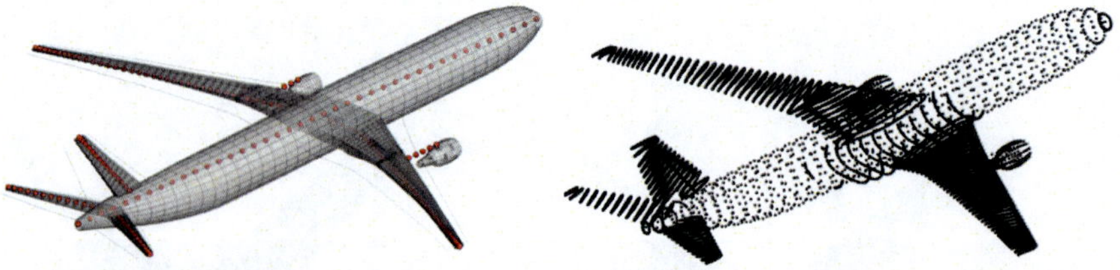

Fig. 2. Structural model. Detailed model including the reduced LRA model (left) and interface nodes (right).

3.3 UltraFLoads

UltraFLoads is a Python framework for high-fidelity multidisciplinary loads analysis, developed at the Institute of Aeroelasticity. It is an object oriented approach, allowing for modularization and combination of disciplines to set up multidisciplinary analysis. As a centralized data scheme is used, modularization can be achieved easily as it decouples the data flow from the process flow.

Simulation scenarios like trimming an aircraft or solving the fluid-structure problem of any aerodynamic method or structural method are provided. As the framework is developed driven by the requirements of a research environment, it needs to allow for quick modification. So, the provided process flows can be altered to some extent by the users on the fly. They can develop their own analysis classes and register them in UltraFLoads. This approach allows for quick prototyping without altering the code.

Since multidisciplinary analysis for full aircraft configurations is complex, multiple features, solvers etc. need to be available. Thus it is tried to use as much as possible of the already developed code for high-fidelity flow simulations at DLR. Multiprocessing, data models, mesh objects, relations model, mesh deformation and much more are provided by the FlowSimulator [9]. On top of the many interfaces to the FlowSimulator objects, multiple convenient Python layers like the TAU-control interface have been developed by the FlowSimulator community. UltraFLoads uses the capabilities of the FlowSimulator and many of the control layers, which are used by an increasing number of researchers.

4 Results

One pull-up (PU), one initial roll (IR) and one roll reversal (RR) are investigated at transonic flight speeds, defined in Table 1. The maneuvers have been selected such that the flight points are still in the flight envelope and provide sufficient dynamic pressure. Also lighter mass configurations are used, compared to the 245 t maximum take off mass [2]. The transonic flow over the aircraft is visible due to the shock in the pressure distribution in Fig. 3.

Table 1. Load case description

ID	Maneuver	Mass (t)	Ma	Altitude (m)	n_z	p (°s^{-1})	$\dot{p}$ (°s^{-2})
PU	Pull-up	192.3	0.73	5500	2.5	0	0
IR	Initialize roll	209.5	0.8	8000	1.67	0	8
RR	Revert roll	209.5	0.8	8000	1.67	16	−16

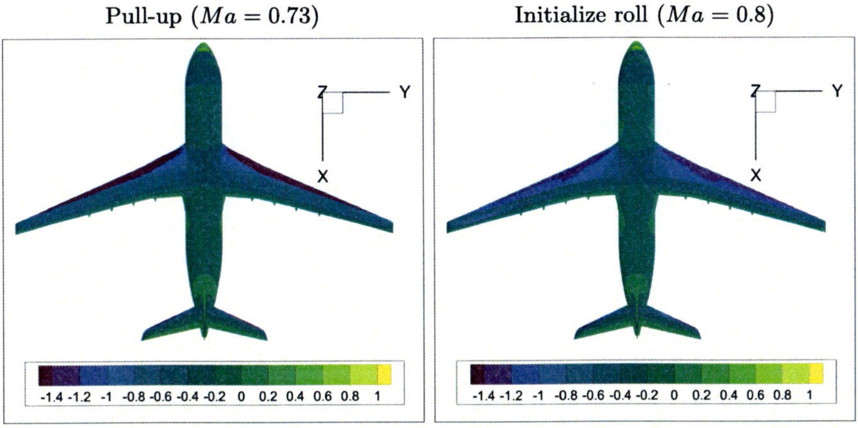

Fig. 3. Stretched top view of the pressure coefficient for the pull-up and the initialize roll maneuver.

The trim variables for the pull-up maneuver are the pitch angle Θ and the scaling factor for the external force field on the horizontal tail plane s_{HTP}. For the rolling maneuvers, the variables are the pitch angle, the aileron deflection (outer and inner) ξ and the scaling factor s_{HTP}.

The trim problem is solved by a Newton approach, where the jacobian is only computed once relatively to the initial trim variables. The columns of the jacobian are estimated by forward finite differencing of the rigid aerodynamics, relatively to the fully converged fluid-structure coupling problem.

Table 2. Trim results for the different maneuvers based on RANS. Percentile difference of loads (RANS-VLM) evaluated at the wing root.

ID	$\Theta(°)$	$\xi(°)$	n_z	$\dot{q}$(°s^{-2})	$\dot{p}$(°s^{-2})	$\delta F_z(\eta = 0.09)$	$\delta M_x(\eta = 0.09)$	$\delta M_y(\eta = 0.09)$
PU	6.24	–	2.5	0.012	0	1.21%	0.89%	2.62%
IR	4.34	−3.09	1.67	0.04	8.05	2.13%	2.69%	3.49%
RR	4.33	−8.43	1.67	0.05	−15.89	1.05%	1.89%	2.31%

Table 3. Percentile contribution of each component to the total vertical force F_z for different maneuvers and aerodynamic methods.

Component	Pull-up		Initial roll		Revert roll	
	VLM	RANS	VLM	RANS	VLM	RANS
Fuselage	16.06%	15.25%	15.69%	14.83%	15.69%	14.88%
Wing	83.87%	85.74%	86.04%	88.56%	86.04%	88.36%
HTP	0.08%	−1.12%	−1.74%	−3.56%	−1.73%	−3.41%

The trim results are summarized in Table 2. The difference for the aerodynamic cutloads (Shear force F_z, bending moment M_x, torsional moment M_y) at the wing's root (relative spanwise coordinate $\eta = 0.09$) show, that the aerodynamic forces based on the RANS equations are larger in all cases. Although 1% or 2% appear to be small, it still influences the structural mass of the load carrying structures. In a design study for a similar transport aircraft it was observed, that a reduction of the root bending moment by 1.05% can reduce the structural mass of the wingbox by 1.17% [10]. The larger aerodynamic bending moments, shear forces and torsional moments can be explained, when looking at the component-wise contribution to the total vertical force in Table 3. The fuselage of the VLM model contributes more to the total force than it is the case for the CFD solution. This holds for all of the investigated maneuvers. Since the fuselage contributes less, the wing has to carry more, explaining the

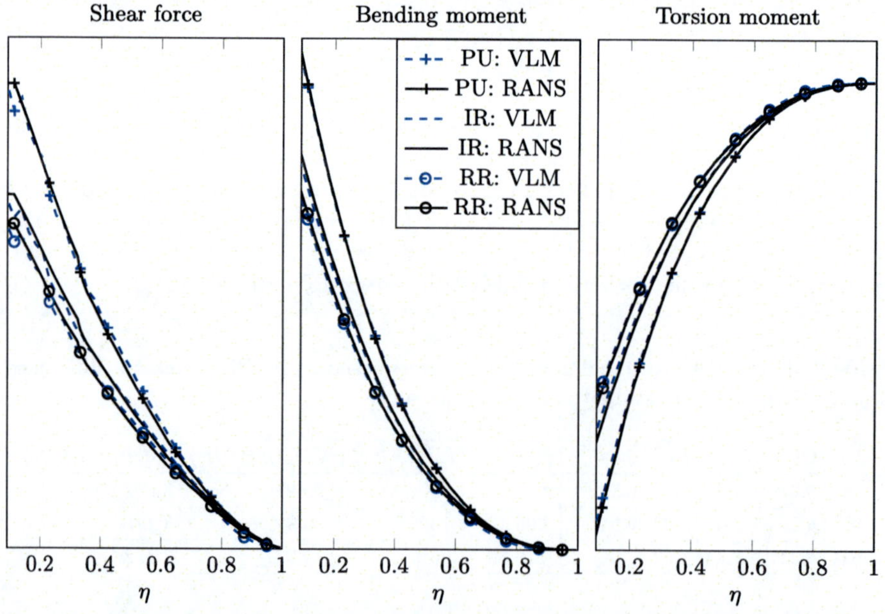

Fig. 4. Aerodynamic loads for the pull up, initial roll and reverse roll over the halfspan of the wing computed with VLM and RANS in the global coordinate system.

higher aerodynamic loads at the wing's root. Also, the larger negative balancing forces on the HTP have to be compensated by wing. The distribution of the shear force, bending moment and torsional moment are shown as cutting loads in Fig. 4. Due to proprietary reasons, no quantitative results are presented. For the inner section and the outer section, the results based on the RANS equations show larger loads. Note, the forces and moments are integral values. Thus differences close to the wing-fuselage intersection include all the differences over the entire wing. In the mid-section the aerodynamic loads based on the VLM generate higher cut-loads. Thus, one cannot conclude that the VLM or the RANS is more conservative. The implications on the structural design would be a stiffer inboard section for loads based on RANS and a softer structure on the outboard compared to the VLM loads. However, as loads and structural sizing pose an implicit problem itself, it is hard to predict the actual outcome of this finding. Further investigation including the structural design is still necessary.

5 Robustness

Even though it is possible to compute quasi-steady rolling and pull-up maneuvers based on solving the RANS equations, the loads computation is still far from being robust. It was observed that for some flight speeds, the flow solution could not be initialized neither by an upwind nor a central flux discretization type. Or it was possible only with an upwind scheme, while for other cases it only worked for central scheme. Also, at some flight points one mesh showed good convergence, but a different mesh with a higher cell density could not even start, while both meshes show the same results for again another flight speed. Thus, the higher complexity comes at the cost of robustness. At the moment, the used meshes are defined to yield good behavior at the cruise conditions. When exploring the boundaries of the envelope, multiple meshes may become necessary to deal with the off design conditions. Another challenge comes from the mesh deformation for large control surfaces. A steady-roll maneuver analysis could not reach the target acceleration, as the deflection caused either the mesh to fail or the flow to separate. Deflecting an aileron upward moves the recompression shock towards the leading edge, where it can cause flow separated regions. Then, converging the flow solution, the fluid-structure problem and the trim problem might not be possible at all. Thus simply applying steady RANS simulation to all flight points for standard loads bookcases does not always work, as for example the quasi-steady assumption may be invalid.

6 Conclusion

The aerodynamic loads based on the RANS equations for a free-flying, flexible transport aircraft have been shown for a pull-up, an initialize roll and a roll reversal maneuver accounting for pitch and roll rates.

Although all maneuvers are in the transonic flow regime, no large difference compared to loads based on the VLM could be found. The aerodynamic loads at the wing's inner section are larger, based on the RANS equations and smaller in the mid-section. Thus, one cannot generally say VLM or RANS is more conservative. However, this might lead to a softer structure at the inboard of the wing if sized by CFD based loads. To elaborate on this finding, structural re-design would be required and will be addressed in future studies. Also, the contribution to the total lift by the fuselage is 5% smaller, when solving the RANS equations for the presented models. Hence the VLM fuselage aerodynamic model should be corrected as well if a cross-fuselage model is used.

Even though quasi-steady pull and rolling maneuvers can be computed using high fidelity methods to account for viscous and transonic flow, the robustness remains a challenge.

This work contributes to high fidelity loads computation. Future studies will elaborate on mixed-fidelity approaches and implications on structural design. Also, the steady flow assumption needs to be addressed.

Acknowledgment. The authors would like to thank Airbus for providing the XRF1 test case as a mechanism for demonstration of the approaches presented in this study.

References

1. European Union Aviation Safety Agency: Certification Specifications and Acceptable Means of Compliance for Large Aeroplanes CS-25. Amendment 25 (2020)
2. Klimmek, T., Schulze, M., Abu-Zurayk, M., Ilic, C., Merle A.: cpacs-MONA – an independent and in high fidelity based MDO tasks integrated process for the structural and aeroelastic design for aircraft configurations. IFASD 2019 - International Forum on Aeroelasticity and Structural Dynamics, Savannah (2019)
3. Voß, A.: Comparing VLM and CFD maneuver loads calculations for a flying wing configuration. In: IFASD 2019 - International Forum on Aeroelasticity and Structural Dynamics, 10–13 June 2019, Savannah (2019)
4. Brooks, T.R., Kenway, G.K.W., Martins, J.R.R.A.: Benchmark aerostructural models for the study of transonic aircraft wings. AIAA J. **56**(7), 2840–2855 (2018). https://doi.org/10.2514/1.J056603
5. De Marco, A., Duke, E., Berndt, J.: A general solution to the aircraft trim problem. In: AIAA Modeling and Simulation Technologies Conference and Exhibit, Hilton Head (2007). https://doi.org/10.2514/6.2007-6703
6. Lomax, T.L.: Structural loads analysis for commercial transport aircraft: theory and practice. In: AIAA Education Series. AIAA, Virginia (1996). https://doi.org/10.2514/4.862465
7. MSC Software, MSC Nastran 2017 Aeroelastic Analysis User's Guide, USA (2016)
8. Gerhold, T.: Overview of the hybrid RANS code TAU. In: Kroll, N., Fassbender, J.K. (eds.) MEGAFLOW- Numerical Flow Simulation for Aircraft Design, pp. 81–92. Springer, Heidelberg (2005). https://doi.org/10.1007/3-540-32382-1_5
9. Reimer, L.: The FlowSimulator—a software framework for CFD-related multidisciplinary simulations. In: European NAFEMS Conference Computational Fluid Dynamics (CFD) – Beyond the Solve, München (2015)
10. Feldwisch, J.M., Klimmek, T., Keye, S.: High-fidelity aero-structural optimization of the wing twist using the NASA common research model. In: DLRK 2018 - Deutscher Luft- und Raumfahrtkongress, 04–06 September 2018, Friedrichshafen (2018)

Check for updates

Deformation Measurements of a Full Span Model with Adaptive Elasto-Flexible Membrane Wings

Jonathan Pflüger(✉), Yuting Chen, and Christian Breitsamter

Chair of Aerodynamics and Fluid Mechanics, Technical University of Munich, Boltzmannstr. 15, 85748 Garching, Germany
jonathan.pflueger@tum.de

Abstract. Unmanned or micro aerial vehicles are designed to perform well in a variety of flight conditions. Therefore, the morphing wing technology aims to constantly adapt the aerodynamics to different flight stages. This work presents experimental investigations of an elasto-flexible membrane wing at a Reynolds number of 264000. The investigated concept enables wing folding over a wide range and it allows the wing to adapt to changing aerodynamic loads. The present study focuses on the deformation of the elasto-flexible membrane and its advantages for the aerodynamic performance of the examined model. The measurements show an increasing camber and thickness of the airfoil for higher angles of attack, which results in a more smooth and delayed stall, in particular for the highly swept configuration.

Keywords: Morphing wing · Flexible wing surface · Camber morphing · Optical deformation measurements · Experimental aerodynamics

1 Introduction

The aerodynamic design of a common aircraft geometry is based on optimizing the flight performance under certain flight conditions. Airliners for example are mainly designed for cruise, since this flight stage heavily dominates their typical mission. However, other aircraft, such as unmanned aerial vehicles (UAV) or micro aerial vehicles (MAV), should perform well within a wide range of flight conditions. Morphing wing configurations may be a solution to this problem, as they are able to adapt their shape to produce optimum performance at various flight stages. The development of morphing wing structures goes back to the origins of aviation [1,2]. The benefits of morphing wings are accompanied by the complicated technical realization concerning the structural design, actuators, materials and the flight controls. The aerodynamic improvements compete against the penalties by additional structures, the increased weight and the energy consumption [3,4].

The morphing methods are subdivided into active and passive means. Frequently used concepts of active morphing are high-lift devices and variable swept

© The Author(s), under exclusive license to Springer Nature Switzerland AG 2021
A. Dillmann et al. (Eds.): STAB/DGLR Symposium 2020, NNFM 151, pp. 537–546, 2021.
https://doi.org/10.1007/978-3-030-79561-0_51

wings. In addition, the change of the wing planform, wing camber and new designs for control surfaces, especially trailing-edge (TE) devices, are investigated. Passive morphing methods allow the wing structure to deform under the influence of aerodynamic loads, which in turn affects the aerodynamic characteristics. One approach is to use an elastic membrane as a lifting surface, where the deformation of the airfoil shape follows from the equilibrium between the inner elastic forces of the membrane and the aerodynamic forces.

The low weight, low cost and structural simplicity represent only some of the benefits regarding membrane wings. Especially, for applications at low Reynolds numbers and small payloads, like for MAVs, the thin and elastic membrane offers further advantages. One unique difference for MAVs, compared to large aircraft, is the focus on agility vs. stability [5]. On the one hand, their small inertias and light weight allow MAVs to operate in a wide range of speeds and flight regimes. On the other hand, these missions implicate more unstable flight conditions, which requires high agility. This can be achieved by an elastic membrane. Possible features are superior stall characteristics, because the shape of the airfoil passively adapts to the inflow at high angles of attack, and the reduction of adverse effects regarding gusts and unsteady aerodynamics, because the elastic membrane is less sensitive to oscillations in the freestream velocity. Song et al. [6] analyzed membrane wings of varying aspect ratio, compliancy, and pre-strain values. Piquee et al. [7] investigated the capacity of a flexible membrane to adapt itself to the incoming flow by experimental and numerical investigations.

The overall scientific aim of the project is the extension and deepening of the understanding of concepts for adaptive aircraft. The concept for the elasto-flexible morphing wing at the Technical University of Munich (TUM) was originally developed by Beguin et al. [8]. The new morphing wing, which was recently developed, is part of a common research project with the Beihang University [9,10]. Both project partners investigate morphing wind tunnel (W/T) models in their facilities.

2 The Elasto-Flexible Morphing Wing

2.1 Description of the Concept

Wings consisting of thin, elastic skin membranes are used in the natural world by several species. Maybe the most common one is the bat, whose flight is investigated in detail by Breuer [11]. Particularly interesting about bats is that they have a high degree of articulation in the wing and the lifting surface is composed of a highly elastic skin membrane. Hence, they can change their wing planform and airfoil to perform various flight maneuvers. The bio inspired concept for the membrane wing of the present investigations is oriented to the "wing design" of the Pterosaurs [12]. The structure of their wings consists of a rigid skeleton, while an elastic skin serves as the aerodynamic wing surface.

Figure 1 shows a sketch illustrating the concept of the investigated configuration within this paper. The position of the leading-edge (LE) spar actively defines the shape of the wing by the folding angle Ω. The wing half span b, the aspect ratio AR and the quarter line sweep angle $\Phi_{(1/4)}$ are actively adjusted. Both wings can rotate individually, which enables asymmetric configurations. The elastic membrane serves as the actual aerodynamic surface and guarantees a smooth and continuous wing surface, while folding the wing in and out.

In addition, the elasticity of the membrane allows the airfoil to passively adapt to aerodynamic loads. This mechanism changes primarily the camber and the thickness of the airfoil, which is investigated within this paper. The membrane is mounted with a certain pre-stress, which is shown in Sect. 4.1. For a predecessor model, the influence of the pre-stress at different flow velocities and wing positions was already investigated in detail by Beguin and Breitsamter [13].

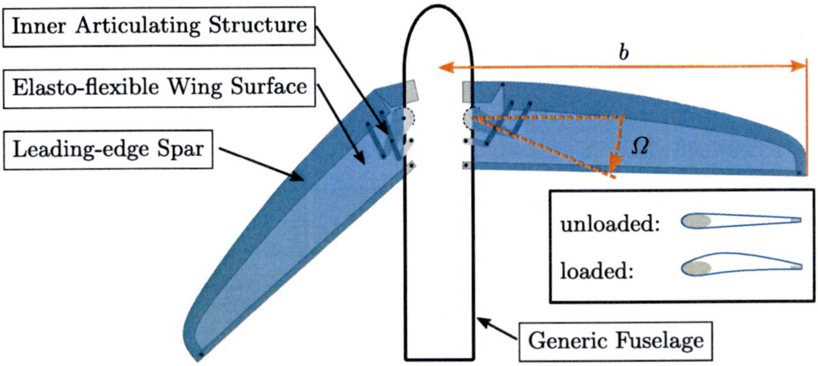

Fig. 1. Illustration of the morphing wing concept.

2.2 Model Design

The model consists of two individually foldable wings, which provide an aspect ratio of $5 \leq AR \leq 10$ and a quarter-chord sweep angle of $5° \leq \Phi_{(1/4)} \leq 45°$. An elasto-flexible membrane serves as the aerodynamic wing surface.

The actuation of the wing is done by two stepper motors and a folding mechanism inside the wing. The position of the wing is determined by the folding angle Ω, see Fig. 1. Figure 2 shows the right side of the model and the schematic cross section of the wing with its components, including the spanned membrane in blue. The main components of the wing structure are given by the elliptical LE spar and the rectangular TE spar. The LE has an elliptical planform and tapers in spanwise direction. The TE spar is straight, because its length must adapt to the current wing position in order to provide a rigid closed frame to stretch the membrane.

The material of the membrane is an elasto-flexible commercial polyurethane (PU) foil provided by the PU-manufacture Pharetra (Pharetra Textile Kunststoffanwendung GmbH & Co. KG). The membrane has a thickness of $t = 65$ µm and an isotropic stiffness with a Young's modulus of about $E_{mem} = 1.03$ MPa. The high elasticity produces a measurable surface deformation by the expected aerodynamic loads. The membrane is welded along the LE, pulled over the wing and fixed under the fuselage cover by a wire.

The deformation of the wing surface depends strongly on the pre-stress of the membrane. The pre-stress can be adjusted by shifting the attachment of the TE at the fuselage in axial direction. For the measurements within this paper, the wing root chord is fixed to $c_r = 0.14$ m.

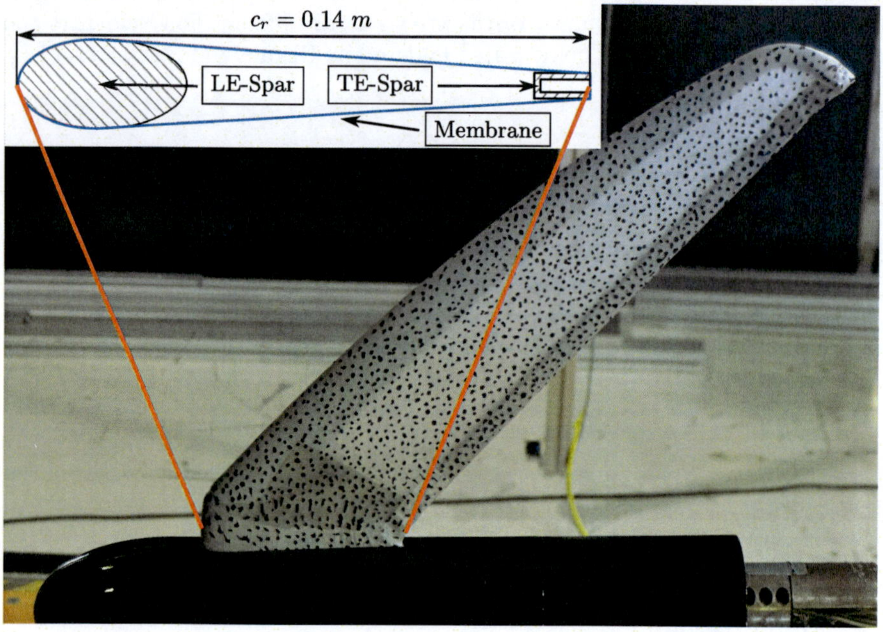

Fig. 2. The elastic membrane on the wing structure with dots for the deformation measurement system, the fuselage and the schematic airfoil cross section.

3 Experimental Setup

The experiments are conducted in the Göttingen-type, closed-loop, low-speed W/T of the Chair of Aerodynamics and Fluid Mechanics at TUM. The dimensions of the test section are 1.8 m × 2.4 m × 4.8 m (*height* × *width* × *length*) and the W/T has a turbulence level below 0.4% and a flow uniformity of less then 0.6% of the freestream velocity. The model is mounted in the W/T test section using a rear sting of a three-axis support. The experiments are performed at a Reynolds number, based on the wing root chord, of $Re_\infty = 264000$ and a Mach number of $Ma_\infty = 0.09$, which corresponds to a dynamic pressure of $q_\infty = 530$ Pa. The measurements comprise angles of attack of $0° \leq \alpha \leq 30°$. All measurements are conducted for three symmetric wing positions: "The High

Aspect ratio Configuration" (HAC, $\Omega = 23°$), the "Intermediate Configuration" (INC2, $\Omega = 43°$) and the "Highly Swept Configuration" (HSC, $\Omega = 63°$).

The deformation of the aerodynamically loaded membrane is measured by a digital image correlation system (DIC). Figure 3 shows the experimental setup on the bottom side of the wing. The same system is mounted on the top side, which results in four cameras to measure the deformation on both sides. The complementary metal oxide semiconductor (CMOS) cameras are placed outside the W/T test section with an angle between the respective optical axes of approximately $\beta = 40°$. The cameras have a maximum frame rate of 10 Hz and a resolution of 1944 × 2592 pixels, which, in conjunction with the imaging optics (Pentax 16 mm CCTV lenses), provide a spatial resolution of 0.34 mm per pixel. For this series of experiments, the deformation of the membrane is recorded with a single snapshot for an exposure time of 0.18 s.

A cylindrical calibration target consisting of a 3-D grid of markers is used to generate the calibration coefficients needed to reconstruct the coordinates in the object space via direct linear transformation (DLT). The randomly distributed black points on the grey membrane are tracked and the open source MATLAB toolbox MultiDIC is used to recover the coordinates of the wing surface in the 3-D object space from two 2-D images [14]. By using a test object, which is moved in z-direction in controlled manner, the uncertainty of the DIC system for out-of-plane displacements is determined to be less than $\Delta z = 0.5$ mm ($\Delta z/c_r = 0.0036$).

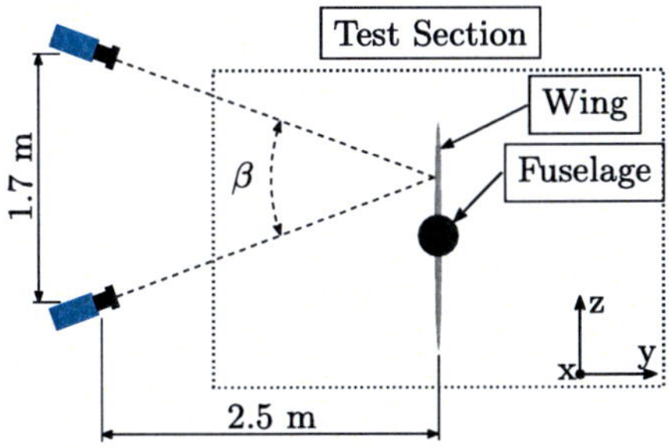

Fig. 3. Front view of the experimental setup for the deformation measurements on the bottom side of the wing.

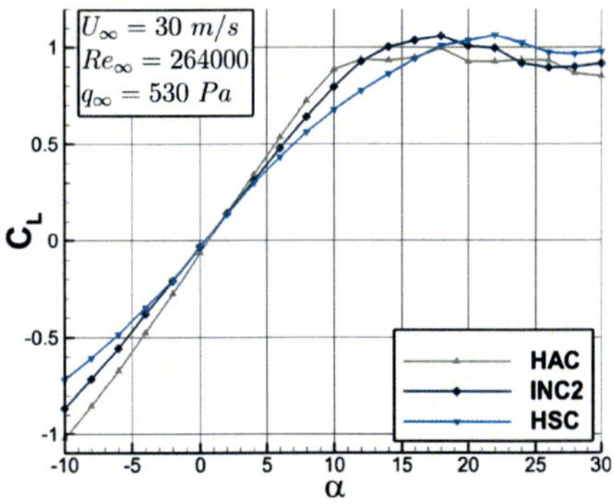

Fig. 4. The lift coefficient as a function of the angle of attack for different configurations.

4 Results

Previous acquired force and moment measurements form the starting point for the discussion of the results of the deformation measurements [15]. Figure 4 shows the comparison of the lift characteristics for the three configurations at a Reynolds number of $Re_\infty = 264000$. The HAC indicates a higher gradient in the linear region, while the HSC shows the highest $C_{L,max}$. Within this paper, the influence of the flexible membrane on the aerodynamic characteristics is investigated.

4.1 Pre-strain of the Membrane

As already stated by Beguin and Breitsamter [13], the pre-strain of the membrane heavily affects the extent of deformation. In detail, the pre-strain depends on the final shape after welding, the length of the wing root, the operation time and the folding angle Ω of the wing. While the contribution of the first three mechanisms to the pre-strain has yet to be measured, the additional strain caused by the folding can be determined by the DIC system. Figure 5 shows the 2nd principle strain ε_2 for all three configurations, which represents the deformation along the direction of the main load. The HSC serves as a reference case, because its shape is closest to the initial membrane cut.

The pre-strain increases significantly with a forward folding of the wing, which mainly affects the inner part between $y = 0.1$ m and $y = 0.3$ m. For the HAC, the membrane is partially stretched by more then 40%. Most of the deformation is caused by a stretching in spanwise direction. The strain is not evenly distributed along the wing span, because the basic shape of the membrane tapers in spanwise direction and the membrane does not slide perfectly on the structure below.

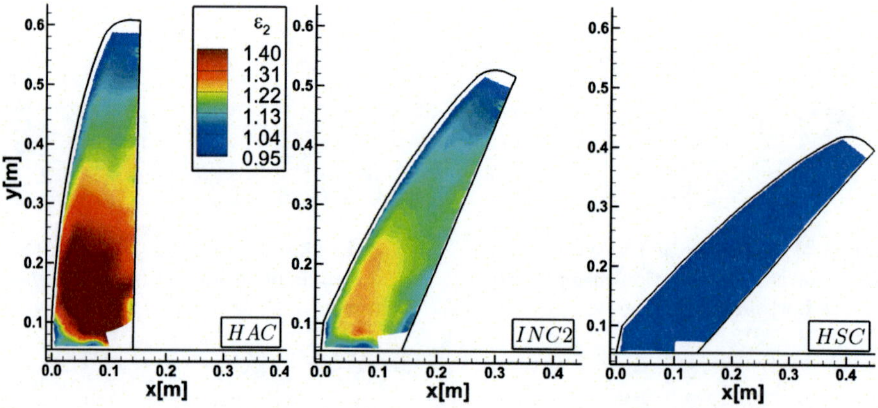

Fig. 5. Second principle strain ε_2 on the membrane with the HSC as a reference.

4.2 Deformation of the Wing Planform

At positive angles of attack, the accelerated airflow on the upper side induces a suction force deforming the membrane upwards. On the lower side, the increased pressure deforms the membrane upwards as well, but with a lower amplitude. The aerodynamics on the suction side is more critical. Therefore, Fig. 6 shows the deformation of the membrane on the upper side. The results are given for an inflow velocity of $U_\infty = 30\,\mathrm{m/s}$ and an angle of attack of $\alpha = 10\,°$. The deformation in z-direction is calculated by the difference between the case with and without inflow and it is made dimensionless by the wing root $c_r = 0.14\,\mathrm{m}$. Because of the inhomogeneous illumination of round edges, the areas with imprecise values at the LE, TE and at the wing tip are blanked. In addition, the inflow bends the whole wing structure upwards, which has to be corrected during the post processing by a rotation of the wing around an axis parallel to the roll axis.

Regarding all three configurations, the highest deformations occur close to the fuselage, which indicates a high aerodynamic load in this section and which is beneficial with respect to the wing root bending moment. Towards the wing tip, the local chord length decreases and so does the deformation of the membrane. The HSC shows the strongest deformation, because it has the longest local wing root and the lowest pre-stress of the membrane, see Fig. 5.

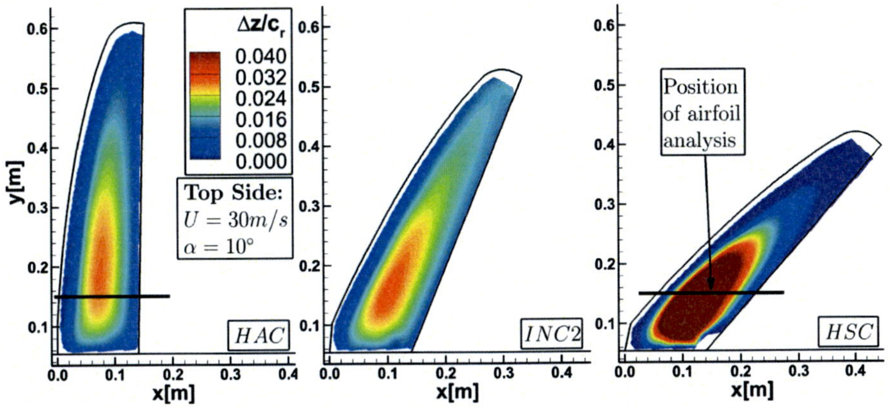

Fig. 6. Deformation of the membrane on the upper side of the wing at $U_\infty = 30\,\mathrm{m/s}$ and $\alpha = 10\,°$ for HAC, INC2 and HSC.

4.3 Deformation of the Wing Airfoil

The deformation measurements performed on the upper and on the lower wing surfaces are combined to analyze the airfoil shapes resulting from different angles of attack. Figure 7 and 8 show a cross section of the HSC and the HAC at $y/b = 0.25$, which is located in the region of the maximum deformation. The red line indicates the static airfoil, while the other ones represent the membrane at an inflow velocity of $U_\infty = 30\,\mathrm{m/s}$ and angles of attack of $\alpha = [5°, 15°, 25°]$ as well as $\alpha = [5°, 14°, 25°]$, respectively. The associated mean camber lines are marked by the dashed line style.

With an increasing angle of attack, the membrane on both configurations and wing sides deforms stronger upwards, which leads to an increased thickness and camber of the airfoil. The HAC reaches the maximum deformation at $\alpha = 14°$, while the HSC does not show a significant drop in the strength of the deformation. Table 1 contains the relative maximum thickness and camber, and their relative positions for the measured angles of attack. At the HSC, the thickness increases by 5% and the camber by around 6% from $\alpha = 0°$ to $\alpha = 25°$. The position of the maximum thickness shifts more downstream, while the position of the maximum camber stays around $x = 0.5 \cdot c_{r,loc}$.

The increasing camber with higher angles of attack results in a negative zero-lift angle of attack α_0, which increases the maximum lift coefficient $C_{l,max}$ [16]. The stronger deformation of the HSC, compared to the HAC, is one reason for the higher $C_{L,max}$ which can be seen in the lift characteristics shown in Fig. 4. Especially on the upper side, the higher thickness alleviates the contour gradient - seen by the airflow - which leads to a weaker pressure gradient at the suction peak.

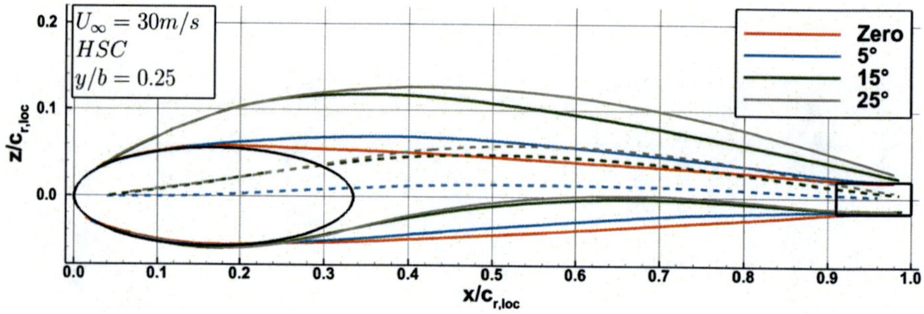

Fig. 7. Cross section of the HSC at $U_\infty = 30\,\mathrm{m/s}$ and $y/b = 0.25$ for different angles of attack (dashed: camber lines).

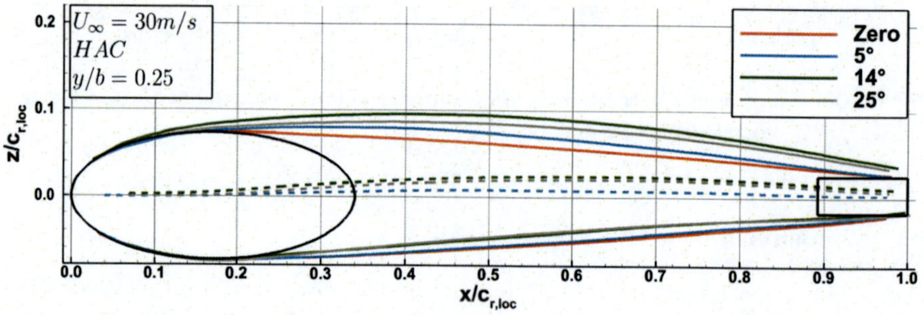

Fig. 8. Cross section of the HAC at $U_\infty = 30\,\mathrm{m/s}$ and $y/b = 0.25$ for different angles of attack (dashed: camber lines).

Table 1. Geometric parameters of the HSC and the HAC as a function of the angle of attack.

α [°]	max. Thickness ($t/c_{r,loc}$) [%]		Position of max. Thickness ($x_t/c_{r,loc}$) [%]		max. Camber ($c/c_{r,loc}$) [%]		Position of max. Camber ($x_c/c_{r,loc}$) [%]	
	HSC	HAC	HSC	HAC	HSC	HAC	HSC	HAC
0	11.5	14.9	18.6	17.1	0.0	0.0	0.0	0.0
5	12.1	15.1	25.2	23.7	1.4	0.8	54.6	55.8
15	16.8	15.5	25.5	21.3	4.9	2.3	49.0	62.0
25	16.2	15.5	27.1	20.6	5.9	2.0	53.7	63.2

5 Conclusion and Outlook

This paper presents experimental investigations of an elasto-flexible membrane wing at a Reynolds number of 264000, based on the wing root chord. The investigated concept not only enables the wing to fold over a wide range, but also allows the wing to adapt to changing aerodynamic loads. The present study focuses on the deformation of the elasto-flexible membrane. Therefore, the deformation on both sides is measured by a digital image correlation system. Advantages for the aerodynamic performance of the examined model are discussed by a comparison of the deformation at different wing positions and angles of attack. The measurements show an increasing deformation with higher angles of attack. Especially, the area close to the fuselage is affected. The airfoil shapes in this area show an increasing camber and thickness. The highly swept configuration deforms stronger than the high aspect ratio configuration due to the lower pre-strain and the longer local wing root, which is one reason for the smoother and delayed stall region of the swept one.

Future work will target at the investigation of the membrane wing under unsteady inflow conditions. Therefore, the cameras should be replaced by high speed cameras to increase the sampling frequency. In addition, measurements behind a gust generator and numerical simulations of a rigid and a flexible configuration will be performed.

Acknowledgement. The project is funded by the Deutsche Forschungsgemeinschaft (DFG, German Research Foundation) grant number BR 1511/12-1.

References

1. Weisshaar, T.: Morphing aircraft systems: historical perspectives and future challenges. J. Aircr. **50**(2), 337–353 (2013). https://doi.org/10.2514/1.C031456
2. Rodriguez, A.: Morphing aircraft technology survey. In: 45th AIAA Aerospace Sciences Meeting and Exhibit, Reno, vol. 1258, pp. 1–9 (2007). https://doi.org/10.2514/6.2007-1258

3. Moorhouse, D., Sanders, B., Spakovsky, M., Butt, J.: Benefits and design challenges of adaptive structures for morphing aircraft. Aeronaut. J. **110**(1105), 157–162 (2006). https://doi.org/10.1017/S0001924000001135
4. Vasista, S., Tong, L., Wong, K.C.: Realization of morphing wings: a multidisciplinary challenge. J. Aircr. **49**(1), 11–28 (2012). https://doi.org/10.2514/1.C031060
5. Valasek, J.: Morphing Aerospace Vehicles and Structures, pp. 11–53. Wiley, Chichester (2012). https://doi.org/10.1002/9781119964032
6. Song, A., et al.: Aeromechanics of membrane wings with implications for animal flight. AIAA J. **46**(8), 2096–2106 (2008). https://doi.org/10.2514/1.36694
7. Piquee, J., Breitsamter, C.: Numerical and experimental investigations of an elasto-flexible membrane wing at a Reynolds number of 280,000. Aerospace **4**(39) (2017). https://doi.org/10.3390/aerospace4030039
8. Béguin, B.: Development and analysis of an elasto-flexible morphing wing. Ph.D. thesis, Technical University of Munich (2014)
9. He, X., Guo, Q., Wang, J.: Extended flexible trailing-edge on the flow structures of an airfoil at high angle of attack. Exp. Fluids **60**(8) (2019). https://doi.org/10.1007/s00348-019-2767-5
10. Guo, Q., He, X., Wang, Z., Wang, J.: Effects of wing flexibility on the aerodynamic performance of an aircraft model. Chin. J. Aeronaut. (in press)
11. Breuer, K., et al.: Biologically-inspired flight for micro air vehicles. Final report, Brown University, School of Engineering, Providence (2012)
12. Wilkinson, M.T.: Sailing the skies: the improbable aeronautical success of the pterosaurs. J. Exp. Biol. **210**, 1663–1671 (2007). https://doi.org/10.1242/jeb.000307
13. Béguin, B., Breitsamter, C.: Effects of membrane pre-stress on the aerodynamic characteristics of an elasto-flexible morphing wing. Aerosp. Sci. Technol. **37**(2014), 138–150 (2014). https://doi.org/10.1016/j.ast.2014.05.00
14. Solav, D., Moerman, K., Jaeger, A., Genovese, K., Hugh, H.: MultiDIC: an open-source toolbox for multi-view 3D digital image correlation. IEEE Access **6**(2018), 30520–30535 (2018). https://doi.org/10.1109/ACCESS.2018.2843725
15. Pflueger, J., Breitsamter, C.: Experimental investigations of a full model with adaptive elasto-flexible membrane wings. Chin. J. Aeronaut. (2020). https://doi.org/10.1016/j.cja.2020.03.037
16. Anderson, J.D.: Fundamentals of Aerodynamics. McGraw-Hill Education, New York (2010)

Numerical Investigation of Flexibility Effects on the CFD-CSM Coupling of Deflected Delta Wing Control Surfaces

Christopher Reinbold[1(✉)], Kaare Sørensen[2], and Christian Breitsamter[1]

[1] Chair of Aerodynamics and Fluid Mechanics, Technical University of Munich, Boltzmannstr. 15, 85748 Garching, Germany
christopher.reinbold@tum.de

[2] Airbus Defence and Space, Aerodynamic Department, Methods and Tools, Rechlinger Str., 85077 Manching, Germany

Abstract. Numerical aeroelastic investigations are conducted for the Model53 delta wing with a deployed slat and two trailing-edge control surfaces at transonic speed and high dynamic pressure. The numerical method is based on coupled high-fidelity CFD-CSM simulations, which is implemented in the multi-disciplinary simulation environment SimServer. The DLR Tau Code is utilized to solve the Reynolds-Averaged Navier-Stokes (RANS) equations and a modal solver is employed to calculate the structural displacements. Control surface regions are modeled with a Chimera approach for hybrid grids. The main focus of the analysis is set on the appropriate handling of control surface deflections for coupled CFD-CSM simulations. Two cases are investigated: In the first case, the control surfaces are deflected for both, the aerodynamic and the structural grid. In the second case, only the control surfaces of the aerodynamic grid are deflected and the aerodynamic forces are transferred to the undeflected structural grid. Both cases are compared in terms of the resulting structural deformation and flow field. The differences between both approaches are minor.

Keywords: Aeroelasticity · Delta wing · Control surfaces · Chimera

1 Introduction

The impact of control surface deflections on the aerodynamic characteristics of an aircraft is of general high importance. Especially, the design process enforces such data in detail. For instance, the generation of aerodynamic data sets or the execution of maneuver simulations through virtual flights require the handling of control surfaces. The movement of control surfaces in high-fidelity Computational Fluid Dynamics (CFD) simulations based on the Reynolds-Averaged Navier-Stokes equations can be modeled with the Chimera technique [1]. The Chimera, or overset, approach offers the possibility to model the deployment of

© The Author(s), under exclusive license to Springer Nature Switzerland AG 2021
A. Dillmann et al. (Eds.): STAB/DGLR Symposium 2020, NNFM 151, pp. 547–556, 2021.
https://doi.org/10.1007/978-3-030-79561-0_52

control surfaces realistically and efficiently, since control surface gaps are captured and a re-meshing of the computational grid for different deflections is not necessary. In order to further increase the accuracy of CFD computations with control surface deflections, structural mass and stiffness needs to be considered due to their effect on control surface efficiency and the aerodynamic characteristics of the aircraft in general.

Over the last years, the modeling of control surfaces with CFD solvers has been subject to several projects, such as ComFlite [2], Digital-X [3], VicToria and Mephisto [4]. Whereas these projects focus on extending the capabilities of the DLR flow solver TAU [5] to perform simulations including control surface deflections with overset or sliding mesh approaches, this work targets to extend the existing implementation of the simulation environment SimServer [6] in order to combine its Chimera [7] and aeroelastic capabilities. Initially, the numerical methods which are involved in the aeroelastic simulations are described. Subsequently, the Model53 wing geometry as well as the aerodynamic and structural model is presented. Finally, the results obtained with the flexible Chimera method are compared for the case where the control surfaces are only deflected in the aerodynamic model and for the case where the control surfaces are deflected in both, the structural and aerodynamic model.

2 Numerical Methods

The numerical methods are all implemented into the SimServer multidisciplinary simulation environment [6]. Emphasis is put on the integration of the Chimera method into the partitioned approach of SimServer for solving aeroelastic problems.

2.1 Computational Fluid Dynamics Solver

The DLR-Tau flow code is used to solve the compressible, three-dimensional Reynolds-Averaged Navier Stokes (RANS) equations [5]. It uses a vertex-centered dual cell method for hybrid, unstructured grids. Convergence acceleration is achieved by utilizing multigrid technique using point-implicit relaxation. A finite-volume formulation with second order upwind scheme is applied for spatial discretization of the steady simulations. The Spalart-Allmaras one-equation turbulence model [8] in the Edward's modified version [9] is employed with enabled rotation correction.

2.2 Fluid-Structure Coupling

For the computation of coupled aeroelastic simulations with movable control surfaces, an important requirement to the fluid-structure coupling is the compatibility with the Chimera implementation. A newly developed interpolation method is based on the point-element-relationship between the CFD- and the

Computational Structural Mechanics (CSM)-surface grid and has been validated with the BACT [10] and HIRENASD [11] test case.

In terms of efficiency and pre-processing effort, it is favorable to only consider the deflection of control surfaces in the CFD model, while keeping the control surfaces in the FE model undeflected. This way, only one FE model is required for various control surface deflections. However, the accuracy of this approach needs to be investigated, which is subject to Sect. 4. In case of considering control surface deflections solely in the CFD model, the forces of each CFD node are evaluated on the deflected flaps of the CFD grid and then transferred to the undeflected FE grid by maintaining the direction of the force vectors, illustrated in Fig. 1. Similarly, displacements on the undeflected FE model are interpolated to the deflected CFD grid by maintaining the direction of the displacement vectors.

Another important aspect to the fluid-structure coupling in combination with moveable control surfaces is the adjustment of the hinge line axis according to the deformations of the wing. To take this into account, the coordinates of the two points defining the hinge axis of each flap are displaced by employing the same algorithm used for the transfer of displacements from the CSM to the CFD grid.

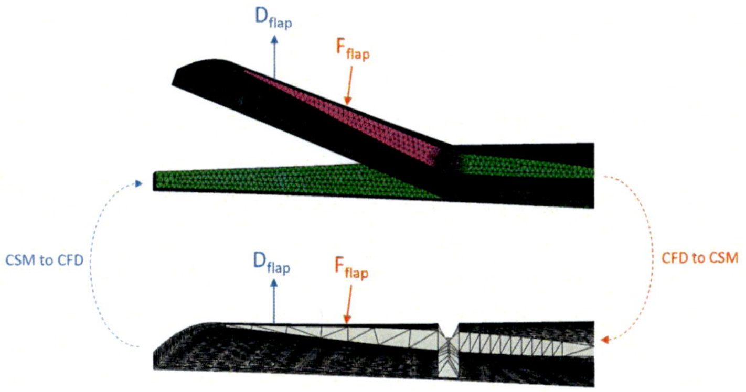

Fig. 1. Illustration of the transfer of an exemplary resulting flap force F_{flap} and displacement D_{flap} between the deflected CFD and undeflected CSM grid.

2.3 CFD Volume Grid Deformation

The surface displacements are propagated to the CFD volume grid with a parallel mesh deformation method based on an advancing front algorithm. To support larger local deformations, the repair of badly distorted elements has been implemented according to [12]. The repair procedure marks regions in the grid containing collapsed elements surrounded by valid cells only. Subsequently, the collapsed cells are rebuilt by employing a volume spline interpolation approach. More information on the mesh deformation method can be found in [6] and [12].

2.4 Chimera Implementation

The Chimera technique is an efficient way to compute multiple body problems, such as control surface deflections, for hybrid grids [13]. In general, it assembles two or more grids into one computational domain through the application of interpolation boundary conditions. To apply the Chimera method in SimServer, one or more submeshes need to be contained within a background mesh. In case of a wing configuration with control surfaces, the wing mesh corresponds to the background mesh whereas each control surface mesh corresponds to a submesh. Geometries belonging to the background mesh or submesh are allowed to intersect. A detailed explanation of the interpolation method and the hole-cutting procedure can be found in [7].

2.5 Structural Solver

For the static aeroelastic simulations in this work, the structural deformations are computed with a linear modal solver. The linear approach is justifiable for the investigated configuration, since the overall elastic deformations remain small. The structure of the wing is designed to have a wing tip deformation of approx. 4% of the semi-span at load corresponding to a load-factor of 9-g.

3 Model53 Delta Wing

The Model53 is a cropped delta wing with a leading-edge sweep angle of $\phi = 53°$ [14]. The root chord length measures $c_r = 8.4$ m and the semi-span is $s = 5.1$ m in its full-scale configuration. The wing is twisted linearly up to $-4°$ at the wing tip and the leading edge slat is deployed to 20°. An outer and inner trailing-edge flap can each be deflected from $-30°$ to 30°. Both flap's leading-edges are constantly positioned at $x/c_r = 0.88$ in chord-wise direction over the span, while the trailing-edge has a sweep angle of 3°. Each flap has a span of $0.49\,\mathrm{s} = 2.5$ m. The pitching moment reference point is located at $x/c_r = 0.5$. An overview of the geometric dimensions of the Model53 is depicted in Fig. 2(a).

3.1 Computational Fluid Dynamics Model

The CFD grid of the Model53 is shown in Fig. 2(b). The inner and outer flap are treated as independent Chimera blocks which intersect with the main wing. A hybrid-unstructured grid is created with the meshing tool CENTAUR, where prismatic cells are used in boundary layer regions and unstructured cells in the remaining domain. A boundary layer mesh resolution with $y^+ < 1$ is used. The computational grid is refined around the wing edges and especially in the vicinity of the leading edge as well as on the upper side of the wing, where the formation of leading-edge vortices is expected, see Fig. 2(b). Compared to the wing edges, intersecting regions between the flaps and the main wing are further refined, to keep the Chimera interpolation error small. The grid used for the numerical simulations has approximately 6 million nodes. A sufficiently large hemispherical farfield boundary with a diameter of 100 reference chord lengths is chosen.

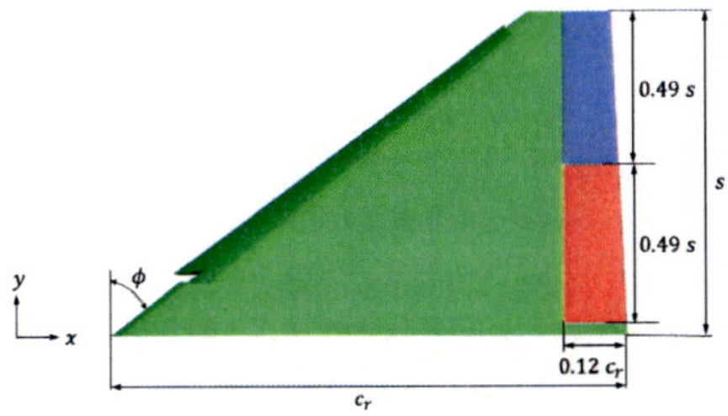

(a) Dimensions, top view.

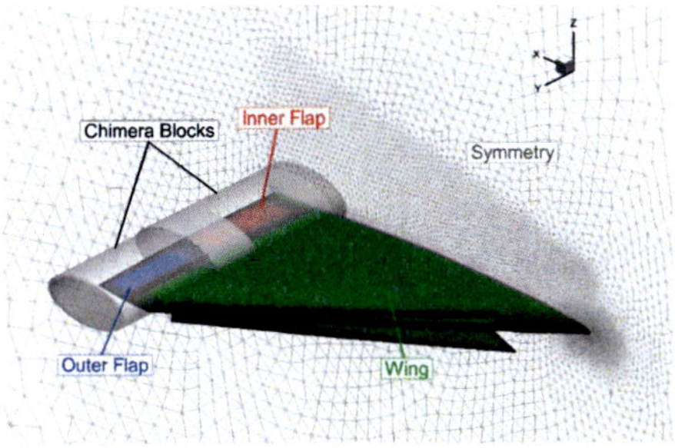

(b) CFD grid, 6 million nodes.

Fig. 2. Dimensions and CFD grid of the Model53.

3.2 Finite Element Model

The FE model of the Model53, Fig. 3, consists of 7530 grid points, 16252 triangle elements and 24 bar elements. For the modeling of the ribs, spars (shown in turquoise in Fig. 3(a)) and skin, shell elements are used which are equipped with stiffening elements to keep buckling fields sufficiently small and to reduce local eigenmodes. The inner and outer flap are structurally modeled and attached to the wing with a hinge concept introduced in [15], which is shown in Fig. 3(b) for the Model53. Each attachment consists of two triangle elements, four bar elements and a double node connecting the bar and triangle elements of the wing and flap side with a rigid body element. Figure 3(c) depicts the FE Model of the Model53 with upward flap deflections $\delta_{FEM} = -30°$, whereas Fig. 3(d) illustrates downward flap deflections $\delta_{FEM} = 30°$. The stiffness of the hinge attachment can be controlled with the bar's properties. This hinge concept enables a realistic behavior and physically meaningful analysis of control surface loads [4]. For the coupling, the 25 lowest mode shapes are used for the modal basis. Modes containing the individual movements of the inner and outer flap with respect to the main wing are contained in this set of 25 modes. The wing is clamped at the root and inertial forces are not considered in this analysis. For all structural components aluminum has been chosen as material. A mesh independence study for the structural grid has been performed.

4 Results

Aeroelastic simulations are performed at transonic speed with a Mach number of $Ma = 0.8$ at sea level altitude, resulting in a dynamic pressure and Reynolds number of $q = 45\ kPa$, $Re = 15.1 \cdot 10^7$ with the root chord as reference length. For each angle of attack $\alpha = [10°, 15°, 20°, 25°]$, two flap deflection angles at the upper and lower deflection limit are considered $\delta = [-30°, 30°]$. This section targets to analyze how the simplification of considering control surface deflections only in the CFD model effects the resulting structural displacements as well as the flow field of the Model53 delta wing.

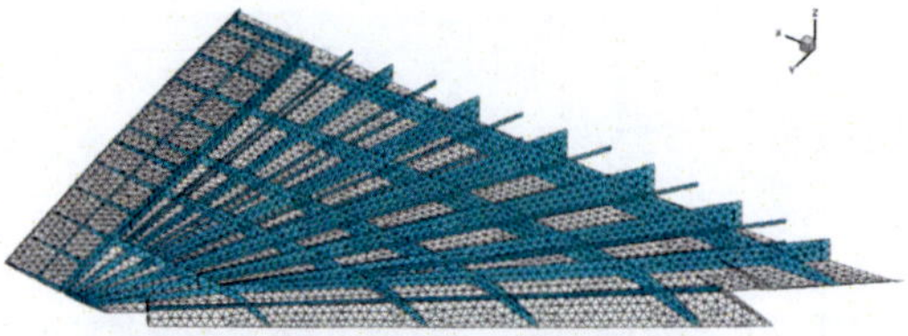

(a) FE model, $\delta_{FEM} = 0°$.

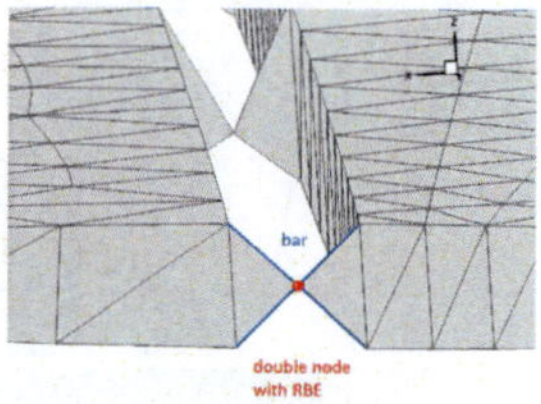

(b) Hinge concept.

(c) FE model, $\delta_{FEM} = -30°$.

(d) FE model, $\delta_{FEM} = 30°$.

Fig. 3. FE model of the Model53 wing for three flap deflection angles $\delta_{FEM} = -30°$, $\delta_{FEM} = 0°$ and $\delta_{FEM} = 30°$.

In Fig. 4, the lift and pitching moment coefficient is plotted over the angle of attack for two flap deflection angles $\delta_{CFD} = -30°$ and $\delta_{CFD} = 30°$, which can each be subdivided in a case where the structural flap is deflected to $\delta_{FEM} = -30°$ or $\delta_{FEM} = 30°$ respectively and in a case where the structural flap is undeflected $\delta_{FEM} = 0°$. For all angles of attack and each flap deflection angle $\delta_{CFD} = -30°$ and $\delta_{CFD} = 30°$, the lift and pitching moment coefficient for the deflected and undeflected structural flaps are in very good agreement, see Fig. 4(a). The deviation for the lift and pitching moment $\Delta C_L = C_{L,\delta_{FEM}=0°} - C_{L,\delta_{FEM}\neq 0°}$ and $\Delta C_{My} = C_{My,\delta_{FEM}=0°} - C_{My,\delta_{FEM}\neq 0°}$ is visualized in Fig. 4(b). It can be seen that the overall deviations are very small, with a mean deviation of $|\Delta C_L| = 0.0019$ and $|\Delta C_{My}| = 0.0012$ in absolute values or $|\Delta C_L/C_{L,\delta_{FEM}\neq 0°}| = 0.4\%$ and $|\Delta C_{My}/C_{My,\delta_{FEM}\neq 0°}| = 0.74\%$ relative to the deflected structural flap case. The largest deviation occurs at $\alpha = 10°$ and $\delta_{CFD} = 30°$, where the highest flap loading relative to the overall wing loading is present with $|\Delta C_L| = 0.0089$, $|\Delta C_L/C_{L,\delta_{FEM}\neq 0°}| = 1\%$ and $|\Delta C_{My}| = 0.00207$, $|\Delta C_{My}/C_{My,\delta_{FEM}\neq 0°}| = 0.98\%$. However, these differences are still considered to be small given the improvements in pre-processing and computational effort. In summary, no considerable reduction in accuracy in terms of the resulting forces and moments can be denoted if the control surfaces of the structural grid are treated undeflected with the coupling strategy introduced in Sect. 2.2.

Figure 5 illustrates the surface pressure coefficient distribution C_P of the Model53 at $\alpha = 20°$ and two flap deflection angles $\delta_{CFD} = -30°$ and $\delta_{CFD} = 30°$. The flow field is characterized by the presence of leading-edge vortices and compressibility effects. Near the apex, the flow separates and forms the apex vortex, which can be identified through the suction footprint in the pressure distribution. In addition, a weaker slat vortex is observable in the pressure con-

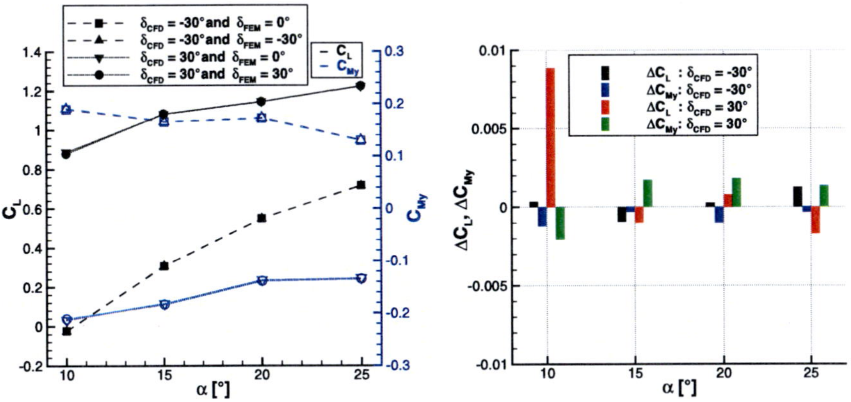

(a) C_L and C_{My}. (b) $\varDelta C_L$ and $\varDelta C_{My}$.

Fig. 4. Lift coefficient C_L and pitching moment coefficient C_{My} as a function of the angle of attack for two flap deflection angles at $Ma = 0.8$, $Re = 15.1 \cdot 10^7$.

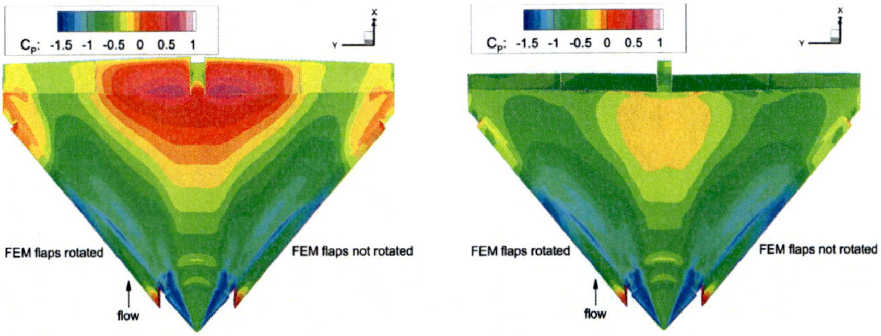

(a) $\delta_{CFD} = -30°$; left: $\delta_{FEM} = -30°$; right: $\delta_{FEM} = 0°$. (b) $\delta_{CFD} = 30°$; left: $\delta_{FEM} = 30°$; right: $\delta_{FEM} = 0°$.

Fig. 5. Surface pressure distribution C_P of the Model53 at $Ma = 0.8$, $Re = 15.1 \cdot 10^7$, $\alpha = 20°$ and $\delta_{CFD} = [-30°, 30°]$.

tours with a lower suction footprint, due to the reduction in local angle of attack. In the left view, the flaps are rotated in both the aerodynamic and structural model, whereas in the mirrored right view, the rotation is only considered in the aerodynamic model. Similar to the lift and pitching moment characteristics, no significant differences are observable between both cases from the pressure coefficient contours.

In line with the observations based on the surface pressure distributions, the z-displacement contours are almost identical for both structural flap treatments: the isolines as well as the maximum displacement values coincide very well, Fig. 6(a) and (b). Figure 7(a) to (d) highlight the differences in the resulting flap positions for both approaches $\delta_{FEM} = 0°$ and $\delta_{FEM} \neq 0°$ com-

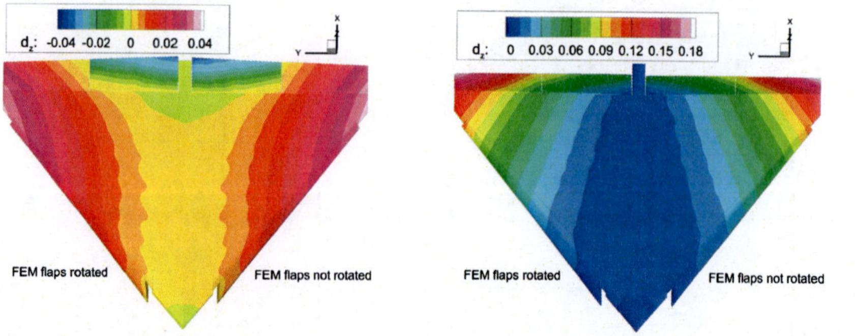

(a) displacement d_z $[m]$ at $\delta_{CFD} = -30°$; left: $\delta_{FEM} = -30°$; right: $\delta_{FEM} = 0°$.

(b) displacement d_z $[m]$ at $\delta_{CFD} = 30°$; left: $\delta_{FEM} = 30°$; right: $\delta_{FEM} = 0°$.

Fig. 6. Displacements of the Model53 at $Ma = 0.8$, $Re = 15.1 \cdot 10^7$, $\alpha = 20°$ and $\delta_{CFD} = [-30°, 30°]$.

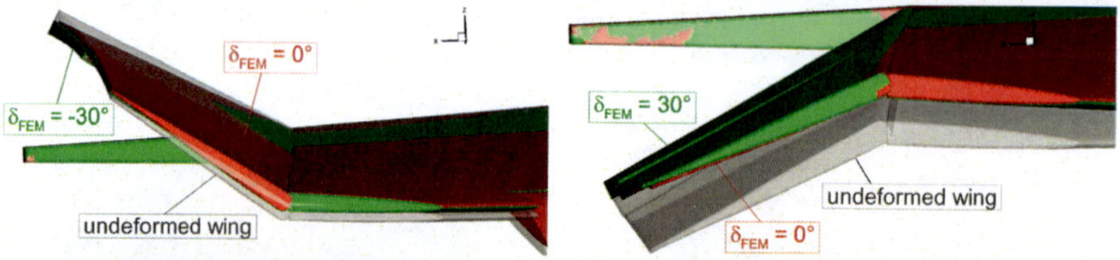

(a) $\delta_{CFD} = -30°$.

(b) $\delta_{CFD} = 30°$.

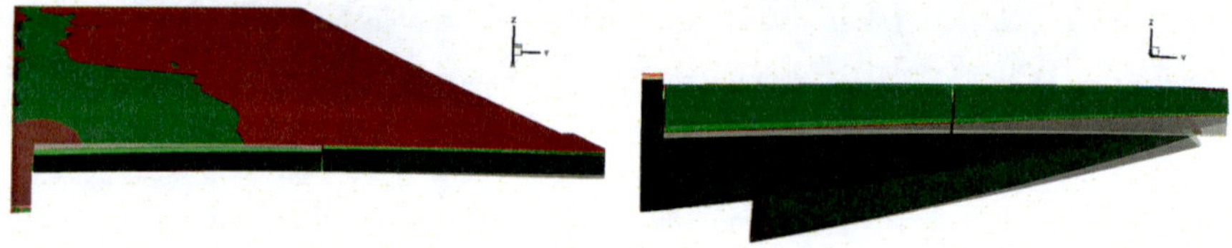

(c) $\delta_{CFD} = -30°$.

(d) $\delta_{CFD} = 30°$.

Fig. 7. Flap deformation of the Model53 at $Ma = 0.8$, $Re = 15.1 \cdot 10^7$, $\alpha = 20°$ and $\delta_{CFD} = [-30°, 30°]$.

pared to the undeformed wing. The main difference between both approaches can be attributed to the rotation of the displacement vectors in case of the structurally deflected flap, that is not captured if the structural flap is considered undeflected. The effective deflection angles for both approaches are $\delta_{eff,FEM=0°} = -28.6°$ and $\delta_{eff,FEM\neq 0°} = -28.1°$ for an initial upward deflection of $-30°$ and $\delta_{eff,FEM=0°} = 26.6°$ and $\delta_{eff,FEM\neq 0°} = 26.1°$ for an initial downward deflection of 30°.

5 Conclusion

Aeroelastic computations involving the Model53 delta wing configuration are performed at transonic freestream condition and high dynamic pressure. The simulations are based on the simulation environment SimServer, which has been extended with the capability to conduct aeroelastic simulations in combination with deformable control surfaces. The Model53 features a deployed slat and an inboard and outboard trailing-edge flap, which can be deflected from $-30°$ to $30°$ degrees using a Chimera mesh in the CFD model. The analysis focused on the appropriate handling of control surfaces deflections for coupled CFD-CSM simulations. A coupling strategy is introduced, where the forces and displacements are transferred between deflected aerodynamic and undeflected structural control surfaces. This approach enables a lower pre-processing and computational effort in aeroelastic simulations with control surface deflections. Concerning the accuracy of this coupling approach, the analysis of the lift and pitching moment coefficient, the surface pressure distribution and the structural deformation showed that the results are almost identical to those, where control surface deflections have also been considered in the structural FE model. The deviation of the lift and pitching moment coefficient of the introduced coupling method is smaller than 1% compared to the structurally deflected flap cases. This deviation can be attributed to a slight rotation of the displacement vectors of the flaps in case of structurally deflected flaps, which is not captured if the structural flap is considered undeflected. However, the forces from the deflected flaps are transferred correctly to the main wing and only small deviations can be observed with regard to the flap displacements. Thus, the presented approach is recommended, if efficient aeroelastic simulations with control surface deflections are targeted and is also expected to work in combination with more flexible wing structures.

Acknowledgments. The fruitful collaboration with Airbus Defence and Space and its partial financial research support is gratefully acknowledged. The authors also thank the Gauss Centre for Supercomputing e.V. (www.gauss-centre.eu) for supporting this project by providing computing time on the GCS Supercomputer SuperMUC-NG at Leibniz Supercomputing Centre (www.lrz.de).

References

1. Capsada, L.A., Heinrich, R.: Development of the DLR tau code for modelling of control surfaces. In Deutscher Luft- und Raumfahrtkongress. Deutsche Gesellschaft für Luft- und Raumfahrt - Lilienthal-Oberth e.V. (2019)
2. Heinrich, R., Reimer, L., Michler, A., Ernst, B., Ritter, M., Neumann, J.: DLR pilot applications in ComFlite. In: Computational Flight Testing, pp. 235–247. Springer, Berlin (2013)
3. Kroll, N., et al.: DLR project Digital-X: towards virtual aircraft design and flight testing based on high-fidelity methods. CEAS Aeronaut. J. **7**(1), 3–27 (2016)
4. Bramsiepe, K., Voß, A., Klimmek, T.: Design and sizing of an aeroelastic composite model for a flying wing configuration with maneuver, gust, and landing loads. CEAS Aeronaut. J. **11**(3), 677–691 (2020)

5. Kroll, N., Langer, S., Schwöppe, A.: The DLR flow solver TAU - status and recent algorithmic developments. In: 52th AIAA Aerospace Sciences Meeting, Maryland, USA (2014)
6. Tremel, U., Deister, F., Sørensen, K., Rieger, H., Weatherill, N.P.: The SimServer–a parallel multidisciplinary simulation environment. Konferenzband DGLR Jahrestagung, Beitrag DGLR-2003-088, Munich, Germany (2003)
7. Sørensen, K., Tremel, U., Rieger, H., Hitzel, S.: Simulation of manoeuvring aircraft with the unstructured chimera approach. In: 45th AIAA Aerospace Sciences Meeting and Exhibit, Reno, Nevada (2007)
8. Spalart, P., Allmaras, S.: A one-equation turbulence model for aerodynamic flows. In: 30th Aerospace Sciences Meeting and Exhibit (1992)
9. Edwards, J., Chandra, S.: Comparison of eddy viscosity-transport turbulence models for three-dimensional, shock-separated FlowFields. In: Fluid Dynamics Conference (1994)
10. Scott, R.C., Hoadley, S.T., Wieseman, C.D., Durham, M.H.: Benchmark active controls technology model aerodynamic data. J. Guidance Control Dyn. **23**(5), 914–921 (2000)
11. Ballmann, J., et al.: Aero-structural wind tunnel experiments with elastic wing models at high Reynolds numbers (HIRENASD-ASDMAD). In: 49th AIAA Aerospace Sciences Meeting Including the New Horizons Forum and Aerospace Exposition, p. 882 (2011)
12. Gerhold, T., Neumann, J.: The parallel mesh deformation of the DLR TAU Code. In: Notes on Numerical Fluid Mechanics and Multidisciplinary Design, vol. 96 (2007)
13. Burggraf, U., Kuntz, M., Schöning, B.: Implementation of the Chimera method in the unstructured DLR finite volume code tau. IN: New Results in Numerical and Experimental Fluid Mechanics II: Contributions to the 11th AG STAB/DGLR Symposium Berlin, pp. 93–100. Vieweg+Teubner Verlag, Wiesbaden (1999)
14. Moioli, M., Reinbold, C., Sørensen, K., Breitsamter, C.: Investigation of additively manufactured wind tunnel models with integrated pressure taps for vortex flow analysis. Aerospace **6**(10), 113 (2019)
15. Maierl, R., Petersson, O., Daoud, F., Bletzinger, K.U.: Automatically generated aeroelastic analysis models including physics based control surface representation. In: 4th Aircraft Structural Design Conference (2014)

Fluid and Thermodynamics

Validation of a Dynamic Simulation Approach for Transient Performance Using the Example of a Turbojet Engine

Jan Göing(✉), Johannes Hogrefe, Sebastian Lück, and Jens Friedrichs

Institute of Jet Propulsion and Turbomachinery, Technische Universität Braunschweig, 38108 Braunschweig, Lower Saxony, Germany
j.goeing@ifas.tu-braunschweig.de
http://www.ifas.tu-bs.de/

Abstract. This investigation aims to provide a validation framework for the dynamic simulation tool ASTOR (AircraftEngine Simulation for Transient Operation Research) for gas turbines. ASTOR is based on a system of ordinary equations which is able to simulate transient performance and dynamic response. The numerical setup of ASTOR is compared to state of the art transient performance simulation techniques for the validation. It is shown that ASTOR is able to calculate the performance of the single-spool turbojet engine under investigation to a level of detail that is in no way inferior than other approaches. Furthermore, an experimental validation is carried out using the in-house Jet-Cat P200SX engine. Measurements of steady state operating points and transient maneuvers show a high degree of consistency with performance simulations conducted in ASTOR.

Keywords: Gas turbine · Transient performance · Dynamic simulation

1 Introduction

Transient maneuvers are an essential part of design process of single components as well as the overall engine in the field of aircraft gas turbines. Especially in an emergency (e.g. go-around, engine failure), a precise prediction and an accurate simulation of the transient performance is required to define the operating range [1]. Due to engine deterioration, these analyses become even more complex and important as stability margins decrease and transient loads increase [2].

In several studies, miscellaneous methods for transient performance simulations are developed [3,4] and [5]. Basically, two iterative methods are used most often, namely, the engine matching (EM) [6,7] technique and the intercomponent volume (ICV) method [8]. Furthermore, a hybrid procedure was developed, which combines the advantages of both methods [9]. In contrast to the EM, the ICV and the hybrid method considers the impact of air volume fillings (volume dynamics (VD)). In general, the volume dynamics have a second order impact on the

© The Author(s), under exclusive license to Springer Nature Switzerland AG 2021
A. Dillmann et al. (Eds.): STAB/DGLR Symposium 2020, NNFM 151, pp. 559–568, 2021.
https://doi.org/10.1007/978-3-030-79561-0_53

transient performance in contrast to rotor inertia, tip clearance, heat transfer and burner time delay [10]. However, for a precise observation of transient performance and for the simulation of further dynamic instabilities e.g. rotational stall or surge [11], the consideration of volume dynamics is indispensable. Unfortunately, the ICV and hybrid method (EM-VD) technique only approximate the influence of VD, due to missing interactions and dynamics of the system of differential equations. Subsequently, unstable effects like rotational stall cannot be simulated.

Therefore, the IFAS (Institute of Jet Propulsion and Turbo Machinery) inhouse performance simulation tool ASTOR (AircraftEngine Simulation for Transient Operation Research) is developed. In ASTOR, the complete system of ordinary differential equations (ODE) is solved to simulate the transient performance in a higher accuracy. In addition to the advantage of more accurate calculation of transient maneuvers and dynamic phenomena can also be calculated, such as rotating stall and surge [12].

In this investigation, ASTOR is validated numerically and experimentally. Therefore, a small single shaft turbojet engine is used. Therein, effects of volumes dynamics are small so that all algorithms are able to compute the transient manoeuvres and an exact comparison is possible. At the beginning of this paper, the jet engine and the test equipment is described in detail. Second, the different algorithms for transient performance calculation are shown and compared. Next, the results of the steady state model are benchmarked with the measurement data for different operating points. Finally, the transient results of multiple numerical models are compared among each other and with test data.

2 Materials and Methods

The experimental setup of the turbojet engine and the numerical algorithm are explained in the following chapter.

2.1 Experimental Setup

In this investigation the experimental data are obtained from the JetCat P200SX, which is a single spool turbojet with a centrifugal compressor and an axial turbine. It has a maximum rotational speed of 112,000 RPM and reaches a maximum thrust of 230 N [13]. Further technical specifications for the design point are listed in Table 1.

The engine is equipped with total pressure (p_t) probes and thermocouples (T_t) at the compressor exit (station 3), turbine inlet (station 4) and behind the exhaust nozzle (station 8). Furthermore the JetCat is attached to a sliding carriage, which is hold back by a force transducer to measure the thrust (F_N). In addition, an optical displacement transducer (ODT) is directed on the shaft, to measure a periodical signal, from which the rotational speed (N) can be computed by a discreet short-time Fourier transformation (DSTFT). General performance parameters like the fuel flow, are read out from the engine's control unit (ECU).

Table 1. Technical specifications of the JetCat P200SX [13]

Max. RPM	112,000
Max. thrust in N	230
Pressure ratio	4
Mass flow in kg/s	0.45
Fuel flow @maxRpm in kg/s	0.0096

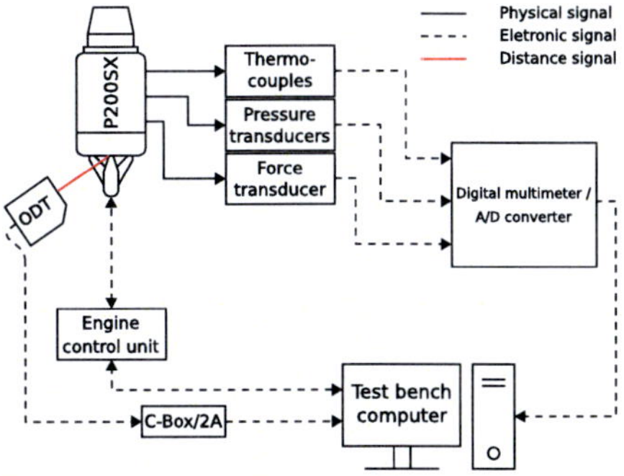

Fig. 1. Sketch of the experimental setup

A schematic sketch of the experimental setup is shown in Fig. 1. The main parts are the P200SX engine, the measuring equipment and the test bench computer. The physical signals for temperature, pressure and thrust are transformed to analog electroncial signals by the transducers and thermocouples, respectively. The transducers and thermocouples are connected to a digital multimeter, where the signals are amplified and digitised. Finally, a LabVIEW software tool is used to sample the measuring data from the thermocouples, the pressure transducers, the force transducer and the ECU, with a sampling rate of approximately $F_s \approx 1.7\,\mathrm{Hz}$. Moreover, the LabVIEW software tool transfers RS232-commands via the RS232 interface of the ECU, to control the operating points according to the profile shown in Fig. 2. Besides, the distance signal is sampled by the C-Box tool software, provided by the manufacturer, with $F_s = 30\,\mathrm{kHz}$. However, the computed rotational speed signal has a sampling rate of $F_s = 50\,\mathrm{Hz}$.

The pressure transducers Model 204 by Setra Systems Inc., with a full span pressure of 345 kPa, are used for the pressure measuring. The accuracy of the pressure transducers by the root of the sum squared is $\pm 0.11\% = \pm 0.38\,\mathrm{kPa}$ [14]. The force transducer type U2B by HBM Inc. has a nominal force of $F_{nom} = 500\,\mathrm{N}$. The sensitivity error for compression is $d_{ZD} = 1.5\%$, to non-linearity $d_{lin} < 0.2\%$ and temperature effect $T_{KC} = 0.02\%$ [15], which results in total accuracy of $\pm 1.72\% = \pm 8.62\,\mathrm{N}$. The accuracy of the temperature measurement is depending on the polynomials, which are used in the digital multimeter Keithley 2750, to convert the induced voltage from the thermocouples, to corresponding temperatures. Therefore the temperatures in the range between 0–773,15 K have an accuracy of $-0.05/+0.06\,\mathrm{K}$ while 773.15–1645.15 K have an accuracy of $-0.05/+0.04\,\mathrm{K}$ [16]. The accuracy of the rotational speed measurement is $\pm 48\,\mathrm{RPM}$.

The operating points (OP) 70%, 80%, 90% and 100% of the maximum rotational speed are investigated in this paper and are changed by step inputs at the times shown in Fig. 2. Prior to the measurement campaign, a test run with the predefined profile is conducted, to ensure comparable conditions for the measurements, with the already warmed up engine. Afterwards, the engine runs 10

Table 2. Deviation between JetCat P200SX and performance twin in different steady state operating points

N in %	Δp_{t3} in %	ΔT_{t3} in %	Δp_{t4} in %	ΔT_{t4} in %	ΔF_N in %
70	0.1	−0.3	1.0	0.1	1.5
80	−1.5	−0.4	−1.4	0.6	0.1
90	−1.8	−0.6	−1.8	−1.3	−1.3
100	0.7	−0.3	0.7	−0.5	2.4

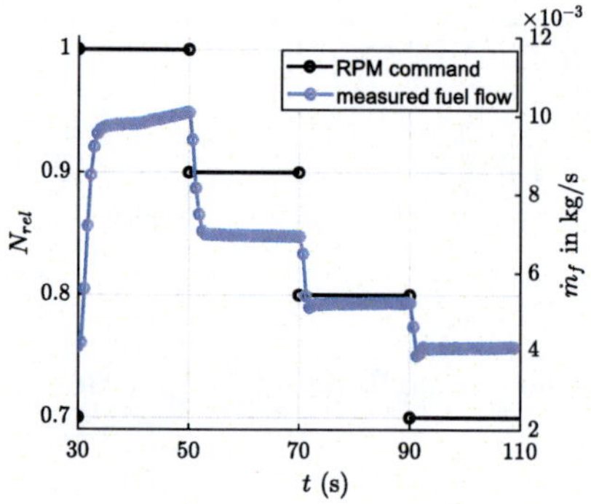

Fig. 2. RPM command signal and average measured fuel flow

measurements, with constant cool down phases. Lastly, for N, p_t, T_t and F_N averaged curves are calculated from the 10 measurements.

In addition, Fig. 2 presents the average measured fuel flow in relation to the rotational speed input command. It can be observed, that the fuel flow rapidly increases respectively decreases during the transient maneuvers, before it has stabilised at a constant level. However, at maximum speed $N_{rel} = 1$ an increase of the fuel flow at $t = 40.59\,\text{s}$ from $\dot{m}_f = 0.0098\,\text{kg/s}$ to $\dot{m}_f = 0.0101\,\text{kg/s}$ at $t = 49.41\,\text{s}$ is noticeable. Although the turbine inlet temperature T_{t4} in Fig. 5 still increases, the measured rotational speed N and the thrust F_N slightly decrease. Therefore it is assumed, that the apparently increasing fuel flow is a sham-effect, caused by losses in the fuel system (e.g. clogged filter [13]), since the ECU only estimates the fuel flow from the applied fuel pump voltage.

2.2 Numerical Setup

Based on long term measurements, a steady state performance twin of the deteriorated IFAS JetCat P200SX is developed. For this purpose, CAD models based on the geometrical dimensions of the inlet, compressor, combustor and turbine were created, to determine the size of control volumes. Furthermore, a CAD model of the engine shaft was created to estimate the inertia ($J = 1.15 \cdot 10^{-4} \text{kg} \cdot \text{m}^2$). Additionally the performance maps of the centrifugal compressor and axial turbine are taken form the literature [17] and [18] and a model-based performance analysis [10] is applied to match the steady state test data. The temperature, pressure and thrust deviation between the real and virtual engine are listed in Table 2 in many rotational speeds. The largest mismatch is visible in F_N for N = 100% with 2.4%.

Starting from the steady state operating point at N = 70%, the initially mentioned methods are outlined in the chapter below. In all methods the fuel flow $\dot{m}_f$ is used as the input parameter.

The main focus of the work lies on the consideration of the equations of motion (see Eqs. 1–3). Further transient effects, e.g. rotor inertia, heat soakage or time delays, are also included with the same conditions in the models (see [19]):

$$\frac{\mathrm{d}\rho_o}{\mathrm{d}t} = \frac{1}{C} \cdot (\dot{m}_i - \dot{m}_o) \tag{1}$$

$$\frac{\mathrm{d}T_{to}}{\mathrm{d}t} = \frac{1}{C_T} \cdot \left(\dot{m}_i \cdot c_{pi} \cdot (T_{ti} + \Delta T_t) - \dot{m}_o \cdot c_{po} \cdot T_{to} - \dot{Q}_i\right) \tag{2}$$

$$\frac{\mathrm{d}\dot{m}_i}{\mathrm{d}t} = \frac{1}{I} \cdot (p_{ti} + \Delta p - p_{to})\,. \tag{3}$$

In ASTOR, the equations of motion are included in a system of ordinary differential equations (see Eq. 4). The efforts e (e.g. pressure, temperature and torque) and flows f (e.g. mass flow, energy flow and rotational speed) are included in the vector $\bar{x}(t)$. In the coefficient matrix A and B the functions of material, geometry, volume dynamics and pressure ratios π are stored. The fuel flow signal is deposited in vector $\bar{u}(t)$ and added with the coefficient matrix B into the ODE system. Here, the pressure ratios are boundary conditions and depend on mass flow and rotational speed [19]. The ODE system is solved by a Runge-Kutta third order method

$$\frac{d}{dt}\bar{x} = A\bar{x} + B\bar{u}. \tag{4}$$

The flow chart of the iterative solvers are depicted in Fig. 3. In the left diagram, the engine matching (EM) algorithm with and without VD is shown. Hence, the global thermodynamic cycle of the engine is computed in each iteration until the module mass flows and the nozzle pressure have converged. Therefore, the auxiliary coordinates, here namely GL-lines (see Fig. 4), are iterated to find out the correct compressor and turbine pressure ratio and mass flow. Next, the turbine inlet temperature is determined with the fuel to air ratio FAR for the next time step iteratively. Finally, the rotational speed is integrated for the next operating point. The iteration is based on the Newton-Raphson method.

Additionally, the impact of volume dynamics on the performance can be computed afterwards (dashed box). These updates of pressure, temperature and mass flow lead to a consideration of the effect but also to a mismatch of the overall thermodynamic cycle [9].

The local intercomponent volume (ICV) solver is represented on the right diagram of Fig. 3. In contrast to EM and EM-VD, the mass flow is computed locally. Starting with the combustion chamber, mass flow and energy equation (see Eq. 1, 2) are used to calculated the pressure p_{t4} in dependence of the fuel flow. Based on the resulting pressure ratio and rotational speed, the mass flow of the turbine is determined from the performance map. This process is repeated until it is converged. The same procedure is performed with the compressor mass flow. Similar to ASTOR, no auxiliary GL-lines are necessary. Next, the nozzle mass flow is computed based on the pressure ratio. This results in different component mass flows, by which the differential equations can be solved. Similar to EM-VD, the interaction of the system components among each other are neglected and the update of the efforts e leads to a small mismatch in the thermodynamic cycle.

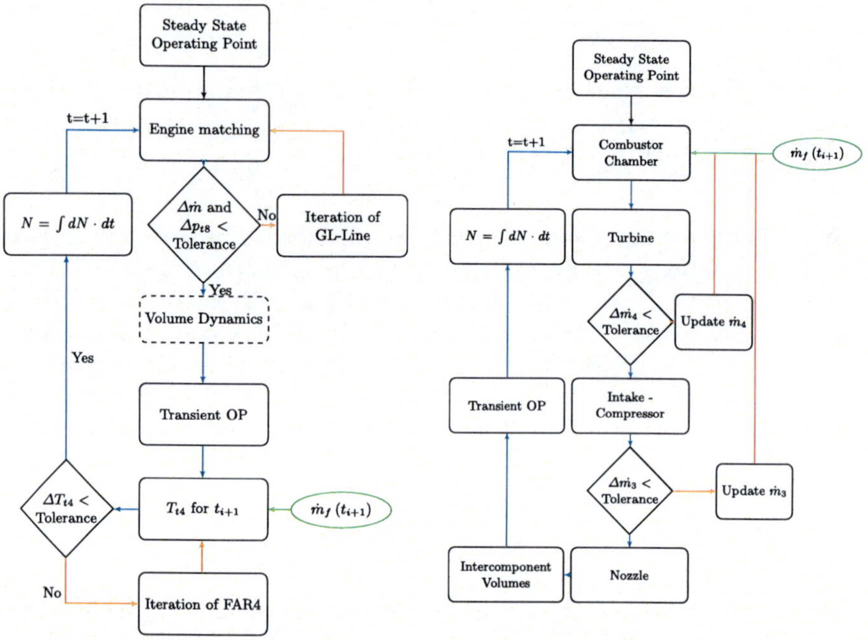

Fig. 3. Flow chart of the three iterative solvers. Left: Engine matching (EM) method and hybrid engine matching (EM-VD) method with volume dynamic (based on [5]). Right: Intercomponent volumes (ICV) method (based on [8]).

3 Results and Discussion

In this chapter, the digital performance twin of the JetCat P200SX is validated numerically and experimentally.

In Fig. 4, the transient operating line of the different performance tools are shown inside of the compressor (left) and turbine (right) performance map. The reduction in the top left-hand corner represents the operating line in the overall performance map. In general, all results describe the acceleration between the operation points $N = 70\%$ to 100% and the deceleration between 100% to 90%, 90% to 80% and 80% to 70% (see Fig. 2). ASTOR is shown in orange, EM in blue, EM-VD in light blue and ICV in dotted yellow. The corrected rotational speeds $N_{corr,rel}$ are represented with the red lines and the GL-lines for the iteration are shown with black lines.

In the left diagram, the compressor π is plotted over the $\dot{m}_{corr}$. In general, the qualitative and quantitative trends are mutually consistent. It is noticeable that in ASTOR the pressure is rising fastest, followed by the ICV and EM-VD. The most significant discrepancy between ASTOR and the iterative methods occurs at the acceleration and lies at $\Delta\dot{m} = 0.04$ kg/s and $\Delta\pi = 0.36$. The mean differences between ASTOR and EM is $\overline{\Delta\dot{m}} = 0.002$ kg/s and $\overline{\Delta\pi} = 0.02$. In the right diagram of Fig. 4, the turbine pressure ratio π is plotted over the product of corrected mass flow $\dot{m}_{corr}$ and rotational speed $N_{corr,rel}$. Similar to the compressor, pressure changes fastest in ASTOR, followed by the ICV and EM-VD. The most significant discrepancy is around $\Delta\dot{m} = 0.0025$ kg/s and $\Delta 1/\pi = 0.14$ during the acceleration. The mean differences between ASTOR and EM is $\overline{\Delta\dot{m}} = 0.0001$ kg/s and $\overline{\Delta 1/\pi} = 0.0072$.

As expected, the volume dynamics belong to a less important phenomenon in these load cases for the small turbojet engine under investigation. However, the small differences can be attributed to this effect [20]. Due to the clearness, only the results of ASTOR will be shown and discussed in the following results.

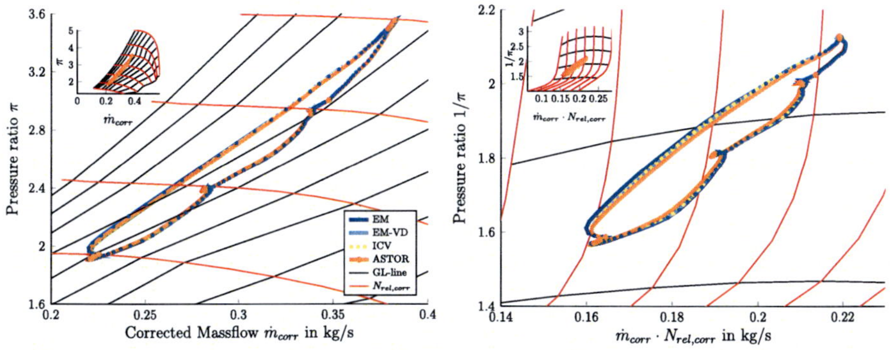

Fig. 4. Comparison of the numeric transient operating line. ASTOR in orange, EM in blue, EM-VD in light blue, ICV in yellow. Left: Pressure ratio π over mass flow $\dot{m}_{corr}$ of the compressor. Right: Pressure ratio $1/\pi$ over the product of $\dot{m}_{corr}$ and $N_{corr,rel}$ of the turbine.

In Fig. 5, important results of the transient maneuver are depicted. The results of the measurement are shown in purple and the ones of ASTOR in orange. In general, the curves show a high level of agreement.

In the top left diagram of Fig. 5, the compressor corrected rotational speed N_{rel} over the time t is plotted. The mean deviation is around $\overline{\Delta N_{rel}} = 0.0062$. Especially the curves of decreasing load between 80% and 70% are in good correspondence (see magnification).

The top right diagram shows the thrust F_N over the time t. A significant deviation between measurement and simulation can be seen during acceleration. The calculated thrust is significantly lower than the measured results. The deviation increases up to 14 N. Possible reasons for this are the behaviour of the fuel flow signal (see Fig. 2) and the large measurement uncertainty of the thrust, which have also an impact on the defined numeric steady state operating point. However, the mean deviation is up to 3 N.

The temperature at station 3 and 4 has a mean deviation of 3.8 K and 10 K. In the bottom left diagram, the temperature T_{t3} is shown and the temperature T_{t4} in the bottom right diagram. Hence, the quantitative deviations can be attributed to the estimated materials and geometries for heat transfer or time delays. However, the qualitative development is in agreement with the test data. Especially the transient system response in the last deceleration is simulated in the same temperature and time range.

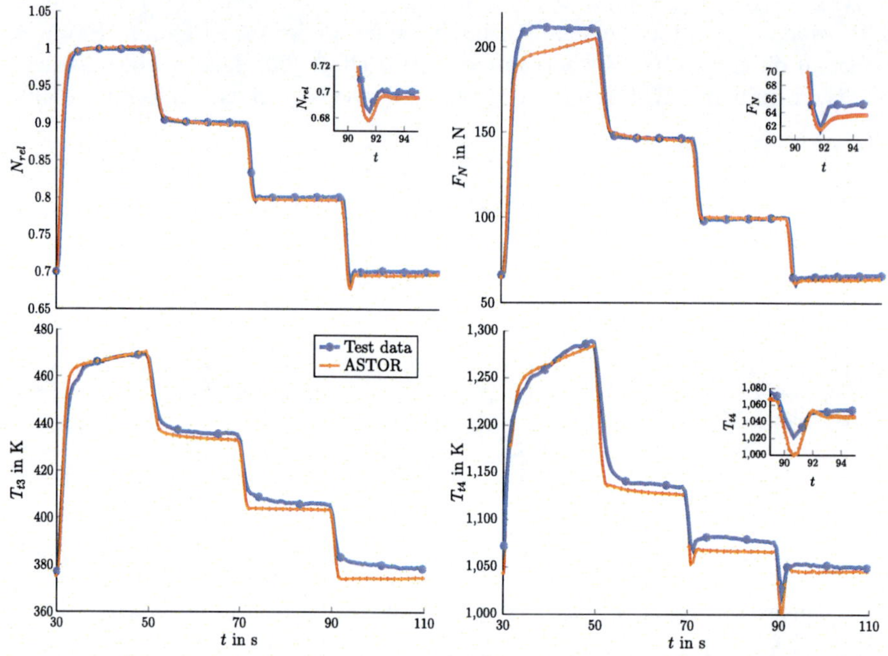

Fig. 5. Transient maneuver over the t. ASTOR in orange, in purple results of the test data. Top left: Compressor corrected rotational speed N_{rel}. Top right: Thrust F_N. Bottom left: Temperature T_{t3} at station 3. Bottom right: Temperature T_{t4} at station 4.

4 Conclusion and Outlook

In this investigation, the performance simulation tool ASTOR has been validated numerically and experimentally using a turbojet engine. Therefore, the performance of the single spool turbojet engine JetCat P200SX was measured and simulated by means of different transient performance calculation methods.

At the beginning an initial framework of the test run was presented. In addition, the steady state virtual performance twin and the different algorithms were described. Afterwards, ASTOR was compared to the engine matching methods (EM), to the hybrid engine matching methods (EM-VD) and to the intercomponent volumes technique (ICV). In general, the results correspond among each other with small deviation due to the impact of volume dynamics. This was to be expected due to the small engine. Next, ASTOR was validated with the test data with high consistency, especially in the load rejection curves.

Particularly in small machines, the influence of volume dynamics can be negligible and estimated very well by the usual algorithms. In further studies, this influence will be examined on large two-spool turbofan engines in more detail.

Nomenclature

ASTOR	AircraftEngine Simulation for Transient Operation Research	$\dot{m}, \dot{m}_f$	Mass flow, fuel flow
		N	Rotational speed
C,C_T	Capacity or thermal storage	ODE	Ordinary differential equations
c_p	Heat capacity	ODT	Optical displacement transducer
d_{lin}	Non-lineariry error	OP	Operating point
d_{ZD}	Compression error	π	Pressure ratio
DSTFT	Discreet Short Time Fourier Transform	p_t	Total pressure
EM	Engine matching	$\dot{Q}$	Heat flow
F_s	Sampling rate	$rel, corr$	Relative, corrected
F_N	Thrust	T_{KC}	Temperature effect error
i, o	in, out coming	T_t	Total temperature
ICV	Intercomponent volumes	$T_{t3/4}$	3/4 behind compressor/burner
IFAS	Institute of Jet Propulsion and Turbo Machinery	t	time and total
		ΔT,Δp	Deviation $(x_1/x_2 - 1) \cdot 100$
I, J	General and moment of inertia	VD	Volume dynamic

Acknowledgment. The present work has been carried out in the subproject D6 within the Collaborative Research Center (CRC) 871 "Regeneration of Complex Capital Goods" which is funded by the Deutsche Forschungsgemeinschaft (DFG, German Research Foundation) – SFB 871/3 – 119193472. Special thanks to Leroy Benjamin and Lennart Harjes from IFAS and André Weingärtner for assistance in setting up measurements.

References

1. Saravanamuttoo, H.I.H., MacIsaac, B.D.: An overview of engine dynamic response and mathematical modeling concepts. In: AGARD Conference Proceedings, no. 324 (1983)

2. Goeing, J., Seehausen, H., Pak, V., Lueck, S., Seume, J.R., Friedrichs, J.: Influence of combined compressor and turbine deterioration on the overall performance of a jet engine using RANS simulation and Pseudo Bond Graph approach. J. Global Power Propul. Soc. **4**, 296–308 (2020). https://doi.org/10.33737/jgpps/131109
3. Fawke, A.J., Saravanamuttoo, H.I.H.: Experimental investigation of methods for improving the dynamic response of a twin-spool turbojet engine (1971)
4. Seldner, K.: Generalized simulation technique for turbojet engine system analysis. National Aeronautics and Space Administration (1972)
5. Numerical Propulsion Simulation Software Consortium: NPSS User Guide. Wolverine Ventures Inc., Jupiter, FL (2006)
6. Kurzke, J.: Transient simulations during preliminary conceptual engine design. ISABE Paper, 2011, Nr. 2011-1321 (2011)
7. Sellers, J.F., Daniele, C.J.: DynGEN: a program for calculating steady-state and transient performance of turbojet and turbofan engines. National Aeronautics and Space Administration (1975)
8. Walsh, P.P., Fletcher, P.: Gas Turbine Performance. Wiley, Chichester (2004)
9. Rahman, N.U., Whidborne, J.F.: Real-time transient three spool turbofan engine simulation: a hybrid approach. J. Eng. Gas Turb. Power **131**(5) (2009)
10. Kurzke, J., Halliwell, I.: An Exploration of Gas Turbine Performance Modeling. Propulsion and Power. Springer (2018)
11. Greitzer, E.M.: Surge and rotating stall in axial flow compressors–Part I: Theoretical compression system model, 190–198 (1976)
12. Garrard, G.D.: ATEC: The Aerodyanmic Turbine Engine Code for the Analysis of Transient and Dynamic Gas Turbine Engine System Operations (1995)
13. Ing. Büro CAT, M. Zipperer GmbH: Bedienungsanleitung Strahlturbinen. https://www.jetcat.de/de/service/Bedienungsanleitungen
14. Setra Systems Inc.: Model 204 Data Sheet. https://www.setra.com/hubfs/Product_Data_Sheets/Setra_Model_204_Data_Sheet.pdf
15. HBM: U2B Force Transducer. https://www.hbm.com/de/2407/u2b-kraftaufnehmer-zum-messen-von-zug-druckkraeften/
16. Keithley: Model 2750 Multimeter/Switch System User's Manual . https://download.tek.com/manual/2750-900-01(F-Aug2011)(User).pdf
17. Grannan, Nicholas, D., Hoke, J.: Trends in JetCAT Microturbojet-Compressor Efficiency. AIAA SciTech Forum, Grapevine, Texas, 9–13 January 2017
18. Montazeri-Gh, M., Fashandi, S.A.M.: Bond graph modeling of a jet engine with electric starter. Proc. Inst. Mech. Eng. Part G J. Aerosp. Eng. **233**(9), 3193–3210 (2019)
19. Goeing, J., Lueck, S., Bode, C., Friedrichs, J. Simulation of the impact of a deteriorated high-pressure compressor on the performance of a turbofan engine using a pseudo bond graph modelling approach. Global Power and Propulsion Society GPPS-BJ-2019-0160 (2019)
20. Fawke, A.J., Saravanamuttoo, H.I.H.: Digital computer methods for prediction of gas turbine dynamic response. SAE Trans. 1805–1813 (1971)

LES of N2/H2 Transcritical Injection in a Rocket Combustion Chamber

Raffaele Olmeda(✉) and Christian Stemmer

Technische Universität München, Boltzmannstr. 15, 85748 Munich, Germany
raffaele.olmeda@tum.de
https://www.mw.tum.de/aer/

Abstract. In modern Launcher Rocket Engines (LRE), the study of the behavior of the injector in combustion chamber has been complicated by the fact that lower fuel temperature and higher pressure are used to achieve larger enthalpy values. In addition to this, methane has become more and more common in aerospace application, therefore transcritical states are often reached by the mixing of fuel and oxidizer. The goal of this work is to correctly characterize the transcritical mixing of coaxial N_2/H_2 injector. Real-gas thermodynamic is used for this purpose in combination with appropriate inlet turbulent boundary conditions. In particular, an effort has been made to use realistic inlet turbulent conditions and a high order scheme has been employed for the calculation of the convective fluxes.

Keywords: Transcritical injection · Binary mixtures · Mixing

1 Introduction

In modern rocket engines, a thorough analysis of the injection process must be done to correctly predict the behavior of the overall combustion chamber. The injection influences several factors such as the nominal pressure, the flame front or resonance phenomena. The analysis is extremely difficult due to the large pressures and low temperatures of the injected fuel, which often comes close to its critical point. Experiments for these scenarios can give only limited information, mainly about the topology of the jet and its density composition, whereas CFD studies can deliver full time- and space-resolved data. The injection of a specie in a warmer environment filled with the same fluid represents a simplified setup which is used to investigate the behavior with a rocket-engine injector. Thermodynamic states in the transcritical regime are generally chosen. The most typical case is injection of cryogenic nitrogen in a warm nitrogen reservoir. Mayer et al. [1] investigated the injection of liquid nitrogen in an ambient temperature chamber, studying the influence of pressure on the spray composition. Shadowgraphy technology was used for the study of the jets. Oschwald and Schick also studied the injection of liquid nitrogen [3]. They noticed how,

© The Author(s), under exclusive license to Springer Nature Switzerland AG 2021
A. Dillmann et al. (Eds.): STAB/DGLR Symposium 2020, NNFM 151, pp. 569–578, 2021.
https://doi.org/10.1007/978-3-030-79561-0_54

passing from a sub- to a supercritical pressure, the jet showed only small disturbances on the surface to an atomization to a spray form. Of more interest are the experiments performed on binary mixtures, which reflects more closely what happens in LREs. At the injector, fuel and oxidizer consist of different species. In these kind of experiments, non-reactive fluids are chosen in order to focus the study on the mixing. Experiments of these kind are described for example by the work of Mayer and Smith [2] and Oschwald et al. [5]. Particularly interesting is how, close to the injector, some thermodynamic states show lower temperatures of the two species, with consequently higher densities than could be expected.

As for the numerical studies, the group of Josette Bellan was among the first to perform simulations of high-pressure mixing. Non-reacting mixing layers were studied with Direct Numerical Simulations [6,7], where temperature and pressure were chosen such that only a single phase existed. Lacaze et al. [8] have investigated binary mixtures with a fully compressible framework. Their work underlined how real-gas effects influence speed of the sound and the specific heat in the cases of high-pressure injection. Very good results have recently been achieved by Matheis and Hickel [9], by Müller et al. [10] and Matheis et al. [11]. Large-Eddy Simulations were used to resolve the jet of a binary mixture. Cubic equations of state were used in combination with the Vapor Liquid Equilibrium model, in order to correctly represent the thermodynamic states in the transition region from cryogenic to hot gas. Good agreement was reached with experiment for the density profiles in longitudinal direction. In the direction perpendicular to the flow, an underestimation of the mixing causes a difference from the experimental reference data of Oschwald et al. [4]. In the present work, three-dimensional LES will be performed for the coaxial injection of transcritical N_2/H_2 for the setup of Oschwald et al. [4]. Previous works focused largely on accurately modeling real-gas thermodynamics, neglecting the effect on the mixing of turbulent fluctuations [9,10], where laminar inlet conditions were used in combination with low-order schemes for stability reasons. In this work, attention is paid to the turbulence inlet boundary condition and their influence on the mixing of the binary mixture. Also, schemes of different accuracy are successfully used in combination with a Ducros sensor. The results will be compared to the experimental data available [4].

2 Governing Equations

The unsteady, three-dimensional, compressible Navier-Stokes equations (NSE) are solved,

$$\frac{D\rho}{Dt} = 0, \tag{1}$$

$$\frac{D(\rho \bar{U})}{Dt} = -\nabla P + \nabla \cdot (\mu \nabla \bar{U}), \tag{2}$$

$$\frac{D(\rho E_{tot})}{Dt} = -\nabla \cdot (P \bar{U}) + \nabla \cdot (\bar{U}\, \bar{\bar{\tau}}) + \lambda \nabla^2(T), \tag{3}$$

where ρ is the density, $\bar{U}$ is the velocity vector, μ is the viscosity, $\bar{\bar{\tau}}$ is the viscous stress tensor, E_{tot} is the total energy, λ is the thermal conductivity. The Lagrangian derivative is defined as $\frac{D}{Dt} = \frac{\partial}{\partial t} + U_i \frac{\partial}{\partial x_i}$, with $i = 1, 2, 3$. In this work, for a general variable n, we write the scalar, vector and matrix of that variable as $n, \bar{n}$ and $\bar{\bar{n}}$. The NSE transport mass, momentum and total energy of the mixture. The transport equation of the mass fraction is

$$\frac{D(\rho X_i)}{Dt} = \nabla \cdot (\bar{J}_i), \tag{4}$$

$$\bar{J}_i = \rho D_i \nabla X_i - X_i \sum_j^N \rho D_j \nabla X_j, \tag{5}$$

where $\bar{J}$ are the diffusion fluxes, N is the number of species (2 for our setup) and D_i are the binary species diffusion coefficients. Only one mass fraction is transported ($i = 1$), whereas the one of the other specie is found with the condition $\sum_i^N X_i = 1$. The Schmidt number is in the range $S_c = 20 - 140$. To calculate the pressure, the Peng-Robinson cubic equation of state is used [14]

$$P = -\frac{a\,\alpha(T)}{v^2 + 2\,b\,v - b^2} + \frac{R\,T}{v - b}, \tag{6}$$

where v is the mixture specific volume, R is the gas constant, T is the temperature, a and b are parameters which depend on the local mixture composition.

3 Numerical Methods

3.1 Solver Description

The CFD code CATUM is used to perform the simulations [12]. It solves the fully compressible Navier-Stokes equations with a finite volume method. The convective fluxes are solved with a linear fourth-order central scheme (4CC). A Ducros sensor changes the scheme to a lower-order upwind scheme in case of instability (in this setup it often necessary). A second-order centered scheme is used for the viscous fluxes. The time integration is computed with a fourth-order Runge-Kutta method. The turbulence Sub-Grid Scales are modeled with implicit LES by means of the advection local deconvolution method by Hickel et al. [13].

3.2 Equation of State

In Eq. 6 the parameters are defined as $a = 0.4572\,(R^2\,T_c^2/\,P_c)$ and $b = 0.0778\,(R\,T_c/\,P_c)$, where the subscript $_c$ characterizes the quantities at the critical point. The temperature dependent function α is defined as

$$\alpha(T) = (1 + c\,(1 - \sqrt{T/T_c})), \tag{7}$$

Table 1. Species parameters.

Specie	$P_c\,[Pa]$	$T_c\,[K]$	ω
H_2	$1.2964\,10^6$	33.145	−0.2160
N_2	$3.3958\,10^6$	126.192	0.0372

with $c = 0.37464 + 1.54226\,\omega + 0.2699\,\omega^2$. The function α accounts for the eccentricity of the specie. The acentric factors ω and critical quantities of nitrogen and hydrogen are listed in Table 1. In the cells, where the thermodynamic state of a mixture must be evaluated, the following mixing rules are used

$$a\,\alpha = \sum_i^N \sum_j^N \chi_i\,\chi_j\,a_{i,j}\,\alpha_{i,j}, \qquad b = \sum_i^N \chi_i\,b_i,$$

$$T_{c(i,j)} = \sqrt{T_{c(i)}\,T_{c(j)}}, \qquad v_{c(i,j)} = 1/8\,\big(\sqrt[3]{v_{c(i)}} + \sqrt[3]{v_{c(j)}}\big)^3,$$

$$P_{c(i,j)} = Z_{(i,j)}\,(R\,T_{c(i,j)}/v_{c(i,j)}), \quad Z_{(i,j)} = 0.5\,(Z_i + Z_j), \quad \omega_{(i,j)} = 0.5\,(\omega_i + \omega_j).$$

Z is the compressibility factor. The double subscript $_{i,j}$ denotes the mixed terms of the parameters, calculated from the values of the single species. In this study we apply a pure version of the Peng Robinson Equation of state. The largest deviation from the NIST database of the Peng Robinson equation occurs for a mass fraction of nitrogen $\chi_{N2} = 1$ with the nitrogen injection thermodynamic conditions (P = 4 MP, T= 118 K). The maximal deviation is 4% of the density value, which has been considered an acceptable error in the framework of the current study. The transport properties are calculated with the approach of Chung et al. [15].

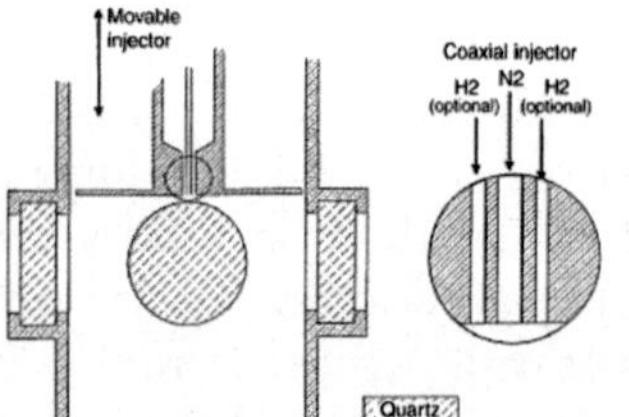

(a)

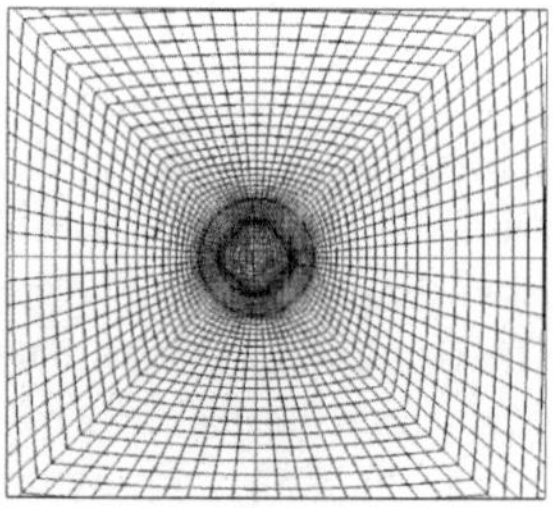

(b)

Fig. 1. Sketch of the experimental setup (a) and YZ plane of the G3 mesh (b).

4 Setup

The experimental setup is the one described by Oschwald et al. [4]. A coaxial injector is composed by two concentric parts. The inner nitrogen injector has a diameter of $D_1 = 1.9$ mm. The outer hydrogen annulus has an inner diameter of $D_2 = 2.4$ mm and an outer diameter of $D_3 = 3.7$ mm. The setup of the experiment is shown in Fig. 1a. The nominal temperature for nitrogen is 118 K, for hydrogen is 270 K. The mass flows are measured with a Coriolis sensor and are $\dot{m}_{N2} = 8.63$ g/s and $\dot{m}_{H2} = 1.92$ g/s. The injected fluids are shot in a chamber filled with warm nitrogen at 298 K. The pressure of the chamber is 4 MPa. Looking at Table 1, it is clear that the inlet conditions would set the problem in a transcritical regime. The work of Matheis et al. [11] proved that at the location of the injection an inlet temperature of nitrogen of 118 K leads to an error of more than 20% with respect to the experiment when calculating the nitrogen partial density. Müller et al.[10] make the same point, noting how during the experiment the sampling was done during a transient phase in which the feeding lines were changing temperatures. Moreover, the group of Oschwald mentioned issues with beam steering and reflection, and problems in calculating the inlet temperature of nitrogen. The first point sampled in the experiment lies directly downstream of the injection point, and can safely be considered a faithful representation of the real thermodynamic state of the injected nitrogen. Following the lead of the past working groups [9,10], the nitrogen inlet temperature has been therefore adjusted to 128 K when studying the nitrogen partial density profiles, fitting the first sampled partial density point, moving the problem to the supercritical regime. An analysis has been made for the setup with $T_c = 128$ K, to assess if any subcritical state appears. For this case, the mixture remains supercritical for the all domain. This assessment is in agreement with the past literature. In the case with $T_c = 118$ K subcritical states do appear, and the numerical scheme treats them as a single phase. At the inlet, temperature and dynamic pressure are fixed as boundary conditions. For the two inlet temperatures $T_c = 118, 128$ K, the Reynolds number at the central injector is $Re = 116 \cdot 10^3, 221 \cdot 10^3$. The mean inlet velocity is calculated from the values of the local pressure and temperature. A time-dependent velocity field in $v(y, z, t)$ and $w(y, z, t)$ with a prescribed amplitude is generated from a set of pseudo-random numbers. The random numbers are generated with the random-number generator of Fortran 90. At every time-step, different random numbers were used. This method is based on the *turbulentInletFvPatchField* of OpenFoam [16]. Inflow fluctuations based on random numbers are not perfect, but due to the lack of scale and frequency information, any more sophisticated method would have been equally imprecise. For one of the cases, a simulation was performed with a more realistic velocity inlet profile for the external annulus. The averaged inlet profile was generated with a precursor LES. Due to the insufficient information on turbulent length and time scales from the precursor simulation, only the mean information was used for the definition of the inflow conditions. This case is defined as Precursor Boundary Condition Turbulent Simulation (PBCTS). A Neumann condition for the pressure is chosen. At the outlet, non-reflective boundary conditions are imposed. For the walls of the chamber, adiabatic conditions are enforced. In Fig. 1b a zoomed image of the mesh on the plane perpendicular to the flow direction is shown. A three-dimensional structured mesh is used. Three different grids with $G1 \approx 2 \cdot 10^5$, $G2 \approx 8 \cdot 10^5, G3 \approx 2 \cdot 10^6$ grid points were used. At the injection point, the resolution of the mesh is $\Delta x_1 = 2 \cdot 10^{-1}$ mm, $\Delta x_2 = 1 \cdot 10^{-1}$ mm, $\Delta x_3 = 5 \cdot 10^{-2}$ mm and $\Delta y_{min} = \Delta z_{min} = 3.5 \cdot 10^{-2}$ mm . Different simulations were performed in order to assess the influence of two different factors: the numerical order of the scheme of the convective fluxes and the type of inlet boundary condition for the velocity. The convective fluxes are calculated with a 4-cell cell-central scheme (4CC) and with a second order upwind scheme (2UP). The simulations have been performed firstly on coarse meshes, then they have been interpolated to finer meshes and finally they have been averaged for seven Flow Through Times (FTT). The experimental data available in [4] are the partial density of nitrogen along the center line of the injector, and the partial density profiles of hydrogen and nitrogen along the radial direction.

5 Results

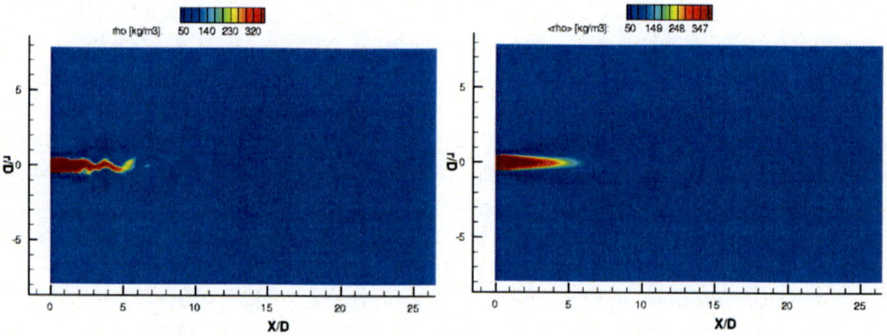

(a) Instantaneous. (b) Averaged.

Fig. 2. Snapshot of the density field for the case with $T_c = 128$ K and the cell-central scheme (4CC).

Snapshots of the density field for the case with $T_{N2} = 128$ K and the cell-central scheme are shown in Fig. 2. The X coordinate was normalized by the internal diameter $D = D_1 = 1.9$ mm. The N_2/H_2 mixture is close to its critical point in the layer between the two species. In this situation, large density changes can occur very rapidly. For the case of $T_{N2} = 128$ K, the density ratio at the moment of injection is $\rho_{N2}/\rho_{H2} = 110$. This causes large gradients of all thermodynamic quantities in the flow field. The shear force between the two fluids produces intense mixing of the two species and jet breakup. After the break up $(X/D > 8)$, a region of intense mixing follows, where we can still recognize eddies with a nitrogen core which have been released by the break up (Fig. 2a). In Fig. 2b the averaged density field is shown. The jet appears to reach density values similar to the nitrogen reservoir right downstream of $X/D = 16$.

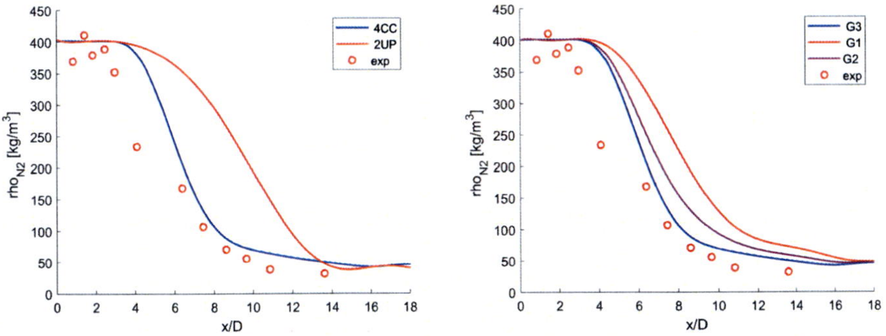

(a) Study on the numerical scheme.

(b) Grid refinement study.

Fig. 3. Mean partial density profile of nitrogen in flow direction along the center line.

A grid refinement study has been performed in order to assess the quality of the LES simulations (Fig. 3b). For this study, the cell-central scheme was used. The grid refinement study shows the tendency of the LES results to converge towards the measured values. As various models with their described limitations are employed in order to tackle the problem of rocket combustion-chamber injection and mixing on an industrial scale, a final convergence has not been reached. The current procedure explores the capabilities of the thermodynamic modeling rather than a full convergence for the LES modeling. A final evaluation of the question of accuracy would have to be done involving a DNS study for the case which is currently not available. A further refinement of the grid would be necessary to show grid convergence and the proper resolution of the turbulent length scales. The high pressure and low temperature of the injected nitrogen causes the speed of sound to be very high mandating a very large amount of time-steps necessary for full convergence of the turbulent scales.

In Fig. 3a and Fig. 4b, we can observe the partial density profile of nitrogen in direction parallel and perpendicular to the flow. All longitudinal measurements were made along the center line of the central jet. All transversal measurements were made in radial direction perpendicularly to the axial center line on a plane 4 mm from the injectors. In the figures presented in these results, distances both in longitudinal and radial direction have been normalized with the inner injector diameter. Two different simulations are compared with the experimental data in Fig. 3a. One was performed with a second-order scheme for the convective fluxes, one with a fourth-order cell-central scheme. This comparison was deemed useful as in previous numerical investigations from other authors with this setup, dissipative schemes were used for numerical stability [11]. In those numerical studies, convergence with the experimental data was reached by using a coarser grid and therefore introducing additional numerical viscosity in the simulation. From Fig. 3a it is clear how large the influence of the order of accuracy of the convective scheme is. The upwind scheme can not properly describe the turbulent fluctuations in direction perpendicular to the flow at the given grid resolution. The mixing of nitrogen and hydrogen is therefore largely delayed, causing the deviation in the partial density profile. On the other hand, the scheme has no influence on the partial density profile in radial direction. The divergence between the two schemes starts after $X/D = 5$, and all the radial profiles were measured at $X/D \approx 2$. The transversal nitrogen partial density shows a good qualitative agreement with the experiment (Fig. 4b) for the case with $T_c = 128$K. The overestimation for $r/D < 0.5$ is caused my an underestimation of the mixing of hydrogen and methane, and it is discussed later more in detail. In Fig. 4a, the partial density profile of hydrogen in radial direction is shown. Radial density profiles close to injection at $X/D \approx 2$ are shown since this location was the only one available from the experiment. An analysis of the radial profiles further downstream would also require additional computational effort. The simulations can not exactly reproduce the peak of the profile nor the asymptotic values for $r/D > 1.5$ as it underestimates the mixing between nitrogen and hydrogen as previously described. The difference in density between hot hydrogen and cold nitrogen is so large, that a small error in the mass fraction would cause a large difference in the partial density profile of hydrogen.

The peak of the partial density lies in the mixing layer between the two species. We can as first thing observe the influence of a different inlet temperature. The lower temperature causes an asymmetric profile, with a higher maximal value and generally higher values over the whole span of r/D with respect to the profile calculated with the higher inlet temperature. Moreover, the profile is shifted to the left for the case with lower temperature, closer to the experimental data. This behavior is caused by the higher mixing rate of the two species due to their higher density ratio and the reduced velocity of the injected nitrogen. The effect of the two different numerical schemes can be clearly seen in the maximal value of the partial density. For the second-order scheme, the peak is overestimated by $0.7\,\mathrm{kg/m^3}$, when compared to the experimental data. The dark blue line is similar to the one obtained by Matheis et al. [11]. In the case of the cell-central scheme instead, the error is reduced by one third. In both cases though, the profiles show a shift to the right of $0.25\,r/D$ compared to the experimental data.

A further case has been analyzed. The described results where obtained using a laminar profile with superimposed random fluctuations as inlet boundary conditions. In order to have a more realistic turbulent inflow behavior, a precursor simulation was made for the external hydrogen annulus. The averaged velocity profiles obtained by the precursor simulation have been then used as inlet boundary condition for the hydrogen annulus. For the turquoise line in Fig. 4a can be seen how this new boundary condition improved the partial density profile with respect to the experimental data. The peak of the profile now lies at $r/D = 0.4$. Also, the flow shows traces of hydrogen in the inner part of the jet, for $r/D < 0.4$. The values are still too low, but an improvement from the previous tests has been achieved. The key seems to be the mixing rate of nitrogen and hydrogen. The underestimated mixing causes a difference of $0.5\,\mathrm{kg/m^3}$

between the experimental data and the best results of this study (PBCTS). Even more realistic inlet conditions (which account for the precursor simulation fluctuations) would probably further improve the quality of the simulations.

These results have shown that in setups where transcritical injection is involved, multiple aspects must be considered when performing a simulation. The correct modeling of real-gas effects is a key factor, but it is not the only one to consider. As it was to be expected, realistic boundary conditions are essential to correctly represent injection setups in rocket engines, in particular close to the injection point.

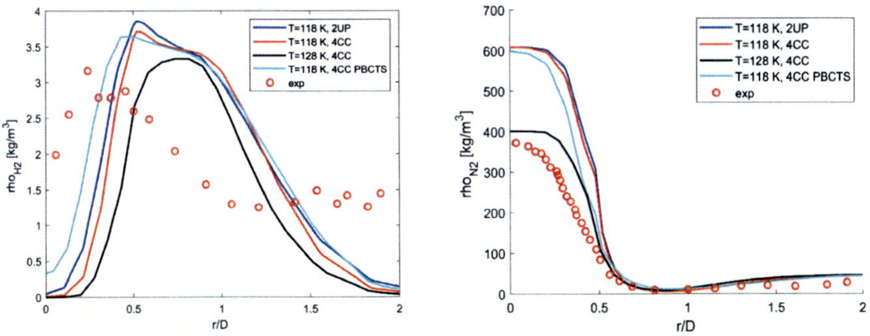

(a) Hydrogen. (b) Nitrogen.

Fig. 4. Mean partial density profiles in radial direction.

6 Conclusion

A framework for the simulation of a binary mixture under transcritical condition has been created. The development of the partial density profiles in direction parallel and perpendicular to the inflow has been studied, and the results have been compared to the experimental data. The focus of this work has been on the influence of two different factor. The first is the inlet temperature, which largely influences the density profiles and the mixing velocity of the two injected species at the inflow boundary. The second is the level of turbulence fluctuations, which intensify the mixing and facilitate the breakup of the jet. The third is the kind of inlet boundary condition used, which has a large effect in the area close to the injector. The results overall have shown good agreement with the experimental data.

For future work, a further analysis of the influence of the turbulent structures and velocity inlet conditions should be performed. A DNS would be quite useful for this purpose, especially if also used to generate realistic inlet data.

The authors gratefully acknowledge the Gauss Center for Supercomputing e.V. (www.gauss-center.eu) by providing computing time on the GCS Supercomputer SuperMUC at Leibniz Supercomputing Center (www.lrz.de) and the DFG for providing financial support in the Sonderforschungsbereich SFB TRR40.

References

1. Mayer, W., Schik, A., Schweitzer, C., Schaffler, M.: Injection and mixing processes in high pressure LOX/GH2 rocket combustors . In: 32nd Joint Propulsion Conference and Exhibit (1996)
2. Mayer, W., Schik, A.: Fundamentals of supercritical mixing and combustion of cryogenic propellants. In: Progress In Astronautics and Aeronautics: Liquid Rocket Thrust Chambers, pp. 339–367. AIAA (2004)
3. Oschwald, M., Schik, A.: Supercritical nitrogen free jet investigated by spontaneous Raman scattering. Exp. Fluids **27**(6), 497–506 (1999)
4. Oschwald, M., Schik, A., Klar, M., Mayer, W.: Investigation of coaxial LN2/GH2-injection at supercritical pressure by spontaneous Raman scattering. In: 35th AIAA/ASME/SAE/ASEE Joint Propulsion Conference and Exhibit 20–24 June 1999, Los Angeles, California (1999)
5. Oschwald, M., Smith, J.J., Branam, R., Hussong, J., Schik, A., Chehroudi, B., Talley, D.: Injection of fluids into supercritical environments. Comb. Sci. Technol. **178**(1–3), 49–100 (2006)
6. Bellan, J.: Supercritical (and subcritical) fluid behavior and modeling: drops, streams, shear and mixing layers, jets and sprays. Progress Energy Combust. Sci. **26**, 329–366 (2000)
7. Bellan, J.: Theory, modeling and analysis of turbulent supercritical mixing. Progress Energy Combust. Sci. **178**, 253–281 (2006)
8. Lacaze, G., Misdariis, A., Ruiz, A., Oefelein, J.C.: Analysis of high-pressure diesel fuel injection processes using LES with real-fluid thermodynamics and transport. Proc. Combust. Inst. **35**(2), 1603–1611 (2015)
9. Matheis, J., Hickel, S.: Multi-component vapor-liquid equilibrium model for LES of high-pressure fuel injection and application to ECN spray A. Int. J. Multiphase Flow **99**, 294–311 (2018)
10. Müller, H., Pfitzner, M., Matheis, J., Hickel, S.: Large-eddy simulations of coaxial LN2/GH2 injection at trans-and supercritical conditions. J. Propul. Power **32**, 46–56 (2015)
11. Matheis, J., Müller, H., Pfitzner, M., Hickel, S.: Large-eddy simulation of cryogenic coaxial LN2/gh2 injection under supercritical pressures. In: Ninth International Symposium on Turbulence and Shear Flow Phenomena (TSFP-9), Melbourne, Australia (2015)
12. Egerer, C.P., Schmidt, S.J., Hickel, S., Adams, N.: Efficient implicit LES method for the simulation of turbulent cavitating flows. J. Comput. Phys. **316**, 453–469 (2016)
13. Hickel, S., Adams, N.A., Domaradzki, J.A.: An adaptive local deconvolution method for implicit LES. J. Comput. Phys. **213**, 413–436 (2016)
14. Peng, D., Robinson, D.B.: A new two-constant equation of state. Ind. Eng. Chem. Fund. **15**, 59–64 (1976)
15. Chung, T.H., Ajlan, M., Lee, L.L., Starling, K.E.: Generalized multiparameter correlation for nonpolar and polar fluid transport properties. Ind. Eng. Chem. Res. **27**(4), 671–679 (1988)
16. Iwamoto, Y., Kusuzaki, R., Sogo, M., Yasuda, K., Yamano, H., Tanaka, M.: Effect of deflected inflow on flows in a strongly-curved 90 degree elbow. Int. J. Fluid Mach. Syst. **10**, 78–85 (2017)

Computational Aerodynamics

Towards a 3D Galerkin-Type High-Order Panel Method: A 2D Prototype

Karsten Bock(✉)

German Aerospace Center (DLR), Institute of Software Methods for Product Virtualization (SP), Zwickauer Straße 46, 01069 Dresden, Germany
karsten.bock@dlr.de

Abstract. Aiming to enable the use of mid-fidelity panel methods over low-fidelity potential methods in the context of mass simulations for aircraft design, a two-dimensional steady-state prototype was implemented to demonstrate the feasibility of panel methods for aerodynamic simulations. In contrast to most state-of-the-art panel codes, this method is based on high-order singularity distributions and a *Galerkin* formulation. This paper demonstrates several advantages of the method over the more traditional low-order collocation-based methods.

Keywords: High-order panel method · Potential aerodynamics

1 Introduction

Modern aircraft design processes require a huge number of aerodynamic simulations, in particular in the loads process including static and dynamic load cases. Although high-fidelity numerical methods such as finite volume methods or spectral element methods are readily available and commonly applied, these tools are usually very demanding in terms of computational cost and process integration. Thus, processes with requirements on simulation turnaround times usually refer to lower fidelity solvers. These apply simpler flow physics models than the *Navier-Stokes* equations, which the high-fidelity tools are usually based on. Boundary element methods on the basis of potential theory are one class of such low- and mid-fidelity methods. Instead of solving field equations, a boundary integral equation is solved on the surface of the body in the flow. Today, low-fidelity methods like the vortex-lattice (VLM) or doublet-lattice method (DLM) are commonly employed, which both neglect the thickness of bodies. In contrast, panel methods use the actual surface of a body and, hence, are labeled as mid-fidelity tools, since they model additional effects in comparison to VLM or DLM. Therefore, using mid-fidelity panel methods instead of the current low-fidelity approaches increases the prediction accuracy [9].

In this paper, panel methods will be categorized into low- and high-order methods by the polynomial order with which the singularities are defined on each panel. Low-order methods use constant source and doublet distributions,

© The Author(s), under exclusive license to Springer Nature Switzerland AG 2021
A. Dillmann et al. (Eds.): STAB/DGLR Symposium 2020, NNFM 151, pp. 581–591, 2021.
https://doi.org/10.1007/978-3-030-79561-0_55

whereas high-order methods use linear or higher degrees. Panel-wise constant doublet singularities implicate doublet jumps at element boundaries, which introduce spurious effects into the potential and flow field [3,6]. This effect decreases the accuracy with which the boundary condition is fulfilled. Furthermore, the evaluation of velocities by superposition tends to be inaccurate for low-order methods on the surface itself or in the immediate vicinity of the surface. Traditional low-order methods struggle with lifting bodies in immediate vicinity to each other as well as internal flows and fail for supersonic flows [3].

Even though high-order panel methods have been developed, e.g. [5,7,13], low-order panel methods remain very popular and state-of-the-art [1,4,10] mainly for two technical reasons beyond availability and habit: ease of implementation and speed of computation. In order to demonstrate the feasibility of high-order panel methods for aerodynamic simulations, a two-dimensional steady-state prototype panel method was implemented. Unlike most state-of-the-art panel methods the method presented here uses high-order singularity distributions in conjunction with a *Galerkin* formulation. This paper describes the chosen method and demonstrates the advantages of the high-order *Galerkin*-type method over the widely employed low-order collocation-based methods. Ultimately, the prototype and study presented here serve as a first step towards the development of a new 3D unsteady, high-order panel code to enhance accuracy, robustness and usability for future mass simulations in aircraft load analyses.

2 Galerkin-Type High-Order Panel Method

2.1 Governing Equations

The *Laplace* equation

$$\nabla^2 \Phi = 0 \tag{1}$$

for the total potential Φ can be derived from the *Euler* equations, which describe a inviscid flow field, under the additional assumption of an irrotational, incompressible, steady-state flow. The perturbation potential ϕ, which represents the influence of a body in the flow, is defined as $\Phi = \phi + \phi_\infty$, i.e. as a small perturbation to the free-stream potential ϕ_∞. *Green's* method yields fundamental solutions for (1), which are solely located on the surface of the body in the flow, e.g. [8]. Therefore, the field problem turns into a boundary problem and the potential at any position $\boldsymbol{p}$ in the field is calculated by

$$\Phi(\boldsymbol{p}) = -\frac{1}{4\pi} \int\limits_{\boldsymbol{x}\in S} \mu\, \boldsymbol{n} \cdot \nabla \left(\frac{1}{\|\boldsymbol{p} - \boldsymbol{x}\|} \right) \mathrm{d}S - \frac{1}{4\pi} \int\limits_{\boldsymbol{x}\in S} \sigma \left(\frac{1}{\|\boldsymbol{p} - \boldsymbol{x}\|} \right) \mathrm{d}S + \phi_\infty \,, \tag{2}$$

where $\boldsymbol{x}$ and $\boldsymbol{n}$ denote a position and the corresponding normal vector on the surface S of the body. Furthermore, the source strength σ and dipole strength μ account for a jump in the normal derivative of potential and a jump of the potential itself across boundary S, respectively. The potential inside the closed body is constant and following *Morino* [8,12] the panel method presented in his paper sets the internal potential to the free-stream potential ϕ_∞. Therefore, the perturbation potential inside the body vanishes and the source distribution is defined by the geometry of the submerged body and the free-stream velocity $\boldsymbol{U}_\infty$ as $\sigma = \boldsymbol{n} \cdot \boldsymbol{U}_\infty$. To fully define the flow around a body, the boundary integral equation (2) has to be solved for the unknown doublet distribution, which means that the perturbation potential ϕ is expected to vanish on the surface in accordance with the body internal potential.

2.2 Discretization

Panel methods are a means to solve the boundary integral equation (2) numerically. To this end, these methods approximate the surface by a set of N elements, called panels. For simplicity, the method presented here utilizes panels of linear shape, though higher order geometry descriptions could be incorporated in principle. These panels carry distributions of the fundamental solutions, namely the source and doublet singularities. In this paper, panel methods are categorized into low- and high-order methods by the polynomial order with which these distributions are defined on each panel. Low-order methods use constant source and doublet distributions, while high-order methods apply linear or higher order polynomials. Most common panel methods employ the low-order approach, e.g. [1,4,10]. In contrast, the present panel method uses up to quadratic order and different orders can be chosen for the source and doublet distributions. These representations are designed to be continuous at the element boundaries, when linear or higher orders are chosen. As in common finite element methods, *Jacobi* polynomials of order n are applied as local basis functions, which then form global basis functions Ψ^n of equal order for the distributions

$$\sigma(\boldsymbol{x}) = \sum_{j=0}^{N_\sigma} \sigma_j \, \Psi_j^{p_\sigma}(\boldsymbol{x}) \qquad \text{and} \qquad \mu(\boldsymbol{x}) = \sum_{j=0}^{N_\mu} \mu_j \, \Psi_j^{p_\mu}(\boldsymbol{x}) \, , \tag{3}$$

where p_σ and p_μ refer to the polynomial orders of the source and doublet distributions. Additionally, the number of global basis functions, i.e. degrees of freedom, are denoted by N_σ and N_μ, while σ_j and μ_j denote discrete nodal values of source and doublet distributions, respectively.

Applying the *Morino* boundary condition, as explicated in Sect. 2.1, the perturbation potential ϕ should vanish on the body surface, which is enforced in a weighted residual sense as

$$0 = \int_{\boldsymbol{x} \in S} \theta(\boldsymbol{x}) \, \phi(\boldsymbol{x}) \mathrm{d}S \, , \tag{4}$$

where θ is a test function. Most commonly, panel methods employ a *Dirac* function $\delta(\boldsymbol{x} - \boldsymbol{x_i})$ as this testing function, resulting in a collocation method, which enforces that the equation is fulfilled in the chosen collocation points. However, the method presented here adopts the global basis functions $\Psi^{p_\mu}(\boldsymbol{x})$ as the test functions instead and, thus, yields a *Galerkin* method, which leads to a fulfillment of Eq. (4) in an integral sense on each panel.

The application of the *Galerkin* method as described, yields an equation like

$$\sum_{j=1}^{N_\mu} \mu_j \underbrace{\int\limits_{\boldsymbol{x}\in S} \left[\Psi_i^{p_\mu}(\boldsymbol{x}) \int\limits_{\boldsymbol{x}_0\in S} \Psi_j^{p_\mu}(\boldsymbol{x}_0)\, \boldsymbol{n}(\boldsymbol{x}_0) \cdot \nabla \left(\frac{1}{\|\boldsymbol{x}_0 - \boldsymbol{x}\|} \right) \mathrm{d}S \right] \mathrm{d}S}_{=c_{ij}}$$

$$= -\sum_{j=1}^{N_\sigma} \sigma_j \underbrace{\int\limits_{\boldsymbol{x}\in S} \left[\Psi_i^{p_\mu}(\boldsymbol{x}) \int\limits_{\boldsymbol{x}_0\in S} \Psi_j^{p_\sigma}(\boldsymbol{x}_0) \left(\frac{1}{\|\boldsymbol{x}_0 - \boldsymbol{x}\|} \right) \mathrm{d}S \right] \mathrm{d}S}_{=b_{ij}} \quad (5)$$

for each global basis function $\Psi_i^{p_\mu}$ representing the unknown doublet distribution. Subsequently, an equation system in the form

$$\boldsymbol{C}\boldsymbol{\mu} = -\boldsymbol{B}\boldsymbol{\sigma} \quad (6)$$

results, where the system matrices $\boldsymbol{C}$ and $\boldsymbol{B}$ are comprised of all the influence factors c_{ij} and b_{ij}. Two integrations are required for the calculation of each influence factor. While the inner integration is performed analytically, the outer integrals are approximated via *Gauß* quadrature. The required analytic integrals over the product of basis functions and fundamental kernels are derived or taken from the literature [6,8]. Finally, a *Kutta* vortex is introduced at the trailing edge of lifting bodies to enforce the *Kutta* condition. This vortex acts equivalently to an infinite doublet panel with one vertex at the trailing edge and one at infinity. The vortex strength is determined by the jump in doublet strength at the trailing edge. A slight modification of the matrices $\boldsymbol{C}$, $\boldsymbol{B}$ and vector $\boldsymbol{\mu}$ incorporates this condition and vortex into Eq. (6). Solving the system yields the vector of unknown doublet strengths, which fully defines the flow field.

2.3 Comparison to Traditional Panel Methods

In contrast to traditional collocation point methods, the *Galerkin* approach offers a general formulation for arbitrary polynomial orders of the singularity distributions, provided the required kernel integrals can be solved. Hence, it lends itself to the implementation of variable- or high-order panel methods. The surface potential distributions are continuous by design, while continuity of the slope is not enforced. With collocation methods, the construction of higher order numerical schemes is possible, but much less generic. It is worth noting, that the *Galerkin* method with constant singularities using an one point *Gauß* quadrature is equivalent to the traditional low-order, collocation point panel method. However, the flexibility comes at a cost: With the *Galerkin* method the inner integrals have to be evaluated once for each quadrature point and, thus, the computational cost increase by about that factor.

The main disadvantages of low-order panel methods arise from the fact that their doublet distributions feature discontinuities at the element borders. Doublet panels can be represented by a combination of a vortex distribution and discrete vortices at the panel boundaries [6]. Therefore, doublet panels feature a velocity influence equal to the one induced by these vortices. In two dimensions these are vortices at the panel vertices and for the three-dimensional case it is a vortex ring around the panel boundary. In case of continuous doublet distributions these boundary vortices of adjacent panels have equal strength but are counter-rotating and, hence, counter each other. However, the jumps present with constant doublet distributions act as vortices effectively and introduce spurious, non-physical velocities into the flow field. This effect decays rapidly with the distance but significantly decreases the accuracy with which velocities in the immediate vicinity of the boundary are modeled. In principle, the desired boundary condition is one of impermeable panels to ensure a flow around the body. However, for the reason described earlier, this impermeability condition does not hold for low-order panel methods and the literature regularly states so, e.g.: "Actually, for constant-strength panels, the panels themselves leak rather badly everywhere except at the control points" [3]. Nevertheless, low-order panel methods are successfully employed given a sufficient number of panels.

Aerodynamic panel simulations are mostly targeted at surface pressures and, hence, surface velocities from which these are calculated. Flow velocities at arbitrary points in the field can be determined by superposition of all potential influences. However, the superposition of low-order doublet panels yields inaccurate results on the surface itself and in the immediate vicinity of it due to the spurious velocities described previously. Since superposition is a costly N^2-problem, it is rarely used for the determination of surface velocities anyway. Instead, surface velocities are calculated from the derivative of the panel doublet strength for *Morino* type panels [8,12]. Low-order panel methods utilize finite-difference schemes to reconstruct the doublet derivatives from the constant basis functions. While this is feasible in two dimensions, it is more complicated and less accurate for three-dimensional geometries. In contrast, for high-order methods it is not necessary to reconstruct the derivatives, they can be calculated directly and strictly local. Linear doublets yield panel-wise constant surface velocities and the linear velocities obtained from quadratic doublet panels are continuous at the panel boundaries in addition.

3 Numerical Results

3.1 Test Cases with Analytic Solutions

The two-dimensional high-order *Galerkin*-type panel method described in Sect. 2.2 was implemented in *Python*. This prototype was applied to two test cases, for which conformal mappings provide analytic solutions for error evaluation: i) the flow around a circle under an angle of attack of $\alpha = 0°$ and ii) a *Kármán-Trefftz* airfoil [2] emerged in a flow at $\alpha = 5°$. A *Kármán-Trefftz* airfoil is similar to the more common *Joukowski* airfoil but features a finite trailing edge angle, which renders the panel method applicable to this case. Figure 1 depicts the paneled surfaces of both test cases and the solutions resulting from the panel method. Both computations utilized 32 panels and linear basis functions for the sources and quadratic ones for the doublets. The flow field is pictured by streamlines, which are colored by the absolute value of the flow velocity $\boldsymbol{u}$ normalized by the free-stream velocity $\boldsymbol{U}_\infty$. As expected, the closed bodies exhibit the free-stream velocity vectors internally, since the flow inside the bodies is governed by the internal potential, which is enforced to be equal to the free-stream potential. Additionally, Fig. 1a pictures the potential field by equipotential lines printed in grey, which meet the streamlines orthogonally. For the airfoil, a lifting body, the *Kutta* condition was enforced at the trailing edge vertex.

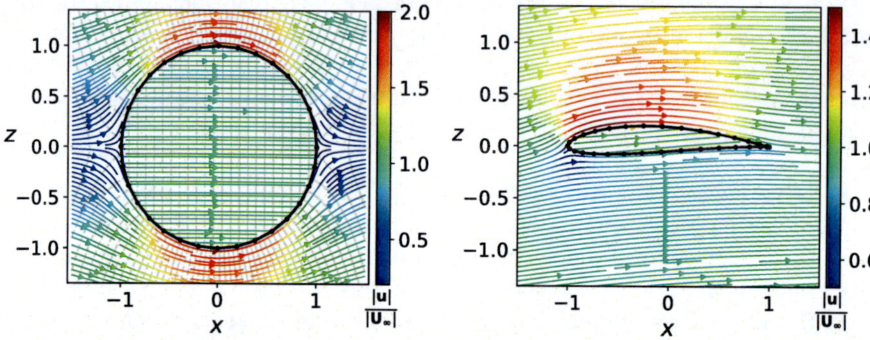

a) Circle at $\alpha = 0^o$.

b) *Kármán-Trefftz* airfoil at $\alpha = 5^o$.

Fig. 1. Streamline plots resulting from computations with 32 panels using linear sources ($p_\sigma = 1$) and quadratic doublets ($p_\mu = 2$).

Figure 2 compares the fulfillment of the boundary condition at the topmost two panels of the circle test case for low- and high-order methods. The low-order solution illustrated in Fig. 2a clearly exhibits the influence of the three effective vortices located at the vertices of the two panels. Additionally, the body internal and external flow does not represent the desired body geometry in the vicinity of the boundary. Streamlines cross panels over large portions of the body, thus the impermeability condition is violated on large portions of the surface of the body. However, at a certain distance to the body, the streamlines mimic the bodies shape, as intended. For this reason, low-order methods are capable to model the flow around a body. However, one has to take care in how to reconstruct the surface velocities and pressures as will be detailed later. In comparison, Fig. 2b depicts the solution with a linear source and quadratic doublet distribution. As described in Sect. 2.2 the doublet distribution is continuous by design and, hence, no spurious vortices are introduced to the potential field. The resulting velocity field conforms well with the body geometry.

Furthermore, Fig. 3 compares the pressure distributions on the surface of the *Kármán-Trefftz* airfoil at an angle of attack of $\alpha = 5°$, where the airfoil

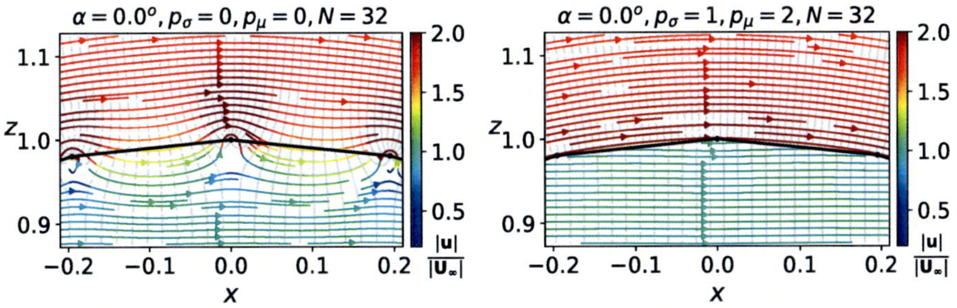

a) Low-Order solution. b) High-order solution.

Fig. 2. Comparison of the flow field at the topmost two panels for the circle case.

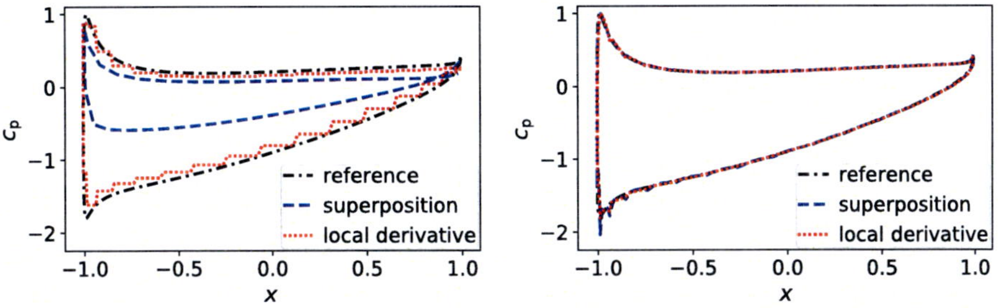

a) Low-order solutions, $p_\mu = p_\sigma = 0$. b) High-order solutions, $p_\mu = 2$, $p_\sigma = 1$.

Fig. 3. Pressure coefficient on surface of *Kármán-Trefftz* airfoil at $\alpha = 5°$ using 32 panels.

is represented by 32 panels. The surface pressure coefficient is calculated as $c_\mathrm{p} = 1 - \boldsymbol{u}^2/\boldsymbol{U}_\infty^2$ directly. The pressure computed via conformal mapping serves as an analytic reference solution. Pressures resulting from the panel method were calculated from superposition of all panel influences as well as the local derivative of the surface doublet strength. Figure 3a shows the solution obtained with a low-order panel method applying panel-wise constant fundamental solutions. As stated in Sect. 2.3 the direct superposition fails to represent the surface pressure for low-order methods. Utilizing a finite-difference approximations of the doublet derivatives yields more accurate results. Yet, the pressure distribution is panel-wise constant. In comparison, the high-order panel method results pictured in Fig. 3b are more accurate. The local derivatives of the quadratic doublet basis functions can be determined strictly local and yield panel-wise linear velocities and pressures. Additionally, successful approximations of the velocities on and near the surface can be achieved by superposition.

Finally, the convergence of the potential and velocities for panel methods of different orders for the circle example are compared. The errors are expressed in terms of L^2-norms, which were determined integrating over all panels by *Gauß* quadrature. At first, Fig. 4a presents the convergence of the error in sur-

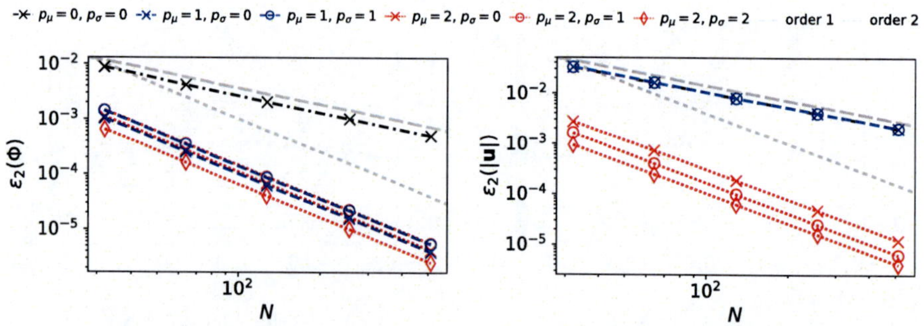

a) Surface potential convergence.

b) Surface velocity convergence.

Fig. 4. Convergence of L^2-norm of the error in potential $\varepsilon_2(\phi)$ and in absolute value of velocities $\varepsilon_2(\|\boldsymbol{u}\|)$ on the surface of the circle represented by N panels.

face potential, denoted as $\varepsilon_2(\phi)$, with respect to the number of panels N. The low-order method exhibits a linearly decreasing error in surface potential. All high-order methods feature a quadratic decay of the potential error, since the error is dominated by the linear geometry representation. Similarly, Fig. 4b contains the results for the errors of absolute values of surface velocities $\varepsilon_2(\|\boldsymbol{u}\|)$. The velocities were calculated from local derivatives of doublet strength. Zeroth and first order methods result in linear order error reduction, because both produce equivalent, panel-wise constant velocity distributions. Note, that for non-equidistant, more complicated meshes the finite-difference approach of low-order panel methods might be less successful in reconstructing the velocities. Applying quadratic doublets increases the accuracy of surface velocities, especially since the velocity distribution is no longer panel-wise constant.

3.2 Miranda Test Case

An additional test case is inspired by and was derived from a three-dimensional nacelle defined by a rotated *NACA* profile [11]. This example is especially difficult to solve for low-order panel codes, because internal flows are sensitive to leakage through the panels. The two-dimensional variant consists of two *NACA 0010* airfoils of chord length $L = 2$ with a passage measuring $0.1\,L$ between them. Figure 5a pictures the paneled surfaces using $N = 32$ elements per airfoil. Since the flow field is symmetric about the center line, both solutions are compared in a single plot. The top half of the picture, above the dash-dotted line, illustrates the low-order solution, while the bottom half shows the high-order solution for comparison. In this plot streamlines are colored by absolute value of normalized velocity. In particular on the side of the airfoil facing the other airfoil, the low-order solution exhibits considerable flow through the panels, the streamlines do not follow the geometry. Contrary, the streamlines of the high-order solution represents the geometric boundary much smoother and more accurate. Figure 5b compares the surface pressure coefficients of solutions with constant, linear and

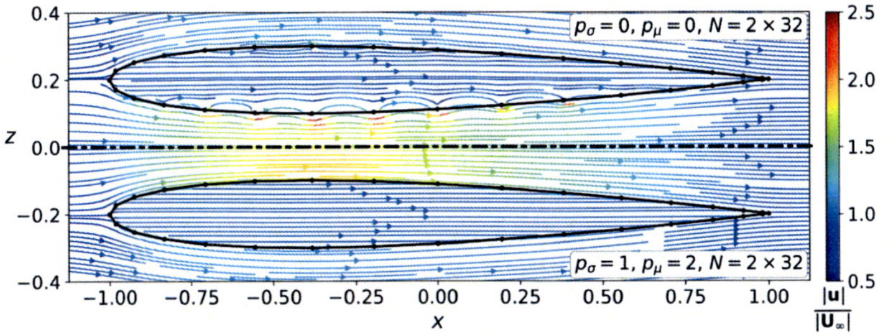

a) Streamline plot comparing the low- and high-order solutions using the same mesh (top half: low-order method, bottom half: high-order).

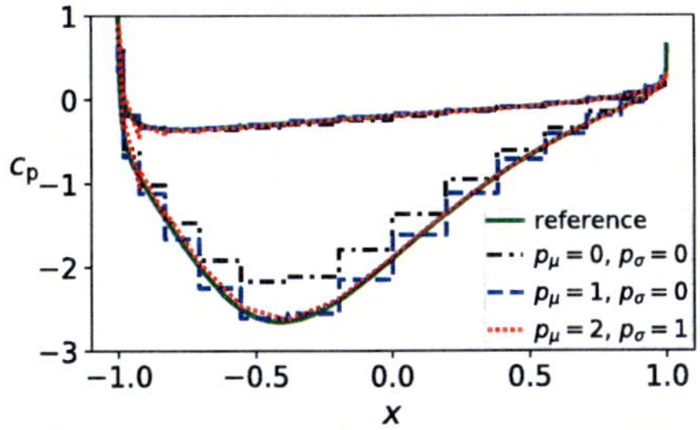

b) Pressure at surface of one profile.

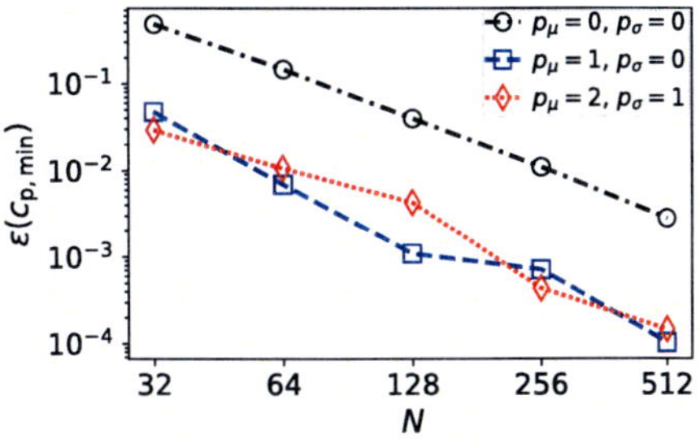

c) Error of pressure minimum.

Fig. 5. Panel method representation of the flow between two *NACA 0010* airfoils.

quadratic doublets to a reference solution. Here, a high-order ($p_\sigma = 1$ and $p_\mu = 2$) panel method solution using 4096 panels serves as the reference solution. The velocities underlying these pressure coefficients were calculated from the surface derivative of the doublets. Similarly to the the three-dimensional case [11], the low-order solution undervalues the pressure drop between the airfoils substantially. Solutions with continuous doublet distributions, i.e. at least linear doublets, yield far more accurate results even with a panel count as low as 32 per airfoil. Eventually, Fig. 5c depicts the error between the minimum pressure coefficient of a solution the reference solution. High-order solutions represent the minimum surface pressure significantly more accurate than the solution obtained with panel-wise constant singularities. Furthermore, the pressure minimum of high-order solutions converges faster towards the reference value with increasing panel counts, than the low-order solution.

4 Conclusions

A two-dimensional, steady-state high-order panel method was implemented using *Morino*-type boundary conditions and a *Galerkin* formulation. This prototype was used to conduct simulations with high- and low-order numerical schemes, since the *Galerkin* approach is independent of the orders of singularities employed. The results presented in this paper confirmed the advantages for which the aforementioned methods have been chosen. High-order methods offer faster convergence of potential and velocities. Moreover, the surface velocities can either be evaluated by superposition or strictly local, which improves the applicability of the method. In comparison low-order solutions are not able to calculate surface velocities accurately by superposition. For local reconstruction of surface velocities, the low-order methods refer to finite-differences schemes, which incorporate at least one layer of neighbor panels. Additionally, high-order methods represent the boundary condition accurately, while traditional low-order methods allow for flow through the surface on large portions of each panel. Extensions to three dimensions will face challenges linked to increased implementation effort and higher computational cost per panel of high-order panel methods. However, the studied two-dimensional high-order prototype performed promisingly overall. Hence, future work will continue towards the development of an high-order unsteady three-dimensional panel method to showcase the potential of advanced, robust mid-fidelity methods for dynamic loads analyses and aeroelastics.

References

1. Bauer, M., Abdel-Maksoud, M.: A 3-D potential based boundary element method for modelling and simulation of marine propeller flows. IFAC Proc. Vol. **45**(2), 1179–1184 (2012)
2. Blom, J.J.H.: Some characteristic quantities of Karman-Trefftz profiles. Techreport NASA-TM-77013, NASA (1983)
3. Erickson, L.L.: Panel methods - an introduction. Techreport NASA-TP-2995, A-89266, NAS 1.60:2995, NASA Ames Research Center (1990)
4. Fiddes, S.P., Burkett, C.W., Kier, T.: An advanced panel method for compressible subsonic unsteady flow past complex geometries. In: International Forum on Aeroelasticity and Structural Dynamics (IFASD) (2015)
5. Hess, J.: A higher-order panel method for three-dimensional potential flow. In: 7th Australasian Conference on Hydraulics and Fluid Mechanics, pp. 517–520 (1980)
6. Hess, J.L.: Calculation of potential flow about arbitrary three-dimensional lifting bodies. Technical report MDCJ 5679-01, McDonnell Douglas Corporation (1972)
7. Johnson, F.T.: A general panel method for the analysis and design of arbitrary configurations in incompressible flows. Technical report NASA-CR-3079, NASA (1980)
8. Katz, J., Plotkin, A.: Low-Speed Aerodynamics. Cambridge Univ. Press, Cambridge (2001)
9. Kier, T.M.: Integrated flexible dynamic maneuver loads models based on aerodynamic influence coefficients of a 3D panel method. In: 56th AIAA/ASCE/AHS/ASC Structures, Structural Dynamics, and Materials Conference. American Institute of Aeronautics and Astronautics (2015)
10. Kunze, P.: Evaluation of an unsteady panel method for the prediction of rotor-rotor and rotor-body interactions in preliminary design. In: 41st European Rotorcraft Forum (2015)
11. Miranda, L.R.: Application of computational aerodynamics to airplane design. J. Aircraft **21**(6), 355–370 (1984)

12. Morino, L.: A General Theory of Unsteady Compressible Potential Aerodynamics. Technical report CR-2464, NASA (1974)
13. Willis, D., Peraire, J., White, J.: A quadratic basis function, quadratic geometry, high order panel method. In: 44th AIAA Aerospace Sciences Meeting and Exhibit. American Institute of Aeronautics and Astronautics (2006)

Check for updates

Numerical Investigation of the Unsteady Flow Field Past an Offshore Wind Turbine in Maintenance Operations

Marion Cormier(✉) and Thorsten Lutz

Faculty 6: Aerospace Engineering and Geodesy, University of Stuttgart,
Institute of Aerodynamics and Gas Dynamics, Pfaffenwaldring 21,
70569 Stuttgart, Germany
{cormier,lutz}@iag.uni-stuttgart.de
http://www.iag.uni-stuttgart.de/en/

Abstract. In the present investigation, the unsteady flow field around a standstill offshore wind turbine is analysed in the hovering and approach regions of maintenance helicopters in order to identify critical flight regions. Three different inflow conditions are analysed, at the rated and cut-off wind speeds. One fictive hovering region and two fictive flight regions for maintenance helicopters are defined and the flow field is separately analysed in these regions. The turbulence kinetic energy and the vertical inclination angle are evaluated. These give indirectly information about the expected blade loads and possible stall occurrence, and a turbulence criterion based on the standard deviation of the vertical velocity component is applied in order to identify possible critical flight regions. The analyses based on the turbulence kinetic energy and on the turbulence criterion suggest the presence of critical areas in the wake of the nacelle for the high wind speed only. Moreover, the analysis of the vertical inclination angle suggests high fluctuations of the velocity direction in the nacelle wake even for the rated wind speed, with possible stall occurrence in the rotor inner part.

Keywords: CFD · Wind turbine aerodynamics · Offshore wind · Maintenance helicopter operations

1 Introduction

Modern offshore wind farms represent one approach to increase the renewable energy's market shares through the high and steady energy yields and optimized energy production costs. Hereby, the maritime inflow conditions with higher mean wind speeds and more constant winds than onshore contribute to the high energy production [13]. Offshore wind farms are not free from regular maintenance activities that necessarily affect the total energy output. In order to reduce off-production time during maintenance operations, helicopter transport of crew

© The Author(s), under exclusive license to Springer Nature Switzerland AG 2021
A. Dillmann et al. (Eds.): STAB/DGLR Symposium 2020, NNFM 151, pp. 592–603, 2021.
https://doi.org/10.1007/978-3-030-79561-0_56

members and material can be opted for rather than ship transport. Yet, helicopter operations are sensitive to the weather conditions and the local flow field over the rotor area. These conditions can, in extreme cases, endanger the flight safety. In the research project HeliOW, a systematic approach has been chosen to study the effects of the wind farm flow on helicopter flight dynamics during maintenance operations. A simulation chain has been built to this extend, in which flow fields from highly-resolved CFD simulations are performed at the University of Stuttgart and used in two helicopter flight simulators: piloted flights through the flow field are conducted within a research flight simulator at the German Aerospace Center (DLR) and desktop-simulations with a coupled pilot model are performed at the University of Munich (TUM).

Wind turbine wakes have increasingly received attention over the last decades, as suggest the numerous numerical and experimental studies related in review articles [12,15,20]. The wakes are mainly characterized by the wind deficit caused by the energy extraction from the rotor, local changes in the wind direction and an increased turbulence level. It can be distinguished between the near-wake region and the far-wake region. In the near-wake region, features related to the blade aerodynamics like concentrated blade tip vortices are present. These can be seen by an approaching helicopter as abrupt changes in inflow velocity and direction that impact the unsteady aerodynamic forces and thus the flight dynamics. Further downstream the wind turbine, the tip vortex helix breaks down into small-scale turbulence in the far-wake region, increasing the momentum of the flow field and modifying the inflow conditions of a downstream located wind turbine.

One can distinct two scenarios which can be hazardous for helicopter flight dynamics during offshore maintenance operations: the perturbations which result from the wake of operating wind turbines when crossing the wind park to reach the malfunctioning wind turbine, and the hovering above the nacelle, at more than 100 m above the sea level and under strong and turbulent winds. Previous studies have addressed the first scenario [1,3,4,18] The present study focuses on the second one, the hovering above the nacelle. The helicopter is exposed to high altitude winds, eventually with partial ground effects if the helicopter blades stick out of the nacelle surface, and should meanwhile keep a constant position for roping crew and material. To simulate flight dynamics under these conditions, time-resolved CFD flow fields of the wind turbine in maintenance configuration are generated, which will be used in a further step in helicopter flight simulators by the project partners TUM and DLR, as described above. To better understand the helicopter reactions observed in the simulators, a detailed analysis of the flow physics around the turbine is mandatory.

In the present study, Detached Eddy Simulations (DES) of a wind turbine in stand still conditions are performed and analyzed. Three cases are considered in total: a wind turbine at both the rated wind speed and the cut-off wind speed, and a wind turbine in full-wake inflow conditions at the cut-off wind speed. The unsteady flow field around the stand still wind turbine is analysed in the hovering and approach regions of maintenance helicopters in order to identify critical regions for the helicopter.

In the next section, the methods applied to simulate the flow field around the turbine are outlined. The compressible flow solver is first introduced and the numerical setup of the wind turbine for the different computational cases is then described. In Sect. 3, the results of the simulations and the flow field analyses are presented. The hovering and approach region are hereby separately analyzed, allowing a clear identification of the flow mechanisms in each phase of maintenance operations. Finally, the conclusions of the present study are summarized and an outlook of research ideas is proposed.

2 Methods

In this section, the numerical methods to simulate the unsteady flow fields and the computational cases are presented. The flow solver FLOWer is first introduced, then the considered wind turbine. It follows a description of the numerical setup used for the simulations and the three different cases investigated in the present study are presented.

2.1 The Flow Solver FLOWer

In the current study, the finite-volume flow solver FLOWer, originally developed at the German Aerospace Center (DLR) [8], is applied to simulate the unsteady flow field around the wind turbine. The unsteady compressible Navier-Stokes equations are solved block-wise with a dual time-stepping scheme and a multigrid algorithm in order to accelerate the convergence of the solution. Each turbine component is meshed separately and the structured grids are embedded within a Cartesian background mesh thanks to the Chimera technique [2].

To capture the unsteady flow separation behind the square-edged nacelle (see Sect. 2.3), Delayed Detached Eddy Simulations (DDES) are performed with the Menter SST turbulence model [10]. Distinction is made between the boundary layers of the wind turbine, where the 2nd order Jameson-Schmidt-Turkel (JST) [5] spatial scheme is used, and the wake region, where the 5th order Weighted Essentially Non-Oscillatory (WENO) [16] scheme is used in order to capture well vortical structures.

The actuator-line method (ACL) [17], that enables consideration of the impact of the rotor on the flow field without need for detailed discretisation of the blade surfaces, is also available in FLOWer and applied in the present study. The blade aerodynamic forces are estimated using the blade-element theory for each blade segment as a function of the local angle of attack, the local undisturbed wind velocity and the local lift and drag coefficients. The aerodynamic force is added in the flow field as a source term along a rotating lifting line, whereby a Gaussian projection function is used for numerical stability reasons. The local angle of attack and inflow velocity are estimated from the flow field at the beginning of each time step.

To mimic the atmospheric turbulence, synthetic turbulent inflow is realised as a superposition of a mean wind profile and turbulent fluctuations. The mean wind profile is imposed as a Dirichlet boundary condition at the inlet of the flow domain and the synthetic turbulent fluctuations, generated with the Mann model [9], are added as source terms downstream the inflow plane.

2.2 The NREL 5-MW Wind Turbine

The generic NREL 5-MW wind turbine is considered, which presents a rotor radius of $D = 63$ m and a hub height of $z_h = 90$ m [6]. Two reasons motivated the choice of this wind turbine: first, it is a public access reference wind turbine, second, the geometry of the nacelle, although very simple, is similar to modern square section nacelles for offshore wind turbines. Although a lot of different designs exist, this bluff body form represents a conservative case with higher perturbations expected in the nacelle wake than real modern wind turbines actually generate. A slightly modified operating point of the NREL 5-MW wind turbine [3, 7] is investigated in the present study.

Fully-Resolved Wind Turbine. For maintenance operations, the wind turbine is placed in the so-called "L-position": the rotor plane is oriented parallel to the wind and one blade points horizontally against the wind direction. This way the helicopter can approach against the wind, from the wake of the turbine, and hover over the nacelle with a reduced collision risk with one of the turbine blades. The NREL 5-MW wind turbine has already been investigated with FLOWer in the past, coupled with a structure solver [7]. The same blade, tower and blade connector grids are used in the present study, but the nacelle grid was adapted to the requirements of the current simulation case. The structured grids were generated with the commercial software Pointwise and with in-house automated scripts.

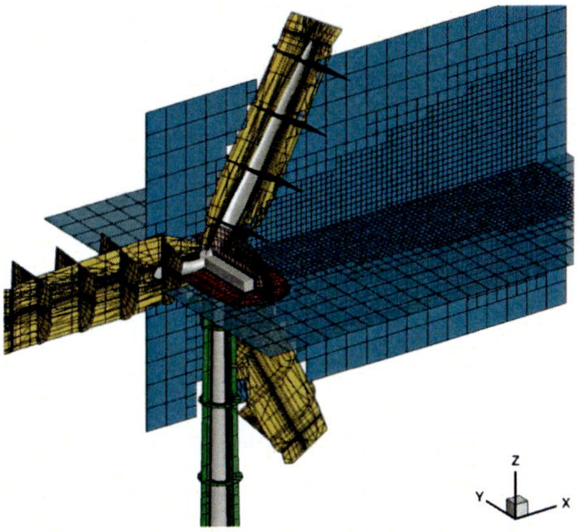

Fig. 1. Structured grids of the fully-resolved wind turbine: blades (yellow), nacelle (red), tower (green) and background with local refinement (blue). Only every $8th$ line is represented here.

For all components, the wall distance fulfills the condition $y^+ \approx 1$. All the turbine grids are embedded in a Cartesian background mesh which features a smallest cell size of $\Delta_x = 1$ m around the turbine and coarsens towards the top, lateral and downstream boundaries using hanging grid nodes. A refined Cartesian region with isotropic cells of size $\Delta_x = 0.2$ m is added in the wake of the nacelle. The setup is represented in Fig. 1, whereby the background mesh is not shown for more clarity.

Actuator Line Rotor. While the turbine of interest is fully resolved, the ACL approach was used to capture the far-wake effects of a neighbouring wind turbine. Previous studies have shown that, with this method, the far wake dynamics are well reproduced at reduced computational costs, compared to an equivalent fully-resolved simulation [19]. As the present study focuses on the detailed aerodynamics of the maintenance wind turbine, the ACL is considered suitable to model the far-wake effects of the upstream wind turbine and optimize computational costs. 99 unevenly distributed blade elements are used for each blade, with a higher density at the blade root and blade tip, where higher gradients in radial direction are expected. A comparison of the line loads for the ACL model and the fully-resolved setup of the current turbine can be found in [3].

2.3 Computational Cases

Three cases are considered in total: an isolated wind turbine in atmospheric inflow at both the rated wind speed and the cut-off wind speed, and a wind turbine in full-wake operating conditions at the cut-off wind speed. The cut-off wind speed is also the typical maximum wind speed for helicopter approaches, representing thus an extreme operation case. The non-isolated case was introduced in order to investigate the effects of a neighbouring wind turbine on the target nacelle to be hovered. The upstream wind turbine is aligned in the inflow direction, creating full-wake inflow conditions for the turbine of interest. The atmospheric inflow wind profile and turbulence characteristics were determined from large offshore databases available in the literature [11]. The main characteristics of the three different simulation setups are summed up in Table 1. The more than 100 Mio. extra cells in Case III are dedicated to the transport of the highly resolved wake of the upstream wind turbine to the turbine of interest.

Table 1. Overview of the simulation cases: wind velocity at hub height V_W, turbulence intensity I and length scale L used in the Mann model, rotational speed Ω of the upstream wind turbine, dimensions of the computational box, time step and total cell number of the setup

Case	V_W [m/s]	I [%]	L [m]	Ω [rpm]	Computational box L_xxL_yxL_z [m]	Time step [ms]	Cells [Mio.]
Case I	11.3	5	45	–	700 m × 700 m × 640 m	5.2	116
Case II	25	6.84	60	–	700 m × 700 m × 640 m	3.4	116
Case III	25	6.84	60	12.1	1820 m × 700 m × 640 m	3.4	247

3 Results

In this sections, the resulting flow fields are presented and analyzed. First, an overview of the 3D unsteady flow field around a stand-still offshore wind turbine is given and the analysis regions are defined. Then, these regions are separately analyzed, allowing a clear identification of the flow mechanisms in each phase of maintenance operations. The turbulence kinetic energy (TKE) and angles are quantitatively analyzed. The results are finally compared to a turbulence criterion for safe helicopter operations.

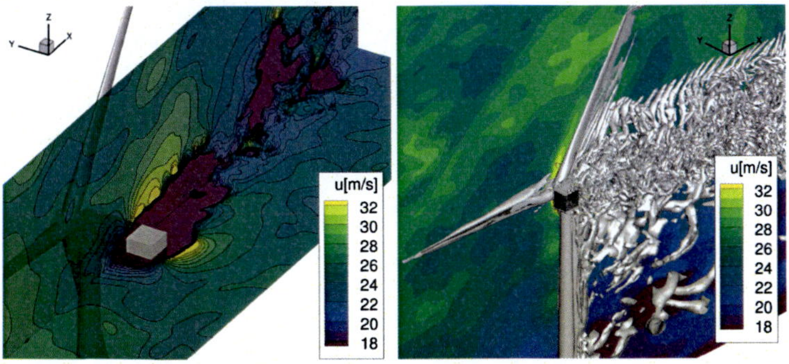

Fig. 2. Instantaneous flow field around the wind turbine in Case II ($V_W = 25\,\mathrm{m/s}$). Streamwise velocity component in a L-cut through the nacelle (left) and isosurface of the λ_2 vortex identification criterion (right).

3.1 Structure of the Time-Resolved Flow Field

Figure 2 (left) shows exemplarily the instantaneous distribution of the streamwise velocity component around the wind turbine for the Case II ($V_W = 25$ m/s). The flow is deflected by the nacelle and a recirculation area is formed in its wake. It forms a von Kármán vortex street in the wake, characterized by strong velocity fluctuations at the separation frequency. Figure 2 (right) shows the vortical structures with help of the λ_2 vortex identification criterion. Highly turbulent structures in the wake of the nacelle can be observed, that are expected to affect the helicopter reaction when entering this region.

3.2 Definition of the Helicopter Flight Regions

In order to separately analyze the flow field in the approach and hovering phases of the helicopter, corresponding regions of the flow field have been defined. Figure 3 shows the contour plot of the streamwise velocity component in a horizontal slice of the flow field at the altitude of 101 m. This is estimated to be the helicopter rotor plane from the macro-data of an EC-135 helicopter hovering 5 m above the nacelle upper surface. The black circular area represents the helicopter rotor during hover, and the black and red parallel lines two fictive approach paths. Usually, pilots approach the nacelle with a lateral angle and move laterally towards the hover position at the end of the approach phase. The black parallel lines represent an approach at an angle of 15° to the wind turbine rotor plane while the red parallel lines represent a more exceptional approach, parallel to the wind turbine blades.

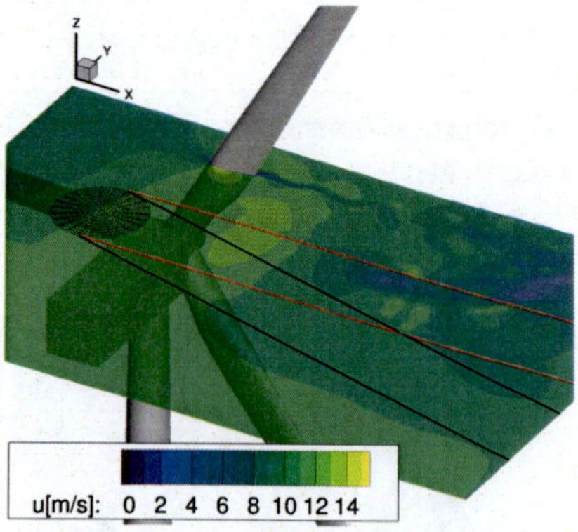

Fig. 3. The helicopter hovering region over the nacelle (black circle) and flight approach paths (red and black parallel lines) at the altitude of 101 m.

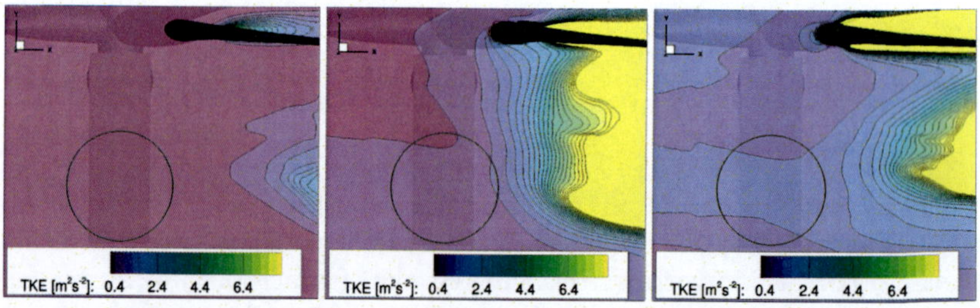

Fig. 4. Turbulence kinetic energy from a 12 s long signal of 300 snapshots at the altitude $z = 101$ m. The black circle represents the helicopter rotor area. The Hann windowing function and a high-pass filter at 0.5 Hz have been used. From left to right: Case I, Case II and Case III.

3.3 Turbulence Kinetic Energy

Hover Region. Figure 4 shows in a top view the turbulence kinetic energy (TKE) in the helicopter rotor plane when hovering over the wind turbine nacelle.

The analysis of TKE shows a maximum value of 2 m^2s^{-2} in all considered cases over the rotor area. It remains in a range that can be met in turbulent atmospheric conditions, suggesting no considerably higher perturbations on the blade loads than in a free-stream flight.

Approach Region. The TKE has been evaluated along the two approach paths defined in Sect. 3.2. The mean value of the TKE over the 2D rotor area has been extracted for each position along the flight path and plotted on Fig. 5. There is a consistent increase of turbulence kinetic energy with increasing distance from the hover area, caused by the turbulent wake of the nacelle. The parallel approach path, in red, features for all considered cases about 5 times higher TKE values than the transverse approach path, in black. Although Case III presents a higher TKE than Case II directly over the nacelle, in agreement with Fig. 4, the opposite trend is observed in both approach paths. This could be explained by an increased turbulent mixing in the nacelle wake in the full-wake event, due to the higher inflow turbulence, that leads to lower fluctuation amplitude in the approach flight in Case III. Remarkably, the TKE level in the transverse approach is for all three cases below 6 m^2s^{-2} and close to the Case I level from the parallel flight. It suggests that even at high wind speeds, choosing a transverse approach path of at least 30° angle from the wind turbine rotor plane reduces strongly the fluctuations amplitude as it draws away from the von Kármán vortex street fluctuations.

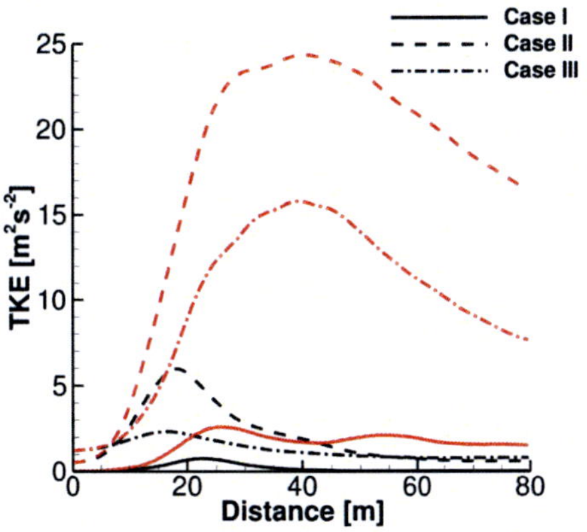

Fig. 5. Mean value of the turbulence kinetic energy over the rotor area along the approach paths at $z = 101$ m, from a 12 s long signal of 300 snapshots. The Hann windowing function and a high-pass filter at 0.5 Hz have been used. Red: parallel flight path, black: transverse approach path.

3.4 Inclination Angle

The vertical inclination angle α corresponds to the angle between the wind velocity vector in the inertial frame of reference and its projection in a horizontal plane. Its analysis gives information about the inflow wind vertical orientation and indirectly about possible occurrence of stall on the helicopter blades. Figure 6 shows the mean and instantaneous evolution of α along the final approach flight paths. The time-average inclination angle is presented in Fig. 6 (left), where the mean value has been extracted over the 2D helicopter rotor area for each position along the flight path. The values of α_{mean} are quite close to each other for all cases and for both approach paths, as it characterizes the mean flow

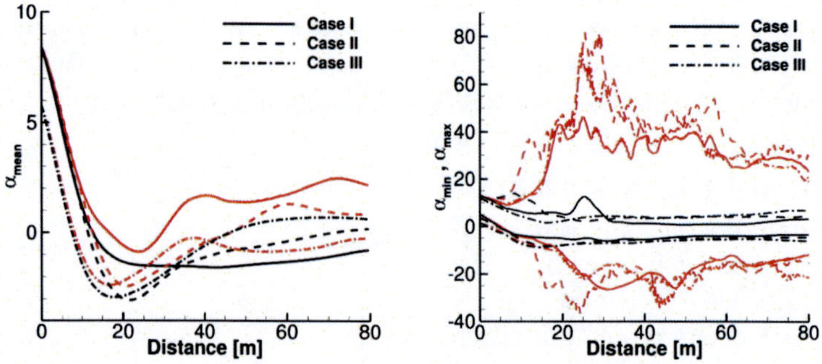

Fig. 6. Average of the mean inclination angle over the helicopter rotor area along the approach paths (left) and envelope of the instantaneous inclination angle at the centerline of the approach paths (right). Red: parallel path, black: oblique path.

field orientation around the nacelle. Directly above the nacelle, the upwards displacement of the mean flow is characterized by positive α_{mean} values, followed by the downwash directly downstream the nacelle with negative inclination angle. Figure 6 (right) shows the instantaneous minimal and maximal values of α along the centerline of both approach paths. Unlike the mean inclination angle, the magnitude of the instantaneous inclination angle features a strong dependency on the chosen flight path. The extreme inclination angles vary from $-40°$ up to more than $80°$ for the approach path through the von Kármán vortex street, while it stays in a range between $-10°$ and $15°$ for the oblique flight path. The resulting variation of the local angle of attack can be approximated as $atan(\frac{\sigma_w}{v_{loc}})$, with σ_w the standard deviation of the vertical wind velocity component and v_{loc} the streamwise component of the local inflow velocity at the blade. A rotational frequency of the EC-135 rotor of 395 RPM and an approach velocity of 4 m/s are assumed here. Considering that $\sigma_w \leq 4$ m/s in the approach region, it would induce on the retreating blade a maximal variation of the angles of attack of 1.7° at 80% of the helicopter blade radius and 4.3° at 40% of the helicopter blade radius. These variations could lead to strong and unpredictable stall events in the inner blade region of the helicopter rotor, where the contribution of the rotational blade velocity is the least. This could, in the worse case, cause critical situations concerning the flight stability.

3.5 Turbulence Criterion for Safe Helicopter Operations

A turbulence criterion for safe helicopter operations has been developed in [14] using piloted flight simulations in a research flight simulator together with data from wind tunnel tests on offshore platforms. They identified a relationship between pilot workload rating and the standard deviation of the vertical velocity component σ_w, from which they derived a turbulence criterion. According to this criterion, σ_w should remain below 2.4 m/s to ensure safe operation conditions. Figure 7 shows the standard deviation σ_w in the helicopter rotor plane for the three cases investigated here. It appears that for the rated wind speed case (Case I), a certain uncritical area can be identified in the observed region. For the high velocity cases II and III, critical state is observed in the regions marked out by the red isoline. This concerns only the zone downstream the hover area, in the von Kármán vortex street of the nacelle. For practical applications it means that, according to that criterion, the helicopter would experience a critical flight state when drifting 5 to 10 m downstream the hover area or if approaching the nacelle parallel to the wind turbine rotor plane.

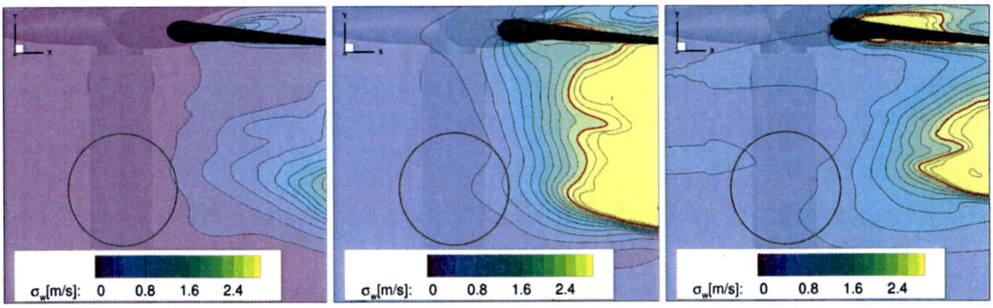

Fig. 7. Standard deviation of the vertical velocity component from a 12 s long signal of 300 snapshots at the altitude $z = 101$ m. The black circle represents the helicopter rotor area. The Hann windowing function and a high-pass filter at 0.5 Hz have been used. The red line represents the threshold value of 2.4 m/s. From left to right: Case I, Case II and Case III.

4 Conclusions

In the present study, unsteady flow fields around a wind turbine were analysed in order to identify possibly critical areas for maintenance helicopters. The high-resolution flow fields were computed for the rated and cut-off wind speed, and an additional case in full-wake inflow conditions was considered. Two fictive helicopter flight regions were assigned for the analysis of the helicopter inflow conditions: a static region over the nacelle for the hovering operations of a fictive EC-135 helicopter and two flight paths to this hovering region for the final helicopter approach. The analysis of the turbulence kinetic energy shows maximal values of 2 m^2s^{-2} in the hover area but up to 25 m^2s^{-2} in the approach area for the worse case. There is a dependency both on the chosen approach path and on the inflow velocity. The result is that at high wind speeds the final approach must be performed from the side, in order to avoid the highly fluctuating von Kármán vortex street. The application of a turbulence criterion, developed specifically for offshore helicopter operations, supports this statement. At high wind fields, the critical value of the standard deviation of the vertical velocity component is exceeded in the nacelle wake region. The analysis of the instantaneous and average vertical inclination angle shows that, independently from the inflow wind speed, flying through the vortex street could lead to unpredictable stall events in the rotor inner region. Theses findings suggest that, for all investigated cases, any drift of the helicopter in the nacelle turbulent wake could strongly affect the helicopter flight dynamics.

It should be noted that the present approach does not consider the interaction of the helicopter rotor aerodynamics, in particular the rotor downwash. Although it is supposed that these interaction effects could minimize the wind direction fluctuations observed at the blade, further analysis in a helicopter simulator or in a simulation frame accounting for the helicopter aerodynamics, for example with help of the ACL technique as seen in [3], shall be performed to confirm these results.

Acknowledgements. This research is funded by the German Federal Ministry for Economic Affairs and Energy (BMWi) within the framework of the German joint research project HeliOW (Grant number: 0324121). The authors gratefully acknowledge the HLRS Stuttgart for providing computational resources.

References

1. Bakker, R., et al.: Wind turbine wakes and helicopter operations. In: 44th European Rotorcraft Forum, Delft, The Netherlands, September 2018
2. Benek, J.A., Steger, J.L., Dougherty, F.C., Buning, P.G.: Chimera. A grid-embedding technique, Arnold Engineering Development Center Arnold AFB TN (1986)
3. Bühler, M., Weihing, P., Klein, L., Lutz, T., Krämer, E.: Actuator line method simulations for the analysis of wind turbine wakes acting on helicopters. J. Phys. Conf. Ser. **1037–6**, 062004 (2018)
4. Horvat, B., Hajek, M., Rauleder, J.: Analysing rotorcraft vortex encounter methods with a Lattice-Boltzmann method based GPU framework. AIAA Scitech Forum 0539 (2020)
5. Jameson, A., Schmidt, W., Turkel, E.: Numerical solution of the Euler equations by finite volume methods using Runge Kutta time stepping schemes. In: 14th Fluid and Plasma Dynamics Conference, vol. 1259 (1981)
6. Jonkman, J., Butterfield, S., Musial, W., Scott, G.: Definition of a 5-MW reference wind turbine for offshore system development. NREL, Golden, CO US (2009)
7. Klein, L., Gude, J., Wenz, F., Lutz, T., Krämer, E.: Advanced computational fluid dynamics (CFD)-multi-body simulation (MBS) coupling to assess low-frequency emissions from wind turbines. Wind Energy Sci. **3**, 713–728 (2018)
8. Kroll, N., Fassbender, J.K.: MEGAFLOW - Numerical Flow Simulation for Aircraft Design. Springer, Heidelberg (2005)
9. Mann, J.: The spatial structure of neutral atmospheric surface-layer turbulence. J. Fluid Mech. **273**, 141–168 (1994)
10. Menter, F.R.: Two-equation eddy-viscosity turbulence models for engineering applications. AIAA J. **32**, 1598–1605 (1994)
11. Peña, A., Gryning, S.-E., Mann, J.: On the length-scale of the wind profile. Q. J. R. Meteorol. Soc. **136**, 2119–2131 (2010)
12. Porté-Agel, F., Bastankhah, M., Shamsoddin, S.: Wind-turbine and wind-farm flows: a review. Bound. Layer Meteorol. **174**(1), 1–59 (2019)
13. Pryor, S.C., Barthelmie, R.J.: Comparison of potential power production at on-and offshore sites. Wind Energy **4**, 173–181 (2001)

14. Rowe, S.J., Howson, D., Turner, G.P.: A turbulence criterion for safe helicopter operations to offshore installations. Aeronaut. J. **110**, 749–758 (2004)
15. Sanderse, B., Van der Pijl, S.P., Koren, B.: Review of computational fluid dynamics for wind turbine wake aerodynamics. Wind Energy **14**, 799–819 (2011)
16. Schäferlein, U., Keßler, M., Krämer, E.: High order CFD-simulation of the rotor-fuselage interaction. 44th ERF, Moscow, Russia (2013)
17. Sørensen, J.N., Shen, W.Z.: Numerical modeling of wind turbine wakes. J. Fluids Eng. **124**, 393–399 (2002)
18. Strbac, A., Martini, T., Greiwe, D., Hoffmann, F., Jones, M. : Analysis of Rotorcraft Wind Turbine Wake Encounters using Piloted Simulation. 45th ERF, Warsaw, Poland (2019)
19. Troldborg, N.: Actuator line modeling of wind turbine wakes (2009)
20. Vermeer, L.J., Sørensen, J.N., Crespo, A.: Review of computational fluid dynamics for wind turbine wake aerodynamics. Wind Energy **14**, 799–819 (2003)

Check for updates

A Discontinuous Galerkin Chimera Method for Unsteady Flow Problems on Moving Grids

Fabian Genuit(✉), Manuel Keßler, and Ewald Krämer

Institute of Aerodynamics and Gas Dynamics (IAG), University of Stuttgart,
Pfaffenwaldring 21, 70569 Stuttgart, Germany
{genuit,kessler,kraemer}@iag.uni-stuttgart.de

Abstract. The discontinuous Galerkin (DG) method is a high-order method in space reducing the amount of cells needed for calculations compared to traditional computational fluid dynamics (CFD) solvers. The present study concerns the implementation of a DG Chimera method and its application to simulations with relative body movements. The chosen test cases of a rotating 2D cylinder and a slowly plunging NACA0012 airfoil deal with body movement and unsteady laminar flow at subsonic Mach numbers. The results obtained agree well with the reference. Furthermore, the flexible application of the DG Chimera method to unstructured grids is demonstrated.

Keywords: Discontinuous Galerkin method · Chimera method · Magnus effect · Slowly plunging NACA0012 airfoil

1 Introduction

In recent years, the application of the DG method has increased and indicates its potential to replace or at least augment the classical CFD methods, like the finite volume (FV) method. Advantageous for the DG method are the less restrictive requirements regarding the spatial resolution of the computational grid compared to FV solvers, since the spatial accuracy order may be chosen arbitrarily and unstructured grids can be used. Moreover, the parallelization of the DG method offers a manageable communication effort which is not dramatically more complex with higher accuracy order.

At the Institute for Aerodynamics and Gas Dynamics (IAG) the DG solver SUNWinT (Stuttgart University Numerical Wind Tunnel) has been developed in the past years. Landmann [11] and Lübon [13] introduced the first basic functionalities and investigated Detached Eddy Simulations. Afterwards, Wurst [18] extended the solver for moving bodies in 2D space. Amongst other things he introduced a Chimera method into the DG framework [19]. The Chimera method is an overlapping grid technique already known from FV solvers [15]. It simplifies the mesh generation especially for complex flow geometries, as they often occur in engineering problems. The entire body is decomposed into an arbitrary number of parts for which the mesh generation is feasible. Then the individual mesh parts are composed to the original assembly and the flow solution can be calculated, taking into account an exchange of flow states between the meshes. Furthermore, the individual mesh parts can perform a movement relative to

© The Author(s), under exclusive license to Springer Nature Switzerland AG 2021
A. Dillmann et al. (Eds.): STAB/DGLR Symposium 2020, NNFM 151, pp. 604–614, 2021.
https://doi.org/10.1007/978-3-030-79561-0_57

the others, if required. Another representative, who introduced the method into the DG context, was Galbraith et al. [3].

While the implementation of Wurst only allowed small movement amplitudes of the Chimera bodies, this work further generalizes the implemented Chimera method. Therefore the hole cutting process and the exchange of the flow solution were improved.

2 Governing Equations

The physics of a fluid in motion is described by the conservation of mass, momentum and energy. When laminar, compressible flows are considered as in this work, the Navier-Stokes equations provide the governing conservation equations

$$\frac{\partial \mathbf{U}}{\partial t} + \nabla \left(\mathbf{F}_{inv}(\mathbf{U}) - \mathbf{F}_{vis}(\mathbf{U}, \nabla \mathbf{U}) \right) = \mathbf{S}(\mathbf{U}, \nabla \mathbf{U}). \tag{1}$$

In the equation the state vector of conservative variables $\mathbf{U}$, the inviscid flux tensor $\mathbf{F}_{inv} = (\mathbf{F}^x_{inv}\, \mathbf{F}^y_{inv}\, \mathbf{F}^z_{inv})^T$ and the viscous flux tensor $\mathbf{F}_{vis} = (\mathbf{F}^x_{vis}\, \mathbf{F}^y_{vis}\, \mathbf{F}^z_{vis})^T$ read as

$$\mathbf{U} = \begin{pmatrix} \rho \\ \rho u_i \\ \rho E \end{pmatrix}, \quad \mathbf{F}^j_{\mathrm{inv}}(\mathbf{U}) = \begin{pmatrix} \rho u_j \\ \rho u_i u_j + p\delta_{ij} \\ (\rho E + p) u_j \end{pmatrix}, \quad \mathbf{F}^j_{\mathrm{vis}}(\mathbf{U}, \nabla \mathbf{U}) = \begin{pmatrix} 0 \\ \tau_{ij} \\ u_i \tau_{ji} + q_j \end{pmatrix}. \tag{2}$$

The density ρ, the components of the velocity vector u_i and the specific total energy E are the unknown state variables. The determination of the pressure p is achieved by an additional equation describing the state of the fluid. In this work, the equation of state for ideal gases is used, since air is the working fluid which is assumed a perfect gas. The equation of state can be written as a function of the specific total energy [11].

The Stokes hypothesis applies to Newtonian fluids, such as air, and the viscous stress tensor τ in (2) becomes symmetrical $\tau_{ij} = \tau_{ji}$. The entries τ_{ij} are a function of the dynamic viscosity μ and the velocity gradients and are well documented in the literature, e.g. [11].

Fourier's law is used to describe the energy fluxes $q_j = -\lambda \partial T / \partial x_j$. The thermal conductivity coefficient λ is assumed isotropic and can be expressed as a function of the Prandtl number Pr, the specific heat capacity c_P and the dynamic viscosity as $\lambda = \frac{\mu c_P}{Pr}$ with $Pr = 0.72$ (for air with $200\ \mathrm{K} \le T \le 700\ \mathrm{K}$).

In (1) the source term $\mathbf{S}$ must be considered in some specific cases, however it is neglected in the present work.

3 Numerical Formulation

Spatial Discretization

The spatial discretization of (1) is achieved with the DG method. A classical finite element Galerkin approach serves as the starting point. The unknown conservative variables $\mathbf{U}$ are approximated by polynomial ansatz functions

$$\mathbf{U}(\mathbf{x}, t) \approx \widetilde{\mathbf{U}}(\mathbf{x}, t) = \sum_{i=1}^{N_b} \mathbf{u}_i(t) \phi_i(\mathbf{x}) \tag{3}$$

with N_b basis functions $\phi_i(\mathbf{x})$ of a polynomial order p only depending on space [10]. In contrast, the unknown degrees of freedom $\mathbf{u}_i(t)$ are time-dependent. The approximated solution (3) is substituted into (1) and multiplied by an arbitrary test function $\mathbf{v}(\mathbf{x}) = \sum_{j=1}^{N_b} \mathbf{a}_j \phi_j(\mathbf{x})$. The Galerkin method uses the same basic functions for the test function and the approximate solution $\widetilde{\mathbf{U}}$, i.e. $\phi_i = \phi_j$. The values a_j are arbitrary but non-zero. Next, the whole domain Ω is approximated with a finite number of elements E and it is demanded for the integral over the whole domain to disappear. An integration by parts yields the so-called semi-discrete weak form of the governing equations on the element level (index $[\cdot]_h$)

$$\begin{aligned}\int_E \mathbf{v}_h \frac{\partial \widetilde{\mathbf{U}}_h}{\partial t}\,\mathrm{d}\Omega + \oint_{\partial E} \mathbf{v}_h \left(\mathbf{F}_{\mathrm{inv}}(\widetilde{\mathbf{U}}_h) - \mathbf{F}_{\mathrm{vis}}(\widetilde{\mathbf{U}}_h, \nabla\widetilde{\mathbf{U}}_h)\right)\cdot \mathbf{n}\,\mathrm{d}\sigma \\ - \int_E \nabla \mathbf{v}_h \cdot \left(\mathbf{F}_{\mathrm{inv}}(\widetilde{\mathbf{U}}_h) - \mathbf{F}_{\mathrm{vis}}(\widetilde{\mathbf{U}}_h, \nabla\widetilde{\mathbf{U}}_h)\right)\mathrm{d}\Omega = 0.\end{aligned} \tag{4}$$

The difference between a classical Galerkin and the DG approach is that the DG method has no requirements for the continuity properties of the solution at the element boundaries. Hence, the resulting possibility of discontinuities in the solution requires a specific treatment of the surface integrals. A Harten, Lax and van Leer (HLL) Riemann solver according to Toro [17], as known from the FV methods, is used for the inviscid fluxes. The viscous flux terms are not only dependent on the conservative variables but also on their gradients and thus are treated with the so-called BR2 method, which was developed by Bassi et al. [1]. The integrals in (4) are calculated numerically by Gaussian integration [16].

Temporal Discretization

The spatially discretized Eq. (4) are also dependent on time and can be written in compact form as $\mathbf{M}\frac{\partial \hat{\mathbf{U}}}{\partial t} = \mathbf{R}(\widetilde{\mathbf{U}}_h)$. The global equation system is composed of the unknown global solution vector $\hat{\mathbf{U}}$, the global mass matrix $\mathbf{M}$ and the residual $\mathbf{R}(\widetilde{\mathbf{U}}_h)$ that consists of the inviscid and viscous fluxes from (4). In order to achieve a high accuracy in time a second order backward differentiation formula (BDF2) [16] is applied [19]. The resulting non-linear system is handled with the Newton-Raphson method which yields the fully discretized Navier-Stokes equations

$$\begin{aligned}\left(\frac{3}{2}\frac{\mathbf{M}}{\Delta t} - \frac{\partial \mathbf{R}\left(\widetilde{\mathbf{U}}^{k-1}\right)}{\partial \hat{\mathbf{U}}}\right)\Delta\hat{\mathbf{U}} = \mathbf{R}\left(\widetilde{\mathbf{U}}^{k-1}\right) - \frac{\mathbf{M}}{\Delta t}\left(\frac{3}{2}\hat{\mathbf{U}}^{k-1} - 2\hat{\mathbf{U}}^{n} + \frac{1}{2}\hat{\mathbf{U}}^{n-1}\right) \\ \text{with } k = 1\ldots m \\ \hat{\mathbf{U}}^{n+1} = \hat{\mathbf{U}}^{k-1} + \Delta\hat{\mathbf{U}}.\end{aligned} \tag{5}$$

Starting with the solution from the prior time step $\hat{\mathbf{U}}^n$ for $k = 1$, at each time step a number of m Newton iterations k are required until the L^2-norm of the update of the solution $\Delta\hat{\mathbf{U}}$ drops below a tolerance ε and the new state $\hat{\mathbf{U}}^{n+1}$ can be determined. At each Newton iteration the linear system has to be constructed and solved. While the evaluation and composition of the left-hand and right-hand side is carried out analytically, the linear system is solved by a GMRES solver, preconditioned with an ILU(0) method, provided by the trilinos package [5]. Note that in (5) $\hat{\mathbf{U}}$ describes a vector composed of the modal values $\mathbf{u}_i$ from the ansatz in (3).

4 Chimera Method

The basic scheme of the Chimera method can be explained with the help of Fig. 1. It illustrates an exemplary Chimera setup containing two overlapping grids: a *body grid* of a 2D circular cylinder (white grid lines) and a *background grid* (black grid lines).

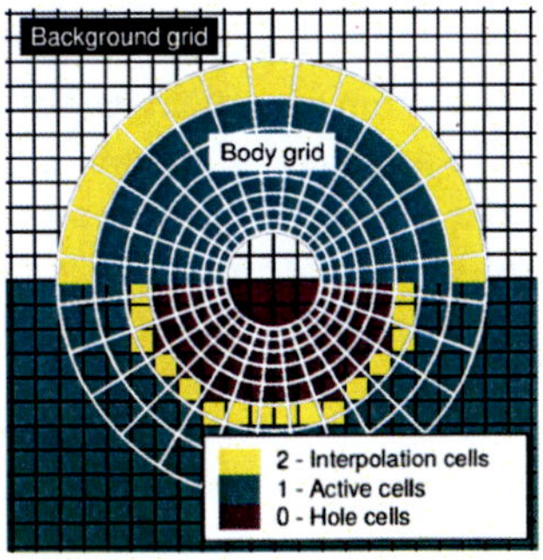

Fig. 1. Exemplary Chimera setup of a 2D cylinder with cell blanking for each of one half of the grid components

Each cell in a Chimera setup is assigned one of three tasks, which is referred to as *cell blanking*. In the upper half of Fig. 1 only the cell blanking for the body grid is colored, whereas the lower half shows the background blanking. Usually, most of the cells are classified as *active cells*. The flow solution of those cells is calculated regularly viz. as if there was only one single grid. Typically, *hole cells* are located in the background grid in the area covered by the body. Since the flow solution in this area is non-physical, these cells are deactivated. The *interpolation cells* (ICs) encircle the hole(s) and the body grid(s). These cells are responsible for the exchange of the solution between the grids which is achieved by interpolation. The ICs in the body grid receive the solution from the active cells in the background grid and vice versa for the ICs of the background grid. Since the ICs of the Chimera grids must not overlap, there exists a small region where the flow solution is calculated twice. Regardless of the chosen spatial accuracy order, the DG Chimera method requires only one row of ICs, since only the surface flux of the adjacent active cell is required. It should be emphasized that the ICs must form a closed surface without gaps. Otherwise, the interpolation of the flow solution between the grids is incomplete and leads to errors.

Mathematically, the exchange of the flow solution between the grids is achieved by a discrete L^2 projection, satisfying the following property

$$\int_{\mathrm{E}} \phi_j^{\mathrm{IC}}\, \widetilde{\mathbf{U}}_h^{\mathrm{IC}}\, d\Omega = \int_{\mathrm{E}} \phi_j^{\mathrm{IC}}\, \widetilde{\mathbf{U}}_h^{\mathrm{DC}}\, d\Omega. \tag{6}$$

ϕ_h^{IC} are the basis functions of the IC and $\widetilde{\mathbf{U}}_h^{\mathrm{IC}}$ describes the unknown solution of the IC by a polynomial function according to (3). On the right-hand side $\widetilde{\mathbf{U}}_h^{\mathrm{DC}}$ is also a polynomial function according to the latter equation and describes the solution of the donor cell (DC), the active cell in the donor grid from which the solution is interpolated. By reformulating (6) as $\mathbf{M}_{IC}\hat{\mathbf{u}}_{IC} = \mathbf{b}$, the vector of the modal coefficients $\hat{\mathbf{u}}_{IC}$ can be determined. $\mathbf{M}_{IC}$ represents the mass matrix of the IC and $\mathbf{b}$ the integral of the right-hand side of (6). Since the calculation of the integrals is conducted numerically using Gaussian integration, it is mandatory to determine the solution in the DCs at all Gaussian points of the IC. Accordingly, the coordinates of all Gaussian points in the IC must be found and evaluated in the corresponding DC. Therefore the inverse of the transformation from the reference space to the physical space is calculated and the appropriate coordinates in the donor cell's reference space are searched in the donor grid. For this purpose a Newton method is used, which usually converges within a few iterations. However, since the correct DC must be identified first during this search process, it can become a bottleneck, especially when parallelization is involved. To this end, a point-to-point communication was implemented in which a preselection reduces the number of cells (or points) to be communicated. Furthermore, in [4] and [19] it has been shown that the evaluation of the projection must also be treated implicitly, if an implicit time discretization is chosen.

The process of identifying the hole and interpolation cells in the background grid is referred to as *hole cutting*. This process is essential for the success of the Chimera method and becomes even more important, if simulations with a movement of the Chimera bodies are taken into account. In the latter case hole cutting usually has to be executed in every single time step and should therefore be implemented efficiently. While hole cutting has previously been realized by stationary simple geometric shapes [19], now a method known from FV solvers has been adapted [15]. Although, the method requires an explicitly defined hole mesh for each body grid, the advantage is that the hole can have an arbitrary shape and can easily be moved with the body. This allows any desired body movement.

To conclude the theoretical part, it should be mentioned that the mesh movement shown in this work is realized with an Arbitrary Lagrangian Eulerian (ALE) formulation [14]. Furthermore, high-order grids [6] are used to describe the curvature of body surfaces.

5 Numerical Results

A fundamental validation of the implementation of the Chimera method in SUNWinT has already been provided by Wurst [18, 19]. In this paper the application of the Chimera method is focused on unsteady flow problems with moving flow bodies.

Magnus Effect

When a body moves through a fluid while spinning around its body axis, a phenomenon called the Magnus effect can be observed. In addition to a drag force the body also experiences a lift force, due to the rotation. Moreover, the drag decreases with increasing rotational speed.

Kang et al. [8] investigated the Magnus effect for a 2D cylinder and provided extensive data which was used for the validation of the Chimera method with a moving body grid.

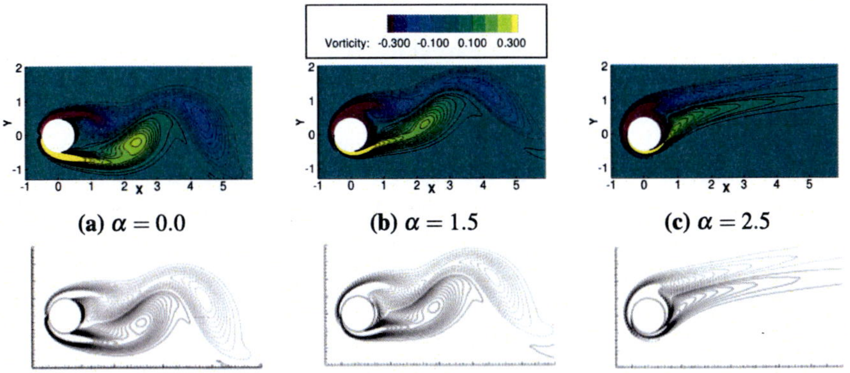

Fig. 2. Contour plots of the vorticity for the Magnus effect (Re = 100) at different rotational speeds α – top: DG Chimera with fifth order ($p4$) – bottom: Kang et al. [9]

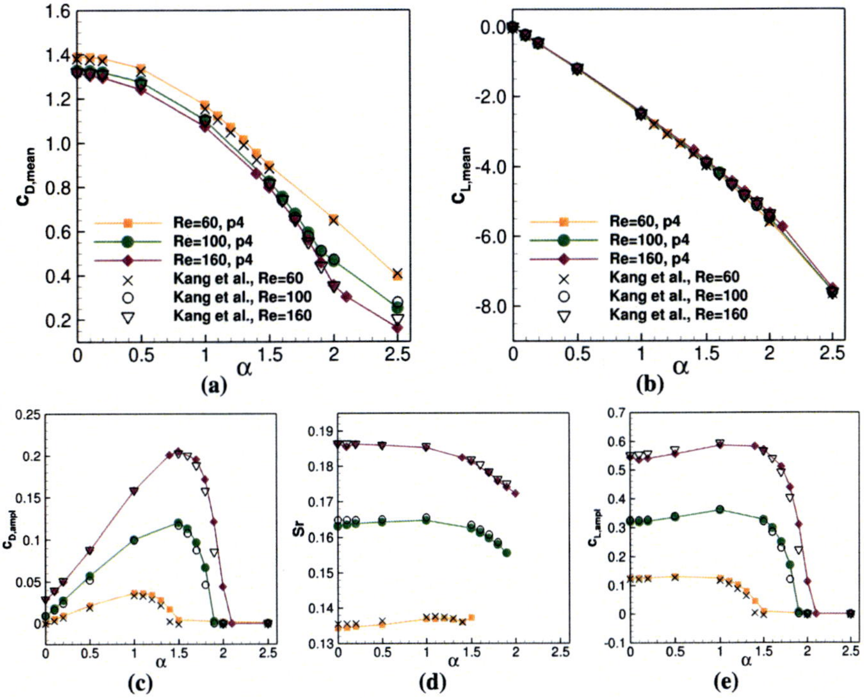

Fig. 3. Flow coefficients for the Magnus effect

Figure 1 shows a part of the numerical setup for this test case. The O-grid of the cylinder represents the body grid that overlaps the background grid. The boundary condition for the cylinder surface is a no-slip wall and the Chimera boundary is located approximately one cylinder diameter D away from the surface. In the background grid the farfield is located $100D$ away in each direction of the cylinder. Instead of translating the body through the fluid, a flow past the cylinder with a Mach number $Ma = 0.1$ was chosen and the body grid additionally performed a counter clockwise rotation around the z-axis. The rotational speed $\alpha = \frac{D\omega}{2u_\infty}$ was varied and different Reynolds numbers of $Re = \{60, 100, 160\}$ were investigated. The numerical mesh consists of 6488 cells in total distributed on the body grid with 224 curvilinear cells ($y^+ \approx 2, p_{\text{geo}} = 3$) and 6264 uncurved cells ($p_{\text{geo}} = 1$) in the background. In comparison the grid used by Kang et al. had 57600 cells [8]. A time step Δt corresponding to 1° of the time needed for one full rotation[1] was chosen with the BDF2 scheme. The simulations were conducted with an ascending spatial order up to five ($p4$), starting from a steady flow state at the cylinder with an order of four ($p3$). In the following only $p4$ results are shown.

Due to the chosen Reynolds numbers, a Kármán vortex street develops for a static cylinder ($\alpha = 0.0$) shown in Fig. 2a for the vorticity at $Re = 100$. With increasing rotational speed α the vortical wake drifts upwards (Fig. 2b). This results from the downwards directed lift which increases with the rotational speed. At an $\alpha = 2.0$ or higher ($Re = 100$) the wake becomes quasi-stationary and has a separation bubble that is located in the upper region of the cylinder wake (Fig. 2c). The qualitative results of the vorticity in Fig. 2 match well with the results provided by Kang et al. [9].

The time-averaged values of the drag coefficient $c_D = 2F_D/\rho u_\infty^2 D$ and the lift coefficient $c_L = 2F_L/\rho u_\infty^2 D$ are plotted over the rotational speed for different Re numbers in Fig. 3a and 3b and agree well with the reference. Although the amplitudes of the force coefficients (Fig. 3c and 3e) and the Strouhal numbers (Fig. 3d) show small deviations to the values provided by Kang et al., they are still in an acceptable range. Since these small deviations mainly occur for higher α, they might be reduced by lowering the time step size and thus improving the temporal accuracy.

Plunging NACA0012 Airfoil

The validation of the Chimera method is extended for real relative displacements of the body grid within the background, which ideally requires a movement of the cell blanking. For this purpose the chosen test case is a plunging airfoil. In the literature [7, 12] this problem is described as the flow around an up and down oscillating 2D NACA0012 airfoil. The motion is prescribed by $y_h(t) = -h_0 \sin(\omega t)$ with $h_0 = 0.08c$ and $\omega = 1.15$ which is referred to as slowly plunging. Wurst [18] has already examined this case to validate the Chimera method, but only with a static hole definition. In this work the cell blanking moves with the body.

Three different numerical setups have been used: two Chimera setups to show the flexibility of the method and a single grid as a further reference. The first Chimera setup is built with rectangular cells and is shown in Fig. 4a. It consists of three grids: the body grid with 1054 curvilinear cells ($y^+ \approx 2, p_{\text{geo}} = 3$), an embedded wake grid with 3072 cells ($p_{\text{geo}} = 1$) and the background grid with 6324 cells ($p_{\text{geo}} = 1$). The second Chimera setup is shown in Fig. 4b. The body grid is the same as in setup 1 and the background grid now has 11810 triangular cells ($p_{\text{geo}} = 1$). The single grid setup is based on the body grid with the Chimera boundary replaced by a farfield condition and extended by

[1] In the $\alpha = 0.0$ case Δt was one twentieth of the overflow duration.

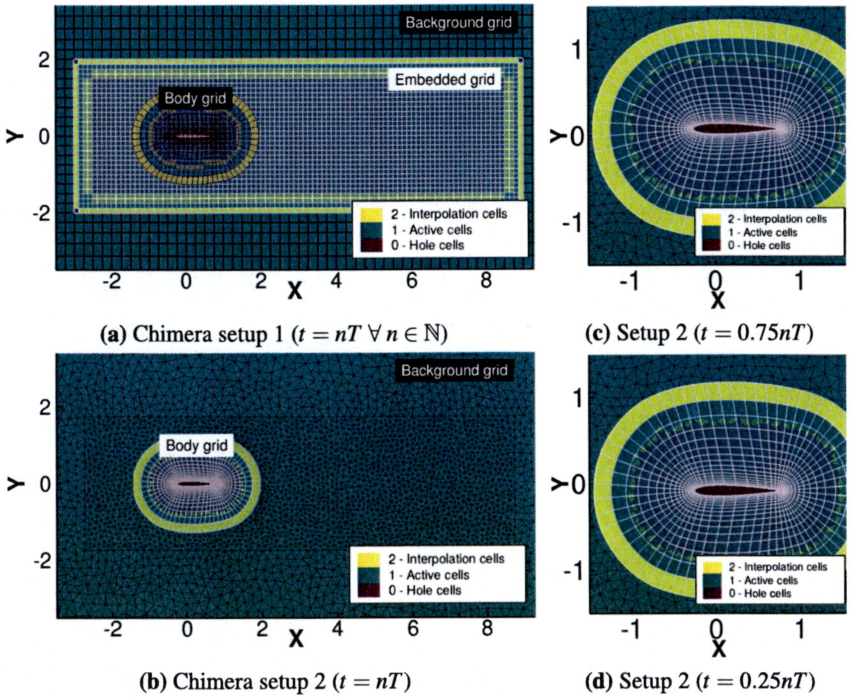

(a) Chimera setup 1 ($t = nT\ \forall\, n \in \mathbb{N}$)

(b) Chimera setup 2 ($t = nT$)

(c) Setup 2 ($t = 0.75nT$)

(d) Setup 2 ($t = 0.25nT$)

Fig. 4. Chimera grids with cell blanking for the slowly plunging NACA0012 airfoil

more cell rows. Thus, setup 3 consists of 2542 curvilinear cells ($y^+ \approx 2, p_{\text{geo}} = 3$). The farfield boundary is located 100 chords away from the body in each direction and the surface of the airfoil is a no-slip wall. At the farfield the Mach number was $Ma = 0.2$ and $Re = 1850$ was used. Starting from a fourth order ($p3$) steady flow state around the airfoil, the simulations were performed using the BDF2 scheme with a time step of $\Delta t = T/100$ in dependence of the plunging period T. The spatial order for the results presented in the following was five ($p4$). The time-dependent cell blanking is illustrated

Table 1. Force coefficients for the slowly plunging NACA0012 airfoil

	$c_{D,\text{mean}}$	$c_{D,\text{max}}$	$c_{D,\text{min}}$	$c_{L,\text{mean}}$	$c_{L,\text{max}}$	$c_{L,\text{min}}$
Single grid ($p4$)	−0.0339	0.1082	−0.1702	−0.0003	4.6897	−4.6820
Chim. setup 1 ($p4$)	−0.0340	0.1085	−0.1707	−0.0004	4.6828	−4.6837
Chim. setup 2 ($p4$)	−0.0340	0.1087	−0.1705	−0.0014	4.6837	−4.6884
Liang et al. [12]	−0.0436	0.1	−0.181	−0.0042	4.77	−4.78
Wurst [18]	−0.0370[a]	–	–	0.0135	–	–
Fries [2]	−0.0370	–	–	0.0013	–	–

[a]corrected by a factor of 1/10 due to typo according to the author

in Fig. 4c and 4d for the hybrid chimera setup at the highest and the lowest position of the airfoil in order to demonstrate the movement of the hole.

Due to the prescribed plunging motion the airfoil generates thrust. In contrast, the averaged c_L is almost exactly zero. Table 1 shows the time-averaged and the minimum and maximum values of the force coefficients, which are almost identical for all setups used. Thus, the influence of a highly resolved wake seems to have little impact on the results, in this case. The comparison to the values reported in the literature [2, 12, 18] only shows minor deviations. The deviation from c_D given by Liang et al. could be explained by a deviation in the static drag coefficient. Liang et al. reported a value of $c_D = 0.054$ for the given Reynolds number, while in this work a value of $c_D = 0.087$ was obtained. This has already been noticed by Wurst and Fries. In this context, the dynamic drag values of Liang et al. match well. Furthermore, by using the moving cell blanking, the average lift coefficient is even closer to zero especially than the value specified by Wurst who used static cell blanking.

The flow field is visualized for the vorticity in Fig. 5a–5c for the three setups used and reveals almost identical wake patterns for both Chimera setups (Fig. 5b and 5c). Due to the cell growth the wake of the single grid setup is quickly under-resolved and vanishes accordingly (Fig. 5a). In comparison to the numerical (Fig. 5d) and experimental (Fig. 5e) results reported in the literature the wake patterns match very well for the Chimera setups. This demonstrates the possibility to easily increase the flow resolution by applying the Chimera method to the area of interest. Moreover, by the usage of triangular cells the cell distribution is more flexible and the resolution of the wake is even sightly better than in the Chimera setup 1.

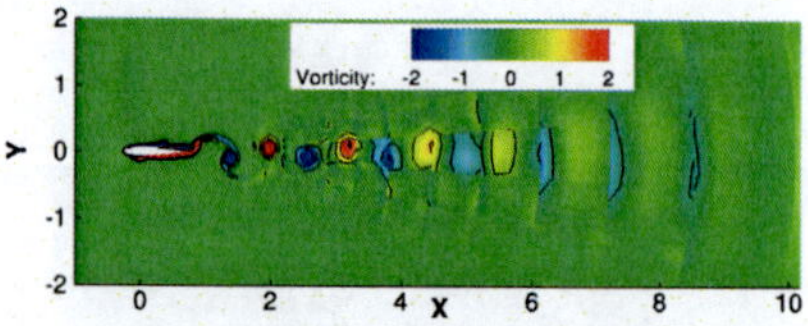

(a) Single grid (*p*4)

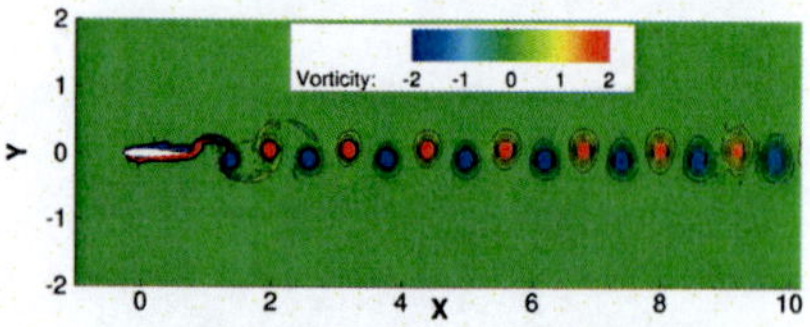

(b) Chimera setup 1 (*p*4)

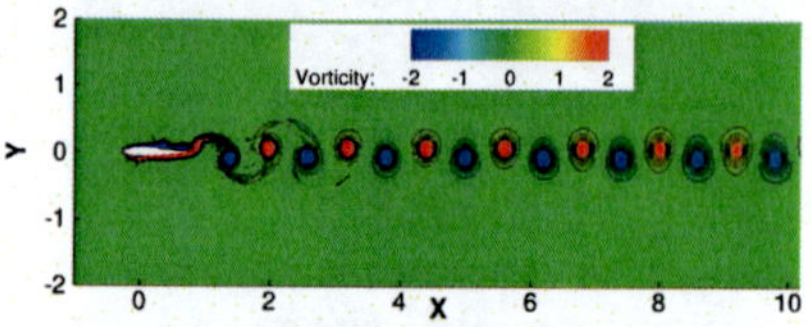

(c) Chimera setup 2 (*p*4)

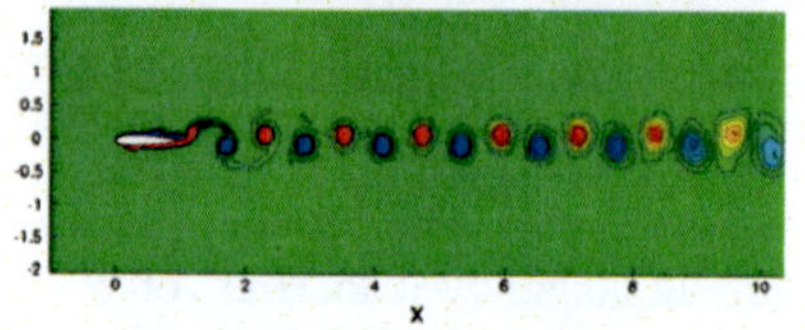

(d) Liang et al. [12]

(e) Jones et al. [7]

Fig. 5. Comparison of the contour plots of the vorticity

6 Conclusion

The Navier-Stokes equations have been discretized with a DG method in space and an implicit BDF2 scheme in time. Furthermore, a DG Chimera method was introduced to allow numerical setups to consist of multiple overlapping grids and enable relative movements of the individual grid parts to each other. An extensive study of the Magnus effect was conducted which successfully validated the method for a trivial relative grid motion.

An oscillating translation movement of a NACA0012 airfoil was the second test case presented. Thus, the hole movement was verified and furthermore the flexibility of the DG Chimera method in using multiple overlapping grids as well as different cell types for each grid was highlighted.

To address real engineering problems at preliminarily subsonic Mach numbers, simulations with lager Chimera body movements, even in 3D, and an efficient adaptation of existing turbulence models to the Chimera method with moving grids will be necessary.

References

1. Bassi, F., Crivellini, A., Rebay, S., Savini, M.: Discontinuous Galerkin solution of the Reynolds-averaged-Navier-Stokes and k-w turbulence model equations. Computers Fluids **34**, 507–540 (2005)
2. Fries, D.: Fluid struktur-interaktion auf bewegten gittern mit einem discontinuous galerkin verfahren. Master's thesis, Universität Stuttgart - Institut für Aerodynamik und Gasdynamik (2014)
3. Galbraith, M.C., Benek, J.A., Orkwis, P.D., Turner, M.G.: A discontinuous Galerkin scheme for Chimera overset viscous meshes on curved geometries. Comput. Fluids **119**, 176–196 (2015)
4. Galbraith, M.C., Knapke, R.D., Orkwis, P.D., Benek, J.A.: A Discontinuous Galerkin Chimera scheme with implicit artificial boundaries. In: 51st AIAA Aerospace Sciences Meeting including the New Horizons Forum and Aerospace Exposition, American Institute of Aeronautics and Astronautics (AIAA), vol. 23 (2013)
5. Heroux, M.A., Bartlett, R.A., Howle, V.E., Hoekstra, R.J., Hu, J.J., Kolda, T.G., Lehoucq, R.B., Long, K.R., Pawlowski, R.P., Phipps, E.T., et al.: An overview of the Trilinos project. ACM Trans. Math. Softw. (TOMS) **31**(3), 397–423 (2005)
6. Hindenlang, F.: Mesh curving techniques for high order parallel simulations on unstructured meshes. PhD thesis, Institut für Aerodynamik und Gasdynamik, Universität Stuttgart (2014)
7. Jones, K., Dohring, C.M., Platzer, M.: Experimental and computational investigation of the Knoller-Betz effect. AIAA J. **36**, 1240–1246 (1998)
8. Kang, S., Choi, H., Lee, S.: Laminar flow past a rotating circular cylinder. Phys. Fluids **11**, 3312–3321 (1999)
9. Kang, S., Choi, H., Lee, S.: Erratum: "laminar flow past a rotating circular cylinder" [phys. fluids 11, 3312 (1999)]. Phys. Fluids **12**(1), 239–240 (2000)
10. Karniadakis, G.E., Sherwin, S.J. (eds.): Spectral/*hp* Element Methods in CFD. Oxford University Press, Oxford (1999)
11. Landmann, B.: A parallel discontinuous Galerkin code for the Navier-Stokes and Reynolds-averaged Navier-Stokes equations. Ph.D. thesis, Insitut für Aerodynamik und Gasdynamik, Universität Stuttgart (2008)

12. Liang, C., Ou, K., Premasuthan, S., Jameson, A., Wang, Z.: High-order accurate simulations of unsteady flow past plunging and pitching airfoils. Comput. Fluids **40**(1), 236–248 (2011)
13. Lübon, C.: Turbulenzmodellierung und Detached Eddy Simulation mit einem Discontinuous Galerkin Verfahren von hoher Ordnung. Ph.D. thesis, Institut für Aerodynamik und Gasdynamik, Universität Stuttgart (2009)
14. Persson, P.O., Bonet, J., Peraire, J.: Discontinuous Galerkin solution of the Navier-Stokes equations on deformable domains. Comput. Methods Appl. Mech. Eng. **198**(17–20), 1585–1595 (2009)
15. Schwarz, T.: Ein blockstrukturiertes Verfahren zur Simulation der Umströmung komplexer Konfigurationen. Ph.D. thesis, DLR research report 2005-20 (2005)
16. Süli, E., Mayers, D.F.: An Introduction to Numerical Analysis. Cambridge University Press, Cambridge (2003)
17. Toro, E.F.: Riemann Solvers and Numerical Methods for Fluid Dynamics: A Practical Introduction, 2nd edn. Springer, Hiedelberg (1999). https://doi.org/10.1007/978-3-662-03915-1
18. Wurst, M.: Development of a high-order Discontinuous Galerkin CFD solver for moving bodies. Ph.D. thesis, Institut für Aerodynamik und Gasdynamik, Universität Stuttgart (2016)
19. Wurst, M., Keßler, M., Krämer, E.: A high-order discontinuous Galerkin Chimera method for laminar and turbulent flows. Comput. Fluids **121**, 102–113 (2015)

Efficient Modelling of Near-Wall Turbulence in Hybrid RANS-LES Simulations

Marius Herr[1,2(✉)] and Axel Probst[2]

[1] TU Braunschweig, Hermann-Blenk-Str. 37, 38108 Braunschweig, Germany
m.herr@tu-bs.de

[2] DLR (German Aerospace Center), Bunsenstr. 10, 37073 Goettingen, Germany
axel.probst@dlr.de

Abstract. This study investigates the applicability and efficiency potential of analytical wall functions in conjunction with hybrid RANS-LES methods in two unstructured flow solvers, i.e. the compressible DLR-TAU code and the incompressible DLR-THETA code. Wall-modelled LES simulations of a periodic channel flow using IDDES confirm the validity of the wall-function approach over a broad range of grids with $12.5 \leq y^+(1) \leq 210$. For the more complex separated flows over a backward facing step and a wall-mounted hump, DDES simulations show that $y^+(1)$ has to be limited to 25 regardless of the flow solver, in order to achieve agreement with results using full near-wall resolution. Despite that restriction, the computing time can be reduced by up to $2/3$ compared to the fully-resolved case.

Keywords: Hybrid RANS-LES · Wall functions · DLR-TAU · DLR-THETA

1 Introduction

Combining the Large-Eddy Simulation (LES) methodology with the statistical RANS modelling approach enables to perform locally scale-resolving simulations of aeronautical flows for high, flight-relevant Reynolds numbers. Despite the efficiency advantage of such hybrid RANS-LES methods (HRLM) compared to pure LES, this approach remains computational-intensive and time-consuming for the specified scope of application [9]. This is due to the high demand on the wall-normal grid resolution for no-slip walls requiring a so-called *low-Re* grid with $y^+(1) \leq 1$, where $y^+(1)$ denotes the wall-adjacent cell-height normalized in wall units.

Coupling the HRLM approach with wall-functions (HRLM-WF), however, allows to avoid the highly-resolved near-wall region by replacing the no-slip condition with a more sophisticated *high-Re* boundary condition and modelling the effects of the unresolved near-wall flow by analytical wall laws. Gritskevich et al. [3] investigated a HRLM approach with common RANS wall-functions by means of various internal flow test cases and confirmed the basic validity of this simulation method. In this work, we adopt the universal wall-function method of Knopp et al. [5] in numerical studies with the compressible flow solver TAU and the incompressible solver THETA in order to systematically explore the applicability of different HRLM-WF models, based on

© The Author(s), under exclusive license to Springer Nature Switzerland AG 2021
A. Dillmann et al. (Eds.): STAB/DGLR Symposium 2020, NNFM 151, pp. 615–624, 2021.
https://doi.org/10.1007/978-3-030-79561-0_58

variants of Detached-Eddy Simulation (DES). Despite similarities of TAU and THETA with respect to the underlying data structures and numerical schemes, fundamental differences are present with regard to the near-wall grid metric (cf. Sect. 2). Hence an applicability of HRLM-WF for both flow solvers is non-trivial and might depend on different criteria for each code. Therefore, after a fundamental exploration of HRLM-WF by means of a periodic channel flow (Sect. 3.1), the approach is analysed separately by TAU and THETA on the basis of the backward facing step (Sect. 3.2) and the NASA wall mounted hump (Sect. 3.3).

2 Numerical Method

The compressible TAU code and the incompressible THETA code are unstructured 2nd-order finite-volume solvers developed by DLR. Both solvers use specialized numerical methods for their respective Mach-number regimes, but they share the same *dual-cell* approach to support identical so-called *primary* grids with mixed cells (e.g. tetrahedra, hexahedra). In scale-resolving simulations such as hybrid RANS-LES, both codes use low-dissipative central schemes for the convective fluxes, which were shown to yield satisfying accuracy for various flow cases [8]. Note that the low-Mach preconditioning available in TAU allows for solving low-speed flow problems, too, despite lower computational efficiency compared to THETA.

Another important difference between TAU and THETA concerns the location of the first off-wall control point, where the flow variables and also the wall function are evaluated: while in TAU this point simply corresponds to the first off-wall cell vertex of the primary grid, THETA introduces an additional *shifted* point at about 0.27 times the height of the first grid cell. This actually increases the near-wall resolution on a given primary grid and reduces the *effective* value of $y^+(1)$, which is therefore denoted as $y_e^+(1)$ for THETA in the subsequent comparisons with TAU.

2.1 Hybrid RANS-LES Modelling

To assess the combination of wall functions with hybrid RANS-LES we consider two well-known variants of Detached-Eddy Simulation (DES): first, the Delayed DES (DDES) [10] which discerns between RANS mode in attached flow and LES mode in detached areas based on the local grid resolution and a boundary-layer (*RANS shielding*) sensor. Second, the Improved DDES (IDDES) [9] which extends the DDES by wall-modelled LES capabilities for partly-resolved attached boundary layers, e.g. downstream of separation or synthetic-turbulence injection.

While IDDES is needed to simulate the periodic channel, DDES is considered sufficient for flows with local resolved separations, i.e. the backward-facing step and the wall-mounted hump. In those test cases, however, the common approach to build 3D meshes by extruding 2D meshes in lateral direction, can delay the onset of resolved turbulence in the initial free shear layers [7]. As a remedy, we employ a vorticity-sensitive LES-filter width, $\tilde{\Delta}_\omega$ [7], instead of the classic maximum-length filter of DES. The Spalart-Allmaras one-equation model serves as RANS background model.

2.2 Analytical Wall Function Approach

The low-Re approach, which is usually applied to the near-wall RANS regions in HRLM, requires a complete resolution of the boundary layer up to the viscous sublayer with $y^+(1) = 1$. In order to relax this strict requirement on the wall-normal grid resolution, we combine the HRLM-models with the analytical wall-function approach by Knopp et al. [5]. The applied wall-functions have been derived from the incompressible RANS momentum equation. This method blends well-known wall laws, i.e. Spalding (F_{Sp}^{-1}), Reichhardt (F_{Rei}) and the logarithmic law of the wall (F_{Log}), such that corresponding RANS results are widely independent of the first cell height $y^+(1)$:

$$F_{\text{Sp}}^{-1} = u^+ + 8.43\left(e^{0.41u^+} - \sum_{n=0}^{5}\frac{(0.41u^+)^n}{n!}\right), \quad F_{\text{Log}} = \frac{\ln(y^+)}{0.41} + 5.1 \tag{1}$$

$$F_{\text{Rei}} = \frac{\ln(1+0.4y^+)}{0.41} + 7.8\left(1 - e^{-\frac{y^+}{11}} - \frac{y^+}{11}e^{-\frac{y^+}{3}}\right) \tag{2}$$

The blending between these laws is designed in line with solutions of the respective turbulence model on low-Re meshes, so that a model-consistent universal wall function formulation is achieved, see [5] for details. On this basis the wall shear stress τ_w is computed for a given $y^+(1)$ and is being prescribed at solid walls. Thus, the previous no-slip boundary condition ($u_w = 0$) is being replaced by a Neumann boundary condition ($\tau_w \propto \frac{\partial u}{\partial y}\big|_{y=0}$). Since such wall functions were originally designed for steady RANS computations, it is non trivial that this approach works for instationary HRLM simulations, too. Nevertheless, the mostly stationary character of the near-wall RANS region in HRLM, as well as the successful application of similar wall functions by [3] justify this approach at least for low values of $y^+(1)$.

Numerical Correction for Wall-Function Computations in THETA. Despite the universal formulation of the wall functions one observes significant skin-friction errors in high-Re computations with TAU or THETA on grids with the first off-wall grid point located in the buffer layer around $y^+(1) \approx 5 \ldots 15$, similar to the findings in [4]. The reason lies in numerical errors in the momentum diffusion (relying on second velocity derivatives) at the first off-wall cell face of such coarse grids which, however, become only relevant in the lower, high-curvature buffer region, where diffusion plays a dominant role.

As an ad-hoc remedy, the relative error of the discretized diffusion term can be quantified as a function of $y_e^+(1)$ by comparison with a fine (*low-Re*) grid and transformed into an empirical correction function ϕ_{corr} for the diffusion. So far, such a correction has only been derived for THETA, whose special near-wall metric prevents a direct transfer to other solvers. It is realized via an additional momentum source at the first off-wall cell face, using the *thin-shear layer approximation* of the wall-normal diffusion and an interpolated look-up table for ϕ_{corr}.

3 Results

3.1 Periodic Channel

For a basic validation of the wall function approach in conjunction with HRLM-based wall-modelled LES, a periodic channel flow at $Re_\tau = u_\tau \delta / \nu \approx 4179$ [6] is computed with TAU using IDDES. The channel geometry comprises the height 2δ, length $2\pi\delta$ and width $\pi\delta$. Periodic boundary conditions are applied in stream- and spanwise directions, and a momentum source term is applied to keep the corresponding target bulk Reynolds number constant at $Re_\delta = U_0\delta/\nu = 98300$, as given by DNS data [6]. For this setting the wall-adjacent RANS areas exhibits a wall normal thickness of about $0.05\,\delta$. The LES mode is active in the remaining region. A structured grid with hexahedral cells is applied, with grid spacings of $\Delta x = 0.1\delta$ (streamwise) and $\Delta z = 0.05\delta$ (lateral). For the temporal discretisation a normalized time step of $\Delta t^+ = 0.4$ is used, overall corresponding to the setup given in [8].

The low-Re mesh with $y^+(1) \approx 1$ serves as reference mesh for the wall-function simulations. For the high-Re meshes $y^+(1)$ is varied systematically over a range from 12.5 to 210. Except for $y^+(1) = 210$, where a homogeneous meshing strategy is applied, the meshes exhibit a wall-normal growth factor of 1.14. For the high-Re meshes the total cell number can be reduced from 39.6% for $y^+(1) = 12.5$ up to 61.4% for $y^+(1) = 52.5$ (cf. Fig. 1 right). The total simulations time amounts to $30l/U_0$ with the channel length l and the bulk velocity U_0, where a time span of $20l/U_0$ is used to calculate temporal statistics. Figure 1 left presents TAU-IDDES results in terms of mean velocity profiles, where good agreement between the low- and high-Re solutions over the entire range of $y^+(1)$ can be stated. The largest, but still moderate deviation with respect to Re_τ is present for $y^+(1) = 12.5$ (cf. Fig. 1 right), which may be attributed to the discretisation errors in the momentum diffusion, as described in Sect. 2.2. Besides, all present simulations share small deviations from the DNS around $y^+ \approx 500$, i.e. above the RANS-LES interface, indicating a remainder of the well-known *log-layer mismatch* issue of wall-modelled LES [9].

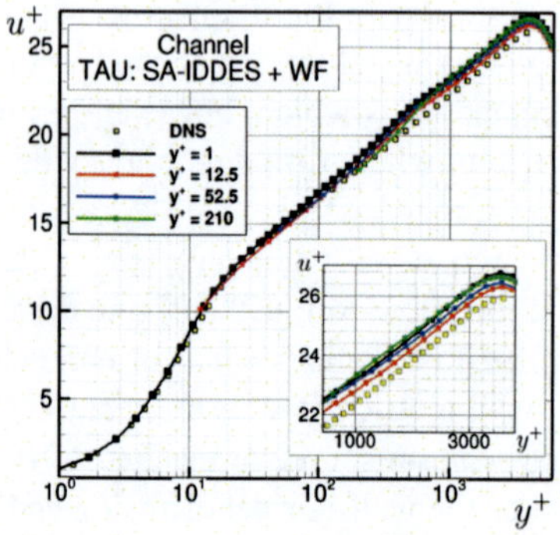

$y^+(1)$	# cells	red. of cells	Re_τ	dev. Re_τ
1	$4.27 \cdot 10^5$	-	4026	-
12.5	$2.58 \cdot 10^5$	39.6%	4092	+1.6%
52.5	$1.65 \cdot 10^5$	61.4%	4069	+1.1%
210	$1.73 \cdot 10^5$	59.4%	4046	+0.5%

Fig. 1. DLR-TAU IDDES results of the periodic channel flow for $Re_\delta = 98.300$ [6]: Velocity profiles for different $y^+(1)$ (left) and overview of mesh properties with averaged values of Re_τ (right).

3.2 Backward-Facing Step

The backward-facing step (BFS) represents a more complex test case including geometry-induced flow separation and reattachment. We consider the experimental flow studies by Driver and Seegmiller [1] with a Reynold number of $Re_h = 37,500$ based on the step height h and a Mach number of 0.13, using DDES in a similar numerical setup as [8]. Periodic boundary conditions are applied at the spanwise domain borders which are separated by a distance of 4h. At the inflow a partially-developed turbulent boundary layer with a thickness as in the experiment is prescribed. The RANS mode, which is active for attached boundary layers, exhibits a thickness of about 1 h for $x < 0$ and reduces to approx. 0.1 h behind the separation.

We employ hybrid meshes with a mostly unchanged, hex-dominated unstructured area located in the LES region downstream of the step, whereas structured areas with varying $y^+(1)$, but constant wall-normal expansion factor of $r = 1.15$, are located close to the walls. In all grids 68 extrusion layers are used to resolve the spanwise domain. As reference for the DDES-WF simulations, a low-Re mesh is created, which fulfils $y^+(1) < 1$ at all wall-adjacent cells. To analyse the impact of $y^+(1)$ on the flow solution, meshes with $y^+(1)$ values of 12.5, 25 and 50, referring to the streamwise position $x/h = -1$, were built. The corresponding wall distances $y(1)$ at this position were applied for all wall-adjacent cells which leads to a suboptimal design of the high-Re grids in the vicinity of the step, exhibiting large cell-volume ratios at the hybrid interface (cf. Fig. 2). Compared to the low-Re grid, the total cell reduction ranges from 24.5 % for $y^+(1) = 12.5$ (resp. 34.1 % for $y_e^+(1) = 13.5$ using THETA) to 34.1 % for $y^+(1) = 50$ (resp. 62.4 % for $y_e^+(1) = 50$).

The physical time step of $dt = 0.07\,h/U_0$ ensures $\mathrm{CFL}_{conv} < 1$ (i.e. CFL-number based on the convection velocity) within the whole LES region. The total simulation time is $300\,h/U_0$, where $200\,c/U_0$ are used to obtain time-averaged results.

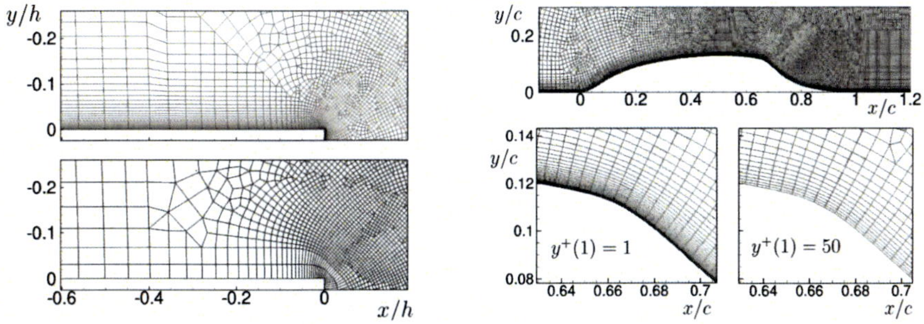

Fig. 2. Left: Mesh comparison of the 2-D backward-facing step. Top view: low-Re mesh with $y^+(1) = 1$. Bottom view: high-Re mesh with $y^+(1) = 50$. Right: 2D-cut of NASA wall-mounted hump discretised with hybrid meshing strategy.

Figure 3 shows skin-friction distributions along the step computed with TAU and THETA using DDES-(WF) on the described grids, together with reference data from the experiment [1] and RANS (Spalart-Allmaras model). Be reminded that for a given grid, the internal THETA grid metric leads to lower *effective* first cell heights than TAU, thus denoted as $y_e^+(1)$. Also note that along the entire x-axes very good agreement between the low-Re solutions of both flow solvers and the experimental data is achieved, which demonstrates the benefit of scale-resolving approaches compared to RANS in massively-separated flow.

Concerning the high-Re results from TAU in Fig. 3 (left), smaller values of $y^+(1)$ lead to a better agreement with the low-Re solution in the recirculation area ($x/h > 0$ and $c_f < 0$), pointing out the limits of DDES-WF for $y^+(1) > 25$ in this challenging flow. Concerning the recovery region downstream of reattachment ($x/h \approx 5.5$), all results become more similar and nearly align at the outflow plane. For $x/h < 0$ the best results are obtained for $y^+ = 25$, whereas the skin friction is overestimated for $y^+(1) = 12.5$ (due to the discretisation errors described in Sect. 2.2), but surprisingly also not matched for $y^+(1) = 50$. The latter remains an open issue which was not further analysed because of the overall large deviations for this mesh.

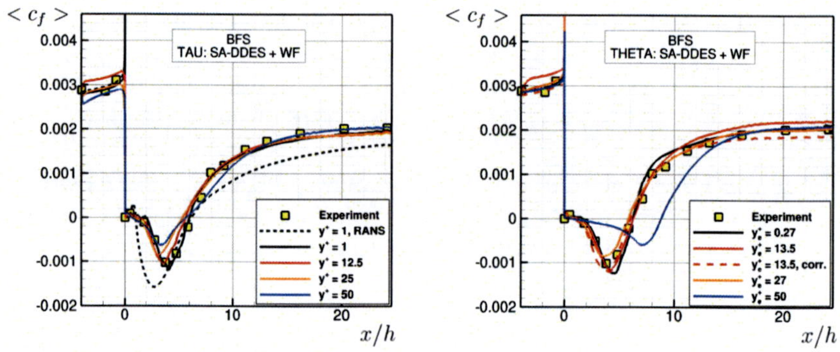

Fig. 3. Mean skin-friction results for the backward-facing step at Re = 37,500: DLR-TAU solutions for different values of $y^+(1)$ (left) and results computed with DLR-THETA and a corrected verion of THETA (right).

Analysing the c_f-curves computed by THETA in Fig. 3 (right), similar trends as for TAU are observed within the recirculation, i.e. better agreement with the low-Re solution is achieved for decreasing values of $y_e^+(1)$. The results for $y_e^+(1) = 50$, however, which corresponds to a *primary* grid with $y^+(1) = 185$, demonstrate the limits of the high-Re approach on very coarse meshes, as the separated region is drastically overpredicted. Figure 3 (right) also illustrates that the deviations for $y_e^+(1) = 13.5$ and $x < 0$ can indeed be traced back to near-wall discretisation errors, since a much better agreement with $y^+(1)$ is achieved with the diffusion correction (*corr.*) mentioned in Sect. 2.2, without interfering with the already good prediction in the recirculation region. As an intermediate summary for this challenging separated flow case, $y^+(1)$ for TAU or $y_e^+(1)$

for THETA should not exceed values of 25 in order to obtain satisfying agreement with the low-Re solutions.

3.3 NASA Wall-Mounted Hump

Another typical and challenging HRLM test case is the NASA wall-mounted hump [2], cf. Fig. 2 (right). Except for the slip-wall condition at the upper wall, the same boundary conditions as for the backward-facing step are applied. At the lower wall, the prescribed inflow matches a boundary layer with $Re_\theta \approx 7200$ from a precursor RANS simulation. The domain length, height, and crest height equal $6.14c$, $0.905c$ and $0.128c$, respectively, where c is the hump chord length. The spanwise extension amounts to $0.4c$ which is considered large enough to obtain span-independent flow solutions. With $Re_c = 936.000$ (based on c and the inflow velocity U_0) and $Ma = 0.1$ both compressible and incompressible simulations are possible. The thickness of the wall-adjacent RANS layer is $0.03c$ before the separation and reduces to $0.003c$ behind it. As for the BFS, a hybrid grid with structured near-wall regions and a hex-dominated unstructured mesh in the separation region is applied. This allows a high local resolution in the latter area, where Δx and Δy are approximately $0.005c$. The low-Re mesh fulfils $y^+(1) < 1$ at all wall-adjacent cells and is stretched by $r = 1.15$ in wall-normal direction. For the high-Re meshes (also $r = 1.15$), $y^+(1)$ equals 12.5, 25 and 50 at the crest position at $x = 0.5$, cf. Fig. 2 (right). Refer to Table 1 for the corresponding grid-point savings, as well as the $y_e^+(1)$-values for THETA. The total simulation time equals $30c/U_0$ ($20c/U_0$ are used for temporal statistics) with a time step size of $dt = c/500U_0$, so that $\mathrm{CFL}_{conv} < 1$ is fulfilled in the LES region.

Figure 4 compares DDES-(WF) skin-friction profiles for the two flow solvers, TAU (left) and THETA (right), with varying $y_{(e)}^+(1)$. For both solvers, high deviations between the low-Re solution and experimental data exist in the area of flow separation. This is due to the grey area at the RANS-LES interface, which remains critical in this flow despite using a modified filter width as remedy, cf. Sect. 2.1. In such regions the production of resolved turbulence is delayed resulting in an unphysical stable shear layer with too low initial turbulent shear stress $< u'v' >$, cf. Fig. 5 (right).

Considering the TAU skin friction profiles, similar findings as for the BFS are made: First, the trend towards better mean skin-friction agreement with the low-Re results for decreasing $y^+(1)$ within the recirculation area. Second, the alignment of profiles in the area of reattachment for increasing x/c, and third, the effect of numerical errors for $y^+(1) = 12.5$, yielding an overestimation of c_f at $x/c < 0.66$. Analogue to the previous test case, overall satisfying agreement with the low-Re reference is obtained, if $y^+(1)$ is limited to a value of 25.

Considering the THETA results in Fig. 4 (right), we observe an unexpected effect in the high-Re results that is in contrast to TAU: on the coarsest near-wall grid, i.e. $y_e^+(1) = 27$, the recirculation area is shifted by $0.15c$ in upstream direction and agrees remarkably well with the experimental data. This apparent reduction of the grey-area problem is supported by plots of the velocity profiles and the turbulent shear stress at the $x/c = 0.8$, i.e. inside the initial separating shear layer (cf. Fig. 5), where a stronger onset of turbulent stress compared to the low-Re results is visible. Note that TAU's high-Re result shows this behaviour, too, but to a lesser extent that is not sufficient to alter the mean skin friction. Despite this apparent improvement by using a coarse grid with wall functions, there is no physical reasoning to infer a consistent advantage over the higher-resolved low-Re approach. At this point, we believe it may be triggered by spurious numerical disturbances arising from the wall function method, calling for a more detailed analysis.

Moreover, Fig. 4 (right) includes THETA results for $y_e^+(1) = 13.5$ using the diffusion correction according to Sect. 2.2. As for the BFS, much better agreement with the low-Re result is obtained, without interfering with the prediction of separation.

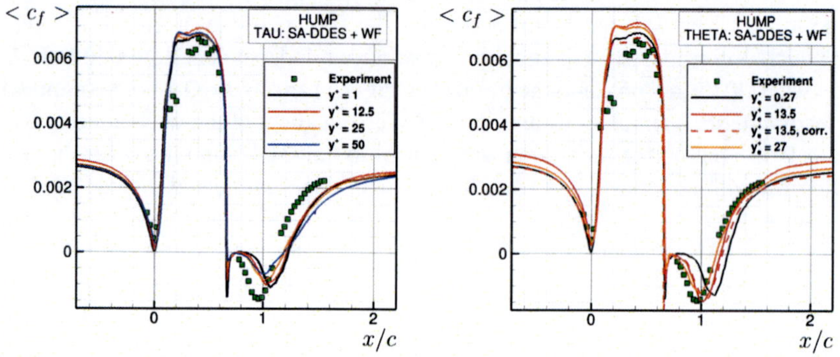

Fig. 4. Mean skin-friction results for the NASA wall-mounted hump at Re = 936.000: DLR-TAU solutions for different values of $y^+(1)$ (left) and results computed with DLR-THETA and a corrected version of THETA (right).

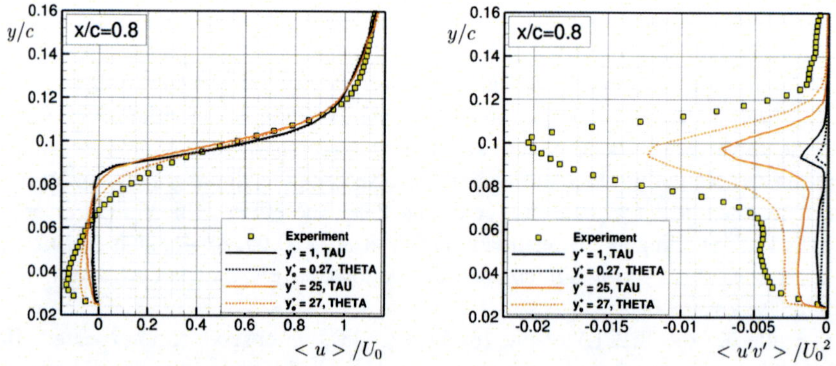

Fig. 5. Profiles of the velocity component in streamwise direction (left) and the normalised shear stress component $< u'w' >$ (right) at position $x/c = 0.8$ for both DLR-TAU and DLR-THETA with experimental data from [2].

3.4 Efficiency Considerations

The reduction of total computing time of high-Re simulations is analysed for the NASA wall-mounted hump. For time integration, TAU utilises a dual-time stepping scheme, i.e. for every physical time step a certain number of inner iterations is calculated. This number is controlled by a set of Cauchy convergence criteria which is kept constant for all simulations. THETA employs a projection scheme with inner linear-solver iterations which are controlled via a fixed residual-reduction threshold.

For TAU, the average time of a single inner iteration can be evaluated separately. Table 1 shows a reduction compared to the low-Re computation up to 36.8 % for $y^+(1) = 50$. In addition, though, the average number of inner iterations is also significantly reduced for increasing values of $y^+(1)$. This is due to the reduced near-wall cell aspect ratio on high-Re meshes, leading to lower numerical stiffness and consequently faster inner convergence of the physical time steps. Considering both effects in combination, one obtains a drastic reduction of total computing time, amounting to 60.4 % for $y^+(1) = 12.5$ and increasing to 72.6 % for $y^+(1) = 50$. For THETA, despite using even coarser meshes for a given $y_e^+(1)$, the overall reduction in computing time is just comparable to TAU. Although the individual effects could not be analysed separately, we assume that the linear-solver convergence in THETA benefits less than TAU from the reduced aspect ratios.

Table 1. Overview of cell and simulation-time reductions on high-Re meshes for the NASA hump.

	$y^+_{(e)}(1)$	red. cells	red. comp. time/it.	red. it./phys. dt	Total red. comp. time
TAU	12.5	28.1 %	25.7 %	46.7 %	60.4 %
	25	33.8 %	31.6 %	51.4 %	66.7 %
	50	38.9 %	36.8 %	56.7 %	72.6 %
THETA	13.5	38.9 %	–	–	61.5 %
	27	44.1 %	–	–	64.4 %

4 Conclusions

The combination of the analytical wall-function approach of Knopp et al. [5] with hybrid RANS-LES methods has been assessed using two different unstructured flow solvers, DLR-TAU and DLR-THETA. With TAU a basic validation of IDDES-WF acting as wall-modelled LES for the periodic channel flow at $Re_\tau \approx 4200$ has shown consistent results with the low-Re solution over a large range of $y^+(1) = 12.5 \ldots 210$. For the separated flows over a backward-facing step and a wall-mounted hump, TAU computations with DDES-WF yield higher demands on the near-wall resolution, i.e. $y^+(1) \leq 25$, in order to match the low-Re reference results in the recirculation regions.

Although this demand is confirmed with THETA, it can be fulfilled with coarser *primary* grids due to THETA's additional internal near-wall point. Moreover, the high-Re results of THETA for the hump show a surprising, yet inexplicable reduction of the grey area, yielding a better agreement with the experiment than the low-Re reference. Finally, a well-known [4] systematic overestimation of the skin friction on high-Re grids with $y^+_{(e)}(1) \approx 12.5$, which was found in all test cases with both flow solvers, could be cured for THETA by an ad-hoc correction of the near-wall momentum diffusion.

With regard to computational efficiency it has been confirmed that wall functions combined with HRLM may provide enormous savings of up to $2/3$ in overall computing time compared to the low-Re HRLM, demonstrating a large potential for future application in high-Reynolds aeronautical flows.

Acknowledgements. We thank Dr. Roland Kessler (DLR Göttingen, now retired) for providing the numerical diffusion correction in THETA.

References

1. Driver, D.M., Seegmiller, H.L.: Features of a reattaching turbulent shear layer in divergent channel flow. AIAA J. **23**(2), 163–171 (1985)
2. Greenblatt, D., Paschal, K.B., Yao, C., Harris, J., Schaeffler, N.W., Washburn, A.E.: Experimental investigation of separation control part 1: baseline and steady suction, AIAA J. **44**(12), 2820–2830 (2006)
3. Gritskevich, M.S., Garbaruk, A.V., Menter, F.R.: A comprehensive study of improved delayed detached eddy simulation with wall functions. Flow Turb Combust **98**(2), 461–479 (2017)
4. Kalitzin, G., Medic, G., Iaccarino, G., Durbin, P.: Near-wall behaviour of RANS turbulence models and implications for wall functions. J. Comp. Phys. **204**(1), 265–291 (2005)
5. Knopp, T., Alrutz, T., Schwamborn, D.: A grid and flow adaptive wall-function method for RANS turbulence modelling. J. Comput. Phys. **220**(1), 19–40 (2006)
6. Lozano-Duran, A., Jimenez, J.: Effect of the computational domain on direct simulations of turbulent channels up to $Re_\tau = 4200$. Phys. Fluids **26**(1), 011702 (2014)
7. Mockett, C., et al.: Two non-zonal approaches to accelerate RANS to LES transition of free shear layers in DES. In: Progress in Hybrid RANS-LES Modelling, Notes on Numerical Fluid Mechanics and Multidisciplinary Design vol. 130, pp. 187–201 (2015)
8. Probst, A., Löwe, J., Reuß, S., Knopp, T., Kessler, R.: Scale-resolving simulations with a low-dissipation low-dispersion second-order scheme for unstructured flow solvers. AIAA J. **54**(10), 2972–2987 (2016)
9. Shur, M.L., Spalart, P.R., Strelets, M.K., Travin, A.K.: A Hybrid RANS-LES approach with delayed-DES and wall-modelled LES capabilities. Int. J. Heat Fluid Flow **29**(6), 406–417 (2008)
10. Spalart, P.R., Deck, S., Shur, M.L., Squires, K.D., Strelets, M.K., Travin, A.K.: A new version of detached-eddy simulation, resistant to ambiguous grid densities. Theor. Comput. Fluid Dyn. **20**(3), 181–195 (2008)

Sensitivity Analysis of Discrepancy Terms Introduced in Turbulence Models Using Field Inversion

Florian Jäckel(✉)

Institute of Aerodynamics and Flow Technology, Center for Computer Applications in AeroSpace Science and Engineering, Bunsenstraße 10, 37073 Göttingen, Germany
florian.jaeckel@dlr.de

Abstract. RANS simulations with the Spalart-Allmaras turbulence model are improved for cases with flow separation using the Field Inversion and Machine Learning approach. A compensatory discrepancy term is introduced into the turbulence model and optimized using high-fidelity reference data from experiments. Influences on the optimization results with respect to regularization, grid resolution and areas in which the optimization is active are investigated. Finally, a neural network is trained and used to augment simulations on a test case.

Keywords: Data-driven turbulence modeling · Field inversion · Machine-learning

1 Introduction

Turbulent flows often play a key role in engineering and in science, hence it is important to be able to predict these flows accurately by numerical simulation. Methods which can resolve turbulent scales like LES and DNS are often still too expensive to be used regularly. Hence, Reynolds-Averaged Navier Stokes (RANS) simulations, which model the influence of turbulence on the mean flow, are the standard method for daily use. One difficulty however is that RANS simulations often fail to predict flow separation accurately e.g. in high-lift flows.

Over the last decades, many approaches aiming to improve turbulence modeling have been tried with varying degrees of success, such as extending existing models, for example with rotation corrections, or using more complex turbulence models, e.g. two equation models or Reynolds-stress models. One approach that has gained increasing interest in the recent years is the Field Inversion and Machine Learning (FIML) approach [2]. For this approach, it is argued that RANS models have always been largely empirical, and hence, with the increasing maturity of machine learning techniques, they can further be improved using machine learning models trained on datasets derived from high-fidelity simulations like DNS and LES or from experiments.

In this paper, an implementation of the FIML approach based on the negative Spalart-Allmaras turbulence model [7] in the DLR TAU-Code is described.

© The Author(s), under exclusive license to Springer Nature Switzerland AG 2021
A. Dillmann et al. (Eds.): STAB/DGLR Symposium 2020, NNFM 151, pp. 625–634, 2021.
https://doi.org/10.1007/978-3-030-79561-0_59

In particular, findings on the influence of the regularization, grid resolutions and regions of the computational domain where the field inversion is active on the field inversion results are presented. Finally, the turbulence model is augmented with a machine learning model and improved results computed with the augmented model are shown.

2 Methodology

2.1 Field Inversion and Machine Learning

The basic idea of the FIML approach is to introduce a discrepancy term β into the turbulence model, here the negative Spalart-Allmaras model:

$$\frac{D\tilde{\nu}}{Dt} = \beta(\mathbf{U},\tilde{\nu})P(\mathbf{U},\tilde{\nu}) - D(\mathbf{U},\tilde{\nu}) + T(\mathbf{U},\tilde{\nu}) \tag{1}$$

Here, P, D and T denote the turbulent production, destruction and transport terms respectively and $\mathbf{U}$ and $\tilde{\nu}$ denote the flow state and the Spalart-Allmaras transport variable. The correction term β is an unknown, spatially varying variable which is assumed to be a function of $\mathbf{U}$ and $\tilde{\nu}$. Since β is unknown, it must be modeled as well. The modeling is done in two separate steps: During a *field inversion* step training data is generated which is then used in a *machine learning* step to train a model relating β and $(\mathbf{U},\tilde{\nu})$.

Field Inversion. The training data needed for machine learning consists of pairs of β and the corresponding flow state $(\mathbf{U},\tilde{\nu})$, one pair per control volume. These pairs are obtained from high-fidelity reference data d_{ref} by solving the inverse problem

$$d_{\text{RANS}}(\beta) = d_{\text{ref}}, \tag{2}$$

where d_{RANS} is the RANS solution depending on β, for β. Solving Eq. (2) is an optimization problem, for which a cost function, here

$$\mathcal{I} = \underbrace{\frac{1}{2}\sum_{i}^{N_i}\sum_{j}^{N_{j,i}}\left[d^j_{i,\text{ref}} - d^j_{i,\text{RANS}}(\beta,\mathbf{U},\tilde{\nu})\right]^2}_{\mathcal{I}_1} + \underbrace{\lambda\frac{1}{2}\sum_{k}^{N_k}(\beta_k - 1)^2}_{\mathcal{I}_2} \tag{3}$$

needs to be minimized. The first term of the cost function, $\mathcal{I}_1$, measures the deviation between d_{ref} and d_{RANS} and is calculated as the squared difference of d_{ref} and d_{RANS}, summed over all $N_{j,i}$ cells where reference data is given and over all N_i reference quantities. Reference quantities can be field variables, such as velocities u, surface variables, such as the skin friction c_f, or integral values such as the lift coefficient c_l.

The second term, $\mathcal{I}_2$, measures and penalizes the magnitude of the turbulence model modification, acting as a Tikhonov regularization. Regularization is needed as the optimization problem typically is ill-posed due to noise in the reference data and the degrees of freedom, i.e. number of cells where β must be optimized, being much higher than the number of cells where reference data is given. The regularization parameter λ is selected according to the *L-Curve* criterion [3].

Because of the non-linearities in the RANS equations which are contained in $\mathcal{I}$, minimization of $\mathcal{I}$ with respect to β must be done iteratively. Hence, a gradient descent method is employed for which the gradient can be computed efficiently using the adjoint method.

Machine Learning. The field inversion step returns discrete values of β at the grid nodes, which can't be transferred directly to simulations on different geometries or different flow conditions. Hence, a relation between β and $(\mathbf{U}, \tilde{\nu})$ must be found first. Therefore, dimensionless flow features $\eta_i(\mathbf{U}, \tilde{\nu})$ are derived and then machine learning is used to find a model

$$f_\beta : \eta_0(\mathbf{U}, \tilde{\nu}), \ldots, \eta_n \mapsto \beta \tag{4}$$

approximating β. In particular, neural networks are used for this regression task. Due to the limited space here, the interested reader is referred to dedicated publications such as [1] for a detailed description of the principles of neural networks.

2.2 Implementation

The Field Inversion and Machine Learning approach has been implemented using the DLR TAU code [8], a highly optimized, parallel, state of the art CFD solver for unstructured grids, using and extending its adjoint capabilities. TensorFlow [9] is used for training and evaluating the machine learning model. The negative Spalart-Allmaras (SA-neg) [7] model is used as turbulence model due to its focus on aeronautical boundary layer flow.

3 Sensitivity Analysis

For the following sections, field inversions are conducted on the S809 airfoil. This airfoil was developed at the National Renewable Energy Laboratory (NREL) for wind turbines, and pressure distributions from wind tunnel measurements are available for a wide range of angles of attack and for multiple Reynolds numbers [5]. At an angle of attack of $\alpha \approx 6°$, flow separation starts to develop at the trailing edge and the predictions of the RANS simulations become unreliable, hence making this case suitable for the FIML approach, where it has been used successfully before [2]. The field inversions in this section are based on a case with an angle of attack of $\alpha = 12.2°$, a Mach number of $\mathrm{Ma} = 0.1$ and a Reynolds number of $\mathrm{Re} = 2 \times 10^6$ with the experimentally measured surface distribution of c_p from [5] as reference.

3.1 Regularization

As described in Sect. 2, the present optimization problem is ill-posed and therefore requires regularization. For an optimal choice of the parameter λ, the *L-Curve* criterion [3] is considered. Hence, field inversions are conducted for a range of values for λ and the resulting cost function terms $\mathcal{I}_1$, $\mathcal{I}_2$ are plotted against each other in the log-log plot in Fig. 1, on the left. The optimal λ_{opt} is found at the curve's inflection point, indicating the point beyond which $\mathcal{I}_1$ does not decrease significantly anymore but the magnitude of the model modification represented by $\mathcal{I}_2$ increases fast. For the current case, the optimal value for the regularization parameter is $\lambda = 5 \times 10^{-4}$ as indicated by the arrow in Fig. 1.

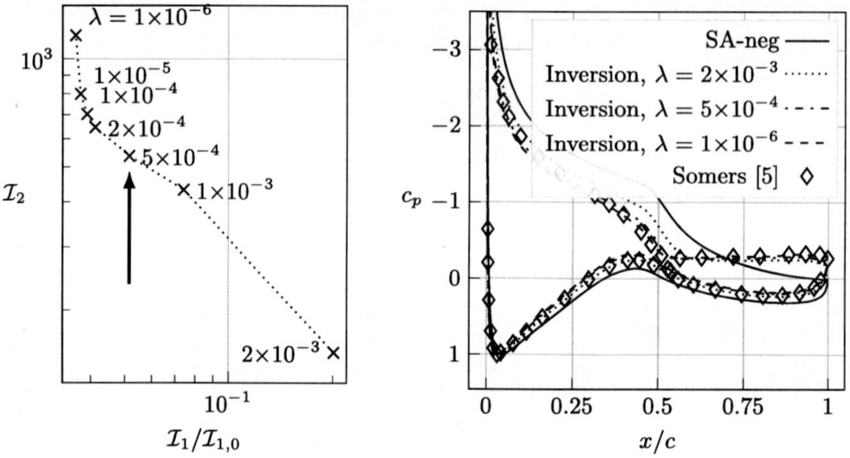

Fig. 1. Left: The L-Curve for the field inversion results for the S809. The optimal λ is marked by the arrow. Right: c_p-distributions from field inversions for different λ.

The right side of Fig. 1 shows the c_p distributions of the baseline turbulence model, the field inversion results for $\lambda = \lambda_{\text{opt}}$, a $\lambda < \lambda_{\text{opt}}$ and a $\lambda > \lambda_{\text{opt}}$ and the reference data. The reference data as well as the field inversion results exhibit a pressure plateau beginning at $x/c \approx 0.5$, indicating flow separation at the trailing edge which is not predicted by the baseline model. For $\lambda = \lambda_{\text{opt}} = 5 \times 10^{-4}$, the c_p-distribution (solid line) is close to the reference data and the value of cost function part $\mathcal{I}_1$ has decreased to 5.16% of its initial value. For stronger regularization, here e.g. $\lambda = 2 \times 10^{-3}$, the c_p-distribution (dash-dotted line) deviates further from the reference data with a decrease of cost function $\mathcal{I}_1$ to 20.1% of its initial value. A pressure plateau is still developed, but is predicted to begin further downstream compared to the results for $\lambda = \lambda_{\text{opt}}$ and the reference data. For weaker regularization, here e.g. $\lambda = 1 \times 10^{-6}$, the c_p-distribution (dashed line) matches the reference data very closely with a decrease of cost function $\mathcal{I}_1$ down to 3.62% of its initial value. However, overfitting can be observed: In the pressure plateau, the reference data include slight fluctuations, to which the inversion result with the weak regularization tries to fit as well. These fluctuations most probably stem from imprecisions during the process of injecting the reference data and should hence not be incorporated.

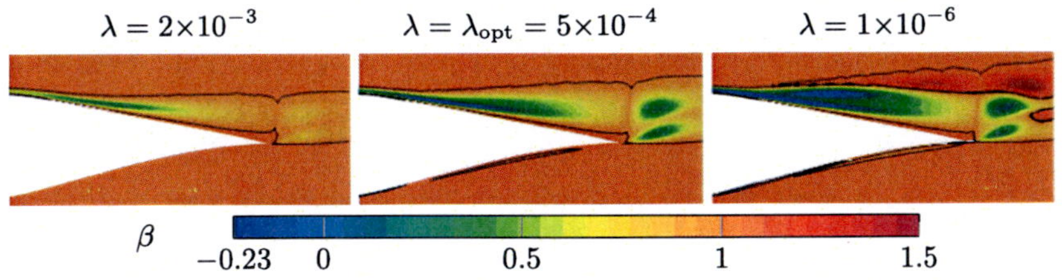

Fig. 2. Resulting fields for β for different values of the regularization parameter λ. The black contour marks $\beta = 1$.

Resulting β-fields for all three cases are shown in Fig. 2. While the β-field for the stronger regularization ($\lambda = 2 \times 10^{-3}$) appears to be equivalent to a down-scaled version of the β-field for $\lambda = \lambda_{\text{opt}}$, the β-field for the weak regularization ($\lambda = 1 \times 10^{-6}$) shows signs of overfitting by exhibiting more pronounced and additional extrema. These findings confirm the relevance of choosing an optimum λ and the L-Curve criterion as a method to do so.

3.2 Grid Resolution

In the next step, the influence of the grid resolution on the field inversion result is investigated. Therefore, field inversions are conducted on six different grids at the same flow conditions as before. The grids were obtained from [6] and range from resolutions of 353×49 grid points to 2113×289 grid points, with y^+_{max} ranging from 4.185 to 0.799. The outer boundary of the computational domain has a distance of $1000c$ from the airfoil, where c is the airfoil chord length. Grid convergence with the baseline turbulence model is reached for the 1057×145 grid.

For each grid, the optimal regularization parameter λ_{opt} was determined separately as described in the previous section. The corresponding results for β are plotted in Fig. 3. On all grids, the field inversion process was successful and the cost function could be reduced adequately. In general, the magnitude of the correction β appears to increase with the grid resolution. For the 1057×145 and the 1409×193 grid, almost no differences in magnitude for the β-field can be made out in the region of flow separation. For the 2113×289 grid, additional maxima can be seen, qualitatively similar to those appearing due to over-fitting as found in the chapter before.

Since the field inversion shows good results on all grids, including under-resolved grids, it must be assumed that the obtained β-fields on the under-resolved grids also account for discretization errors. Currently, we are interested

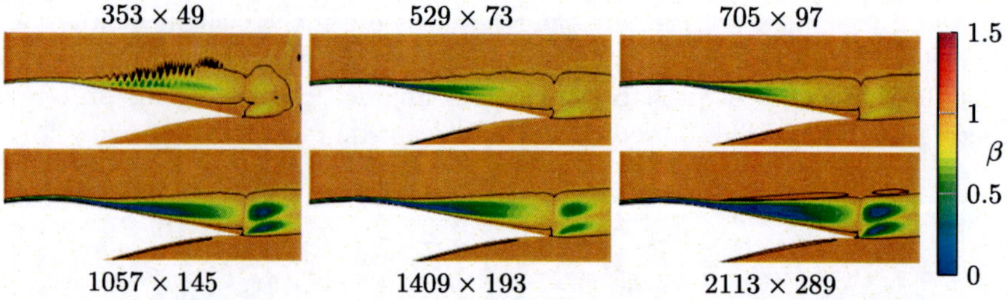

Fig. 3. Field inversion results β for different grids.

only in improving the turbulence model itself and hence use only results from converged grids, but it can be reasoned that a FIML approach could be used to improve simulations on under-resolved grids in general as well.

3.3 Area of Optimization

Finally, the influence of the correction term β depending on the area where it is active is investigated. As seen on the left in Fig. 4, three areas where β has changed noticeably can be distinguished, marked with A, B and C. In region A, at the upper surface close to the leading edge, turbulent production was increased by increasing β. In region B, spanning almost the entire upper surface except for the leading edge, turbulent production was reduced by decreasing β. Region C, which encompasses the two spots corresponding to the vortices downstream of the trailing edge, again sees a reduction in turbulent production.

The influence of the correction term depending on the area where it is active is investigated by running RANS simulations with the augmented model starting with the full β-field as seen in Fig. 4 and turning off the correction in the different areas one by one by setting β to 1. The different tries are declared as R_{ABC}, where A, B and C are either 1 or 0, depending on whether the correction is active in the respective region or not.

The results are shown on the right in Fig. 4. Turning the correction term off in all regions (R_{000}) corresponds to the baseline turbulence model, hence the cost function part $\mathcal{I}_1$ remains unchanged at 100% of its initial value. Activating the correction term in regions A and C decreases $\mathcal{I}_1$ only slightly to 98.0% (R_{100}), 95.8% (R_{001}) and 93.7% (R_{101}) of its initial value. The correction term in region B has by far the largest influence by reducing $\mathcal{I}_1$ to 4.53% (R_{010}) of the initial value alone and to 4.35% (R_{110}), 3.88% (R_{011}) and 3.84% (R_{111}) when active in conjunction with regions A and C. This behaviour was expected however since region B covers the largest area and especially the area where flow separation occurs.

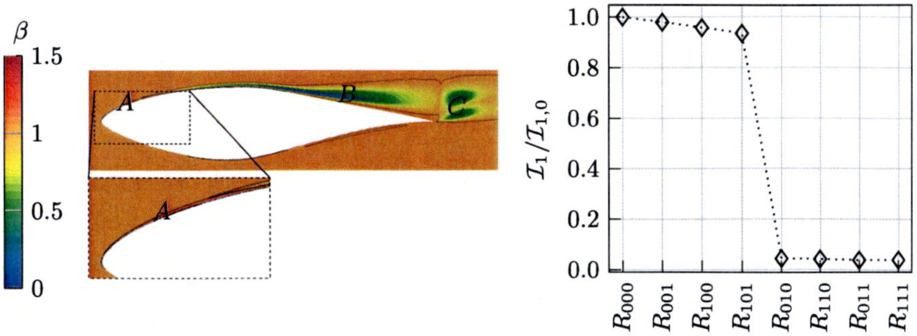

Fig. 4. Left: The three areas of interest marked by A, B and C. Right: Results after activating β in the different regions.

4 Application

4.1 Training

In this section the entire FIML approach is applied. First, training data is generated by conducting field inversions on the S809 airfoil for angles of attack $\alpha = 6.2°, 10.2°$ and $20.1°$. As discussed in the previous chapter, for each case the regularization parameter λ is determined separately and the field inversions are conducted only on the fully resolved 1057×145 grid to ensure optimal training data. From the complete set of data, the samples in the freestream, where the turbulence model is not active, are discarded together with 99.5% of the remaining samples where $\beta = 1$. This is done to reduce the frequency at which $\beta = 1$ appears in the training data as otherwise the machine learning model would simply learn to always predict $\beta = 1$. The remaining samples are split into a training set totaling approximately 38000 samples and a testing set totaling approximately 4000 samples.

Second, a machine learning model is trained to model β according to Eq. (4). The neural network used in this case is a fully connected network with two layers with 200 neurons each. Dropout layers are applied after each hidden layer with a drop-out rate of 0.2. The mean squared error is used as loss function and the network is trained for 500 epochs using the Adam optimizer. The dimensionless flow features $\eta_i(\mathbf{U}, \tilde{\nu})$ used as input are

$$\eta_1 = \log(\frac{P}{D}), \eta_2 = \log(\chi), \eta_3 = \log(\frac{|S|}{|\Omega|}) \text{ and } \eta_4 = \log(\frac{\mu_t|S|}{\tau_w}), \tag{5}$$

similar to the features presented in [4]. The features represent the ratio of the production and the destruction term of the SA model, the non-dimensionalized SA transport variable, the ratio of the strain and vorticity magnitudes and the ratio of the local turbulent shear stress to the shear stress at the wall respectively. The logarithms are applied due to the fact that each features' values span multiple orders of magnitudes.

4.2 Numerical Results

Finally, RANS simulations with the ML-augmented turbulence model are conducted for multiple angles of attack from $\alpha = 0.0°$ up to $\alpha = 20.1°$.

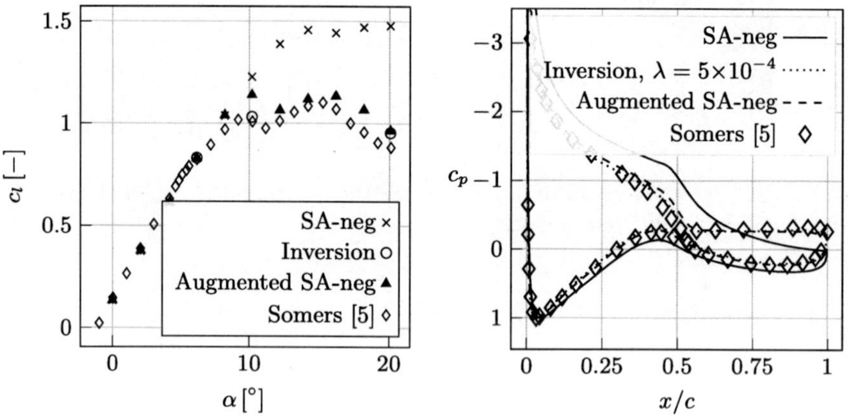

Fig. 5. Left: Lift coefficients for the S809 airfoil at Re $= 2 \times 10^6$. Right: c_p-distributions for $\alpha = 12.2°$

Figure 5 compares results from the augmented RANS model, the original turbulence model, the field inversion and the reference data. For $\alpha \lessapprox 6.2°$, the agreement between the measured lift and the predicted lift using the original RANS model is good, while it quickly worsens for larger α as soon as flow separation occurs. The lift predictions of the field inversion results agree much better, matching the reference lift with deviations of only 0.8% ($\alpha = 6.2°$), 2.3% ($\alpha = 10.2°$) and 8% ($\alpha = 20.1°$). The predictions of the ML-augmented turbulence model remain unchanged in the linear area of the lift curve, which is intended since the results of the original turbulence model are already satisfying here. For higher α, the augmentation improves the turbulence model noticeably as the lift is predicted far more accurately. The maximum deviation of the augmented model is considerably lower with 13% (at $\alpha = 10.2°$) as opposed to a maximum deviation of $> 50\%$ (at $\alpha = 20.1°$) of the baseline model.

On the right in Fig. 5, the improvement of the predictions due to the ML-augmentation is illustrated based on the c_p-distribution exemplary for $\alpha = 12.2°$, for which the field inversion was done but not included in the training data for the ML model. While the c_p-distribution of the augmented model (dashed line) doesn't come as close to the experimental values as the field inversion result (dotted), it still represents a large improvement with respect to the baseline model (solid). Figure 6 compares the results for the β-field as computed during field inversion and as predicted by the ML-augmentation. While the inversion result is smoother in general and shows stronger modifications, the presumably most import area leading up to the point of flow separation shortly after the

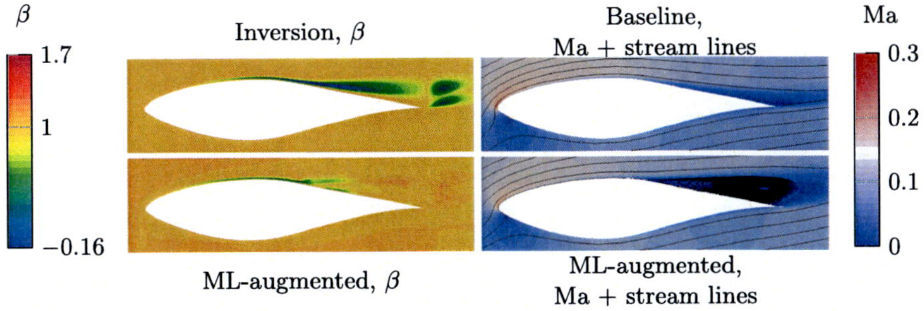

Fig. 6. Left: β-field from field inversion and as predicted by the ML-model. Right: Mach number and stream lines from the baseline turbulence model and the ML-augmented model.

thickest airfoil section, is predicted similarly by the ML-model. In the right half of Fig. 6, the Ma-number and the streamlines of the solution are plotted for the original turbulence model and the ML-augmented model. While the original model shows only a very small region of flow separation, the ML-augmented model develops a much larger separation bubble as expected.

5 Conclusion

The field inversion and machine learning approach was successfully reproduced based on the DLR TAU-code and the negative Spalart-Allmaras model. The influence of the regularization parameter, the grid resolution and the regions where β is optimized during field inversion on the field inversion results were investigated based on the flow around the S809 airfoil with flow separation at the trailing edge. For the regularization, a method to determine the optimal regularization parameter was demonstrated and the influence a suboptimal choice can have was discussed. It was shown and discussed that the field inversion can also compensate for spatial discretizaton, returning correct results on coarse grids as well. Different regions where the optimization is active were detected and their influence on the end result was investigated. Under consideration of these information, a machine learning model was trained using data from field inversions of the S809 at multiple angles of attack. The RANS turbulence model was then augmented by the machine learning model and improved results were shown on the S809 airfoil.

A future step in the context of the FIML approach is to include inversion results from additional test cases with different geometries and at different flow conditions in the training data, which is expected to be essential to be able to provide a robust and reliable augmentation. With growing experience with the FIML approach it is also planned to investigate its application to more complex turbulence models such as the k-ω-SST model and, potentially, Reynolds-Stress models like the SSG/LRR-ω model.

References

1. LeCun, Y., Bengio, Y., Hinton, G.: Deep learning. Nature **521**, 436–444 (2015)
2. Singh, A.P., Duraisamy, K.: Using field inversion to quantify functional errors in turbulence closures. Phys. Fluids **28**, 045110 (2016)
3. Hansen, P.C.: The L-curve and its use in the numerical treatment of inverse problems. In: Johnston, P., (ed.) Computational Inverse Problems in Electrocardiology, Advances in Computational Bioengineering, pp. 119–142, WIT Press (2000)
4. Holland, J.R., Baeder, J.D., Duraisamy, K.: Field inversion and machine learning with embedded neural networks: physics-consistent neural network training. In: AIAA Aviation, Forum, 17–21 June 2019. Dallas, Texas (2019)
5. Somers, D.M.: Design and experimental results for the S809 airfoil, Technical Report. N. p, United States (1997)
6. Coder, J.G.: Standard test cases for transition model verification and validation in computational fluid dynamics. In: AIAA Aerospace Sciences Meeting, 8–12 January 2018. Kissimmee, Florida (2018)
7. Allmaras, S.R., Johnson, F.T., Spalart, P.R.: Modifcations and clarifications for the implementation of the spalart-allmaras turbulence model. In: Seventh International Conference on Computational Fluid Dynamics (ICCFD7), Big Island, Hawaii, 9–13 July 2012
8. Schwamborn, D., Gerhold, T., Heinrich, R.: The DLR TAU-Code: recent applications in research and industry. In: European Conference on Computational Fluid Dynamics, ECCOMAS CDF (2006)
9. Abadi, M., et al.: TensorFlow: Large-Scale Machine Learning on Heterogeneous Systems. White paper (2015)

Spliss: A Sparse Linear System Solver for Transparent Integration of Emerging HPC Technologies into CFD Solvers and Applications

Olaf Krzikalla[1(✉)], Arne Rempke[1,2], Alexander Bleh[3], Michael Wagner[1], and Thomas Gerhold[1]

[1] Institute of Software Methods for Product Virtualization, German Aerospace Center (DLR), Dresden, Germany
olaf.krzikalla@dlr.de
[2] Institute of Aerodynamics and Flow Technology, German Aerospace Center (DLR), Göttingen, Germany
[3] Institute of Propulsion Technology, German Aerospace Center (DLR), Cologne, Germany

Abstract. We introduce Spliss, a novel block sparse linear solver library, which is designed to meet current and upcoming challenges in the CFD development. Spliss supports a wide range of linear operators typically used in CFD applications. This includes sparse block matrices of variable block sizes and different scalar types as well as matrix-free operators. The provided solving methods comprise standard methods such as Jacobi or Gauss-Seidel smoothers, Krylov subspace methods, and domain-specific methods such as direct block-tridiagonal solvers. Spliss uses contemporary and emerging HPC technologies such as one-sided communication, hybrid and heterogeneous parallelization. The C++ template interface allows for extensions of the library and specialization of its internals for optimization. We explain the integration of Spliss in the two CFD solvers CODA and TRACE. Finally, we show how CODA benefits from using Spliss by offloading the linear solver to GPUs.

Keywords: Sparse linear solver · Sparse block structure · CFD solver

1 Introduction

The efficient solving of large linear equation systems that result from the discretization of the Reynolds-averaged Navier-Stokes (RANS) equations in CFD methods requires algorithms that are well adapted to the specific numerical problems. Existing libraries do not cover all required features such as variable-sized blocks or mixed scalar types in a matrix (e.g. Trilinos [17]). For some libraries optimizations regarding HPC challenges such as GPU-aware computing or hybrid parallelization are yet to come (e.g. PETSc [3]). Therefore, we have decided to develop the novel linear solver library Spliss (Sparse Linear System Solver). Spliss is tailored to the wide

© The Author(s), under exclusive license to Springer Nature Switzerland AG 2021
A. Dillmann et al. (Eds.): STAB/DGLR Symposium 2020, NNFM 151, pp. 635–645, 2021.
https://doi.org/10.1007/978-3-030-79561-0_60

requirements of CFD applications while at the same time being independent of the particular CFD solver. Moreover, Spliss addresses the needs of CFD developers targeting to utilize today's supercomputing systems.

In this paper we first introduce the rich feature set of Spliss, which specifically targets CFD solvers. Then we explain the technical design of Spliss, by which we can achieve computational efficiency and parallel scalability on HPC platforms. We follow up with a detailed description about the integration of Spliss in the CFD applications CODA [7,10–12] and TRACE [6]. Finally we present first results of the integration process. These results show, that Spliss enables the usage of formerly unemployed HPC resources, which reduces the time-to-solution of CFD applications.

1.1 Matrix Structure

The linear system of equations, for which a method such as (among others) the implicit RANS method requires an approximated solution, is characterized by a sparse structure resulting from the connectivity of the geometric mesh. For each geometrical element of the mesh, e.g. volumes for the finite volume method (FVM) or nodes for the finite element method (FEM), the local degrees of freedom (DoF) are represented in a vector and their derivatives in a matrix. Since the number of degrees of freedom per element in general is larger than one in CFD, this results in a sparse block matrix. Figure 1 shows typical examples for such a sparse matrix with dense blocks.

Fig. 1. Selected matrix block structures covered by Spliss. Left: constant block sizes. Center: variable block sizes. Right: complex values in main diagonal blocks.

The size of the dense blocks depends on mainly two ingredients. First, the physical model involved, e.g. 5 for the conservative variables $\rho, \rho\mathbf{v}$, ρE of 3D Euler equations, 6, 7 or up to 12 when additionally utilizing one-equation, two-equation or Reynolds stress turbulence models for the RANS equations. Second, the discretization utilized, e.g. one DoF per physical variable for FVM, 4 or 10 for second or third order of a modal Discontinuous Galerkin method for 3D elements. Therefore, depending on the chosen method, the nonzero matrix block can be quite small (e.g. 5×5 for FVM Euler) or quite large (e.g. 392×392 for sixth-order method with 56 degrees of freedom per element for each of the 7 equations for a two-equation turbulence model). More importantly, the block sizes can vary within a single matrix. This happens, for instance, when choosing different discretization orders for different areas of the mesh using adaptive p-refinement [12]. A schematic illustration for such a matrix can be seen in the center of Fig. 1. In contrast to this, the left side of Fig. 1 shows a sparse matrix of fixed-sized blocks, which results from a homogeneous approach for all elements.

For time-spectral methods, such as linear frequency domain (LFD) or harmonic balance (HB), the matrix has mixed scalar types, e.g. with complex entries for the diagonal blocks and real values for other blocks [2,5]. Spliss respects this property and does not force the user to use a matrix with all complex entries, most of them having an imaginary part of zero. Instead, Spliss allows matrices to contain more than a single scalar type, as illustrated in the right part of Fig. 1. Here the matrix consists of complex diagonal and real-valued off-diagonal entries.

Furthermore, no explicit representation of the matrix might be present, only a method to evaluate the effect of the operator on a given vector ("matrix-free") [13]. This is especially interesting for a 2nd order Jacobian [9], for which no representation is available due to memory restrictions. Instead, the user may only provide a method to compute the matrix vector product, which is then used within a solver method to solve the equation system.

1.2 Solution Methods

The supported solving methods comprise standard methods such as Jacobi or (multi-color) Gauss-Seidel smoothers [1], direct methods to invert dense blocks, and Krylov subspace methods [16,18]. Unlike other standard libraries, Spliss is able to combine these methods in a highly flexible way, rather than just a solver-preconditioner combination. It is possible to chain multiple solver components and, therefore, experiment with different combinations as illustrated in Fig. 2. Note that although this allows a very flexible construction of specialized solver chains, the solver components are still able to implement optimizations in case they are combined in a way where this brings a benefit, e.g. because they should invert the same linear operator.

Additionally to those "standard" solver methods, domain-specific solver components are available such as a line-implicit method, which is especially advantageous for solving the equation systems that emerge from resolved boundary layers

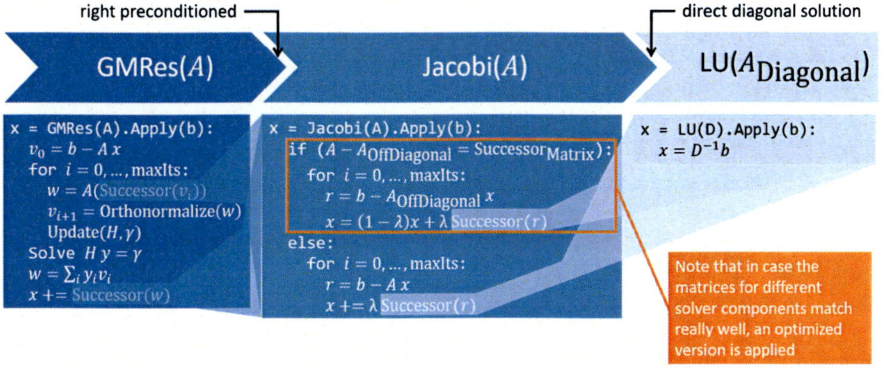

Fig. 2. GMRES method preconditioned by a Jacobi method which uses the direct block diagonal inversion as block-local preconditioner.

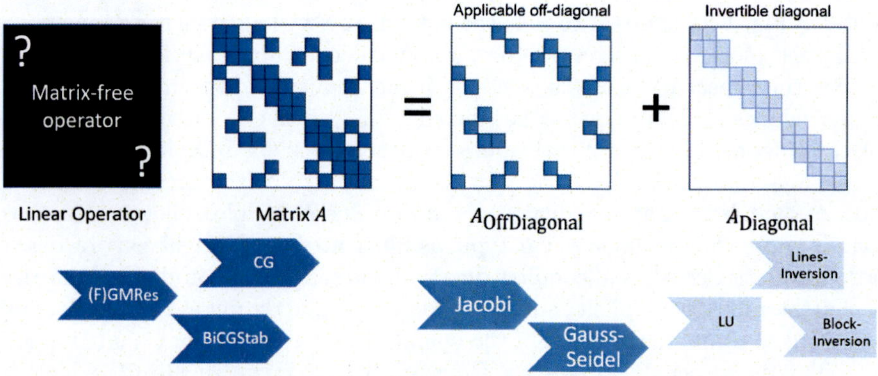

Fig. 3. Different solver components demand for different parts of the matrix.

in viscous CFD meshes [8]. It implements the Thomas algorithm [4] for solving a block-tridiagonal system exactly. This can be interpreted as extending the block-diagonal part of the matrix into a generalized diagonal part. Figure 3 gives a visual overview on different solver components, which can be combined to the user's wish, their requirements regarding the matrix or linear operator, and the concept of splitting the matrix into an invertible generalized diagonal and an applicable off-diagonal matrix part. Note that in case of a wider diagonal such as the block-tridiagonal system, $A_{\text{OffDiagonal}}$ contains significantly less nonzeros than A.

2 The HPC-Aware Implementation of Spliss

A key design goal of Spliss is computational efficiency and parallel scalability for current and emerging HPC technologies. Spliss not only leverages all available parallelization levels of contemporary HPC platforms, but also offers different parallelization techniques for each of these levels. The use of the C++ template mechanism shifts decisions such as the usage of a GPU and the selection of an appropriate memory layout to the compile time. This saves runtime and gives the compiler the possibility for more aggressive optimizations.

The parallelization strategy of Spliss is based on the fact, that large amounts of the solution process are spent in the matrix-vector multiplication. In such a multiplication each entry of the result vector can be computed independently of all other entries. Hence, those entries can be computed in parallel. The decomposition of matrix and vector data across processes reflects this parallelization strategy. For a given matrix each process holds a matrix band consisting of block rows. Band borders are always located at block borders. The distribution of the matrix also defines the distribution of the result vector. Figure 4 illustrates this concept for one of the processes. The square depicts a matrix, the orange-blue rectangle above depicts the source vector and the blue rectangle to the right depicts the result vector. The middle matrix band is colored in saturated blue and orange. This band is stored locally for the illustrated process. The result

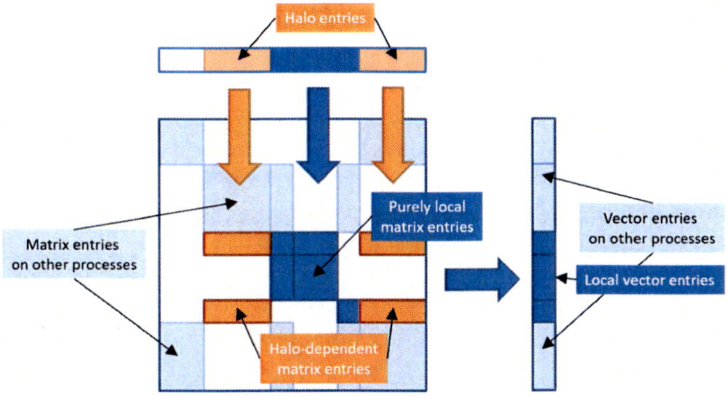

Fig. 4. Process distribution of data structures in Spliss.

vector is distributed in the same way. Result entries that belong to the band are stored on the same process – the local vector entries in the figure.

The matrix blocks in the local band are separated in purely local and halo-dependent blocks. That separation depends on the distribution of the source vector. For purely local matrix blocks also the respective source vector entries are stored locally. Source vector entries stored on other processes form the halo, which must be transferred during a multiplication operation. The number of halo entries depends on the sparsity pattern of the matrix. Only those source vector entries are part of the halo, for which there is a matrix block in the locally stored band, since only such entries contribute to the computation of the local result vector. E.g. the left-most source vector block in the figure is not part of the halo as there is no entry in the local matrix band for this block.

Spliss provides implementations of the described parallelization approach for MPI as well as for GASPI. The MPI implementation uses the classical two-sided communication model. Since this model requires a process handshaking during the halo exchange, it makes the execution of the individual processes tightly coupled. In contrast, GASPI provides a one-sided communication model. With this model a double buffering approach has been implemented in Spliss. This approach loosens the tight coupling of the processes, which leads to a more asynchronous execution flow and hence to a better scalability.

The shared-memory parallelization of Spliss employs the Alpaka library [14] and basically uses the same approach that is used for distributed parallelization. Every thread process a complete block row of the matrix. Thus no write conflicts on the result vector can occur. On GPUs with their large number of threads block rows are further partitioned into scalar rows to leverage all available GPU resources.

Spliss implements various memory layouts for matrices. These layouts define the storage order of the individual matrix entries in the memory. By this means Spliss can automatically adapt the underlying matrix implementation to the specific needs of the respective SIMD or GPU target architecture.

3 The Integration of Spliss in CFD Applications

Spliss is currently available in version 1.4.0 and provides all the features described in Sect. 1. It is already integrated and used in CFD applications, among others in CODA and TRACE. These applications can thus take advantage of Spliss' advanced HPC technologies while being spared complex HPC programming tasks such as those described in the following paragraphs.

For instance, Spliss is built in such a way, that the shared library object can use GPUs without exposing this fact to the library interface. An application is not even required to use the GPU framework during its compilation or link process. It is sufficient to link against a GPU-enabled Spliss library in order to benefit from possibly available GPUs. By that means, CODA as well as TRACE already leverage GPUs without the necessity of any specific code adaptions.

Another example of modern HPC approaches provided by Spliss is the handling of hybrid parallelization. Other libraries may expose a fork-join model for multi-threading support. That is, a library routine is called by s single thread, which then internally spawns worker threads. Those threads are joined again before the program returns from the routine. This model puts an implicit barrier at each library call and is, therefore, prone to load imbalances.

Spliss can be used in this classical way too. In addition, it also offers the possibility to be used with all threads of an application. In that case it doesn't fork or join threads internally, but uses the existing application threads (see Fig. 5). There are no barriers at library calls. Therefore threads run asynchronously for a longer time period, which can mitigate the effects of load imbalances. By that means a better scalability is achieved. Moreover, the application programmer can call library routines in a natural way. It is not necessary to create single-threaded regions for library calls. This approach enabled CODA – which has its own flexible threading model – to seamlessly integrate Spliss.

In contrast to CODA, the CFD solver TRACE rather uses the classical threading approach thus demonstrating the flexibility of Spliss regarding different threading models. Furthermore TRACE imposes several requirements that had been implemented in Spliss and successfully adapted in TRACE. First, it

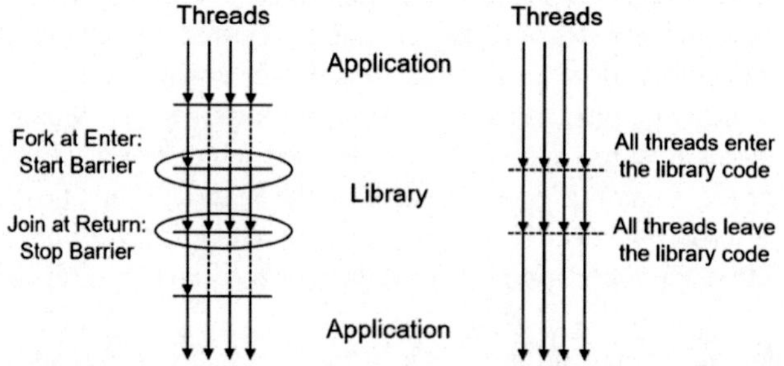

Fig. 5. Left: execution flow of threads during a library call in the fork-join model. Right: execution flow of threads during a library call with reused application threads as supported by Spliss.

should be noted that TRACE is mainly formulated in C hence an adaption module had to be implemented to couple Spliss. We decided against a generic C-Interface for Spliss since an individual coupling strategy can benefit more from specific optimizations especially when setting up the block structured matrix. Second, the legacy solver implementation in TRACE is using a domain decomposition consisting of structured and unstructured blocks. The former is featuring a hyperplaned coloring and an according permutation of the rows. With the support for row permutations in Spliss, the specific algorithm in TRACE was fully reimplemented while allowing for an HPC-optimal computation. Third, TRACE takes advantage of the unique ability of Spliss to define matrices in which some entries are complex-valued while other entries are real-valued, thus saving memory and reducing computational overhead. In conclusion the coupling with TRACE has successfully demonstrated the combined usage of a variety of Spliss features even in the context of a legacy C-Code.

4 Evaluation

In this section we evaluate Spliss in comparison to the well-established library PETSc and highlight the benefits of Spliss for a CFD solver such as CODA by offloading the computationally intensive linear solver to GPUs.

4.1 Evaluation of Spliss in Comparison to PETSc

In order to verify the correctness of the implemented methods, we compare the results obtained by Spliss to those that can be obtained by using the open

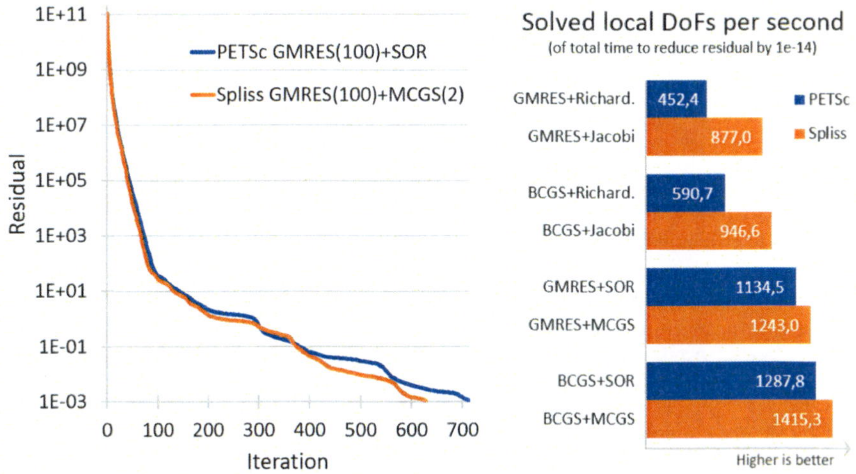

Fig. 6. Left: Comparison of the residual over the iterations obtained by Spliss and PETSc using GMRES with a Gauss-Seidel preconditioner. Right: Comparison of the performance of similar solver methods implemented by PETSc and Spliss.

source library PETSc [3]. The left part of Fig. 6 shows some residual results for using the GMRES method, preconditioned (from right) by a Block-Gauss-Seidel smoother. The PETSc implementation uses the successive over-relaxation (SOR) preconditioner, which means a forward and a backward sweep on each local domain. The Spliss implementation uses two iterations of a multi-color Gauss-Seidel solver component [1], which enables more parallelization on the domain e.g. by using threads or GPUs. Although the smoothers are different on the algorithmical basis, they reduce the residual quite similar.

The right part of Fig. 6 shows the results of a performance comparison of PETSc and Spliss for different solver methods. We compared four different methods: BiCGStab and GMRES, both preconditioned from right by either a Block-Jacobi method (Richardson scheme in PETSc) or a Block-Gauss-Seidel method (SOR in PETSc). The full linear equation system consists of 10.8 million degrees of freedom. The matrix is a sparse block matrix with a constant block size of 3×3 scalar entries and was obtained by using a mesh deformation method utilizing the elasticity-analogy [15]. The residual was reduced by 14 orders of magnitude on 128 MPI domains in parallel using no threads or GPUs.

The most significant aspect is the massive deviation for the GMRES+Richardson/Jacobi method. Here Spliss is almost twice as fast as PETSc. The causes for this are probably the design decision explained in Subsect. 1.2 to split diagonal and off-diagonal matrix parts and the good parallel implementation as described in Sect. 2. This combination of solver method was one of the main drivers in the Spliss development, so it seems likely Spliss achieves an outstanding performance here. The deviation is a little bit smaller, but still massive for the BiCGStab method preconditioned by Richardson (PETSc)/Jacobi (Spliss) method for probably the same reasons.

When switching from Jacobi to SOR respective multi-color Gauss-Seidel (MCGS), the performance increase for PETSc is larger than for Spliss. This originates in the different algorithmical Gauss-Seidel approach explained above, which involves more synchronizations and halo exchanges for the multi-color Gauss-Seidel than for the per-domain-sequential Gauss-Seidel. However, the multi-color approach offers more parallelization possibilities by using threads or GPUs, which are not tested here for better comparability to PETSc.

4.2 Evaluation of CODA with Spliss on GPUs

Focusing on the specific task of linear-system solving allows for integrating more advanced, hardware-adapted optimizations. One example is the usage of GPUs, where Spliss enables the execution of the computationally intensive linear solver on GPUs with no changes to the CFD solver CODA. In the following we evaluate first results of CODA using Spliss to run on a GPU cluster with NASA's Onera M6 test case "2308"[1]. This test case is well studied and provides experimental data and numerical solution by other CFD applications for comparison. It solves

[1] https://turbmodels.larc.nasa.gov/onerawingnumerics_val.html.

Table 1. Runtime comparison between CPU and GPU versions.

		Iterate (in min.)			Linear solver (in min.)		
		CPU	GPU	Ratio	CPU	GPU	Ratio
Mesh 4 (extra coarse)	1 × 20	4.69	4.02	85.7%	2.37	1.73	73.0%
	4 × 20	1.62	1.68	103.7%	1.00	1.08	108.0%
Mesh 3 (coarse)	1 × 20	58.0	40.8	70.3%	41.8	24.7	59.1%
	4 × 20	15.5	16.4	105.8%	11.2	12.0	107.1%

the Reynolds-averaged Navier-Stokes equations (RANS) with a Spalart-Allmaras one-equation turbulence model in its negative form (SAneg).

First, we compare numerical results of CODA with Spliss running on CPUs and GPUs. We verified characteristic residuals of the CPU version and GPU version and found that the deviation between the computed residuals for CPU and GPU for each iteration are in the same range or smaller than deviations between CPU measurements with different decompositions. The computed solution for lift and drag coefficients is identical for the CPU and GPU version.

Second, we compare the runtimes of the CPU and GPU version for the coarse (Mesh 3) and extra coarse (Mesh 4) mesh sizes with 1.08M and 135k elements, respectively, in two setups: a) single node: 1 Intel Xeon 6230 CPU with 20 cores vs. 1 Nvidia Tesla V100 GPU and b) four nodes: 4 CPUs vs 4 GPUs (1 GPU per CPU/node). Table 1 shows the runtime results for these setups for the entire iteration phase (300 iterations) and the time of the iteration phase spent in the linear solver, i.e. Spliss; whereas all times include all transfers and synchronizations between CPU and GPU. For the single node setups the GPU version provides a significant runtime improvement up to 41%. For the multi-node setups the GPU version performs slightly worse than the CPU version since data exchanges between GPUs are currently executed via the host CPU resulting in a significant overhead. We expect this issue to resolve with the integration of CUDA-aware MPI communication, which allows direct transfers between GPUs without host involvement.

5 Conclusion

The introduced sparse linear solver library Spliss makes a significant contribution to the development of CFD solvers. First, it enables solving linear systems with increased performance by transparently integrating more advanced, hardware-adapted optimizations, such as off-loading to GPUs. Second, Spliss supports a variety of algorithms and data structures that are required, e.g. line-implicit methods for the boundary layer of viscous CFD meshes or different scalar types for time-spectral methods. Third, Spliss relieves the developer of CFD software in dealing with the hardware layer and is easy to use, e.g. due to the adoption of the specific threading model of CODA. Finally, we show that Spliss allows the transparent use of GPUs within the CFD solver CODA, which can provide a significant runtime improvement without affecting numerical stability.

Acknowledgements. The authors would like to thank those people who enabled, planned and contributed to Spliss: Jan Backhaus, Ralf Hartmann, Johannes Holke, Jens Jä-gersküpper, Norbert Kroll, Martin Joachim Kühn, Stefan Langer, Tobias Leicht, Arthur Stück, Jonas Thies and Daniel Vollmer. The work was funded by the DLR programs VicToria, KonSENS and the project TOSCANA, which is part of the german government's aviation research program under funding code 20X1704A. It has been supported by the EXCELLERAT project, which has received funding from the European Union's Horizon 2020 research and innovation programme under grant agreement No. 823691.

References

1. Adams, L., Ortega, J.M.: A multi-color SOR method for parallel computation. In: ICASE 82.9 (1982)
2. Ashcroft, G., Frey, C., Kersken, H.-P.: On the development of a harmonic balance method for aeroelastic analysis. In: Proceedings of the 6th European Conference on Computational Fluid Dynamics (2014)
3. Balay, S., et al.: PETSc Web page (2020). https://www.mcs.anl.gov/petsc
4. Conte, S.D., De Boor, C.: Elementary numerical analysis: an algorithmic approach. Society for Industrial and Applied Mathematics (2017)
5. Da Ronch, A., McCracken, A.J., Badcock, J.J., Widhalm, M., Campobasso, M.S.: Linear frequency domain and harmonic balance predictions of dynamic derivatives. J. Aircr. **50**(3), 694–707 (2013)
6. Franke, M., Kügeler, E., Nürnberger, D.: Das DLR-Verfahren TRACE: Moderne Simulationstechniken für Turbomaschinenströmungen. In: DGLR-Jahrbuch. Deutscher Luft- und Raumfahrtkongress 2005 (2005)
7. Huismann, I., Fechter, S., Leicht, T.: HyperCODA – extension of flow solver CODA towards hypersonic flows. In: Jahresbericht 2020 zum 22. DGLR-Fach-Symposium der STAB (2020)
8. Langer, S.: Application of a line implicit method to fully coupled system of equations for turbulent flow problems. Int. J. Comput. Fluid Dyn. **27**(3), 131–150 (2013)
9. Langer, S., Schwöppe, A., Kroll, N.: The DLR flow solver TAU - status and recent algorithmic developments. In: 52nd Aerospace Sciences Meeting. AIAA (2014)
10. Leicht, T., et al.: DLR-project digital-X: next generation CFD solver 'Flucs'. Deutscher Luft- und Raumfahrtkongress 2016 (2017). http://www.dglr.de/publikationen/2017/420027.pdf
11. Leicht, T., et al.: Towards a unified platform for massively parallel numerical flow simulation. Deutscher Luft- und Raumfahrtkongress 2018 (2018)
12. Löwe, J.: p-adaptive discontinuous Galerkin method in CODA. In: Jahresbericht 2020 zum 22. DGLR-Fach-Symposium der STAB (2020)
13. Luo, H., Baum, J.D., Löhner, R.: A fast, matrix-free implicit method for compressible flows on unstructured grids. J. Comput. Phys. **146**(2), 664–690 (1998)
14. Matthes, A., et al.: Tuning and optimization for a variety of many-core architectures without changing a single line of implementation code using the Alpaka library (2017). https://arxiv.org/abs/1706.10086

15. Rempke, A.: Netzdeformation mit Elastizitätsanalogie in multidisziplinärer FlowSimulator-Umgebung. In: Jahresbericht 2016 zum 20. DGLR-Fach-Symposium der STAB (2016)
16. Saad, Y.: A flexible inner-outer preconditioned GMRES algorithm. SIAM J. Sci. Comput. **14**(2), 461–469 (1993)
17. The Trilinos Project Team: The Trilinos Project Website (2020). https://trilinos.github.io
18. Van der Vorst, H.A.: Bi-CGSTAB: a fast and smoothly converging variant of Bi-CG for the solution of nonsymmetric linear systems. SIAM J. Sci. Stat. Comput. **13**(2), 631–644 (1992)

Check for updates

Convection Treatment for RANS Turbulence Model Equations

Sebastian Braun, Stefan Langer, and Tobias Leicht(✉)

German Aerospace Center (DLR), Institute of Aerodynamics and Flow Technology, Lilienthalplatz 7, 38108 Braunschweig, Germany
{sebastian.braun,stefan.langer,tobias.leicht}@dlr.de

Abstract. In the context of finite-volume methods it is generally accepted that at least second order discretization schemes should be considered for the mean flow equations due to their superior accuracy compared to first order schemes. For turbulence model equations, the situation is not so clear when the Boussinesq approximation is applied. Since the coupling of the turbulence model into the mean flow equations is rather weak (via the eddy viscosity and a rather small contribution to the turbulent kinetic energy to the total energy conservation), one can hope for a minor effect on solution accuracy. The goal of this article is to identify the relevant details required in a careful implementation of a first order treatment of the convective part of the turbulence model, such that the obtained results remain comparable to a fully second order discretization. Compelling arguments are given that a first order treatment of turbulence does not necessarily lead to a relevant loss of accuracy in the investigated cases, while yielding a significant gain in robustness.

Keywords: RANS equations · Convection treatment of turbulence · Finite volume method

1 Introduction

The compressible Reynolds-averaged Navier-Stokes (RANS) equations are a well-established model for high Reynolds-number viscous flows. The Reynolds stresses – additional terms appearing in the RANS equations – can be modeled via the Boussinesq approximation in which the Reynolds stress tensor is proportional to the mean strain rate tensor. The proportionality factor is known as eddy viscosity and can be modeled via partial differential or integral equations of a transport type. Such an approach results in a system of equations for the mean flow, coupled with another system of equations modeling the effects of turbulence.

Converging steady-state solutions for CFD cases based on RANS equations with a second-order scheme can be very challenging, in particular when the number of degrees of freedom is systematically increased. Improvements in accuracy of second-order over lower order schemes are of little relevance if the convergence of the scheme cannot be reliably established.

© The Author(s), under exclusive license to Springer Nature Switzerland AG 2021
A. Dillmann et al. (Eds.): STAB/DGLR Symposium 2020, NNFM 151, pp. 646–655, 2021.
https://doi.org/10.1007/978-3-030-79561-0_61

Therefore, one of the common propositions is to treat the convection term(s) of the turbulence model with a first order approach only, aiming at a good compromise between accuracy and robustness. Indeed, reducing the convection treatment of turbulence equations to first order often has a tremendous positive effect on robustness, so that it becomes much more feasible to obtain converged solutions. The mesh asymptotic values of outputs are not affected, as the first order treatment leads to an overall scheme that remains a consistent approximation of the continuous problem.

However, when naively applying this on an otherwise unmodified scheme, the adverse effect on a given mesh can be surprisingly strong and a much higher mesh resolution is required to achieve acceptable accuracy. In the following, we demonstrate this effect and how to overcome it with a suitable modification of the underlying scheme.

2 Governing Equations of Fluid Flow

The Navier-Stokes equations represent a conservation law for mass, momentum and energy. Using conservative variables $\mathbf{W} := (\rho, \rho u_1, \rho u_2, \rho u_3, \rho E)$ the equations are given in integral form by

$$\frac{\mathrm{d}}{\mathrm{d}t}\int_{\Omega} \mathbf{W}\,\mathrm{d}\mathbf{x} + \int_{\partial\Omega} (\mathbf{f}_c \cdot \mathbf{n} - \mathbf{f}_v \cdot \mathbf{n})\,\mathrm{d}s = 0, \qquad \Omega \subset \mathbb{R}^3, \tag{1}$$

where $\mathbf{n} = (n_1, n_2, n_3)$ denotes the outer unit normal vector and $\mathbf{f}_{c,v}$ the convective and viscous fluxes respectively. Turbulence is modeled using either the one-equation turbulence model introduced by Spalart and Allmaras (SA) [1,2],

$$\frac{\mathrm{d}}{\mathrm{d}t}\int_{\Omega} \tilde{\nu}\,\mathrm{d}\mathbf{x} + \int_{\partial\Omega} \left(\mathbf{f}^{\mathrm{SA}}_{c,turb} \cdot \mathbf{n} - \mathbf{f}^{\mathrm{SA}}_{v,turb} \cdot \mathbf{n}\right)\mathrm{d}s(y) = \int_{\Omega} Q^{\mathrm{SA}}\,\mathrm{d}\mathbf{x}. \tag{2}$$

or a two-equation type $k - \omega$ model, originally introduced by Wilcox [3]

$$\frac{\mathrm{d}}{\mathrm{d}t}\int_{\Omega} (k,\omega)\,\mathrm{d}\mathbf{x} + \int_{\partial\Omega} \left(\mathbf{f}^{k\omega}_{c,turb} \cdot \mathbf{n} - \mathbf{f}^{k,\omega}_{v,turb} \cdot \mathbf{n}\right)\mathrm{d}s(y) = \int_{\Omega} Q^{k,\omega}\,\mathrm{d}\mathbf{x}. \tag{3}$$

To shorten notation we denote the additional variables for the turbulence model by W_t, that is $W_t = \tilde{\nu}$ for the Spalart-Allmaras model and $W_t = (k, \omega)$ for the $k - \omega$ model.

2.1 Upwind Schemes

The total convective flux $(\mathbf{f}_c \cdot \mathbf{n}, \mathbf{f}_{c,turb} \cdot \mathbf{n})$, including mean flow and turbulence model, is realized by some numerical flux function $\mathcal{H} = \mathcal{H}(W_L, W_R, n)$ for stability, where W_L and W_R denote the left and right states across some edge in the computational mesh. Examples for a second order upwind biased scheme $\mathcal{H}$ are a Roe scheme [4] or a matrix valued artificial dissipation scheme. These schemes represent a good compromise between robustness and accuracy. They share the requirement for constructing the Roe-matrix

$$\left|\mathbf{A}^{\mathrm{Roe,fc}}\right| := \left|\begin{pmatrix} \frac{\partial(\mathbf{f}_c\cdot\mathbf{n})\left(W^{\mathrm{Roe}},W_t^{\mathrm{Roe}}\right)}{\partial W} & \frac{\partial(\mathbf{f}_c\cdot\mathbf{n})\left(W^{\mathrm{Roe}},W_t^{\mathrm{Roe}}\right)}{\partial W_t} \\ \hline \frac{\partial\left(\mathbf{f}_{c,turb}\cdot\mathbf{n}\right)\left(W^{\mathrm{Roe}},W_t^{\mathrm{Roe}}\right)}{\partial W} & \frac{\partial\left(\mathbf{f}_{c,turb}\cdot\mathbf{n}\right)\left(W^{\mathrm{Roe}},W_t^{\mathrm{Roe}}\right)}{\partial W_t} \end{pmatrix}\right|.$$

Then, in general, the local linearized Riemann solver is given by

$$\mathcal{H}(W_L, W_R, n) := \frac{1}{2}\left[\langle f_c(W_L), n\rangle + \langle f_c(W_R), n\rangle\right] - \frac{1}{2}\left|\mathbf{A}^{\mathrm{Roe,fc}}\right|(W_R - W_L). \quad (4)$$

The definition of the states W_L and W_R depends on the actual scheme. For example, for a second order Roe scheme the states are obtained via linear reconstruction, exploiting gradient information together with some limiting process, while for a second order matrix-valued artificial dissipation an averaging based on an undivided Laplacian together with a blending function is used (see e.g. [5,6] for details). Roe-averaged states $\left(W^{\mathrm{Roe}}, W_t^{\mathrm{Roe}}\right)$ are used for the construction of the Roe-matrix.

To compute $\left|\mathbf{A}^{\mathrm{Roe,fc}}\right|$, the eigendecomposition (see e.g. [7]) of this operator is required. In order to obtain a stable second order scheme, which can deal with stagnation points and shock waves, the local Riemann solver needs to be adapted, such that it avoids vanishing dissipation. To this end, a procedure referred to as an *entropy fix* enforces a lower limit on the (absolute) eigenvalues. This can be done via a simple cut-off or in a differentiable smoothed form as proposed by Harten and Hyman [8]. In the former case, the eigenvalues are replaced by

$$|\Lambda|_{\mathrm{ef}} := \mathrm{diag}\left(|V|_{\mathrm{ef},1}, |V|_{\mathrm{ef},1}, |V|_{\mathrm{ef},1}, |V+a|_{\mathrm{ef},2}, |V-a|_{\mathrm{ef},3}, |V|_{\mathrm{ef},1}^{\mathrm{turb}}, |V|_{\mathrm{ef},1}^{\mathrm{turb}}\right)$$

where for some $0 \leq \delta_{\mathrm{ef}} \leq 1$ and $0 \leq \delta_{\mathrm{ef}}^{\mathrm{turb}} \leq 1$

$$|V|_{\mathrm{ef},1} := |\lambda_i|_{\mathrm{ef},1} := \max\left\{|V|, \delta_{\mathrm{ef}}(|V|+a)\right\}, \qquad i = 1,2,3, \quad (5a)$$

$$|V+a|_{\mathrm{ef},2} := |\lambda_4|_{\mathrm{ef},2} := \max\left\{|V+a|, \delta_{\mathrm{ef}}(|V|+a)\right\}, \quad (5b)$$

$$|V-a|_{\mathrm{ef},3} := |\lambda_5|_{\mathrm{ef},3} := \max\left\{|V-a|, \delta_{\mathrm{ef}}(|V|+a)\right\}, \quad (5c)$$

$$|V|_{\mathrm{ef},1}^{\mathrm{turb}} := |\lambda_i|_{\mathrm{ef},1} := \max\left\{|V|, \delta_{\mathrm{ef}}^{\mathrm{turb}}(|V|+a)\right\}, \qquad i = 6,7. \quad (5d)$$

Understanding the local Riemann solver (4) presented in this section as a prototype, we consider three different implementations in following.

2.2 Fully Coupled Discretization

Equations (4) and (5) define a consistent, fully coupled discretization of the total convective flux $(\mathbf{f}_c \cdot \mathbf{n}, \mathbf{f}_{c,turb} \cdot \mathbf{n})$. As it is possible, in principle, to choose different values for the entropy fix parameter δ_{ef} for the various eigenvalues, $\mathcal{H}$ given in (4) defines a class of Riemann solvers with regard to the entropy fix. Combined with the above expressions for mean flow states W and turbulence states W_t we obtain a second order scheme. Instead, using simply the cell-average states $(W_t)_R = (W_t)_j$ and $(W_t)_L = (W_t)_i$ yields a first order treatment of the turbulence part. For the fully coupled discretization we employ a single value $\delta_{\mathrm{ef}} = \delta_{\mathrm{ef}}^{\mathrm{turb}}$ for all eigenvalues.

2.3 Loosely Coupled Discretization

In implementations it is not uncommon to neglect the off-diagonal terms in $\left|\mathbf{A}^{\mathrm{Roe,fc}}\right|$ appearing in (4), that is we set

$$\frac{\partial\left(\mathbf{f}_c \cdot \mathbf{n}\right)\left(W^{\mathrm{Roe}}, W_t^{\mathrm{Roe}}\right)}{\partial W_t} = \frac{\partial\left(\mathbf{f}_{c,turb} \cdot \mathbf{n}\right)\left(W^{\mathrm{Roe}}, W_t^{\mathrm{Roe}}\right)}{\partial W} = 0.$$

For the Spalart-Allmaras model these terms are generally negligible in size compared to other terms [9]. This may be interpreted as treating the discretization of the total convective flux $(\mathbf{f}_c \cdot \mathbf{n}, \mathbf{f}_{c,turb} \cdot \mathbf{n})$ in a loosely coupled manner, by applying the idea of the Roe decomposition to two separate sub-systems of equations. Then, the local Riemann solver for the turbulence flow equation takes the form

$$\mathcal{H}^{turb,1}\left((W_t)_L, (W_t)_R, n\right) := \frac{1}{2}\left[\langle f_{c,turb}\left((W_t)_L\right), n\rangle + \langle f_{c,turb}\left((W_t)_R\right), n\rangle\right] \\ -\frac{1}{2}|V|^{\text{turb}}_{\text{ef},1}\left((W_t)_R - (W_t)_L\right). \tag{6}$$

As above, a first order scheme can be obtained using the cell-average states. Results show that a first order scheme in combination with a non-zero entropy fix often yields excessively diffusive schemes. Consequently, following the strategy presented in [10,11], we implement the first order scheme without application of an entropy fix, that is we choose $|V|^{\text{turb}}_{\text{ef},1} = |V|$. [1]

2.4 Loosely Coupled, Simplified HLLC Discretization

Considering convection in a loosely coupled sense, we can also take the freedom to apply a different approximate Riemann solver for the turbulence model. For simplicity, we consider (in particular for two-equation turbulence models) the straightforward upwinding

$$\mathcal{H}^{turb,2}\left((W_t)_L, (W_t)_R, n\right) := \begin{cases} \langle u_{e_{ij}}, n\rangle (W_t)_L, & \langle u_{e_{ij}}, n\rangle \geq 0, \\ \langle u_{e_{ij}}, n\rangle (W_t)_R, & \langle u_{e_{ij}}, n\rangle < 0, \end{cases} \tag{7}$$

where $u_{e_{ij}} = \frac{1}{2}(u_j + u_i)$ is the averaged velocity for the edge e_{ij}. This approach that does not require an entropy fix can be considered as a simplified Harten–Lax–van Leer with contact (HLLC) flux [12]. As above, we consider the states $(W_t)_L = (W_t)_i$ and $(W_t)_R = (W_t)_j$ for a first order discretization.

3 Numerical Examples

To demonstrate the possible impact of the discussed discretization strategies, we consider well established simple test cases. Numerical results have been obtained using two different implementations of the finite volume method. The first of these is based on median-dual grids and applies a central discretization scheme with matrix-valued dissipation, while the second implementation in the flow solver CODA [13] uses a linear reconstruction process on a cell-centered metric. All presented results with the Spalart-Allmaras turbulence model are obtained from the former implementation, while all $k-\omega$ results are obtained with CODA. Similar observations were found for both codes on all cases in additional computations not shown here.

[1] Alternatively, this can be interpreted as using the maximum absolute eigenvalue of only the turbulence model equations in the entropy fix procedure. This yields no modification of the eigenvalues for all $0 \leq \delta^{\text{turb}}_{\text{ef}} \leq 1$, since the original eigenvalue and the maximum eigenvalue have the same size.

To assess the data obtained using different discretization strategies, we compute the relative error to reference data (as detailed in the examples below), for example considering the drag coefficient $\text{err}_{\text{rel}}(C_D) = \frac{|C_D^{\text{ref}} - C_D|}{|C_D|}$.

3.1 Turbulent Flat Plate

As a first example we consider the flow over a flat plate at zero pressure gradient with Mach number 0.2 and Reynolds number $Re = 5.0 \cdot 10^6/(\text{unit grid length})$. As reference data we use results from [11] that were obtained on a hexahedral mesh of size 4353 × 3073, that is with more than 12.0×10^6 DoF. First, we examine the Spalart-Allmaras model. Table 1 shows the influence of the discretization strategies. First of all, one major effect is observed. For the discretization order of $p = 2$, the choice of the entropy fix in the investigated range from $0 \leq \delta_{\text{ef}}^{\text{turb}} \leq 0.2$ has minor impact on the overall accuracy, whereas for the order of $p = 1$ the effect is tremendous, even for a small choice of $\delta_{\text{ef}}^{\text{turb}} = 0.05$. On the other hand, we see that the method for $p = 1$ and $\delta_{\text{ef}}^{\text{turb}} = 0.0$ as well as the scheme (7) yield results of an accuracy comparable to the scheme of order $p = 2$.

Similar results were obtained for the Wilcox-$k - \omega$ two equation turbulence model [3]. The left part of Fig. 1 shows the influence of the discretization order and the entropy fix on the drag coefficient as a function of mesh resolution. The overall conclusion remains the same as for the Spalart-Allmaras model. Using an entropy fix $\delta_{ef}^{\text{turb}}$ larger than 0.0 in combination with fully coupled first order discretization tremendously decrease the accuracy on a given mesh, up to a point where a quite fine mesh containing almost one million cells is required to achieve an acceptable solution for this simple test case. In contrast, the effect of the discretization order itself is much smaller. Even on very coarse meshes the differences caused by the order of the discretization is small when no additional entropy fix is present in the first order treatment of the turbulence model. In addition to the integrated values shown in the left part, the right part of Fig. 1 shows the distribution of the turbulence kinetic energy (TKE) within the boundary layer at $x/l = 1.0$. The results obtained with a discretization order

Table 1. Data for turbulent flat plate (SA model)

Mesh	Entropy fix	p	c_f at $x/c = 0.97$	$\text{err}_{\text{rel}}(c_f)$	Drag coefficient	err_{rel} (drag)
4353 × 3073	$\delta_{\text{ef}}^{\text{turb}} = 0.0$	1	$2.70431 \cdot 10^{-3}$	0.0	$2.858029 \cdot 10^{-3}$	0.0
545 × 385	$\delta_{\text{ef}}^{\text{turb}} = 0.0$	2	$2.70339 \cdot 10^{-3}$	$3.4 \cdot 10^{-4}$	$2.859159 \cdot 10^{-3}$	$4.0 \cdot 10^{-4}$
545 × 385	$\delta_{\text{ef}}^{\text{turb}} = 0.05$	2	$2.70328 \cdot 10^{-3}$	$3.8 \cdot 10^{-4}$	$2.859054 \cdot 10^{-3}$	$3.6 \cdot 10^{-4}$
545 × 385	$\delta_{\text{ef}}^{\text{turb}} = 0.1$	2	$2.70320 \cdot 10^{-3}$	$4.1 \cdot 10^{-4}$	$2.858956 \cdot 10^{-3}$	$3.2 \cdot 10^{-4}$
545 × 385	$\delta_{\text{ef}}^{\text{turb}} = 0.2$	2	$2.70308 \cdot 10^{-3}$	$4.5 \cdot 10^{-4}$	$2.858522 \cdot 10^{-3}$	$1.7 \cdot 10^{-4}$
545 × 385	$\delta_{\text{ef}}^{\text{turb}} = 0.0$	1	$2.70313 \cdot 10^{-3}$	$4.4 \cdot 10^{-4}$	$2.858661 \cdot 10^{-3}$	$2.2 \cdot 10^{-4}$
545 × 385	$\delta_{\text{ef}}^{\text{turb}} = 0.05$	1	$2.64010 \cdot 10^{-3}$	$2.4 \cdot 10^{-2}$	$2.790916 \cdot 10^{-3}$	$2.3 \cdot 10^{-2}$
545 × 385	$\delta_{\text{ef}}^{\text{turb}} = 0.1$	1	$2.58410 \cdot 10^{-3}$	$4.4 \cdot 10^{-2}$	$2.733813 \cdot 10^{-3}$	$4.3 \cdot 10^{-2}$
545 × 385	Simpl. HLLC	1	$2.70311 \cdot 10^{-3}$	$4.4 \cdot 10^{-4}$	$2.858697 \cdot 10^{-3}$	$2.3 \cdot 10^{-4}$

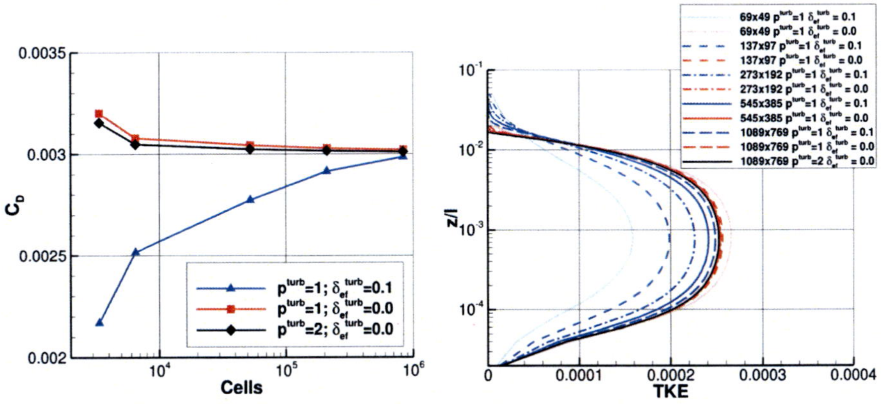

Fig. 1. Mesh convergence for the turbulent flat plate ($k - \omega$ model)

of 1 and an entropy fix of $\delta_{ef}^{\text{turb}} = 0.0$ quickly converge towards the reference solution computed with a second order discretization. In contrast, large offsets can be observed when an entropy fix of $\delta_{ef}^{\text{turb}} = 0.1$ is used. The adverse effect is noticeable even on the finest mesh used here.

3.2 Turbulent Flow over RAE 2822 Airfoil

As a second test case we consider turbulent flow over the RAE 2822 airfoil at $Ma = 0.73$, angle of attack $\alpha = 2.79°$ and $Re = 6.5 \cdot 10^6$. This case features a shock on the upper part of the airfoil. Here, we take the result from the finest mesh of size 1280×256 and $p = 2$ as reference. From the results obtained with the SA model in Table 2, we see that the relative difference for $p = 1$ on the grid of size 1280×256 in drag and lift coefficient is below 0.5% compared to the reference, and the absolute difference less than 1 drag count. For the coarse mesh, we have a relative error of about 1.2% and 2 drag counts difference for

Table 2. Data for RAE 2822 airfoil

Mesh	Entropy fix	p	Drag coeff.	err_{rel} (drag)	Lift coeff.	err_{rel} (lift)
1280×256	$\delta_{\text{ef}}^{\text{turb}} = 0.0$	2	$1.751034 \cdot 10^{-2}$	0.0	$8.001134 \cdot 10^{-1}$	0.0
1280×256	$\delta_{\text{ef}}^{\text{turb}} = 0.0$	1	$1.742627 \cdot 10^{-2}$	$4.8 \cdot 10^{-3}$	$7.982642 \cdot 10^{-1}$	$2.3 \cdot 10^{-3}$
1280×256	Simpl. HLLC	1	$1.742553 \cdot 10^{-2}$	$4.8 \cdot 10^{-3}$	$7.982481 \cdot 10^{-1}$	$2.3 \cdot 10^{-3}$
640×128	$\delta_{\text{ef}}^{\text{turb}} = 0.0$	1	$1.730268 \cdot 10^{-2}$	$1.2 \cdot 10^{-3}$	$7.935826 \cdot 10^{-1}$	$8.2 \cdot 10^{-3}$
640×128	Simpl. HLLC	1	$1.729945 \cdot 10^{-2}$	$1.2 \cdot 10^{-2}$	$7.935116 \cdot 10^{-1}$	$8.3 \cdot 10^{-3}$
320×64	$\delta_{\text{ef}}^{\text{turb}} = 0.0$	2	$1.730002 \cdot 10^{-2}$	$1.2 \cdot 10^{-2}$	$7.843892 \cdot 10^{-1}$	$2.0 \cdot 10^{-2}$
320×64	$\delta_{\text{ef}}^{\text{turb}} = 0.0$	1	$1.709501 \cdot 10^{-2}$	$2.3 \cdot 10^{-2}$	$7.796433 \cdot 10^{-1}$	$2.6 \cdot 10^{-2}$
320×64	Simpl. HLLC	1	$1.708240 \cdot 10^{-2}$	$2.4 \cdot 10^{-2}$	$7.793550 \cdot 10^{-1}$	$2.6 \cdot 10^{-2}$

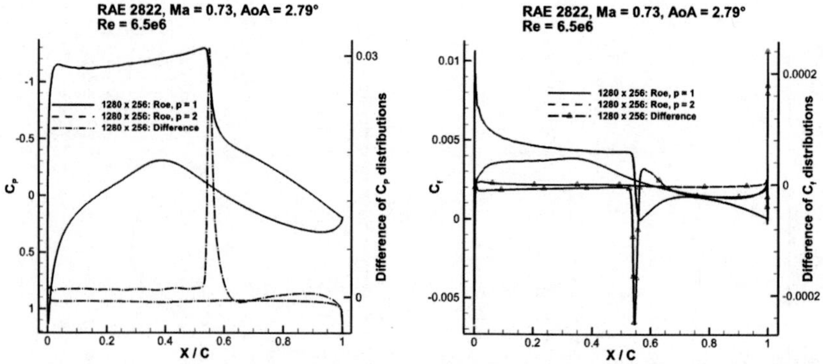

Fig. 2. Comparison of C_P- and C_f-distribution for mesh 1280 × 256

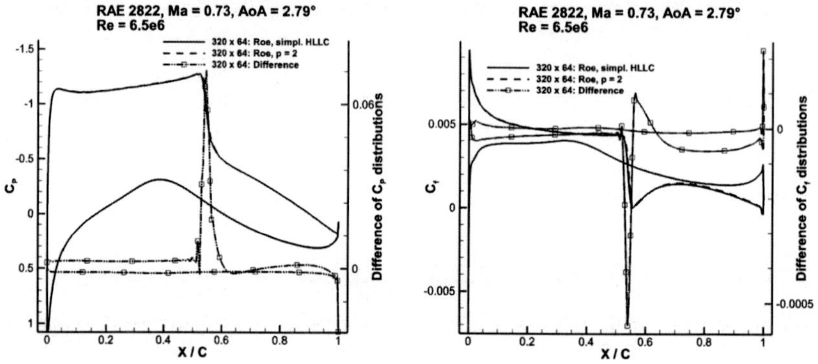

Fig. 3. Comparison of C_P- and C_f-distribution for mesh 320 × 64

$p = 2$, whereas we obtain larger but still comparable values of about 2.3% and 4 drag counts for $p = 1$.

As a further indication of the effect on aerodynamically relevant quantities Figs. 2 and 3 illustrate the distribution of pressure and skin friction coefficients C_p and C_f. Relevant features such as the suction peak, shock position and magnitude are almost identical between $p = 1$ and $p = 2$ results. The additionally plotted difference supports this observation and is in fact more than an order of magnitude smaller than the value it refers to, that is

$$\|C_p^{p=2} - C_p^{p=1}\|_\infty < \|C_p^{p=2}\|_\infty \cdot 10^{-1}, \qquad \|C_f^{p=2} - C_f^{p=1}\|_\infty < \|C_f^{p=2}\|_\infty \cdot 10^{-1}.$$

As illustrated in Fig. 4, showing convergence histories, the solver applied here did not converge well on all cases. In fact, with settings that worked successfully for all other cases, the computation stalled early for $p = 2$ on the medium mesh of size 640×128. This in an indication of the improved robustness obtained for first

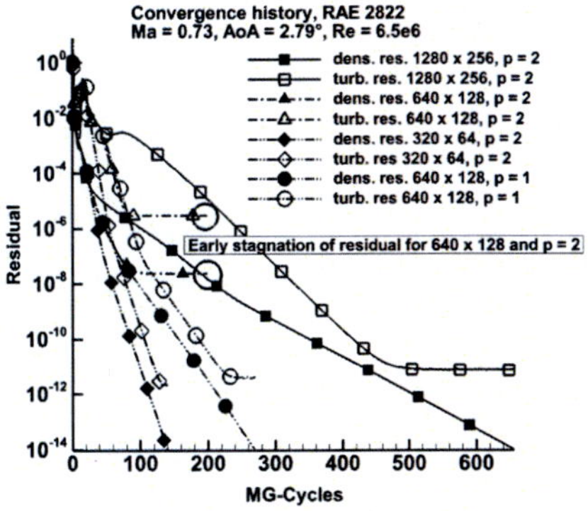

Fig. 4. Convergence histories for RAE 2822 test case

order discretization of the turbulence convective part, where such an irregular behavior has not been observed. This finding is supported by experience from many other cases not reported here.

3.3 Turbulent Flow over NACA 0012 Airfoil

As a final example we consider turbulent flow over the NACA 0012 airfoil, at $Ma = 0.15$, angle of attack $\alpha = 10°$ and $Re = 6 \cdot 10^6$. Here we consider the $k - \omega$ model of Wilcox.

Figure 5 illustrates the eddy viscosity computed for this case. The solution on a fine mesh with 1792×512 grid points using a loosely coupled scheme with no entropy fix applied in the lower right represents the reference. Compared to this, the result obtained on a coarse mesh with 224×64 grid points is so far off that it could be considered totally useless when an entropy fix of $\delta_{\text{ef}} = \delta_{\text{ef}}^{\text{turb}} = 0.1$ is used in a fully coupled discretization scheme, cf. the upper left part of Fig. 5. Reducing the entropy fix to $\delta_{\text{ef}} = \delta_{\text{ef}}^{\text{turb}} = 0.0$ yields reasonable results on the same mesh, cf. the upper right part of the figure. Due to reducing the entropy fix for the mean flow equation in the coupled scheme, convergence is rather difficult, however. This improves considerably when changing to the loosely coupled scheme, keeping $\delta_{\text{ef}}^{\text{turb}} = 0.0$ but increasing the value in the mean flow part back to $\delta_{\text{ef}} = 0.1$. The resulting eddy viscosity in the lower left part of Fig. 5 is rather similar. Using the simplified HLLC flux for the turbulence equations yields results that are visually indistinguishable and thus not shown here. Furthermore, it should be noted that the resolution of the 224×64-mesh is already in a range typically found in industrial 3D applications w. r. t. the geometrical complexity of the case.

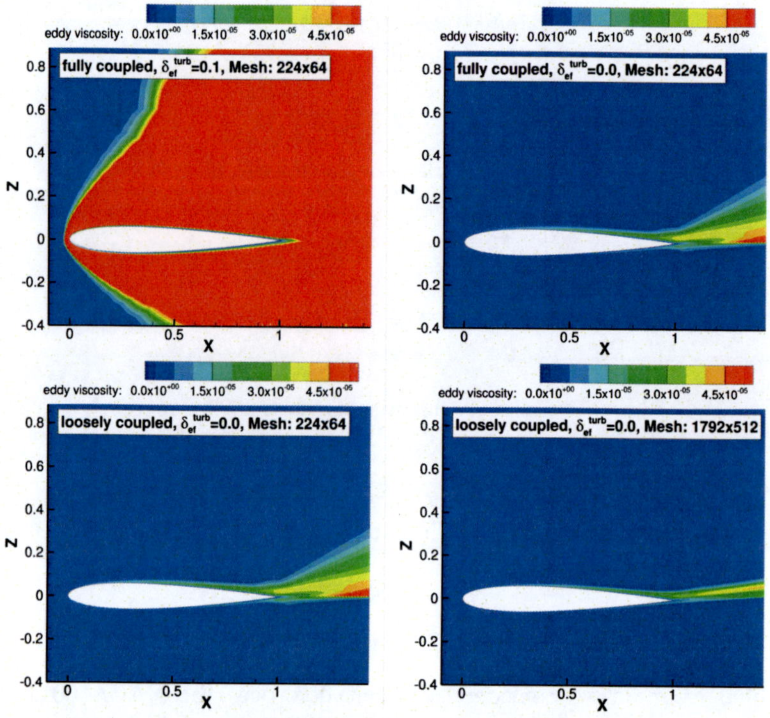

Fig. 5. Effect of the entropy fix value on the eddy viscosity distribution around the NACA 0012 airfoil at $\alpha = 10°$, $Ma = 0.15$, $Re = 6 \cdot 10^6$.

4 Conclusion

We have demonstrated that in principle it is possible to discretize turbulence model equations along with the mean flow equations with second order schemes. At least for simple cases, fully converged results could be obtained. Convergence gets more problematic with increasing mesh size, complexity of the case and the turbulence model applied - what might be feasible for the Spalart-Allmaras model becomes more difficult for the Wilcox-$k-\omega$ model and almost unfeasible for Menter's SST model in many cases (not shown here).

While the robustness of the solver can readily be improved tremendously via switching the treatment of the convection part of the turbulence model to first order, strongly adverse effects of potential entropy fixes can be observed in fully coupled schemes. A loosely coupled point of view gives more freedom to choose a suitable form for discretizing of the additional turbulence convective parts, avoiding the need for entropy fixes in the first order treatment. Exploiting this freedom allows to recover an acceptable accuracy on reasonably resolved meshes.

Two specific formulations of the scheme for first order turbulence convection have been considered – one based on the Roe flux and another one based on a pure scalar upwinding. The effect of this choice differs for the two implementations considered here, with either of the scheme yielding better robustness in combination with the specifics of one of the two codes. Differences in results are comparatively small, however, so the effect on accuracy is negligible.

Modelling Reynolds stresses via eddy viscosity models yields diffusion terms with second order velocity derivatives in the mean flow equations. In Reynolds Stress Models (RSM), the Reynolds stresses are independent variables that appear directly as first order derivatives in the mean flow equations. This leads to the question whether they should be considered as additional convective terms and how to treat them. Thus, the influence of discretization accuracy for RSM is an interesting topic for future research.

References

1. Spalart, P.R., Allmaras, S.R.: A one-equation turbulence model for aerodynamic flows. In: AIAA Computational Fluid Dynamics Conference, number 1992–439 in Conference Proceeding Series. AIAA (1992)
2. Allmaras, S.R., Johnson, F.T., Spalart, P.R.: Modifications and clarifications for the implementation of the Spalart-Allmaras turbulence model. In: International Conference on Computational Fluid Dynamics 7, number ICCFD7-1902 in Conference Proceeding Series, Hawaii (2012)
3. Wilcox, D.C.: Reassessment of the scale-determining equation for advanced turbulence models. AIAA J. **26**(11), 1299–1310 (1988)
4. Roe, P.L.: Approximate Riemann solvers, parameter vectors, and difference schemes. J. Comput. Phys. **43**, 357–372 (1981)
5. Langer, S.: Agglomeration multigrid methods with implicit Runge-Kutta smoothers applied to aerodynamic simulations on unstructured grids. J. Comput. Phys. **277**, 72–100 (2014)
6. Langer, S.: Preconditioned Newton methods to approximate solutions of the Reynolds averaged Navier-Stokes equations. Technical report, Institut für Aerodynamik und Strömungstechnik (2018)
7. Langer, S.: Application of a line implicit method to fully coupled system of equations for turbulent flow problems. Int. J. Comput. Fluid Dyn. **27**(3), 131–150 (2013)
8. Harten, A., Hyman, J.M.: Self adjusting grid methods for one-dimensional hyperbolic conservation laws. J. Comput. Phys. **50**, 235–269 (1983)
9. Suárez Martínez, G.: Coupling strategies for solving the RANS equations. Technical report, Institut für Aerodynamik und Strömungstechnik, February 2020
10. Langer, S.: Investigation and application of point implicit Runge-Kutta methods to inviscid flow problems. Int. J. Numer. Methods Fluids **69**(2), 332–352 (2012)
11. Langer, S.: Investigations of a compressible second order finite volume code towards the incompressible limit. Comput. Fluids **149**, 119–137 (2017)
12. Toro, E.F., Spruce, M., Speares, W.: Restoration of the contact surface in the HLL-Riemann solver. Shock Waves **4**(1), 25–34 (1994)
13. Leicht, T., et al.: DLR-project Digital-X - next generation CFD solver 'Flucs.' In: Deutscher Luft- und Raumfahrtkongress, Braunschweig, Germany (2016)

DES of Weapon Bay in Fighter Aircraft Under High-Subsonic and Supersonic Conditions

Karthick Rajkumar[1](✉), Eike Tangermann[1], Markus Klein[1], Sebastian Ketterl[2], and Andreas Winkler[2]

[1] Bundeswehr Universität München, Werner-Heisenberg-Weg 39, 85577 Neubiberg, Germany
karthick.rajkumar@unibw.de

[2] Division of Flight Physics - Aerodynamics - Aerodynamic Design, Airbus Defence and Space, 80577 Manching, Germany

Abstract. A cavity configuration features a highly unsteady separated flow, which is characterised by an intense aero-acoustic coupling mechanism. Strong pressure oscillations produced in and around the cavity have the potential to cause structural fatigue and induce resonance phenomena inside the cavity. In this study, a novel open cavity configuration with doors attached on the sides and a length to depth ratio of 5.7 has been studied numerically using the TAU code developed by German Aerospace Center for high subsonic and supersonic flows. The turbulence is resolved partially using a hybrid RANS-LES turbulence model. The study comprises the Mach numbers (Ma) 0.8 and 1.2 with Reynolds number (Re) 12×10^6. The Rossiter modes occurring in the cavity due to the feedback mechanism have been numerically computed and validated for both flow conditions using reference data from experimental measurements.

Keywords: Cavity flow · The Rossiter modes · Hybrid RANS-LES

1 Introduction

A cavity configuration can be seen in many real-world applications. In military and civil aircraft, the cavity configurations find its usage in areas such as spoilers, weapon bays, landing gears etc. The flow behaviour over the cavity configuration exhibits highly unsteady characteristics and is oscillatory in nature [1]. The flow oscillates predominantly at higher frequencies which potentially cause structural fatigue problems and limits the cavity life. With the support of experimental investigations, high fidelity CFD simulations could help to understand these complex flow physics and to accelerate the design stage of cavity development.

A cavity, even as simple as a rectangular cuboid, exhibits a highly separated shear flow originating from the leading edge of the cavity. Based on the flow over the cavity, it can be divided into three different types: open, closed and transitional cavity [2]. In the open cavity type, the shear flow that develops from

© The Author(s), under exclusive license to Springer Nature Switzerland AG 2021
A. Dillmann et al. (Eds.): STAB/DGLR Symposium 2020, NNFM 151, pp. 656–665, 2021.
https://doi.org/10.1007/978-3-030-79561-0_62

the front edge impinges directly on the rear wall. In the closed cavity, the shear flow reattaches to the floor of the cavity and separates again before impinging on the rear wall. In the transitional cavity, the flow reattachment to the floor of the cavity is intermittent and unstable. Several factors, such as length to depth ratio (L/D) and Mach number (Ma), can determine the type of cavity flow.

Rossiter [3] described the mechanism of distinct modes appearing in the open cavity. He devised the oscillation model (Eq. 1) based on three observations inside the cavity. 1. Downstream convection of vortices from the shear layer, 2. Impingement of vortices at the downstream edge generating acoustic waves and 3. Acoustic waves travelling upstream and exciting further disturbances in the shear layer, leading to a self-sustained oscillation process.

$$f = \frac{U_\infty}{L}\frac{m-\alpha}{Ma+1/\kappa} \tag{1}$$

Here f is the frequency at which the modes appear, U_∞ is the freestream velocity, L is the length of the cavity, m is the mode number, α is the phase delay constant, Ma is the Mach number and κ is the ratio of convection velocity of vortical structure to the freestream velocity. The Rossiter model is still widely used to determine the frequencies in the cavity with good accuracy, but it does not perform well in supersonic flow conditions. Heller et al. [4] modified the Rossiter model by assuming that the speed of sound inside the cavity is equal to that in the stagnating freestream, which holds valid in supersonic conditions as well.

Several attempts have been made to study the cavity flows numerically, particularly on the M219 cavity [5] to describe the fluid mechanics and determine the frequency components of the wall pressure spectra. Henderson et al. [6] and Inagaki et al. [7] have successfully utilized URANS methods for modelling the narrowband of the acoustic spectra, but it has been shown that URANS models are unable to predict the broadband spectra. To characterize both the narrowband and broadband spectra, advanced turbulence modelling such as DES methods based on Spalart-Allmaras and k-ω SST have been employed and obtained better results in predicting the broadband spectra than by the URANS models in the M219 configuration [8]. The current study employs the improved delayed DES (IDDES) variant of the DES method based on the Spalart-Allmaras model on a novel cavity NX-CB1 configuration [9]. Handa et al. [13] have studied the generation and propagation of pressure waves in the cavity and it has been reported that vortices in the shear layer generate pressure waves which come in contact with already reflected pressure waves off the rear wall.

In this study, a novel open cavity NX-CB1 with opened doors at the sides [9] has been investigated numerically to predict the Rossiter modes at high subsonic and supersonic flow conditions using a hybrid RANS-LES method. The numerical simulations have been performed using the DLR-TAU Computational Fluid Dynamics (CFD) code [10] under the flow conditions of Mach numbers 0.8 and 1.2. The wall pressure fluctuations have been post-processed to analyse the root-mean-square (RMS) values and acoustic spectra using the Fast Fourier transform (FFT). The numerically computed RMS values and wall spectra have been validated against the experimental data, which have been made available for this study by Airbus Defence and Space [9].

1.1 Description of Geometry

The geometry NX-CB1 as shown in Fig. 1 has a close similarity to the well known M219 cavity configuration [5]. A rectangular cavity with length to depth ratio (L/D) 5.7 and length to width ratio (L/W) 4.16 as seen from the Fig. 1 is cut onto a flat side at some distance from the rig's sharp leading edge and it is placed in the centre line of the cavity rig. The cavity flow is analysed for two different Mach numbers of 0.8 and 1.2 under Reynolds number 12×10^6 with the angle of attack and the side slip angle set to 0. In the wind tunnel experiments conducted by Mayer et al. [9], the wall pressure fluctuations at locations L1, L2, L3, L4, L7 and L8 have been recorded using pressure transducers and they have been made available for this study for validation of the numerical simulations.

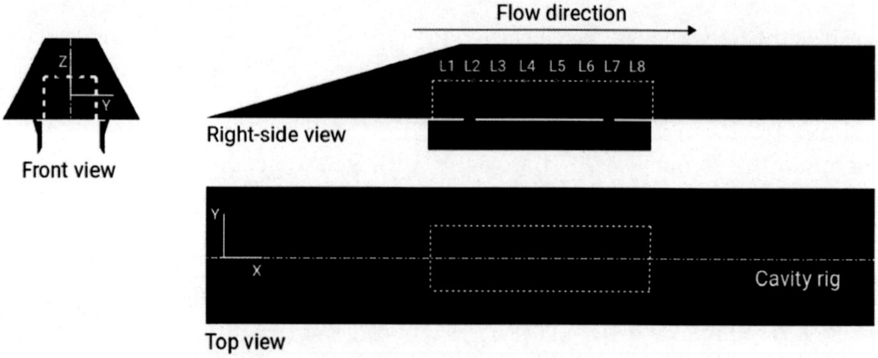

Fig. 1. Model of the NX-CB1 cavity

2 Numerical Method

In this study, the DLR-TAU code has been used to carry out all the numerical simulations. The DLR-TAU code is a finite volume (FV) flow solver based on the compressible Navier-Stokes formulation developed by German Aerospace Center (DLR) [10]. Solving all the time and length scales of a turbulent flow involves a substantial computing power even for a moderate Reynolds number application. Therefore, a concept of turbulence modelling is required to model the fluctuating turbulent components of the flow field. Depending on the application and availability of computing resources, an appropriate turbulence model can be employed.

2.1 Hybrid RANS-LES Method

In industrial applications, the commonly used approach of turbulence treatment is solving the Reynolds Averaged Navier-Stokes equations (RANS), which is based on a statistical averaging procedure. Unsteady RANS is often preferred when there is a slow evolution of flow with time and it loses a lot of information when it is applied to the cavity flow due to the highly unsteady nature of the flow behaviour. Another alternative is an LES based model, which stands as a compromise between DNS and RANS model. LES resolves all the larger eddies that carry major energy content of the flow and models the smaller eddies. The LES method requires that the grid is fine enough to capture the energy carrying eddies and that its cut-off length lies in the inertial range of the Kolmogorov energy spectrum [11]. The bottleneck in employing LES is that there will be a strict requirement of using very fine cells to resolve the energy carrying eddies in the near-wall region. Therefore, the Improved Delayed DES (IDDES) model based on Spalart-Allmaras (S-A), a variant under the category of hybrid RANS-LES methods [12] has been used in this study. In the SA-IDDES model, the core flow field away from the wall is treated through the scale resolving formulations, whereas in the near-wall region, the method employs the Spalart-Allmaras (SA) model with additional shielding function to prevent the activation of the LES mode within the boundary layer.

2.2 Spatial and Numerical Discretization

The spatial discretization of the geometry (Fig. 2) consisting of unstructured elements with tetrahedral, prism, hexahedral and pyramid cells has been produced with 23 Million grid nodes. Using RANS results employing the k-w model, the integral turbulent length scale has been estimated and accordingly, the regions in and around the cavity have been refined sufficiently such that a minimum of 1 cell is used to resolve the turbulent length scale and a valid LES grid is produced. As discussed in the *Introduction* section, shear layer vortices are one of the major driving forces behind the oscillation process, particular attention has been given to resolve the shear layer region where the vortices are generated with utmost accuracy while maintaining an acceptable computing time. Additionally, as the prime objective of the study is to analyse the pressure fluctuations on the cavity, only the flow inside the cavity and below the flat plate has been considered for the refinement. A slice of mesh at midplane y = 0 showing the refinement levels near the cavity has been shown on the right of the Fig. 2. The shear layer in the cavity has been completely refined due to high unsteadiness of flow. Because the angles of attack and sideslip are zero, the model has been meshed in half and mirrored about the symmetry axis so that the asymmetric grid effects are effectively avoided. Moreover, a non-dimensional wall coordinate (y^+) less than 1.0 has been set along the walls of the cavity in both the flow cases. 35 prism layers of cells have been used to refine the boundary layers on the cavity walls. In the boundary layer along the ceiling, 10 to 12 cells are placed within the y^+ of 5.0 in Ma = 0.8, whereas 5 to 6 cells are placed within the y^+ of 5.0 in Ma = 1.2.

A second-order central scheme and backward Euler scheme have been used to discretize spatial flux and temporal terms, respectively. In both the flow conditions of Ma = 0.8 and Ma = 1.2, the time step size has been chosen such that the convective Courant-Friedrichs-Lewy number (CFL) is well below 2.0. Convergence criteria are based on Cauchy convergence control of the variables

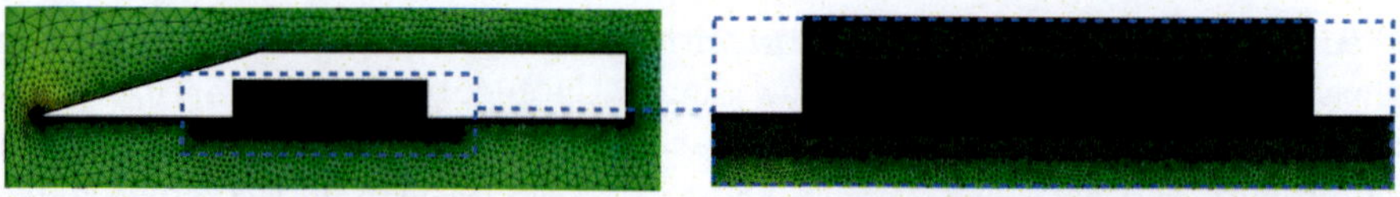

Fig. 2. Mesh resolution at midplane y = 0: refinement levels around the cavity (left); inside the cavity (right)

volume averaged turbulent kinetic energy, maximum eddy viscosity, total vorticity and maximum Mach number with tolerance values of 1e-05, 1e-05, 1e-05 and 1e-04 respectively.

3 Results and Discussion

In this section, some of the flow features of the NX-CB1 cavity in terms of turbulent structures, velocity fields and root mean square (RMS) of pressure field and the pressure spectra along the ceiling will be discussed.

3.1 Qualitative Description of the Flow Field

The resolution of vortical structures is crucial to obtain the acoustic spectra at a reasonable accuracy. At an instant of time, vortical structures in the shear layer and inside the cavity detected through Q-criterion (vorticity magnitude as contour color) are shown on the left of Fig. 3. The ability to resolve those vortical structures as seen on the left of Fig. 3 is attributed to the usage of the SA-IDDES turbulence model. Due to proprietary reasons, only the relative magnitude levels of contour values are presented and the same is followed for all the results shown. The position of front and rear edges of the cavity are marked with a dotted white line for a better visibility. It can be seen that the vortical structures have a high magnitude of vorticity on the doors of the cavity near the front wall. The evolution of a typical vortex structure in the shear layer with time is that it develops from the front edge of the cavity and combines with other turbulent structures as they are convected downstream and dissipates after impinging on the rear wall of the cavity. On the impingement of the vortex structure on the rear wall, a transverse travelling pressure wave is generated and travels upstream to reflect off on other walls of the cavity [13]. Highly turbulent behaviour is apparent on the downstream corner of the cavity near the rear wall with the redirected flow stream interacting with the oncoming shear flow components (see right of Fig. 3 and Fig. 4). A RANS model would not be able to capture these behaviours in the cavity, which are the typical underlying physics of open cavity flows.

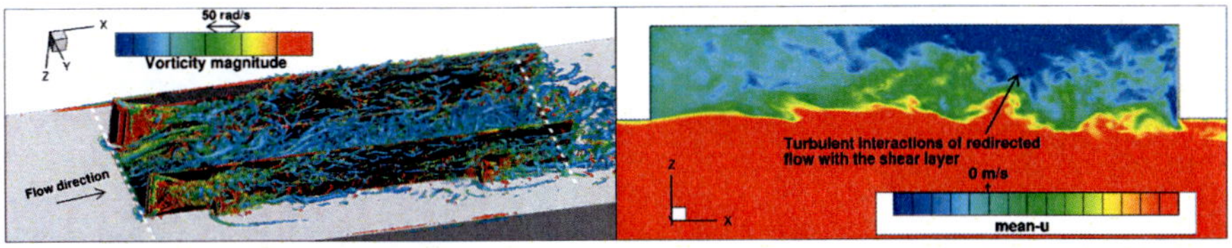

Fig. 3. Snapshots of instantaneous fields for Ma = 1.2: Vortical structures in the cavity (left); streamwise velocity at an instant of time (right)

3.2 Mean-u and p-RMS Fields

Figure 4 shows the time-averaged streamwise velocity contours at midplane y = 0 for Ma = 0.8 and Ma = 1.2. It can be seen that the shear layer develops from the front edge of the cavity wall and it impinges directly on the rear wall, which is an expected behaviour from the open cavity configuration. Few components of the flow which impinge on the edge of the rear wall get either redirected inside the cavity or travels further downstream depending on the angle of flow impingement on the rear wall edge. The flow components that get redirected inside the cavity form a large recirculation region. The negative contour levels have been marked with dotted lines for a better visibility of upstream travelling components. The shear layer in Ma = 1.2 deflects by a little margin into the cavity near the front edge, whereas the shear layer in Ma = 0.8 shows no such visible deflection, which makes the recirculation region in the cavity of Ma = 1.2 more slender than in Ma = 0.8.

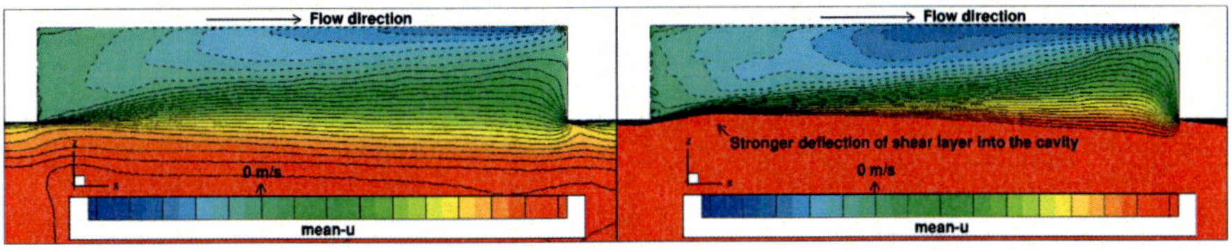

Fig. 4. Time averaged streamwise velocity contour at mid-plane y = 0: Ma = 0.8(left); Ma = 1.2(right)

Figure 5 shows the plot of root mean square (RMS) of pressure along the centerline of the ceiling of the cavity compared with the measured data. The simulated RMS data fit the experimental data reasonably well with a slight under prediction of values near 80% of the cavity length. It is evident from the surface RMS pressure distribution on the right of the Fig. 5 that the pressure fluctuations are substantially higher in the second half of the cavity than in the first half with maximum RMS values on the rear wall due to complex interactions of shedding vortices with it.

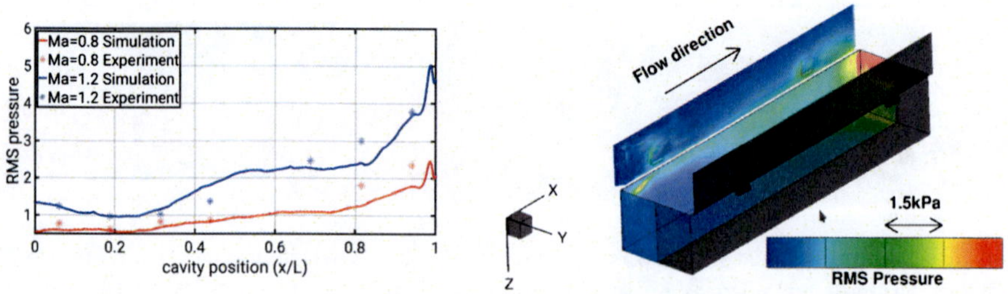

Fig. 5. RMS pressure: along the center line of the cavity ceiling in Ma = 0.8 and Ma = 1.2 (left); over the cavity walls in Ma = 1.2 (right)

3.3 Spectral Analysis

Power spectral density (PSD) is a measure of energy contained in the pressure signal as a function of frequencies. Figures 6 and 7 show the PSD spectra for some of the probe locations along the cavity ceiling for Ma = 0.8 and Ma = 1.2, respectively. It can be seen that there exist narrowband and broadband spectra as observed in other cavity related studies [8]. These tones are the so-called Rossiter modes [3]. The frequencies at which the Rossiter modes occur can be computed through the modified Rossiter oscillation model (Eq. 2).

$$f = \frac{U_\infty}{L} \frac{m - \alpha}{Ma/(\sqrt{1 + (\gamma - 1)Ma^2/2}) + 1/\kappa} \tag{2}$$

The Modified Rossiter model has an additional variable γ as the adiabatic exponent in its formulation, derived by assuming the speed of sound in the cavity is equal to the stagnation sound speed. Table 1 shows the frequencies of the modes computed from modified Rossiter model, measured data together with the simulation results.

Table 1. Frequencies of Rossiter modes

Mode	Ma = 0.8			Ma = 1.2		
	Theory (Hz)	Exp. (Hz)	CFD (Hz)	Theory (Hz)	Exp. (Hz)	CFD (Hz)
1	263	272	266	352	352	345
2	670	755	752	896	888	856
3	1076	1160	1144	1440	1464	1444
4	1484	1600	1622	1985	2640	2533

A Fast Fourier Transform (FFT) has been performed based on Welch's method to decompose the pressure fluctuations into its frequency components. A total of 0.125 s of physical time has been processed for the FFT analysis using the Hamming window function with the maximum offset length of FFT windows corresponding to the integral time scale computed through the autocorrelation function. For comparison with the experimental data, results of the numerical simulations have been plotted together in Fig. 6. The lowest frequency that the simulated data can resolve is kept 30 Hz. The modal shapes and their frequencies of the experimental results have been captured reasonably well by the numerical simulations. At probe location L1, the modal shapes have been captured well with moderate fluctuations near the first mode. The second, third and the fourth mode are predicted well by the simulation. At probe location L3, the 4th Rossiter mode does not occur in the measured data, which is predicted as well by the simulation results. At probe location L4, a strong contrast in the magnitude of 3rd mode relative to 2nd and 4th mode can be seen and the simulation data fits this pattern as well with a slightly under-predicted 3rd mode in the simulation data. At probe location L8, all the modes have been captured well with the moderate under-prediction in the spectral magnitudes. However, the relative magnitudes between the modes are correctly captured.

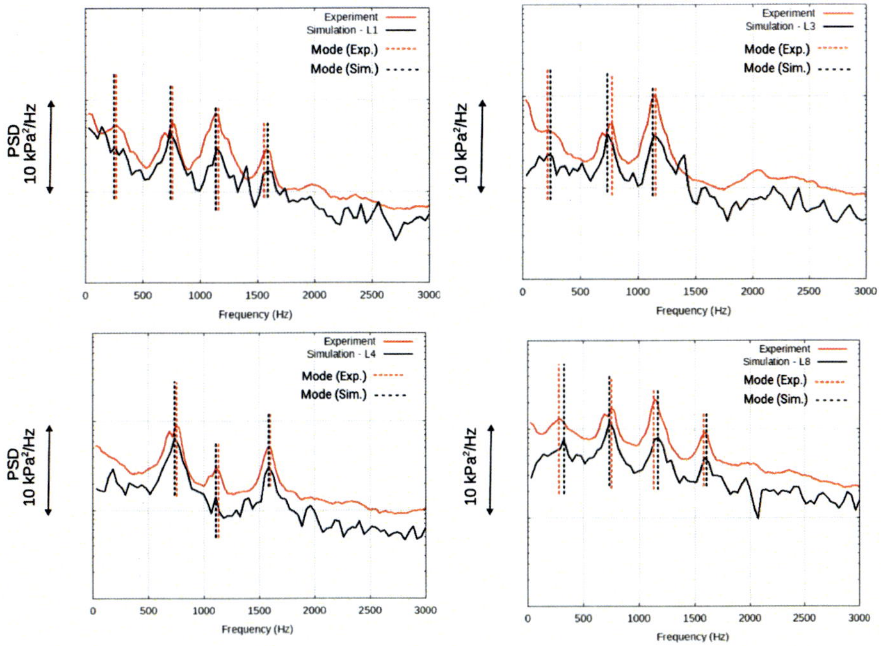

Fig. 6. PSD of NX-CB1 at probe locations L1, L3, L4 and L8 (Ma = 0.8)

In Ma = 1.2, the modal shapes are captured quite well along the probe locations (L1 and L8 are shown in Fig. 7). It is evident from the Table 1 that the increase in Mach number has shifted the frequencies of the modes to higher frequencies and the numerical simulation has effectively captured this behaviour. In particular, the distinct modes appearing in the experimental data have been captured better than in Ma = 0.8 case. It is also seen that as the frequencies increase, the Rossiter modes computed in the simulations slightly shift to the left relative to the experimental data.

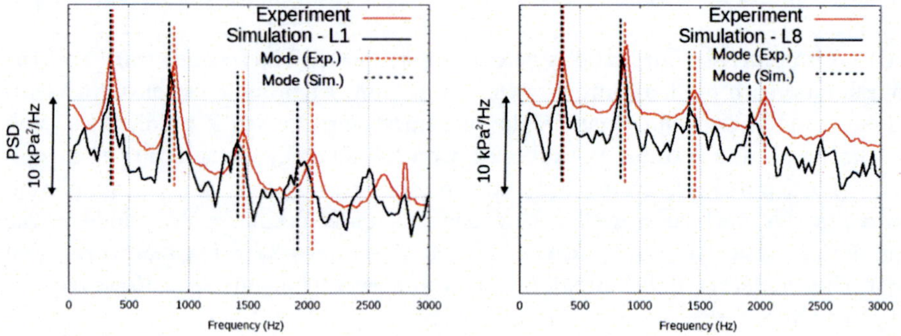

Fig. 7. PSD of NX-CB1 at probe locations L1 and L8 (Ma = 1.2)

4 Conclusion and Outlook

In this study, a novel cavity configuration with sidewise doors NX-CB1 has been studied numerically with use of the DLR-TAU CFD code. The narrowband and broadband of the acoustic spectra occurring in the cavity has been investigated for two different flow conditions of Ma = 0.8 and Ma = 1.2 using a hybrid RANS-LES model and it has been proven that the SA-IDDES model can quite reasonably capture the shapes and frequencies of the Rossiter modes in both the flow conditions with a slight under-prediction in the magnitude levels. In particular, in Ma = 1.2, there is an excellent demonstration of distinct Rossiter modes in the simulated data with a slight frequency shift in the higher frequency range. Additionally, compared to most of the cavity studies such as Allen et al. [8], the present study has acquired 2.5 times shorter series of unsteady data in terms of the characteristic time scale of the first mode. By taking this factor into account, the simulation results are in considerably good agreement with the experimental data. Future work will address the shift in the modal frequencies as well as possibilities to compute the Rossiter modes more effciently while maintaining good accuracy.

Acknowledgements. This work has been carried out with the funding support from Airbus Defence and Space under the project "Analysis of Unsteady Effects in Fighter Aircraft Aerodynamics". The authors would like to acknowledge the German Aerospace Center (DLR) for providing the TAU code and valuable recommendations for carrying out the simulations.

References

1. Aradag, S.: CFD for High Speed Flows in Engineering. VDM Dr. Muller, Saarbrucken, Germany (2008)
2. Chappell, P.D.: Aerodynamics and aero-acoustics of rectangular planform cavities - Part I: Time-averaged flow. ESDU Item **02008** (2009)
3. Rossiter, J.E.: Wind-Tunnel Experiments on the flow over rectangular cavities at Subsonic and transonic speeds. Report and Memoranda 3438, Aeronautical Research Council, London, UK (1966)
4. Heller, H., Bliss, D.: The physical mechanism of flow-induced pressure fluctuations in cavities and concepts for their suppression. In: 2nd Aeroacoustics Conference (1975)
5. Henshaw, M.J. de C.: M219 cavity case in verification and validation data for computational unsteady aerodynamics, RTO-TR-26, AC/323(AVT)TP/19, October 2000
6. Henderson, J., Babcock, K., Richards, B.E.: Subsonic and transonic transitional cavity flows. In: 6th AIAA/CEAS Aeroacoustics Conference and Exhibit, Paper No. AIAA-2000-1966, June 2000
7. Inagaki, M., Murata, O., Kondoh, T., Abe, K.: Numerical prediction of fluid-resonant oscillation at low Mach number. AIAA J. **40**(9), 1823–1908 (2002)
8. Allen, R., Mendonca, F., Kirkham, D.: RANS and DES turbulence model predictions on the M219 cavity at M = 0.85. In: RTO AVT Symposium on Functional and Mechanical Integration of Weapons and Land and Air Vehicles, VA, USA, 7–9 June 2004
9. Mayer, F., Mancini, S., Kolb, A.: Experimental Investigation of installation effects on the aeroacoustic behavior of rectangular cavities at high subsonic and supersonic speed. Presented at Deutscher Luft- und Raumfahrtkongress, Germany (2020)
10. Langer, S., Schwöppe, A., Kroll, N.: The DLR Flow Solver TAU - status and recent algorithmic developments. In: 52nd Aerospace Sciences Meeting, National Harbor, Maryland, USA (2014)
11. Chaouat, B.: The state of the art of hybrid RANS/LES modeling for the simulation of turbulent flows. Flow Turbul. Combust. **99**(2), 279–327 (2017). https://doi.org/10.1007/s10494-017-9828-8
12. Spalart, P.R., Jou, W.H., Strelets, M., Allmaras, S.R.: Comments on the feasibility of LES for wings, and on a hybrid RANS/LES approach. In: Liu, C., Liu, Z. (eds.) First AFOSR International Conference on DNS/LES, Ruston, LA, Advances in DNS/LES 4–8 August 1997
13. Handa, T., Miyachi, H., Kakuno, H., Ozaki, T.: Generation and propagation of pressure waves in supersonic deep-cavity flows. Exp. Fluids **53**, 1855–1866 (2012)

Check for updates

Numerical Investigation of a Porous Trailing Edge by a Zonal RANS/LES Simulation

Sutharsan Satcunanathan(✉), Matthias Meinke, and Wolfgang Schröder

Institute of Aerodynamics, RWTH Aachen University,
Wüllnerstraße 5a, 52062 Aachen, Germany
s.satcunanathan@aia.rwth-aachen.de

Abstract. A hybrid Reynolds-averaged Navier-Stokes (RANS)/large-eddy simulation (LES) approach for flows with porous surfaces is presented. For the two-dimensional geometry considered, a computationally cheap 2D-RANS simulation in the entire domain is performed first, followed by a LES in regions, where more accurate unsteady flow data are required. For the RANS part a two-equation low Reynolds number k-ε turbulence model is modified to include the porous treatment. The boundary conditions for the LES, which is fully embedded inside the RANS domain, are derived from the RANS solution. To enforce a physically realistic transition from an averaged RANS solution towards a resolved turbulent flow field, at the inflow of the LES coherent structures are generated by means of the reformulated synthetic turbulence generation (RSTG) method. The methodology is applied to a NACA 0012 airfoil with a solid and porous trailing edge in a subsonic flow. The simulation is set up with the broadband turbulent boundary-layer trailing-edge (TBL-TE) noise prediction as a future objective in mind, i.e., the noise sources in the trailing-edge region are captured by the LES.

Keywords: CFD · Porous media · Hybrid RANS/LES

1 Introduction

Experimental and numerical investigations suggest an airframe noise mitigation potential of porous surface treatments applied to aerodynamic bodies. Acoustic prediction methods usually require a preceding high-fidelity flow simulation of the turbulent flow field as an input for a subsequent determination of the sound field. Additionally, an overall evaluation of the noise mitigation concept requires to take the aerodynamic penalty in terms of, e.g., the lift-to-drag ratio into account. One way to treat the flow through a porous media is to conduct a spatially fully resolved simulation of the porous microstructures without any modeling. However, due to the computational expense, except for validation or benchmark tests, it is more common to utilize the method of volume-averaging [1], which assumes a separation of scales between the pores and the flow features of interest. Compared to a direct pore structure resolving simulation it

© The Author(s), under exclusive license to Springer Nature Switzerland AG 2021
A. Dillmann et al. (Eds.): STAB/DGLR Symposium 2020, NNFM 151, pp. 666–676, 2021.
https://doi.org/10.1007/978-3-030-79561-0_63

enables the efficient prediction of the integral effects. The air-saturated porous medium is treated as a homogeneous phase with spatially filtered continuously varying field variables in space and time, which are then governed by the volume-averaged Navier-Stokes equations (VANS). This constitutes a closure problem, where the impact of the microscopic scales is substituted by surface filter terms, which need to be modeled. Among the different works Ref. [4] is worthwhile to mention for the current investigations. A parametric study of the flow and acoustic fields of various shaped blunt trailing-edges of flat plates with different porous media have been conducted. For an aerodynamic and acoustic evaluation the skin friction and the surface resistance to quantify the acoustic impedance have been investigated. Recommendations about the range of values of the two major parameters characterizing a porous material, the porosity and the pore size, have been suggested for a good trade-off, from which an optimal fictitious porous material was tailored for the present study.

For an evaluation of the effectiveness of porous media to mitigate TBL-TE noise, a 2D NACA 0012 airfoil with a rectangular porous segment, covering the last 15% of the airfoil to allow for a redistribution of the near-wall turbulent structures before passing over the trailing edge, is considered. The airfoil is in a uniform flow at $0°$ angle-of-attack, a Reynolds number based on the chord length c of $Re = 1 \times 10^6$ and a Mach number of $Ma = 0.15$. The focus in the present study is on the unsteady flow field. At these conditions the flow remains attached, such that we take advantage of a computationally efficient zonal approach, in which the RANS equations are solved in the entire domain first, followed by a LES in the noise source region, which is fully embedded inside the RANS domain. It starts at $0.6c$ to allow the turbulent boundary layer to fully evolve before advecting over the porous surface. The overall RANS-LES setup is sketched in Fig. 1. The RANS equations in free fluid are obtained by Reynolds averaging the instantaneous Navier-Stokes equations. In the presence of porous media, however, there is no agreement on whether to apply volume-averaging first followed by Reynolds averaging or vice versa. It is argued that Reynolds averaging first will preserve microscale turbulence. Here we adopt the formulation by Mößner et al. [5], who use volume-averaging before Reynolds averaging, which they justify by the results of Breugem [6], who obtained a good match between a volume averaged DNS and a geometry resolved DNS of a generic porous channel flow. One key element in the hybrid RANS/LES method is the generation of a physically meaningful inflow for the LES satisfying the underlying conservation equations. All the turbulent statistics encoded in the RANS turbulence model need to be transferred into resolved large scale eddies carrying the turbulent kinetic energy and moments as predicted by the upstream RANS solution. This is accomplished by the RSTG method by Roidl et al. [7]. From the RANS solution virtual eddy cores are determined and injected at the RANS-LES interface and their induced velocity fluctuations at the inlet plane are superimposed to the mean flow field.

In this paper the modifications to the governing equations of the RANS and the LES part for porous media are presented and the zonal coupling strategy is outlined. The methodology is applied to a NACA 0012 airfoil with a porous trailing edge. Results are compared against a solid reference configuration.

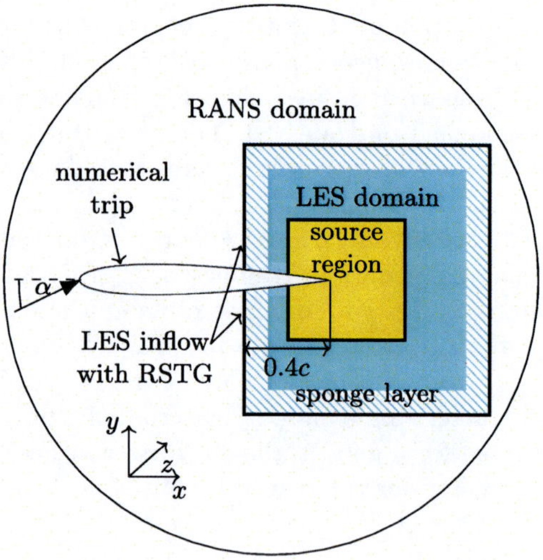

Fig. 1. Schematic of the RANS and LES domains around the NACA 0012 airfoil.

2 Problem Formulation

The modified compressible macroscopic equations in the presence of porosity are briefly stated. It is worth starting with the equations for the LES, since the RANS equations are built upon the volume averaged equations. In this section all variables are non-dimensionalized by appropriate reference quantities.

2.1 LES

In the following all variables are intrinsic volume averaged quantities, but for the sake of clarity no special notation is used. The velocity is Favre-averaged. For technical details regarding the averaging operator or a derivation the reader is referred to [1,8]. The conservation equations in non-dimensional form read

$$\frac{\partial \rho}{\partial t} + \frac{\partial \rho u_j}{\partial x_j} + \rho u_j \frac{\partial \ln \phi}{\partial x_j} = 0, \tag{1}$$

$$\frac{\partial \rho u_i}{\partial t} + \frac{\partial \rho u_i u_j}{\partial x_j} + \frac{\partial p}{\partial x_i} - \frac{1}{Re}\frac{\partial \tau_{ij}}{\partial x_j} + \rho u_i u_j \frac{\partial \ln \phi}{\partial x_j} + \frac{\mathcal{F}_i}{\phi} = 0, \tag{2}$$

$$\begin{aligned} &\frac{\partial \rho E}{\partial t} + \frac{\partial u_k(\rho E + p)}{\partial x_k} - \frac{1}{Re}\frac{\partial u_j \tau_{jk}}{\partial x_k} - \frac{1}{Re Pr}\frac{\partial k_{d,k}}{\partial x_k} \\ &+ \left(u_k(\rho E + p) - \frac{1}{Re} u_j \tau_{jk} - \frac{1}{Re Pr} k_{d,k} \right) \frac{\partial \ln \phi}{\partial x_k} = 0. \end{aligned} \tag{3}$$

For repeated indices the Einstein summation convention is to be used. It is t the time, x_i ($i \in [x, y, z]$) are the Cartesian coordinates, ρ is the density, p is the pressure, u_i are the velocity components, E is the total energy, $k_{d,i}$ and τ_{ij} ($j \in [x, y, z]$) are the components of the thermal diffusion vector and the viscous shear stress tensor. The porosity for a rigid porous material is $\phi \in [0,1] \neq f(t)$. The Prandtl number is denoted by Pr. The scale-up from a micro to a macro perspective introduces spatial filter terms embodying the integral effects of the solid phase, exerting a force on the fluid. These are expressed in terms of macroscopic quantities by Darcy's law and the Forchheimer correction term [2,3], which in non-dimensional form for an isotropic medium read

$$\mathcal{F}_i = \underbrace{\frac{1}{Re_{d_p}\sqrt{Da}}\phi\mu u_i}_{Darcy} + \underbrace{\frac{1}{\sqrt{Da}}\phi^2 c_F \rho \sqrt{u_j u_j} u_i}_{Forchheimer} . \tag{4}$$

The quantity μ is the dynamic viscosity, $d_p \sim \sqrt{\kappa}$ is a measure for the effective pore diameter, c_F is the Forchheimer coefficient, $Da = \kappa/L^2$ is the Darcy number, where $\kappa[m^2]$ is the permeability, quantifying the flow resistivity and L being some problem specific length scale and Re_{d_p} is the pore Reynolds number classifying the flow regime. The Darcy drag is related to the viscous Stokes drag in creeping flows ($Re_{d_p} < 1$), whereas the Forchheimer contribution accounts for the inertia effects, i.e., the form drag at higher Re_{d_p}. To arrive at Eqs. (1)–(4), the no-slip condition and the adiabatic condition at the pore surfaces have been assumed. Following the discussion in [6], the spatial correlations in the unresolved structures, the subgrid-scale stresses in an LES, are neglected. For the homogeneous material considered hereafter gradients in ϕ only occur at the fluid-porous interface. The present modeling does not include surface roughness effects. One of the challenges in porous media modeling is the interface, where flow quantities and material properties undergo rapid changes, rendering the equations to be very stiff and length scale constraints underlying the derivation of the volume averaged equations and the closure may be violated. Here we treat the interface by matching the solutions in the pure fluid and pure porous regions by means of jump conditions for the primitive flow variables, which in simple words account for the reduced cross-sectional area, dictated by the porosity, the flow faces, when entering the porous material. The whole interface treatment is beyond the scope of this paper, but some references are [5,8,9].

2.2 RANS

The Reynolds averaging of Eqs. (1)–(4) will result in a closure problem for the Reynolds stresses because of the non-linearity of the inertia terms. Here we choose to use the low Reynolds number k-ε model by Chien [10]. The advantage of using the (isotropic) rate of dissipation ε as the second turbulence variable over other two-equation models is the fact, that equations for the turbulent kinetic energy k, ε or the Reynolds stresses can be derived by analytical manipulations of the Navier-Stokes equations and therefore starting from Eqs. (1)–(4) model terms for k and ε can be derived. The additional terms for the porous media are adopted from Mößner et al. [5], who proposed the modifications for a Reynolds stress model. The closure (4) becomes

$$\mathcal{F}_i = \frac{\phi\mu\tilde{u}_i}{Re_{d_p}\sqrt{Da}} + \frac{\phi^2 c_F \bar{\rho}}{\sqrt{Da}} \left(|\tilde{u}|\tilde{u}_i + \frac{k}{|\tilde{u}|}\tilde{u}_i + \frac{\tilde{u}_j}{|\tilde{u}|}\widetilde{u_i' u_j'} - \frac{1}{2}\frac{\tilde{u}_i\tilde{u}_j\tilde{u}_k}{|\tilde{u}|^3}\widetilde{u_j' u_k'} \right), \tag{5}$$

where $\bar{()}$ denotes Reynolds averaging, $\tilde{()}$ denotes Favre averaging and $()' = () - \tilde{()}$. The porous terms in the k equation are obtained by taking half the contraction of the model terms in the Reynolds stress model. The non-dimensional transport equations for k and ε read

$$\frac{\partial \rho u_j k}{\partial x_j} = \frac{1}{Re} \frac{\partial \mu_k \frac{\partial k}{\partial x_j}}{\partial x_j} + P - \rho\varepsilon(1 + M_t^2) - \frac{1}{Re} \frac{2\mu k}{y_w^2} + \mathcal{D}_k - \mathcal{P}_k, \tag{6}$$

$$\frac{\partial \rho u_j \varepsilon}{\partial x_j} = \frac{1}{Re} \frac{\partial \mu_\varepsilon \frac{\partial \varepsilon}{\partial x_j}}{\partial x_j} + C_1 \frac{\varepsilon}{k} P - \frac{\varepsilon}{k} \left(c_2 f_2 \rho\varepsilon + \frac{1}{Re} \frac{2\mu k e^{-c_4 y_w^+}}{y_w^2} \right) - \mathcal{P}_\varepsilon, \tag{7}$$

where P is the classical production term, $\mu_k = \mu + \mu_t/\sigma_k$ and $\mu_\varepsilon = \mu + \mu_t/\sigma_\varepsilon$ with σ_k and σ_ε being the turbulent Prandtl numbers, the compressibility correction $M_t^2 = 1.5k/a^2$ with a the speed of sound, y_w (y_w^+) the distance to the wall (in viscous units) and $\mathcal{D}$, $\mathcal{P}_k$ and $\mathcal{P}_\varepsilon$ the porous model terms. The turbulent eddy viscosity is computed from $\mu_t = c_\mu f_\mu \rho k^2/\varepsilon$. The auxiliary functions are $f_2 = 1 - \frac{0.4}{1.8} e^{-Re_T^2/36}$ and $f_\mu = 1 - e^{-c_3 y_w^+}$ with $Re_T = \rho k^2/\mu\varepsilon$. An additional diffusion term

$$\mathcal{D} = \frac{\partial}{\partial x_j} \left(c_{D,p,eff} \frac{k^2}{\varepsilon} \frac{\partial k}{\partial x_j} \right) \tag{8}$$

was added inside the porous media, where $c_{D,p,eff} \in [0, c_{D,p})$ depends on a smooth indicator function as given in [5] and was argued to be necessary for low permeability materials only. The porous terms in Eqs. (6) and (7) are

$$\mathcal{P}_k = \frac{2\phi\mu k}{Re_{d_p}\sqrt{Da}} + \frac{\phi^2 c_F \bar{\rho}}{2\sqrt{Da}} \left(4k|\tilde{u}| + 2\frac{\tilde{u}_i \tilde{u}_k}{|\tilde{u}|} \widetilde{u_i' u_k'} + 3\frac{\tilde{u}_k}{|\tilde{u}|} \widetilde{u_i' u_i' u_k'} - \frac{\tilde{u}_i \tilde{u}_m \tilde{u}_k}{|\tilde{u}|^3} \widetilde{u_i' u_m' u_k'} \right), \tag{9}$$

$$\mathcal{P}_\varepsilon = \frac{2\phi\mu\varepsilon}{Re_{d_p}\sqrt{Da}} + \frac{\phi^2 c_F \bar{\rho}}{\sqrt{Da}} \left(\frac{8}{3}|\tilde{u}|\varepsilon + \frac{2\nu}{Re} \frac{\partial |\tilde{u}|}{\partial x_j} \frac{\partial k}{\partial x_j} - 2c_\varepsilon \frac{k}{\varepsilon} \frac{\tilde{u}_k}{|\tilde{u}|} \widetilde{u_k' u_i'} \frac{\partial \varepsilon}{\partial x_i} \right. \tag{10}$$

$$\left. + \frac{\nu}{Re} \frac{\partial}{\partial x_j} \left(\frac{\tilde{u}_i \tilde{u}_k}{|\tilde{u}|} \frac{\partial \widetilde{u_i' u_k'}}{\partial x_j} \right) \right).$$

In [5] a standard parameter set was given for use of the terms in their Reynolds stress model. To begin with, they set some parameters to zero due to a lack of reference data for calibration. This includes the triple correlation, such that here this term will not be taken into account either. Low Reynolds number models incorporate the wall distance to express a near wall damping effect. At a fluid-porous interface these distances are modified to account for the relaxed blocking effect compared to a solid wall as also suggested in [5]. Values for all remaining constants can be found in the original works. For the turbulence quantities k and ε additional interface conditions need to be prescribed.

2.3 Numerical Solution

The RANS and LES equations are solved by a finite-volume method. The former is discretized on a structured body-fitted mesh for efficiency reasons, while the latter is discretized on a locally refined unstructured Cartesian mesh, where a fully conservative cut-cell approach is used for the immersed boundaries [11]. In both solvers the inviscid fluxes are computed based on an advective upstream splitting method (AUSM) in conjunction with a second-order accurate MUSCL (Monotone Upstream-centered Scheme for Conservation Laws) type reconstruction of the cell surface values. The viscous fluxes are discretized by a second-order centered scheme. In the LES the semi-discretized equations are integrated in time using a low-storage explicit 5-stage Runge-Kutta method also of second-order accuracy. A monotone integrated LES (MILES) ansatz, in which the low dissipation of the numerical scheme replaces an explicit subgrid scale model of the high wave number spectrum is used. The RANS and LES solutions are coupled by a one-way coupling, i.e., the LES solution does not affect the RANS solution.

3 Computational Setup

Two simulations of a 2D NACA 0012 airfoil with a solid ('s') and a porous ('p') trailing edge in a uniform flow at $Ma = 0.15$, $Re = 1.0\times10^6$ and $\alpha = 0°$ have been conducted. The RANS equations are solved on a 2D structured multi-block mesh of O-type with a finite trailing edge thickness of 0.26%c. In case 'p' it consists of 860×10^3 cells outside the airfoil and 7×10^3 cells inside the porous region, distributed across 4 and 5 blocks, respectively. The mesh extends 50c away from the airfoil and the finest resolution in inner coordinates in wall-normal direction is $\Delta y^+_{wall,FP} < 1$ at the solid wall and the fluid-porous (FP) interface. In the streamwise direction the cells are clustered near the trailing edge and near the transition from the solid boundary to the FP-interface with a minimum spacing of $\Delta x^+ \approx 10\Delta y^+_{wall,FP}$. Figure 2 (left) shows a magnification of the trailing edge region. The mesh for the solid setup was derived from the porous mesh by making as little changes as possible to exclude any impacts on the comparison coming from the mesh. The fully embedded LES domain covers the last 40% of the airfoil and extends 4c into the wake, 6c in the flow-normal direction and 0.1c in the spanwise direction. The mesh is locally refined near the airfoil wall and the FP-interface with a minimum isotropic grid spacing of $\Delta^+ \approx 7$, which results in a total of 230×10^6 cells in case 'p'. A snapshot is given in Fig. 2 (right). The LES domain is surrounded by a sponge layer to absorb reflections from the eddies leaving the domain and to smoothly impose the far-field conditions from the preceding RANS solution. At the inflow only pressure sponges are applied to minimize spurious noise from the RSTG. The LES is initialized with the mean flow field obtained beforehand with the RANS solver to shorten the transient phase. In the spanwise direction periodic boundary conditions are applied.

The material parameters for the Darcy-Forchheimer model follow from Ref. [4] for the porosity $\phi \approx 0.5$ and the mean particle diameter $d_p^+ \approx 7$ and correspond to a fictitious porous material with a good trade-off between aerodynamic and acoustic performance. For widely used materials the Ergun Eq. [12] and the Kozeny-Carman formulation [13] relate those quantities to the permeability and the Forchheimer coefficient, resulting in $Da = 9.3 \times 10^{-10}$ and $c_f = 0.38$.

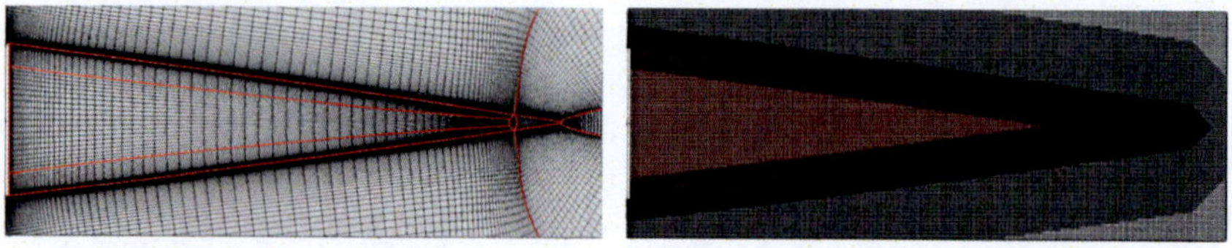

Fig. 2. Detailed views of the structured RANS mesh (left) and the unstructured Cartesian mesh of the LES (right) for the porous configuration. The red shadowed area in the left figure indicates the porous region.

4 Results

In a first step the pure RANS solution and the LES solution downstream of $x/c = 0.6$ are compared against viscous XFOIL predictions in terms of the mean pressure coefficient c_p and the wall friction coefficient c_f for case 's' in Fig. 3. The distributions lie on top of each other. The c_f-distributions contain more information about the state of the boundary layer. The k-ε model was numerically forced to transition at $\approx 0.1c$ and attains a turbulent boundary layer shortly downstream. The transition in XFOIL, with a shorter transition length, was set at $0.12c$. Strictly speaking, the turbulence model has its range of validity in the fully turbulent regime, i.e., the solution around the laminar-turbulent transition region is to be interpreted with caution. The flow remains attached along the whole airfoil. Downstream of the LES inflow plane c_f initially drops below the value of the pure RANS solution, before recovering after a transition length of $\approx 4\delta_0$, where δ_0 is the boundary layer thickness at the inlet, indicating that the boundary layer attains a fully turbulent state. Upstream of the LES domain the c_p and c_f distributions of case 'p', not shown here, are identical, which evidences that for $\alpha = 0°$ the low permeability porous trailing edge has a negligible influence on the LES inflow condition. Only for $x/c > 0.85$ the velocity shearing was observed to be lower at the FP-interface and so was c_f, since the flow is able to penetrate the porous media. In Fig. 4

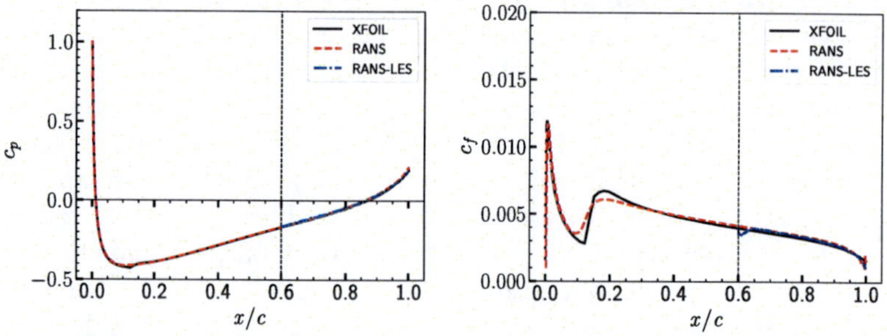

Fig. 3. Distributions of c_p (left) and c_f (right) of case 's'. Dashed line indicates the inflow plane of the LES domain.

(left) the profiles of the turbulent isotropic integral length scale Λ computed from the RANS solutions are plotted in the near wake at $x/c = 1.0038$. As part of the BANC-II workshop a comparable configuration at $Ma = 0.1118$ was investigated, where measurement data by IAG Stuttgart [14] was taken as a reference. These are plotted alongside, where the same definition $\Lambda = 0.4k^{3/2}/\varepsilon$ has been adopted to post-process the simulation data. Despite the differences in the profiles the prediction of the peak level is satisfactory and the overall quality is comparable to other participants of the BANC-II workshop. For the present very low permeable material the Λ only slightly deviates near the symmetry axis. From the LES solution unsteady flow data was acquired for $2.65c/U_\infty$, where U_∞ is the freestream velocity. Figure 4 (right) visualizes the turbulent kinetic energy k in the trailing edge region. Overall for the flow over the porous surface the level of k in the boundary layer is higher. This can be explained with a view on Fig. 5, where the normal components of the Reynolds stress tensor, contributing to k, are plotted alongside k at $x/c = 0.99$. The profiles start diverging for $y/c < 0.015$. For $y \to 0$ the increase in the profiles flattens earlier in case 's' with the plateau lying well below those near the FP-interface. The peak in the streamwise velocity fluctuation vanishes in case 'p'. At the solid wall the velocity induced by the eddies passing by experience a more severe blockage effect with the wall-normal velocity component strictly forced to zero. A view on profiles plotted more upstream reveals, that the relative dominance of the normal and spanwise components $u^2_{y,rms}$ and $u^2_{z,rms}$ in case 'p', because of the 'softer' boundary condition allowing the eddies to partially penetrate the surface, contribute to a higher k. The observations are in accordance with those in [15] about the disappearance of elongated streaky structures near the permeable surface, leading to the reduction of the peak streamwise velocity fluctuations and the increase in $u^2_{y,rms}$ and $u^2_{z,rms}$, due to the weakening of the wall-blocking and wall-induced viscous effects. The distribution of the root-mean-square of the pressure fluctuation p'_{rms} in Fig. 6 (left) shows the opposite trend. Very near the airfoil surface the overall level is reduced in case 'p'. To satisfy the no-

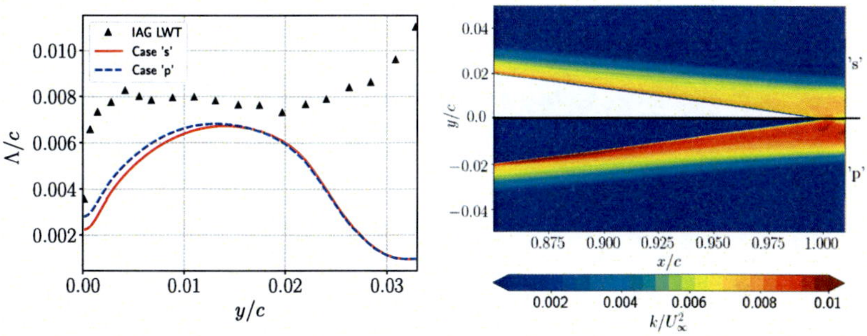

Fig. 4. Longitudinal integral length scale Λ from the RANS solutions (left) and turbulent kinetic energy k in $x/c \in [0.85, 1.01]$ from the LES solutions (right), where the upper half belongs to case 's' and the lower half to case 'p'.

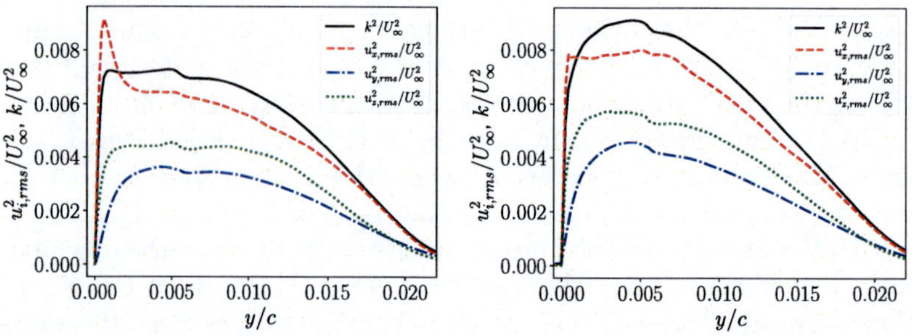

Fig. 5. Profiles of the normal Reynolds stresses and k at $x/c = 0.99$ from the LES solutions for case 's' (left) and case 'p' (right).

slip condition hydrodynamic pressure fluctuations can be thought to induce an image velocity field, that cancels the velocity exactly to zero at the wall. The permeable wall more effectively resists to local disturbances. The model predicts pressure fluctuations inside the porous media, which gradually dissipate deep inside the material. Finally, snapshots of the dominating noise source component of the instantaneous perturbed Lamb vector $\boldsymbol{L}' = [\boldsymbol{\omega} \times \boldsymbol{u}]'$ are shown in Fig. 6 (right), which would be the input to a subsequent acoustic solver. Qualitatively, hardly any differences in the distribution of L_y' can be observed. The maximum levels $\max(|L_y'|)$ are comparable. To complete the study on the noise mitigation potential of porous surfaces, Computational Aeroacoustics (CAA) simulations based on these noise sources will be performed to evaluate the far-field noise in the future. Inside the low permeable porous material all kind of fluctuations, thus vortical structures, are rapidly damped out by the Darcy-Forchheimer model terms, so that no noise sources are located in there.

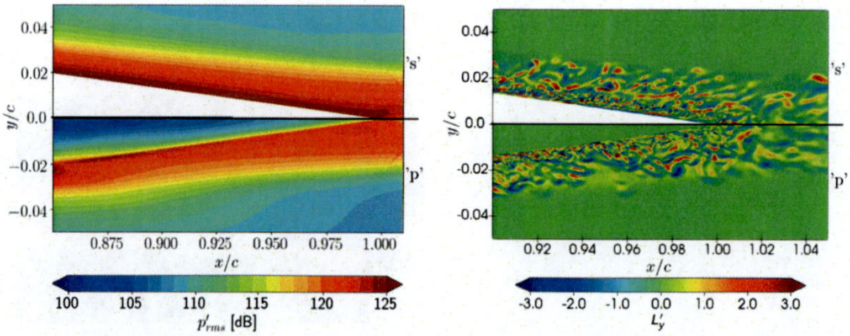

Fig. 6. Distributions of p'_{rms} (left) and L_y' (right) from the LES solutions, where upper half belongs to case 's' and lower half to case 'p'.

5 Conclusions

The feasibility of a hybrid RANS/LES approach was presented with particular focus on flow over and through porous surfaces. Therefore the method of volume averaging was applied to adopt a macroscopic point of view. For the RANS part a low Reynolds number k-ε model was augmented by model terms as proposed in [5]. The methodology was applied to a NACA 0012 airfoil with the objective to capture the turbulent eddies, which will contribute to TBL-TE noise, by a higher-fidelity LES. For the symmetric flow at $\alpha = 0°$ a negligible impact of the porous trailing edge on the potential flow field around the airfoil was observed. Near wall turbulence is altered slightly in that streaky structures with a preferential streamwise orientation disappear, but wall-normal and spanwise fluctuations increase, due to the less blocking permeable wall. Future work will include angle-of-attack variations to evaluate the performance of porous materials on lifting bodies.

Acknowledgments. The support of this research by the Deutsche Forschungsgemeinschaft DFG is gratefully acknowledged. The authors also gratefully acknowledge the Gauss Centre for Supercomputing e.V. (www.gauss-centre.eu) for funding this project by providing computing time on the GCS Supercomputer HAWK at Höchstleistungsrechenzentrum Stuttgart (www.hlrs.de).

References

1. Whitaker, S.: Theory and Applications of Transport in Porous Media: The Method of Volume Averaging. Kluwer Academic Publishers (1999)
2. Whitaker, S.: Flow in porous media I: A theoretical derivation of Darcy's law. Transp. Porous Media **1**(1), 3–25 (1986)
3. Whitaker, S.: The Forchheimer equation: a theoretical development. Transp. Porous Media **25**(1), 27–61 (1996)
4. Koh, S.R., Meinke, M., Schröder, W.: Numerical analysis of the impact of permeability on trailing-edge noise. J. Sound Vib. **421**, 348–376 (2018)
5. Mößner, M., Radespiel, R.: Modelling of turbulent flow over porous media using a volume averaging approach and a Reynolds stress model. Comput. Fluids **108**, 25–42 (2015)
6. Beugem, W.: The influence of wall permeability on laminar and turbulent flows: theory and simulations. Ph.D. thesis (2004)
7. Roidl, B., Meinke, M., Schröder, W.: Boundary layers affected by different pressure gradients investigated computationally by a zonal RANS-LES method. Int. J. Heat Fluid Flow **45**, 1–13 (2014)
8. Satcunanathan, S., Meinke, M.H., Schröder, W.: Prediction of noise mitigation by porous media based on a direct-hybrid CFD/CAA method. In: 25th AIAA/CEAS Aeroacoustics Conference, pp. 2019–2696 (2019)
9. Ochoa-Tapia, J.A., Whitaker, S.: Momentum transfer at the boundary between a porous medium and a homogeneous fluid—I. Theoretical development. Int. J. Heat Mass Transfer **38**(14), 2635–2646 (1995)
10. Chien, K.Y.: Predictions of channel and boundary-layer flows with a low-Reynolds-number turbulence model. AIAA J. **20**(1), 33–38 (1982)

11. Schneiders, L., Hartmann, D., Meinke, M., Schröder, W.: An accurate moving boundary formulation in cut-cell methods. J. Comput. Phys. **235**, 786–809 (2013)
12. Macdonald, I.F., El-Sayed, M.S., Mow, K., Dullien, F.A.L.: Flow through porous media-the Ergun equation revisited. Ind. Eng. Chem. Fundam. **18**(3), 199–208 (1979)
13. Clennell, M.B.: Tortuosity: a guide through the maze. Geolog. Soc. London Spec. Publ. **122**(1), 299–344 (1997)
14. Herrig, A.: Validation and application of a hot-wire based method for trailing-edge noise measurements on airfoils. Ph.D. thesis (2011)
15. Breugem, W.P., Boersma, B.J., Uittenbogaard, R.E.: The influence of wall permeability on turbulent channel flow. J. Fluid Mech. **562**, 35 (2006)

Meshing Strategy for Movable Control Surfaces: Towards High-Fidelity Flight Maneuver Simulations

Larissa Bruna Streher[1,2](✉) and Ralf Heinrich[1]

[1] Deutsches Zentrum für Luft-und Raumfahrt, Institute of Aerodynamics and Flow Technology, Lilienthalplatz 7, 38108 Braunschweig, Germany
{larissa.streher,ralf.heinrich}@dlr.de

[2] Bernoulli Institute for Mathematics, Computer Science and Artificial Intelligence, University of Groningen, Nijenborgh 9, 9747 Groningen, AG, The Netherlands

Abstract. High-fidelity flight maneuver simulations are crucial for the development of realistic digital aircraft models. However, such simulations are still hampered by difficulties in modeling the relative body motion between control and lifting surfaces when using realistic configurations. The presence of spanwise gaps between lifting and control surfaces impedes the application of concepts such as mesh deformation, and hampers the usage of mesh deformation combined with the overset method since the mesh generation process is particularly cumbersome. To reduce the user effort to create overset meshes, we have developed a methodology to automatically create overlapping regions for matching block interfaces. Hence, the usage of the overset method combined with mesh deformation for modeling moving control surfaces is facilitated, and a significant advance towards the computation of high-fidelity flight maneuvers is achieved.

Keywords: CFD · Mesh generation · Control surface modeling

1 Introduction

Aircraft control surfaces are movable aerodynamic devices designed to maintain the aircraft stable, and yet responsive to any control surface deployment. The deflection of such surfaces changes the aerodynamic forces and moments acting on the aircraft and are, therefore, used for trimming and maneuvering purposes. The accurate prediction of the flow around maneuvering aircraft is of major importance for the development of realistic digital models of aircraft, which would enable a reduction of time and costs involved in the design and certification of aircraft [5]. The simulation of high-fidelity flight maneuvers throughout the entire flight envelope is, however, still hampered due to difficulties in the dynamical modeling of movable control surfaces.

Many approaches for modeling moving control surfaces have already been proposed, such as mesh regeneration, mesh deformation, and combinations of overset (Chimera) grids with either rigid body rotation or mesh deformation

© The Author(s), under exclusive license to Springer Nature Switzerland AG 2021
A. Dillmann et al. (Eds.): STAB/DGLR Symposium 2020, NNFM 151, pp. 677–686, 2021.
https://doi.org/10.1007/978-3-030-79561-0_64

[3]. Mesh regeneration consists of generating new high-quality meshes for each deflection angle of the control surfaces. This alternative is, however, very time consuming and only suitable for steady control surface deflections since it does not preserve the mesh connectivity. Mesh deformation, in contrast, deforms the mesh to fit the new configuration without modifying the grid connectivity and is, therefore, suitable for steady and unsteady control surface deflections. This approach is, however, not capable of modeling the spanwise gaps present between the lifting and control surfaces, since highly sheared grid elements appear in the gap region already at low deflection angles [6]. The combination of overset meshes with rigid body rotation, on the other hand, is capable of modeling the spanwise gaps, but also requires the introduction of a streamwise gap. Albeit such a gap is usually present in real configurations, it is normally sealed. Thus, a wing configuration that considers the streamwise gap is usually not realistic and therefore not suitable for high-fidelity simulations. Conversely, the combination of Chimera meshes with mesh deformation enables the usage of realistic wing configurations without changing the mesh connectivity for each deflection angle. This approach is, therefore, applied in this work.

The combination of overset meshes with mesh deformation is based on the creation of overset grid blocks for the control surfaces, which are deployed through mesh deformation algorithms. Although this approach has been successfully tested, e.g., by Reimer and Heinrich [7], it has a major drawback: the process of generating valid, high-quality Chimera meshes is cumbersome (especially for the narrow gap regions) and time consuming.

Aiming at reducing the user effort required to generate high-quality overset meshes, we have developed a methodology that extends matching block interfaces. This methodology has been implemented on an add-on of the finite volume software DLR TAU [9], yielding the so-called *AutoLap* software. With *AutoLap*, overlapping regions with user-defined number of cell layers can, then, be easily and automatically generated for TAU mesh files based on the NetCDF interface [8]. Thereby, computations of aircraft with movable control surfaces using the overset method combined with mesh deformation are facilitated.

2 Automatic Overlapping Region Generation for Matching Block Interfaces: The *AutoLap* Software

The *AutoLap* software is based on the idea of Blades and Marcum [2]: extruding the faces along the grid block interface into the adjacent subdomain based on information of this subdomain. Here, we apply this idea exclusively to blocks that have matching interfaces, i.e., when the element faces and nodes on the interfaces of both subdomains are located at the same place. Consequently, elements on the adjacent subdomain can be simply added to the connectivity list of the grid block being considered.

As a starting point for *AutoLap*, a multi-block grid with matching block interfaces is to be imported into the software. The user then defines the number of cell layers to be added either to a specific matching interface or to all available matching interfaces. Next, *AutoLap* identifies the matching interface pairs of the input grid using bounding boxes that enclose all points of each boundary previously defined as a matching interface. If the definition of bounding boxes is not sufficient to determine the interface pairs, arbitrary points on the considered interface are tested to determine whether there is a point located on the same position that lies on the candidate interface pair and, if needed, on other matching interfaces or on the grid block of the candidate interface pair.

Once the matching interface pairs are defined, the layers of elements that are to be added to the (target) interface can be identified and flagged based on the neighboring elements of the partner (donor) interface elements. First, the points on the target matching interface are located and the corresponding donor points, i.e., points that are located at the same position as the target points but are part of the donor grid block, are flagged. The donor points are, then, divided into so-called *Chimera* and *non-Chimera points*. The former are the points that are connected to any Chimera boundary (on either the target or the donor grid block), or the points that are no longer on both partner matching interfaces. The latter, i.e., the *non-Chimera points*, are the points that do not meet the criteria to be *Chimera* points.

With the flagged points divided into *Chimera* and *non-Chimera points*, the volume elements can be flagged. All volume elements connected to the *non-Chimera points* are flagged, as these elements are never present on the target grid block. For the *Chimera points*, the flagging process is carried out carefully in order to avoid the generation of duplicate elements. First, all volume elements connected to the donor *Chimera points* that were not already flagged with the *non-Chimera points* are identified, and the underlying donor points are further separated into so-called *valid* and *invalid* (donor) *points*. The *valid points* are either those that have no corresponding points on the target grid block, or those that have such corresponding points and meet the following criteria: they do not lie on a Chimera boundary on the donor grid block, or they lie on such a boundary, provided that they simultaneously lie on a matching interface on the donor block, or their corresponding target point belongs to a Chimera surface on the target block. The *invalid points* are those that do not meet the criteria to be considered *valid points*. If the previously identified donor element consists exclusively of *valid points* that are not located on domain boundaries, it is directly flagged. Otherwise, the element is only flagged if it is not already present on the target grid block.

Finally, the faces of all previously flagged volume elements are analyzed in order to accommodate the extended boundary. If the face is connected to a volume element, *AutoLap* first verifies whether a corresponding surface element is available on the target grid block. In case such a surface is indeed available, the surface element on the target block is marked for later deletion as the underlying boundary is shifted by one cell layer. Otherwise, the points constituting the face are stored and the new surface element, which accommodates the shifted boundary, is flagged. In case the considered face is connected to a surface element, *AutoLap* verifies whether this element is located on a matching interface with the target block. If the surface element does not lie on such an interface, it is flagged in order to accommodate the new boundary. Otherwise, the donor surface element is tagged for a later change of the boundary type to Chimera, whereas the corresponding target surface element, if available, is marked for later deletion.

If the specified matching interface must be extended by more than one cell layer, the entire flagging process is repeated until the user-defined number of cell layers is marked. In case the cell layer being flagged is not the first one, the point flagging process that forms the basis for the volume element search is, however, based on the surface elements marked on the previously flagged layer.

As soon as the flagging process is completed, the flagged elements can be added to the target grid block. First, the points that are not present on the target grid block and that are required to construct the flagged elements, are created at the same location as the underlying donor points. Second, volume elements that are topologically the same as the flagged ones are generated. Next, the surface elements that have been flagged for deletion are removed from the grid connectivity. Subsequently, the surface elements flagged on the last cell layer are created. Finally, the boundary type of the newly created elements is set to Chimera, and the boundary type of the donor surface elements that were flagged for later modification are updated to Chimera as well. The overlapping region is, then, successfully created for the specified matching interface. If the partner matching interface or other matching interfaces are also to be treated, the same process is to be applied, bearing in mind that the input grid for the former is the grid with the original matching interface and for the latter the grid with the extended patched surface. If no further matching interfaces are to be treated, the grid with the automatically generated overlapping region can be written and used for CFD simulations with the overset meshes technique.

3 Test Cases

AutoLap is tested here for three different test cases. First, the solution of a Prandtl-Meyer expansion fan is computed for single-block reference grids and two-block Chimera grids, which overlapping regions are generated by *AutoLap*. The solution obtained with all grids are compared, ensuring that the overlapping region was successfully created. Next, a test case in which a Chimera boundary of the expansion fan slides in the stream-wise direction according to a sinusoidal function is investigated. The grid is gradually deformed and the concept that we aim to apply to control surfaces, i.e., mesh deformation combined with overset meshes with automatically generated overlapping regions, is proven. Finally, the deflection of an arbitrary aileron of a NACA wing is tested for a reference grid fully generated by a mesh generation software and a grid in which the overlapping region in the gaps between control and lifting surfaces are generated with the *AutoLap* software. The results are compared and the applicability of *AutoLap* for realistic configurations with moving control surfaces is verified.

All computations are carried out with the cell-vertex DLR TAU Code [9]. The governing equations are discretized in space with a second-order accurate finite volume method. The convection term is discretized with a second-order accurate central scheme with added Matrix dissipation. To deform the meshes, surface deflections are prescribed and a thin plate spline [4] is used to interpolate the deflections into the volume grid.

3.1 Prandtl-Meyer Expansion Fan

Here, we investigate the supersonic flow of air about a concave corner, which leads to the so-called Prandtl-Meyer expansion fan. The flow is herein considered inviscid and has the following free-stream properties: pressure of $p_\infty = 101325\,\text{Pa}$, temperature of $T_\infty = 273.15\,\text{K}$, and Mach number of $Ma_\infty = 1.5$. The two-dimensional expansion fan geometry is characterized by a turn angle of $\theta = 7.6°$, an inlet height of $H_{\text{in}} = 2.3\,\text{m}$, an outlet height of $H_{\text{out}} = 2.5\,\text{m}$, and a width of $L_{\text{x}} = 2\,\text{m}$, whereby the expansion corner lies at $0.25\,L_{\text{x}}$.

Two single-block reference grids are first generated for the above described geometry: *ref. structured* (single-block structured grid with 6954 mesh nodes) and *ref. unstructured* (single-block unstructured grid with 9662 grid nodes). In addition to the reference grids, two other two-block grids are created with the matching interface located at $z_{\text{int}} = 0.3\,\text{m}$, as illustrated in Fig. 1. Overlapping regions consisting of two cell layers for the multi-block grids are, then, automatically generated by the *AutoLap* software and two Chimera grids are obtained: *Chim. structured* with 7564 mesh nodes, and *Chim. unstructured* with 10278 mesh nodes. Except from the overlapping region and the grid block structure, these overset meshes are essentially the same as the reference grids (see Fig. 1).

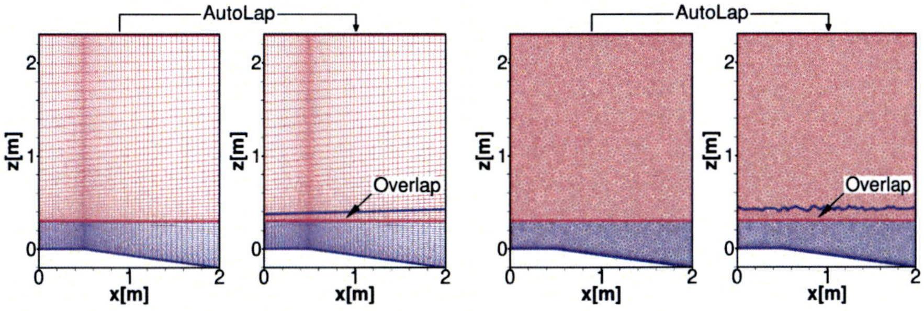

(a) Structured grid. (b) Unstructured grid.

Fig. 1. Expansion fan grid before (left) and after (right) the generation of overlapping regions with *AutoLap*. The bottom and top grid blocks are depicted in purple and pink, respectively.

Steady-State Test Case. The steady-state solution of the expansion fan is computed with the reference and Chimera grids. As expected, donor cells are found for all Chimera interpolation points. Figure 2 shows the Mach number distribution for the reference and Chimera grids, as well as the location of the forward and rearward Mach lines obtained with the inviscid compressible flow theory (analytical solution). The dimensionless flow properties downstream of the rearward Mach line are also summarized in Table 1. The results with the reference and Chimera grids agree really well for both structured and unstructured grids. The mean values of the flow variables downstream of the rearward Mach line are approximately the same for the Chimera and reference grids, whereas the standard deviation for the Chimera grids are lower due the presence of the overlapping regions where underlying interpolations are performed. The results also agree relatively well when compared to the analytical solution. *AutoLap* is thus able to generate valid overlapping regions for the overset method.

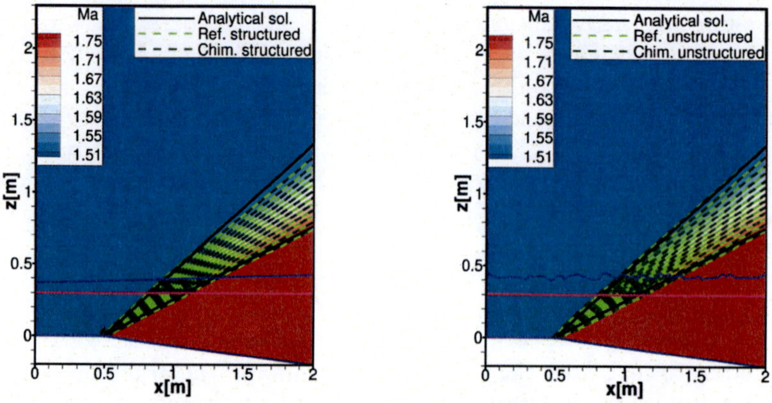

(a) Structured grid.

(b) Unstructured grid.

Fig. 2. Mach distribution of the steady-state expansion fan. The overlapping region of the overset meshes is located between the purple and pink lines.

Sliding Chimera Boundary Test Case. A time dependent case, in which the chimera surface of the top grid block (depicted in pink in Fig. 1) moves in the streamwise direction is investigated here. An explicit backward Euler time-stepping scheme is applied and the grid is gradually deformed in each time step according to a sinusoidal function with a maximal amplitude of deformation of $\Delta x = 0.1\,\text{m}$. Figure 3 illustrates the structured and unstructured overset grids after being deformed by the maximal amplitude. The elements on the overlapping region are, therefore, not anymore the same for both top and bottom grid blocks, and the Chimera boundaries become non-matching. Figure 4 illustrates the Mach number distribution in relation to the geodesic coordinates for the structured and unstructured overset grids and Table 2 summarizes the dimensionless flow properties downstream of the rearward Mach line for the maximal deformations. There is a good agreement between the solutions with no deformation and with the maximum deformations, even if the deformed and non-deformed meshes are not the same anymore. Hence, the developed methodology is also suitable to generate overlapping regions for overset grids used in cases where the position of the mesh nodes changes in time, i.e., cases with relative body motion as in the deployment of control surfaces.

Table 1. Summary of the dimensionless flow properties downstream of the rearward Mach line for the steady-state test case. The mean values and standard deviations of the numerical solutions are computed with regard to the grid points located downstream of the rearward Mach line.

	Ma	p	T
Analytical sol.	1.7578	0.6814	0.8962
Ref. structured	$1.7573 \pm 7.42 \times 10^{-3}$	$0.6822 \pm 7.68 \times 10^{-3}$	$0.8964 \pm 2.93 \times 10^{-3}$
Chim. structured	$1.7573 \pm 7.15 \times 10^{-3}$	$0.6821 \pm 7.40 \times 10^{-3}$	$0.8964 \pm 2.83 \times 10^{-3}$
Ref. unstructured	$1.7580 \pm 5.84 \times 10^{-3}$	$0.6816 \pm 6.22 \times 10^{-3}$	$0.8961 \pm 2.32 \times 10^{-3}$
Chim. unstructured	$1.7580 \pm 5.66 \times 10^{-3}$	$0.6816 \pm 6.02 \times 10^{-3}$	$0.8961 \pm 2.25 \times 10^{-3}$

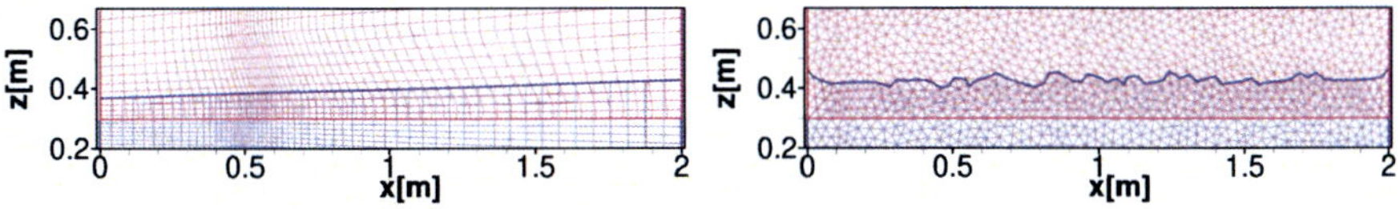

(a) Structured grid. (b) Unstructured grid.

Fig. 3. Expansion fan grids after a mesh deformation of Δx. Close-up on the deformed overlapping region.

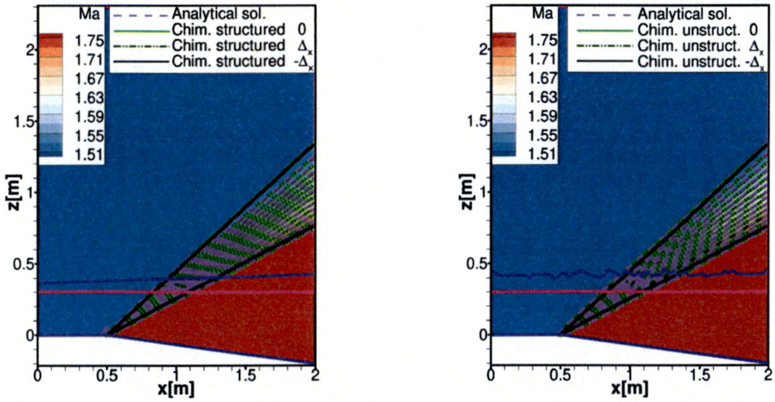

(a) Structured grid. (b) Unstructured grid.

Fig. 4. Mach distribution of the expansion fan with a sliding Chimera boundary. Results are shown for the undeformed mesh, and meshes deformed by Δx and $-\Delta x$. The overlapping region is located between the purple and pink lines.

3.2 NACA0012 Wing with a Generic Aileron

A NACA0012 finite wing with a generic aileron is studied here. The wing is characterized by a chord length of $c = 1\,\mathrm{m}$ and a spanwise length of $L_{\mathrm{w}} = 3.05\,\mathrm{m}$. The generic aileron is located at $0.69\,L_{\mathrm{w}}$ and has a spanwise length of $L_{\mathrm{ail}} = 0.8\,\mathrm{m}$. Gaps of 0.01 m separate the control and lifting surfaces.

Table 2. Summary of dimensionless flow properties downstream of the rearward Mach line for the sliding Chimera boundary test case. The mean values and standard deviations are computed with respect to the grid points located downstream of the rearward Mach line of the deformed grids.

Mesh	Ma	p	T
Chim. struct. Δx	$1.7573 \pm 7.21 \times 10^{-3}$	$0.6821 \pm 7.46 \times 10^{-3}$	$0.8964 \pm 2.85 \times 10^{-3}$
Chim. struct. $-\Delta x$	$1.7571 \pm 7.20 \times 10^{-3}$	$0.6823 \pm 7.44 \times 10^{-3}$	$0.8964 \pm 2.84 \times 10^{-3}$
Chim. unstruct. Δx	$1.7578 \pm 5.63 \times 10^{-3}$	$0.6817 \pm 5.97 \times 10^{-3}$	$0.8961 \pm 2.24 \times 10^{-3}$
Chim. unstruct. $-\Delta x$	$1.7578 \pm 5.80 \times 10^{-3}$	$0.6818 \pm 6.14 \times 10^{-3}$	$0.8961 \pm 2.30 \times 10^{-3}$

All computations are performed with an angle of attack of $\alpha = 5°$ and with the following free stream flow properties: $Ma_\infty = 0.7$, Reynolds number $Re_\infty = 7 \times 10^6$, and Prandtl number $Pr_\infty = 0.72$. The one-equation Spalart-Allmaras turbulence model in its negative form [1] is used to model the Reynolds stress tensor. Upwind spatial discretization schemes are used to initialize the simulation, switching to a second-order accurate central scheme with added matrix dissipation once a reasonable physical solution is obtained.

Here, a mesh strategy based on the creation of four grid blocks and one hole geometry is utilized, as illustrated in Fig. 5. One unstructured grid block is created for the plain wing, whereas three structured grids blocks are created for the aileron and spanwise gaps. The plain wing grid block overlaps with the structured grid blocks, so that a hole geometry is defined to blank the surface of the wing, where the aileron and the gaps are located. The gaps and aileron grid blocks are, here, structured in order to account for the vortices caused by the presence of the gaps.

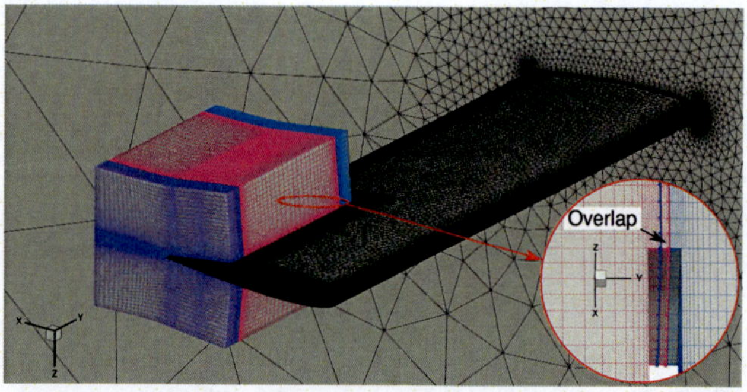

Fig. 5. Mesh for the NACA wing with a generic aileron with a close-up view on the overlapping region generated by the *AutoLap* software.

Two grids composed of two-cell layer overlapping regions between the structured grid blocks are analyzed in this work: *reference Chimera* and *automatic Chimera*. For the former grid, the overlapping regions between the structured grid blocks are created directly by the mesh generation software, whereas for the latter, matching interfaces between the structured grid blocks are created at first and are automatically extended with the methodology developed here. Both grids are characterized by the exactly same number of mesh nodes and elements: 4,046,039 mesh nodes, 8,333,501 volume elements, and 278,682 surface elements.

Here, we compute the steady flow solution for both grids with the aileron deployed by angles of $\delta = -5°$ and $\delta = -20°$ in order to ensure the functionality of *AutoLap* for moving control surfaces. During the mesh deformation process, the gap blocks (displayed in blue and purple in Fig. 5) are maintained fixed, whereas the aileron grid block (illustrated in pink in Fig. 5) is deformed. Donor cells are found for all Chimera interpolation points.

The pressure coefficient distribution and the eddy viscosity for the control surface wake (solved with the structured grid blocks) are illustrated in Fig. 6 for both meshes. The presence of the spanwise gaps leads to the formation of vortices, which influence the overall flow circulation. Thus, high-fidelity simulations of moving control surfaces must indeed consider these gaps. As the results obtained with both grids are in perfect agreement, the applicability of the *AutoLap* software for moving control surfaces is hereby proven.

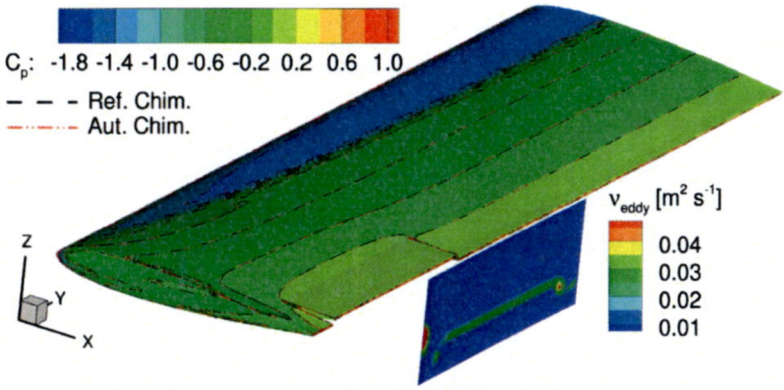

(a) Aileron deployed by $\delta = -5°$.

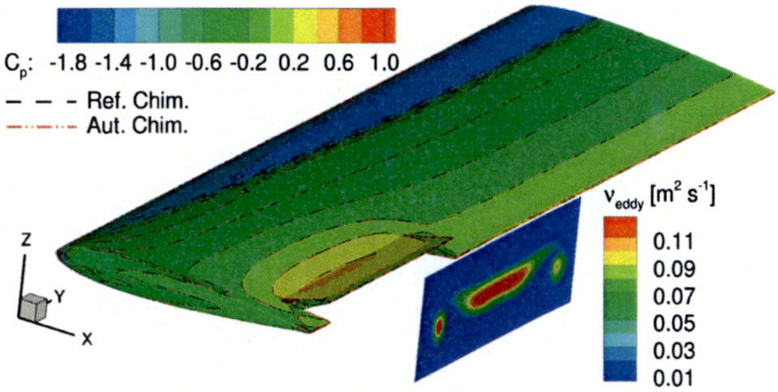

(b) Aileron deployed by $\delta = -20°$.

Fig. 6. Pressure coefficient C_p and eddy viscosity ν_{eddy} distributions for the *automatic Chimera* mesh. Black dashed lines and red dashed dotted lines are the isolines of the pressure coefficient distribution for the *reference Chimera* and *automatic Chimera* grids, respectively.

4 Conclusions

The simulation of moving control surfaces is particularly challenging for realistic wing configurations. The presence of small spanwise gaps between lifting and control surfaces hampers the deployment of the control surfaces with alternatives other than mesh regeneration. Here, we focus on one approach that is successful even for these challenging configurations: the combination of the overset method with mesh deformation. The mesh generation process of overset meshes is, however, often cumbersome and time consuming. Aiming at reducing the user effort required to generate high-quality Chimera meshes, we have developed a methodology to automatically create the overlapping regions, whereby a grid with matching block interfaces is used as a starting point. This methodology has been implemented on an add-on of the DLR TAU solver, which yielded the so-called *AutoLap* software. This software is particularly interesting because it enables the extension of matching surfaces with a user-defined number of cell layers, so that even a trial-and-error approach is viable in order to generate overlapping regions within which all Chimera boundary points can be properly interpolated. *AutoLap* has been tested for the Prandtl-Meyer expansion fan with and without mesh deformation, as well as a NACA wing with a generic aileron, which was deflected by -5 and $-20°$ using a mesh deformation algorithm. Through comparisons with reference grids, the methodology developed here is verified. Thus, an important tool to assist the simulation of dynamically moving control surfaces with the combination of overset meshes and mesh deformation is provided, which can also ease the realization of high-fidelity flight maneuver simulations.

References

1. Allmaras, S.R., Johnson, F.T.: Modifications and clarifications for the implementation of the spalart-allmaras turbulence model. In: ICCFD7, pp. 1–11 (2012)
2. Blades, E.L., Marcum, D.L.: A sliding interface method for unsteady unstructured flow simulations. Int. J. Numer. Meth. Fl. **53**(3), 507–529 (2007)
3. Capsada, L.A., Heinrich, R.: Development of the DLR TAU code for modeling of control surfaces. In: DLRK 2018, p. 480018 (2018)
4. Duchon, J.: Splines minimizing rotation-invariant semi-norms in Sobolev spaces. In: Schempp, W., Zeller, K. (eds.) Constructive Theory of Functions of Several Variables, pp. 85–100. Springer, Heidelberg (1977)
5. Kroll, N., et al.: DLR project digital-X: towards virtual aircraft design and flight testing based on high-fidelity methods. CEAS Aeronaut. J. **7**(1), 3–27 (2015). https://doi.org/10.1007/s13272-015-0179-7
6. Rampurawala, A.M.: Aeroelastic analysis of aircraft with control surfaces using CFD. Ph.D. thesis, University of Glasgow (2005)
7. Reimer, L., Heinrich, R.: Modeling of movable control surfaces and atmospheric effects. In: Notes on Numerical Fluid Mechanics and Multidisciplinary Design: Computational Flight Testing, vol. 123, pp. 183–206. Springer, Heidelberg (2013)
8. Rew, R., Davis, G.: NetCDF: an interface for scientific data access. IEEE Comput. Graph. **10**(4), 76–82 (1990)
9. Schwamborn, D., et al.: The DLR-TAU-code: recent applications in research and industry. In: ECCOMAS CFD 2006, p. 1109 (2006)

Aerodynamic Shape Design of a Powered Helicopter Cell Using an Adjoint RANS Method with Actuator Disk Modelling

Jan-Hendrik Wendisch[1(✉)], Marc Wentrup[1], and Arthur Stück[2]

[1] Institut für Aerodynamik und Strömungstechnik, Lilienthalplatz 7, 38108 Braunschweig, Germany
{jan-hendrik.wendisch,marc.wentrup}@dlr.de
[2] Institut für Softwaremethoden zur Produkt-Virtualisierung, Zwickauer Str. 46, 01069 Dresden, Germany
arthur.stueck@dlr.de

Abstract. A shape optimization approach backed by adjoint RANS is suggested, which allows for helicopter cell-shape design in powered flight conditions and is realizable in context of industrial design. To this end the DLR TAU code was extended to support an actuator-disk rotor modelling for the discrete-adjoint gradient computation. The steady-state approach accounts for rotor effects in the shape optimization and reliefs from the burden associated with time-accurate adjoint computations for transient rotor analyses. The adjoint CFD method was included in a tool-chain for gradient-based helicopter cell-shape optimization in conjunction with a free-form parametrization of the cell shape and a radial-basis function approach that propagates surface deformations into the CFD volume mesh. The suggested approach was successfully demonstrated in a drag minimization study for the modified helicopter fuselage of the ROtor-Body-INteraction (ROBIN) configuration combined with a realistic load distribution for the 7AD main rotor. The shape optimization was conducted for forward flight conditions at full-scale using unstructured grids. The study confirmed the practical value of the suggested approach: accounting for the rotor downwash in the cell-shape design optimization turned out to be a critical factor for success; the cell shape obtained from an optimization in which rotor effects were neglected deteriorated the aerodynamic performance in powered flight conditions.

Keywords: Aerodynamic helicopter design · Shape optimization · Adjoint CFD · Sensitivity analysis · Actuator disk

1 Introduction

The aerodynamic performance of blunt helicopter afterbody configurations is often dominated by flow recirculation and vortex separation. In such conditions, the aerodynamic forces and moments on the helicopter cell are expected to be particularly sensitive to the influence of the main rotor downwash the cell is exposed to. In the context of industrial

© The Author(s), under exclusive license to Springer Nature Switzerland AG 2021
A. Dillmann et al. (Eds.): STAB/DGLR Symposium 2020, NNFM 151, pp. 687–696, 2021.
https://doi.org/10.1007/978-3-030-79561-0_65

design optimization, this immediately calls for viscous CFD evaluations accounting for rotor effects. Recently, different adjoint CFD methods have been presented in the literature for gradient-based helicopter design in conjunction with many design parameters [e.g. 1–3]. However, unsteady adjoint rotor analyses with resolved [3] or modeled blades [10] quickly become prohibitive in the framework of industrial design optimization. In spite of their limitations, steady-state Reynolds-averaged Navier-Stokes (RANS) analyses with actuator-disk rotor modelling are deemed a natural choice for adjoint-based performance optimization of helicopter cells: the approach captures steady-state rotor effects while it reliefs from the burden of tracing back the transient time trajectory in reverse in the adjoint sensitivity analysis.

The RANS method including the discrete-adjoint sensitivity analysis is described in Sect. 2 focusing on the actuator-disk rotor modelling. The adjoint RANS method is used for gradient-based shape optimization of a helicopter cell which is presented in Sect. 3 followed by the conclusions drawn in Sect. 4.

2 Navier-Stokes Method and Actuator Disk Modelling

The discrete-adjoint of the DLR RANS solver TAU has been extended to support the adjoint-based computation of sensitivity derivatives for shape optimization in conjunction with the actuator-disc modelling suggested by Raichle [4].

2.1 Finite-Volume Discretization

A standard second-order accurate, edge-based finite-volume discretization is used to approximate the compressible, steady-state RANS equations on unstructured, median-dual grids [9]. The convective fluxes were approximated by a central scheme, which is augmented by a matrix-valued numerical diffusion term based on blended first and third undivided differences, referred to as matrix dissipation. A central finite-volume discretization was used for the viscous fluxes. The latter were based on a Green-Gauss gradient approximation in combination with a compact gradient face-correction to yield the spatial gradients on the cell faces. We used the negative extension of the Spalart-Allmaras turbulence model in this study. The discretized RANS equations in residual form are denoted $\boldsymbol{R}(\boldsymbol{U}(\boldsymbol{X}), \boldsymbol{X}) = \boldsymbol{0}$, with $\boldsymbol{U}$ representing the vector of the conserved state variables and $\boldsymbol{X}$ being the position vector of the volume nodes.

2.2 Discrete-Adjoint Method

The discrete-adjoint approach was pursued to efficiently compute the total sensitivity derivative of the objective functional J for a large number of design variables in the design vector $\boldsymbol{D}$:

$$\frac{dJ}{d\boldsymbol{D}} = \underline{\frac{\partial J}{\partial \boldsymbol{X}}\frac{d\boldsymbol{X}}{d\boldsymbol{D}} + (\boldsymbol{U}^*)^T\left(\frac{\partial \boldsymbol{R}}{\partial \boldsymbol{X}}\frac{d\boldsymbol{X}}{d\boldsymbol{D}}\right)} \quad \text{with} \quad \left(\frac{\partial \boldsymbol{R}}{\partial \boldsymbol{U}}\right)^T \boldsymbol{U}^* = -\frac{dJ}{d\boldsymbol{U}} \tag{1}$$

The set of adjoint state equations on the right-hand side of (1) is solved for the adjoint of the state vector $\boldsymbol{U}^*$, which is then used to evaluate in the sensitivity-derivative equation on the left-hand side of (1) from a matrix-vector product. Since the adjoint state equations are independent of the mesh perturbation dX/dD, they need to be solved only once per objective functional for an arbitrary number of design variables. In the current implementation, the underlined terms in the sensitivity derivative evaluation are computed by means of central finite differences at the cost of two (infinitesimal) volume-mesh deformations per entry in the design vector $\boldsymbol{D}$ plus two TAU residual computations. In the operator of the adjoint state equations, the non-linear residual vector $\boldsymbol{R}$ is fully differentiated with respect to the state $\boldsymbol{U}$ including the complete Spalart-Allmaras turbulence model. Contributions to the adjoint state equations stemming from the non-compact second-order accurate finite-volume discretization – such as Green-Gauss gradients or undivided Laplace operators in the numerical dissipation operator – are reconstructed in a matrix-free multi-pass approach as described by Stück [5, 6]. To this end, the adjoint multipliers corresponding to the Green-Gauss gradients and undivided differences are computed in a first face loop before they are fed into the computation of the adjoint residual in a second face loop. By means of this on-the-fly evaluation of the adjoint residual, we avoid to explicitly store the complete Jacobian matrix $\partial \boldsymbol{R}/\partial \boldsymbol{U}$ and get around associated artificial data structures and potential memory and run-time overheads.

The discrete-adjoint TAU code was extended in this study to support the actuator-disk modelling as described in the following.

2.3 Actuator-Disk Extension for the Discrete-Adjoint Method

According to Raichle [4] the actuator disk is considered as an internal boundary at the upstream (index 1) and downstream side (index 2) of which two dual half-cells virtually merge, cf. Fig. 1 (left). For comparison, an interior dual cell surrounding node *i* is plotted on the right-hand side as a reference.

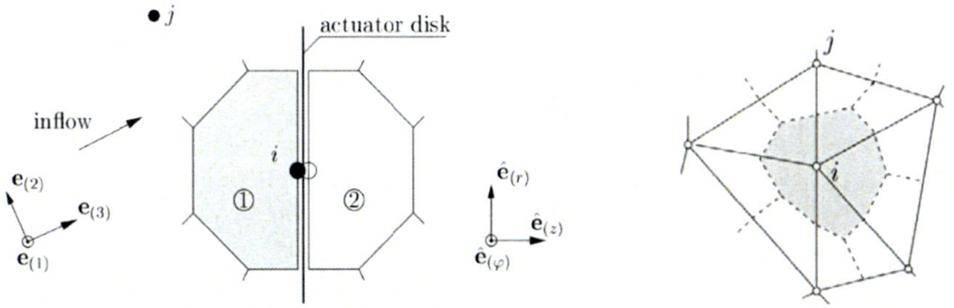

Fig. 1. Left: actuator disk considered as an internal boundary with up- and downstream dual half cells 1 and 2. The local Cartesian coordinate system of the actuator disk is defined by the unit base vectors $\hat{e}_{(r)}$, $\hat{e}_{(\varphi)}$ und $\hat{e}_{(z)}$ with the origin located on the actuator disk. The unit base vectors $e_{(1)}$, $e_{(2)}$ und $e_{(3)}$ specify the global Cartesian coordinate system which is used in the CFD method. Right: interior control volume around node *i* with neighbor node *j*

The finite-volume balance of the "merged" dual cell includes the numerical fluxes through its regular cell faces. The cell balance is completed by a momentum source contribution, $Q_i = f_i\, dA$, imposed on the intersection cell face dA on the actuator disk.

The source of momentum induces a contribution to the energy cell balance, $Q = v_i f_i \, dA$, wherein v_i denote the coordinates of the flow velocity vector. In order to devise the discrete-adjoint system, all flux terms being the central flux scheme augmented by the numerical diffusion, viscous fluxes and source terms were differentiated with respect to the conserved variables in the up- and downstream half-cells, $\boldsymbol{U}_1$ und $\boldsymbol{U}_2$, and those of the surrounding cells of the numerical second-order stencil. The actuation source density of momentum f_i was frozen in the differentiation. The up- and downstream half-cells 1 and 2 can be added to give one discrete residual balance per dual cell cut by the actuator disk; however, in line with the interior-boundary perspective, we formally introduce separate up- and downstream states, $\boldsymbol{U}_1$ und $\boldsymbol{U}_2$, and the corresponding adjoint variables $\boldsymbol{U}_1^*$ and $\boldsymbol{U}_2^*$. This requires for a dedicated closing condition. Here the non-linear jump relations for the actuator disk come into play, defining an algebraic, non-linear relationship between the up- and downstream half-cells states over the actuator disk, $\boldsymbol{U}_2\Big(\boldsymbol{V}_2\Big(\hat{\boldsymbol{V}}_2\Big(\hat{\boldsymbol{V}}_1(\boldsymbol{V}_1(\boldsymbol{U}_1))\Big)\Big)\Big)$. The primitive variables in the global coordinate system are represented by $\boldsymbol{V}^T = (\rho, v_1, v_2, v_3, p)$, their counterpart in the local coordinate system aligned with the actuator disk is denoted $\hat{\boldsymbol{V}}^T = \big(\rho,\ \hat{v}_r, \hat{v}_\varphi, \hat{v}_z,\ p\big)$. The local coordinate system of the actuator disk is specified by the unit base vectors $\hat{\boldsymbol{e}}_{(r)}, \hat{\boldsymbol{e}}_{(\varphi)}$ und $\hat{\boldsymbol{e}}_{(z)}$. A vector $\boldsymbol{r}$ expressed in the global Cartesian coordinate system $\boldsymbol{e}_{(1)}, \boldsymbol{e}_{(2)}$ und $\boldsymbol{e}_{(3)}$ and the local coordinate system of the actuator disk reads

$$\boldsymbol{r} = r_i \boldsymbol{e}_{(i)} = r_1 \boldsymbol{e}_{(1)} + r_2 \boldsymbol{e}_{(2)} + r_3 \boldsymbol{e}_{(3)} = \hat{r}_r \hat{\boldsymbol{e}}_{(r)} + \hat{r}_\varphi \hat{\boldsymbol{e}}_{(\varphi)} + \hat{r}_z \hat{\boldsymbol{e}}_{(z)} = \hat{r}_i \hat{\boldsymbol{e}}_{(i)} \,,$$

with the coordinate transformation $r_i = T_{ij} \hat{r}_j$. As T_{ij} is orthogonal, matrix inversions can be avoided in the adjoint implementation. The constitutive equations for the nonlinear jump conditions $\boldsymbol{H}\Big(\hat{\boldsymbol{V}}_1,\ \hat{\boldsymbol{V}}_2\Big) = 0$ expressed in the local coordinate system of the actuator disk are

$$\begin{aligned}
0 &= \hat{v}_{\varphi 2} - \left[\hat{v}_{\varphi 1} + \frac{\hat{f}_\varphi}{\rho_1 \left| \hat{v}_{z1} \right|} \right] \\
0 &= \hat{v}_{r2} - \hat{v}_{r1} \\
0 &= \rho_2 \hat{v}_{z2} - \rho_1 \hat{v}_{z1} \\
0 &= \rho_2 \hat{v}_{z2}^2 + p_2 - \left[\rho_1 \hat{v}_{z1}^2 + p_1 + \hat{f}_z \right] \\
0 &= \frac{\hat{v}_{z2}^2}{2} + h_2 - \left[\frac{\hat{v}_{z1}^2}{2} + h_1 + \frac{\hat{f}_z}{\rho_1} \right] .
\end{aligned} \tag{2}$$

The jump conditions (2) represent a balance for a degenerated, flattened control volume surrounding the internal cut, cf. Raichle [4]. Equations (2) relate the up- and the downstream state, $\hat{\boldsymbol{V}}_1$ and $\hat{\boldsymbol{V}}_2$ in the merged dual-grid cell.

In order to formulate the discrete-adjoint system for the actuator disk, the elements of the dependency chain $\boldsymbol{U}_2\Big(\boldsymbol{V}_2\Big(\hat{\boldsymbol{V}}_2\Big(\hat{\boldsymbol{V}}_1(\boldsymbol{V}_1(\boldsymbol{U}_1))\Big)\Big)\Big)$ are systematically differentiated in the following:

$$\delta \boldsymbol{U}_2 = \frac{\partial \boldsymbol{U}_2}{\partial \boldsymbol{V}_2}\,\frac{\partial \boldsymbol{V}_2}{\partial \hat{\boldsymbol{V}}_2}\,\frac{\partial \hat{\boldsymbol{V}}_2}{\partial \hat{\boldsymbol{V}}_1}\,\frac{\partial \hat{\boldsymbol{V}}_1}{\partial \boldsymbol{V}_1}\,\frac{\partial \boldsymbol{V}_1}{\partial \boldsymbol{U}_1}\,\delta \boldsymbol{U}_1\,. \tag{3}$$

Differentiation of (2) with respect to the state $\hat{\boldsymbol{V}}^T = (\rho,\ \hat{v}_r, \hat{v}_\varphi, \hat{v}_z,\ p)$ in the up- and downstream cell implicitly defines the expression $\partial\hat{\boldsymbol{V}}_2/\partial\hat{\boldsymbol{V}}_1$, viz.

$$\frac{\partial \boldsymbol{H}}{\partial \hat{\boldsymbol{V}}_2}\delta\hat{\boldsymbol{V}}_2 = -\frac{\partial \boldsymbol{H}}{\partial \hat{\boldsymbol{V}}_1}\delta\hat{\boldsymbol{V}}_1\ . \tag{4}$$

The remaining operations in (3) are the linearized expression between the conserved and the primitive variables $(\partial\boldsymbol{U}/\partial\boldsymbol{V})$, its inverse $(\partial\boldsymbol{V}/\partial\boldsymbol{U})$ and the coordinate transformation $\partial v_i/\partial\hat{v}_j = T_{ij}$. By means of the linearized closure (3), the up- and downstream variations $\delta\boldsymbol{U}_1$ and $\delta\boldsymbol{U}_2$ are linked in the primal case or the corresponding adjoint multipliers $\boldsymbol{U}_1^*$ and $\boldsymbol{U}_2^*$ are related in the adjoint case, either evaluated from the right or from the left:

primal (evaluated from the right, ←)

$$(\boldsymbol{U}_2^*)^T\left(\frac{\partial \boldsymbol{U}_2}{\partial \boldsymbol{V}_2}\,\frac{\partial \boldsymbol{V}_2}{\partial \hat{\boldsymbol{V}}_2}\left(\frac{\partial \boldsymbol{H}}{\partial \hat{\boldsymbol{V}}_2}\right)^{-1}\left(-\frac{\partial \boldsymbol{H}}{\partial \hat{\boldsymbol{V}}_1}\right)\frac{\partial \hat{\boldsymbol{V}}_1}{\partial \boldsymbol{V}_1}\,\frac{\partial \boldsymbol{V}_1}{\partial \boldsymbol{U}_1}\right)\delta\boldsymbol{U}_1\,. \tag{5}$$

adjoint (evaluated from the left, →)

This immediately yields the following forward and reverse jump closure sequences of operations in the primal and the adjoint case:

primal (↑, bottom to top) / adjoint (↓, top to bottom)

$$\begin{aligned}
\delta\boldsymbol{U}_2 &= \left(\tfrac{\partial \boldsymbol{U}_2}{\partial \boldsymbol{V}_2}\right)\delta\boldsymbol{V}_2 & \boldsymbol{V}_2^* &= \left(\tfrac{\partial \boldsymbol{U}_2}{\partial \boldsymbol{V}_2}\right)^T\boldsymbol{U}_2^*\\
\delta\boldsymbol{V}_2 &= \left(\tfrac{\partial \boldsymbol{V}_2}{\partial \hat{\boldsymbol{V}}_2}\right)\delta\hat{\boldsymbol{V}}_2 & \hat{\boldsymbol{V}}_2^* &= \left(\tfrac{\partial \boldsymbol{V}_2}{\partial \hat{\boldsymbol{V}}_2}\right)^T\boldsymbol{V}_2^*\\
\left(\tfrac{\partial \boldsymbol{H}}{\partial \hat{\boldsymbol{V}}_2}\right)\delta\hat{\boldsymbol{V}}_2 &= \delta\boldsymbol{H} & \left(\tfrac{\partial \boldsymbol{H}}{\partial \hat{\boldsymbol{V}}_2}\right)^T\boldsymbol{H}^* &= \hat{\boldsymbol{V}}_2^*\\
\delta\boldsymbol{H} &= \left(-\tfrac{\partial \boldsymbol{H}}{\partial \hat{\boldsymbol{V}}_1}\right)\delta\hat{\boldsymbol{V}}_1 & \hat{\boldsymbol{V}}_1^* &= \left(-\tfrac{\partial \boldsymbol{H}}{\partial \hat{\boldsymbol{V}}_1}\right)^T\boldsymbol{H}^*\\
\delta\hat{\boldsymbol{V}}_1 &= \left(\tfrac{\partial \hat{\boldsymbol{V}}_1}{\partial \boldsymbol{V}_1}\right)\delta\boldsymbol{V}_1 & \boldsymbol{V}_1^* &= \left(\tfrac{\partial \hat{\boldsymbol{V}}_1}{\partial \boldsymbol{V}_1}\right)^T\hat{\boldsymbol{V}}_1^*\\
\delta\boldsymbol{V}_1 &= \left(\tfrac{\partial \boldsymbol{V}_1}{\partial \boldsymbol{U}_1}\right)\delta\boldsymbol{U}_1 & \boldsymbol{U}_1^* &= \left(\tfrac{\partial \boldsymbol{V}_1}{\partial \boldsymbol{U}_1}\right)^T\boldsymbol{V}_1^*\,.
\end{aligned} \tag{6}$$

To support the adjoint-based actuator-disk modelling for shape design optimization, the individual primal and adjoint operators (6) of the linearized closure chain (3) were implemented in the linearized DLR TAU code to consistently compute the linear half-cell residuals in forward (propagation from upstream half-cell 1 to downstream half-cell 2) and reverse mode (from 2 to 1). In this way, forward-reverse duality is conserved on the level of instructions for the actuator disk. The evaluation of the sensitivity derivative based on finite-differencing of the residual, cf. left-hand side of (1), is automatically supported by including the actuator-disk modelling in the non-linear residual.

2.4 Solution Algorithms

The non-linear steady-state RANS problem was solved by means of the implicit Euler method, in which a lower-upper symmetric Gauss-Seidel (LUSGS) sweep was performed per numerical time step. The iteration matrix of the backward-Euler LUSGS method approximates the eigenvalues of the face-flux Jacobian by its spectral radius which reduces the computational effort in the preconditioner.

The corresponding discrete-adjoint solution algorithm consistently used the reverse of the backward-Euler method, in which the transposed of the primal LUSGS operator was applied. In the discrete-adjoint solution process, the backward-Euler LUSGS operated as a left-preconditioner to a restarted GMRes method. A specified number of implicit Euler sweeps was performed per Krylov vector. The geometric multigrid method was not used in this study, since the agglomeration algorithm currently does not preserve the half-cell wise coupling over the actuator disk surfaces on coarser grid levels.

2.5 Verification Studies

To verify the discrete-adjoint implementation, the individual residuals for the cells located in the second-order numerical stencil surrounding the actuator-disk were numerically differentiated and compared against the discrete differentiation implemented in forward and reverse (adjoint) mode. Details have to be omitted here for the sake of brevity. Second-order convergence was observed in conjunction with central finite differences. In addition, strict duality of the forward and reverse linear implementation was verified for all cells in the numerical stencil surrounding the actuator disk, $(\boldsymbol{U}^{*})^{T}(\partial \boldsymbol{R}/\partial \boldsymbol{U})\delta \boldsymbol{U} = (\delta \boldsymbol{U})^{T}(\partial \boldsymbol{R}/\partial \boldsymbol{U})^{T}\boldsymbol{U}^{*}$.

3 Drag Minimization for a Helicopter Cell in Powered Conditions

3.1 ROBIN-Mod7 Helicopter Configuration

The ROBIN [8] mod 7 configuration [7] was considered at full-scale at a Reynolds-Number of $Re = 4 \times 10^7$ and a Mach-Number of $Ma = 0.206$ based on the helicopter length and free-stream inflow conditions in forward flight. The configuration consists of a blunt fuselage with a length of *8.4* m and a main rotor radius of *4.2* m.

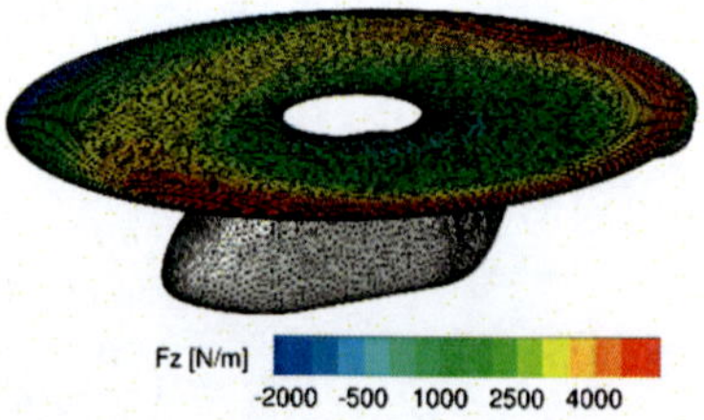

Fig. 2. Computational mesh with sectional load distribution on actuator disk

The unstructured mixed-element mesh consists of 2.41 million nodes with a first wall spacing giving a y + between 0.7 and 1.01 for roughly 90% of the fuselage surface. The mesh is considered a fair tradeoff between accuracy and computational costs in the context of shape optimization. All of the optimization runs were carried out based on the mesh which is depicted in Fig. 2 along with the load distribution of the rotor disk. The load distribution for the main rotor was computed by the comprehensive rotor code HOST for a 7AD main rotor in cruise flight conditions and scaled to the ROBIN rotor dimensions.

3.2 Optimization Workflow and Components

The helicopter optimization tool chain used here is illustrated in Fig. 3, cf. [1, 2]. The toolchain consists of two main sections highlighted in grey (left) and light blue (right), respectively. A gradient-based optimizer (conjugate gradient method) and the parametrization of the cell-shape based on the free-form deformation method are found in the grey section. The python environment was locally executed on a workstation. The blue section, which includes the CFD objective evaluation, the adjoint-based gradient computation and the volume mesh deformation according to the radial basis function (RBF) method, was carried out in a parallel workflow on the HPC cluster. Objective functional values and gradients were exchanged between the blue and the grey box. In the given workflow, 30 free-form parameters were used to describe the symmetric cell shape in the drag minimization. The freeform deformation tool acted on the positions of the 1481 surface control points of the RBF which, in turn, governed the movements of the surface and the volume nodes of the CFD grid. To evaluate the sensitivity derivatives of the objective functional, cf. left-hand side of Eq. (1), the RBF method was consistently used to deform the volume mesh in the process of finite differencing of the residual.

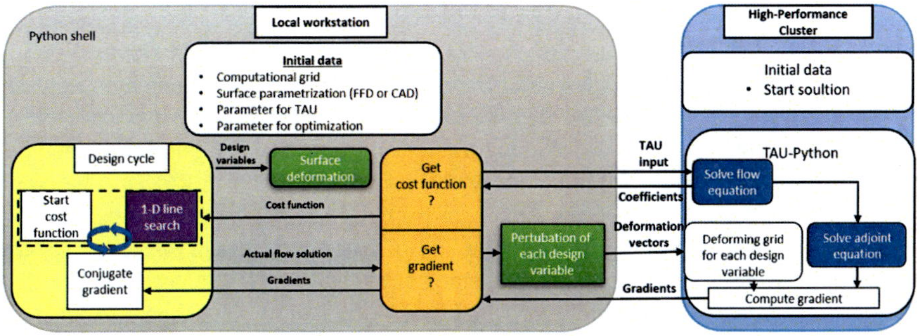

Fig. 3. Schematic of the coupling tool chain [1] that involves the discrete-adjoint TAU method

3.3 Optimization Study

Different optimization runs were performed with the proposed workflow. In the first optimization, the actuator disk influence was deactivated by setting the distribution of sectional loads in the actuator disk to zero.

To evaluate the objective functional, the RANS equations were solved using the implicit-Euler iteration in conjunction with the LUSGS method. The convergence of the (non-linear) RANS solution algorithm is plotted in Fig. 4 for the baseline shape with a powered actuator disk (ACD) and without the actuator disk (no ACD). Additionally, the convergence of the RANS solution of the first optimization cycle are shown. The gradient-based optimization process is driven by the discrete-adjoint sensitivity analysis. Five cycles of the restarted GMRes method were carried out here. 80 Krylov vectors were computed in each GMRes cycle to span the Krylov space. In the preconditioner, 500 backward-Euler iterations were performed, each of which containing a forward-backward sweep of the LU-SGS method. The convergence history for the nested iteration scheme for the discrete-adjoint problem defined on the baseline shape with a powered actuator disk is depicted in Fig. 5. The convergence is plotted over the number of restarted GMRes cycles. The optimization history was not strictly monotonous which may be attributed to the line-search algorithm. The drag minimization was finalized when the objective functional leveled out. Further investigations regarding the optimization method including the line-search approach could not be realized in the present work.

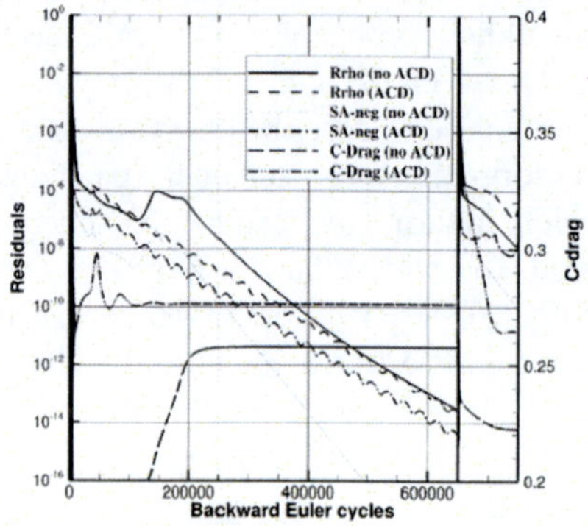

Fig. 4. Convergence of the initial RANS evaluation and of the first optimization cycle

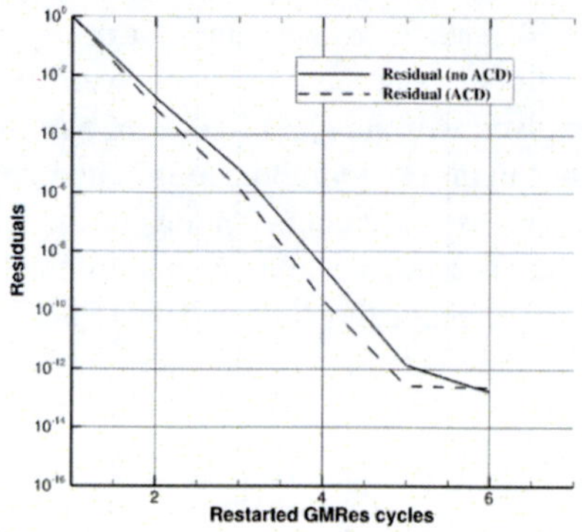

Fig. 5. Convergence of adjoint mass residual plotted over the restarted GMRes cycles

When the downwash effect of the main rotor on the fuselage is missing in inactive rotor conditions, the shape optimization yields a drag reduction of 24.4%. The comparison of the pressure distributions and skin-friction lines between the initial and the optimized fuselage shape is depicted in the Fig. 6. Without the effect of the main rotor downwash, a pressure recovery at the beginning of the afterbody was observed. This led to an earlier onset of flow separation, which was mitigated by the shape optimization and effectively reduced the drag. The fully converged RANS solution obtained on the relatively coarse non-symmetric grid turned out to be non-symmetric for the blunt body configuration.

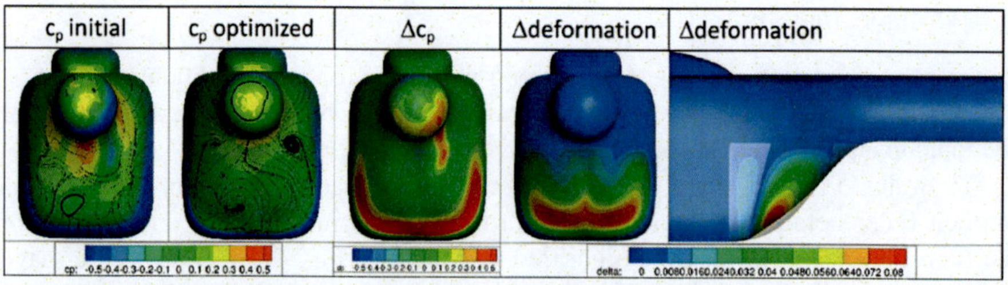

Fig. 6. Comparison of pressure distributions and skin friction lines between initial (transparent) and optimized shape; deformation (inboard = positive) with inactive main rotor

In the next step, the optimized shape obtained for inactive actuator disk condition was recomputed with an active actuator disk representing the powered main rotor. Figure 7 reveals that the area of separated flow broadens for the optimized shape, when the rotor is activated, resulting in a drag increase. It is interesting to note that the shape optimized for inactive conditions performs worse than the initial shape, when operating in powered conditions; compared to the initial shape, the drag of the cell shape optimized for inactive rotor conditions increases by 3.5% in powered conditions.

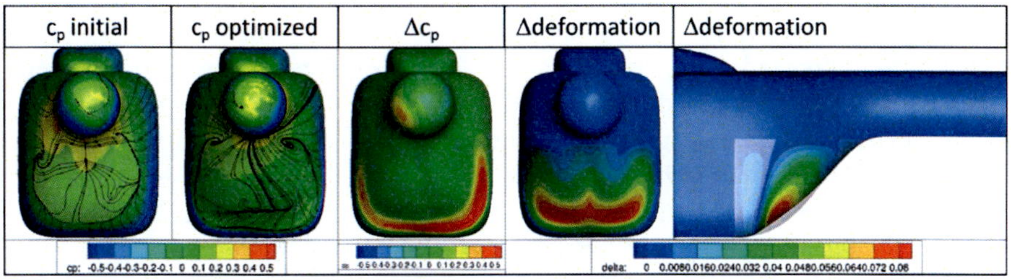

Fig. 7. Comparison of pressure distributions and skin friction lines between initial (transparent) and optimized shape; deformation (inboard = pos.) activating main rotor after optimization

Eventually, the shape optimization was repeated from the initial shape in powered conditions. By means of the presented actuator-disk extension to the discrete-adjoint method, the rotor influences in powered conditions could consistently be accounted for in the adjoint-based computation of the objective functional gradient. In contrast to the initial optimization study with an inactive rotor, a drag reduction of only 5.26% was achieved in conjunction with a powered actuator-disk model for the main rotor. The limited potential for optimization in powered conditions is attributed to the smaller separation zone observed for the initial shape as shown in Fig. 8. In that case, the optimization is more challenging due to the better performance of the initial design. A consistent gradient computation that includes the actuator disk model is deemed to be even more important in such conditions. Further drag reductions in powered conditions under the influence of the main rotor may have been prevented by the symmetry constraint on the fuselage, which was implicitly imposed by the FFD parametrization. A full FFD box that allows for non-symmetric deformations of the fuselage will be investigated in future studies to unblock such improvements.

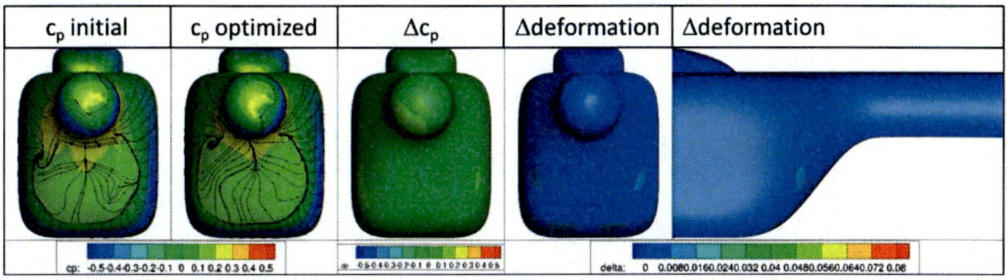

Fig. 8. Comparison of pressure distributions and skin friction lines between initial (transparent) and optimized shape; deformation (inboard = pos.) with main rotor influences.

4 Conclusions

The discrete-adjoint method of the DLR TAU code was extended to support the actuator disk modelling suggested by Raichle [4] for helicopter-cell shape optimization in powered flight conditions, in which the influence of the main rotor is considered. The feasibility and the usability of the suggested advanced design optimization approach backed by viscous adjoint CFD and actuator-disk modelling was successfully demonstrated for a simplified helicopter configuration: the drag of the powered ROBIN mod7 helicopter fuselage was minimized in viscous steady-state flow conditions. For the considered blunt-body test case, the adjoint method exhibited a remarkable solver convergence reducing the non-linear and adjoint residuals for at least 13 orders of magnitude. An optimization study was carried out to investigate the influence of the main-rotor downwash on the shape optimization. The study indicates that the main rotor influences significantly affect the aerodynamic performance of the helicopter cell. For the considered helicopter scenario, it was demonstrated that rotor effects have to be considered in the gradient-based shape optimization to obtain a drag reduction in powered flight conditions. This has confirmed the practical value of the suggested performance optimization approach, which allows to account for the actuator-disk modelling in the discrete-adjoint sensitivity analysis.

Acknowledgement. This research in the project CHARME was funded by the German Federal Ministry for Economic Affairs and Energy (BMWi) based on a resolution of the German Bundestag.

References

1. Wentrup, M.: An adjoint based gradient optimization chain for complex helicopter fuselage parts using a free form deformation Or CAD based parameterization method. In: 41th European Rotorcraft Forum, Munich (2015)
2. Zhang, Q, Wendisch, J.-H., Lee, J.-D.: An adjoint-based optimisation method for helicopter fuselage backdoor geometry. In: 36th European Rotorcraft Forum, Paris (2010)
3. Nielsen, E.J., Diskin, B.: Discrete adjoint-based design for unsteady turbulent flows on dynamic overset unstructured grids. AIAA J. **51**(6), 1355–1373 (2013)
4. Raichle, A.: Flusskonservative Diskretisierung des Wirkscheibenmodells als Unstetigkeitsfläche. PhD Dissertation, Fakultät für Maschinenbau der Technischen Universität Braunschweig, ISSN 1434–8454, ISRN DLR-FB-2017–62 (2017)
5. Stück, A.: Dual-consistency study for Green-Gauss gradient schemes in an unstructured Navier-Stokes method. J. Comput. Phys. **350**, 530–549 (2017)
6. Stück, A.: An adjoint view on flux consistency and strong wall boundary conditions to the Navier-Stokes equations. J. Comput. Phys. **301**, 247–264 (2015)
7. Schaeffler, N.W., et al.: Progress towards fuselage drag reduction via active flow control – a combined CFD and experimental effort. In: 36th European Rotorcraft Forum, Paris (2010)
8. Raymond, E.M., Gorton, S.A.: Steady and Periodic Pressure Measurements on a Generic Helicopter Fuselage Model in the Presence of a Rotor. NASA-TM 2000–210286 (2000)
9. Blazek, J.: Computational Fluid Dynamics, 3rd edn. Butterworth-Heinemann (2015)
10. O'Brien, D.: Analysis of Rotor-Fuselage Interactions using Various Rotor models, 43 Aerospace Science Meeting, Reno (2005)

Neural Network Modeling of Transonic Buffet on the NASA Common Research Model

Rebecca Zahn(✉), Tim Linke, and Christian Breitsamter

Chair of Aerodynamics and Fluid Mechanics, Technical University of Munich, Boltzmannstr. 15, 85748 Garching, Germany
rebecca.zahn@tum.de

Abstract. The application of reduced-order modeling (ROM) techniques in the context of aerodynamic nonlinear system identification of realistic aircraft configurations gained increasing attention in recent years. Therefore, in the present study the application of a recurrent neuro-fuzzy model (NFM) that is serial connected with a multilayer perceptron (MLP) neural network is introduced concerning the computation of transonic buffet aerodynamics. In particular, the intention of the ROM is the prediction of coefficient time-series trends in contrast to a precise resolution of detailed flow effects. Further, a reduction of computational time compared to a full-order reference Computational Fluid Dynamics (CFD) solution is pursued. The training of the ROM is accomplished based on a data set computed by means of unsteady Reynolds-averaged Navier-Stokes (URANS) simulations. The performance of the trained ROM is demonstrated by predicting the buffet flow characteristics of the NASA Common Research Model (CRM) investigated at transonic flow conditions. Therefore, the wing of the configuration is excited by an external pitching motion beyond buffet onset. By comparing the ROM result with a reference URANS solution, a precise prediction capability of the aerodynamic characteristics as well as a reduction in computational time is demonstrated.

Keywords: Buffet aerodynamics · Reduced-order modeling · Multilayer perceptron neural network · Neuro-fuzzy models · Nonlinear system identification

1 Introduction

Transonic wing buffet, also referred to as shock buffet, represents an unsteady aerodynamic phenomenon which is characterized by the appearance of self-sustained shock oscillations and boundary layer separation. These oscillations result in low-amplitude shock motions, typically represented by Strouhal numbers in the range of $Sr = 0.2-0.6$, depending on the wing geometry and flight parameters. As a result of the shock-induced and separated flow, the aircraft structure is susceptible to excitations which lead to structural degradation and reduced passenger comfort during flight [1].

© The Author(s), under exclusive license to Springer Nature Switzerland AG 2021
A. Dillmann et al. (Eds.): STAB/DGLR Symposium 2020, NNFM 151, pp. 697–706, 2021.
https://doi.org/10.1007/978-3-030-79561-0_66

While transonic buffet on airfoils has been in focus of several experimental and numerical studies in the recent past, the investigation of transonic wing buffet is still developing. Numerical studies performed by Iovnovich and Raveh [2] investigated the influence of geometry parameters on the buffet phenomenon of finite and infinite swept wings. Based on their study, an increase in pressure disturbance propagation in spanwise direction with increasing sweep angle was observed. In addition, the movement of local shock oscillations towards the wing tip as so called buffet-cells was shown, resulting in higher buffet frequencies compared to 2D buffet. Crouch et al. [3] investigated finite and infinite swept wings based on a global stability analysis, yielding a different instability mode compared to studies of two-dimensional or unswept wings. Based on experimental studies by Koike et al. [4] and Paladini et al. [5], broadband pressure fluctuations were found to be associated to transonic wing buffet. Further, the characteristic frequency ranges were found to be more broadband compared to 2D buffet cases.

In order to capture the relevant flow physics associated to 3D buffet, hybrid RANS/Large-Eddy simulation (LES) or unsteady Reynolds-averaged Navier-Stokes (URANS) simulations must be performed [2,6]. Therefore, the required computational effort is time and cost consuming, especially when considering industrial applications. In order to solve this issue, nonlinear system identification and reduced-order model (ROM) approaches are under development and have been applied to some extent towards the computation of transonic buffet of industry-relevant configurations.

Data based techniques such as proper orthogonal decomposition (POD) and dynamic mode decomposition (DMD) have been applied by Timme [1] and Ohmichi et al. [7] in order to identify instability mechanisms and dominant flow modes of 3D wing buffet. Further, Candon et al. [8] investigated deterministic and non-deterministic ROM approaches for dynamic bending and torsion load spectra prediction resulting from transonic buffet on a high-agility aircraft configuration. Another study performed by Candon et al. [9] compared three deep learning architectures, namely a standard recurrent neural network (RNN), a long short-term memory (LSTM) neural network and a bidirectional LSTM concerning buffet load prediction. Based on their investigations the bidirectional LSTM yields the best prediction capability regarding structural dynamic responses of a high-agility aircraft during buffet critical maneuvers.

However, none of the aforementioned studies investigated the application of ROM methods for the computation of integral quantities such as force and moment coefficients of a realistic aircraft configuration at buffet condition. Therefore, the following study presents the application of a ROM approach based on a serial connection of a recurrent neuro-fuzzy model (NFM) and a multilayer perceptron (MLP) neural network [10] for the prediction of aerodynamic coefficient time-series associated to transonic wing buffet. However, the resolution of detailed flow effects or local fluctuations is not the intention of the present study. For model calibration, a training data set computed by means of forced-motion URANS simulations is provided. As an application case, the NASA Common Research Model (CRM) configuration is investigated at transonic buffet condition, whereas the pitch degree of freedom of the wing structure is excited by an user-defined signal. For performance evaluation, the ROM results are compared to a full-order CFD reference solution.

2 Computational Setup

In the following section, a short introduction of the applied CFD solver is given. Further, the nonlinear system identification technique including the training and application procedure is introduced.

2.1 CFD Solver

The TAU flow solver developed by DLR [11] is used in the present study in order to compute the necessary training and reference data. TAU solves the URANS equations by means of a shock-capturing finite volume scheme in combination with unstructured and hybrid grids. The spatial discretization is accomplished by a central differencing scheme. In order to reduce numerical dissipation, a ratio of 0.4 between matrix and scalar dissipation in combination with a fourth order dissipation coefficient of 1/128 is chosen. A Kok skew symmetric scheme is applied for the discretization of the meanflow fluxes, whereas the discretization of the turbulence fluxes is accomplished by a first order Roe scheme. A Green Gauss scheme is applied for the reconstruction of the gradients. Time integration is achieved using an implicit backward Euler scheme, whereas a lower-upper symmetric Gauss-Seidel (LU-SGS) scheme is chosen as a linear solver. Further, a 3 v multigrid cycle is applied to the grid in order to accelerate convergence. A dual time stepping approach is applied for all simulations with a physical time step of $6.67 \cdot 10^{-6}$ s and 100 inner iterations per time step. Further, a CFL number of the implicit pseudo time steps of two is chosen. For turbulence modeling, the standard Spalart-Allmaras model with a quadratic constitutive relation (QCR) extension is implemented. The mesh deformation for the respective pitching motion of the CRM wing is accomplished by the application of a radial basis function (RBF) based grid deformation using a TAU-Python interface.

2.2 Nonlinear System Identification Approach

In general, a ROM is computed by processing a data set including known input and output quantities, which represent the characteristic system features. Therefore, full-order CFD simulations are performed in order to capture the flow condition of interest, while selected rigid body degrees of freedom (DoF) are excited using a training signal, covering the amplitude and frequency range of interest.

The combined NFM-MLP network, which is in more detail introduced in Refs. [10,12], is based on a time-discrete external filtering approach. Therefore, the response of a dynamic system is represented by a function F which incorporates (time-delayed) system inputs and outputs [12]. For the following application (see Sect. 3.1), the input of the NFM-MLP ROM is defined by the wing pitching angle θ, whereas the system output is denoted by the lift coefficient

(C_L) and pitching moment coefficient (C_{My}). Therefore, the nonlinear system identification approach can be defined as follows, considering k as the current discrete time step ($t_{k+1} = t_k + \Delta t$):

$$\begin{pmatrix} \tilde{C}_L(k) \\ \tilde{C}_{My}(k) \end{pmatrix} = F \left\{ \left[\begin{pmatrix} \theta_k \\ \theta_{k-1} \\ \vdots \\ \theta_{k-m} \end{pmatrix}, \begin{pmatrix} C_{L,k-1} \\ C_{L,k-2} \\ \vdots \\ C_{L,k-n} \end{pmatrix} \begin{pmatrix} C_{M_y,k-1} \\ C_{M_y,k-2} \\ \vdots \\ C_{M_y,k-n} \end{pmatrix} \right] \right\} \tag{1}$$

In Eq. 1, $m \in N$ and $n \in N$ refer to the maximum dynamic delay-orders for the respective input and output quantities. Based on the available training data, the combined NFM-MLP network is applied in order to determine the unknown functional relationship (defined as F) between the system inputs and outputs. Considering buffet aerodynamics, the main challenge is given by the representation of the stochastic responses resulting from the self-sustained flow field unsteadiness, which appears even without any external excitation. Therefore, the maximum delay orders (m,n) must be larger compared to system identification tasks of e.g. flutter or limit-cycle oscillations (LCOs) and should include at least a single buffet period in the input vector [13]. Based on this requirement, the delay orders are defined as $m = 250$ and $n = 250$ according to [10].

3 Application

3.1 Test Case: Common Research Model

In order to investigate the prediction capability of the ROM approach, the ROM is applied for the computation of the lift coefficient (C_L) and pitching moment (C_{M_y}) coefficient time series of the NASA CRM configuration. According to numerical investigations performed by Ehrle et al. [14], the buffet condition of the CRM configuration is characterized by a Mach number of $Ma_\infty = 0.85$, a Reynolds number of $Re = 30 \cdot 10^6$ and an angle of attack of $\alpha = 5°$. In order to represent the cryogenic wind tunnel conditions defined during experimental studies [15], nitrogen with Sutherland parameters is chosen as the working fluid.

Based on a grid convergence study conducted by Ehrle et al. [14], a computational grid including approximately $36 \cdot 10^6$ elements is applied for the following simulations. The mean aerodynamic chord (MAC) of the configuration is defined as 0.189 m and the boundaries of the farfield are extended by 50 times of the model length. Further, a sufficient boundary resolution with $y^+ < 1$ is accomplished on all surfaces. Due to aeroelastic deformations measured during the wind tunnel campaign [15], twist and upward bending of the wing are accounted for in the numerical simulations by means of a local mesh deformation.

3.2 Nonlinear System Identification

In order to capture the characteristics of the aerodynamic system, a sufficient training data set must be computed which represents all system features. For the following application, the training data is defined such that the overall amplitude and frequency range of interest is excited. Therefore, an amplitude-modulated pseudo-random binary signal (APRBS) is applied in the present study for a sufficient excitation of the pitch degree of freedom. In addition, the APRBS has been scaled to a minimum and maximum pitch amplitude of $\theta_{min,max} = \pm 1.0°$. In Fig. 1, the signal which is chosen for the present study, is visualized.

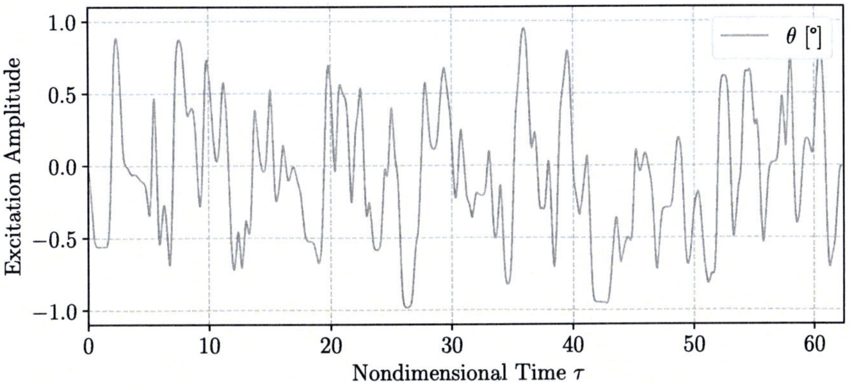

Fig. 1. Time series of the scaled training signal (APRBS) for the excitation of the pitching motion of the CRM wing structure.

For network training purposes, the available CFD data is divided into two data sets. Based on the division, 70% of the data are used for ROM training, whereas the remaining data is applied for ROM validation.

Prior to the final application of the ROM for unknown buffet test cases, the ROM is evaluated based on the training data itself by performing multi-step ahead predictions. Therefore, the ROM is trained based on a user-defined sequence of the training data and applied for the prediction of a certain amount of future training points. In Fig. 2 and Fig. 3, the lift and pitching moment coefficient time series due to the APRBS excitation are shown. Besides the reference CFD solution, the prediction trend of the ROM is visualized. For error quantification between the CFD and ROM results, a fit factor Q is computed [10]. A high prediction capability is indicated by a fit factor close to 100%. The resulting fit factors for the lift and pitching moment coefficient are computed as $Q(C_L) = 92.04\%$ and $Q(C_{My}) = 91.35\%$, respectively.

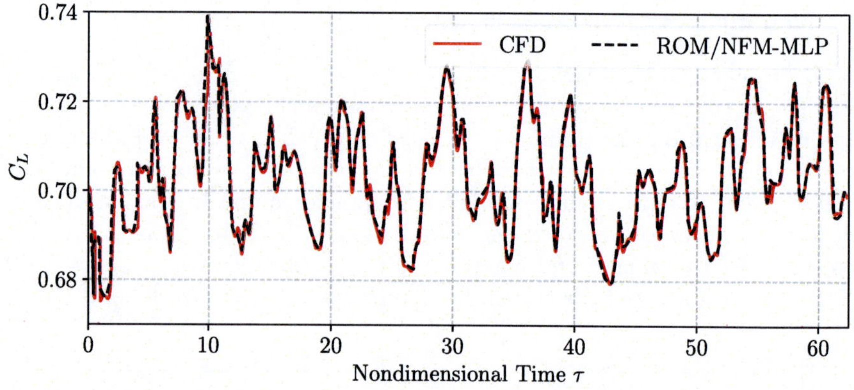

Fig. 2. Lift coefficient response caused by the scaled APRBS excitation (CRM, $Ma_\infty = 0.85$, $Re = 30 \cdot 10^6$, $\alpha = 5°$, $\theta = \pm 1.0°$). Besides the CFD reference solution, the simulation result of the ROM is shown.

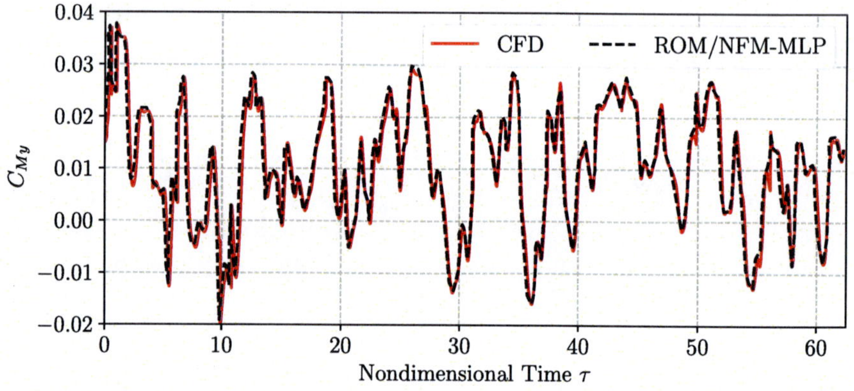

Fig. 3. Pitching moment coefficient response caused by the scaled APRBS excitation (CRM, $Ma_\infty = 0.85$, $Re = 30 \cdot 10^6$, $\alpha = 5°$, $\theta = \pm 1.0°$). Besides the CFD reference solution, the simulation result of the ROM is shown.

4 Validation and Discussion

For the final performance test, the APRBS-based trained ROM is applied to harmonic pitching oscillations with varying frequencies and amplitudes. The buffet condition of the CRM configuration investigated using URANS simulations is characterized by frequencies in the range of $Sr = 0.20 - 0.65$ [14]. Therefore, the applied oscillations are chosen in order to be included in the buffet frequency range, resulting in the following selected reduced frequencies: $k_{red,Ex} = \omega \cdot \text{MAC}/\text{U}_\infty = [1.57, 3.76]$. These frequencies correspond to Strouhal numbers of $Sr = 0.25$ and $Sr = 0.60$, respectively. Further, each reduced frequency is investigated representing two different excitation amplitudes, namely

$\theta_{min,max} = [\pm 0.5°, \pm 1.0°]$. In order to rate the prediction capability of the ROM, full-order reference CFD simulations have been performed for the respective frequencies and amplitudes. Therefore, approximately four excitation periods have been computed with both the ROM and the TAU solver.

In Fig. 4 and Fig. 5, the results of the lift and pitching moment coefficient as obtained by CFD and the ROM are visualized for both frequencies with an excitation amplitude of $\theta = \pm 1.0°$. As it can be seen, the ROM is clearly able to reproduce the time-series trends of both coefficients with sufficient accuracy, which is also emphasized by the corresponding fit factors (see Table 1). Further, Fig. 6 and Fig. 7, visualize the results for an excitation amplitude of $\theta = \pm 0.5°$ as calculated by CFD and the ROM. Analogously to the other excitation amplitude, a good agreement between the CFD and ROM results becomes apparent, which is also represented by the computed fit factors.

Table 1. Fit factors for the sine excitation with varying reduced frequencies and amplitudes

$k_{red,Ex}$	θ	$Q(C_L)$	$Q(C_{M_y})$
1.57	±0.5°	90.46%	89.78%
3.76		90.27%	90.62%
1.57	±1.0°	89.04%	90.78%
3.76		89.23%	90.94%

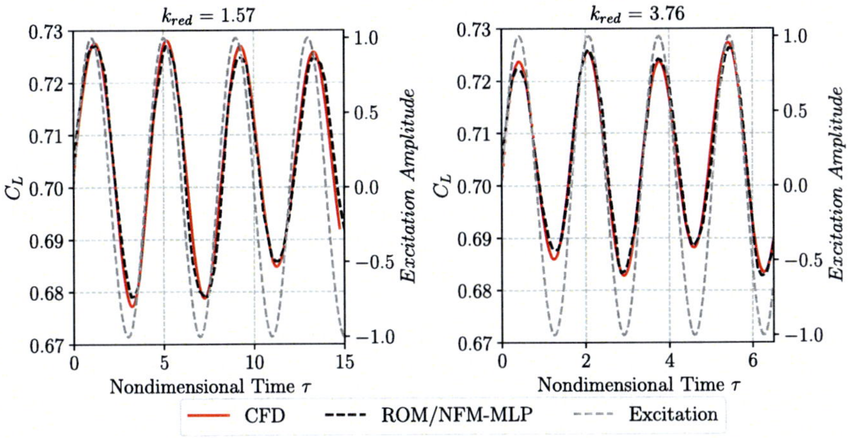

Fig. 4. Time-domain response of the lift coefficient induced by pitch oscillations at selected reduced frequencies and a pitch amplitude of $\theta = \pm 1.0°$ (CRM, $Ma_\infty = 0.85$, $Re = 30 \cdot 10^6$, $\alpha = 5°$). Besides the CFD reference solution, the simulation result of the ROM is shown.

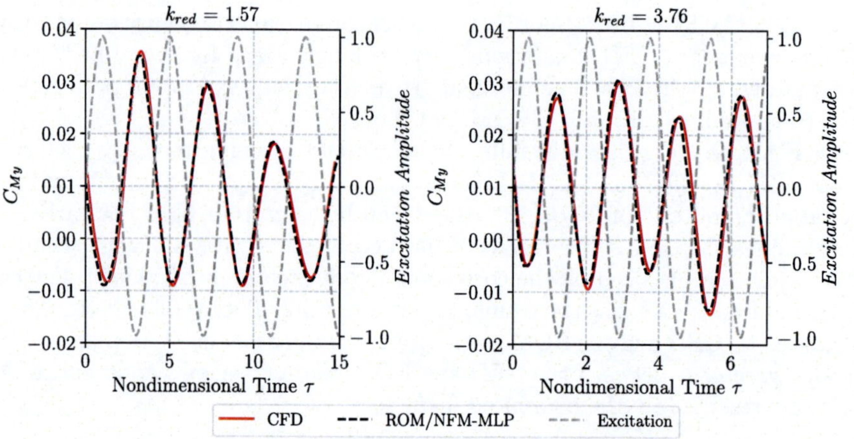

Fig. 5. Time-domain response of the pitching moment coefficient induced by pitch oscillations at selected reduced frequencies and a pitch amplitude of $\theta = \pm 1.0°$ (CRM, $Ma_\infty = 0.85$, $Re = 30 \cdot 10^6$, $\alpha = 5°$). Besides the CFD reference solution, the simulation result of the ROM is shown.

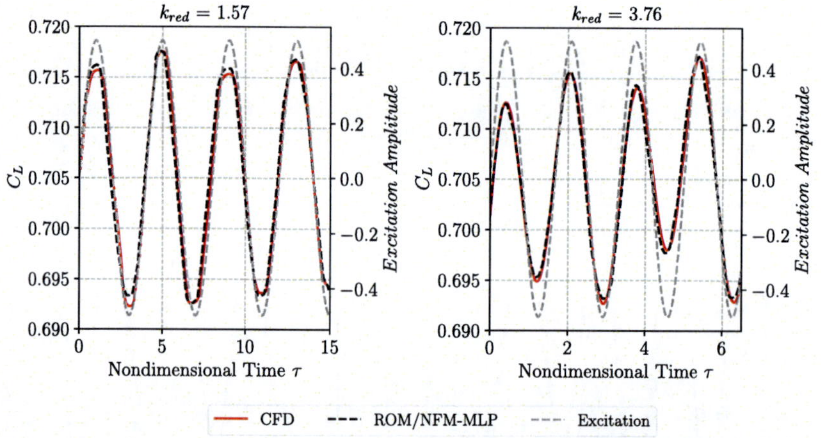

Fig. 6. Time-domain response of the lift coefficient induced by pitch oscillations at selected reduced frequencies and a pitch amplitude of $\theta = \pm 0.5°$ (CRM, $Ma_\infty = 0.85$, $Re = 30 \cdot 10^6$, $\alpha = 5°$). Besides the CFD reference solution, the simulation result of the ROM is shown.

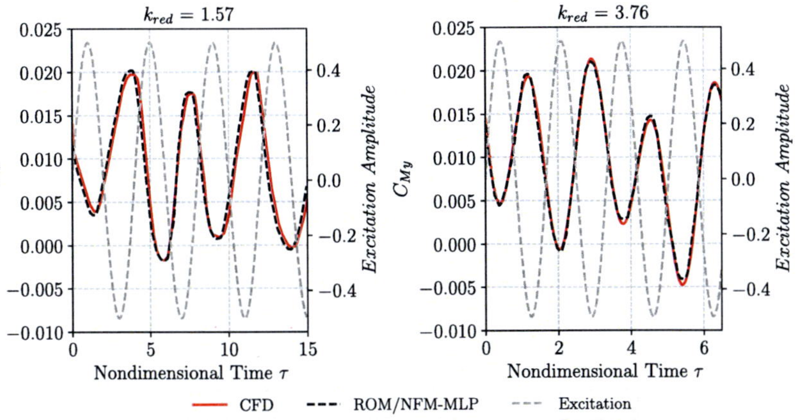

Fig. 7. Time-domain response of the pitching moment coefficient induced by pitch oscillations at selected reduced frequencies and a pitch amplitude of $\theta = \pm 0.5°$ (CRM, $Ma_\infty = 0.85$, $Re = 30 \cdot 10^6$, $\alpha = 5°$). Besides the CFD reference solution, the simulation result of the ROM is shown.

5 Conclusion

A reduced-order modeling (ROM) framework based on a serial connected recurrent neuro-fuzzy model (NFM) and multilayer perceptron (MLP) neural network was applied to compute transonic buffet aerodynamics of an realistic aircraft configuration. In particular, the aim of the study was a sufficient prediction of aerodynamic force and moment time-series trends associated to transonic wing buffet. For performance evaluation, the NASA Common Research Model (CRM) was investigated at buffet condition. A comparison of the ROM results and a reference computational fluid dynamics (CFD) solution yield a good prediction capability of the applied ROM approach. Further, a reduction in computational time by approximately two orders of magnitude was achieved. Future work will include the coupling of the ROM with a structural solver in order to simulate aeroelastic buffeting.

Acknowledgement. The authors gratefully acknowledge the Deutsche Forschungsgemeinschaft (DFG, German Research Foundation) for funding this work in the framework of the research unit FOR 2895 (Unsteady flow and interaction phenomena at high speed stall conditions), subproject TP7, grant number BR1511/14-1. Further, the authors thank the Gauss Centre for Supercomputing e.V. (www.gauss-centre.eu) for funding this project by providing computing licences and computing time on the GCS Supercomputer SuperMUC-NG at Leibniz Supercomputing Center (www.lrz.de). Also, the authors would like to thank the Institute of Aerodynamics and Gas dynamics (IAG) at the University of Stuttgart for providing the computational grid of the CRM configuration.

References

1. Timme, S.: Global instability of wing shock-buffet onset. J. Fluid Mech. **885** (2020)
2. Iovnovich, M., Raveh, D.E.: Numerical study of shock buffet on three-dimensional wings. AIAA J. **53**(2), 449–463 (2014)
3. Crouch, J.D., Garbaruk, A., Strelets, M.: Global instability analysis of unswept- and swept-wing transonic buffet onset. In: Fluid Dynamics Conference, AIAA-2018-3229, Atlanta, GA (2018)
4. Koike, S., Ueno, M., Nakakita, K., Hashimoto, A.: Unsteady pressure measurements of transonic buffet on the NASA common research model. In: 34th AIAA Aplied Aerodynamic Conference, AIAA-2016-4044, Washington, D.C. (2016)
5. Paladini, E., Dandois, J., Sipp, D., Robinet, J.C.: Analysis and comparison of transonic buffet phenomenon over several three-dimensional wings. AIAA J. **57**(1), 1–18 (2018)
6. Sator, F., Timme, S.: Delayed detached-eddy simulation of shock-buffet on half wing-body configuration. AIAA J. **55**(4), 1230–1240 (2016)
7. Ohmichi, Y., Ishida, T., Hashimoto, A.: Modal decomposition analysis of three-dimensional transonic buffet phenomenon on a swept wing. AIAA J. **56**(10), 3938–3950 (2018)
8. Candon, M., Levinski, O., Altaf, A., Carrese, R., Marzocca, P. : Aircraft transonic buffet load prediction using artificial neural networks. In: 59rd AIAA Structures, Structural Dynamics and Materials Conference, AIAA-2019-0763, San Diego, CA (2019)
9. Candon, M., et al.: On the application of a long short-term memory deep learning architecture for aircraft dynamic loads monitoring. In: 59rd AIAA Structures, Structural Dynamics and Materials Conference, AIAA-2020-0702, Orlando, FL (2020)
10. Winter, M., Breitsamter, C.: Nonlinear identification via connected neural networks for unsteady aerodynamic analysis. Aerosp. Sci. Technol. **77**, 802–818 (2018)
11. Schwamborn, D., Gerhold, T., Heinrich, R.: The DLR TAU-code, recent applications in research and industry. In: European Conference on Computational Fluid Dynamics ECCOMAS CFD 2006, Netherlands (2006)
12. Nelles, O.: Nonlinear System Identification - from Classical Approaches to Neural Networks and Fuzzy Models. Springer, Berlin & Heidelberg, Germany (2011)
13. Winter, M., Breitsamter, C.: Reduced-order modeling of transonic buffet aerodynamics. Notes Num. Fluid Mech. **142**, 511–520 (2020)
14. Ehrle, M., Waldmann, A., Lutz, T., Krämer, E.: Simulation of transonic buffet with an automated zone DES approach. CEAS J. **11**, 1025–1036 (2020)
15. Lutz, T., Gansel, P.P.: Going for experimental and numerical unsteady wake analyses combined with wall interference assessment by using the NASA CRM-model in ETW. In: 51st AIAA Aerospace Sciences Meeting, AIAA-2013-871, Grapevine, TX (2013)

Experimental Aerodynamics

Experimental Investigation of a Line-Cavity System Equipped with Fiber-Optic Differential Pressure Sensors in a Shock Tube

Florian M. Heckmeier(✉), Niklas Mooshofer, Thomas Hopfes, Christian Breitsamter, and Nikolaus A. Adams

Chair of Aerodynamics and Fluid Mechanics, Department of Mechanical Engineering, Technical University of Munich, Boltzmannstr. 15, 85748 Garching, Germany
florian.heckmeier@aer.mw.tum.de

Abstract. The aerodynamic behavior inside a line-cavity system is investigated within this work. Acoustic effects, like attenuation and resonance, are mainly dependent on the geometric line-cavity system properties: radius r, line length L and cavity volume V. In order to determine the transfer function from the system entry to the location of the pressure sensor at the cavity end, newly developed fiber-optic differential pressure sensors are used to acquire signals of high bandwidth. In contrast to approaches in the frequency domain, where e.g. a speaker emits signals of dedicated frequencies, in this work, the transfer function is calculated in the time domain. A step pressure change in a shock tube is produced and leads to the excitation of frequencies in a large bandwidth simultaneously. In addition to the fiber-optic pressure sensor at the end of the line-cavity system, a further fiber-optic sensor is flush mounted to the shock tube test section as a reference. By applying system-identification routines, the transfer function can be deduced. Experimental investigations of two line-cavity systems of various lengths show very good results. The signals of the reference pressure signals can be reproduced very accurately.

Keywords: Pneumatic line-cavity system · System identification · Acoustic transfer function · Fiber-optic differential pressure sensor · Shock tube · Unsteady pressure measurement

1 Introduction

The determination of the acoustic line-cavity system (LCS) is crucial when operating fast-response multi-hole probes, because the pressure signal is subject to attenuation and resonance inside the line-cavity system. In order to ensure high spatial resolution, in general, a small probe tip is preferred. Due to the large dimensions of the sensor compared to the probe tip, the sensors can not be flush mounted and a placement within the probe shaft is unavoidable. However, the

© The Author(s), under exclusive license to Springer Nature Switzerland AG 2021
A. Dillmann et al. (Eds.): STAB/DGLR Symposium 2020, NNFM 151, pp. 709–718, 2021.
https://doi.org/10.1007/978-3-030-79561-0_67

resulting transfer function between the pressure at the tip and the location of the sensor can be determined, for example with an analytic model introduced by Bergh and Tijdeman [1]. While their model provides the basic characteristics, it does not account for manufacturing imperfections that cause the actual system to differ from the analytic solution to a certain extent. Therefore, it is essential to determine the transfer function experimentally. This has recently been done in a frequency test rig [2] by means of a speaker emitting sinusoidal signals at varying frequencies. In this study, a different method to determine the transfer function is applied. The shock tube of the Chair of Aerodynamics and Fluid Mechanics of the Technical University of Munich (TUM-AER) [3] generates a shock wave that provides a step change in pressure. Fiber-optic pressure sensors are flush mounted to the wall of the test section for reference, and at the end of a given line-cavity system, respectively. Within a system identification step, the dicrete-time pressure signals acquired due to the shock pressure step are measured and set into relation as described by Paniagua and Dénos [4]. In contrast to the determination in the frequency test rig, which shows a bandwidth restriction due to the selected speaker, the shock allows to also identify high frequencies.

The work is structured into Sect. 2 describing the theoretical background, followed by Sect. 3 depicting the experimental setup and test plan. In the last part, in Sect. 4, dedicated results on the determination of the transfer function in a shock tube are shown.

2 Theoretical Background

In this section, an introduction on the theory of the pneumatic line-cavity system is given. Furthermore, the applied system identification approach is demonstrated. Finally, a brief description of the mounted newly developed fiber-optic differential pressure sensor is outlined.

2.1 Theory on the Line-Cavity System

The line-cavity system is mainly characterized by its length L, its diameter $d = 2r$, and the volume in front of the sensor V, as depicted in Fig. 1. An analytic, recursive solution for small disturbances and for narrow tubes by Bergh and Tijdeman [1] describes the line-cavity system. Here, laminar flow and a fluid governed by the ideal gas law is assumed. The complex ratio of the pressure signal at the sensor location in the cavity to the pressure signal at the reference

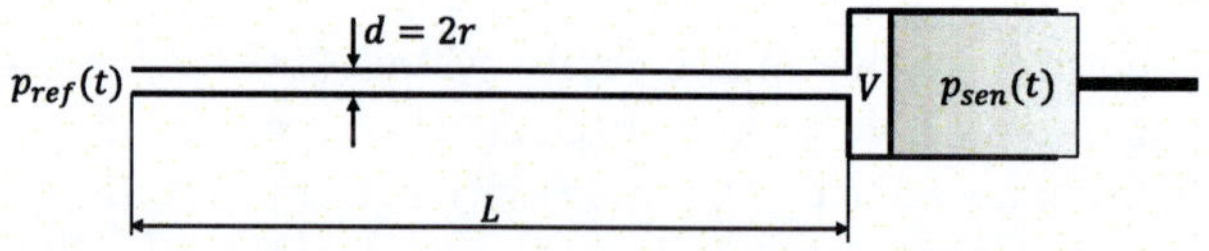

Fig. 1. Line-cavity system sketch

location P_{sen}/P_{ref} describes the attenuation and phase shift for the acoustic wave propagation inside a system of a single tube as a function of the angular frequency ω. It is denoted as transfer function (TF) of the system $H(\omega)$:

$$H(\omega) = \frac{P_{sen}(\omega)}{P_{ref}(\omega)} = \left[\cosh(\psi L) + \frac{V\eta\psi}{\kappa\pi r^2}\sinh(\psi L)\right]^{-1} \quad (1)$$

$$with \qquad \psi = \tfrac{\omega}{c}\sqrt{\tfrac{\kappa}{\eta}\tfrac{J_0(\gamma)}{J_2(\gamma)}} \quad (2)$$

$$\eta = \left[1 + \tfrac{\kappa-1}{\kappa}\left(\frac{J_2(\gamma\sqrt{Pr})}{J_0(\gamma\sqrt{Pr})}\right)\right]^{-1} \quad (3)$$

$$\gamma = i^{3/2} r \sqrt{\tfrac{\rho\omega}{\mu}} \quad (4)$$

Thereby, J_i is the Bessel function of ith order, κ the specific heat ratio and Pr, ρ and μ the Prandtl number, the density and dynamic viscosity of the fluid, respectively.

2.2 System Identification

The time-discrete signals can be Z-transformed (i.e. time-discrete equivalent of the Laplace transform) and the transfer function of the linear system can be expressed by two polynomials of order m and n [4]:

$$\frac{Y(z)}{U(z)} = \frac{b_0 \cdot z^{-d} + b_1 \cdot z^{-1-d} + \cdots + b_m \cdot z^{-m-d}}{1 + a_1 \cdot z^{-1} + a_2 \cdot z^{-2} + \cdots + a_n \cdot z^{-n}} \quad (5)$$

where $Y(z)$ and $U(z)$ are the Z-transformed of the time series data of the line-cavity and the reference sensor signals, respectively, and the exponent d depicts the delay between the input and output signal. After the identification of the unknown parameters a_i and b_i, the amplitude ratio and phase lag corresponding to the transfer function can be computed at any frequency f for a specific sampling rate f_s.

$$\left|\frac{Y(f)}{U(f)}\right| = \sqrt{\frac{B_1^2 + B_2^2}{A_1^2 + A_2^2}} \quad (6)$$

$$\varphi(f) = atan(B_2/B_1) - atan(A_2/A_1) \quad (7)$$

$$A_1 = 1 + \sum_{i=1}^{m} a_i \cdot \cos(2\pi f/f_s \cdot (-i)) \quad (8)$$

$$A_2 = \sum_{i=1}^{m} a_i \cdot \sin(2\pi f/f_s \cdot (-i)) \quad (9)$$

$$B_1 = \sum_{i=0}^{m} b_i \cdot \cos(2\pi f/f_s \cdot (-i-d)) \quad (10)$$

$$B_2 = \sum_{i=0}^{m} b_i \cdot \sin(2\pi f/f_s \cdot (-i-d)) \quad (11)$$

Hence, for following measurements, the transfer function can be applied to correct the pressures attenuated in the line-cavity system, as described in [2]. In this work, the MATLAB built-in system identification functions are used [5]. Before applying the signal identification functions, the acquired signals are divided into a training and validation set. The delay d between the input and output signal is estimated with the function *delayest*. By using the function *tfest*, the free parameters a_i and b_i of the transfer function identification can be calculated depending on the chosen number of poles and zeros. In order to achieve a good fit for the training signal, various combinations of the number of poles and zeros are tested. In addition to the estimated fit of the *tfest*-output, the estimated transfer function is applied on the validation data set. The combination of poles and zeros with the best fit/correlation to the reference pressure signal is chosen and the transfer function parameters are stored.

2.3 Fiber-Optic Pressure Sensor

This section only provides a brief overview of the working principle of the fiber-optic pressure sensors. Further information can be taken from the literature [6, 7]. The sensor measurement is based on the Fabry-Pérot interferometer effect in combination with an edge-filter interrogator. A white light source emits light that travels through the fiber. In the sensor cavity, the light spectrum is modulated depending on the distance between the fiber end and the membrane. When pressure is applied, the deflection of the membrane leads to a changed superposition of optical waves differing from the initial operation condition without any applied pressure. The phase condition of the interfered wavelength and, hence, the reflected modulated light is captured and processed by an edge-filter interrogator [6]. In Fig. 2, a sketch of the 2nd generation fiber-optic pressure sensor is shown. Due to manufacturing benefits, an octagonal shape is chosen. In order to enhance the measurement resolution, the design also includes a reference pressure line to operate the sensor in gauge/differential mode. Furthermore, a buffer is applied around the fiber and the pressure line for improved mechanical stability. The sensor is manufactured out of two fused silica wafers in a selective-laser etching (SLE) process [8]. In the static calibration, the sensor is pressurized at multiple pressure levels. As a result, the correlation between the optic output

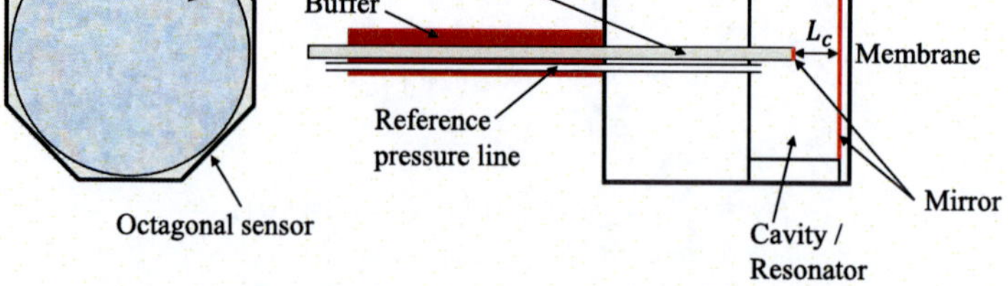

Fig. 2. Sketch of the fiber-optic differential pressure sensor (FOPS)

parameter ρ_{opt} and the applied pressure p is determined. In theory, around the operation point and in the pressure range of the sensor, the output should be linear:

$$p = (\rho_{opt} - \rho_0) \cdot k_{opt}^{-1} \tag{12}$$

Here, ρ_0 is the zero output at 0 Pa gauge pressure, and k_{opt} is the slope fit of the linear regression.

As an example, in Fig. 3, the calibrated points and the linear fit of one of the applied fiber-optic sensors in this experiment (FOPS-19E3) is shown. The linear fit gives a calibration coefficient of $k_{opt,19E3} = -3.4529 \cdot 10^{-6}$ Pa^{-1}. Preliminary shock-tube tests showed that the sensors withstand pressure levels to be expected inside fast-response probes. Using state-of-the-art piezo-electric sensors as reference sensors, gauge pressure levels above 20 kPa could be acquired and a linear behavior up to this level could be confirmed for the fiber-optic sensors.

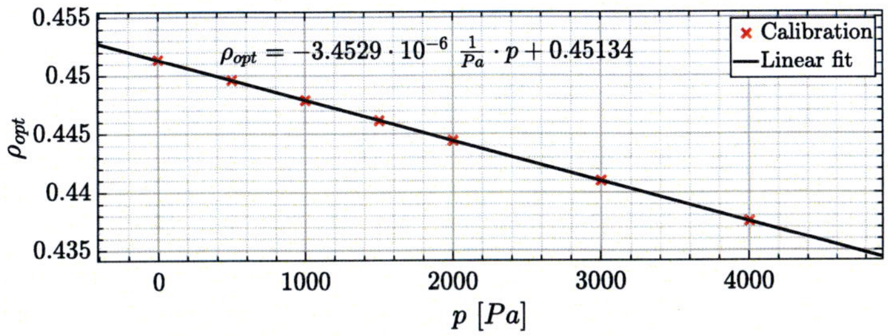

Fig. 3. Static calibration of the differential pressure sensor FOPS-19E3

3 Experimental Setup: Shock Tube

Fig. 4 depicts the layout of the shock tube and connected systems for the current experiments. The tube has an overall length of 24 m, an inner diameter of 290 mm, and is segmented into a driver, a driven, and a test section. A cookie-cutter is installed upstream of the test section to remove boundary flows and to enable the transition from the tube to the 190 mm × 190 mm square test section. Static conditions are monitored with K-type thermocouples for the temperature and with Setra pressure transducer models ASM and 730 for the absolute pressure. Before experiments, a 0.2 mm-thick PVC diaphragm is clamped between the driver and driven sections with a pair of crossed 0.1 mm thick NiCr heating wires in direct contact. While the driven section is kept at atmospheric conditions, the driver section is filled with air to a level of 1.15 bar. When the recording is set up, an electric current is supplied to the heating wires. The locally melted diaphragm ruptures, and a planar shock wave forms, propagates towards the downstream test section, and induces a uniform flow along with a step change in pressure, temperature, and density. Pressure and flow conditions

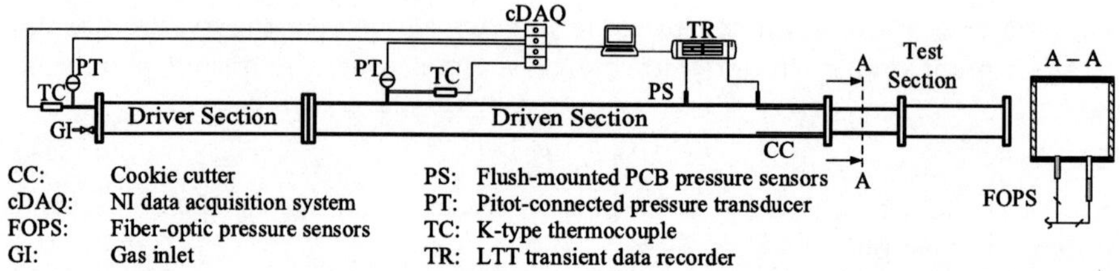

Fig. 4. Schematic layout of the shock tube facility at TUM-AER

remain steady for approximately 2 ms before reflected waves alter the conditions. During experiments, the pressure variation are measured with PCB Piezotronics ICP fast-response pressure sensors at a sampling frequency of 1 MHz. Two sensors are flush mounted along the tube and allow to calculate the shock velocity shortly upstream of the test section based on the time lag between the signals. Along with pre-shock conditions, this allows to calculate post-shock flow properties based on moving shock relations.

Figure 5a depicts one of the fiber-optic differential pressure sensors that are used to determine the transfer function. The octagonal shape mentioned earlier is clearly visible. The sensor housing for the fiber-optic sensors mimics the shape of the regular PCB sensors of the shock-tube facility and the sensor is sealed and fixed into the housing with modeling clay. While the housing is rather big, it enables a fast installation of the fiber-optic sensor at desired locations. For the experiment, two sensors are mounted in parallel in the test section as indicated by the cutting plane A – A in Fig. 4. One of them is flush-mounted and serves as a reference, while the other is connected to the LCS under investigation (see Fig. 5c). The LCS consists of a tube with a diameter of $d = 2r = 0.9$ mm and a length of $L = \{50, 100\}$ mm which is fixed in a housing similar to that of the sensors. At the end of the tube, a brass connector with the same outer diameter as the sensor housing is attached to the LCS (see Fig. 5c). The brass connector is used together with a silicone tube to connect the LCS to the sensor in order to achieve a negligible cavity volume V and enable a fast replacement of various sensors and LCSs. In total, six experiments are conducted for the present study, varying the length of the LCSs L as summarized in Table 1. Here, beside the variation in length, the shock Mach number M_{shock} and the pressures of the driver and the driven section, p_{driver} and p_{driven}, are shown, respectively.

Table 1. Test plan for all conducted measurements in the shock tube

L	M_{shock}	p_{driver}	p_{driven}	# experiments
50 mm	1.0291–1.0389	1.1–1.15 bar	96.5 kPa	3
100 mm	1.0374–1.0399	1.15 bar	96.4 kPa	3

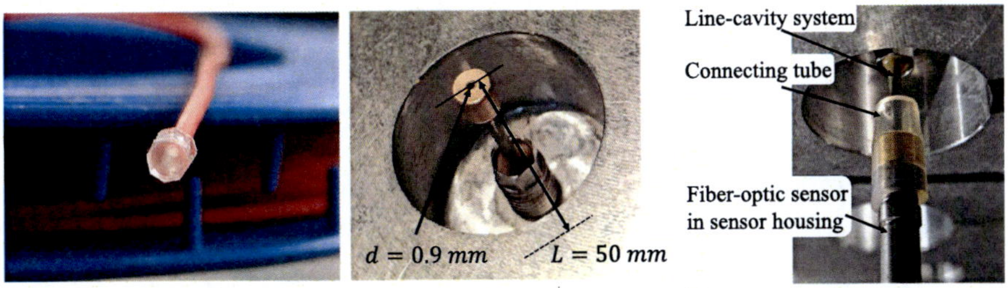

(a) Fiber-optic pressure sensor (b) Line-cavity system (c) Line-cavity system with attached sensor

Fig. 5. Line-cavity system and sensor assembly

4 Results and Discussion

In this section, the results of the shock-tube tests for the determination of the transfer function of the LCS are shown. In Fig. 6a, the acquired pressure signals of the flush-mounted reference sensor p_{ref} and the sensor signal at the end of the LCS p_{sen} are shown for test case #2 (see Table 2). All signals are sampled with a sampling rate of $f_s = 50$ kHz, which is currently the maximum achievable sampling rate of the DAQ system for the fiber-optic pressure sensors. The data is separated into three parts: The first pressure rise, which corresponds to the initial shock (see Fig. 6b), is used for the system identification training as described in Sect. 2.2. Furthermore, in order to identify the best number of poles and zeros for the system identification, a validation data set is chosen independently. For the generalization, the entire signal is reconstructed. With the delay estimation function *delayest*, the delay *d* is estimated to 40 samples. Fitting the free transfer function parameters and cross-checking it with the validation data, the optimal number of poles and zeros is identified for this LCS with $L = 50$ mm: with 19 poles and nine zeros, a fit to estimate the training input data is 97.51% and the correlation of the reconstructed validation data is above 99%. As a result, the amplitude ratio of the discrete-time identified transfer function for the LCS under investigation is shown in Fig. 6c, along with the analytic solution by Bergh and Tijdeman [1]. In contrast to the analytic solution, which shows resonance at very similar frequencies, the attenuation is more pronounced in the actual acquired and post-processed data. In Fig. 7, the transfer function is applied to reconstruct the signal, which is the final generalization step: Apart from the shock moment with not negligible overshoots before the shock (see Fig. 7a), the reconstructed pressure curve and the measured reference signal show a very high agreement, as depicted in Fig. 7b. Furthermore, the transfer function that was identified with test case #2 can be applied on other cases. Figure 8 depicts the reconstructed signals of case #1 and #3 at different signal locations. Again, a very high agreement between the reconstructed signal and the reference signal is clearly visible. In order to quantify the quality of the transfer function determination and to show the reconstruction accuracy, Table 2 depicts

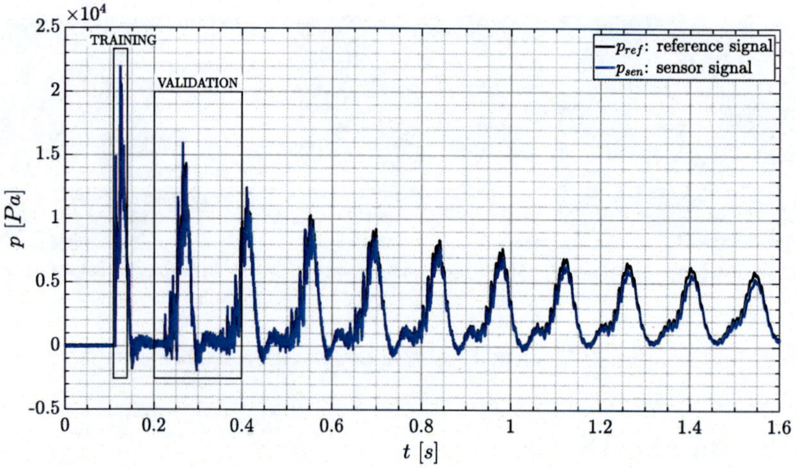

(a) Division of the input/output data: training and validation data sets indicated by the boxes

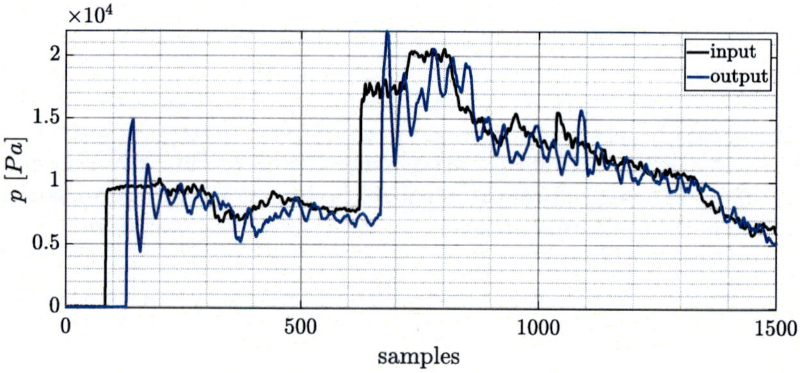

(b) Training input and output data

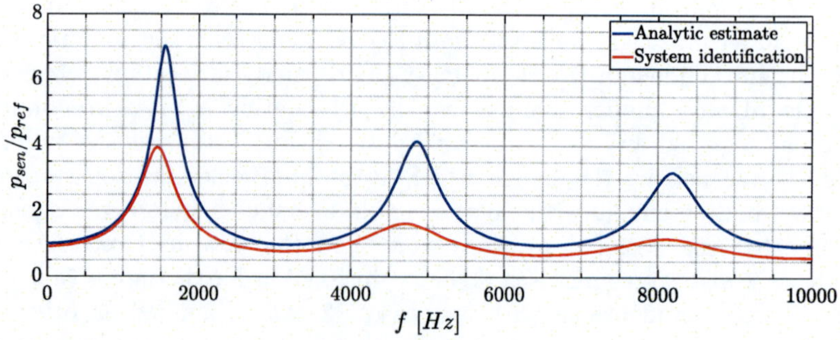

(c) Amplitude ratio of TF

Fig. 6. Post-processing for the transfer function determination of the signal acquired for test case #2

Table 2. Results overview

#	L	M_{shock}	p_{driver}	Usage	Correlation [%]
1	50 mm	1.0291	1.10 bar	Generalization	99.90
2	50 mm	1.0389	1.15 bar	Training + validation	99.90
3	50 mm	1.0370	1.15 bar	Generalization	99.93
4	100 mm	1.0399	1.15 bar	Generalization	98.52
5	100 mm	1.0384	1.15 bar	Generalization	99.28
6	100 mm	1.0374	1.15 bar	Training + validation	99.97

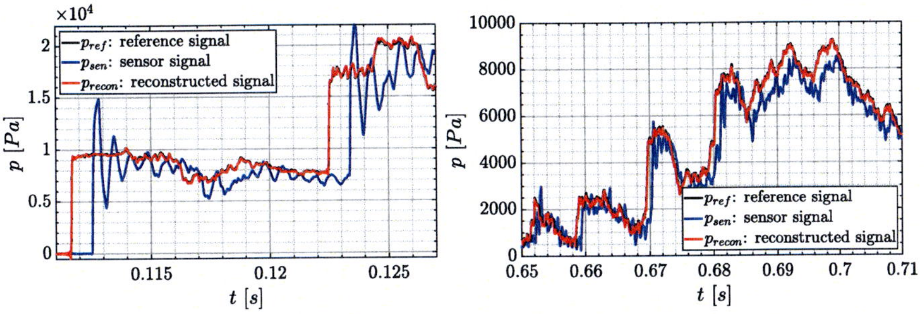

(a) Test case #2: Initial pressure step (b) Test case #2: Generalization

Fig. 7. Reconstruction of the pressure signal acquired in the shock-tube for case #2 with coinciding reference and reconstructed signals

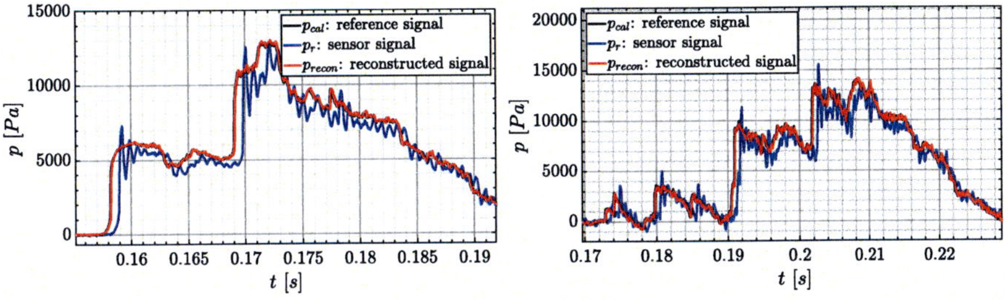

(a) Test case #1: Generalization (b) Test case #3: Generalization

Fig. 8. Generalization for different cases #1 and #3 with coinciding reference and reconstructed signals

the correlation of the reconstructed with the reference pressure signal for all test cases. Furthermore, the usage of the data set, namely training, validation or generalization, is mentioned. For both LCS lengths, a very high reconstruction accuracy is achieved. Hence, the determination of the transfer function via a system identification process shows very promising results and can be used in addition to the rather time consuming determination in the frequency domain [2].

5 Conclusion

In this paper, a discrete-time system identification approach is applied in order to investigate the acoustic transfer function in a pneumatic line-cavity system (LCS). Newly developed fiber-optic differential pressure sensors are used to acquire the high bandwidth pressure signal data resulting from a change in pressure generated in the shock tube facility at TUM-AER. Various LCSs with differing dimensions are tested. Post-processed data show a high agreement and reproducibility when comparing it to reference data: A frequency range up to 10 kHz can be captured. Moreover, a reconstruction accuracy in the application of the estimated transfer function on new pressure data (generalization) is higher than 99 %. Furthermore, the newly developed 2nd generation fiber-optic pressure sensors show a good dynamic behavior. The linear pressure range of the sensor is confirmed up to 20 kPa. As a reasonable next step, a fast-response five-hole probe will be tested in the shock tube to measure the transfer function of the fully assembled probe.

Acknowledgment. The authors would like to thank fos4X GmbH for providing the sensors and the DAQ system.

References

1. Bergh, H., Tijdeman, H.: Theoretical and experimental results for the dynamic response of pressure measuring systems. NLR-TR F.238, p. 21 (1965)
2. Heckmeier, F.M., Iglesias, D., Kreft, S., Kienitz, S., Breitsamter, C.: Development of unsteady multi-hole probes based on fiber-optic pressure sensors. Eng. Res. Expr. **1**, 025023 (2019)
3. Wang, Z., Giglmaier, M., Hopfes, T., Köglmeier, L., Adams, N.A.: Impact of sensor housing geometries on transient stagnation pressure measurements in impulse facilities. Exper. Therm. Fluid Sci. **109**, 109851 (2019)
4. Paniagua, G., Dénos, R.: Digital compensation of pressure sensors in the time domain. Exper. Fluids **32**, 417–424 (2002)
5. Ljung, L.: System Identification Toolbox - User's Guide. The MathWorks Inc, Natick, MA, USA (2020)
6. Kienitz, S., Kreft, S., Schmid, M., Staats, M., and Koch, A.W.: Miniature airworthy fiber-optic pressure sensor for measuring static pressure and acoustics. In: Proceedings Aerospace Europe Conference 2020, 25–28. February 2020, Bordeaux, France (2020)
7. Heckmeier, F.M., Meusel, K., Kienitz, S., Iglesias, D., Breitsamter, C.: Development of fiber-optic pressure sensors for the usage in unsteady multi-hole probes. In: Proceedings 27. GALA Fachtagung - Experimentelle Strömungsmechanik, 3.-5. September 2019, Erlangen, Germany (2019)
8. Gottmann, J., Hermans, M., Repiev, N., Ortmann, J.: Selective laser-induced etching of 3D precision quartz glass components for microfluidic applications - up-scaling of complexity and speed. Micromachines **8**, 110 (2017)

2D Airfoil-Testing in an Open Wind Tunnel Test Section: Test Setup, Instrumentation, Interference -Corrections, Validation

Oliver Meyer(✉)

Universität der Bundeswehr München, 85577 Neubiberg, Germany
oliver.meyer@unibw.de
https://www.unibw.de/mb/institute/we4/we42/startseite

Abstract. The wind tunnel of the faculty of mechanical engineering at the Unviversität der Bundeswehr München is used for the investigation of two-dimensional airfoil characteristics. The measurement results in the open test section showed a relatively low repeatability and additionally larger deviations from reference data after the application of test section interference corrections. Therefore, the testing procedure is improved and the interference corrections are changed to a method, which represents the actual test situation much better. As a result, the measurements are very repeatable and the corrected airfoil polars match very well with reference data from larger wind tunnels and XFOIL calculations.

Keywords: Airfoil testing · Wind tunnel · Open test section · Interference corrections · Airfoil polars

1 Introduction

The low speed wind tunnel of the faculty of mechanical engineering is used for the determination of airfoil polars. The airfoil models are equipped with end plates to realize 2D flow conditions in the open test section. The lift forces are directly measured via the overhead balance and the drag is determined via wake rake measurements and a corresponding momentum loss analysis. Additionally, the angle of attack is required and the geometry of the test setup to plot the polars for the lift-, drag- and moment-coefficients c_l, c_d, c_m. The main topics to ensure a reliable and accurate airfoil polar determination may be summarized as follows:

1. Realize 2D flow conditions around the airfoil model (shown in [1]).
2. Measure and determine lift and drag accurately.
3. Find or develop and apply a suitable test section interference correction.

Especially the second and third topic require some attention as the following examples will show. Figure 1 shows the visible low repeatability of the measurements for one airfoil tested at different dates from different operators. The test

© The Author(s), under exclusive license to Springer Nature Switzerland AG 2021
A. Dillmann et al. (Eds.): STAB/DGLR Symposium 2020, NNFM 151, pp. 719–728, 2021.
https://doi.org/10.1007/978-3-030-79561-0_68

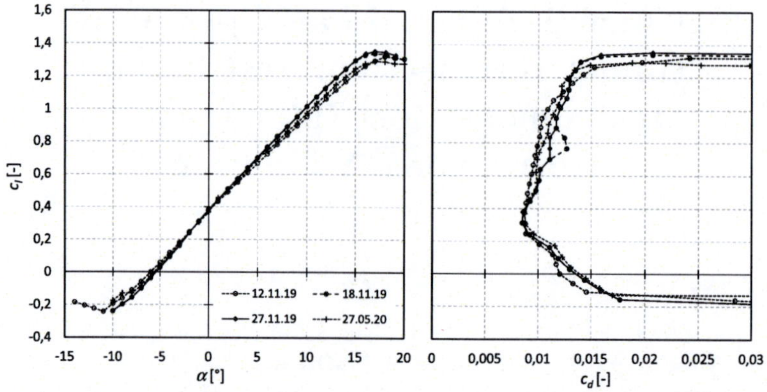

Fig. 1. Uncorrected lift and drag polars of one airfoil before optimization - visible scatter

setup, the used instrumentation and the data analysis have to be investigated to improve the repeatability and to reduce the scatter. A correction for test section interferences must be applied to the derived polars. The correction used so far was based on a procedure taken from Rae and Pope [7] (explained in [2]) and showed generally a good coincidence with reference data in the lift polar but throughout with too high $c_{l,max}$-values and larger deviations in drag, see Fig. 2. In particular the difference in $c_{l,max}$ is suspicious, as the measurements were performed at lower Reynolds numbers than the available references and hence the expected $c_{l,max}$ should have been smaller than the reference $c_{l,max}$. However, the used procedure was not developed for the present boundary conditions and the applied corrections for lift, drag and the geometrical representation of the models in the correction as used in [2] can be improved. Therefore, a more suitable correction should be found and applied correctly.

2 Wind Tunnel and Test Setup

The wind tunnel is a Göttinger type with an open test section. The plane view and technical data are shown in Fig. 3. It is a general purpose wind tunnel with an external overhead balance and an optional moving belt system for ground simulation. The wind tunnel is mainly used for educational purposes and industrial projects. The wind tunnel and the measurement systems have undergone several modifications in the near past to accommodate 2D airfoil models and to determine airfoil polars. The flow quality in the test section was optimized and the test-setup was modified to the given 2D airfoil models to ensure a 2D-flow in the open test section [1].

However, airfoil investigations are often performed in closed test sections with external balance systems and preferably test sections with a large height to model-wingspan ratio to minimize wall interferences. Residual interferences in closed test sections may be determined by wall pressure signature measurements

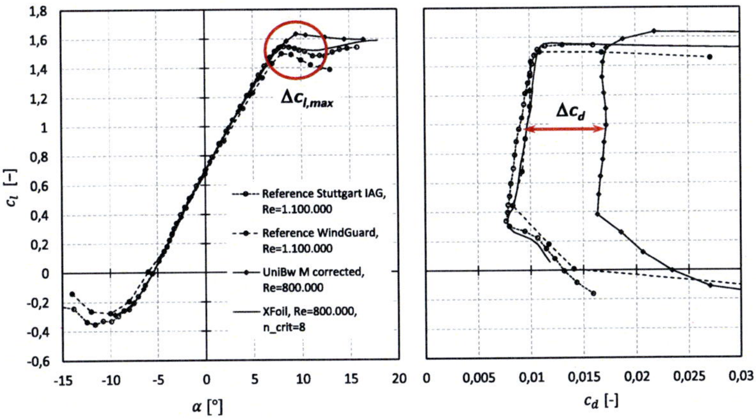

Fig. 2. Lift and drag polars after interference correction before optimization - remaining deviation from reference, data taken from [2]

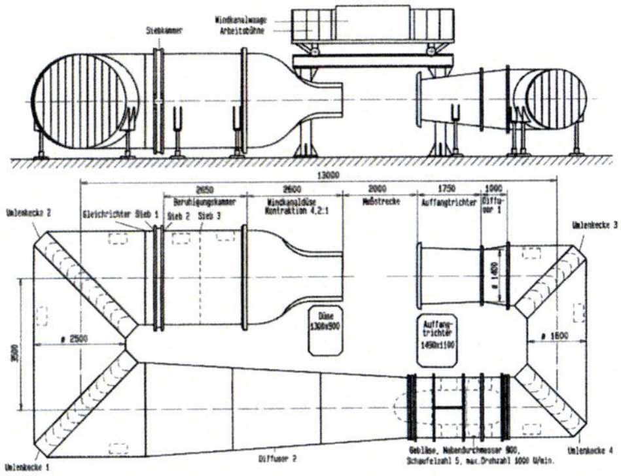

Technische Daten

Test Section (h x w x l)	0,9m x 1,3m x 2m
Velocity	0 - 40 m/s
Test Section Re-nr.:	$2.8\cdot10^6$
Power	75kW
Turbulence	0.35%
Ground Simulation (Belt with BL Removal System)	0-40 m/s
Axial Static Pressure Gradient	$\frac{dc_p}{dx} < \frac{0.002}{0.1m}$

Balance Data

Component	Range [N]	Resolution [N]	Max deviation from value
Drag I	0 - 100	0,1N	0.24%
Drag II	0 - 100	0,1N	0.32%
Lift I	0 - 200	0,1N	0.02%
Lift II	0 - 200	0,1N	0.02%
Lift III	0 - 200	0,1N	0.02%
Lateral	0 - 100	0,1N	0.32%

Fig. 3. The wind tunnel of the department of mechanical engineering

and the measured data can be corrected for the acting inferferences accordingly [3]. Here, the open test section is not converted to a closed test section for the following reasons: first, the models in the resulting test section would generate large interferences due to the given test section and model geometry and second, the resulting test section would be too short to apply interference corrections satisfactory. Therefore, it was decided to apply circular end plates to the models to enforce a local 2D flow at the airfoils and to use the given overhead balance and model mounts. The end plate diameter is twice the airfoil chord length and the resulting flow has a two-dimensional character throughout the linear lift gradient range of the airfoil, [1,2]. An overview of the airfoil in the test section and the flow rakes is shown in Fig. 4a. A more detailed top view of the airfoil setup in the test section is depicted in Fig. 4b.

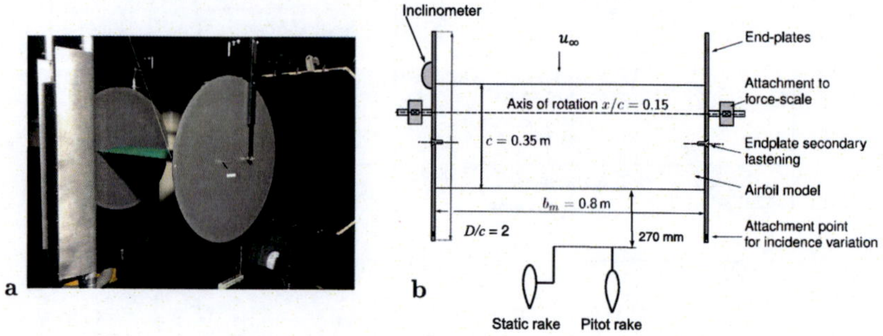

Fig. 4. Test-setup: **a** overview, **b** top view

The main geometrical characteristics in the test setup can be summarized as follows:

1. The airfoil rotating point is positioned at $x = 730$ mm from the nozzle exit, i.e. at a relative x-position of $x/L = 0.365$ with a test section length of $L = 2000$ mm.
2. The airfoil is vertically and laterally centred in the test section.
3. The end plates are circular and have a relative diameter of $D/c = 2$ with a chord length of the model of $c = 350$ mm and an end plate diameter of $D = 700$ mm respectively.
4. The end plates are located entirely within the core flow of the free jet.
5. The static and Pitot rake are positioned 270 mm behind the trailing edge of the airfoil.

The design parameters of the used airfoils are listed in Table 1.

Table 1. Design parameters of airfoil-1 and airfoil-2

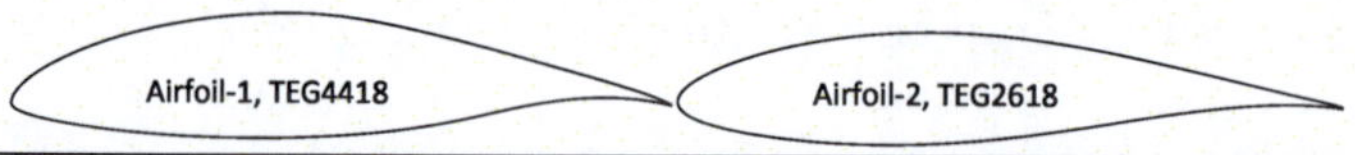

Parameter		Airfoil-1	Airfoil-2
relative thickness	t/c	18 %	18 %
rel. pos. of max. thickness	$x_{t,max}/c$	40.9 %	35.1 %
rel. LE radius	r_{LE}/c	1.42 %	1.96 %
test Reynolds number	Re	800.000	800.000

3 Investigation of Instrumentation and Testing Procedure

The lift polars in Fig. 1 show deviations in the lift gradient and $c_{l,max}$. The difference in $c_{l,max}$ is max. +5% and the lift gradients differ max. 9% from each other. The drag polars in Fig. 1 (right) also show visible deviations from test to test. The drag coefficients deviate from 5% to 20% from each other in the linear lift range. Various parameters were indicated to likely influence the measurement repeatability, such as: model inclination determination, model positioning accuracy, angle and positioning sensitivity of wake rake measurements and the numerical wake drag determination method.

3.1 Angle of Attack Measurement

The inclination angles of the models are indirectly determined via the vertical movement of the rear model suspension and the calculation of a corresponding model angle via the geometry of the setup (trigonometrics). This setup is responsive to errors due to the indirect method and due to the manual measurement of angles for initial setup and calibration. The relatively strong differences in the lift gradients (Fig. 1 left) may be subjected to differences in inclination measurements. Therefore, the inclination measurement is changed to a direct measurement via a calibrated inclinometer which is mounted to a model end plate, see Fig. 4b.

3.2 Flow Rakes in Angular Flow

The total pressure measurement behind the model is mandatory for the drag determination. The static pressure measurements may be required if the wake measurement has to be situated too close behind the model where the static pressure may not be completely equalized across the free stream yet [4]. Here, a static pressure rake with a less amount of pressure heads is used to care for a certain correction, if required. Both rakes are made from stainless steel tubes fixed to streamlined structures where the pressure tubes are internally routed out of the flow. The pressure head tubes have an outer diameter of 1.2 mm and an inner diameter of 0.8 mm. The static pressure tube is designed according to the example in [5]. Both rakes, the total and the static, are vertically fixed and the sensor heads will experience a certain flow angle during a polar measurement. Therefore, flow angle measurements are performed to analyse the angular sensitivity of the rakes and to derive corrections if necessary. The rakes are measured in total, i.e. all pressure heads are measured at once while the rakes are tilted in the test section flow. The total pressure rake results represent the average values of all 106 pressure heads and the static pressure rake results represent the average values of 20 pressure heads.

Figure 5 shows the relative deviation of the total pressures in percentage of the dynamic pressure in yaw (a) and pitch (b). The error markers depict the standard deviation from the averaged values. The results for the yaw-variation (Fig. 5a) show almost constant values in the yaw-range of $-5°$ to $5°$, i.e. the

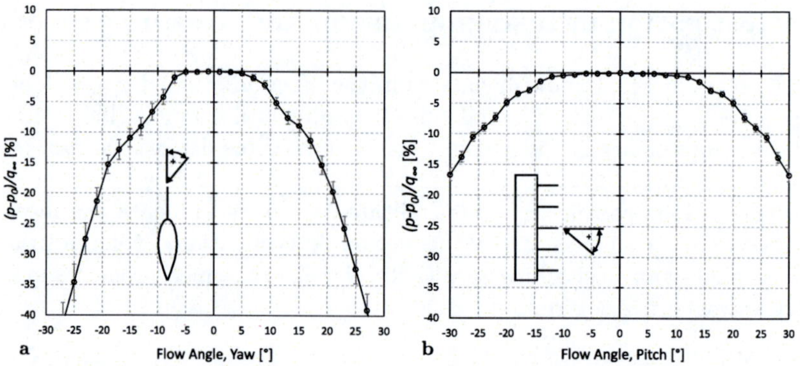

Fig. 5. Angle calibration of total pressure rake: **a** yaw, **b** pitch

rake must be adjusted within this range in the setup. The flow pitch angle may vary much more during test due to the inclination of the model and may produce larger measurement errors. Fortunately, the rake shows a low sensitivity to angle changes under pitch: Fig. 6b shows a minor angular total pressure sensitivity of all 106 sensors between pitch angles of $-10°$ to $10°$, 2.5% deviation at $-15°$ and $15°$ and appr. 5% deviation at $-20°$ and $20°$. The standard deviations are very small for this test case. Tüchsen showed in [6] with CFD-calculations for the given test setup, that the expected flow angle at the rake heads behind the airfoil model is significantly smaller than the inclination angle of the model, namely between $2°$ and $6°$ smaller depending on the assumed boundary conditions. Hence, the expected flow angle error with attached airfoil flow up to $\alpha \approx 15°$ and fully separated flow above $\alpha \approx 18°$ may be very small in the tested model inclination range. Therefore, a flow angle correction for the total pressure measurement may not be required.

The angular sensitivity of the static pressure rake is much higher in the yaw direction, see Fig. 6b, which should result in a careful initial positioning of the rake. The yaw angle during test should be constant due to the 2D nature of the flow in the test setup. The pitch sensitivity shows deviations from the reference static pressure of appr. 3% at $-15°$ and $15°$ and max. 7% at $-20°$ and $20°$. Here, a flow angle correction of the measured static pressure may not be necessary for the same reasons as mentioned for the total pressure measurement above. Additionally, the expected static pressure deviations are small across the wake.

3.3 Miscellaneous

Axial position of flow rakes: The wake rakes are positioned 270 mm behind the models. Tüchsen showed in [6], that the determined polars are independent from position changes within -100 mm and $+150$ mm from the initial position, see Fig. 4b.

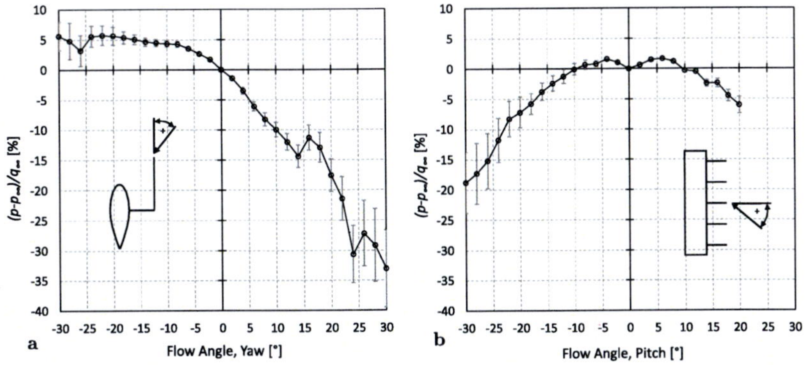

Fig. 6. Angle calibration of static pressure rake: **a** yaw, **b** pitch

Vertical position of flow rakes: The total pressure rake has a high amount of pressure heads in the center of the rake for accurate wake measurements within the linear range of the lift polar. The amount of pressure heads is reduced to the vertical boundaries of the free stream, see [10]. I.e. at higher model inclination angles the wake measurement may have a reduced accuracy. Tüchsen showed in [6] that a vertical repositioning of the rake has a negligible effect on the measured results.

Wind tunnel balance: The wind tunnel balance is investigated with reference weights to check the calibration. The balance system displays the weights in each direction within the specified accuracy in the specified range.

Model positioning: The exact and repeatable model positioning in the test section is very important and variations between different test dates were unfortunately observed. A vertically or laterally shifted and inclined model obviously delivers different results. An exact positioning of the models must be performed by each individual operator.

Numerical drag determination: The 106 total pressures are measured via two 64-channel μDAC modules from PSI and the drag coefficient calculation is performed via an appropriate program. The bump observed in the drag polar in Fig. 1 at $c_l \approx 0.8$ determined on the 18.11.2019 could have been attributed to an error in the program where the wake was measured by the two modules simultaneously, i.e. the wake was transitioning from one module to the other. The numerical integration is corrected in the program accordingly.

Following parameters are finally indicated to influence the observed low measurement accuracy and repeatability: the model positioning accuracy in the test section, the model inclination measurement and the numerical evaluation of drag coefficients. The results show a very good repeatability after optimization of the formentioned items and hence a very low scatter in the derived airfoil polars, Fig. 7. All polars coincide very reliable and accurately now.

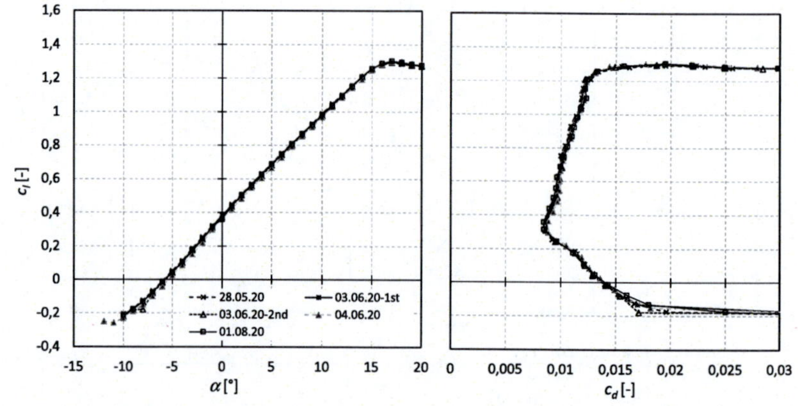

Fig. 7. Airfoil polars after test setup optimization—low scatter

4 Wind Tunnel Interference Correction

The airfoil drag evaluation via wake measurements, the angle of attack measurement and the accurate model positioning deliver very repeatable results now. Unfortunately, most of these parameters are affected by test section interferences and differ from pure 2D free flow conditions. The airfoil polars deviate clearly from pure 2D reference data as a result of these interferences, see Fig. 8. The interference corrections used so far [1,2] are based on the work of Rae and Pope [7]. Unfortunately, the corrections consider other test setup boundary conditions which should not be applied directly to the present test case of 2D-testing with end plates in an open test section. Additionally, potential for improvement was detected in the application of drag correction and the use of geometrical data of the models and test section. Ewald stated 1998 in [3] that the setup of 2D models in open test sections disappeared in wind tunnel testing and this case is therefore not mentioned in [3]. But Ewald referred to Garner [8] for 2D testing in open test sections. Garner described corrections for exactly the given test case in [8] which can be summarized as follows:

The correction is based on geometrical data of the test section (height h, relative geometry factors G_0 and G_1), airfoil dimensions (chord c and thickness t) and flow informations (speed c, Mach number M and Prandtl-Glauert factor β). Usually, the parameter h equals the nozzle height in open test sections by definition but the use of the height of the free stream core flow at the model position appears more reasonable for the current scenario. The used dimension h was determined by Tüchsen in [6]. First, a correction for the angle of attack $\Delta\alpha$ is determined according to [8] Eq. 2.74. This corrects the effect of the test section boundaries and dimensions on the effective angle of attack. The lift coefficient c_l is corrected for two interferences, the altered downwash due to the open and limited test section, [8] Eq. 2.75 and the solid blockage effect, see [8] Chap. 5.9.5. The required correction factor ϵ_B is the sum of the solid blockage factor ϵ_S calculated from [8] Eq. 5.61 and the wake blockage factor ϵ_W which is neglected here because the wake blockage effect is small for an airfoil as long as the flow is attached. The drag coefficient c_d is corrected for the horizontal buoyancy effect according to Glauert [9] due to a present very small horizontal pressure gradient in the test section and a solid blockage correction for the present test case given by Garner in [8] Chap. 5.9.5. The application of the correction and the used formulas are explained in detail and stepwise in [10].

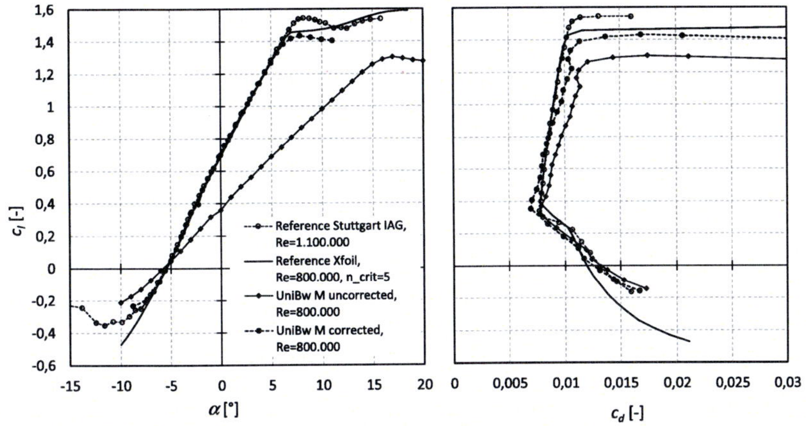

Fig. 8. Airfoil-1: polars before and after correction

5 Validation Results and Conclusion

The measured and corrected polars for two different airfoils are shown in Fig. 8 and 9. The geometrical characteristics of the airfoils are listed in Table 1. The results in Fig. 8 are compared with reference data from the Laminarwindkanal of the IAV Stuttgart. The Reynolds numbers are different and deviations are expected. The $c_{l,max}$-values measured and corrected in the UniBw M are now lower than the reference results tested at a higher Reynolds number. This corresponds with expectation and the XFOIL calculation where the lift gradient also reduces at a comparable α. The lift gradients match also very well. The corrected drag polar shows a very good coincidence with the references and the deviation is much smaller now compared to former results shown in Fig. 2. The second airfoil tested and corrected, Fig. 9, shows also a very good coincidence with the references in the lift and drag polar. Both, the improvements in the measurement procedures as well as the application of a suitable correction result in very repeatable and accurate polars now. In particular the drag determination is highly improved compared to the former test results in Fig. 2.

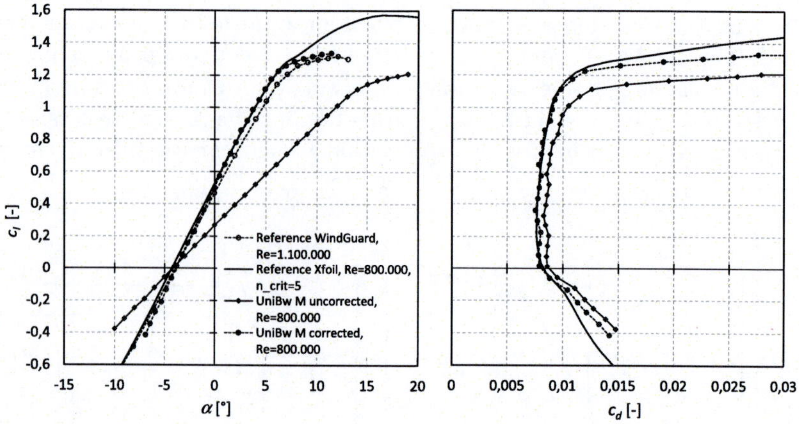

Fig. 9. Airfoil-2: polars before and after correction

References

1. Meyer, O., Terreblanche, T., Klein, M.: Optimized experimental environment for wing profile investigations in the low speed wind tunnel. In: Proceedings of the 5th International Conference on Experimental Fluid Mechanics, Munich, Germany, pp. 728–733 (2018)
2. Terreblanche, T.: Optimization of profile polars for wind turbine rotor blades with the use of leading-edge vortex generators. TU Braunschweig, Master Thesis (2017)
3. Ewald, B.F.R. (ed.) et al.: Wind Tunnel Wall Correction, AGARDograph 336, NATO (1989). ISBN 92-836-1076-8
4. Jones, B.M.: The Measurement of Profile Drag by the Pitot Traverse Method. ARC Rep. and Mem. Nr. 1688 (1936)
5. Nitsche, W., Brunn, A.: Strömungsmesstechnik, vol. 2, Springer (2006) ISBN-10 3-540-20990-5
6. Tüchsen, Th.: Analyse des Göttinger Windkanals zur Bestimmung von Profilpolaren und Validierung der 2D-Windkanalkorrekturen, Master Thesis, Universität der Bundeswehr München (2020)
7. Barlow, J.B., Rae, W.H., Pope, A.: Low Speed Wind Tunnel Testing, 3rd edn., New York (1999). 0-471-55774-9
8. Garner, H.C., Rogers, E.W.E., Acum, W.E.A., Maskell, E.C.: Subsonic Wind Tunnel Wall Corretions, AGARDograph 109, NATO (1966)
9. Glauert, H.: Wind Tunnel Interference on Wings, Bodies and Airscrews, ARC R& M No. 1566 (1933)
10. Meyer, O.: Bestimmung von Profilpolaren (im Windkanal der WE4/2), Universität der Bundeswehr (2020). https://www.unibw.de/mb/institute/we4/we42/downloads/polaren.pdf

On the Effectiveness of Retrofitted Airfoil Leading Edge Devices to Limit Lift

Oliver Meyer(✉)

Universität der Bundeswehr München, 85577 Neubiberg, Germany
oliver.meyer@unibw.de
https://www.unibw.de/mb/institute/we4/we42/startseite

Abstract. Horizontal Axis Wind Turbines are operated with different safety mechanisms which protect the wind turbines from overloading during strong wind situations. However, wind turbines may collapse due to sudden gust events or other overload situations. An additional safety measure may be an aerodynamic load limit of the airfoils themselves by adding passive lift limiting devices to the wing. Ideally, these devices would not add extra drag and the aerodynamic performance will maintain in the normal operation range of the wings. The investigated small radii or 'noses' added to the leading edge of a laminar airfoil shows very promising results: the lift limit can be adjusted by the shape, position and spanwise distribution of the added 'nose'. Additional drag can be kept to a minimum within a certain operation range of the airfoil, depending on the geometrical characteristics of the added nose.

Keywords: Airfoil aerodynamics · Wind turbine airfoils · High lift limitation · Leading edge modifications

1 Introduction

Horizontal Axis Wind Turbines (HAWT) are designed for a certain wind load spectrum to operate near an optimum design point. This applies to the entire system as well as the wings and the airfoils used in each wing section. The HAWT is usually inactive during wind loads below or above the design load range. However, the operation of wind turbines shows that HAWT's may suffer from severe damage due to a probable overload situation.

Therefore, some supplementary aerodynamic safety measures are investigated to restrict the wing load right above the design or operation point by the aerodynamic behaviour of the airfoil. This would act as a direct aerodynamic overload protection for the turbine wing and may contribute to the endurance of the HAWT. Additionally, the used method should not raise the drag of the airfoil in the normal operation range to maintain the aerodynamic performance of the HAWT. The desired effect in the lift and drag polar of a high lift restricted airfoil is depicted in Fig. 1. Ideally, the lift limitation feature is easy to retrofit to existing facilities.

© The Author(s), under exclusive license to Springer Nature Switzerland AG 2021
A. Dillmann et al. (Eds.): STAB/DGLR Symposium 2020, NNFM 151, pp. 729–738, 2021.
https://doi.org/10.1007/978-3-030-79561-0_69

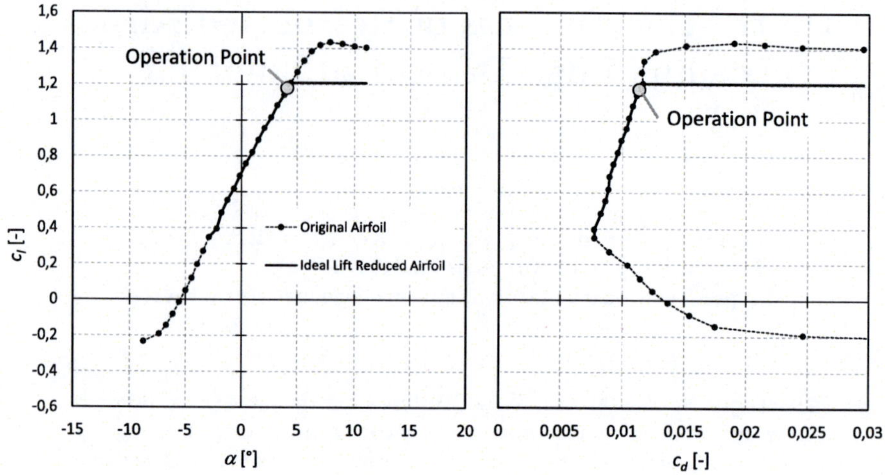

Fig. 1. Principle of lift and drag polars of example baseline profile (dashed) and targeted lift polar for safer wind turbine operation (solid)

Some methods such as tubercles and Leading Edge Vortex Generators (LEVoG's by the DLR), Fig. 2, were investigated in [1] and [2] or tripping devices were suggested, see [3], to limit the high lift of a typical wind turbine airfoil.

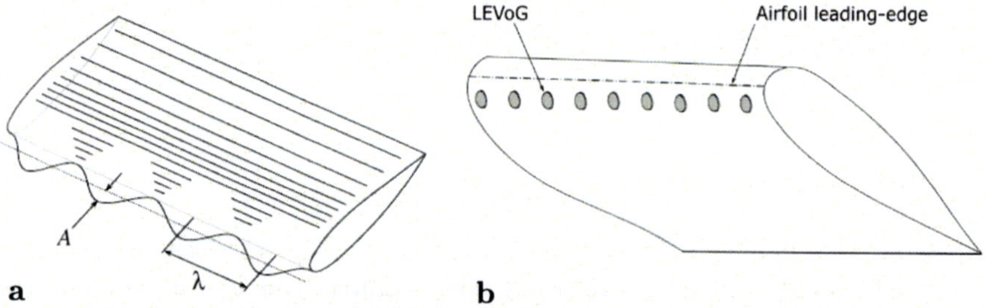

Fig. 2. Principle of Tubercle Leading Edge **a** and LEVOG's **b** [4]

However, the tubercles show some promising results for particular parameters of tubercle amplitude A and wavelength λ. But the design and the realization may be very complex. LEVoG's may be easy to retrofit but the lift limiting effect may be too low and the accompanying raise in drag may be too high, see [4]. A tripping device as suggested in [3] limits the lift but probably produces extra drag and noise. Therefore, some other methods should be investigated, which are simpler to implement and which also show the desired effect.

Here, the relatively old idea of locally extended small leading edge radii is investigated. The small leading edge radius or sharp leading edge at the root of an airplane wing stalls the wing locally to signalize the pilot the lift limit of the airplane. This leading edge modification is realized for example at the wing roots of a Dornier Do 335 (1943), see Fig. 3.

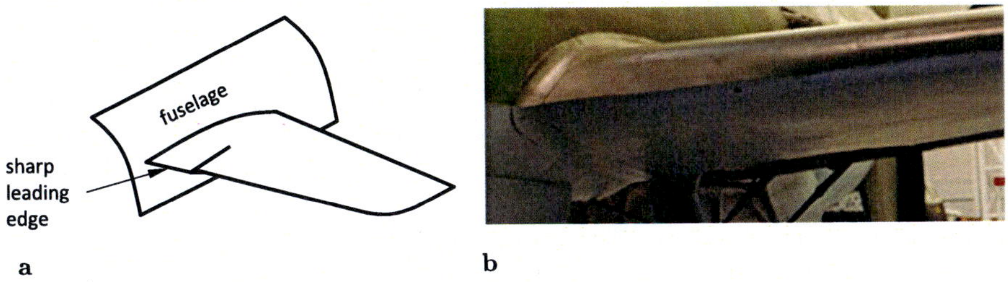

Fig. 3. Principle of sharp leading edge **a** and example at Do 335 **b**

The advantage of a sharper leading edge may be a minimum or even no drag raise if the stagnation point of the flow and the modified leading edge are aligned correctly at a certain angle of attack. The lift limit effect could be designed by the circumferential position, amplitude and radius of the leading edge modification.

2 Wind Tunnel and Test Setup

The wind tunnel is a Göttinger type with an open test section. The plan view and technical data are shown in Fig. 4. It is a general purpose wind tunnel with an external overhead balance and an optional moving belt system for ground simulation. The wind tunnel is mainly used for educational purposes and industrial projects.

The airfoils are equipped with end plates to ensure a 2D flow around the airfoil models in the open test section [1] and the models are mounted to the overhead balance to measure the lift forces. The drag coefficients are determined via wake pressure loss measurements using the method described in [5]. The derived airfoil polars are then corrected with respect to the acting test section interferences using a correctional procedure described by Garner in [6]. The effectiveness of this test setup and validation results are shown in [7]. The wind speed during test is 40 m/s and the airfoil chord length is 0.35 m which leads to a test Reynolds number of appr. $Re = 800.000$ under atmospheric conditions.

However, the airfoil Reynolds numbers during HAWT operation are higher and the airfoil characteristics may change accordingly. It was shown in [4] that at a higher Reynolds number of $Re = 1.1 \cdot 10^6$ the maximum lift $c_{a,max}$ and drag level may differ slightly for the used airfoil, but the aerodynamic behaviour,

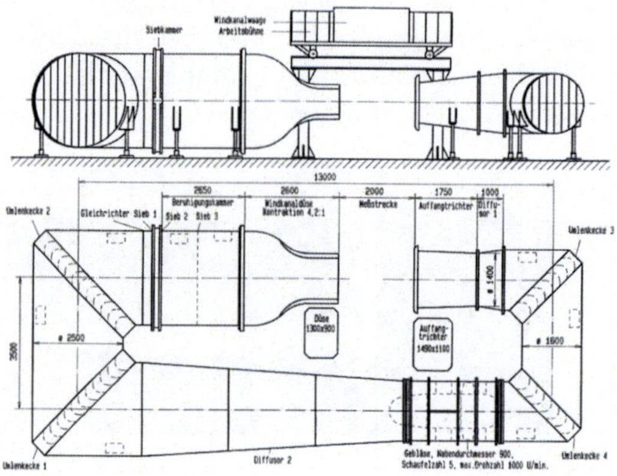

Technische Daten

Test Section (h x w x l)	0,9m x 1,3m x 2m
Velocity	0 - 40 m/s
Test Section Re-nr.:	$2.8 \cdot 10^6$
Power	75kW
Turbulence	0.35%
Ground Simulation (Belt with BL Removal System)	0-40 m/s
Axial Static Pressure Gradient	$\frac{dc_p}{dx} < \frac{0.002}{0.1m}$

Balance Data

Component	Range [N]	Resolution [N]	Max deviation from value
Drag I	0 - 100	0,1N	0.24%
Drag II	0 - 100	0,1N	0.32%
Lift I	0 - 200	0,1N	0.02%
Lift II	0 - 200	0,1N	0.02%
Lift III	0 - 200	0,1N	0.02%
Lateral	0 - 100	0,1N	0.32%

Fig. 4. The wind tunnel of the Department of Mechanical Engineering

i.e. the shape of the airfoil polars remains comparable. Therefore, the observed airfoil characteristics at $Re = 0.8 \cdot 10^6$ should indicate a comparable behaviour at higher Reynolds numbers for HAWT. Nevertheless, the observed effects should be validated during an airfoil design process subsequently at higher Reynolds numbers and in field testing with real HAWT's.

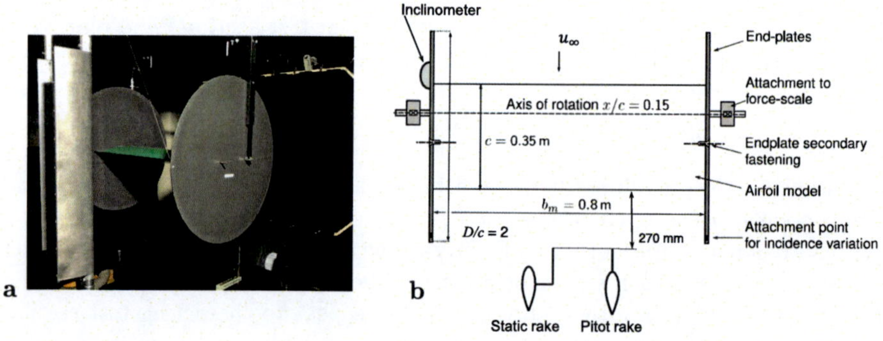

Fig. 5. Test-setup: **a** overview, **b** top view

3 Investigation of Leading Edge Radius Variations

A laminar wind turbine airfoil with the characteristics listed in Table 1 is selected for the first parameter variations. This airfoil was also used in the former investigations with tubercles and LEVoG's. First, some numerical investigations are performed with Xfoil [8] to select promising leading edge variations for subsequent wind tunnel tests.

Table 1. Design parameters of laminar airfoil TEG4418

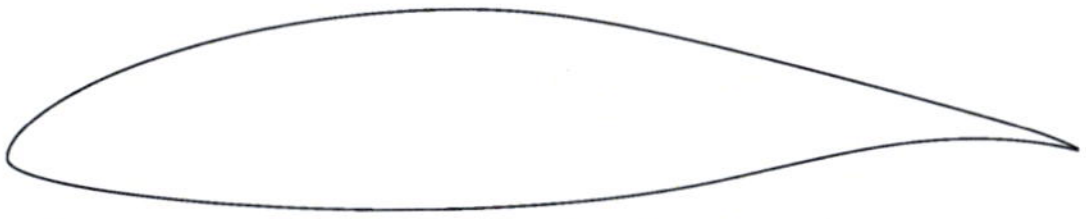

Parameter		Value
relative thickness	t/c	18 %
rel. pos. of max. thickness	$x_{t,max}/c$	40.9 %
rel. LE radius	r_{LE}/c	1.42 %
test Reynolds number	Re	800.000

3.1 Xfoil Results

The Xfoil calculations are performed with the wind tunnel test Reynolds number of $Re = 800.000$ and an 'amplification factor' of $n_{crit} = 5$ which corresponds to the turbulence level Tu in the wind tunnel of appr. $Tu = 0.35\%$, [8]. If predictions are planned for higher Reynolds numbers and/or different turbulence levels, Re and n_{crit} may be adjusted accordingly. Investigations with the present airfoil in [2] and [4] showed, that Xfoil predicts also higher Reynolds numbers ($Re = 1.1 \cdot 10^6$) at lower turbulence levels ($n_{crit} = 8$) very well.

The calculations are first performed stepwise 'by hand' to detect critical regions where flow separation may begin and the inclination range is adjusted accordingly. The calculations are performed in angle of attack (α) steps of $\alpha = 0.1°$. Two leading edge radius variations are selected arbitrarily: first a small radius and second a slightly larger radius in the area of the expected stagnations point at 0° angle of attack, depicted in Fig. 6a as nose 1 and nose 2. Additionally, three circumferential positions of nose 1 are investigated, see Fig. 6b.

The polars in Fig. 7 show the measured and calculated reference base airfoil data and additionally the two leading edge variations. Nose 1 shows no significant effect in the lift polar at $c_{l,max}$ (Fig. 7a), but nose 2 seems to generate the desired lift limiting effect. The drag polar in Fig. 7b shows a very narrow low drag operation range of nose 2, i.e. appr. $2° \leq \alpha \leq 5°$, hence the nose may be

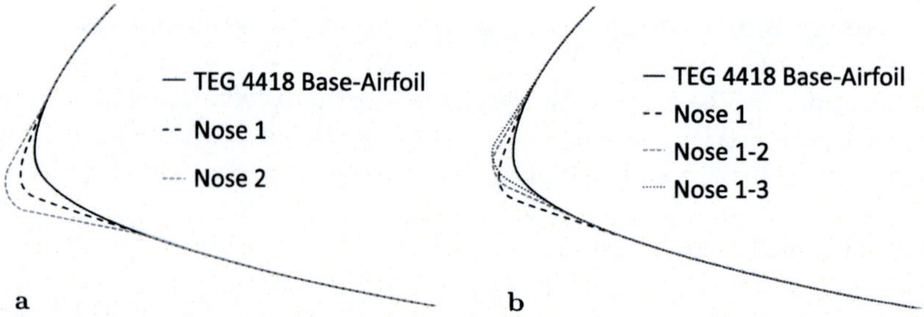

Fig. 6. Two radii added to the baseline airfoil (nose 1 and nose 2) **a**, circumferential position variation of nose 1 **b**

too large in size. The smaller nose 1 shows a much broader operation range of appr. $0^\circ \leq \alpha \leq 7^\circ$ but shifted to too high angles of attack.

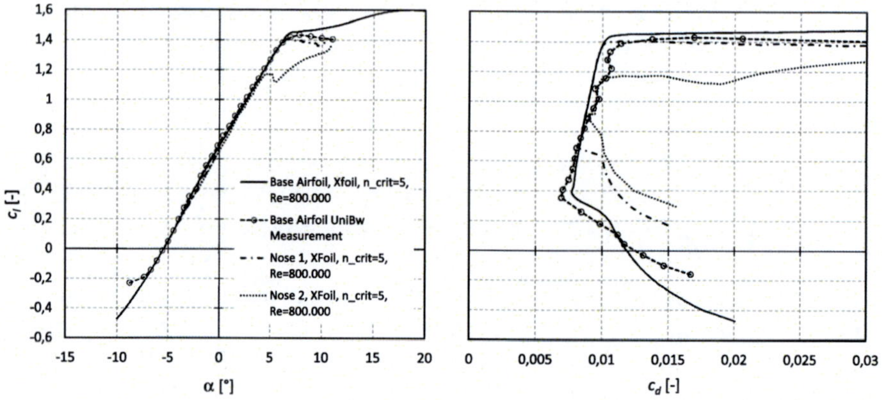

Fig. 7. Xfoil-polars of nose 1 and nose 2 compared to base airfoil data (Xfoil and measurement)

Therefore, the circumferential position of nose 1 is moved upwards, i.e. in the direction of the upper side of the airfoil, see Fig. 6b, to alter the lift limit towards lower angles of attack. The results are shown in the polars in Fig. 8. The lift polar (Fig. 8a) depicts the intended effect and the maximum lift is reduced depending on the position of the nose. The criterion is the beginning of the flow separation on the suction side of the airfoil where the lift polar leaves the linear lift gradient. The drag polar in Fig. 8b shows that the drag in the operation range with the noses attached may not raise compared to the base airfoil. The range of low drag stays large compared to nose 2 (Fig. 7b) and can be shifted to lower angles of attack with the variations nose 1–2 and nose 1–3.

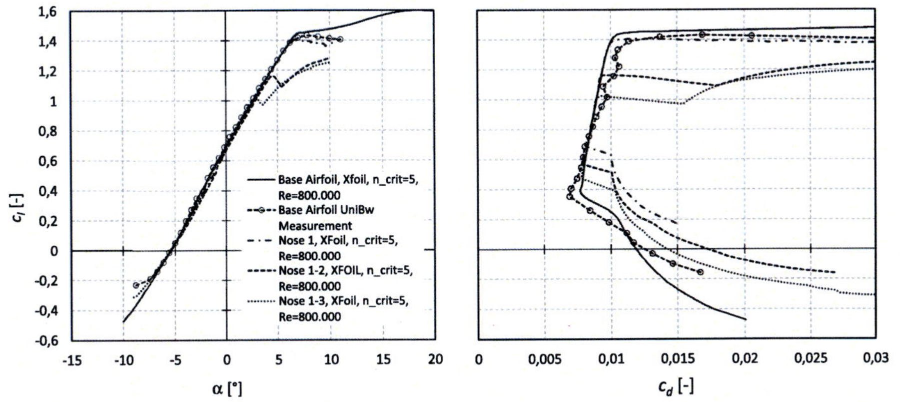

Fig. 8. Xfoil-polars of nose 1 at different circumferential positions compared to base airfoil data (Xfoil and measurement)

The aerodynamic effect of the nose 1 variations is additionally depicted in Fig. 9 with the calculated pressure distributions in the design point of $\alpha = 4°$. The results for the base airfoil and nose 1 are identical for this angle of attack, both in the lift polar in Fig. 8 (left) and in the pressure distribution in Fig. 9. Nose 2 begins to 'work' right at $\alpha = 4°$ which can be observed in the lift polar in Fig. 8 (left) where the lift begins to reduce. This effect can also be seen in the pressure distribution, because the end of the laminar separation bubble (investigated in [2] and shown also in [1]) moves a little upstream on the suction side

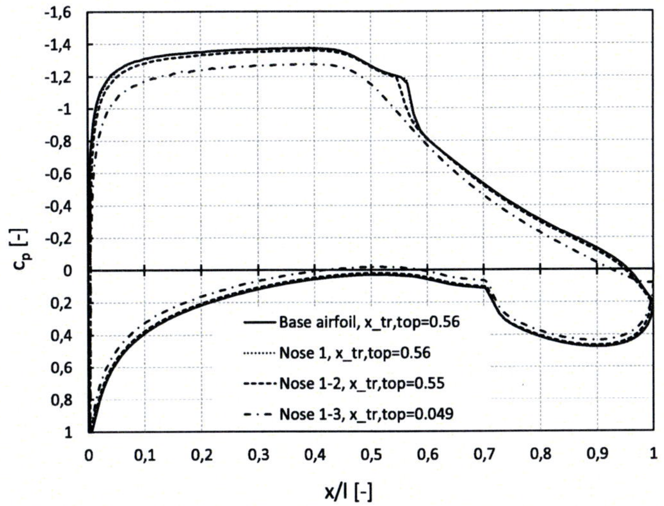

Fig. 9. Xfoil-pressure distributions c_p of nose 1 at different circumferential positions compared to base airfoil at design point $\alpha = 4°$

and the transition position moves upstream accordingly. The calculated positions of the transitions ($x_{tr,top}$) on the upper airfoil surface are named in the graph. Nose 3 begins to reduce the lift at appr. $\alpha = 3°$ with accompanying raise in drag, see Fig. 8. The pressure distribution shows, that the laminar separation bubble disappeared on the upper side and the transition moved further upstream. A reduced lift is obvious in the reduced pressure difference of upper and lower side and a raise in drag may be indicated by the reduction of the trailing edge pressure from $c_{p,te} = 0.2$ to $c_{p,te} = 0.08$.

3.2 Wind Tunnel Results

In a first step, the experimental nose variations are performed with the base airfoil model and individually attached nose shapes. The most promising leading edge modifications are then realized with new designed and manufactured airfoil wing models. Nose 2 is selected for first wind tunnel experiments because the larger size compared to nose 1 is easier to transfer manually to the airfoil. The nose shape is moulded to the airfoil model with silicone and a moulding gauge. The attached silicone nose is then stepwise reduced to 75%, 50% and 25% wingspan as shown in Fig. 10.

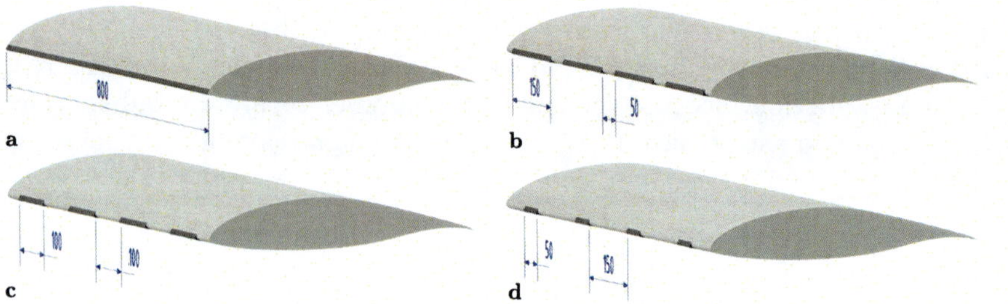

Fig. 10. LE variations with nose 2 added to 100% span **a**, 75% span **b**, 50% span **c** and 25% span **d**

The determined lift behaviour for the different nose 2 variations shows an almost equal behaviour for 100%, 75%, and 50% spanwise distribution of the nose. 25% spanwise distribution shows a maximum lift between 100% modification and base airfoil (no modification). Therefore, the 100% and 25% cases are build as new high-quality wind tunnel models for more accurate measurements. The lift polars in Fig. 11 shows the above mentioned effect of the 100% and 25% spanwise nose-distribution in solid black lines. The other two silicone modifications are shown in grey dashed lines additionally. The drag polar shows that the airfoil with leading edge modifications produces only minimum extra drag within a certain angle of attack range. The range is appr. within $-1° \leq \alpha \leq 3°$ with nose 2 at 100% span and gets larger with reduced spanwise distribution of nose 2. With 25% spanwise distribution there is very little extra drag between appr. $-2° \leq \alpha \leq 7°$ and no measured extra drag between appr. $-6° \leq \alpha \leq -2°$. The low drag α-ranges of the other distributions (grey dashed lines) lie between 25% and 100% spanwise distribution of the nose. It must be kept in mind for these test cases (50% and 75%) that the leading edge model quality was 'hand crafted' and the results may not be of the same quality.

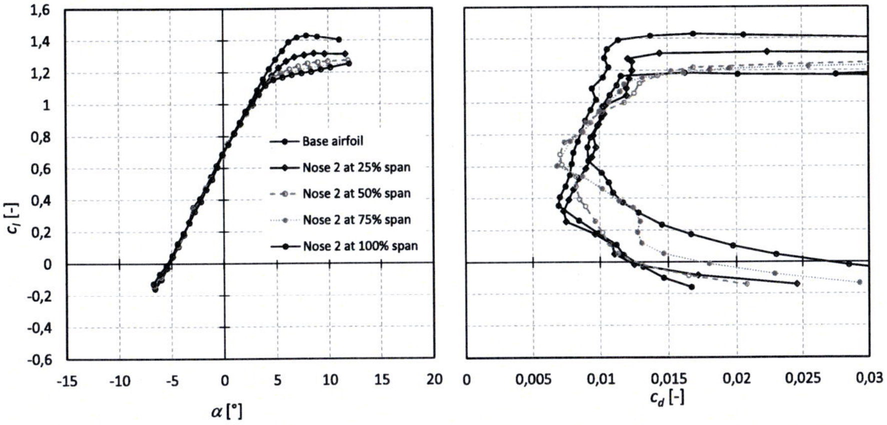

Fig. 11. Lift and drag polars of base airfoil compared to nose 2 leading edge modifications

4 Conclusion and Outlook

Small leading edge radii added to an airfoil can change the characteristics to the desired behaviour of limited lift and minimum added drag in a certain operation range. The effect can be adjusted by the size, shape and position of the radius as well as the spanwise distribution of the radius devices. Additionally, the devices are easy to manufacture and simple to retrofit to existing wind turbine wings.

These investigations and results must be understood as a first design step for further investigations in larger wind tunnels at higher Reynolds numbers and potential field measurements with full size HAWT. However, the results are very promising and show the desired effect. The retrofitted small leading edge radii will be further investigated on an airfoil with a larger initial leading edge radius and a less critical stall behaviour.

Acknowledgements. The author would like to thank M.Eng. Marten Danker, staff member of the authors Fluid Dynamics group, for his support and contribution of the graphics in Fig. 10 and measurement data for Fig. 11.

References

1. Meyer, O., Terreblanche, T., Klein, M.: Optimized experimental environment for wing profile investigations in the low speed wind tunnel. In: Proceedings of the 5th International Conference on Experimental Fluid Mechanics, Munich, Germany, pp. 728–733 (2018)
2. Terreblanche, T.: Optimization of profile polars for wind turbine rotor blades with the use of leading-edge vortex generators. TU Braunschweig, Master Thesis (2017)
3. Sun, Y.: Wind Turbine Blade with Tripping Device and Method Thereof, European Patent Application, EP 3 133 281 A1 (2017)
4. Meyer, O., Terreblanche, T., Klein, M.: Wing Profile Investigations in the Low Speed Wind Tunnel of the UniBwM, Deutscher Luft- und Raumfahrtkongress 2018. In: Conference Proceedings (2018)
5. Jones, B.M.: The Measurement of Profile Drag by the Pitot Traverse Method. ARC Rep. and Mem. Nr. 1688 (1936)
6. Garner, H.C., Rogers, E.W.E., Acum, W.E.A., Maskell, E.C. (ed.) Subsonic Wind Tunnel Wall Corrections, AGARDograph 109, NATO (1966)
7. Tüchsen, Th.: Analyse des Göttinger Windkanals zur Bestimmung von Profilpolaren und Validierung der 2D-Windkanalkorrekturen. Master Thesis, Universität der Bundeswehr München (2020)
8. Drela, M., Youngren, H.: Xfoil 6.9, General Description. MIT, Cambridge (2001)

Wake Vortex Analysis on Transport Aircraft Wing Featuring Dynamic Flap Motion

Johannes Ruhland(✉) and Christian Breitsamter

Chair of Aerodynamics and Fluid Mechanics, Technical University of Munich, 85748 Garching, Germany
johannes.ruhland@tum.de

Abstract. The aim of this work is to impose a disturbance on the velocity field already in the near field of the wake vortex system of a generic high-lift transport aircraft configuration by oscillating flaps. This should result in an excitation of long wave instabilities in the downstream development of the wake vortex system. For aerodynamic force measurements and particle image velocimetry measurements, experiments are carried out in the Göttingen-type low-speed wind tunnel A of the Chair of Aerodynamics and Fluid Mechanics of the Technical University of Munich. Complimentary to the experimental data, unsteady Reynolds-averaged Navier–Stokes flow simulations are conducted. A baseline configuration with statically deflected flaps is compared to a configuration with actuated flaps. It can be shown experimentally as well as numerically that a significant disturbance can be imposed already in the near field of the wake vortex system by oscillating flaps.

Keywords: Dynamic flap motion · Wake vortex analysis

1 Introduction

Due to growing difficulties in airport and runway expansion, major airports around the world are facing a long-term capacity problem. The minimum separation distance between two aircraft during approach, takeoff, and enroute is solely driven by the maximum takeoff weight and therewith by the size of the two involved aircraft. The Federal Aviation Administration in the US and the Civil Aviation Authority in Europe regulate the separation distances, which are conservatively estimated to account for the different behaviour of the vortices under different atmospheric conditions. The aim is to maintain existing passenger capacities and to reduce the separation distance between two aircraft without reducing the level of safety already obtained today and therewith relieve hub airports. In this work the active influence on the wake vortex system of a generic transport aircraft by oscillating flaps is investigated. It has already been shown in several publications that an active excitation of instabilities can lead to a faster decay of the wake vortex system

© The Author(s), under exclusive license to Springer Nature Switzerland AG 2021
A. Dillmann et al. (Eds.): STAB/DGLR Symposium 2020, NNFM 151, pp. 739–748, 2021.
https://doi.org/10.1007/978-3-030-79561-0_70

[3–5,7,10,11]. However, the majority of the methods are referred to largely basic phenomena of multi-vortex systems or require additional systems. Crouch et al. [7] investigated the influence of actively controlled trailing edge flaps. Therefore no additional high lift elements have to be integrated into the wing structure of an aircraft. In this study, a half model of a generic transport aircraft in landing configuration, equipped with movable trailing edge flaps, is examined. In addition to investigations with static flap deflections, the flaps are also subject to periodic oscillations, which are intended to influence the wake vortex system downstream of the configuration. The aim is to impose a certain disturbance on the wake vortex system already in the near field to excite the long wave instability, called Crow instability [9], in the farfield. The reduced frequency k associated with the wavelength λ_{Crow} of the Crow instability is $k \approx (0.056 \pm 0.006) b/b_0$ for $\lambda_{Crow} \approx (9 \pm 1) b_0$ [5]. The goal is to impose this dominant frequency on the wake vortex structures already in the near field. This is to be achieved by the actuation frequency of the flaps corresponding to the frequency of the developing instability.

2 Experimental Technique

2.1 Wind Tunnel Model and Test Facility

The investigated geometry is a generic transport aircraft configuration based on the reference geometry LR-270 (Long Range 270). The reference geometry LR-270 [13] was created within the project BIMOD (Influencing maximum lift and wake vortex instabilities by dynamic flap movement), which is conducted by the Chair of Aerodynamics and Fluid Mechanics (TU Munich), the Institute of Aerospace Systems (RWTH Aachen) and the Institute of Structural Mechanics (RWTH Aachen). The long-range aircraft with a maximum take-off mass of 270 t is designed with the aircraft design software MICADO of the Institute of Aerospace Systems (RWTH Aachen). Based on the LR-270 geometry, a wind tunnel model has been derived. Figure 1 shows an isometric view of the wind tunnel model and informations on essential geometric parameters are summarized in Table 1. An advanced dropped hinge flap (ADHF) system is integrated at the trailing edge and a droop nose device is located at the leading edge along the halfspan s. The model has three flaps. The inboard flap (IF) can be deflected statically, the midboard flap (MF) and outboard flap (OF) can be operated both statically and dynamically. The spoiler deflection angle is kept constant. The midboard flap and outboard flap can be actuated anti-symmetrically to keep the global lift coefficient constant. The amplitude of the flap movement is determined by an excenter of the driving unit. The frequency of the oscillation can be adjusted via the rotational speed of the motor. To force turbulent boundary-layer characteristics, trip dots are installed at the leading edge of the model [6]. This ensures the comparability of the experimental data to numerical results obtained by unsteady Reynolds-averaged Navier-Stokes (URANS) simulations associated with fully-turbulent boundary layers. The reduced frequency k is defined as:

$$k = \frac{f(b/2)}{U_{ref}} \tag{1}$$

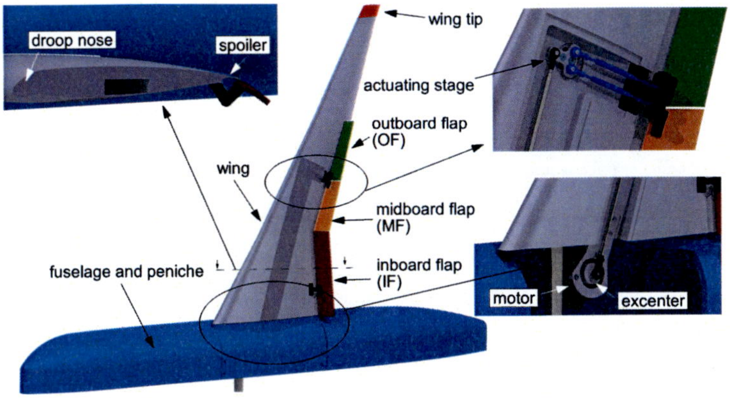

Fig. 1. Isometric and detaild view of the LR-270 model.

For a span of $b = 2.2\,\text{m}$, a reduced frequency of $k = 0.08$ [5] and a freestream velocity of $U_{ref} = 25\,\text{m/s}$, an actuation frequency of $f = 1.8$ Hz results. The actuation amplitude is set to $\Delta\delta = \pm 5^{\circ}$ to generate a noticeable varying circulation distribution along the half span. Experiments are carried out in the Göttingen-type low-speed wind tunnel (W/T) A of the Chair of Aerodynamics and Fluid Mechanics of the Technical University of Munich (TUM-AER) to conduct aerodynamic force measurements and particle image velocimetry measurements (PIV). The W/T A is operated with an open test section, which has a size of $1.8 \times 2.4 \times 4.8\,\text{m}^3$ (height $\times$ width $\times$ length). The experiments are performed at low subsonic speeds at a Reynolds number of $Re = 0.5 \cdot 10^6$ ($Re = \frac{U_{ref} \cdot c_{ref}}{\nu}$) and a Mach number of $Ma = 0.07$.

2.2 Particle Image Velocimetry

A stereo-PIV measurement system is used to measure the velocity field downstream of the model in several cross-flow sections. The Stereo-PIV system is mounted on a three-axis traversing system. A double pulsed Nd:Yag laser with a maximum power of 325 mJ per pulse and a wave length of 532 nm illuminate the measurement plane. Two sCMOS cameras with a resolution of 2560×2160 pixel were placed up- and downstream of the measurement plane. Seeding particles with a diameter of $d_{part} \approx 1\ \mu\text{m}$ were fed into the flow. The cameras recorded 400 image pairs per cross-flow section with a sampling frequency of $f_{meas} =$ 15 Hz for measurements with non actuated flaps. For triggered measurements with oscillating flaps, 200 image pairs per triggered phase (overall 8 phases) were recorded with a sampling frequency of 1.8 Hz matching the flap oscillation frequency. The presented quantities in this study are the mean velocity values determined from the 400 or 200 acquired samples. Information on the applied Stereo-PIV setup is summarized in Table 2.

Table 1. Essential geometrical data of the half model LR-270.

Scale			1:27.7
Wing			
	Half span	s	1.1 m
	Reference area	S_{ref}	0.233 m^2
	Aspect ratio	Λ	8.53
	Taper ratio	λ	0.12
	Reference chord	c_{ref}	0.312 m
	Sweep (25%-line) (inboard/outboard)	φ	30.08°/34.25°
	Dihedral	ν	$\sim 9°$
	Twist	ϵ	$-1.5°$ to $-4.7°$
	Thickness ratio (inboard/midboard/outboard)	c/t	0.132/0.11/0.094
High-Lift Devices			
	Droop nose deflection	DND	25°
	Spoiler deflection	SPD	6.8°
	Inboard flap deflection	IFD	20° – 40° (static)
	Midboard flap deflection	MFD	20° – 40° (actuable)
	Outboard flap deflection	OFD	20° – 40° (actuable)
Fuselage			
	Length	L	2.032 m
	Diameter	D	0.220 m
	Peniche height	H	0.075 m

Table 2. Stereo-PIV setup.

Angle between cameras	$\Theta = 60°$
Image pairs per section	400 (no flap actuation)/200 (flap actuation)
Final field of view	$d_y \times d_z \approx 0.45 \times 0.210\,m^2$
Number of vectors per section	$\approx$4500
Spatial resolution	$\Delta d = 1.5 \cdot 10^{-3}$ m
Non-dimensional spatial resolution	$\Delta d/s = 1.36 \cdot 10^{-3}$

The wing tip of the model at an angle of attack of $\alpha = 0°$ is defined as the zero point of the x-axis (flow direction). Measurements are conducted at the non-dimensional downstream positions $x/b = x^* = [0.064; 0.109; 0.2; 0.273]$ for experiments without actuated flaps and $x^*= [0.064; 0.273]$ for experiments with actuated flaps.

3 Numerical Setup

3.1 Computational Mesh

The generation of the computational mesh (see Fig. 2) is conducted with ANSYS ICEMCFD. A block structured 3D mesh featuring hexahedral elements is created. The boundary layer is resolved by choosing a dimensionless wall distance of $y^+ \leq 1$ on the entire half model geometry. Convergence and mesh independency studies were conducted to determine a reasonable size of the computational mesh and to ensure low numerical errors caused by the spatial discretization. A 40,000,000, 80,000,000 and 120,000,000 grid is used. Forces and moments have a deviation of < 0.6% for 40,000,000 compared to the 80,000,000 case. However, the 80,000,000 case was used, since a high grid resolution downstream had to be achieved. The geometric dimensions of the numerical domain coincides with the test section of the wind tunnel.

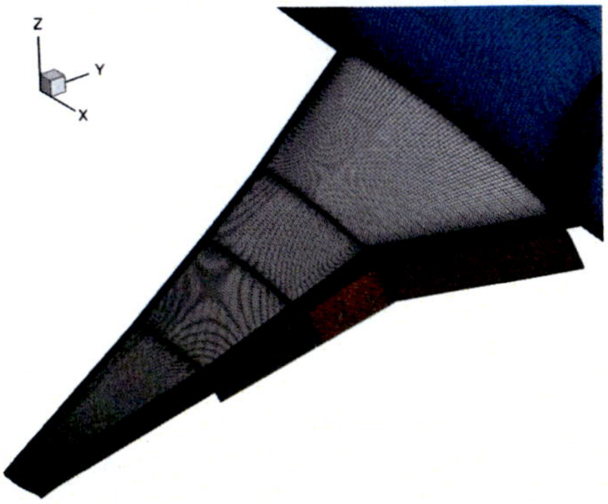

Fig. 2. Surface mesh of the wing section of the LR-270 numerical model.

3.2 Flow Solver

The numerical flow simulations are conducted with ANSYS FLUENT. The (unsteady) Reynolds-averaged Navier-Stokes equations (U/RANS) are solved. Turbulence modelling is performed by the shear stress transport $k-\omega$ SST turbulence model [12] and the turbulence properties at the boundaries are set in order to provide a turbulence intensity of $Tu = 0.2\%$ at the front of the geometry matching wind tunnel conditions. The COUPLED algorithm [1] treats the pressure-velocity coupling. The second-order pressure scheme is employed for the pressure interpolation and second-order upwind schemes are chosen for the spatial discretization of momentum, turbulent kinetic energy and specific dissipation rate. Moreover, a least squares cell-based formulation is used for the gradient calculation. A bounded second-order implicit scheme is selected for the temporal discretization [1]. The time step size Δt is selected by means of convergence studies, flow field analysis and frequency f of the flap actuation:

$$\Delta t = \frac{1}{f} \cdot \frac{1}{10^{\mathrm{N}}} \text{ with N} = 1, 2, 3 \tag{2}$$

In order to define the appropriate time step size, simulations with different time step sizes are carried out for N = [1; 2; 3]. A sufficient time step size resolution is achieved for N = 2. To model the motion of the midboard and outboard flap, user defined functions (UDFs) are implemented. The motion function is defined as:

$$g(t) = \Delta\delta \cdot sin(2\pi f \cdot t). \quad (3)$$

If the equation is differentiated with respect to time, the velocity equation is obtained:

$$g'(t) = \Delta\delta \cdot 2\pi f \cdot cos(2\pi f \cdot t). \quad (4)$$

Based on Eq. 4 the UDFs are created. As already mentioned, in this study the amplitude is set to $\Delta\delta = \pm 5°$ and the reduced frequency is set to $k = 0.08$. A negative $\Delta\delta$ corresponds to a reduction of the flap deflection angle and a positive $\Delta\delta$ corresponds to an increase of the flap deflection angle. Figure 3 shows the plotted function $g(t/T_P)$ with $T_P = 1/f$ and the maximum/minimum flap deflection settings for $k = 0.08$, $U_{ref} = 25$ m/s, $c_{ref} = 0.312$ m and the inboard flap deflected at 25° as well as midboard and outboard flap, the latter with a superimposed oscillation with ±5°. The computational mesh must be adapted for each time step applying dynamic mesh motion methods, which include layering and smoothing [1,2]. To achieve the specified convergence criteria (conservatives $< 10^{-6}$), 1200 CPUh were required for the baseline configuration and 13800 CPUh for the oscillating flap configuration.

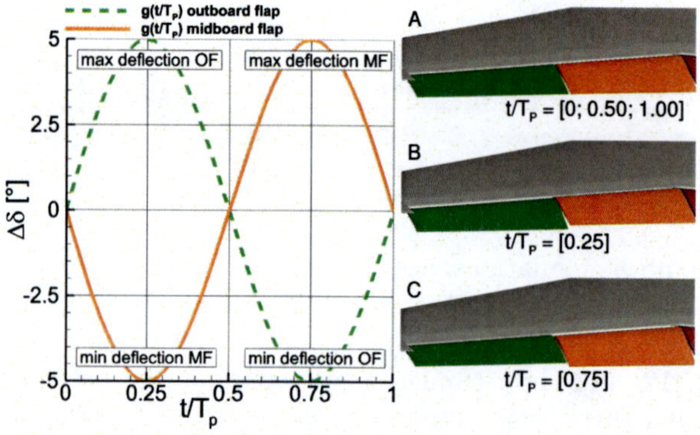

Fig. 3. Motion function for midboard and outboard flap (left). Flap deflections at t/T_p= [0; 0.5; 1.0] (A), t/T_p= [0.25] (B) and t/T_p= [0.75] (C) (right).

4 Results

The following chapter presents experimental and numerical results of the baseline high-lift configuration with statically flap deflections at IFD = MFD = OFD = 25° (approach setting) and the configuration with oscillating midboard and outboard flap (f = 1.8 Hz, $\Delta\delta = \pm 5°$). The discussion comprises the non-dimensional axial vorticity, the non-dimensional vorticity magnitude $\xi_{(x)} = \frac{\omega_{(x)} b}{2U_{ref}}$ ($\omega \equiv$ vorticity) and the non-dimensional lateral amplitude.

4.1 Baseline Configuration

The flow field is analyzed for the LR-270 configuration at $\alpha = 12°$ (angle of attack), $C_L = 1.4$ (lift coefficient), $Re = 0.5 \cdot 10^6$ and $Ma = 0.07$ in four cross-flow sections $x/b = x^* = [0.064; 0.109; 0.2; 0.273]$. The following designations are introduced for the discrete vortices: Wing-tip vortex (WTV), outboard flap vortex (OFV), midboard flap vortex (MFV), inboard flap vortex (IFV) and inboard flap fuselage vortex (IFFV). Figure 4 shows an isometric view of the cross-flow fields of the non-dimensional vorticity magnitude ξ of PIV measurements (left) and U/RANS simulations (right). There is a good agreement of the vortex patterns between experimental and numerical data. The cross-flow sections $x^* = [0.064; 0.273]$ are examined in more detail in Fig. 5.

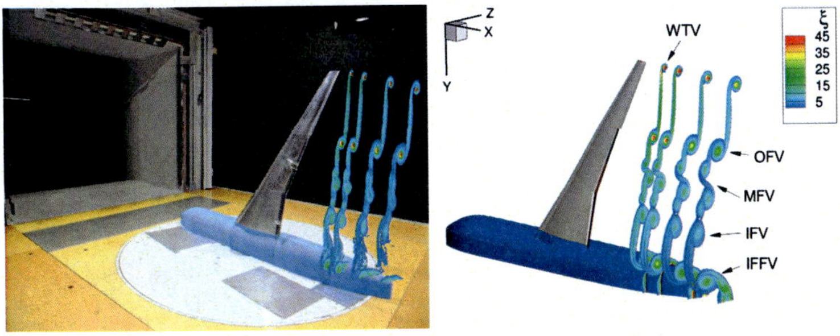

Fig. 4. Non-dimensional vorticity magnitude ξ (PIV-measurements (left) and U/RANS simulation (right)) for non-dimensional downstream distances $x^* = [0.064, 0.109, 0.2, 0.273]$ at $Re = 0.5 \cdot 10^6$, $Ma = 0.07$, $\alpha = 12°$, $C_L = 1.4$, IFD = MFD = OFD = 25°, k = 0 and $\Delta\delta = 0°$.

The structures around the MFV differ due to the unresolved wing-flap levers in the numerical simulation. The beginning of the wake vortex roll-up process can already be seen in the near field at the WTV for experimental and numerical data. The WTV moves along the vortex sheet towards the OFV. The two strongest vortices based on peak vorticity levels are the OFV and the WTV with the latter clearly being the strongest of all such distinct vortices. The maximum peak vorticity level of the OFV is associated with the highly loaded outboard flap. The high vorticity level of the WTV is caused by the wing tip geometry (no winglet). The MFV and IFV each result from a superposition of the discrete vortices that are created at the respective side edges of the IF, MF and OF. Comparing the position of the vortex structures of the experimental data with the numerical data, a good agreement can be seen. However, the axial vorticity in the numerical data is underestimated compared to the experimental data for $x^* = 0.273$. One reason for this behavior is the dissipation rate of the numerical turbulence model. However, this is not necessary for the further course of the project, from $x^* = 0.064$ onward the downstream development of the vortex system is solved by a large eddy simulation (LES) setup. For the initialization of

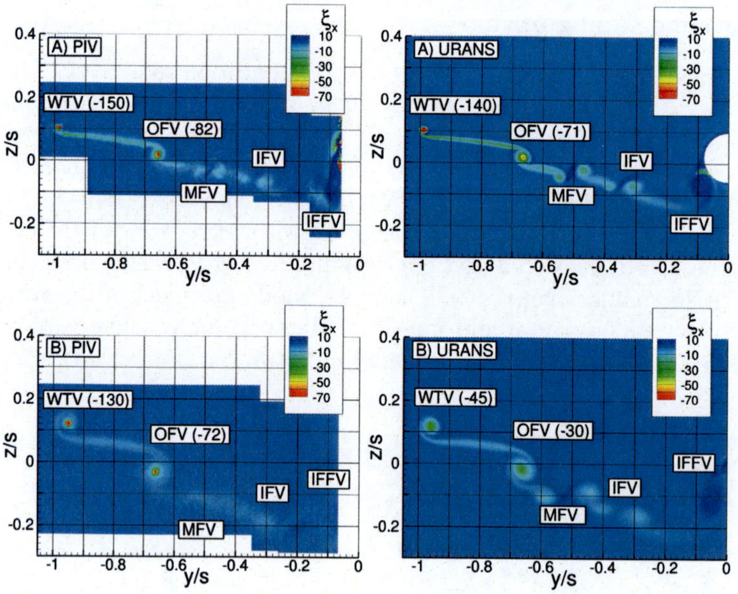

Fig. 5. Non-dimensional axial vorticity ξ_x (PIV-measurements (left) and U/RANS simulation (right)) for non-dimensional downstream distances $x^* = [0.064\ (A), 0.273\ (B)]$ at $Re = 0.5 \cdot 10^6$, $Ma = 0.07$, $\alpha = 12°$, $C_L = 1.4$, $IFD = MFD = OFD = 25°$, $k = 0$ and $\Delta\delta = 0°$.

the LES the velocity field and the turbulence intensity of the U/RANS solutions are impressed on the LES setup at $x^* = 0.064$.

4.2 Oscillating Flap Configuration

Now, the results of the configuration with actuated flaps are considered. The inboard flap deflection angle is set statically to IFD = 25°. MF and OF oscillate inversely at f = 1.8 Hz (k = 0.08), $\Delta\delta = \pm 5°$ at 25° flap deflection. Eight phases at t/T_p = [0; 0.125; 0.25; 0.375; 0.50; 0.625; 0.750; 0.875; 1.00] are triggered (see Fig. 3). The global pattern of the wake vortex system is comparable to the baseline configuration. The decisive factor is whether a significant disturbance can be imposed on the flow by the flap oscillation. For this reason it is useful to analyze the local oscillation of the OFV compared to the OFV motion of the baseline case due to meandering. Figure 6 shows the transient local position of the OFV core (PIV and URANS) in dependency of the triggered phase at $x^* = 0.064$ and $x^* = 0.273$. There is a good agreement between the numerical and experimental data regarding the spatial movement. Also marked is the range (black circle) in which the OFV core moves for the baseline case (meandering). The range was determined using the 400 individual image pairs of the PIV measurements of the baseline case. For each image pair the position of the vortex core was determined and the circle includes all these vortex core positions. Thus, it can be shown that a strong disturbance can be introduced into the vortex system. Furthermore it can be noticed that the lateral amplitude ($\Delta A^* = \sqrt{(\Delta \frac{y}{s})^2 + (\Delta \frac{z}{s})^2}$) of the vortex motion increases downstream from x* = 0.064 ($\Delta A^* \approx 0.011$) to x* = 0.273 ($\Delta A^* \approx 0.018$) for PIV measurements and URANS simulations compared to the baseline case ($\Delta A^* \leq 0.001$). It has been shown that increasing the disturbance level fosters the Crow instability amplification rate, thus leading to an earlier wake vortex decay [8]. Frequency dependent forcing may markedly enhance this process [3].

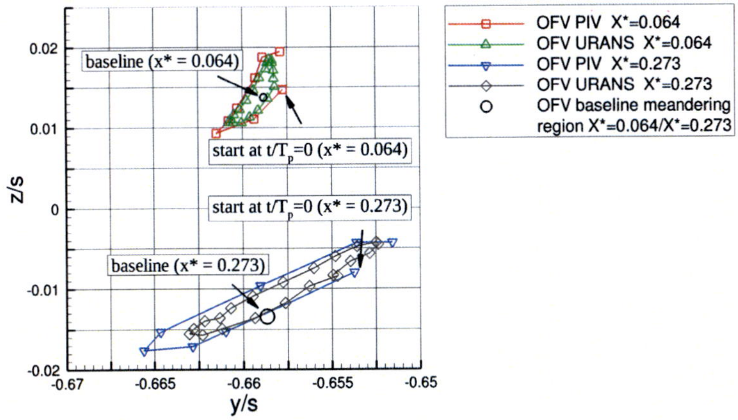

Fig. 6. Comparison of baseline case and oscillating flap case regarding OFV oscillation for non-dimensional downstream distances x* = [0.064; 0.273] at Re = $0.5 \cdot 10^6$, Ma = 0.07, $\alpha = 12°$, $C_L = 1.4$, IFD = MFD = OFD = 25°, k = 0.08 and $\Delta\delta = \pm 5°$.

5 Conclusion

The present investigations concentrate on forced motion disturbances on the wake vortex system of a generic transport aircraft already in the near field by means of oscillating flaps. In this way, a long-wave instability (Crow instability) is supposed to be excited in the further downstream course of the wake vortex system. The investigations are carried out using both an experimental and numerical setup. The generic transport aircraft model is investigated under wind tunnel conditions at Mach number Ma = 0.07 and Reynolds number Re = $0.5 \cdot 10^6$ with approach settings (angle of attack $\alpha = 12°$, lift coefficient $C_L = 1.4$, flap settings 25°, droop nose deflection 25° and spoiler deflection 6.8°). A baseline configuration with static flap deflection is compared to a configuration with oscillating trailing edge flaps. The model has three trailing edge flaps. The inboard flap is at any time statically deflected. The midboard and outboard flaps can be actuated inversely. The examined actuation reduced frequency is 0.08 and the amplitude of the flap movement is set to $\Delta\delta = \pm 5°$. It is shown that a strong disturbance can be introduced into the flow field by means of oscillating flaps. The non-dimensional lateral amplitude ΔA^* of the outboard flap vortex motion increases from $\Delta A^* \approx 0.011$ (non-dimensional distance $x^* = 0.064$) to $\Delta A^* \approx 0.018$ ($x^* = 0.273$) compared to a configuration with non-actuated flaps ($\Delta A^* \leq 0.001$). For the non-dimensional distance $x^* = 0.064$, the experimental and numerical data are in good agreement with regard to the non-dimensional axial vorticity. The discrepancy for the non-dimensional axial vorticity of experimental and numerical data increases downstream, where the numerical setup underestimates the experimental results.

Acknowledgements. The funding of this investigation within the LUFO V3 project BIMOD (FKZ: 20E1702C) by the German Federal Ministry for Economic Affairs and Energy (BMWi), is gratefully acknowledged. Furthermore, the authors want to thank ANSYS for providing the flow simulation software.

References

1. ANSYS Fluent Theory Guide. Release 17.0. ANSYS, Inc., Southpointe Canonsburg USA, January 2016
2. ANSYS Fluent User's Guide. Release 17.0. ANSYS, Inc., Southpointe Canonsburg USA, January 2016
3. Allen, A.: Beeinflussung des Nachlaufwirbelsystems von Transportflugzeugen. Techn. Univ., Diss, München (2008)
4. Bellastrada, C.: Minimization of wake vortex hazard trough differential flap settings. Techn. Univ., Diss, München (2012)
5. Breitsamter, C.: Nachlaufwirbelsysteme großer Transportflugzeuge: Experimentelle Charakterisierung und Beeinflussung. Techn. Univ., Habil.-Schrift, München (2007)
6. Breitsamter, C.: Strömungsphysik und Modellgesetze, Lect. Notes Technische Universität München (2018)
7. Crouch, J.D., Miller, G.D., Spalart, P.R.: Active-control system for breakup of airplane trailing vortices. AIAA J. **39**(12), 2374–2381 (2001)
8. Crow, S.C., Bate, E.J.: Lifespan of trailing vortices in a turbulent atmosphere. AIAA J. **13**(7), 476–482 (1976)
9. Crow, S.C.: Stability theory for a pair of trailing vortices. AIAA J. **8**(12), 2172–2179 (1970)
10. Greenblatt, D.: Flap vortex management using active gurney flaps. AIAA J. **47**(12), 2845–2856 (2009)
11. He, Y., Yang, J.W., Bao, F.: Wake vortex control using modified flaps. Appl. Mech. Mater. **365–366**, 827–834 (2013)
12. Menter, F.: Two-equation eddy-viscosity turbulence models for engineering applications. AIAA J. **32**(8), 1598–1605 (1994)
13. Stephan, R., et al.: Influence of dynamic flap movement on maximum lift and wake vortex evolution. Deutscher Luft- und Raumfahrtkongress, DocumentID: 490118 (2019)

Aeroacoustic Noise

Check for updates

Fast Non-empiric Propeller Source Noise Model with Mean Flow

Andrea Franco(✉), Roland Ewert, Malav Mukesh Soni, Michael Mößner, and Jan Werner Delfs

Department of Technical Acoustics German Aerospace Center (DLR), Institute of Aerodynamics and Flow Technology, Lilienthalplatz 7, 38108 Braunschweig, Germany
andrea.franco@dlr.de

Abstract. A fast and physical-principles-based method to compute the noise radiation for installed propellers' applications is introduced. The simulation of the sound propagation includes newly defined perturbation equations, with propeller sound sources modelled as rotating singularities. The system of equations proposed is derived from the Linearised Euler Equations, after the application of a split-approach on the velocity perturbations. Two separate coupled systems of equations are obtained, the Acoustic Perturbation Equations, governing the acoustic velocity and pressure perturbations, and the Vortical Convection Equations, describing the convection of non-acoustic velocity perturbations. The moving sources are regularised with a Gaussian kernel over the computational mesh, allowing their description within an unstructured quadrature-free Discontinuous Galerkin experimental solver for Computational Aeroacoustics applications. The prediction capabilities of the method proposed have been successfully validated with a simple test case.

Keywords: CAA · Propeller noise · APE+VCE

1 Introduction

The demand for novel sustainable aerial vehicles' concepts satisfying the EU Flightpath 2050 guidelines, opens new challenges for aircraft manufacturers concerning re-evaluation of the available propulsion technologies and aircraft re-design by means of unconventional propulsion integration configurations. Regardless of the configuration chosen, a requirement to fulfill for future aircraft concepts is a perceived reduction of 65% of noise emissions by 2050. Recent advances in aircraft propulsion configurations include distributed electric propulsion systems, and buried or semi-buried engine integrations on the airframe [1]. The noise abatement milestones are guiding research of propeller aeroacoustics in providing manufacturers with new design methods and tools capable of accurately and reliably estimating propeller noise emissions for any arbitrary arrangement. This work presents a fast and accurate method for computations of highly integrated propellers, developed from previous models of Gutin [2] and Lowson [3], who replaced the propeller blades by rotating source singularities.

© The Author(s), under exclusive license to Springer Nature Switzerland AG 2021
A. Dillmann et al. (Eds.): STAB/DGLR Symposium 2020, NNFM 151, pp. 751–760, 2021.
https://doi.org/10.1007/978-3-030-79561-0_71

The method introduced is implemented in a Computational Aeroacoustics (CAA) framework to address highly non-uniform mean flows, as well as installation noise predictions. Source singularities avoid the resolution of the propeller geometry, increasing the method flexibility. A similar approach was considered by Dierke et al. [9] in the finite difference block-structured CAA solver PIANO of DLR. In this work, the method is implemented within a quadrature-free Discontinuous Galerkin (DG) scheme, solving the governing equations on unstructured grids, thus, reducing the effort to generate computational meshes for highly complex geometries.

Alongside the proposed propeller model, a suitable robust system of equations for acoustic computations is required, which is capable of appropriately representing acoustic and vortical eigenmodes. Linearised Euler Equations (LEE) include the aforementioned eigenmodes and have been extensively used in the past for CAA applications; however, they would not be free from instabilities, such as the unbounded Kelvin-Helmholtz type. Motivated by the work of Goldstein [4], and further developments by Atassi et al. [5,6], a newly defined system of equations is presented, based on a velocity-split approach, separating acoustic and vortical modes as introduced in [7].

The propeller noise prediction method of this contribution is at the same time fast and physical-principles-based, and able to address both conventional and unconventional propulsion configurations, without requiring a priori empirical knowledge. Initial results of the implementation in the DG CAA experimental solver DISCO++ of DLR are reported. A rotating singularity is considered as a test case, to validate the concept against known analytical solutions. To address future installation noise predictions applications with arbitrary non-uniform inflow conditions, the rotating singularities model will be extended with line-distributed rotating source singularities of strengths determined from actuator disk RANS simulations.

2 Propeller Source Noise Model Implementation

The propeller source noise model is based on a rotating source singularity, representative either of loading or thickness noise contributions, or a combination of these two self-noise source components. This modelling choice allows to simulate a propeller without requiring moving surfaces or the exact blade geometry in the CAA simulations.

2.1 Calculation of the Time Dependent Source Singularity Location

Each propeller blade is replaced by a single source singularity, rotating around the propeller axis as time advances. The location of the singularities is calculated in a local cylindrical coordinate system, centered at the hub of the physical propeller to be modelled, with initial location $\mathbf{X_{init}} = (r_{init}, \phi_{init}, z_{init})^T$ and updated circumferential coordinate ϕ_{time} computed from the source's Revolutions Per Second (RPS) prescribed in input, as seen in Fig. 1.

Multiple propeller blades N_{blade} are modelled through a circumferential offset $2\pi/N_{blade}$ applied to ϕ_{time}. The sources' values would be defined either by the integrated load vector $\mathbf{f} = (f_r, f_\phi, f_z)^T$ or the displacement $\dot{\theta}$ of the propeller on the air, and are included appropriately in the sources vector $\mathbf{S} = (S_i)^T$, where i depends on the perturbation equation of interest.

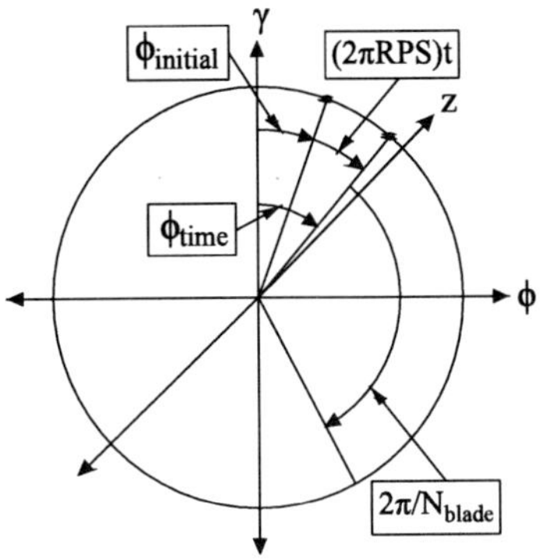

Fig. 1. Propeller cylindrical coordinate system

2.2 Regularisation of Sources

The definition of a singularity in space-discretised schemes, as the DG-method considered in this work, is a difficult task to consider, and is frequently the cause for numerical instabilities. Rather, the source must be approximated with a distributed value, which can be incorporated in the computational mesh. This is accomplished by regularising the values of the sources on the grid using a 3D isotropic Gaussian kernel $f(r, \phi, z)$:

$$f(r, \phi, z) = \left(\frac{ln(2)}{\pi}\right)^{\frac{3}{2}} \frac{1}{\varepsilon} \exp\left(\frac{ln(\frac{1}{2})d^2}{\varepsilon^2}\right) \quad (1)$$

where d is the distance between the current position of the singularity in motion and a grid point, and ε is the half-width at half-maximum of the Gaussian probability density function, representing the distance over which the Gaussian distribution drops from the maximum value to half its maximum. The integrated source values prescribed do not correspond to the required fluctuating source components to be included in perturbation equations. Instead they are described as the sum of a steady mean and a fluctuating component, where only the second component acts as a source in the perturbation equations.

For this reason, the steady mean value $\mathbf{S_{mean}}$ to be subtracted from the input source values is obtained by integrating over a circle the Gaussian regularised source. The resulting fluctuating component is obtained as $\mathbf{S'} = \mathbf{S} - \mathbf{S_{mean}}$, and after being scaled for the perturbation equation of interest, it is assigned to the appropriate high order point determined by the DG discretisation.

3 System of Perturbation Equations

In this work, a newly defined system of equations combined with the propeller model is presented, which is suited to represent appropriately the interaction of acoustic and vortical eigenmodes, and is robust enough to address applications including complex geometries. The newly proposed system of equations is derived from the LEE, shown here in their primitive variables formulation:

$$\frac{\partial \rho'}{\partial t} + \nabla \cdot (\rho_0 \mathbf{u}' + \rho' \mathbf{u_0}) = \dot{\theta}' \tag{2}$$

$$\frac{\partial \mathbf{u}'}{\partial t} + (\mathbf{u_0} \cdot \nabla)\mathbf{u}' + (\mathbf{u}' \cdot \nabla)\mathbf{u_0} + \frac{\nabla p'}{\rho_0} - \rho' \frac{\nabla p_0}{\rho_0^2} = \frac{\mathbf{f}'}{\rho_0} \tag{3}$$

$$\frac{\partial p'}{\partial t} + \mathbf{u_0} \cdot \nabla p' + \mathbf{u}' \cdot \nabla p_0 + \gamma p_0 \nabla \cdot \mathbf{u}' + \gamma p' \nabla \cdot \mathbf{u_0} = \frac{c_0^2}{\gamma} \dot{\theta}' \tag{4}$$

where ρ, u, p, t, and γ are respectively the density, velocity, pressure, time, and specific heat ratio of air, with subscript 0 indicating mean flow quantities and primed symbols indicating perturbation quantities. Assuming isentropic flow conditions described by the linearised expression $p' = \rho' c_0^2$, with c_0 as the speed of sound computed from the mean flow, Eq. (2) can be rewritten:

$$\frac{\partial p'}{\partial t} + c_0^2 \nabla \cdot \left(\mathbf{u_0} \frac{p'}{c_0^2} + \rho_0 \mathbf{u}' \right) = c_0^2 \dot{\theta}' \tag{5}$$

As a consequence of introducing Eq. (5), Eq. (4) can be neglected and, including the isentropic relation written as $\rho' = p'/c_0^2$, the isentropic formulation of the LEE (LEE-s) reads:

$$\frac{\partial p'}{\partial t} + c_0^2 \nabla \cdot \left(\mathbf{u_0} \frac{p'}{c_0^2} + \rho_0 \mathbf{u}' \right) = c_0^2 \dot{\theta}' \tag{6}$$

$$\frac{\partial \mathbf{u}'}{\partial t} + (\mathbf{u_0} \cdot \nabla)\mathbf{u}' + (\mathbf{u}' \cdot \nabla)\mathbf{u_0} + \nabla \left(\frac{p'}{\rho_0} \right) = \frac{\mathbf{f}'}{\rho_0} \tag{7}$$

$$\rho' = \frac{p'}{c_0^2} \tag{8}$$

The proposed newly defined system of equations is obtained from Eq. (6), Eq. (7), and Eq. (8), by rewriting Eq. (7) using the mathematical identity:

$$(\mathbf{u_0} \cdot \nabla)\mathbf{u}' + (\mathbf{u}' \cdot \nabla)\mathbf{u_0} = \nabla(\mathbf{u}' \cdot \mathbf{u_0}) + \omega' \times \mathbf{u_0} + \omega_\mathbf{0} \times \mathbf{u}' \tag{9}$$

with $\omega' = \nabla \times \mathbf{u}'$ and $\omega_\mathbf{0} = \nabla \times \mathbf{u_0}$. The perturbation velocities $\mathbf{u}'$ of Eq. (7) are then split as $\mathbf{u}' = \mathbf{u^a} + \mathbf{u^r}$, where $\mathbf{u^a} = \nabla \varphi$ is the potential (acoustic) component and $\mathbf{u^r}$ is the non-acoustic component including the information of the initial solenoidal field of the LEE as well. In the resulting split momentum equation for $\mathbf{u^a}$, the term $\omega_\mathbf{a} \times \mathbf{u_0}$ is identically null, since $\omega_\mathbf{a} = \nabla \times \mathbf{u_a} = \mathbf{0}$.

Re-arranging the terms among the split momentum equations, the system of equations reads:

$$\frac{\partial p'}{\partial t} + c_0^2 \nabla \cdot \left(\mathbf{u_0} \frac{p'}{c_0^2} + \rho_0 \mathbf{u^a} \right) = -c_0^2 \nabla \cdot (\rho_0 \mathbf{u^r}) + c_0^2 \dot{\theta}' \quad (10)$$

$$\frac{\partial \mathbf{u^a}}{\partial t} + \nabla (\mathbf{u^a} \cdot \mathbf{u_0}) + \nabla \left(\frac{p'}{\rho_0} \right) = -\mathbf{X} \quad (11)$$

$$\frac{\partial \mathbf{u^r}}{\partial t} + (\mathbf{u_0} \cdot \nabla)\mathbf{u^r} + (\mathbf{u^r} \cdot \nabla)\mathbf{u_0} = -\omega_0 \times \mathbf{u^a} + \mathbf{X} + \frac{\mathbf{f}'}{\rho_0} \quad (12)$$

Equations (10) and (11), without the source term $-c_0^2 \nabla \cdot (\rho_0 \mathbf{u^r})$, are the APE-4 [8] as being used in CAA to predict the propagation of acoustic modes over a non-uniform medium. Equation (12) is an additional transport equation, hereafter referred to as Vortical Convection Equation (VCE), representing the convection of non-acoustic velocity perturbations on a background mean flow. The system of equations composed of Eq. (10), Eq. (11), and Eq. (12) is named APE+VCE, and is equivalent to the LEE-s, as it can be verified by adding Eq. (11) with Eq. (12). Furthermore, the velocity splitting is not required to provide exactly distinct potential and solenoidal components of the velocity, as described in [7]. To formulate a mathematically equivalent system of equations to the LEE-s, the vector field $\mathbf{X}$ was introduced, without any assumption being made on its nature with respect to the eigenmodes of the equation. $\mathbf{X}$ can be freely defined to suit a specific velocity-split definition, hence, the splitting is general and can be flexibly used for installation noise prediction purposes, in which vortical and acoustic eigensolutions are non-distinct. In the remainder of this work, it was decided to set $\mathbf{X} = 0$, in order to avoid the introduction of additional sources in Eq. (11), thus, verifying the possible application of the proposed equations to simplified test cases. The original fluctuating source terms of the LEE can be assigned to the RHS of Eq. (10) for externally prescribed mass sources, and to the RHS of Eq. (12) for externally prescribed fluctuating forces.

4 Validation

The proposed method, combining the propeller model with the APE+VCE, has been validated, and the results are shown in the form of polar directivities and a narrow band spectrum of the fluctuating pressure signal. Tests have been conducted for both a rotating force and mass singularity, in a uniform flow aligned with the axis of rotation of the sources. A sensitivity study for the appropriate choice of ε was considered, and the results are presented as a comparison of polar directivities for different chosen frequency modes. Additional comparisons involve the solution of various perturbation equations, viz. LEE (Eq. (2), Eq. (3), Eq. (4)), LEE-s (Eq. (6), Eq. (7), Eq. (8)), APE+VCE (Eq. (10), Eq. (11), Eq. (12)), and APE (Eq. (10), Eq. (11) without $-c_0^2 \nabla \cdot (\rho_0 \mathbf{u^r})$).

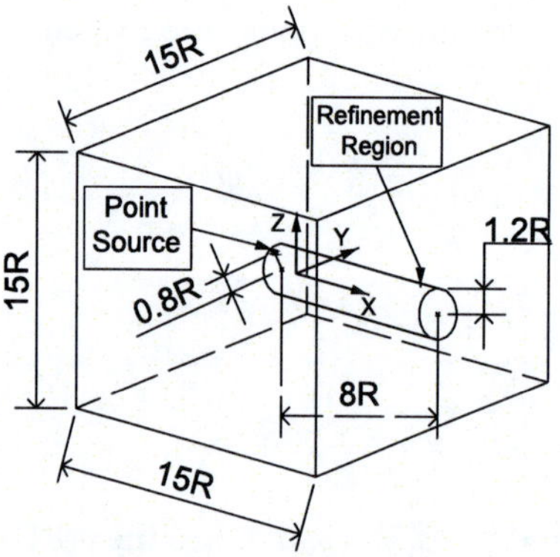

Fig. 2. Computational domain

Table 1. Simulations input parameters

Input Parameter	Value
Free Stream Mach nr. M_∞	0.15
Revolutions Per Second	119.09 s^{-1}
Refer. Propeller Blade Length	0.33 m
Thrust	152.09 N
Blade Volume	0.00072 m^3
Farf. Tetra Cell Size Δx_{Far}	0.4326
Refin. Tetra Cell Size Δx_{Ref}	0.04326
Number of Time Steps	400,000
Time Step Size	0.0001

The computational domain is sketched in Fig. 2. Table 1 summarises the simulation input parameters, including the strengths of the source singularities. To resolve the source singularities regularised with Eq. (1), a cylindrical refinement region is generated, with a radial extension chosen to prevent a sharp transition of grid cells' size in the area of vortex propagation. Within the refinement region, each type of singularity is located at 0.8R in the radial direction to best represent loading and blade thickness effects for a propeller. A non-reflecting boundary condition, based on an upwind flux vector splitting approach, with prescribed null incoming eigenmodes, is applied on the domain outer surfaces. The computational time was approximately equivalent to four complete revolutions of the rotating singularity. Δx_{Far} was chosen to resolve the first ten harmonics. All the results are shown in terms of Sound Pressure Level (SPL), considering a ratio of $\varepsilon/\Delta x_{Ref} = 2$, to appropriately resolve the regularised source on the computational grid. To better appreciate the energy content of the solution, each SPL result is described considering the 1st, 3rd, and 5th frequency mode of the fluctuating pressure signals, captured on the z-x coordinate plane. Microphones are placed at a distance of 10R from the origin of the cylindrical coordinate system of the propeller. The results obtained from CAA simulations are compared with related results from known analytical equations for moving source singularities, formulated either in presence or absence of flow. Equations (13a) and (13b) describe respectively the pressure perturbation field for a given mass θ_p or force singularity $\mathbf{f_p}$.

$$p'_{\theta_p}(\mathbf{x},t) = \frac{1}{4\pi}\left\{\frac{\frac{\partial \theta_p}{\partial \tau} + \mathbf{e_R}\cdot\frac{\partial \mathbf{M_q}}{\partial \tau}(1-M_{qR})^{-1}\theta_p}{R_i(1-M_{qR})^2} + \frac{a_\infty(M_{qR}-\mathbf{M}_q^2)}{R_i^2(1-M_{qR})^3}\theta_p\right\} \tag{13a}$$

$$\begin{aligned} p'_{f_p}(\mathbf{x},t) = \frac{1}{4\pi}&\left\{\frac{\frac{\partial \mathbf{f_p}}{\partial \tau}\cdot\mathbf{e_R} + (\mathbf{f_p}\cdot\mathbf{e_R})\frac{\partial \mathbf{M_q}}{\partial \tau}\cdot\mathbf{e_R}(1-M_{qR})^{-1}}{a_\infty R_i(1-M_{qR})^2}\right\} \\ +\frac{1}{4\pi}&\left\{\frac{-\mathbf{f_p}\cdot\mathbf{M_q} + (1-\mathbf{M_q}^2)\mathbf{f_p}\cdot\mathbf{e_R}(1-M_{qR})^{-1}}{R_i^2(1-M_{qR})^2}\right\} \end{aligned} \tag{13b}$$

$\mathbf{M_q}$ represents the Mach number of the source motion, $\mathbf{x}$ the observer position, $\mathbf{y}(\tau)$ the singularity position dependent on the retarded time τ, $R_i = |\mathbf{R_i}| = |\mathbf{x} - \mathbf{y}(\tau)|$ the distance between source and observer, $\mathbf{e_R} = \mathbf{R_i}/R_i$ the unit distance vector between source and observer, $\mathbf{M_{qR}} = \mathbf{M_q} \cdot \mathbf{e_R}$ the momentary Mach number component in the direction of the observer, and a_∞ the speed of sound. If a uniform flow is considered, the distance R_i will change as $R_i = |\mathbf{R_i}| = |\mathbf{x} - \mathbf{y}(\tau) - \mathbf{U}_\infty(\tau - t)|$, and the Mach number will change as well as $\mathbf{M_q} = \mathbf{M_q} + \mathbf{U}_\infty$, with $\mathbf{U}_\infty$ representing the free stream velocity vector.

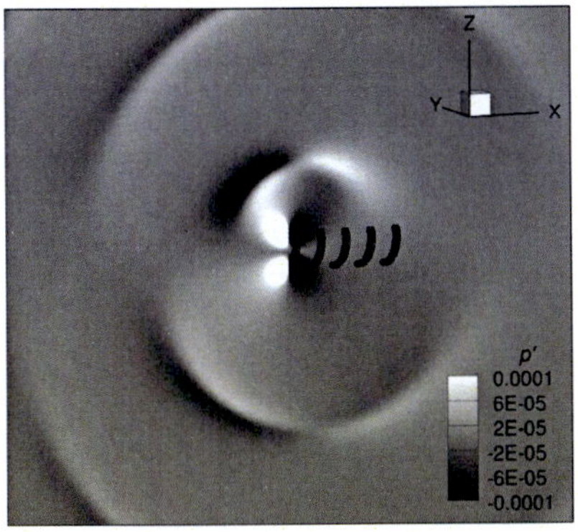

Fig. 3. Contour plot of the fluctuating pressure for simulated thrust singularity, with Q-Criterion superimposed

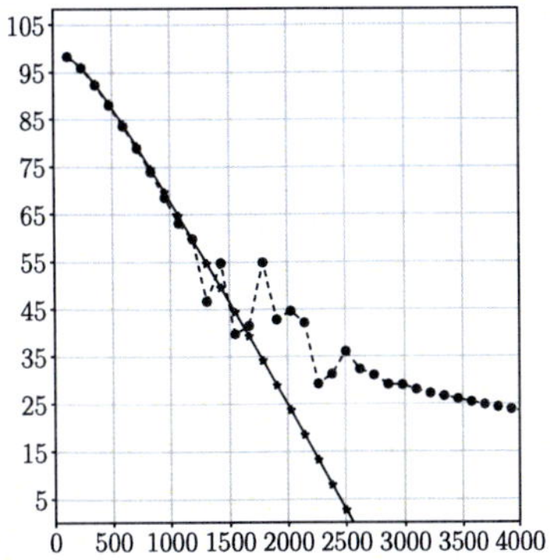

Fig. 4. Narrow band spectrum of p', simulated thrust, $M_\infty = 0.15$. Horizontal axis: frequency (Hz), Vertical axis: SPL (dB). The lines in the plot correspond to Analytical results (*solid*), APE+VCE results (*dash*)

A qualitative inspection of the results of a simulated thrust force with the APE+VCE system of equations exhibits the fundamental features to be expected from the simplified test case investigated. Figure (3) presents the pressure fluctuations contour, with Q-Criterion iso-surfaces superimposed, computed from $\mathbf{u^r}$. The proposed system of equations describes qualitatively well the radiated sound pattern. Quadrupole-type pressure fluctuation patterns can be observed. A ring vortex pattern, generated due to the rotation of the force singularity, results from the Q-Criterion iso-surfaces. An important feature of propeller noise prediction-methods is the ability to capture the logarithmic decreasing trend of the harmonics, as highlighted in Fig. 4, where the narrow band spectrum of the fluctuating pressure signal at a location of 45° is shown. The results are obtained simulating a thrust force with $M_\infty = 0.15$. Each marker in the plot refers to a single harmonic. The SPL of the first 10 harmonics presents the expected rapid decreasing trend in dB values, as higher harmonics are investigated.

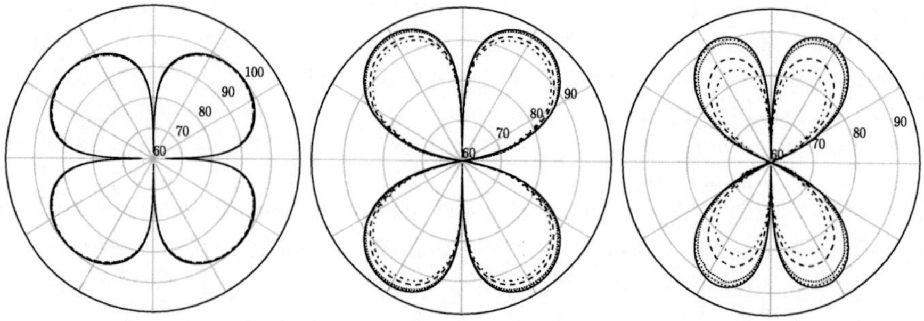

Fig. 5. SPL (dB) for different $\varepsilon/\Delta x_{Ref}$. Simulated thrust with $M_\infty = 0$. From left to right are shown the 1st, 3rd and 5th frequency mode. The lines in each plot correspond to: Analytical results (*solid*), $\varepsilon/\Delta x_{Ref} = 2$ (*loose dot*), $\varepsilon/\Delta x_{Ref} = 4$ (*dot*), $\varepsilon/\Delta x_{Ref} = 8$ (*dash*), and $\varepsilon/\Delta x_{Ref} = 10$ (*dashdot*)

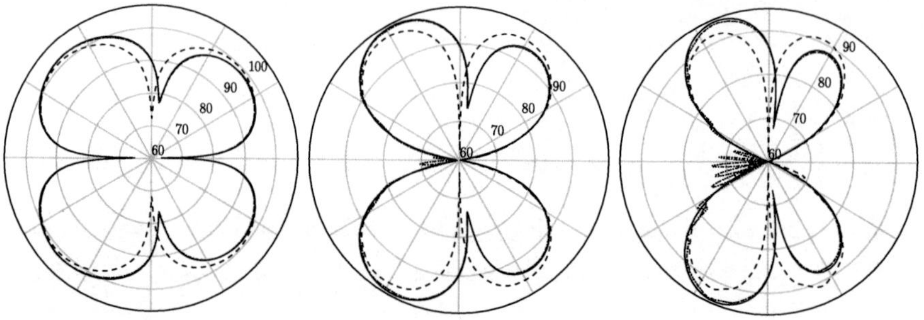

Fig. 6. SPL (dB) equations comparison. Results of a simulated thrust with $M_\infty = 0.15$. From left to right are shown the 1st, 3rd and 5th frequency mode. The lines in each plot correspond to: Analytical results (*solid*), LEE (*loose dot*), LEE-s (*dot*), APE+VCE (*dashdot*), and APE (*dash*)

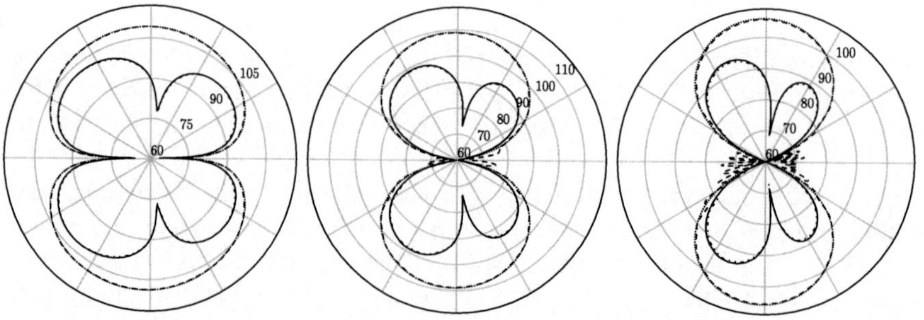

Fig. 7. SPL (dB) Thrust and mass singularities comparison, with $M_\infty = 0.15$. From left to right are shown the 1st, 3rd and 5th frequency mode. The lines in each plot correspond to: Analytical Thrust results (*solid*), Analytical Mass results (*dot*), APE+VCE Mass results (*dash*), APE+VCE Thrust results (*dashdotdot*)

For frequencies higher 1100 Hz, large SPL values are seen, which might be due to a limited mesh resolution or spurious numerical noise. In Fig. 5, the results of a sensitivity study for different ratios $\varepsilon/\Delta x_{Ref}$ are shown, with a single thrust force source in the APE+VCE equations and with $M_\infty = 0$. For the 1st mode, the louder levels are observed at around 45° in the first quadrant with symmetrical values in other quadrants. Higher modes show a shift of the maximum levels location towards 90° and 270°. The energy content of the 1st mode is seen not to depend on the choice of $\varepsilon/\Delta x_{Ref}$. For the 3rd mode, the SPL becomes sensitive towards $\varepsilon/\Delta x_{Ref}$ ratios, showing that larger values of $\varepsilon/\Delta x_{Ref}$ influence the higher harmonics. When analyzing the results of the 5th mode, the differences obtained with different $\varepsilon/\Delta x_{Ref}$ are clearly visible: values up to $\varepsilon/\Delta x_{Ref} = 4$ show similar levels, whereas, higher $\varepsilon/\Delta x_{Ref}$ produce a reduction in SPL. This indicates how a singularity, regularised over several computational grid points, would exhibit a modification of the radiated sound wave with a considerable reduction in its energy content. Microphones have been placed at a distance 10R to avoid the nearfield effects. Axial noise emission, as seen from the SPL of the 5th mode, indicates the presence of numerical noise otherwise not present from analytically predicted results. The equivalence of the APE+VCE with the LEE-s and LEE, as stated in Sect. (3), is verified with a simplified test case simulating a single thrust force with $M_\infty = 0.15$. The results included in Fig. 6 in terms of SPL, show a comparison of these equations together with analytical and APE solutions. The results of all the equations, except for the case of APE, show a very good agreement of the SPLs for the 1st mode. The expected Doppler shift of the directivity pattern is not seen in the results of APE. Due to the filtering of entropical and vortical modes applied in their formal derivation, the APE would require an appropriate source definition to restore the correct prediction of the pressure signal [8]. In the APE+VCE formulation, this requirement is highlighted by the source term present in Eq. (10). The comparison of different equations for the 3rd and 5th frequency mode maintains overall a very good agreement, APE results aside. Some excess noise is detectable in the vicinity of the axis of rotation of the force singularity. A similar explanation as provided for the excess noise seen in Fig. 5 can be considered also in this case. Other than a thrust force, a mass source singularity was also simulated, to qualitatively verify the prediction capabilities of the method. Figure (7) shows the results in terms of SPL, obtained with $M_\infty = 0.15$. The directivity patterns of all the frequency modes are represented as expected. As the mode number increases, excess noise is visible at locations close to the rotational axis of the singularity, as seen also in Fig. 5.

5 Conclusions and Future Outlook

A new propeller noise prediction method was developed in a CAA framework, including regularised force and mass sources singularities, as well as the newly proposed APE+VCE system of perturbation equations.

The proposed method aims to reduce, at an acceptable level of accuracy, the computational cost of CAA simulations for propeller noise estimation, in order to investigate installation noise. Initial results of this approach show the expected trends and patterns of the predicted fluctuating pressure signal. SPLs of simulated thrust and mass singularities, in a uniform background medium, capture the essential noise prediction features. Some excess numerical noise is seen to be present near the rotational axis of the singularities. As a future development, numerical noise mitigation techniques will be included, and the model will be extended to line-distributed rotating source singularities.

Acknowledgements. We would like to acknowledge the funding by the Deutsche Forschungsgemeinschaft (DFG, German Research Foundation) under Germany's Excellence Strategy – EXC 2163/1 – Sustainable and Energy Efficient Aviation – Project-ID 390881007

References

1. Gohardani, A.S., Doulgeris, G., Singh, R.: Challenges of future aircraft propulsion: a review of distributed propulsion technology and its potential application for the all electric commercial aircraft. Prog. Aerosp. Sci. **47**, 369–391 (2011). https://doi.org/10.1016/j.paerosci.2010.09.001
2. Gutin, L.: On the sound field of a rotating propeller, NACA TM-1195 (1948)
3. Lowson, M. V., Lighthill, M. J.: The sound field for singularities in motion. In: Proceedings of the Royal Society of London, vol. 286, pp. 559–572, 3 August 1965. https://doi.org/10.1098/rspa.1965.0164
4. Goldstein, M.E.: Unsteady vortical and entropic distortions of potential flows round arbitrary obstacles. J. Fluid Mech. **89**, 433–468 (1978). https://doi.org/10.1017/S0022112078002682
5. Atassi, H.M., Golubev, V.V.: Unsteady disturbances in swirling turbomachinery flows. In: Fransson, T.H. (ed.) Unsteady Aerodynamics and Aeroelasticity of Turbomachines, pp. 147–164. Springer, Dordrecht (1998). https://doi.org/10.1007/978-94-011-5040-8_10
6. Logue, M., Atassi, H.: Scattering of acoustic waves by a rotor. In: 14th AIAA/CEAS Aeroacoustics Conference, AIAA, Vancouver, CA, p. 2989 (2008). https://doi.org/10.2514/6.2008-2989
7. Ewert, R., Kreuzinger, J.: Hydrodynamic/acoustic splitting approach with flow-acoustic feedback for universal subsonic noise computation, Preprint submitted to Journal of Computational Physics (2020). arXiv:2009.07155
8. Ewert, R., Schroeder, W.: Acoustic perturbation equations based on flow decomposition via source filtering. J. Comput. Phys. **188**, 365–398 (2003). https://doi.org/10.1016/S0021-9991(03)00168-2
9. Dierke, J., Akkermans, R.A., Delfs, J. W., Ewert, R.: Installation effects of a propeller mounted on a wing with Coanda flap. Part II: numerical investigation and experimental validation. In: 20th AIAA/CEAS Aeroacoustics Conference, Atlanta, GA (2014). https://doi.org/10.2514/6.2014-3189

Investigation of Lighthill-Source Components in a Planar Jet-Wing-Flap Configuration

Daniel Schütz(✉) and Holger Foysi

Chair of Fluid Mechanics, University of Siegen, Paul-Bonatz-Str. 9-11, 57068 Siegen, Germany
{daniel.schuetz,holger.foysi}@uni-siegen.de

Abstract. A round jet with $Re = 3600$ and $M = 0.9$ as well as two planar nozzle-wing flap configurations were simulated by using LES. An evaluation was proposed to predict the effect of the instantaneous Lighthill sources, S, their linear components, S^{l}, and non-linear components, S^{n}. The simulations indicate that conclusions can be made regarding the directional characteristic of the emitted sound field. These findings were used in investigations of a nozzle-wing-flap configuration. The wing-nozzle configuration showed increased values of the mean Lighthill source terms on the bottom of the wing and flap, contrary to the case of an isolated wing.

1 Introduction

Installation effects cause the wings and flaps to get into the area of influence of the engine jet, which can lead to aerodynamic and aeroacoustic interference. For the latest aircraft generations the trend is towards a reduction in the number of flaps and towards smaller deflection angles of the take-off flaps in the range of 9° to 22° [10], which initially reduces the jet flap interaction (JFI). However, since the engines have to be placed closer to the wing due to the high bypass ratios, significant aeroacoustic interaction effects can occur even if no high-lift configurations are used and the rear flap is not directly impacted by the jet. Because the engine is mounted closer to the wing, interaction effects between jet and wing are possible, too. Regarding such jet wing-flap configurations there is still not enough data available. Recent publications [1,4,7,8,13] are beginning to deal with this issue due to the application-technical relevance described above. Even if the problem is simply composed of the well investigated individual problems, that of a jet and wing noise, the interaction creates a problem that cannot be represented a priori by looking and simply combining the individual problems alone. However, for a rough estimate of the installation effect, the noise generated by the complete configuration is approximated as a linear superposition of independent sources, i.e. JFI noise is noise from the complete configuration substracting isolated jet and isolated wing noise [4]. The idea of this paper is not to isolate the sound field due to the JFI, but to investigate the sound sources occuring in the JFI. For this purpose the source term, $\partial_i \partial_j T_{ij}$, of the Lighthill equation is considered,

© The Author(s), under exclusive license to Springer Nature Switzerland AG 2021
A. Dillmann et al. (Eds.): STAB/DGLR Symposium 2020, NNFM 151, pp. 761–772, 2021.
https://doi.org/10.1007/978-3-030-79561-0_72

$$\frac{\partial^2 \rho'}{\partial t^2} - a_\infty^2 \frac{\partial^2 \rho'}{\partial x_j^2} = \frac{\partial^2 T_{ij}}{\partial x_i \partial x_j}. \tag{1}$$

It is important to note that the double divergence of T_{ij} on the right side of (1) is a nominal but not a priori real acoustic source. One reason is that the Lighthill equation was derived from the basic fluid mechanics equations without further approximations and linearizations and the nominal source term combines and thus implicitly takes into account all effects except for effects like refraction. Another reason is that the major part of $\partial_i \partial_j T_{ij}$ does not radiate into the acoustic far-field and the standard Lighthill equation cannot give a simple statement about what parts of the noise originate from which region of the jet [6].

The aim of the work is to locate relevant sources of sound of the JFI and to investigate them more closely with regard to their radiation characteristics. The article proposes to do this by means of a decomposition of the Lighthill source term and a subsequent directional derivation of the source components. The required data is provided by means of LES simulation.

2 Lighthill Source Componentens

Regarding far-field sound, the contribution of the viscous term $\partial^2 \tau_{ij}/\partial x_i \partial x_j$ can be neglected [3,9] and investigations by Colonius & Freund [2] showed that viscous effects in the case of jet noise do not contribute significantly even if the Reynolds number is low. In the following the source term defined by

$$S := \frac{\partial^2 \rho u_i u_j}{\partial x_i \partial x_j} + \frac{\partial^2}{\partial x_i^2}(p' - a_\infty^2 \rho') \tag{2}$$

is considered. It is known that the largest part of S does not radiate into the acoustic far field. For a deeper insight into the sound-generating mechanisms it is therefore useful to split S further. For this purpose, the Reynolds decomposition is applied to the Lighthill stress tensor, T_{ij}, and it follows

$$T_{ij} = \overline{T}_{ij} + \rho(\overline{u}_i u_j' + u_i' \overline{u}_j) + \rho u_i' u_j' + (p' - a_\infty^2 \rho')\delta_{ij} - \tau_{ij}',$$

where $\overline{T}_{ij} = \rho \overline{u}_i \overline{u}_j + (\overline{p} - a_\infty^2 \overline{\rho})\delta_{ij} - \overline{\tau}_{ij}$. If τ_{ij} is not taken into account, the individual terms can be interpreted as a middle component, T_{ij}^{m}, a linear component, T_{ij}^{l}, a non-linear component, T_{ij}^{n} and an entropy component, T_{ij}^{s}, and one can write

$$T_{ij} = T_{ij}^{\mathrm{m}} + T_{ij}^{\mathrm{l}} + T_{ij}^{\mathrm{n}} + T_{ij}^{\mathrm{s}}, \qquad \text{where}$$

$$T_{ij}^{\mathrm{m}} = \rho \overline{u}_i \overline{u}_j + (\overline{p} - a_\infty^2 \overline{\rho})\delta_{ij}, \quad T_{ij}^{\mathrm{l}} = \rho(\overline{u}_i u_j' + u_i' \overline{u}_j), \quad T_{ij}^{\mathrm{n}} = \rho u_i' u_j', \quad T_{ij}^{\mathrm{s}} = (p' - a_\infty^2 \rho')\delta_{ij}.$$

Since the sound emission of $T_{ij}^{\overline{\rho}} = \overline{\rho} u_i u_j + (p' - a_\infty^2 \rho')\delta_{ij}$ is approximately equal to that of T_{ij} [6], the density is not explicitly split into $\rho = \overline{\rho} + \rho'$. This consideration does not yet allow any statements to be made about the direction in which a source of noise is emitted. Therefore, we introduce a directionally weighted source term S^k, which is intended to describe radiation in the direction of the observer position at $\boldsymbol{x} = (x^o, y^o, z^o)$, as

$$S_r^k = \frac{(x_i - y_i)(x_j - y_j)}{|\boldsymbol{x} - \boldsymbol{y}|^2} \frac{\partial^2 T_{ij}^k}{\partial y_i \partial y_j}.$$

Here, $\boldsymbol{y} = (x, y, z)$ is the position in the LES area, if the distance $r = |\boldsymbol{x} - \boldsymbol{y}|$ between observer and source goes towards infinity, S_r^k depends only on the observer direction $\hat{\boldsymbol{r}}$ and the source components in the x-y-plane at $z = 0$ ($\phi = \pi/2$) projected into direction $\hat{\boldsymbol{r}} = (\sin\phi\cos\theta, \sin\phi\sin\theta, \cos\phi)$ are determined via

$$S_\theta^k = \cos^2\theta \, \frac{\partial^2 T_{xx}^k}{\partial x^2} + \sin^2\theta \, \frac{\partial^2 T_{yy}^k}{\partial y^2} + 2\cos\theta\sin\theta \, \frac{\partial^2 T_{xy}^k}{\partial x \partial y}, \tag{3}$$

where θ is the angle measured from the x-axis and ϕ the angle measured from the z-axis. Freund [6] found out that the contribution of T_{ij}^{s} to the sound field is the smallest of all components for all angles, so the entropy component will not be considered here. We assume in the following, that an approximation to the sound sources resulting from the JFI is obtained by averaging the source components over time for the JFI configuration and the isolated jet and wing field, then combining them to deduce the JFI component from the entire configuration as follows

$$\overline{S_{JFI}^k} = \overline{S_{Jet+Flap}^k} - \overline{S_{Jet}^k} - \overline{S_{Flap}^k} \tag{4}$$

This won't give a perfect representation of the true JFI sound sources as cancellation might occur, for example, but it should point to regions where the relevant sources could be located. These regions are then used to extract time series of the transient sources in the LES in this area, stored and used to determine the resulting sound field following the procedure of [6].

3 Validation of the Source Treatment

To validate the proposed approach for determining the dominant source fractions and their directivity, a round jet with a Mach number of $M = 0.9$ and a Reynolds number of $Re_D = \rho_j U_j D_j/\mu_j = 3600$ was simulated and compared with the results of Freund [6]. The LES and Ffowcs William Hawkins (FWH) procedure used for this purpose has been described and validated in detail in previous works, e.g. [11]. The density and temperature of the jet are $\rho_j/\rho_\infty = 1$ and $T_j/T_\infty = 1$ respectively. The six virtual observers B1 to B6 are located on a radius of $10D_j$ measured from the nozzle and at $\theta = 15°, 30°, 45°, 60°, 75°$ and $90°$ (see Fig. 1 left). The mean values and higher moments of the LES solution were found to be in good agreement with the DNS of Freund [5] and measurements from the literature.

Sound Sources. The overall sound pressure level (OASPL) along an arc at $R = 60r_0$ from the nozzle is shown in Fig. 1 and compared to data from [12] and [6]. Compared to [12] our FWH solution is closely following the LES simulation results without showing the spurious increase at high radiation angles. S_r^k and S_θ^k were calculated for all six observers and observer angles for the total, linear and non-linear source parts, visualised in Figs. 2, 3, 4, 5, 6, 7 and 8. The time, plane and the contour levels are identical to Fig. 2 for all illustrations. The L_2 norm can be interpreted as an integral value for the source strength. The illustration of the total source S in Fig. 2 indicates that the nominal source part directed to

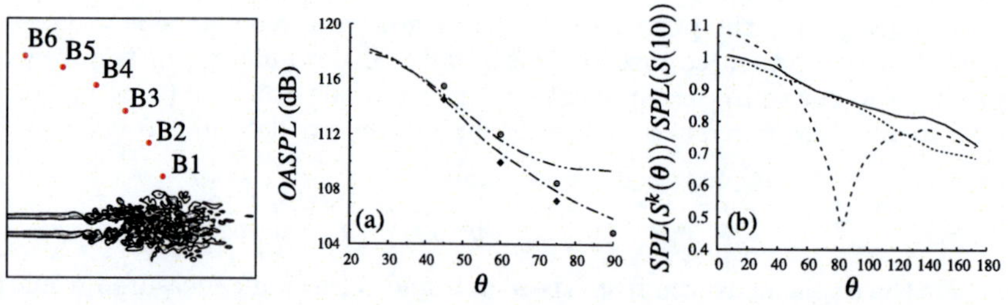

Fig. 1. Observer positions (left). (a) OASPL along an arc at $R = 60r_0$ from the nozzle and with θ measured from the jet axis: ◇ FWH solution; ∘ LES solution, (each shifted by -9.54 dB from radius $20r_0$ to $60r_0$). Comparison with simulation data: -·- LES; -··- FWH of [12]. (b) Normalised sound pressure level of the total source S (—) and the components S^{l} (- - -) and S^{n} $(\cdots)$. The data was taken from [6].

observer B6 seems to be larger than that for the observers near the jet axis. Additionally, the L_2 norm increases with radiation angle. However, this observation is in contrast to the total sound pressure level in Fig. 1 (a), which decreases with increasing angle. If the source in Fig. 3, weighted by the distance to the observer (taking into account the $1/r$ distance law), $S_r^k/r \equiv \frac{(x_i-y_i)(x_j-y_j)}{|\boldsymbol{x}-\boldsymbol{y}|^3}\frac{\partial^2 T_{ij}^k}{\partial y_i \partial y_j}$, is considered, the source strengths match with the calculated SPL at the observer position. We observe that the scaled source in the direction of the observer B1 is the most intensive, that for B6 the weakest. For the linear sources a similar effect can be observed. From Fig. 4 it is seen that the sources radiating in the direction of observer B6 are stronger in total, but the shear noise arriving at B6 is less than the noise at B1 due to the larger distance. It is also noticeable that the source regions to which the observer is positioned orthogonally make only negligible contributions. From this it is assumed that shear noise sources have a strong directional effect and hardly radiate orthogonally to the mean flow. In contrast, no significant preferred direction is visible for the non-linear sources, as shown in the Fig. 5. They seem to radiate in all directions, but more intensively with increasing angle. The strong noise impact to B1 is again due to its close location to the source. These observations agree well with the SPLs of the components from the simulations of [6], shown in Fig. 1(b), as discussed in the following. For a more general consideration the source area was differentiated globally in the direction of θ and visualised in Figs. 6, 7 and 8. Here, the directional character is shown for an observer at an angle θ, measured from the jet axis, far away from the nozzle. For $\theta = 15°$ $L_2(S_\theta)$ is maximal and decreases with increasing angle (Fig. 6). The shear sources for $\theta = 15°$ are maximal, too, and about 3% stronger in the L_2 norm than the total source S. The area of high activity is limited to the area at the end of the mixing layer, where eddies pair and the flow is characterised by coherent structures and at the end of the potential core. The L_2-norm decreases with increasing θ and around $\theta = 90°$ a significant decrease of the source strength is visible. An opposite behavior is observed for the non-linear sources. $L_2(S_\theta^{\mathrm{n}})$ increases slightly with radiation

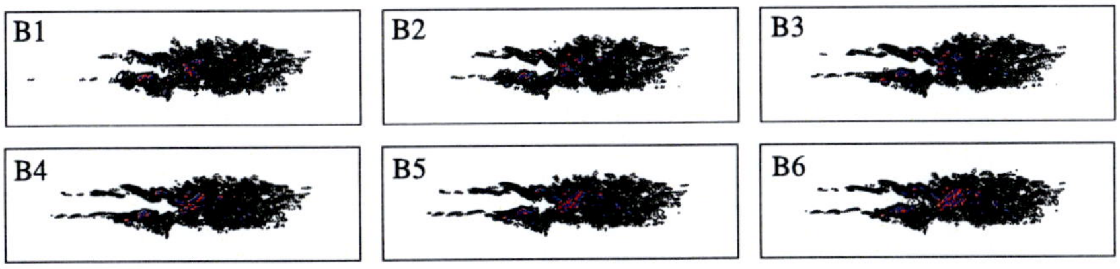

Fig. 2. Instantaneous visualization of S_r in the x-y plane at $z = 0$. Contours: 8 levels at $Sr_0^2/(\rho_j U_j^2) = \pm 0.5, 1.5, 3, 6$. Positive and negative values are shown solid resp. dashed, the maximum plane is also coloured red and the minimum blue. The normalised L_2 norm in the x-y plane for increasing angles is $L_2(S_r)/L_2(S) = [0.62, 0.60, 0.65, 0.72, 0.82, 0.95]$.

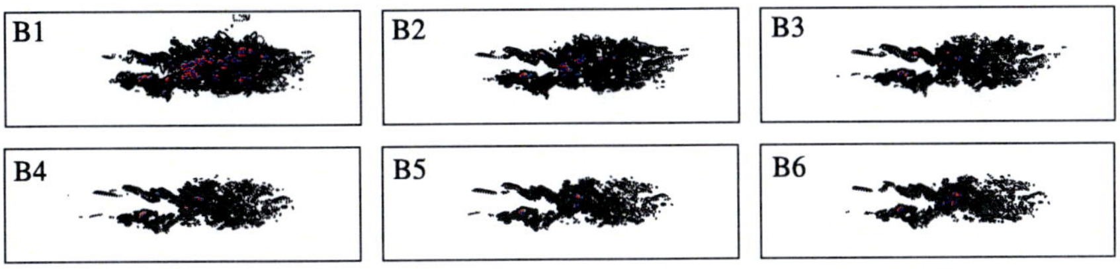

Fig. 3. Instantaneous visualisation of S_r/r. Contours (positive: solid, negative:dashed): 8 levels at $S_r/r \times r_0^3/(\rho_j U_j^2) = \pm 0.05, 0.15, 0.3, 0.6$. The maximum level is coloured red, the minimum blue.

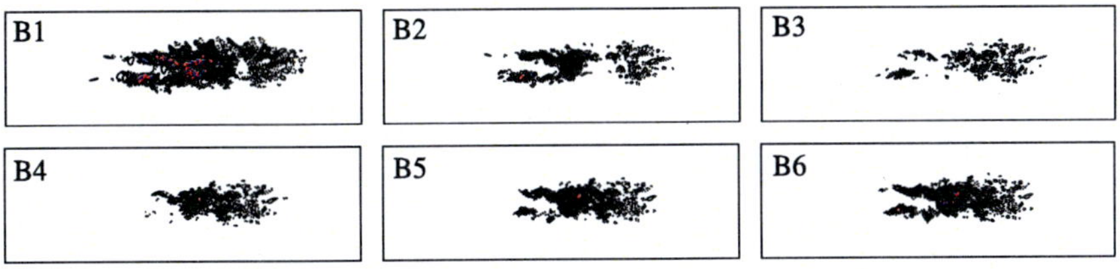

Fig. 4. Instantaneous visualization of S_r^{l}/r. $L_2(S_r^{\mathrm{l}})/L_2(S) = [0.55, 0.31, 0.21, 0.35, 0.58, 0.81]$.

angle. S^{n} therefore radiates more strongly in the direction orthogonal to the main flow. It should be noted again that the sources considered here are nominal sources, meaning that a strong source weighted in a certain direction does not necessarily result in strong acoustic radiation in that direction, too.

The instantaneous characteristics of the source components observed here with regard to their strength and direction, however, agree surprisingly well with the averaged directional characteristic from Ref. [6] (see Fig. 1 (b)) and Ref. [12]. These references split the total source term as described above and calculated the sound field of the respective source term components. As in the visualisations here, the sound pressure levels of the total source S at the reference slightly drop monotonically with increasing radiation angle; the linear component decreases

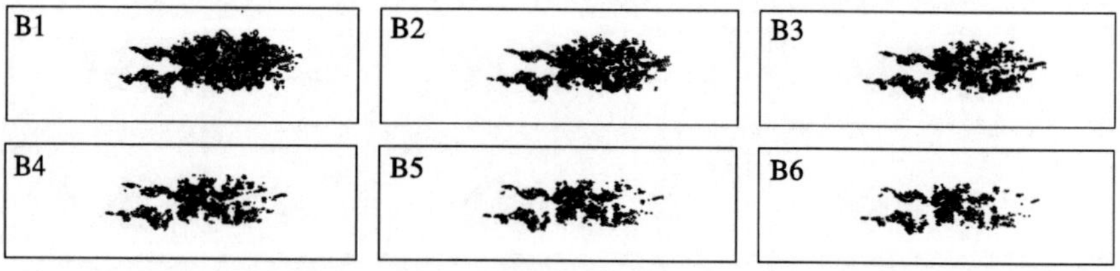

Fig. 5. Instantaneous visualization of S_r^{n}/r. $L_2(S_r^{\mathrm{n}})/L_2(S) = [0.38, 0.48, 0.55, 0.57, 0.57, 0.55]$.

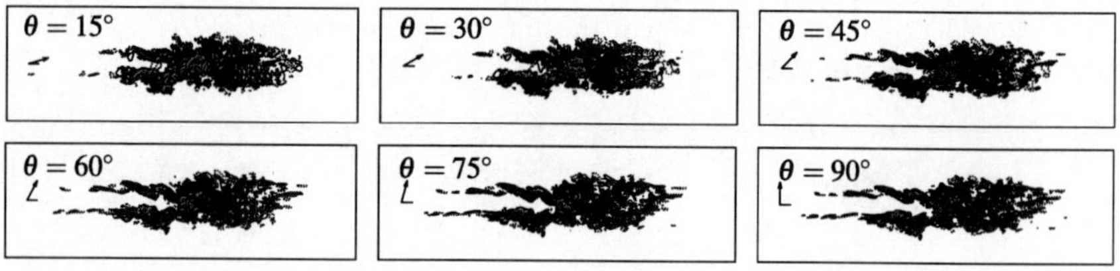

Fig. 6. Instantaneous visualization of S_θ. $L_2(S_\theta)/L_2(S) = [0.97, 0.74, 0.60, 0.59, 0.63, 0.65]$.

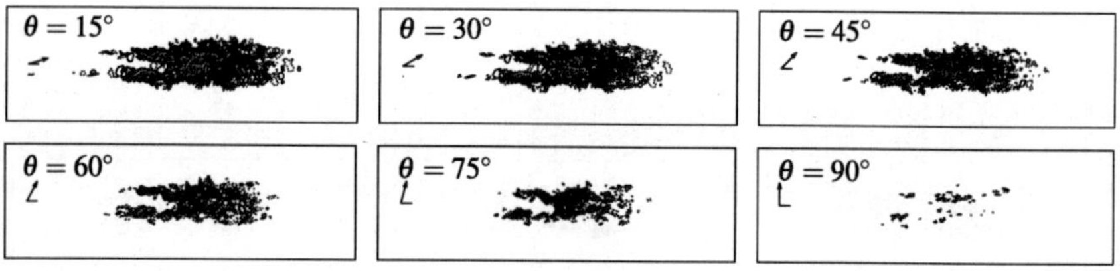

Fig. 7. Instantaneous visualization of S_θ^{l}. $L_2(S_\theta^{\mathrm{l}})/L_2(S) = [1.03, 0.78, 0.56, 0.40, 0.25, 0.07]$

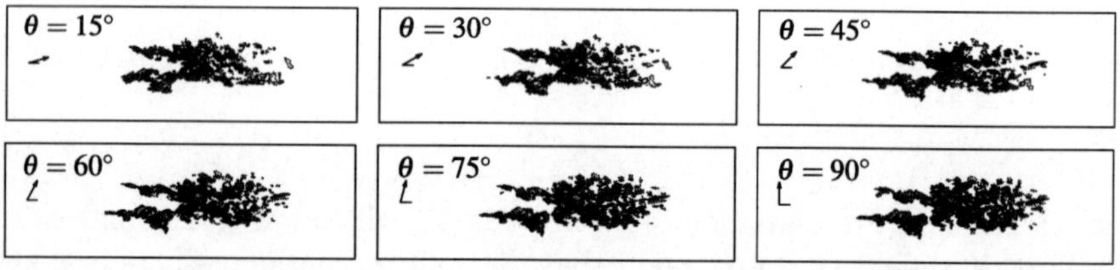

Fig. 8. Instantaneous visualization of S_θ^{n}. $L_2(S_\theta^{\mathrm{n}})/L_2(S) = [0.37, 0.32, 0.34, 0.42, 0.50, 0.56]$.

much more strongly towards its minimum at $\theta \approx 90°$, for $90° < \theta < 140°$ the level rises and for angles greater than 140° it begins to drop again. The non-linear component in the visualisations here increases slightly with radiation angle, in contrast to the average acoustic pressure in the reference. This could be due to the instantaneous view, here. Low-frequency sound, primarily due to S^l, dominates the area at small angles along the x axis, the noise of S^n is low here. But due to the high negative correlation of S^s and S^l for $\theta < 90°$, a large part of the radiated sound from linear sources is eliminated by that of entropy sound [6,12]. High frequency noise, primarily radiated by S^n, is dominant for angles between 30° and 90° and reduced near the jet axis. The procedure proposed here encourages us to get insight into the radiation properties on the basis of instantaneous data, avoiding the expensive extraction of the source fractions over a long time periods and subsequent solution of the Lighthill equation for each individual source term, for a first indication of what is happening inside the flow.

4 Jet-Wing-Flap Configuration

Two different geometry configurations (Refδ8 and P1δ22) were simulated. The varied geometry parameters were the wing and flap position and the flap angle δ. As a reference, the isolated nozzle was also simulated for both flow configurations. Nozzle, wing and flap are uniformly extended in span direction z. In a first effort, to get inside into the physics involved here, the wing nozzle configuration was not simulated using a real nozzle mounted to the wing, but first a 2D cut through the 3D nozzle configuration was used and extended into the third dimension, treated as a homogeneous direction in the following. This resulted in a 3D plane jet-wing interaction, intended get an idea of some physical mechanisms involved (shear layer flap interaction, scattering) and guide the extension to the real nozzle-wing configuration, intended to be done in a later phase of the project.

Flow Parameters and Numerical Setup. The flow velocity of the secondary nozzle is $U_s = 50\text{m/s}$, the Mach number $M_s = 0.146$ and $Re_{D_s} \simeq 5 \times 10^5$. The speed difference to the co-flow is negative, $\Delta U = U_s - U_\infty \simeq -0.03a_\infty$ ($Re_{\Delta U} \simeq 10^5$), and the convection speed of the outer shear layer is $U_c \simeq 0.16a_\infty$ and that of the inner $U_c \simeq 0.23a_\infty$. The flow of the primary nozzle is faster relative to the secondary flow and the speed difference is $U_p - U_s = 0.17a_\infty$. The jets are isothermal with $T_s = T_p = T_\infty = 293.15\text{K}$. The reference speed is $U_{\text{ref}} = 128.5\text{m/s}$ and the diameter of the secondary nozzle is $D_s = 0.15548\text{m}$. The Reynolds number, based on the chord length of the main wing $C = 0.3\text{m}$, is $Re_C = \rho_\infty U_\infty C/\mu \simeq 1.19 \times 10^6$.

Figure 9 shows the grids of the two configurations. In configuration P1δ22 the wing is mounted closer to the nozzle in vertical direction and the flap is further extended.

The simulation area is discretisized with eight overlapping grids. The simulation area is $-2 \geq x/D_s \leq 10$, $-2.5 \geq y/D_s \leq 2.5$ and $-0.05 \geq z/D_s \leq 0.05$. The boundary layer resolution with $\delta x^+ \approx 35$, $\delta y^+ \approx 0.7$ and $\delta z^+ \approx 14$ is very good regarding the LES approach. The LES procedure was described in [11] in more detail.

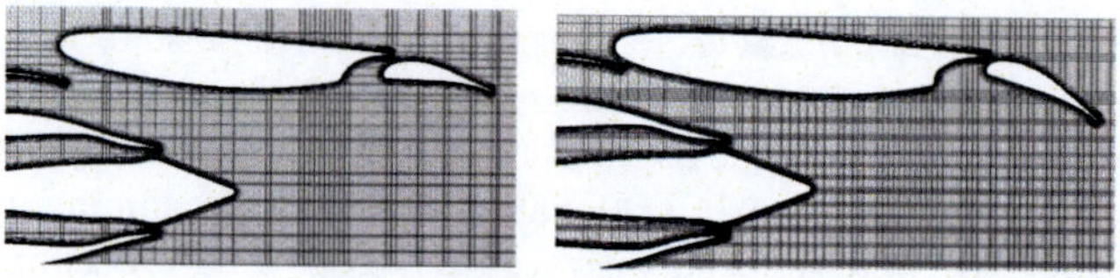

Fig. 9. Grids: Refδ8 with $\delta = 8°$ (*left*); P1δ22 with $\delta = 22°$ (*right*). Shown is every third grid line.

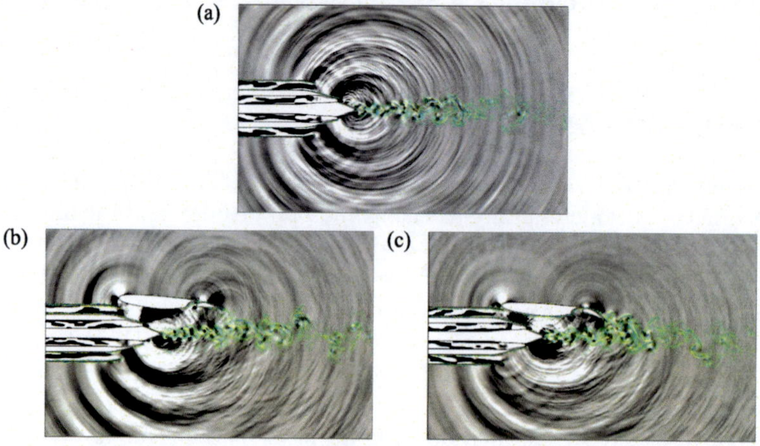

Fig. 10. Instantaneous visualization of the non-linear component of the Lighthill source term, $S^{\mathrm{n}} = \partial^2(\rho u_i' u_j')/\partial x_i \partial x_j$, and the dilatation, $\Theta = \nabla \cdot \boldsymbol{u}$: (a) nozzle, (b) Ref$\delta$8 and (c) P1$\delta$22. Colour scales: $S^{\mathrm{n}} = \pm 160 \rho_s U_{\mathrm{ref}}^2 / D_s^2$ and $\Theta = \pm 0.01 U_{\mathrm{ref}}/D_s$.

Lighthill Source Components. Figure 10 shows an instantaneous visualisation of the non-linear component of the Lighthill source term and the divergence of the velocity field. The following three phenomena can be observed: (i) The bypass jet excites the boundary layer on the bottom side of the wing, which grows downstream. If the nozzle is mounted close to the wing, the turbulent boundary layer interacts with Kelvin-Helmholtz type structures from the bypass mixing layer. (ii) In addition, the turbulent boundary layer on the bottom of the wing may interact with the lower trailing edge of the main wing. (iii) The reflection of jet noise at the wing and flap is a major factor influencing the noise level below the jet if the flap angles are moderate and the distance between jet and wing is not too small. This effect is particularly clear in the Refδ8 configuration, observed in Fig. 10 (b)and less with the P1δ22 configuration, as in the case of the P1 configuration, the acoustic waves from the primary jet have not enough space to develop before hitting the wing surface.

The linear component S^{l} is shown in Fig. 11. The same preferential radiation direction of S^{l} along the flow axis is seen. At $\theta = -90°$, S^{l} is minimal. Here, a clear difference between the two JFI configurations can be observed: Due to the

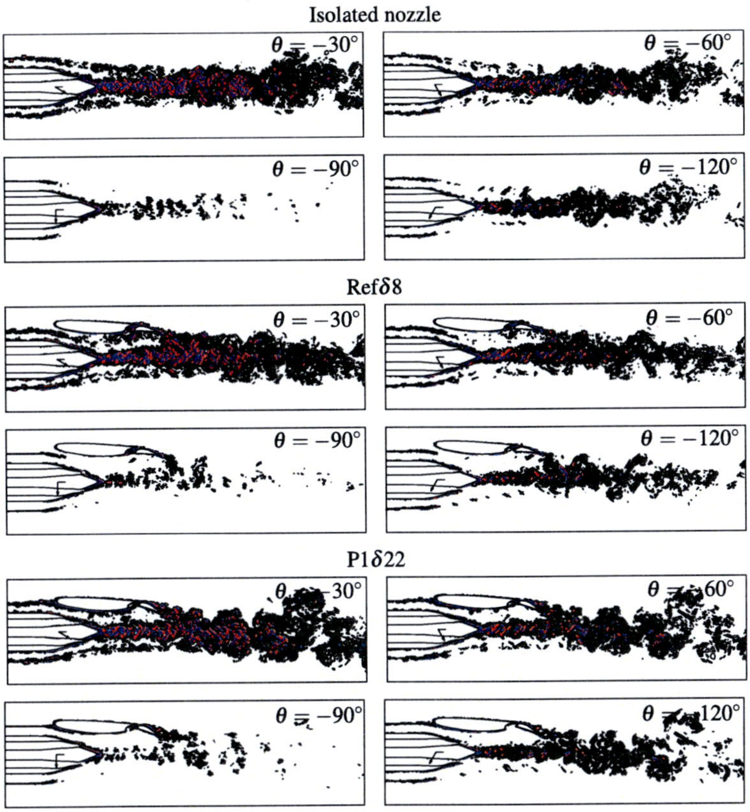

Fig. 11. Instantaneous visualization of the Lighthill source S^{l}_{θ} in the x-y plane at $z = 0$. Contours: 8 levels at $SD_s^2/(\rho_s U_{\mathrm{ref}}^2) = \pm 0.5, 1.5, 3, 6$. Positive and negative values are solid respectively dashed, the maximum level is also coloured red and the minimum blue.

fact that at P1δ22 the wing is closer to the nozzle, the nozzle boundary layer interacts with the wing leading edge and S^{l} is significantly increased in this area.

For the non-linear component, S^{n}, (Fig. 12) no preferred direction of radiation is shown, as expected. At P1δ2 the source activity above the flap and in the boundary layer of the bottom side of the wing is increased compared to Refδ8.

From the observation of these illustrations it is assumed that configuration P1δ2 causes more leading edge noise but is not necessarily more intensive in terms of JFI noise than configuration Refδ8. It is also clear that the visualisation of the source terms does not take into account reflections of sound waves at the wing and flap.

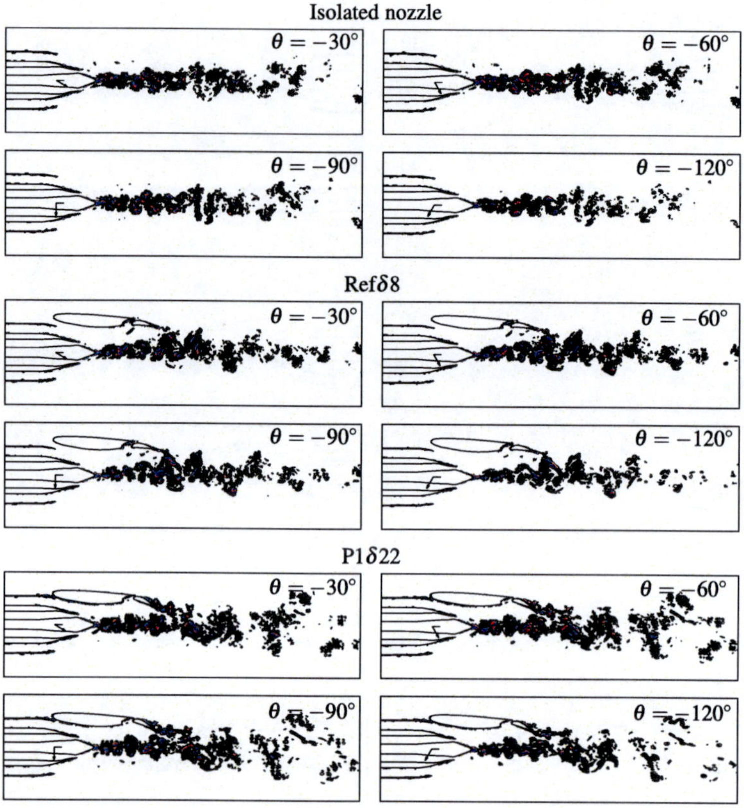

Fig. 12. Instant visualisation of S_θ^{n}. The contours are identical with Fig. 11.

5 Conclusion and Outlook

Motivated by the task of being able to assess as many geometries and parameter variations as possible with regard to their acoustic radiation at acceptable costs, an alternative to the evaluation via sound pressure level (SPL) was tested. LES of a turbulent jet was used to investigate first, to what extent an examination of instantaneous Lighthill sources and their components (S^{l}, S^{n}, S^{n}, S^{s}) allow conclusions to be drawn about the directivity of the emitted sound field. The results of [6] served as a basis for comparison, where the total Lighthill source was split into linear, non-linear and entropy terms, their emitted sound fields calculated and the sound pressure level plotted over the radiation angle θ. Here, the above-mentioned components were extracted from the LES and weighted in the direction of the six observers and uniformly in different θ directions using differentiation. Whereas, similarly to the SPL profile of [6], the linear component has a clear preferred direction along the jet axis, no such direction is discernible for the non-linear component. We could show with our approach, that the L_2 norm of the instantaneous source fields behaved exactly as predicted by [6], in that the SPL from S^{l} for radiation angles of $\theta < 30°$ is larger than the total source S SPL and decreases for $\theta \approx 90^o$ to a distinct minimum. Furthermore, the linear source fraction of source regions for a source-observer vector, $\boldsymbol{r} = \boldsymbol{x} - \boldsymbol{y}$ at 90° to the jet axis almost disappears, whereas sound from non-linear sources S^{n} dominates. If $\boldsymbol{r}$ is oriented along the jet axis, S^{l} is the dominant SPL contribution at the observer point $\boldsymbol{x}$.

In the second part of the work, the insulated wing, the insulated nozzle and the entire nozzle-wing-flap configuration were simulated for two different flap angles. The non-linear component of the Lighthill source, S^{n}, is suitable as an indicator of acoustic emission in the boundary layer region. From regions of elevated S^{n} values, an emission of sound can also be observed. For S^{l} and S^{s} this unambiguousness is not available. In a next step, the localisation of the JFI sources according to equation (4) will be started and the full configuration is to be simulated and analysed.

References

1. Belyaev, I.V., Zaytsev, M.Y., Kopiev, V.F., Ostrikov, N.N., Faranosov, G.A.: Studying the effect of flap angle on the noise of interaction of a high-bypass jet with a swept wing in a co-flow. Acoust. Phys. **63**(1), 14–25 (2017). https://doi.org/10.1134/S1063771016060026
2. Colonius, T., Freund, J.B.: Application of lighthill's equation to a mach 1.92 turbulent jet. AIAA J. **38**, 368–370 (2000)
3. Crighton, D.G.: Basic principles of aerodynamic noise generation. Prog. Aerosp. Sci. **16**, 31–96 (1975)
4. Faranosov, G., Kopiev, V., Ostrikov, N., and Kopiev, V.: The effect of pylon on the excess jet-flap interaction noise. In: Proceedings of the 22nd AIAA/CEAS Aeroacoustics Conference, AIAA Paper 2016-3043, Lyon, France (2016)
5. Freund, J.B.: Noise sources in a low-reynolds-number turbulent jet at mach 0.9. J. Fluid Mech. **438**, 277–305 (2001)
6. Freund, J.B.: Noise-source turbulence statistics and the noise from a mach 0.9 jet. Phys. Fluids **15**, 1788–1799 (2003)
7. Jordan, P., et al.: Jet-flap interaction tones. J. Fluid Mech. **853**, 333–358 (2018)
8. Lawrence, J.L.T., Self, R.H.: Installed jet-flap impingement tonal noise. In: Proceedings of the 21st AIAA/CEAS Aeroacoustics Conference, AIAA Paper 2015-3118, Dallas, Texas (2015)
9. Lilley, G.M.: Jet noise: classical theory and experiments. In: Aeroac. of Flight Vehicles: Theory and Practice, NASA Reference Publication 1258, vol. 1, pp. 211–289, NASA Langley Research Center, Hampton, Virginia (1991)
10. Rudolph, P.K.C.: High-lift systems on commercial subsonic airliners. Technical report, NASA, NASA Ames Research Center, Moffett Field, California (1996)
11. Schütz, D., Foysi, H.: Towards the prediction of flow and acoustic fields of a jet-wing-flap configuration. In: Dillmann, A., et al. (eds.) New Results in Numerical and Experimental Fluid Mechanics XI. Notes on Numerical Fluid Mechanics and Multidisciplinary Design, vol. 136, pp. 659–669. Springer, Cham (2018). https://doi.org/10.1007/978-3-319-64519-3_59

12. Uzun, A., Lyrintzis, A.S., Blaisdell, G.A.: Coupling of integral acoustics methods with les for jet noise prediction. In: Proceedings of the 42th Aerospace Sciences Meeting & Exhibit, AIAA Paper 2004-0517, Reno, Nevada (2004)
13. Vera, J., Self, R.H., Kingan, M.J.: The prediction of the radiated pressure spectrum produced by jet-wing interaction. In: Proceedings of the 21st AIAA/CEAS Aeroacoustics Conference, AIAA Paper 2015-2216, Dallas, Texas (2015)

A FWH Method for Aeroacoustic Prediction in Presence of Vorticity and Convection

Daniel Schütz(✉) and Holger Foysi

Chair of Fluid Mechanics, University of Siegen, Paul-Bonatz-Str. 9-11, 57068 Siegen, Germany
holger.foysi@uni-siegen.de

Abstract. A far-field solver based on the acoustic analogy of Ffowcs Williams and Hawkings for acoustic problems with convection and vortical waves was implemented in the open source package OVERTURE. The numerical calculations of CAA test cases showed good agreement with analytical solutions or results from measurements and simulation data from the literature. This paper presents the implemented solver and its verification. In futher work, the developed framework will be used to investigate jet-flap interaction and the emitted sound field, numerically.

1 Introduction

As a result of efficiency-optimising measures, a new generation of turbofan engines with a very high bypass ratio (so-called UHBR engines) is increasingly being installed on aircraft. Installation effects cause the wings and flaps to get into the area of influence of the engine jet, which can lead to aerodynamic and aeroacoustic interference (Jet-Flap Interaction (JFI)). The configuration is complex and its investigation expensive. In order to reduce the costs of a simulation of the aerodynamic and acoustic field, some hybrid methods have been presented in the past where the calculation area is divided into an aerodynamic region, Ω_{CFD}, which includes all relevant acoustic sources S, and an acoustic region, Ω_{CAA} [4]. The Ffowcs Williams and Hawkings [5] (FWH) equation is commonly used to predict sound in the far field. Efforts to integrate it into a numerical framework, which allows to efficiently solve the near and far field of the above described configuration in the presence of a main flow in a simulation, however, are rarer to be found yet. Most publications in this field consider acoustic phenomena in flows without the presence of a background flow, which, in a JFI under real conditions, is always present. In the context of this paper, a LES-FWH method was developed and integrated into the open-source package OVERTURE (https://www.overtureframework.org), which allows to calculate the acoustic pressure at discrete observer points simultaneously to the CFD simulation and in the presence of a background flow. The flow variables in the near-field and the acoustic pressure in the far-field are computed by means of the simultaneous solution of the compressible Navier-Stokes and FWH equations in one simulation.

© The Author(s), under exclusive license to Springer Nature Switzerland AG 2021
A. Dillmann et al. (Eds.): STAB/DGLR Symposium 2020, NNFM 151, pp. 773–784, 2021.
https://doi.org/10.1007/978-3-030-79561-0_73

2 Governig Equations in CFD-Domain

In the CFD area the Navier-Stokes equations are solved in the primitive form:

$$\begin{aligned}
&\frac{\partial \rho}{\partial t} + (\mathbf{u}\cdot\nabla)\rho + \rho(\nabla\cdot\mathbf{u}) = 0,\\
&\frac{\partial \mathbf{u}}{\partial t} + (\mathbf{u}\cdot\nabla)\mathbf{u} + \frac{1}{\rho}\nabla p = \frac{\mu}{\rho}\left(\nabla^2\mathbf{u} + \frac{1}{3}\nabla(\nabla\cdot\mathbf{u})\right),\\
&\frac{\partial T}{\partial t} + (\mathbf{u}\cdot\nabla)T + (\gamma-1)T(\nabla\cdot\mathbf{u}) = (\gamma-1)\left(\frac{\tilde{k}}{\rho}\nabla^2 T + \frac{\mu}{\mathscr{R}_g\rho}\Phi\right),
\end{aligned} \tag{1}$$

where $\mathbf{u} = [u, v, w]$ is the velocity vector in Cartesian coordinates, ρ is the density, p is the pressure and T is the total temperature of the fluid. The thermal equation of state provides the pressure $p = \rho\mathscr{R}_g T$, $\gamma = c_p/c_v$ is the ratio of the specific heat capacities, $\mathscr{R}_g$ is the specific gas constant and $\tilde{k} = k/\mathscr{R}_g$ is the dimensionless thermal conductivity, with Φ being defined as

$$\Phi = \frac{\partial u_i}{\partial x_j}\left(\frac{\partial u_j}{\partial x_i} + \frac{\partial u_i}{\partial x_j}\right) - \frac{2}{3}(\nabla\cdot\mathbf{u})^2.$$

The approximation of the spatial derivatives of the equation system (1) is carried out in the inner simulation area with a central 7-point finite difference scheme of sixth order. At the boundaries the radiation and outflow boundary conditions of Tam and Dong [12] are utilized to avoid reflections.

3 Governing Equations in CAA-Domain

The integration surface S (also called control surface) is mathematically defined by the function $f(\mathbf{x}, t)$. The control surface is defined via $f = 0$ and $|\nabla f| = 1$, within this control area $f < 0$, outside of the control volume we have $f > 0$; $\nabla f = \hat{\mathbf{n}}$ is the outwardly directed unit normal vector.

Following Najafi-Yazdin *et al.* [10], we consider a medium moved uniformly with $\mathbf{U}_\infty$ in the formulation of the FWH analogy and split the velocity into $\mathbf{u} = \mathbf{u}' + \mathbf{U}_\infty$. The discrete FWH equation for this case are written as follows:

$$\left[\left(\frac{\partial}{\partial t} + \mathbf{U}_\infty\cdot\nabla\right)^2 - a_\infty^2\nabla^2\right]p'(\mathbf{x},t)\,H(f) = \left[\frac{\partial}{\partial t} + \mathbf{U}_\infty\cdot\nabla\right][\rho_0 U_n\delta(f)] - \frac{\partial\,[L_i\delta(f)]}{\partial x_i} + \frac{\partial^2\,[T_{ij}H(f)]}{\partial x_i\partial x_j}, \tag{2}$$

with the Heaviside function $H(f)$ and the terms on the right

$$U_i = \frac{\rho}{\rho_0}\left(u_i' + U_{\infty i} - v_i\right) + v_i - U_{\infty i}, \quad L_i = \rho u_i'\left(u_j'\hat{n}_j + U_{\infty j}\hat{n}_j - v_j\hat{n}_j\right) + P_{ij}\hat{n}_j,$$

where $U_n = U_i\hat{n}_i$. The terms for thickness-noise and loading noise now include the main flow velocity, the Lighthill stress tensor T_{ij} remains unchanged from the classical formulation. For $\mathbf{U}_\infty = \mathbf{0}$ you get the classical FWH equation. For the solution of (2) one needs Green's convective function (Najafi-Yazdin *et al.* [10]):

$$G(\mathbf{x}, t; \mathbf{y}, \tau) = \frac{\delta(g)}{4\pi\mathscr{R}} = \frac{\delta(\tau - t + R/a_\infty)}{4\pi\mathscr{R}} \tag{3}$$

where $\mathbf{x}$ is the location and t is the time of the observer, whereas $\mathbf{y}$ and τ are the source location and time, respectively, where $\mathbf{y} = \mathbf{y}(\tau)$. R is the effective acoustic distance between the observer and the source. For a uniform main flow in any direction the quantities $\mathscr{R}$ and R are defined as (Zhong and Zhang [14]):

$$\mathscr{R} = \frac{1}{\alpha}\sqrt{r^2 + (\alpha \mathbf{M}_\infty \cdot \mathbf{r})^2}, \quad R = \alpha^2(\mathscr{R} - \mathbf{M}_\infty \cdot \mathbf{r}), \quad \text{with } \alpha^2 = \frac{1}{1 - |\mathbf{M}_\infty|^2},$$

where $\mathbf{r} = \mathbf{x} - \mathbf{y}$ and $r = |\mathbf{r}|$ is the geometric distance between observer and source. The derivatives of $\mathscr{R}$ and R with respect to x_i can be calculated directly via

$$\tilde{\mathscr{R}}_i = \frac{\partial \mathscr{R}}{\partial x_i} = \frac{r_i + \alpha^2 M_{\infty i}(\mathbf{M}_\infty \cdot \mathbf{r})}{\alpha^2 \mathscr{R}}, \quad \tilde{R}_i = \frac{\partial R}{\partial x_i} = \alpha^2 \left(\tilde{\mathscr{R}}_i - M_{\infty i}\right). \tag{4}$$

In the investigations here the integration area $f = 0$ is stationary and the acoustic distances $\mathscr{R}$ and R are not dependent on the source time τ_e. In this case, after some analysis not shown here, the pressure terms for a problem with main flow follow as

$$4\pi p'_T(\mathbf{x}, t) = \int_{f=0} \left[(1 - M_{\infty i}\tilde{R}_i)\frac{\rho_0 \dot{U}_n}{\mathscr{R}} - U_{\infty i}\tilde{\mathscr{R}}_i \frac{\rho_0 U_n}{\mathscr{R}^2}\right]_{\tau_e} \mathrm{d}S, \tag{5}$$

$$4\pi p'_L(\mathbf{x}, t) = \int_{f=0} \left[\frac{\dot{L}_i \tilde{R}_i}{a_\infty \mathscr{R}} + \frac{L_i \tilde{\mathscr{R}}_i}{\mathscr{R}^2}\right]_{\tau_e} \mathrm{d}S. \tag{6}$$

Furthermore, we made use of the *advanced time* approach of Casalino [4] to build p'. A great advantage of this approach is that it can be easily parallelized. The FWH surface was chosen to be a closed surface, with the exit plane placed within the sponge zone, described below. This resulted in similar results compared to [3], avoiding the averaging procedure, as only the sponge zone strength needs to be adjusted.

4 Verification and Validation

The solution of the implemented FWH method is compared with analytical solutions and measurements for the following verification cases: A radiating acoustic pulse in (i) a stationary medium, (ii) a uniform flow with $\mathbf{M}_\infty = (0.5, 0, 0)$ and (iii) with $\mathbf{M}_\infty = (0.5, 0.5, 0.5)$. Additionally, a round jet with Reynold number $Re = 3600$ and Mach number $M_j = 0.9$ was simulated. The first three cases serve to verify the numerical implementation, while the fourth will demonstrate the method's ability to predict turbulence-induced noise using a well-known test case.

3-D Pulse

As a simple test case for verification, a three-dimensional acoustic pulse was simulated. The analytical solution for a uniform main flow with $\mathbf{M}_\infty = (M_x, M_y, M_z)$ is given by:

$$p_e = p'(x,y,z,t) = \frac{R-t}{2R} e^{-\alpha(R-t)^2} + \frac{R+t}{2R} e^{-\alpha(R+t)^2}, \qquad (7)$$

where $R = \left[(x - M_x t)^2 + (y - M_y t)^2 + (z - M_z t)^2\right]^{1/2}$, $\alpha = \ln(2)/b^2$ and $b = 3$. Table 1 shows the parameters of the simulated cases, having identical uniform resolution in the three spatial directions except for case C4 that includes stretching. The LES area is defined as $-50 \le x, y, z \le 50$, the control area is formed by a cube of edge length $L = 50$, localized in $-25 \le x, y, z \le 25$. In Table 2 the positions of the observers are given and the respective theoretical simulation intervals required for a converged FWH solution without convection, which was set to $[30, 100]$. Since the solution in the LES area for $t < 0$ is zero, for all cases a time interval from $t = 0$ to 100 was simulated.

Without convection ((i)$|\mathbf{M}_\infty| = 0$): For cases C1 and C2, the additional computational effort for calculating the FWH solution is 1% and scales linearly with the number of observers (see case C5). The performance compared to case C3 is due to the lower relative overhead due to the higher spatial resolution of case 3. The additional computing time for case C4 is dependent on the non-uniform grid stretching. The variables were non-dimensionalized using the density ρ_0, pressure $a_\infty^2 \rho_0$, grid spacing of the reference grid Δ and time Δ/a_∞, where a_∞ is the speed of sound.

Table 1. Parameters of the simulated test cases. exec-time is the execution time of the program on 8 cores. TPSM (time-per-step-per-million-grid-points) and nTPSM is the execution time per time step and 10^6 grid points normalized with the TPSM of the reference case.

Case	N_{obs}	Observer o	N_x	N_{cell}	Δx	Δt	T_{sim}	exec-time	TPSM	nTPSM
Ref	0	–	101	14406	1	0.25	50	227	1.1125	1
C1	1	2	101	14406	1	0.25	50	231	1.125	1.01
C2	1	2	101	14406	1	0.125	50	461	1.125	1.01
C3	1	2	201	58806	0.5	0.125	50	2560	0.7875	0.71
C4	1	2	101	14406	$f(x)$	0.125	50	923	2.25	2.02
C5	6	$1,2,3,4,5,6$	101	14406	1	0.25	100	485	1.175	1.06

Table 2. Position of the observers, $\mathbf{x}^o = (x^o, y^o, z^o)$, and theoretical minimum LES simulation interval, $ta_\infty/\Delta = [t^o_{\text{start}}, t^o_{\text{end}}]$, for a convergent FWH solution in the time interval $ta_\infty/\Delta = [30, 100]$. $\mathcal{T}^o = r^o/\Delta$ is the dimensionless travel time of the acoustic waves from the control surface to the observer o.

Observer o	x^o	y^o	z^o	r^o_{min}	r^o_{max}	$\mathcal{T}^o_{\text{min}}$	$\mathcal{T}^o_{\text{max}}$	$[t^o_{\text{start}}, t^o_{\text{end}}]$
1	0	26	0	1	62.06	1	62.06	$[-32.06, 99]$
2	0	50	0	25	82.92	25	82.92	$[-52.92, 75]$
3	0	75	0	50	106.07	50	106.07	$[-76.07, 50]$
4	26	26	26	1.73	88.33	1.73	88.33	$[-58.33, 98.27]$
5	28.87	28.87	28.87	6.70	93.3	6.70	93.3	$[-63.3, 93.3]$
6	43.3	43.3	43.0	31.70	118.3	31.70	118.3	$[-88.3, 68.3]$

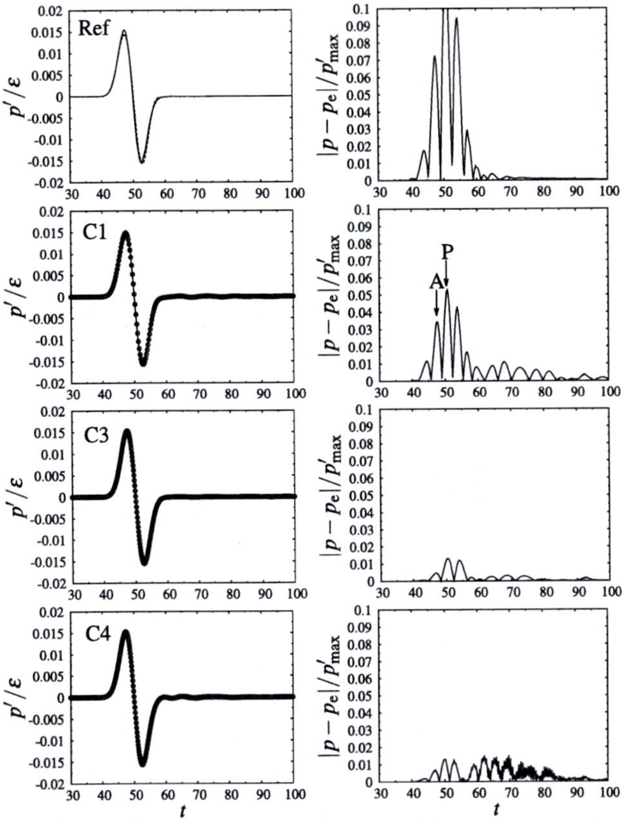

Fig. 1. Acoustic pressure at $\mathbf{x} = (0, 50, 0)$ (*left*): —, analytical solution; - - -, LES solution and ∘, FWH solution. Normalized error $|p - p_e|$ (*right*), where p_e is the analytical solution and $p'_{max} = \max\{p'_e|\}$.

Figure 1 shows the acoustic pressure p' (*left*) and normalized errors (*right*) at point $\mathbf{x} = (0, 50, 0)$ for the cases C1, C3, C4 and the reference solution. The phase (P) and amplitude error (A) is less than 6% of the pressure amplitude for the case with low spatial and temporal resolution, C1. If the resolution is doubled in each case, the error is reduced to below 2%. A similar error reduction can be achieved by refining the grid in the area of the cells relevant to the observer (C4). However, slight oscillations occur in the range $t > 55$. Since the LES method is dispersive and dissipative in contrast to the FWH method, the error increases with distance. The error of the FWH solution is determined by the error of the LES solution on the control surface and the rounding error in the calculation of t_{adv}, as well as by the approximation of the surface integrals. Approximation errors decrease with the distance of the observer from the control surface. The results of the most critical case, C5 (observer 4), are shown in Fig. 2. The observer direction runs exactly through the corner of the control cuboid.

The relative error remains below 8% and is therefore only marginally larger than the error of the reference solution. If the error caused by the LES solution is accounted for, the error, caused by the FWH method alone, is reduced to below 3% (dotted line in Fig. 2) (*right*) (Fig. 3).

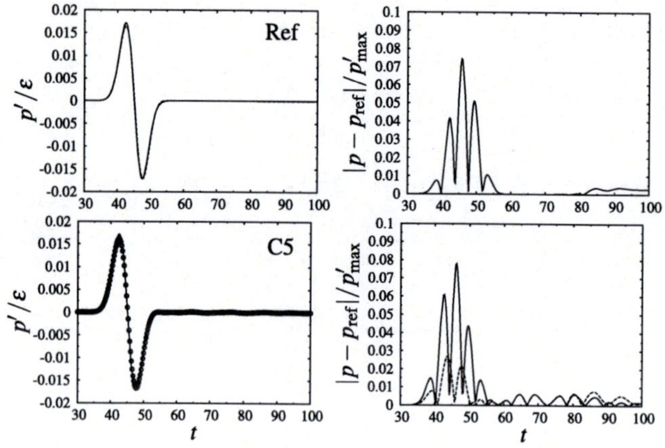

Fig. 2. Acoustic pressure at $\mathbf{x} = (26, 26, 26)$ (*left*): —, analytical solution; - - -, LES solution and ∘, FWH solution. Normalized error (*right*) with $p_{\mathrm{ref}} = p_{\mathrm{e}}$ (—) and $p_{\mathrm{ref}} = p_{\mathrm{LES}}$ (- - -).

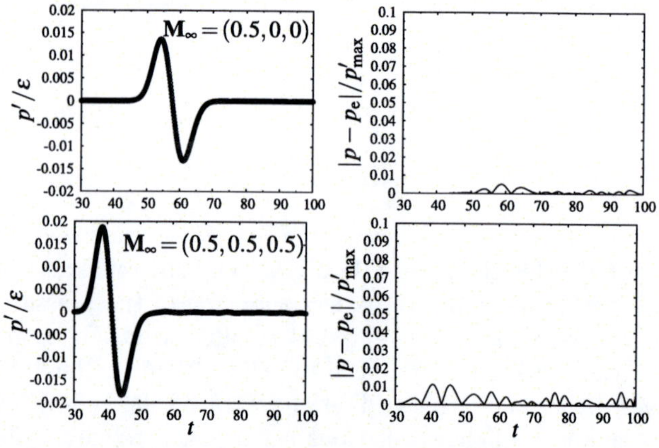

Fig. 3. Acoustic pressure at $\mathbf{x} = (0, 50, 0)$ (*left*): —, analytical solution; and ∘, FWH solution. Normalized error $|p - p_{\mathrm{e}}|$ (*right*), where p_{e} is the analytical solution and $p'_{\max} = \max\{p'_{\mathrm{e}}|\}$.

With Convection ((ii) $M_x = 0.5$ and (iii) $|\mathbf{M}_\infty| \simeq 0.866$): The cases with convection were simulated on the C3 grid with $\delta t = 0.125$ ($|\mathbf{M}_\infty| = 0.5$) and $\delta t = 0.094$ ($|\mathbf{M}_\infty| \simeq 0.866$). The pressure was calculated for point B2 at $\mathbf{x} = (0, 50, 0)$.

The simulation intervals of B2, $[t^2_{\text{start}}, t^2_{\text{end}}]$, required for a convergent solution in the interval $t = [30, 100]$, are $[-50.52, 71.13]$ and $[-20, 79.29]$, respectively. The pressure curves and the error compared to the analytical solution of both cases can be seen in Fig. 2. The agreement with the analytical solution, Eq. 7, is very good in both cases. For $\mathbf{M}_\infty = (0.5, 0, 0)$ the error is at all times less than 1% and for $\mathbf{M}_\infty = (0.5, 0.5, 0.5)$ less than 1.5% of the pressure amplitude. Thus the theory has been validated for the case with and without convection and its implementation has been verified.

Round Jet

In order to validate the method for more realistic cases, a three-dimensional round jet ($M_j = 0.9$ and a Reynolds number of $Re_D = \rho_j U_j D_j / \mu_j = 3600$) was simulated. The inflow axial velocity of the jet ist given by the following tanh-profile: $u_0(r) = \frac{U_j}{2}\left[1 + \tanh\left(\frac{r_0 - r}{2\delta_\theta}\right)\right]$, where r_0 is the jet radius, U_j ist the inflow centerline velocity and $\delta_\theta = 0.05 r_0$ is the initial momentum thickness of the shear layer. The initial inlet temperature profile was determined using the Crocco-Buseman relation for an isothermal jet.

Numerical Specifications. The LES area is defined by $-2 \le x/r_0 \le 30$, $7 \le y/r_0 \le 25$ and $-6 \le z/r_0 \le 6$ and the Navier-Stokes equations are discretisized with $277 \times 187 \times 127$ grid points. The control surface is spanned by a cuboid of edge length $(28, 10, 10) r_0$. A damping zone was introduced after $x = 20 r_0$, by adding a term of the form $-\sigma(U - U_\infty)$ to the governing equations. The sponge zone serves to smooth vortices before they pass the closed outlet FWH surface. The σ strength grows quadratically with x up to $\sigma_{\max} = 0.15$ at the outlet at $y = 0$. To increase the physical acoustic area, a tanh function is used to decrease the damping strength for positive y. The derivatives were calculated using a central difference method of sixth order. The time step was fixed at $\Delta t = 0.007 r_0 / U_j$, the solution filtered at every fifth time step by using an explicit filter of tenth-order, similar to Ref. [8]. This filtering was shown to serve as a replacement for a LES model, too, in under resolved cases [8]. The origin of the effective radiation location (direction vector, $r\mathbf{e}_r$) for the radiation and outflow boundary conditions was set at $x = 15 r_0$. For the excitation of the laminar-turbulent transition a vortex ring forcing similar to Ref. [2] was used.

Aerodynamic Field. The development of the jet is in good agreement with the DNS of Freund [7], as shown in the following. Figure 4 shows the time-averaged Mach number at various downstream locations. Here, too, the results from the simulation presented are very close to the measurements of Stromberg *et al.* [11], too. The Mach number profile changes with downstream position in accordance to the reference data (Fig. 4(a)–(c)). The mean length of the potential core, measured from the position of forcing x_0, can be deduced from Fig. 4(d), and is approximately $14 r_0$. Additionally (not shown here for brevity and lack of space), the second moments of velocity agree nicely with the literature and the energy spectrum shows the typical $k^{-5/3}$ decay.

Aeroacoustic Field. Figure 5 shows an isosurface of the second invariant of the velocity gradient tensor, $II_{\nabla \mathbf{u}}$, and the divergence of the velocity, $\Theta = \nabla \cdot \mathbf{u}$.

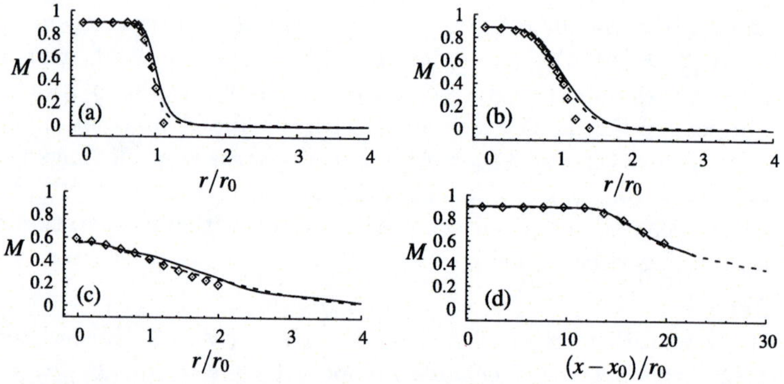

Fig. 4. Mach number distribution: —, present simulation; - - -, Freund [7] (DNS); ◇, Stromberg *et al.* [11] (exp. Data). (a) $x = 2r_0$; (b) $x = 10r_0$; (c) $x = 20r_0$; (d) streamwise development at ($r = 0$).

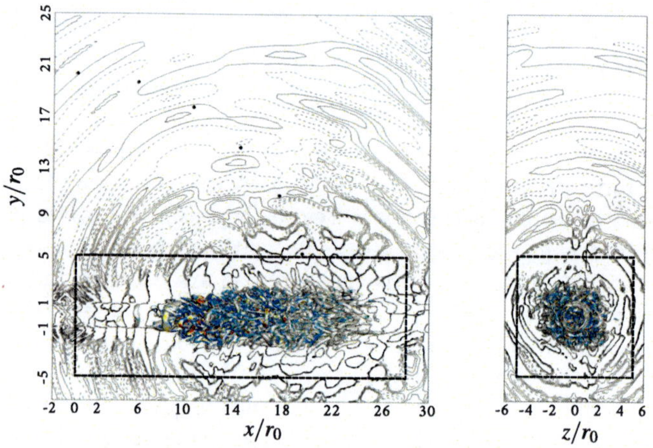

Fig. 5. Instantaneous visualisation of $II_{\nabla \mathbf{u}}$ and the dilatation $\Theta = \nabla \cdot \mathbf{u}$ in the x-y plane at $z = 0$ (*left*) and in the y-z plane at $x = 15r_0$ (*right*). Contour planes of the dilatation $\Theta = \nabla \cdot \mathbf{u}$ between $\pm 5 \times 10^{-4} U_j/r_0$ (increments: $2 \times 10^{-4} U_j/r_0$; —, $\Theta > 0$; - - -, $\Theta < 0$) and isosurface for $Q = 0.25 U_j^2/r_0^2$ coloured with the instantaneous Lighthill source S with colour scale from blue ($S = -6\rho_j U_j^2/r_0^2$) to red ($S = 6\rho_j U_j^2/r_0^2$), values between $-1.5 < Sr_0^2/(\rho_j U_j^2) < 1.5$ are colourized grey. The dotted box marks the control surface and the points • define the position of the 6 observers at $R = 20r_0$ at a distance of 15° between $\theta = 15°$ to 90°.

The black dots in the representation mark the position of the virtual observers on an arc with radius $20r_0$, measured from the coordinate origin. At these locations, the solution obtained from the FWH equation is compared with the LES solution and solutions from the literature. Since the observer B1 ($\theta = 15°$) fulfils neither the condition to be located in the acoustic far field ($|\mathbf{r}| \gg \lambda$), nor the condition for the geometrical far field ($|\mathbf{r}| \gg \Omega_s$), the worst agreement with the reference solution is expected in this case. Therefore, it is assumed that the scaled rms pressure at $R = 60r_0$ for $\theta = 15°$ to $30°$ is overestimated. Small spurious waves in the inlet region are caused by analytically, but not numerically zero divergence from the vortex ring forcing. These waves are orders of magnitude smaller than the relevant acoustics. This effect was investigated in the past by simulations without developing jets, for example, in combination with filtering and grid stretching [6]. In Fig. 6, the pressure fluctuations calculated via FWH solver at the six discrete observer positions are compared with the LES solution. The FWH solution agrees very well in amplitude and phase with the LES solution, as expected, for $\theta \geq 30°$. Suprisingly, for the observer B1 at $\theta = 15°$, the agreement in phase is good, too. The amplitude is a bit underestimated, but this was not expected otherwise due to the very small distance of the observer point to the control surface. Since the mean pressure field in the far field becomes uniform ($\overline{p}(\mathbf{x}) \rightarrow p_\infty$), the shift in the mean pressure also decreases with increasing distance of the observer from the control surface. It is also noticeable that the pressure obtained from the LES varies more smoothly than the FWH solution, which is due to the numerical dissipation and the smoothing of the LES solution by the numerical filter. The overall sound pressure level (OASPL) in Fig. 7 was calculated from the rms value of the pressure in the time interval under consideration. For comparison with reference data, the values were scaled to the value at $60r_0$ using $p_{\mathrm{rms}}(60r_0) = \frac{20}{60}p_{\mathrm{rms}}(20r_0)$ and plotted over the angle θ. For the observers with $\theta > 30$ the agreement with the SAE ARP 876C database and measurements is very good. Along the jet axis the pressure levels are slightly higher. This is because the scaling is only valid for pressure fluctuations of purely acoustic nature. The observer B1, however, is still subject to aerodynamic pressure fluctuations due to the relatively small arc radius (also shown by the dilation in Fig. 5), which is why the rms value is much higher there. For a larger arc this effect should disappear. Similar curves were also found in the FWH simulations by Uzun *et al.* [13] (Fig. 7). Uzun still had the problem that for a closed control surface at the outlet the sound was overestimated for large angles, whereas it was greatly underestimated for small angles in case of an open control surface, there. The former is due to the fact that vortex and entropy modes flowing over the outlet surface represent quadrupoles, which affect the FWH solution in the area opposite the outlet surface (large angles). The results obtained for the impulse test case and the simulation of the three-dimensional jet demonstrate that the FWH solver implemented allows an accurate and efficient prediction of the sound pressure in the far field for the investigations of a JFI configuration planned in the future. The formulation and its implementation in the Overture package was thus validated.

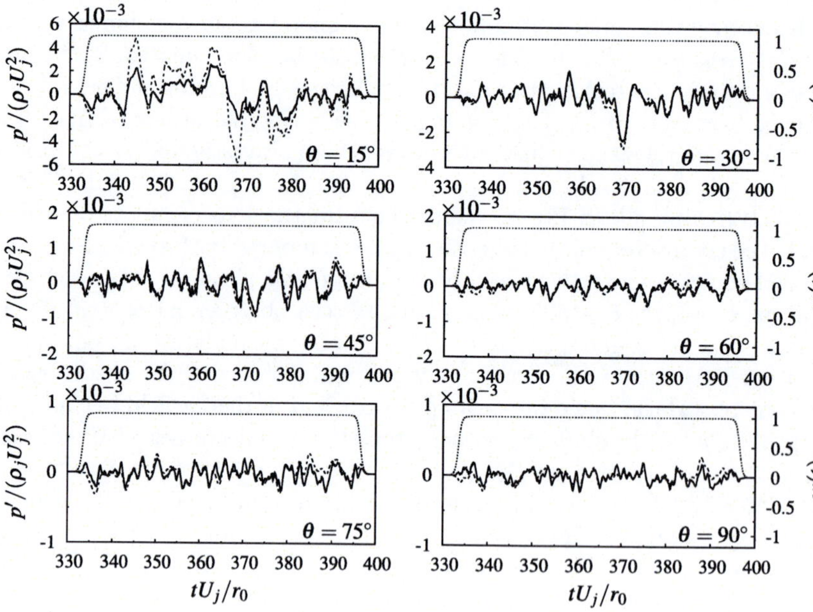

Fig. 6. Windowed acoustic pressure for the 6 observers at $R = 20r_0$: —, FWH solution; - - -, LES solution. $\cdots$, tanh window, $w(t)$. The profiles were moved such that $\overline{p'} = 0$.

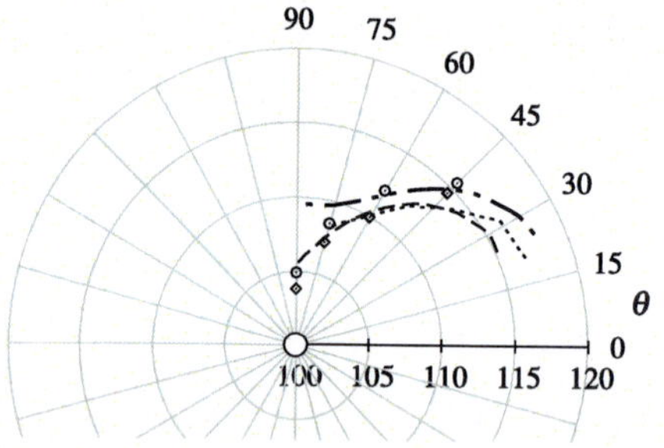

Fig. 7. Overall sound pressure level (OASPL) in dB along an arc at $R = 60r_0$ from the nozzle and with θ measured from the jet axis: $\diamond$, FWH solution and $\circ$, LES solution, (each shifted by -9.54 dB from radius $20r_0$ to $60r_0$). Comparison with simulation data: -·-·-; closed FWH with $Re = 4 \times 10^5$ Uzun [13] and $\cdots$, SAE ARP 876C database [1]. Comparison with experimental data: - - -, $Re = 2 \times 10^5$ Mollo-Christensen [9].

5 Conclusion and Outlook

For three-dimensional compressible flows, a convective Ffowcs Williams and Hawking solver, coded in C++, was presented and compared to literature results. This type of implementation proved to be cheap in terms of computing time and was included in the OVERTURE framework, allowing non-standard high-fidelity simulations. The approach used here allows the calculation of the acoustic pressure at defined observer points simultaneously to LES and is based on the work of Najafi-Yazdin *et al.* [10], in combination with the source-time approach of Casalino [4]. The FWH solver proved to be very efficient with an increase in computing time of only about 1% per observer compared to the standard LES simulation. It was found that the computing time scaled linearly with the number of observers in the investigated area. The solver could be successfully verified by comparing to analytical solutions of selected CAA benchmark cases and by the extensive LES/FWH simulation of a round jet with $Re = 3600$ and $M_j = 0.9$. The agreement with the analytical solution of the three-dimensional pressure pulse was very good even for the hardest case with convection in all directions and observation direction through a corner of the integration cube. For $\mathbf{M}_\infty = (0.5, 0, 0)$ the error at all times was less than 1% and for $\mathbf{M}_\infty = (0.5, 0.5, 0.5)$ less than 1.5% of the pressure amplitude. In the case of the jet, the comparison of the time series of the acoustic pressure obtained from the combination of LES and FWH with that obtained from the direct calculation via pure LES was surprisingly good. For all relevant observation points the instantaneous pressure curves were in excellent agreement within the calculated simulation period. The averaged sound pressure levels agreed well with measurements and data from the literature, too. In a next step, the jet flap interaction, mentioned in the introduction, will be investigated numerically.

References

1. SAE ARP 876C: Gas Turbine Jet Exhaust Noise Prediction. Society of Automotive Engineers (1985)
2. Bogey, C., Bailly, C., Juvé, D.: Noise investigation of a high subsonic, moderate Reynolds number jet using a compressible large eddy simulation. Theor. Comput. Fluid Dyn. **16**, 273–297 (2003)
3. Brès, G.A., Nichols, J.W., Lele, S.K., Ham, F.E.: Towards best practices for jet noise predictions with unstructured large eddy simulations. In: Proceedings of the 42nd AIAA Fluid Dynamics Conference and Exhibit, AIAA Paper 2012, New Orleans, Louisiana, p. 2965 (2012)
4. Casalino, D.: An advanced time approach for acoustic analogy predictions. J. Sound Vib. **261**, 583–612 (2003)
5. Ffowcs Williams, J.E., Hawkings, D.L.: Sound generation by turbulence and surfaces in arbitrary motion. Proc. R. Soc. Lond. Ser. A **264**(1151), 321–342 (1969)
6. Foysi, H., Mellado, M., Sarkar, S.: Simulation and comparison of variable density round and plane jets. Int. J. Heat Fluid Flow **31**, 307–314 (2010)
7. Freund, J.B.: Noise sources in a low-Reynolds-number turbulent jet at Mach 0.9. J. Fluid Mech. **438**, 277–305 (2001)
8. Mathew, J., Lechner, R., Foysi, H., Sesterhenn, J., Friedrich, R.: An explicit filtering method for large eddy simulation of compressible flows. Phys. Fluids **15**, 2279 (2003)
9. Mollo-Christensen, E., Kolpin, M.A., Martucelli, J.R.: Experiments on jet flows and jet noise far-field spectra and directivity patterns. J. Fluid Mech. **18**, 285–301 (1964)

10. Najafi-Yazdi, A., Brès, G.A., Mongeau, L.: An acoustic analogy formulation for moving sources in uniformly moving media. Proc. R. Soc. Lond. A **467**, 144–165 (2011)
11. Stromberg, J.L., McLaughlin, D.K., Troutt, T.R.: Flow field and acoustic properties of a Mach number 0.9 jet at a low Reynolds number. J. Sound Vib. **72**, 159–179 (1980)
12. Tam, C.K.W., Dong, T.Z.: Radiation and outflow boundary conditions for direct computation of acoustic and flow disturbances in a nonuniform mean flow. J. Comput. Acoust. **4**, 175–201 (1996)
13. Uzun, A., Lyrintzis, A.S., Blaisdell, G.A.: Coupling of integral acoustics methods with les for jet noise prediction. Int. J. Aeroacoustics **3**, 297–346 (2005)
14. Zhong, S., Zhang, X.: A sound extrapolation method for aeroacoustics far-field prediction in presence of vortical waves. J. Fluid Mech. **820**, 424–450 (2017)

Author Index

© The Editor(s) (if applicable) and The Author(s), under exclusive license
to Springer Nature Switzerland AG 2021

A. Dillmann et al. (Eds.): STAB/DGLR Symposium 2020, NNFM 151, pp. 785–787, 2021.
https://doi.org/10.1007/978-3-030-79561-0